Infosys Science Foundation Series

Infosys Science Foundation Series in Mathematical Sciences

The **Infosys Science Foundation Series in Mathematical Sciences**, a Scopus-indexed book series, focuses on high-quality content in the domain of pure and applied mathematics, biomathematics, financial mathematics, operations research and theoretical computer science. With this series, Springer hopes to provide readers with monographs, textbooks, edited volume, handbooks and professional books of the highest academic quality on current topics in relevant disciplines. Literature in this series will appeal to a wide audience of researchers, students, educators and professionals across the world.

Aymen Ammar · Aref Jeribi

Spectral Theory of Block Multivalued Operator Matrices

Springer

Aymen Ammar
Department of Mathematics
University of Sfax
Sfax, Tunisia

Aref Jeribi
Department of Mathematics and Statistics,
College of Science
Imam Mohammad Ibn Saud Islamic
University (IMSIU)
Riyadh, Saudi Arabia

ISSN 2363-6149 ISSN 2363-6157 (electronic)
Infosys Science Foundation Series
ISSN 2364-4036 ISSN 2364-4044 (electronic)
Infosys Science Foundation Series in Mathematical Sciences
ISBN 978-981-95-0396-4 ISBN 978-981-95-0397-1 (eBook)
https://doi.org/10.1007/978-981-95-0397-1

Mathematics Subject Classification: 47A06, 47A08, 47A05, 47A25, 47A11, 47A10, 47A53

This Springer imprint is published by the registered company Springer Nature Singapore Pte Ltd.
The registered company address is: 152 Beach Road, #21-01/04 Gateway East, Singapore 189721, Singapore

In memory of my mother Ayada

—Aymen Ammar

To
my mother Sania,
my father Ali,
my wife Fadoua,
my children Adam and Rahma,
my brothers Sofien and Mohamed Amin,
my sister Elhem, and
all members of my extended family

—Aref Jeribi

Preface

This book introduces spectral theory of bounded and unbounded block matrices of linear relations by focusing on the Fredholm theory in both, Banach and Hilbert spaces, the local spectral theory of multivalued linear operator and the block matrix of linear relations. The main goal of this book is to elaborate some results on the spectral properties of block matrix linear relations with unbounded entries. We present some conditions to prove some Frobenius–Schur decompositions for linear relations and characterize the essential spectra stability of these linear relations. Our findings generalize many known ones in the literature.

The work corresponds to a crowning to multiple outstanding papers published in prestigious journals, and this synthesis could be extremely valuable in terms of opening further fruitful lines of investigation as well as, offering promising future directions. It is to some extent a sequel of the authors recent research on multivalued linear operators, both Banach and Hilbert spaces as well as their spectral theory. There are scarce works handling exclusively spectral theory of bounded and unbounded block matrices of linear relations or their variants.

This book stands for the pioneering document that tackles spectral theory of bounded and unbounded block matrices of linear relations. From this perspective, the producers attempt to contribute to the infrequent creative writing on the theme. Indeed, the smallest needed experience material has been gathered in order to allow a relatively pleasant access to the book. Detailed proofs of all theorems are included and exhibited with accuracy and clarity, allowing students to thoroughly familiarize themselves with all the basic concepts.

Sfax, Tunisia
Riyadh, Saudi Arabia
June 2025

Aymen Ammar
Aref Jeribi

Contents

1 Introduction 1
References 7

2 Algebraic and Topological Properties of Block Multivalued Operator Matrices 13
2.1 Algebraic and Topological Properties 13
2.1.1 The Distance Between Two Nonempty Subsets 13
2.1.2 Canonical Isometry 14
2.1.3 Direct Sum 15
2.2 Kuratowski Measure of Noncompactness 17
2.2.1 The Diameter of a Set 17
2.2.2 Definition 18
2.2.3 Gap Between Two Linear Subspaces 19
2.3 Quasi-Banach Spaces 19
2.4 Vector-Valued Analytic Functions 20
2.5 Relations Sets 21
2.5.1 Definitions and Notations 21
2.5.2 Composition of Relations 24
2.6 Multivalued Linear Operators in Vector Space 26
2.6.1 Definitions and Basic Properties 26
2.6.2 Index and Co-Index of Linear Relation 30
2.7 Algebraic Properties of Linear Relations 31
2.7.1 Sum of Multivalued Linear Operators 31
2.7.2 Extensions and Restrictions of Linear Relation 34
2.7.3 Addition of 2×2 Matrices Linear Relations 37
2.8 Product of Linear Relations 39
2.8.1 Product of two Linear Relations 39
2.8.2 Ascent and Descent of a Linear Relation 44
2.8.3 Product of 2×2 Matrices of Linear Relations 46
2.8.4 Product of 3×3 Matrices of Linear Relations 59

2.9 Quotient Map of Linear Relation 63
2.9.1 Quotient Map of 2×2-Block Matrix Linear Relations 64
2.9.2 Quotient Map of 3×3-block matrix 68
2.10 Norm of Linear Relations 72
2.10.1 Norm of Multivalued Linear Operator 72
2.10.2 Continuity and Openness of a Multivalued Linear Operator 75
2.10.3 Minimum Modulus of Multivalued Operator 81
2.11 Closed Linear Relation 87
2.11.1 Closure and Closed Linear Relation 87
2.11.2 Sum of Closed Linear Relations in Banach Space 91
2.11.3 Composition of Closed Linear Relations 98
2.11.4 Closed Linear Relations in Hilbert Spaces 99
2.12 Closable Linear Relation 104
2.12.1 Definition and Some Auxiliary Results 104
2.12.2 Product of Closable Linear Relations 109
2.13 Diagonally and Off-Diagonally Dominant Block Matrix Linear Relation 117
2.13.1 Definitions and Some Results 117
2.13.2 Closability of Diagonally and Off-Diagonally Dominant Block Matrix Linear Relation 120
2.14 Selection of Linear Relations 123
2.15 Adjoint of Linear Relation 137
2.15.1 Definitions and Properties 137
2.15.2 Hermitian Linear Relation 141
2.15.3 Self-adjointness of 2×2 Block Matrix of Linear Relations 145
2.15.4 The Closeness of 3×3 Diagonal Matrices Linear Relations 154
2.15.5 Adjoint of 3×3 Matrices of Linear Relations 154
2.16 Gap Between Two 2×2 Closed Multivalued Linear Operator Matrices 158
2.16.1 Gap Between Two Linear Relations 158
2.16.2 Relationship Between the Gap and Norm of Multivalued Linear Operator 161
References 175

3 Spectra and Local Spectral Theory for Block Multivalued Operator Matrices 177
3.1 Resolvent Set of Linear Relation 177
3.2 Spectrum of Linear Relation 185
3.2.1 Spectrum 186
3.2.2 Approximate Point Spectrum of Linear Relation 188

3.2.3 Riesz Projection and Isolated Point of the Spectrum of a Linear Relation 190
3.2.4 Discrete Spectrum of a Linear Relation 192
3.3 Pencil Resolvent Set of Linear Relation 198
3.4 The Local Resolvent Set of Linear Relation 200
3.4.1 Local Spectrum Set of Linear Relation 205
3.4.2 Analytic Residuum of Multivalued Linear Operators 207
3.5 Compact Linear Relations 210
3.5.1 Demicompact Linear Relations 213
3.5.2 Relatively Demicompact Linear Relation 217
3.6 Linear Relations with Fredholm Property 226
3.6.1 Linear Relations That are Upper and Lower Semi-Fredholm 226
3.6.2 Closed Linear Relations That are Upper and Lower Semi-Fredholm 227
3.6.3 Stability and Perturbation Results Concerning Linear Relations 233
3.6.4 Fredholm of an Upper Triangular Matrix of Linear Relations 236
3.6.5 Fredholmness of 3×3 Matrices Linear Relations 255
3.6.6 Closed Fredholm and Semi-Fredholm Perturbation Classes 258
3.6.7 Perturbation by Demicompact Linear Relation 264
3.7 Essential Spectra of Linear Relation 269
3.7.1 Definitions and Preliminary Results 269
3.7.2 Pencil Essential Spectra of Multivalued Linear Operators 274
3.8 Local Spectrum of a Linear Relation 280
3.8.1 Local Spectral Subspace of Linear Relations 290
3.8.2 The Single-Valued Extension Property 298
3.8.3 Relationship Between the Spectrum and the Defect Spectrum of a Closed Linear Relation on a Hilbert Space Having SVEP 306
3.8.4 The Analytical Core of Closed Multivalued Linear Operator 310
3.8.5 Glocal Spectral Subspace 314
3.9 Some Properties of the Onto Spectrum of a Linear Relation 321
3.9.1 Characterization of the Onto Spectrum 321
3.9.2 Some Results of an Upper Triangular Linear Relation Matrix $\mathscr{M}_C$ Involving SVEP 325
3.9.3 Local Spectral Properties of $\mathscr{L}_0$ 331
3.9.4 Local Spectral of 3×3 Linear Relations Matrices 344
3.9.5 Spectral Properties of 3×3 Linear Relations Matrices Through the SVEP 348

3.10 The Jeribi Essential Spectrum ... 357
3.10.1 Definition ... 357
3.10.2 Open Question ... 357
References ... 358

4 Pseudospectra and Numerical Range of Linear Relations ... 365
4.1 Pseudospectra of Multivalued Linear Operator ... 365
4.1.1 The ε-Pseudospectra of Linear Relations ... 375
4.1.2 The Pseudospectra and the ε-Pseudospectra of a Sequence of Closed Linear Relations ... 394
4.2 Essential Pseudospectra of Linear Relations ... 401
4.2.1 Some Properties of Pseudospectra of a Linear Relation ... 401
4.2.2 Stability of Essential Pseudospectra of Linear Relations ... 412
4.3 The ε-Essential Pseudospectra of Linear Relations ... 419
4.3.1 The Essential ε-Pseudospectra of Linear and a Sequence Linear Relations ... 422
4.3.2 The Essential ε-Pseudospectra of a Sequence of Linear Relations ... 426
4.4 Pencil Pseudospectra and Pencil Essential Pseudospectra of Linear Relations ... 429
4.4.1 Pencil Pseudospectra of Linear Relations ... 429
4.4.2 Pencil Essential Pseudospectra of Linear Relations ... 437
4.5 Numerical Range of a Linear Relation ... 445
4.5.1 The Numerical Range of a Linear Relation T Associated with A ... 446
4.5.2 The Pseudo-Spectral Theory Associated with the Numerical Range ... 451
4.5.3 The Numerical Range of Two Linear Relations ... 459
4.5.4 The Spectral Theory Associated with the Numerical Range ... 471
References ... 480

5 Frobenius–Schur Factorization and Essential Spectra of Block Multivalued Operator Matrices ... 483
5.1 Frobenius–Schur Factorization for 2×2 Matrices Multivalued Linear Operator ... 483
5.1.1 Closability and Closure of $\mathscr{L}_0$ ... 483
5.1.2 Demicompact of the Block Multivalued Matrix ... 501
5.1.3 Essential Spectra of $\mathscr{L}$ ... 506
5.2 Case Where the Domains Consisting of Vectors Satisfying Certain Relations Between Their Components ... 508
5.2.1 Frobenius–Schur's Decomposition ... 508
5.2.2 Essential Spectra ... 524

5.3 Essential Spectra of a 2×2 Matrix of Linear Relations 526
5.3.1 Essential Spectra of an 2×2 Upper-Triangular Matrix of Linear Relations in Hilbert Space 526
5.3.2 Essential Spectra of 2×2-Block Matrices Linear Relations 541
5.4 The Pencils Essential Spectra of Block Matrices of Multivalued Linear Operators 551
5.4.1 Closedness and Closability of Pencils Matrix of Multivalued Linear Operator 552
5.4.2 Frobenius–Schur's Decomposition of the Pencils 2×2 Matrix of Multivalued Linear Operator 555
5.4.3 The Pencils Essential Spectra of $\mathscr{L}$ 570
5.5 Essential Spectra of 3×3-Block Matrices Linear Relations 574
5.5.1 The Closedness and Closability of 3×3 Matrices Linear Relations 574
5.5.2 Frobenius–Schur Factorization for Multivalued 3×3 Matrices Linear Operator 576
5.5.3 Essential Spectra 3×3 Matrices Linear Operator 596
5.5.4 Essential Spectra of Upper-Triangular Matrix 599
References 605

Index 607

Symbols

$\mathbb{N}$	The set of natural numbers
$\mathbb{R}$, $\mathbb{C}$	The fields of real and complex numbers, respectively
$\mathbb{R}^n$	The n-dimensional real space
$\widetilde{X}$	The completion of the normed space X
$\inf(M)$	The infimum of the set M
$\sup(M)$	The supremum of the set M
$\widetilde{U}$	The completion of U
B_r	The ball with radius r
B_X	The unit ball in X
$\text{dist}\,(x, y)$	The distance between x and y
X^*	The dual of X
$\dim(X)$	The dimension of the space X
$\|x\|$	The norm of x
$(X, \|.\|)$	The linear normed space X
$(x_n)_n$	The sequence $(x_n)_n$
$G(U)$	The graph of U
$\mathscr{C}(\Omega, \mathbb{R})$	The space of real continuous functions on Ω
L_p	The L_p space
$\mathscr{L}(X, Y)$	The space of linear operators from X into Y
$\mathscr{L}(X)$	The space of linear operators from X into X
$\rho(U)$	The resolvent set of U
$\sigma(U)$	The spectrum of U
$\rho_V(U)$	The V-resolvent set of U
$\sigma_V(U)$	The V-spectrum of U
$U_{\vert M}$	The restriction of U to M
$\mathscr{W}(X, Y)$	The set of weakly compact linear operator from X into Y
$\mathscr{R}(X, Y)$	The collection of relations from X into Y
$L\mathscr{R}(X, Y)$	The collection of linear relations from X into Y
$B\mathscr{R}(X, Y)$	The collection of bounded linear relations from X into Y
$C\mathscr{R}(X, Y)$	The class of all closed linear relations from X into Y

$BC\mathscr{R}(X, Y)$	relation fromThe class of all bounded and closed linear relations from X into Y
$K\mathscr{R}(X, Y)$	The families of all compact linear relations from X into Y
$WK\mathscr{R}(X, Y)$	relation fromThe families of all weakly compact linear relations from X into Y
$\mathscr{D}\mathscr{C}(X, Y)$	The families of all demicompact linear relations from X into Y
$N(U)$	The null space of U
$R(U)$	The range space of U
$\mathscr{P}\mathscr{R}\mathscr{K}(X, Y)$	The families of all precompact linear relations from X into Y
$\mathscr{S}\mathscr{S}\mathscr{R}(X, Y)$	The set of strictly singular linear relations from X into Y
J_M^X	The injection operator from M into X
$x_n \to x$	x_n converges strongly to x
$\overline{M}$	The closure of the set M
$co(\cdot)$	The convex hull
$\overline{co}(\cdot)$	The closed convex hull
$\mathscr{P}\mathcal{R}(\Phi_+(X, Y))$	The set of upper semi-Fredholm perturbations of linear relations from X into Y
$\mathscr{P}\mathcal{R}(\Phi_-(X, Y))$	The set of lower semi-Fredholm perturbations of linear relations from X into Y
$\mathscr{P}\mathscr{R}(\mathscr{A}_\alpha(X, Y))$	The set of α-Atkinson perturbations linear relations from X into Y
$\mathscr{P}\mathscr{R}(\mathscr{A}_\beta(X, Y))$	The set of β-Atkinson perturbations linear relations from X into Y
$\mathscr{F}(X, Y)$	The set of Fredholm linear relations from X into Y
$\mathscr{F}_+(X, Y)$	The set of upper semi-Fredholm linear relations from X into Y
$\mathscr{F}_-(X, Y)$	The set of lower semi-Fredholm linear relations from X into Y
$\Phi_+(X, Y)$	The set of upper semi-Fredholm closed linear relations from X into Y
$\Phi_-(X, Y)$	The set of lower semi-Fredholm closed linear relations from X into Y
$\mathscr{W}^l(X, Y)$	The set of left Weyl linear relations from X into Y
$\mathscr{W}^r(X, Y)$	The set of right Weyl linear relations from X into Y

Chapter 1
Introduction

The notion of a multivalued operator rests upon generalizing the abstraction of a linear operator (single-valued) to that of a multivalued operator. E. A. Coddington (see [1]) and R. Arens was introduced this idea who provided a scientific treatment in [2]. Since then, it has been recognized as useful in miscellaneous fields and has been the subject of studies in multiple specific contexts, (see [3]). Currently, the evaluation of multivalued linear operator in Hilbert or Banach spaces is of great significance since it displays promising applications in several difficulties in physics and dissimilar areas of applied mathematics.

A. Favini and A. Yagi, in their paper [4], announce the concept of multivalued linear operators or linear relation as an method toward degenerating evolution equations with estimated to the time derivative. The notion of linear relations (or multivalued linear operators) were started into functional analysis by J. Von Neumann as long ago 1950 (see [5]). From a mathematical outlook, it is quite fascinating to tackle differential equations controlled by non-densely defined operators. One of the principal aims of this theory is to explore the closure, the completion and the inverse of linear relations. Besides, linear relations have been a hot area of research. Surely, number of research works have been oriented particularly to this area like [6–35].

The classical single-valued linear operators theory does not involve some cases that include the adjoint of a non-densely defined linear operator and the inverse of a non-invertible operator. Thereupon, a theory of multivalued linear operators, also called linear relations, and non-densely defined Hermitian operators seems to be highly needed. It is clear that the adjoint of a non-densely defined linear operator is multivalued. Hence, it is important to extend certain results of spectral theory to the multivalued linear case as along with to obtain perturbation theorems for multivalued linear operators. As in the situation of linear operators, the spectral theory of linear relations, involving the associated analytic functional calculus, corresponds to a significant tool to study diversified properties of these objects and for deriving diverse of their applications.

A. Ammar and A. Jeribi, *Spectral Theory of Block Multivalued Operator Matrices*, Infosys Science Foundation Series, https://doi.org/10.1007/978-981-95-0397-1_1

Let X and Y be two Banach vector spaces over the complex plane $\mathbb{C}$. A linear relation U from X into Y is a mapping from a subspace

$$\mathscr{D}(U) = \{u \in X \text{ such that } Uu \neq \emptyset\}$$

of X, called the domain of U, into the collection of nonempty subsets of Y such that

$$U(\alpha u + \beta v) = \alpha U(u) + \beta U(v),$$

for all α, β nonzero scalars in $\mathbb{C}$ and u, $v \in \mathscr{D}(U)$. If U maps the points of its domain to singletons, then U is said to be a single-valued linear operator or simply an operator. We denote by $L\mathscr{R}(X, Y)$ the class of linear relations from X into Y and we write $L\mathscr{R}(X) = L\mathscr{R}(X, X)$. A linear relation $U \in L\mathscr{R}(X, Y)$ is uniquely determined by its graph, $G(U)$, which is established by

$$G(U) = \left\{ \begin{pmatrix} u \\ v \end{pmatrix} \in X \times Y : u \in \mathscr{D}(U),\ v \in Uu \right\}.$$

Therefore, we can recognize U with $G(U)$. The opposite of U is the linear relation U^{-1}. The graph of U^{-1} is established as

$$G(U^{-1}) = \left\{ \begin{pmatrix} v \\ u \end{pmatrix} \text{ such that } \begin{pmatrix} u \\ v \end{pmatrix} \in G(U) \right\}.$$

Let K and L be two subspaces of X, then

$$K^{\perp} = \left\{u' \in X^* \text{ such that } u'(k) = 0,\ \forall k \in K\right\}$$

and

$$L^{\top} = \left\{v \in X \ : \ l'(v) = 0,\ \forall\, l' \in L\right\},$$

where X^* is the dual of X. The adjoint U^* of the linear relation U is defined as follows:

$$G(U^*) = G(-U^{-1})^{\perp}.$$

Thus, we means that $(v', u') \in G(U^*)$ if, and only if, $v'(v) - u'(u) = 0$ for all $(u, v) \in G(U)$. Hence, $u' \in U^* v'$ if, and only if, $v'(Uu) = u'(u)$ for all $u \in \mathscr{D}(U)$, where

$$\begin{aligned} \langle (v, u), (v', u') \rangle &= \langle u, u' \rangle + \langle v, v' \rangle \\ &= u'(u) + v'(v). \end{aligned}$$

We can therefore characterize the adjoint by

$$G(U^*) = \left\{ \begin{pmatrix} v' \\ u' \end{pmatrix} \in Y^* \times X^* \ : \ u' \text{ is an extension of } v'U \right\}.$$

In 2013, A. Ammar, T. Àlvarez, and A. Jeribi (see [7]) studied the essential spectra of a 2×2-block matrix of linear relations. Additionally, A. Ammar, A. Jeribi, and B. Saadaoui integrated some concepts for a 2×2-block relations matrix in a Banach space. In particular, they established the Frobenius–Schur of relations matrices under certain conditions (see [36]). In [37], A. Ammar, S. Fakhfakh, and A. Jeribi extended initially the main results of Jeribi [38] and Tretter [39–41] to a large class that is the class of multivalued Block matrices operator in Banach spaces. For a given block matrix $\mathscr{A}$,

$$\mathscr{A} := \begin{pmatrix} A & B \\ C & D \end{pmatrix},$$

where the entries are linear operators. C. Tretter characterized diagonally dominant and off-diagonally dominant block operator matrices and invested these notions to establish criteria for an operator matrix to be closed and closable. Departing from this definition, we shall extend this result in the first place to entries linear relations A, B, C and D. We are chiefly concerned with investigation of the closedness and closability of $\mathscr{A}$ by applying the definition of the diagonally dominant and off-diagonally dominant block matrix $\mathscr{A}$. In the second place, we need to apply perturbation theorems to upper semi-Fredholm relations so as to assess the stability properties of the essential spectra under small perturbations (see [7, 8, 10, 42–66]). T. Álvarez, A. Ammar, and A. Jeribi in their work [7] attempted to extend these pertinent results for block operator matrix to the multivalued operators as well as the unbounded block operator matrix. Our basic objective lies in using information about the entries A, B, C, and D to investigate various spectral properties of the block operator matrix $\mathscr{A}$.

In [36], A. Ammar, A. Jeribi, and B. Saadaoui tackled the algebraic and topological properties of block multivalued operator matrices. In the theory of unbounded matrices of linear relations, the Frobenius–Schur factorization corresponds to a basic tool serving to handle the spectrum and various spectral properties. This was first recognized by Álvarez et al. in [7]. Independently and under slightly different assumptions, it was elaborated later in [36]. In fact, Ammar et al. in [51] imposed some conditions on $F(\mu) := C(A - \mu I)^{-1}$, $E(\mu) := \overline{(A - \mu I)^{-1} B}$ and $M(\mu) := D - C(A - \mu I)^{-1} B$ with $\mu \in \rho(A)$, under which we get the following factorization:

$$\overline{\mathscr{A}} = \mu I + \begin{pmatrix} I & 0 \\ F(\mu) & I \end{pmatrix} \begin{pmatrix} A - \mu I & 0 \\ 0 & \overline{M(\mu)} - \mu I \end{pmatrix} \begin{pmatrix} I & E(\mu) \\ 0 & I \end{pmatrix},$$

where $\overline{\mathscr{A}}$ is the closure of $\mathscr{A}$ and $\rho(A)$ is the resolvent set of A. As matter of fact, our reasoning shall be mainly focused on the specific properties of the first entry A, allowing us therefore to get a closable block relation matrix $\mathscr{A}$.

In 1952, N. Dunford, has introduced the notion of single-valued extension property (SVEP) and some results of local spectral theory for bounded operators in Banach spaces [67]. For a survey findings related to local spectral theory of linear operators, the reader is referred to [68–82]. Let λ_0, be a complex number. An operator U is said to have the single valued property at λ_0, if the only analytic function in a neighborhood of λ_0 that verifies

$$(\lambda - U)f(\lambda) = 0$$

is $f = 0$ (see [83]). This notion has a fundamental in local spectral theory. After, it is contented by a great variety of spectral decomposition problem of linear bounded operators. Furthermore, the single-valued extension property plays an intrinsic role in Fredholm theory. We can cite P. Aiena and her co-authors [84–93]. We refer the reader to multiple works which have handled different notions and several aspects of local spectral theory, such as [94–97]. In 2020, A. Ammar, A. Bouchekoua, and A. Jeribi extended this concept to the closed linear relation [98]. More precisely, he establish several results of Fredholm theory, based on the SVEP. Further, Ammar et al. [98], Mnif and Ouled-Hmed [99–101], extended this concept to multivalued linear operator. Further, let x be a vector in X and U be a bounded linear relation, we define the local resolvent of U at x as the set of all complex number λ for which there exist an open neighborhood $\mathcal{V}$ of λ and an analytic function $f : \mathcal{V} \longrightarrow X$ such that the following equation:

$$(\lambda_0 - U)f(\lambda_0) = x + U(0),$$

holds for all $\lambda_0 \in \mathcal{V}$, when $U(0)$ is the multivalued part of $\mathcal{V}$. We denote by $\rho_U(x)$ the local resolvent of U at x. The complement of $\rho_U(x)$ in $\mathbb{C}$ is called the local spectrum of U at x and is denoted by $\sigma_U(x)$. In [70], A. Ammar, S. Fakhfakh, and D. Kouas investigated some properties of local spectral theory for upper triangular 2×2-block matrix of linear relations. Also, in [68–70] A. Ammar et al. checked up on a few properties of the local spectra of a 2×2-block matrix of linear relations.

Over the last years, the study of the spectrum of an operator was elaborated by means of the local analysis of the spectrum. Numerous works were devoted to the investigation of the local spectral theory of bounded or closed operators on Banach spaces (see [102, 103]). One of the most outstanding tools in the latter theory is called the single-valued extension property (abbreviated SVEP). This notion was firstly coined by Dunford [67] and explored by several mathematicians. We can state Finch [83], Laursen and Neumann [103] and Aiena and her co-authors [84–93]. We refer the reader to Finch's research [83] for the definition of the single-valued extension property of the closed linear operator U on Banach space. The operator U is said to have the SVEP, if U has the SVEP at every element in $\mathbb{C}$. This property has been investigated by multiple mathematicians in the case of bounded operators defined on a Banach space (see [104]). The basic significance of the SVEP resides in its connection with some basic spaces of the local spectral theory. One of the main objectives of this book is to explore the local spectral theory of multivalued linear operators. Motivated by the methods invested in the study of the local spectral theory of linear operators, we introduce and analyze some basic concepts of the local spectral theory, namely the local spectral, the analytic spectral subspace, the glocal space, and the analytic core, for closed linear relations (or multivalued linear operator) on a Banach space. Subsequently, we characterize them by means of the single-valued extension property.

In this book, resting upon [105, 106], we investigate certain properties of local spectral theory of upper triangular linear relations matrices $\mathscr{M}_C$ which are defined by

$$\mathscr{M}_C = \begin{pmatrix} A & C \\ 0 & B \end{pmatrix},$$

on $X \bigoplus Y$, where A, B, and C are linear relations. Subsequently, enlightened by the definitions and the notions used in the study of Ammar et al. [98], we shall investigate and characterize the local spectral properties of $\mathscr{M}_C$. Specifically, we have given a characterization of $\mathscr{M}_C$, when one will the relation B has the SVEP as indicated by the following equality:

$$\sigma(\mathscr{M}_C) = \sigma(A) \bigcup \sigma(B), \quad \text{for all } C \in L\mathscr{R}(Y, X),$$

where $\sigma(\mathscr{M}_C)$ is the spectrum of $\mathscr{M}_C$ and $\sigma(B)$ is the spectrum of B. After that, we shall provide some new conditions under which the matrix $\mathscr{M}_C$ has the SVEP.

The concept of numerical range of an operator on Banach and Hilbert spaces treated by several researchers (see [107]). There are many ways to define the last concept of closed linear operators in a Hilbert space. Among them we are interested of the numerical range of a closed linear operator A which is the range of the restriction to the unit sphere of the quadratic form associated with A. This is a convex subset of the complex plane whose closure contains its spectrum. In this book, we introduce the new concept of the numerical range of a linear relation and a linear operator on Hilbert spaces and study its properties. After that, we establish some results describing the relationship between the various concepts of spectral theory and the numerical range. The classical definition of the numerical range of a closed linear operator in a Hilbert space is the range of the restriction to the unit sphere of the quadratic form associated with it. Among the most important properties of the numerical range proved by a few authors are that it is convex and that its closure contains the spectrum of the operator.

In [36, 108], A. Ammar et al. examined a 3×3-block matrices of linear relations on a product of Banach space. It was demonstrated that, under certain conditions, this block matrix defines a closable operator and its essential spectra are determined. In [109], A. Ammar, T. Diagana, and A. Jeribi explored the spectral properties of a 3×3 block matrices of linear relations

$$\mathscr{L} := \begin{pmatrix} A & B & C \\ D & E & F \\ G & H & L \end{pmatrix},$$

with unbounded entries and with a domain consisting of vectors satisfying certain relations between their components. This book is devoted to study some properties and characterizations of spectra (resp. essential spectra) of the block matrix of linear relations $\mathscr{L}$. Specifically, we considered some properties of local spectral theory that can be used to develop in case of block matrix of multivalued linear operators. In particular, we provide some sufficient conditions that ensure that the spectra of $\mathscr{L}$ is equal to the union of the spectra of the linear relations A and D. What's more, we scrutinize the relationship between the essential spectra about the diagonal relations A and D and the new sets $\Xi_{\pm}(\mathscr{L})$, $\Xi_{+}(\mathscr{L})$, $\Xi_{-}(\mathscr{L})$, and $\Xi_{0}(\mathscr{L})$. In this note, we prove equality up to these sets between the essential spectra of $\mathscr{L}$ and the essential spectra of A and D. Finally, we determined the adjoint of $\mathscr{L}$ and we characterize its spectrum.

Basically, our book involves four chapters.

This chapter outlines a selection of known properties of linear relations acting on normed spaces. Multiple results from the theory of block matrices of linear relations are set forward. In particular, results concerning the sum, the product, the selection, the perturbation results, and the gap of 2×2 closed multivalued linear operator matrices are displayed. Also, the product of 3×3-block operators matrices of linear relations are studied. Relations are provided in terms of their induced linear operators. In Chap. 2, we depict and analyze some concepts of the local spectral theory: the local spectral, the analytic spectral subspace, the glocal space and the analytic core for closed linear relations (or multivalued linear operator) on a Banach space. Afterward, we characterize them by means of the single valued extension property. After that, some perturbation results of unbounded 2×2-block matrices of linear relations are reported. Initially, we extend the main results of C. Tretter in [39–41] to linear relations. Additionally, we set forward a matrix linear relation for diagonally dominant and off-diagonally dominant of block matrix linear relations $\mathscr{A}$, we provide a necessary and sufficient condition for $\mathscr{A}$ to become closed and closable and we also attempt to determine the conditions down which a block matrix linear relations happen a demicompact block matrix linear relation. Basically, the operators occurring in the representation of $\mathscr{A}$ are unbounded. In Chap. 3, we study pseudospectra of linear relation, and we introduce the new concept of the numerical range of two linear relations on Hilbert spaces. After that, we examine the aspects of the coupled numerical range for a linear relation and a linear operator on Hilbert spaces. Besides, we study its properties. After that, we establish some results describing the relationship between the various concepts of spectral theory and the numerical range. The intrinsic target of Chap. 4 resides in identifying some hypotheses which should allow the block matrix linear relations $\mathscr{A}$ to be closable. From this perspective, it is interesting to determine to present the suitable conditions for the entries A, B, C, and D. These conditions rely mainly upon the Frobenius–Schur decomposition of $\mathscr{A}$. Subsequently, we shall present the essential spectra of the closure of block matrix linear relations $\mathscr{A}$. Eventually, in this chapter we study the essential spectra of 3×3 block matrices of linear relations.

References

1. Coddington, E.A.: Extension theory of formally normal and symmetric subspaces. In: Memoirs of the American Mathematical Society, vol. 134. American Mathematical Society, Providence, R.I (1973)
2. Arens, R.: Operational calculus of linear relations. Pac. J. Math. **11**, 9–23 (1961)
3. Alaarabiou, El H., Benilan, P.: Sur le carré d'un operateur non linéaire. Arch. Math. **66**, 335–343 (1996)
4. Favini, A., Yagi, A.: Multivalued operators and degenerate evolution equations. Ann. Math. Pura. Appl. **163**(4), 353–384 (1993)
5. Von Neumann, J.: Functional Operators. II. The geometry of orthogonal spaces. In: Annals of Mathematics Studies, vol. 22. Princeton University Press, Princeton, N.J. (1950); Rend. Circ. Mat. Palermo. (27), 373–392 (1909)
6. Abdmouleh, F., Àlvarez, T., Jeribi, A.: On a characterization of the essential spectra. Filomat **36**(9), 2921–2938 (2022)
7. Álvarez, T., Ammar, A., Jeribi, A.: On the essential spectra of some matrix of linear relations. Math. Methods Appl. Sci. **37**(5), 620–644 (2014)
8. Ammar, A., Boutaf, F.Z., Jeribi, A.: Essential spectra in non-archimedean fields. Mat. Stud. **58**(1), 82–93 (2022)
9. Ammar, A., Boutaf, F.Z., Jeribi, A.: Some results of essential spectra of sum of two bounded linear operators in non-Archimedean Banach space. Bol. Soc. Mat. Mex. (3), 29; (2), Paper No. 33 (2023)
10. Ammar, A., Daoud, H., Jeribi, A.: Pseudospectra and essential pseudospectra of multivalued linear relations. Mediterr. J. Math. **12**(4), 1377–1395 (2015)
11. Ammar, A., Ezzadam, A., Jeribi, A.: A characterization of the essential spectra of 2×2-block matrices of linear relations. Bull. Iranian Math. Soc. **48**(5), 2463–2485 (2022)
12. Ammar, A., Fakhfakh, S., Jeribi, A.: Some results on the stability of $\sigma_{eap}(\cdot)$ and $\sigma_{edelta}(\cdot)$ of linear relations. Funct. Anal. Approx. Comput. **14**(2), 51–62 (2022)
13. Ammar, A., Jeribi, A., Saadaoui, B.: Some decomposition for multivalued matrix pencils linear operator. Bol. Soc. Mat. Mex. (3) 27; (1), Paper No. 17, 23 (2021)
14. Ellouz, H., Feki, I., Jeribi, A.: On the closedness and self-adjointness of 3×3 operator matrices and application to a non-relativistic three-channel potential scattering model. Asymptot. Anal. **130**(3–4), 409–426 (2022)
15. Jeribi, A., Krichen, B., Salhi, M.: Characterization of essential spectra involving weakly demicompact operators on Banach spaces. Bull. Iranian Math. Soc. **48**(4), 1513–1538 (2022)
16. Jeribi, A., Moalla, N., Yengui, S.: Perturbation theory, M-essential spectra of 2×2 operator matrices and application to transport operators. J. Math. Appl. **44**, 35–55 (2021)
17. Latrach, K., Jeribi, A.: A nonlinear boundary value problem arising in growing cell populations. Nonlinear Anal. **36**(7), Ser. A: Theory Methods, 843–862 (1999)
18. Latrach, K., Jeribi, A.: Some results on Fredholm operators, essential spectra, and application. J. Math. Anal. Appl. **225**(2), 461–485 (1998)
19. Latrach, K., Jeribi, A.: Sur une équation de transport intervenant en dynamique des populations. (French) [On a transport equation arising in growing cell populations] C. R. Acad. Sci. Paris Sér. I Math. **325**(10), 1087–1090 (1997)
20. Latrach, K., Jeribi, A.: On the essential spectrum of transport operators on L1-spaces. J. Math. Phys. **37**(12), 6486–6494 (1996)
21. Jeribi, A., Krichen, B., Zitouni, A.: Spectral properties for γ-diagonally dominant operator matrices using demicompactness classes and applications. Rev. R. Acad. Cienc. Exactas Fís. Nat. Ser. A Mat. RACSAM **113**(3), 2391–2406 (2019)
22. Jeribi, A., Krichen, B., Zitouni, A.: Properties of demicompact operators, essential spectra and some perturbation results for block operator matrices with applications. Linear Multilinear Algebra **68**(12), 2506–2522 (2020)

23. Jeribi, A., Latrach, K.: Quelques remarques sur le spectre essentiel et application à l'équation de transport. (French) [Some remarks on the essential spectrum and application to the transport equation] C. R. Acad. Sci. Paris Sér. I Math. **323**(5), 469–474 (1996)
24. Jeribi, A., Latrach, K., Megdiche, H.: Time asymptotic behavior of the solution to a cauchy problem governed by a transport operator. J. Integral Equ. Appl. **17**(2), 121–139 (2005)
25. Jeribi, A., Mahfoudhi, K.: Generalized Drazin-meromorphic pseudospectrum for a bounded linear operator on a Banach space. Rend. Circ. Mat. Palermo (2) 72; (2), 1303–1314 (2023)
26. Jeribi, A., Megdiche, H., Moalla, N.: On a transport operator arising in growing cell populations. II. Cauchy problem. Math. Methods Appl. Sci. **28**(2), 121–145 (2005)
27. Jeribi, A., Mnif, M.: Fredholm operators, essential spectra and application to transport equations. Acta Appl. Math. **89**(1–3), 155–176 (2006)
28. Jeribi, A., Moalla, N.: Fredholm operators and Riesz theory for polynomially compact operators. Acta Appl. Math. **90**(3), 227–245 (2006)
29. Jeribi, A., Moalla, N.: A characterization of some subsets of Schechter's essential spectrum and application to singular transport equation. J. Math. Anal. Appl. **358**(2), 434–444 (2009)
30. Jeribi, A., Moalla, R.: Nonlinear alternatives of Leray-Schauder type in Banach algebra involving four operators with application. Numer. Funct. Anal. Optim. **34**(10), 1097–1114 (2013)
31. Jeribi, A., Moalla, N., Walha, I.: Spectra of some block operator matrices and application to transport operators. J. Math. Anal. Appl. **351**(1), 315–325 (2009)
32. Jeribi, A., Moalla, N., Yengui, S.: S-essential spectra and application to an example of transport operators. Math. Methods Appl. Sci. **37**(16), 2341–2353 (2014)
33. Jeribi, A., Ould Ahmed Mahmoud, S.A., Sfaxi, R.: Time asymptotic behaviour for a one-velocity transport operator with Maxwell boundary condition. Acta Appl. Math. **98**(3), 163–179 (2007)
34. Jeribi, A.: Résolution de divers problèmes elliptiques par des méthodes d'éléments finis - Cours, exercices, et problèmes corrigés. Edition Ellipses, p. 444 (2021). ISBN 978-2-340-06117-0
35. Jeribi, A., Walha, I., Gustafson, W., Kato, W.: Schechter and Browder essential spectra of some matrix operator and application to two-group transport equation. Math. Nachr. **284**(1), 67–86 (2011)
36. Ammar, A., Dhahri, M.Z., Jeribi, A.: Some properties of upper triangular 3×3-block matrices of linear relations. Boll. Unione Mat. Ital. **8**(3), 189–204 (2015)
37. Ammar, A., Fakhfakh, S., Jeribi, A.: Stability of the essential spectrum of the diagonally and off-diagonally dominant block matrix linear relations. J. Pseudo-Differ. Oper. Appl. **7**(4), 493–509 (2016)
38. Jeribi, A.: Spectral Theory and Applications of Linear Operators and Block Operator Matrices. Springer, New York (2015)
39. Tretter, C.: Spectral issues for block operator matrices. In: Differential equations and mathematical physics (Birmingham, al, 1999), vol. 16 of AMS-IP Stud. Adv. Math., 407–423. Amer. Math. Soc., Providence, RI (2000)
40. Tretter, C.: Spectral Theory of Block Operator Matrices and Applications. Imperial College Press, London (2008)
41. Tretter, C.: Spectral inclusion for unbounded block operator matrices. J. Func. Anal. **11**, 3806–3829 (2009)
42. Álvarez, T.: On almost semi-Fredholm linear relations in normed spaces. Glasg. Math. J. **47**(1), 187–193 (2005)
43. Álvarez, T., Ammar, A., Jeribi, A.: A characterization of some subsets of S-essential spectra of a multivalued linear operator. Colloq. Math. **135**(2), 171–186 (2014)
44. Álvarez, T., Benharrat, M.: Relationship between the Kato spectrum and the Goldberg spectrum of a linear relations. Mediterr. J. Math. **13**, 365–378 (2015)
45. Álvarez, T., Cross, R.W., Danie, W.: Multivalued Fredholm type operators with abstract generalised inverses. J. Math. Anal. Appl. **261**(1), 403–417 (2001)

46. Álvarez, T., Gonzalez, M.: Paracomplete normed spaces and Fredholm theory. Rend. Circ. Mat. Palermo **2**(28), 257–264 (1999)
47. Álvarez, T., Wilcox, D.: Perturbation theory of multivalued Atkinson operators in normed spaces. Bull. Austral. Math. Soc. **76**(2), 195–204 (2007)
48. Ambrozie, C.-G., Vasilescu, F.-H.: Banach Space Complexes, Mathematics and Its Applications, vol. 334. Kluwer Academic Publishers Group, Dordrecht (1995)
49. Ammar, A., Ben Ezzadam, A., Jeribi, A.: Some results of Fredholm perturbations of multivalued linear operator in normed spaces. Filomat **37**(5), 1399–1408 (2023)
50. Benhamadou, M., Jeribi, A.: Analyse numérique matricielle - Méthodes et algorithmes, exercices et problèmes corrigés, Edition Ellipses, p. 680 (2020). ISBN 978-2-340-03803-5
51. Ammar, A., Jeribi, A., Saadaoui, B.: Fredholm linear relation and some results of demicompact for multivalued matrix linear operator. Filomat **37**(5), 1507–1522 (2023)
52. Belabbaci, C.: Jeribi essential spectrum. Lib. Math. (N.S.) **37**(1), 65–73 (2017)
53. Athmouni, N., Baloudi, H., Jeribi, A., Kacem, G.: On weighted and pseudo-weighted spectra of bounded operators. Commun. Korean Math. Soc. **33**(3), 809–821 (2018)
54. Belabbaci, C.: The S-Jeribi essential spectrum. Ukraïn. Mat. Zh. **73**(3), 308–313 (2021)
55. Benkhaled, H., Elleuch, A., Jeribi, A.: Demicompactness results for strongly continuous semigroups, generators and resolvents. Mediterr. J. Math. **15**(2), Paper No. 71, 19 (2018)
56. Benkhaled, H., Elleuch, A., Jeribi, A.: The class of order weakly demicompact operators. Rev. R. Acad. Cienc. Exactas Fís. Nat. Ser. A Mat. RACSAM **114**(2), Paper No. 76, 8 (2020)
57. Benkhaled, H., Elleuch, A., Jeribi, A.: Uniformly continuous cosine families properties around weak demicompactness concept. Hacet. J. Math. Stat. **51**(6), 1674–1679 (2022)
58. Benkhaled, H., Elleuch, A., Jeribi, A.: Demicompactness properties for uniformly continuous cosine families. Mediterr. J. Math. **19**(4), 157 (2022)
59. Benkhaled, H., Hajji, M., Jeribi, A.: L-weakly and M-weakly demicompact operators on Banach lattices. Filomat **36**(13), 4319–4329 (2022)
60. Benkhaled, H., Hajji, M., Jeribi, A.: On the class of demi Dunford-Pettis operators. Rend. Circ. Mat. Palermo (2) **72**(2), 901–911 (2023)
61. Benkhaled, H., Jeribi, A.: The class of demi KB-operators on Banach lattices. Turkish J. Math. **47**(1), 387–396 (2023)
62. Ben Amar, A., Jeribi, A., Krichen, B.: Essential spectra of a 3×3 operator matrix and an application to three-group transport equations. Integral Equ. Oper. Theory **68**(1), 1–21 (2010)
63. Ben Amar, A., Jeribi, A., Mnif, M.: Some results on Fredholm and semi-Fredholm perturbations. Arab. J. Math. (Springer) **3**(3), 313–323 (2014)
64. Ben Brahim, F., Jeribi, A., Krichen, B.: Polynomially demicompact operators and spectral theory for operator matrices involving demicompactness classes. Bull. Korean Math. Soc. **55**(5), 1351–1370 (2018)
65. Ben Brahim, F., Jeribi, A., Krichen, B.: Spectral theory for polynomially demicompact operators. Filomat **33**(7), 2017–2030 (2019)
66. Ben Brahim, F., Jeribi, A., Krichen, B.: Spectral properties for polynomial and matrix operators involving demicompactness classes. Extracta Math. **33**(1), 11–32 (2018)
67. Dunford, N.: Spectral theory. II. Resolutions of the identity. Pac. J. Math. **2**, 559–614 (1952)
68. Ammar, A., Bouchekoua, A., Jeribi, A.: The local spectral theory for linear relations involving SVEP. Mediterr. J. Math. **18**(2), Paper No. 77, 27 (2021)
69. Ammar, A., Bouchekoua, A., Lazrag, N.: Aiena's local spectral theory for a block matrix linear relations through localized SVEP. Rendiconti del Circolo Matematico di Palermo Series **2** (2022)
70. Ammar, A., Fakhfakh, S., Kouas, D.: The local spectral theory for upper triangular matrix linear relations involving SVEP. Rendiconti del Circolo Matematico di Palermo Series **2**, 1–14 (2021)
71. Chamkha, Y., Mnif, M.: Browder spectra of upper triangular matrix linear relations. Publ. Math. Debrecen **82**(3–4), 569–590 (2013)
72. Lajnef, M., Mnif, M.: Isolated spectral points of a linear relation. Monatsh. Math. **191**, 595–614 (2020)

73. Lajnef, M., Mnif, M.: On generalized Drazin invertible linear relations. Rocky Mountain J. Math. **50**(4), 1387–1408 (2020)
74. Lajnef, M., Mnif, M.: Upper and lower generalized Drazin invertible linear relations. Anal. Mathematica **48**, 779–802 (2022)
75. Lajnef, M., Mnif, M.: On isolated points of the approximate point spectrum of a closed linear relation. Extracta Mathematica **37**, 75–90 (2022)
76. Jeribi, A.: Quelques remarques sur les opérateurs de Fredholm et application à l'équation de transport. (French) [Some remarks on Fredholm operators and application to the transport equation] C. R. Acad. Sci. Paris Sér. I Math. **325**(1), 43–48 (1997)
77. Jeribi, A.: A characterization of the essential spectrum and applications. J. Boll. dell. Union. Mat. Ital. J. Boll. dell. Union. Mat. Ital. (8) 5-B, 805–825 (2002)
78. Jeribi, A.: A characterization of the Schechter essential spectrum on Banach spaces and applications. J. Math. Anal. Appl. **271**, 343–358 (2002)
79. Jeribi, A.: On the Schechter essential spectrum on Banach spaces and application. Facta Univ. Ser. Math. Inform. **17**, 35–55 (2002)
80. Jeribi, A.: A nonlinear problem arising in the theory of growing cell populations. Nonlinear Anal. Real World Appl. **3**(1), 85–105 (2002). (Reviewer: Mohamed Boulanouar) 45K05 (47N20 92D20)
81. Jeribi, A.: Time asymptotic behaviour for unbounded linear operator arising in growing cell populations. Nonlinear Anal. Real World Appl. **4**(5), 667–688 (2003)
82. Jeribi, A.: Fredholm operators and essential spectra. Arch. Inequal. Appl. **2**(2–3), 123–140 (2004)
83. Finch, J.K.: The single-valued extension property on a Banach space. Pac. J. Math. **58**(1), 61–69 (1975)
84. Aiena, P.: Fredholm And Local Spectral Theory, with Applications to Multipliers, pp. xiv+444. Kluwer Academic Publishers, Dordrecht (2004). ISBN: 1-4020-1830-4
85. Aiena, P.: Semi-Fredholm operators, perturbation theory and localized SVEP. Caracas, Venezuela (2007)
86. Aiena, P.: Fredholm and local spectral theory II. With application to Weyl-type theorems. Lecture Notes in Mathematics, vol. 2235, pp. xi+544 Springer, Cham (2018). ISBN: 978-3-030-02265-5, 978-3-030-02266-2
87. Aiena, P., Aponte, E., Bazan, E.: Weyl type theorems for left and right polaroid operators. Integral Equ. Oper. Theory **66**, 1–20 (2010)
88. Aiena, P., Guillen, J.R.: Weyl's theorem for perturbations of paranormal operators. Proc. Am. Math. Soc. **35**, 2433–2442 (2007)
89. Aiena, P., Trapani, C., Triolo, S.: SVEP and local spectral radius formula for unbounded operators. Filomat **28**(2), 263–273 (2014)
90. Aiena, P., Trapani, C., Triolo, S.: SVEP and local spectral radius formula for unbounded operators. Filomat **28**(2), 263–273 (2014)
91. Aiena, P., Triolo, S.: Fredholm spectra and Weyl type theorems for Drazin invertible operators. Mediterr. J. Math. **13**(6), 4385–4400 (2016)
92. Aiena, P., Triolo, S.: Local spectral theory for Drazin invertible operators. J. Math. Anal. Appl. **435**(1), 414–424 (2016)
93. Aiena, P., Triolo, S.: Weyl-type theorems on Banach spaces under compact perturbations. Mediterr. J. Math. **15**(3), Paper No. 126, 18 (2018)
94. Finch, J.K.: The single-valued extension property on a Banach space. Pac. J. Math. **58**(1), 61–69 (1975)
95. Laursen, K.B., Neumann, M.M.: On analytic solutions of the equation $(T - \lambda) f(\lambda) = x$. LEU Seminar Notes in Function Analysis PDEs, p. 256265. Louisiana State University, Louisiana (1994)
96. Laursen, K.B., Vrbovà, P.: Some remarks on the surjectivity spectrum of linear operators. Czechoslovak Math. J. **39**(114) (1989)
97. Neumann, M.M.: On local spectral properties of operators on Banach spaces. Int. Workshop Oper. Theory. Rend. Circ. Mat. Palermo (2) Suppl (56), 15–25 (1998)

98. Ammar, A., Bouchekoua, A., Jeribi, A.: The ε-pseudospectra and the essential ε-pseudospectra of linear relations. J. Pseudo-Differ. Oper. Appl. **11**(2), 879–915 (2020)
99. Mnif, M., Ouled-Hmed, A.A.: Analytic core and quasi-nilpotent part of linear relations in Banach spaces. Filomat **32**(7), 2499–2515 (2018)
100. Mnif, M., Ouled-Hmed, A.A.: Local spectral theory and onto spectrum of linear relations. Ukrainian Math. J. **73**, 255–275 (2021)
101. Mnif, M., Ouled-Hmed, A.-A.: Local spectral theory and onto spectrum of linear relations, translated from Ukraïn. Mat. Zh. **73**(2), 222–237 (2021); Ukrainian Math. J. **73**(2), 255–275 (2021)
102. Erdélyi, I., Wang, S.W.: A local spectral theory for closed operators. In: London Mathematical Society Lecture Note Series, vol. 105. Cambridge University Press, Cambridge (1985)
103. Laursen, K.B., Neumann, M.M.: An introduction to local spectral theory. In: London Mathematical Society Monographs. New Series, vol. 20. The Clarendon Press, Oxford University Press, New York (2000)
104. Dales, H.G., Aiena, P., Eschmeier, J., Laursen, K., Willis, G.A.: Introduction to Banach algebras, operators, and harmonic analysis. In: London Mathematical Society Student Texts, vol. 57. Cambridge University Press, Cambridge (2003)
105. Elbjaoui, H., Zerouali, E.: Local spectral theory for 2×2 operator matrices. Int. J. Math. Sci. **42**, 2667–2672 (2003)
106. Houimdi, M., Zguitti, H.: Propriétés spectrales locales d'une matrice carrée des opérateurs. (French) [[Local spectral properties of a square operator matrix]]. Acta Math. Vietnam (2), 137–144, 25 (2000)
107. Kittaneh, F.: Numerical radius inequalities for Hilbert space operators. Studia Math. **168**(1), 73–80 (2005)
108. Ammar, A., Jeribi, A., Saadaoui, B.: A new decomposition for multivalued 3×3 matrices. Random Matri. Theory Appl. 2250028 (2022)
109. Ammar, A., Diagana, T., Jeribi, A.: Perturbations of Fredholm linear relations in Banach spaces with application to 3×3-block matrices of linear relations. Arab J. Math. Sci. **22**(1), 59–76 (2016)

Chapter 2
Algebraic and Topological Properties of Block Multivalued Operator Matrices

The intention of this chapter is to introduce desired results which are used throughout the book. Moreover, we introduced and studied the algebraic and topological properties of bounded and unbounded block matrices of linear relations in both Banach and Hilbert spaces. We mention that several results in this chapter may be found in most standard books dealing with multivalued linear theory and functional analysis (see [1–5]).

2.1 Algebraic and Topological Properties

2.1.1 The Distance Between Two Nonempty Subsets

The distance between two nonempty subsets M and N of a normed space X is defined by the formula

$$\operatorname{dist}(M, N) = \inf \{\|m - n\| \text{ such that } m \in M \text{ and } n \in N\}.$$

In particular, for $m \in M$, the distance between $\{m\}$ and N is given by

$$\begin{aligned}\operatorname{dist}(m, N) &= \operatorname{dist}(N, m)\\ &= \inf \{\|m - n\| \text{ such that } n \in N\}.\end{aligned}$$

The closed and the open disk of $\mathbb{C}$ centered at λ_0 and with radius r is denoted, respectively, by $\overline{\mathbb{D}}(\lambda_0, r)$ and $\mathbb{D}(\lambda_0, r)$. If $\lambda_0 = 0$, then $\mathbb{D}(0, r) = \mathbb{D}_r$ and $\overline{\mathbb{D}}(0, r) = \overline{\mathbb{D}}_r$.

A. Ammar and A. Jeribi, *Spectral Theory of Block Multivalued Operator Matrices*, Infosys Science Foundation Series, https://doi.org/10.1007/978-981-95-0397-1_2

2.1.2 Canonical Isometry

Theorem 2.1.1 [6, Theorem 2.28] *Let X be a vector space and let E and F be vector subspaces of a vector space X over $\mathbb{K}$ ($\mathbb{R}$ or $\mathbb{C}$). Then,*

$$\frac{E+F}{F} \cong \frac{E}{E\bigcap F},$$

where $\cong$ is a canonical isometry. ◇

For subspaces E, F of X we write $E \subset_e F$ (E is essentially contained in F) if there exists a finite-dimensional subspace $U \subset X$ such that $E \subset F + U$. In [7], V. Muller proved that $E \subset_e F$ if, and only if,

$$\dim\left(\frac{E+F}{F}\right) < \infty$$

if, and only if, there is a finite-dimensional subspace $U \subset E$ such that $E \subset F + U$ if, and only if,

$$\dim\left(\frac{E}{E\bigcap F}\right) < \infty.$$

Let X be a vector space and $E \subset X$ be a closed subspace. Let us denote by X/E the quotient space induced by a closed subspace E of X.

Definition 2.1.1 Let $(X, \|\cdot\|)$ be a real normed vector space and $M \subset X$ be a closed subspace. For $x \in X$, we define

$$\|[x]\|_{X/M} = \inf_{m\in M} \|x - m\|. \tag{2.1.1}$$

The quotient space X/M is a real vector space and the formula (2.1.1) defines a norm function on X/M. ◇

The following lemma is from L. E. Labuschagne in [8].

Lemma 2.1.1 *Let E be a normed vector space and let M be a closed subspace of E and let N be a subspace of E containing M. Then,*

(i) N is closed subspace if, and only if, N/M is closed.
(ii) $E/N \cong (E/M)\big/(N/M)$ and $Q_N = Q_{N/M}Q_M$, where $\cong$ is a canonical isometry. ◇

2.1.3 Direct Sum

Let X be a normed vector space. We will denote byX^* the space of all continuous functionals u' defined on X. Therefore X^* is the norm dual of X with norm

$$\|u'\| = \inf\left\{\mu \in \mathbb{K} \ : \ |u'(u)| \leq \mu\|u\| \text{ for all } u \in X\right\}.$$

If U and V are subspaces of X, we define the following

$$U^{\perp} := \left\{u' \in X^* \text{ such that } \ u'(u) = 0 \ \text{ for all } u \in U\right\}$$

and

$$V^{\top} := \left\{u \in X \text{ such that } \ v'(u) = 0 \ \text{ for all } v' \in V\right\}.$$

Remark 2.1.1 Let U be a vector subspace of normed space X and let V be a vector subspace of X^*. Then,

(i) $U^{\perp}$ and $V^{\top}$ are closed linear subspaces of X^* and X, respectively.
(ii) $(U^{\perp})^{\top} = \overline{U}$.
(iii) $(V^{\top})^{\perp} = \overline{V}$. ◇

Theorem 2.1.2 [2, Theorem III.3.9] *Let X be Banach space and let U, V be closed subspaces of X. Then, $U + V$ is closed if, and only if, $U^{\perp} + V^{\perp}$ is closed.* ◇

Theorem 2.1.3 [5, Theorem 2.7] *Let X be a normed space and u be a nonzero element of X. Then, there exists $u' \in X^*$ such that $\|u'\| = 1$ and $u'(u) = \|u\|$.* ◇

Definition 2.1.2 (i) Let X be a vector space and let G be a vector subspace of X. Any vector subspace H is said to be a complement of G, denoted

$$X = H \bigoplus G, \tag{2.1.2}$$

if $X = G + H$ and $G \bigcap H = \{0\}$.

(ii) Let X be a vector space. Every pair of vector subspaces G and H satisfying (2.1.2) determines a linear projection P by means

$$G = R(P)$$

and

$$H = N(P),$$

where $R(P)$ is the range of P and $N(P)$ is the kernel or null space of P. If the projection P is bounded, then G is said to be topologically complemented and H is said to be a topological complement of G. ◇

Remark 2.1.2 (i) [2, Terminology II.4.3] Let X be Banach space and M be a closed subspace of X. If M has a closed complementary subspace N, then M is topologically complemented by N.

(ii) If X is a normed space, then the finite-dimensional subspaces of X, and the closed finite codimensional subspaces of X are topologically complemented.

(iii) If X is an Hilbert space, then each closed subspace M of X is topologically complemented by its orthogonal complement $M^\perp$. ◇

Lemma 2.1.2 (i) *Let U, V be subspaces of H and let E, F be subspaces of K, where H and K are two Hilbert spaces. Then,*

$$(U \times E) \bigoplus (V \times F) = (U \bigoplus V) \times (E \bigoplus F).$$

(ii) *Let U be a subspace of H and let V be a subspace of K. Then,*

$$U^\perp \times V^\perp = (U \times V)^\perp. \qquad \diamond$$

Proof (i) Let $(m, n) \in (U \times E) \bigoplus (V \times F)$. Then, there exist $(x, y) \in U \times E$ and $(z, w) \in V \times F$ such that

$$\begin{aligned}(m, n) &= (x, y) + (z, w)\\ &= (x + z, y + w).\end{aligned}$$

This implies that $m = x + z \in U + V$ and $n = y + w \in E + F$. Hence, we infer that $(m, n) \in (U + V) \times (E + F)$. It remains to prove that

$$U \bigcap V = \{0\} \text{ and } E \bigcap F = \{0\}. \tag{2.1.3}$$

Let $(m, n) \in (U \bigcap V) \times (E \bigcap F)$. Then, $(m, n) \in U \times E$ and $(m, n) \in V \times F$. This implies that $(m, n) \in (U \times E) \bigcap (V \times F) = \{(0, 0)\}$. Hence, (2.1.3) holds. This shows that

$$(U \times E) \bigoplus (V \times F) \subset (U \bigoplus V) \times (E \bigoplus F).$$

Conversely, let $(m, n) \in (U \bigoplus V) \times (E \bigoplus F)$. Then, $m = x + z$ and $n = y + w$ with $x \in U$, $z \in V$, $y \in E$ and $w \in F$. Consequently, $(x, y) \in U \times E$, $(z, w) \in V \times F$. Hence,

$$(m, n) = (x, y) + (z, w) \in (U \times E) + (V \times F).$$

At this level, we shall show that $(U \times E) \bigcap (V \times F) = \{(0, 0)\}$. Let us assume that $(m, n) \in (U \times E) \bigcap (V \times F)$. Then, $(m, n) \in U \times E$ and $(a, b) \in V \times F$. This leads us to deduce that $m \in U \bigcap V = \{0\}$ and $n \in E \bigcap F = \{0\}$.

(ii) Since $\overline{U}$ and $\overline{V}$ are closed subspaces of H and K, respectively, then $\overline{U} \times \overline{V} \subset H \times K$ is so. This implies that

$$\begin{aligned} H \times K &= (\overline{U} \times \overline{V}) \bigoplus (\overline{U} \times \overline{V})^{\perp} \\ &= (\overline{U} \times \overline{V}) \bigoplus (\overline{U \times F})^{\perp} \\ &= (\overline{U} \times \overline{V}) \bigoplus (U \times V)^{\perp}. \end{aligned}$$

On the other hand, we can write

$$\begin{aligned} H \times K &= (\overline{U} \bigoplus U^{\perp}) \times (\overline{U} \bigoplus U^{\perp}). \\ &= (\overline{U} \times \overline{V}) \bigoplus (U^{\perp} \times V^{\perp}) \ \text{(by } (i)). \end{aligned}$$

By the uniqueness of the decomposition, we infer that

$$U^{\perp} \times V^{\perp} = (U \times V)^{\perp}.$$

This completes the proof. Q.E.D.

2.2 Kuratowski Measure of Noncompactness

2.2.1 The Diameter of a Set

The diameter of a set U is the number

$$\text{diam}\,(U) := \begin{cases} \sup\{\text{dist}(u, v) \text{ such that } u \in U \text{ and } v \in U\} & \text{if } U \neq \emptyset \\ 0 & \text{otherwise.} \end{cases}$$

The following properties are satisfied:
(a) $\text{diam}(U) = 0$ if, and only if, U is an empty set or consists of exactly one point.
(b) If X complete space, then

(b_1) If $U_1 \subset U_2$, then $\text{diam}(U_1) \leq \text{diam}(U_2)$.

(b_2) $\text{diam}(\overline{U}) = \text{diam}(U)$, where $\overline{U}$ is the closure of U.

(b_3) [**Cantor's intersection theorem**]: If $(U_n)_n$ is a decreasing sequence of nonempty, closed and bounded subsets of X and $\lim_{n\to+\infty} \text{diam}(U_n) = 0$, then the intersection

$$U_\infty = \bigcap_{n\in\mathbb{N}} U_n$$

of all U_n is nonempty and consists of exactly one point.

(c) If X is a Banach space, then we have the following

(c_1) $\text{diam}(\lambda U) = |\lambda| \, \text{diam}(U)$, $\forall \lambda \in \mathbb{R}$.
(c_2) $\text{diam}(x + U) = \text{diam}(U)$ for any $x \in X$.
(c_3) $\text{diam}(U_1 + U_2) \leq \text{diam}(U_1) + \text{diam}(U_2)$.
(c_4) $\text{diam}(\text{co}(U)) = \text{diam}(U)$, where $\text{co}(U)$ is the convex hull of U.

2.2.2 *Definition*

Definition 2.2.1 Let E be a bounded subset of X. The Kuratowski measure of noncompactness is defined by $\delta(E) = \inf\{d > 0$ such that E can be covered by a finite number of sets of diameter $\leq d\}$. ◇

It is obvious that $0 \leq \delta(F) \leq \text{diam}(F) < +\infty$ for each nonempty bounded subset F of X.

Proposition 2.2.1 [9, Proposition 1] *For random bounded sets E, F, $E_n \subset X$, and $\lambda \in \mathbb{C}$, we have*

(1) $\delta(E) = 0$ *if, and only if,* $\overline{E}$ *is compact, where* $\overline{E}$ *is the closure of* E.
(2) $\delta(\overline{E}) = \delta(E)$ *and* $\delta(\lambda E) = |\lambda| \, \delta(E)$. *Likewise, if* $E \subset F$, *then*

$$\delta(E) \leq \delta(F).$$

(3) *If* $E_{n+1} \subset E_n = \overline{E_n}$, *and* $\lim\limits_{n \to +\infty} \delta(E_n) = 0$, *then*

$$E_\infty := \bigcap_{n=1}^{+\infty} E_n \neq \emptyset$$

and

$$\delta(E_\infty) = 0.$$

(4) $\delta(E \bigcup F) = \max\{\delta(E), \delta(F)\}$ *and* $\delta(E + F) \leq \delta(E) + \delta(F)$.
(5) $\delta(E) = \delta(\overline{co}\,(E))$, *where* $\overline{co}\,(E)$ *is the convex closure of* E.
(6) *For all* $\varepsilon > 0$, *we have*

$$\delta(N_\varepsilon(E)) \leq \delta(E) + 2\varepsilon,$$

where

$$N_\varepsilon(E) = \{x \in X \text{ such that } dist(x, E) < \varepsilon\}.$$ ◇

2.2.3 Gap Between Two Linear Subspaces

Let U and V be two linear subspaces of a normed space X. The gap between U and V is defined, in the case where $U \neq \{0\}$, by the following formula

$$\delta(U, V) = \sup_{u \in U,\ \|u\|=1} \text{dist}(u, V). \tag{2.2.1}$$

Any other way we define $\delta(\{0\}, V) = 0$ for any subspace V. Furthermore, $\delta(U, \{0\}) = 1$ if $U \neq \{0\}$, as demonstrated from the definition (see (2.2.1)). We can additionally define the symmetric or maximal gap between U and V by

$$\widehat{\delta}(U, V) = \max\{\delta(U, V), \delta(V, U)\}.$$

The gap $\delta(U, V)$ can be identified as the smallest number δ such that

$$\text{dist}(u, V) \leq \delta\|u\|,$$

for all $u \in U$. The concept of a gap between linear subspaces and linear operators was presented by M. G. Krein and M. A. Krasnoselski in the 1940s (see [10]). Number among the works in this focus we can state, for example (see [11, 12, 12–18]. Think about that $\text{dist}(\alpha u, V) = |\alpha|\text{dist}(u, V)$, $\alpha \in \mathbb{R}^*$, we deduce that

$$\sup_{u \in U\ \|u\| \leq |\alpha|} \text{dist}(u, V) = |\alpha|\delta(U, V). \tag{2.2.2}$$

Let's consider subspaces U and V, then

(i) $\delta(U, V) = \delta(\overline{U}, \overline{V})$ and $\widehat{\delta}(U, V) = \widehat{\delta}(\overline{U}, \overline{V})$.
(ii) $\delta(U, V) = 0$ if, and only if, $\overline{U} \subset \overline{V}$.
(iii) $\widehat{\delta}(U, V) = 0$ if, and only if, $\overline{U} = \overline{V}$.
(iv) $0 \leq \delta(U, V) \leq \widehat{\delta}(U, V) \leq 1$.

2.3 Quasi-Banach Spaces

Let X be a vector space over the field $\mathbb{K}$ of real or complex numbers. A functional

$$\langle \cdot, \cdot \rangle : X \times X \longrightarrow \mathbb{K}$$

is called an inner product if it satisfies the following conditions:

- $\overline{\langle u, v \rangle} = \langle v, u \rangle$, for all $u, v \in X$.
- $\langle u, u \rangle > 0$ unless $u = 0$, for all $u \in X$.
- $\langle u, \lambda v + \mu w \rangle = \overline{\lambda}\langle u, v \rangle + \overline{\mu}\langle u, w \rangle$, for all $u, v, w \in X$ and $\lambda, \mu \in \mathbb{K}$.

- $\langle \lambda u, v\rangle = \lambda\langle u, v\rangle$, for all $u, v \in X$ and $\lambda \in \mathbb{K}$.
- $\|u\| = \sqrt{\langle u, u\rangle}$, for all $u \in X$.

The pair $(X, \langle\cdot,\cdot\rangle)$ is called inner product space. A Hilbert space is an inner product space, which is complete. By a quasi-norm $\|\cdot\|$ defined on a linear space X, we mean a real-valued function with the following properties:

(i) Let $u \in X$. Then, $\|u\| = 0$ if, and only if, $u = 0$.
(ii) $\|\lambda u\| = |\lambda|\|x\|$ for $u \in X$ and $\lambda \in \mathbb{K}$.
(iii) $\|u + v\| \leq c_X(\|u\| + \|v\|)$ for $u, v \in X$. Here $c_X > 1$ is a constant.

Formula (iii) is called the quasi-triangle inequality. If $c_X = 1$, then $\|\cdot\|$ is said to be a norm. In this case, (iii) passes into the well-known triangle inequality. A normed space (or normed vector space) is a pair $(X, \|\cdot\|)$, where X is a vector space over $\mathbb{K}$ and $\|\cdot\|$ is a norm on X. A norm induces a distance by the formula

$$\mathrm{dist}(u, v) = \|v - u\|, \quad u,\ v \in X.$$

Therefore, a normed space is a metric space, and thus a topological vector space. Every quasi-norm $\|\cdot\|$ given on a linear space X induces a metrizable topology such that the algebraic operations are continuous. A fundamental system of neighborhoods of the zero element is formed by the subsets $\varepsilon\overline{\mathbb{B}}(0, 1)$ with $\varepsilon > 0$, where $\overline{\mathbb{B}}(0, 1)$ is the closure of the open ball with radius 1 around 0 in X, cf.

$$\mathbb{B}(0, 1) := \{u \in X : \|u\| < 1\}.$$

A quasi-Banach space is a linear space X equipped with a quasi-norm $\|\cdot\|$ which becomes complete with respect to the associated metrizable topology. This means that all Cauchy sequences are convergent. In the most important case when $\|\cdot\|$ is a norm we call X a Banach space.

2.4 Vector-Valued Analytic Functions

The aim of this section is to give some concept and handful results with respect to vector-valued analytic functions. Let X be a complex Banach space and let O be an open set in $\mathbb{C}$. A function $f : O \longrightarrow X$ is said to be analytic, if the composition $x' \circ f : O \longrightarrow \mathbb{C}$ is analytic for each $x' \in X^*$. It is well known that analytic functions are holomorphic (or complex differentiable), in the sense that, for arbitrary $\theta \in O$, the norm limit

$$f'(\theta) = \lim_{\eta \to \theta} \frac{f(\theta) - f(\eta)}{\theta - \eta}$$

exists in X. The space of all analytic functions from O into X, where O is open set in $\mathbb{C}$ and X is a complex Banach space, is denoted by $H(\mathscr{U}, X)$.

Lemma 2.4.1 [19, Theorem A.3.1] Let X be a complex Banach space and let O be an open set in $\mathbb{C}$.

(i) Every analytic function $f : O \longrightarrow X$ is infinitely differentiable, and admits, for each $\eta_0 \in O$ and each $\varepsilon > 0$ with $\overline{\mathbb{D}}(\eta_0, \varepsilon) \subseteq O$, the representation

$$f(\eta) = \sum_{n=1}^{\infty} a_n \, (\eta - \eta_0)^n, \text{ for all } \eta \in \overline{\mathbb{D}}(\eta_0, \varepsilon),$$

where $\overline{\mathbb{D}}(\eta_0, \varepsilon) = \{\eta \in \mathbb{C} \text{ such that } |\eta - \eta_0| \leq \varepsilon\}$ and the coefficients $a_n \in X$ are given by

$$a_n = \frac{f^{(n)}(\eta_0)}{n!} \text{ for all } n \geq 0.$$

Moreover, the power series representing f converges absolutely and uniformly on the disk $\overline{\mathbb{D}}(\eta_0, \varepsilon)$.

(ii) Given any sequence of vectors $a_n \in X$ for all $n \geq 0$, the series

$$f(\eta) = \sum_{n=1}^{\infty} a_n \, (\eta - \eta_0)^n$$

converges for $\eta \in \mathbb{D}(\eta_0, \varepsilon) = \{\eta \in \mathbb{C} \text{ such that } |\eta - \eta_0| < \varepsilon\}$, and defines an analytic function $f : \mathbb{D}(\eta_0, \varepsilon) \longrightarrow X$, where ε denotes the radius of convergence given by

$$\frac{1}{\varepsilon} = \limsup_{n \to +\infty} \|a_n\|^{1/n}. \qquad \diamond$$

When $\eta_0 = 0$, the open disk centrad at 0 and with radius ε is simply denoted by $\mathbb{D}_{\varepsilon}$. It follows from Lemma 2.4.1 that analytic functions are locally represented by vector-valued power series.

Lemma 2.4.2 [19, Theorem A.3 .3] *Let X be a complex Banach space, $\mathscr{U}$ be an open set in $\mathbb{C}$ and let $f : \mathscr{U} \longrightarrow X$. If f is a bounded analytic function on the entire complex plane, then f is constant.* $\diamond$

2.5 Relations Sets

2.5.1 Definitions and Notations

Giving M and N be two arbitrary nonempty sets. A multivalued function U from M into N is a mapping outlined on a nonempty subset $\mathscr{D}(U)$ of M (called the domain of

U) which takes on values in $2^N\backslash\emptyset$ (the collection of nonempty subset of N). If U maps the points of its domain to singletons, then U is said to be a single-valued operator, or function. We will denote $\mathscr{R}(M,N)$ by the collection of multivalued function or relations from M into N and as useful we write $\mathscr{R}(M) := \mathscr{R}(M,M)$. A number of examples of multivalued function:, the inverse of function, partial order relations, equivalent relations, the adjoints of linear operators and convex processes,. … For $m \in M$ and $m \notin \mathscr{D}(U)$, we define $Um = \emptyset$. This convention implies that

$$\mathscr{D}(U) = \{m \in M\colon\ Um \neq \emptyset\}.$$

Giving $U \in \mathscr{R}(M,N)$. The graph of U is a subset of $M \times N$ defined by

$$G(U) = \left\{\begin{pmatrix} m \\ n \end{pmatrix} \in M \times N : m \in \mathscr{D}(U) \text{ and } n \in Um\right\}.$$

A relation set is uniquely determined by its graph. Conversely, any nonempty subset of $M \times N$ determines uniquely a relation. The identity and the null relation defined on a nonempty subset Ξ of M are denoted, respectively, by I_Ξ (or simply I) and 0_Ξ, when Ξ is understood, it is the relation in $\mathscr{R}(\Xi)$ whose graph is

$$G(I_\Xi) = \left\{\begin{pmatrix} \xi \\ \xi \end{pmatrix} : \xi \in \Xi\right\}.$$

$$G(0_\Xi) = \left\{\begin{pmatrix} \xi \\ 0 \end{pmatrix} : \xi \in \Xi\right\}.$$

Giving $\Xi \subset M$, the range of Ξ by U is

$$U(\Xi) = \bigcup_{\xi \in \Xi \bigcap \mathscr{D}(U)} U(\xi). \tag{2.5.1}$$

Specifically, $U(M)$, denoted by $R(U)$, is said the range of U. The inverse of U is the relation U^{-1} present with

$$G(U^{-1}) = \left\{\begin{pmatrix} n \\ m \end{pmatrix} \in N \times M : \begin{pmatrix} m \\ n \end{pmatrix} \in G(U)\right\}.$$

It is readily apparent that

$$(U^{-1})^{-1} = U.$$

Let E be a nonempty subset of N, then

$$U^{-1}(E) = \left\{u \in \mathscr{D}(U) \text{ such that } E \bigcap Uu \neq \emptyset\right\}. \tag{2.5.2}$$

Hence, for $v \in R(U)$, we have

$$U^{-1}v = \{u \in \mathscr{D}(U) \text{ such that } v \in Uu\}.$$

Consequently,

$$R(U) = \mathscr{D}(U^{-1})$$

and

$$R(U^{-1}) = \mathscr{D}(U).$$

Definition 2.5.1 Let M, N be two arbitrary nonempty sets and U be a relation from M into N. Then,

(i) A relation U is said to be onto, if $R(U) = N$.
(ii) A relation U is said to be one-to-one, if U^{-1} is an operator.
(iii) A relation U is said to be bijective, if it is both injective and surjective.

Lemma 2.5.1 *Let M, N be two arbitrary nonempty sets and U be a relation from M into N. If U is one-to-one, then*

$$\textit{for all } x, \ y \in \mathscr{D}(U), \ \textit{we have } Ux = Uy \Rightarrow x = y. \tag{2.5.3}$$

◇

Proof Let $z \in Ux = Uy$. Then, $\begin{pmatrix} x \\ z \end{pmatrix} \in G(U)$ and $\begin{pmatrix} y \\ z \end{pmatrix} \in G(U)$. This implies that $\begin{pmatrix} z \\ x \end{pmatrix} \in G(U^{-1})$ and $\begin{pmatrix} z \\ y \end{pmatrix} \in G(U^{-1})$. Hence, from the assumption, U is injective, we infer that U^{-1} is an operator which yields $x = y$. Q.E.D.

The example below shows that the converse of Lemma 2.5.1 is false.

Example 2.5.1 Let $U : \{u, v, w\} \longrightarrow \{u, v, w\}$ be a relation defined by

$$U(u) = \{u, v\},$$

$$U(v) = \{v, w\}$$

and

$$U(w) = \{u, v, w\}.$$

We have $U(u) \neq U(v)$, $U(u) \neq U(w)$, and $U(v) \neq U(w)$. Then, (2.5.3) is holds. But $U^{-1}(u) = \{u, w\}$. Consequently, U^{-1} is not an operator. Signify U is not injective.

Definition 2.5.2 Let M, N be two nonempty sets, U be a relation from M into N and $K \subset M$ such that $\mathscr{D}(U) \bigcap K \neq \emptyset$. The restriction of U to K, denoted by $U_{|K}$, is a relation from M into N which is defined by

$$\left(U_{|K}\right) x = Ux, \quad x \in \mathscr{D}\left(U_{|K}\right) = K \bigcap \mathscr{D}(U).$$

◇

Remark 2.5.1 Note that

$$G\left(U_{|K}\right) = (K \times N) \bigcap G(U) = \left\{ \begin{pmatrix} x \\ y \end{pmatrix} \in G(U) \text{ such that } x \in K \right\}. \qquad \diamond$$

Example 2.5.2 Let E be a subset of U and let J_E^U (abbreviated J_E when U is understood) be denotes the natural injection map of E into U, i.e., $J_E \in \mathscr{R}(E, U)$,

$$\mathscr{D}(J_E) = E$$

and

$$J_E(e) = e \ for \ all \ \ e \in E.$$

Then,

$$T_{|E} = T J_E J_E^{-1}. \tag{2.5.4}$$

Note that $I_E = I_{U|E}$ in particular, both sides bare in $\mathscr{R}(U, U)$. Thus, from (2.5.4), we have

$$I_E = J_E J_E^{-1}.$$

Although I_E and J_E have identical graphs, I_E is not even an extension of J_E.

2.5.2 *Composition of Relations*

Let M, N, K be nonempty sets, U be a relation from M into N and V be a relation from N into K such that $\mathscr{D}(V) \bigcap R(U) \neq \emptyset$. Then, VU is a relation from M into K which is defined by

$$(VU)(u) = V(Uu), \quad \text{for all} \quad u \in \mathscr{D}(VU),$$

where

$$\begin{aligned} \mathscr{D}(VU) &= \{u \in M \text{ such that } V(Uu) \neq \emptyset\} \\ &= \{u \in M \ \text{ such that } \mathscr{D}(V) \bigcap Uu \neq \emptyset\}. \end{aligned}$$

Lemma 2.5.2 [2, Subsection I.1.3] Let M, N, K be nonempty sets, U be a relation from M into N and V be a relation from N into K such that $\mathscr{D}(V) \bigcap R(U) \neq \emptyset$. Then,

(i) $\mathscr{D}(VU) = U^{-1}(\mathscr{D}(V))$.

(ii) $G(VU) = \left\{ \begin{pmatrix} x \\ y \end{pmatrix} \in M \times K : \begin{pmatrix} x \\ z \end{pmatrix} \in G(U) \text{ and } \begin{pmatrix} z \\ y \end{pmatrix} \in G(V) \text{ for some } z \in V \right\}. \qquad \diamond$

Proof (i) In view of (2.5.2) implies that

$$\begin{aligned}\mathcal{D}(VU) &= \{x \in U \text{ such that } Ux \bigcap \mathcal{D}(V) \neq \emptyset\} \\ &= U^{-1}(\mathcal{D}(V)).\end{aligned}$$

(ii) Let $\begin{pmatrix} x \\ y \end{pmatrix} \in G(VU)$. Then, $x \in \mathcal{D}(VU)$ and $y \in VUx$. By using (2.5.1), there exists $z \in Ux$ such that $y \in Vz$. This implies that $z \in N$, $\begin{pmatrix} x \\ z \end{pmatrix} \in G(U)$ and $\begin{pmatrix} z \\ y \end{pmatrix} \in G(V)$. Conversely, let $\begin{pmatrix} x \\ z \end{pmatrix} \in G(U)$ and $\begin{pmatrix} z \\ y \end{pmatrix} \in G(V)$ for some $z \in V$. This implies that $z \in Ux$ and $y \in Vz$. This leads to $y \in VUx$. Q.E.D.

Proposition 2.5.1 *Let M, N, K be nonempty sets, U be a relation from M into N and V be a relation from N into K such that $\mathcal{D}(V) \bigcap R(U) \neq \emptyset$. Then,*

(i) $U = UI_{\mathcal{D}(U)} = I_{R(U)}U$.
(ii) $G(I_{\mathcal{D}(U)}) \subset G(U^{-1}U)$.
(iii) $(VU)^{-1} = U^{-1}V^{-1}$.
(iv) $\mathcal{D}(VU) = U^{-1}V^{-1}(R(V))$.
(v) *U is one-to-one if, and only if,* $U^{-1}U = I_{\mathcal{D}(U)}$.
(vi) *U is an operator if, and only if,* $UU^{-1} = I_{R(U)}$. ◇

Proof (i) Let $\begin{pmatrix} x \\ y \end{pmatrix} \in G(U)$. Then, $x \in \mathcal{D}(U)$ and $y \in Ux$. This implies that $\begin{pmatrix} x \\ x \end{pmatrix} \in G(I_{\mathcal{D}(U)})$ and $\begin{pmatrix} x \\ y \end{pmatrix} \in G(U)$. Hence, by using Lemma 2.5.2 (ii), we infer that $\begin{pmatrix} x \\ y \end{pmatrix} \in G(UI_{\mathcal{D}(U)})$. Conversely, let $\begin{pmatrix} x \\ y \end{pmatrix} \in G(UI_{\mathcal{D}(U)})$. Then, there exists $z \in M$ such that $\begin{pmatrix} x \\ z \end{pmatrix} \in G(I_{\mathcal{D}(U)})$ and $\begin{pmatrix} z \\ y \end{pmatrix} \in G(U)$. Because of this $\begin{pmatrix} x \\ z \end{pmatrix} \in G(I_{\mathcal{D}(U)})$ implies that $z = x$. This permits to concluded that $\begin{pmatrix} x \\ y \end{pmatrix} \in G(U)$. The same argument as above gives $U = I_{R(U)}U$.

(ii) Let us assume that $\begin{pmatrix} x \\ y \end{pmatrix} \in G(I_{\mathcal{D}(U)})$. Hence, $x \in \mathcal{D}(U)$ and $x = y$. This implies that there exists $z \in N$ such that $z \in Ux$ which yields $\begin{pmatrix} x \\ z \end{pmatrix} \in G(U)$ and $\begin{pmatrix} z \\ x \end{pmatrix} \in G(U^{-1})$. Then, we gather that $\begin{pmatrix} x \\ x \end{pmatrix} \in G(U^{-1}U)$.

(iii), we have

$$\begin{aligned}G((VU)^{-1}) &= \left\{\begin{pmatrix} x \\ y \end{pmatrix} \in K \times M : \begin{pmatrix} x \\ y \end{pmatrix} \in G(VU)\right\} \\ &= \left\{\begin{pmatrix} x \\ y \end{pmatrix} \in K \times M : (y, z) \in G(U) \text{ and } \begin{pmatrix} z \\ x \end{pmatrix} \in G(V) \text{ for some } z \in N\right\} \\ &= \left\{\begin{pmatrix} x \\ y \end{pmatrix} \in K \times M : \begin{pmatrix} z \\ y \end{pmatrix} \in G(U^{-1}) \text{ and } \begin{pmatrix} x \\ z \end{pmatrix} \in G(V^{-1}) \text{ for some } z \in N\right\} \\ &= G(U^{-1}V^{-1}).\end{aligned}$$

(iv) It can be inferred from Lemma 2.5.2 (i) that $\mathscr{D}(VU) = U^{-1}(\mathscr{D}(V))$. Since $\mathscr{D}(V) = V^{-1}(R(V))$. This implies that $\mathscr{D}(VU) = U^{-1}V^{-1}(R(V))$.

(v) Let us assume that $\begin{pmatrix} x \\ y \end{pmatrix} \in G(U^{-1}U)$. Hence, there exists $z \in N$ such that $\begin{pmatrix} x \\ z \end{pmatrix} \in G(U)$ and $\begin{pmatrix} z \\ y \end{pmatrix} \in G(U^{-1})$ which yields $\begin{pmatrix} z \\ x \end{pmatrix} \in G(U^{-1})$ and $\begin{pmatrix} z \\ y \end{pmatrix} \in G(U^{-1})$. Based on the assumption U is one-to-one, we infer that $x = U^{-1}z = y$. As a consequence, $\begin{pmatrix} x \\ y \end{pmatrix} \in G(I_{\mathscr{D}(U)})$. Now, using (ii) enables to conclude that $U^{-1}U = I_{\mathscr{D}(U)}$. Conversely, let $\begin{pmatrix} x \\ y \end{pmatrix} \in G(U^{-1})$ and $\begin{pmatrix} x \\ z \end{pmatrix} \in G(U^{-1})$. Then, $\begin{pmatrix} x \\ y \end{pmatrix} \in G(U^{-1})$ and $\begin{pmatrix} z \\ x \end{pmatrix} \in G(U)$. This leads to $\begin{pmatrix} z \\ y \end{pmatrix} \in G(U^{-1}U)$. The fact that $G(U^{-1}U) = G(I_{\mathscr{D}(U)})$ implies that $y = z$. Hence, U^{-1} is an operator. This is equivalent to mean that U is one-to-one. Substituting U for U^{-1} in (v) yields (vi).Q.E.D.

2.6 Multivalued Linear Operators in Vector Space

This section recalls the concept of linear relations, basic properties, and elementary results which will be needed in the sequel.

2.6.1 Definitions and Basic Properties

Let X and Y be two vector spaces over the field $\mathbb{K}$ ($\mathbb{R}$ or $\mathbb{C}$).

A linear relation (or a multivalued linear operator) is a mapping $U \subset X \times Y$ which goes from a subspace

$$\mathscr{D}(U) = \{x \in X \text{ such that } Ux \neq \emptyset\} \subset X,$$

called the domain of U, into the collection of nonempty subsets of Y such that

$$U(\alpha_1 x_1 + \alpha_2 x_2) = \alpha_1 U(x_1) + \alpha_2 U(x_2)$$

for all α_1, $\alpha_2 \in \mathbb{K}^*$, and x_1, $x_2 \in \mathscr{D}(U)$. We define $Ux = \emptyset$ for all $x \in X \backslash \mathscr{D}(U)$. The class of linear relations as defined earlier will be denoted by $L\mathscr{R}(X, Y)$ and as useful we write $L\mathscr{R}(X) := L\mathscr{R}(X, X)$. The range $R(U)$ of U is defined by

$$R(U) = \bigcup_{x \in \mathscr{D}(U)} Ux.$$

If $R(U) = Y$, then we are saying that U is onto, and if $M \subset X$, then the range of M under U is defined by the set

$$U(M) = \bigcup_{x \in M \bigcap \mathscr{D}(U)} Ux.$$

The inverse of $U \in L\mathscr{R}(X, Y)$ is the linear relation $U^{-1} \in L\mathscr{R}(Y, X)$. The graph of U^{-1} is a subset of $Y \times X$ defined by

$$G(U^{-1}) = \left\{ \begin{pmatrix} y \\ x \end{pmatrix} \in Y \times X \text{ such that } \begin{pmatrix} x \\ y \end{pmatrix} \in G(U) \right\}.$$

If $M \subset Y$, then the inverse image of M down U is defined by the set

$$U^{-1}(M) := \left\{ x \in \mathscr{D}(U) \text{ such that } Ux \bigcap M \neq \emptyset \right\}.$$

The subspace

$$U^{-1}(0) = \left\{ x \in X \text{ such that } \begin{pmatrix} x \\ 0 \end{pmatrix} \in G(U) \right\}$$

is called the null space (or kernel) of U, and is denoted by $N(U)$. If U^{-1} is a single-valued operator, then U is called one-to-one. If U is one-to-one and onto we say that U is bijective.

Proposition 2.6.1 [2, Proposition I.2.16] *Let X, Y be two vector spaces and let U be a relation from X into Y. Then, $U \in L\mathscr{R}(X, Y)$ if, and only if, for all scalars α_1, α_2 and for all $x_1, x_2 \in \mathscr{D}(U)$,*

$$\alpha_1 U(x_1) + \alpha_2 U(x_2) \subset U(\alpha_1 x_1 + \alpha_2 x_2). \qquad \diamond$$

Proposition 2.6.2 *Let $U \in \mathscr{R}(X, Y)$. The following properties are equivalent*

(i) $U \in L\mathscr{R}(X, Y)$.
(ii) $G(U)$ is a subspace of $X \times Y$.
(iii) U^{-1} is a linear relation.
(iv) $G(U^{-1})$ is a subspace of $Y \times X$. $\diamond$

Proof $(i) \Longrightarrow (ii)$ Let $\begin{pmatrix} x_1 \\ y_1 \end{pmatrix}$ and $\begin{pmatrix} x_2 \\ y_2 \end{pmatrix} \in G(U)$ and $\alpha \in \mathbb{K}\backslash\{0\}$. Since $\begin{pmatrix} x_1 \\ y_1 \end{pmatrix} \in G(U)$ implies $y_1 \in Ux_1$, then $\alpha y_1 \in \alpha Ux_1 = U(\alpha x_1)$. In the same way, since $\begin{pmatrix} x_2 \\ y_2 \end{pmatrix} \in G(U)$, then $y_2 \in Ux_2$. So, $\alpha y_1 + y_2 \in U(\alpha x_1) + Ux_2 = U(\alpha x_1 + x_2)$. We obtain

$$\begin{pmatrix} \alpha x_1 + x_2 \\ \alpha y_1 + y_2 \end{pmatrix} \in G(U). \tag{2.6.1}$$

Finally, by using (2.6.1), we have $\begin{pmatrix} \alpha 0 \\ \alpha 0 \end{pmatrix} \in G(U)$. So, $G(U)$ is subspace of $X \times Y$.
$(ii) \Longrightarrow (i)$ Assume that $G(U)$ is subspace of $X \times Y$. Then, $\mathcal{D}(U)$ is a linear subspace of X. Let $x_1, x_2 \in \mathcal{D}(U)$. To show that $Ux_1 + Ux_2 \subset U(x_1 + x_2)$. Let $y \in Ux_1 + Ux_2$. Then, $y = y_1 + y_2$, where $y_1 \in Ux_1$ and $y_2 \in Ux_2$. We infer that

$$\begin{aligned} \begin{pmatrix} x_1 + x_2 \\ y \end{pmatrix} &= \begin{pmatrix} x_1 + x_2 \\ y_1 + y_2 \end{pmatrix} \\ &= \begin{pmatrix} x_1 \\ y_1 \end{pmatrix} + \begin{pmatrix} x_2 \\ y_2 \end{pmatrix} \in G(U). \end{aligned}$$

So, $y \in U(x_1 + x_2)$. Next, to show that

$$U(x_1 + x_2) \subset Ux_1 + Ux_2.$$

Let $y \in U(x_1 + x_2)$ and select $y_1 \in Ux_1$ arbitrarily. Then,

$$\begin{pmatrix} x_1 + x_2 \\ y \end{pmatrix} = \begin{pmatrix} x_1 \\ y_1 \end{pmatrix} + \begin{pmatrix} x_2 \\ y - y_1 \end{pmatrix} \in G(U)$$

and hence $\begin{pmatrix} x_2 \\ y - y_1 \end{pmatrix} \in G(U)$ whence $y - y_1 \in Ux_2$. Therefore, $y \in y_1 + Ux_2 \subset Ux_1 + Ux_2$ as required. We have thus shown that

$$U(x_1 + x_2) = Ux_1 + Ux_2.$$

Now, let $\alpha \in \mathbb{K}^*$ and $x \in \mathcal{D}(U)$. If $y \in Ux$, then $\alpha \begin{pmatrix} x \\ y \end{pmatrix} = \begin{pmatrix} \alpha x \\ \alpha y \end{pmatrix} \in G(U)$ so $\alpha y \in U(\alpha x)$. Hence, $\alpha U(x) \subset U(\alpha x)$. On the flip side if $y \in U(\alpha x)$, hence $(x, \alpha^{-1} y) \in G(U)$, so y $\in \alpha U(x)$. Thus, $U(\alpha x) \subset \alpha U(x)$. Hence, $U(\alpha x) = \alpha U(x)$. Therefore, U is a linear relation.
$(ii) \Longleftrightarrow (iii) \Longleftrightarrow (iv)$ The remaining equivalence now follow by symmetry. Q.E.D.

Example 2.6.1 Let $X = C([0,1],\mathbb{R}) = \{f : [0,1] \longrightarrow \mathbb{R} : f \text{ is continuous}\}$ and

$$\begin{aligned} U : X &\longrightarrow \mathscr{P}(X)\backslash\emptyset \\ f &\longrightarrow \left\{\int f\right\}. \end{aligned}$$

Then, U is a linear relation. Indeed, we have $\mathscr{D}(U) = X$. Let $f, g \in X$, then

$$\begin{aligned} U(f+g) &= \left\{\int (f+g)\right\} \\ &= \left\{\int f + \int g\right\} \\ &= \left\{\int f\right\} + \left\{\int g\right\} \\ &= U(f) + U(g). \end{aligned}$$

Let α be nonzero scalar and $f \in X$. Then,

$$\begin{aligned} U(\alpha f) &= \left\{\int (\alpha f)\right\} \\ &= \left\{\alpha \int f\right\} \\ &= \alpha \left\{\int f\right\} \\ &= \alpha U(f). \end{aligned}$$

Hence, U is a linear relation everywhere defined. Moreover, we have

$$U(0) = \left\{\int 0\right\} = \mathbb{R}.$$

Remark 2.6.1 (i) $\mathscr{D}(U^{-1}) = R(U)$ and $\mathscr{D}(U) = R(U^{-1})$.
(ii) $R(U) = \left\{y \in Y \text{ such that } \begin{pmatrix} x \\ y \end{pmatrix} \in G(U)\right\}$.
(iii) $N(U) = \left\{x \in \mathscr{D}(U) \text{ such that } \begin{pmatrix} x \\ 0 \end{pmatrix} \in G(U)\right\}$.
(iv) If U is a linear operator such that $N(U) \neq \{0\}$, then U^{-1} is a linear relation.
(v) $U(0) = \left\{y \in Y \text{ such that } \begin{pmatrix} 0 \\ y \end{pmatrix} \in G(U)\right\}$. ◇

Lemma 2.6.1 *Let X, Y be two vector spaces and let $U \in L\mathscr{R}(X, Y)$. Then,*

(i) *$U(0)$ is a vector subspace of Y.*
(ii) *$U^{-1}(0)$ is a vector subspace of X.* ◇

Proof (i) Let $y \in Ux$. Then, $0 = y - y \in Ux - Ux$. Based on the assumption $U \in L\mathscr{R}(X, Y)$, we infer that $Ux - Ux = Ux + U(-x) = U(x - x) = U(0)$. It is shown that $0 \in U(0)$. Let $x,\ y \in U(0)$. Since

$$U(0) = \left\{ y \in Y \text{ such that } \begin{pmatrix} 0 \\ y \end{pmatrix} \in G(U) \right\}$$

we have $\begin{pmatrix} 0 \\ x \end{pmatrix} \in G(U)$ and $\begin{pmatrix} 0 \\ y \end{pmatrix} \in G(U)$. Since U is a linear relation implies from Proposition 2.6.2 (ii) that $(0, x + y) \in G(U)$. This leads to $x + y \in U(0)$. Let $\alpha \in \mathbb{K}^*$ and $x \in U(0)$. Arguing as above, we show that $\begin{pmatrix} 0 \\ \alpha x \end{pmatrix} \in G(U)$ which yields $\alpha x \in U(0)$. Hence, $U(0)$ is a vector subspace of Y.
(ii) The fact that $U \in L\mathscr{R}(X, Y)$ implies that $U^{-1} \in L\mathscr{R}(Y, X)$ is a linear relation. Now, by applying (i), we obtain $U^{-1}(0)$ is a vector subspace of X. This completes the proof. Q.E.D.

Lemma 2.6.2 [2, Corollary I.2.9] *Let X, Y be two vector spaces and let $U \in L\mathscr{R}(X, Y)$. Then,*

(i) U is an operator or single-valued if, and only if, $U(0) = \{0\}$.
(ii) U is one-to-one if, and only if, $N(U) = \{0\}$. ◇

Lemma 2.6.3 [2, Corollary I.2.10] *Let X, Y be two vector spaces and let $U \in L\mathscr{R}(X, Y)$. Then, $UU^{-1}(0) = U(0)$ and $U^{-1}U(0) = U^{-1}(0)$.* ◇

2.6.2 *Index and Co-Index of Linear Relation*

Let $U \in L\mathscr{R}(X, Y)$, we define the following quantities

$$\alpha(U) = \dim(N(U)),$$

$$\beta(U) = \dim(Y/R(U)),$$

and

$$\overline{\beta}(U) = \dim(Y/\overline{R(U)}).$$

If $\alpha(U)$ and $\beta(U)$ are finite, then the index of U is defined by

$$i(U) = \alpha(U) - \beta(U).$$

The index of U^{-1} is called co-index of U and is denoted by $ci(U)$. It has been observed that

$$ci(U) = \alpha(U^{-1}) - \beta(U^{-1}).$$

Since $\alpha(U^{-1}) = \dim(U(0))$ and $\beta(U^{-1}) = \operatorname{codim}(\mathscr{D}(U))$, it follows that

$$ci(U)) = \dim(U(0)) - \operatorname{codim}(\mathscr{D}(U)).$$

So, if we take U is a single-valued operator everywhere defined, then $ci(U) = 0$.

Remark 2.6.2 Let X, Y be two normed spaces and $U \in L\mathscr{R}(X, Y)$. If $R(U)$ is closed, then $\beta(U) = \overline{\beta}(U)$. ◇

2.7 Algebraic Properties of Linear Relations

2.7.1 *Sum of Multivalued Linear Operators*

Definition 2.7.1 Let $U, V \in L\mathscr{R}(X, Y)$. The linear relation $V + U$ is defined by

$$(V + U)x := Vx + Ux,$$

for $x \in \mathscr{D}(V + U)$. ◇

Remark 2.7.1 Let $U, V, W \in L\mathscr{R}(X, Y)$. Then,

(i) $\mathscr{D}(U + V) = \mathscr{D}(U) \bigcap \mathscr{D}(V)$.
(ii) The algebraic sum $U + V$ is also a linear relation defined by

$$\begin{aligned} G(U+V) &= \left\{ \begin{pmatrix} x \\ y+z \end{pmatrix} \in X \times Y \text{ such that } \begin{pmatrix} x \\ y \end{pmatrix} \in G(U),\ \begin{pmatrix} x \\ z \end{pmatrix} \in G(V) \right\} \\ &= \left\{ \begin{pmatrix} x \\ y \end{pmatrix} \in X \times Y \text{ such that } y = u + v \text{ with } \begin{pmatrix} x \\ u \end{pmatrix} \in G(U),\ \begin{pmatrix} x \\ v \end{pmatrix} \in G(V) \right\}. \end{aligned}$$

More particularly, if $U \in L\mathscr{R}(X)$, then for all $\eta \in \mathbb{C}$, we have

$$G(\eta - U) = \left\{ \begin{pmatrix} x \\ \eta x - y \end{pmatrix} \in X \times X \text{ such that } \begin{pmatrix} x \\ y \end{pmatrix} \in G(U) \right\}.$$

(iii) $U + (W + V) = (U + W) + V$.
(iv) $V + U = U + V$.
(v) $\alpha(\beta U) = (\alpha\beta)U$, for all $\alpha, \beta \in \mathbb{K}$. ◇

(vi) For all $\alpha \in \mathbb{K}$, we may define

$$G(\alpha U) = \left\{ \begin{pmatrix} x \\ \alpha y \end{pmatrix} \in X \times Y \text{ such that } \begin{pmatrix} x \\ y \end{pmatrix} \in G(U) \right\}.$$

Remark 2.7.2 Let X, Y be two vector spaces V, $U \in L\mathcal{R}(X, Y)$ and let α be a scalar. Then,

(i) $V + U$ and αU are linear relations.
(ii) $0\,U$ is a linear operator and $0\,U = 0_{\mathcal{D}(U)}$. Indeed, by using (iv) of Remark 2.7.1, we have

$$\begin{aligned} G(0\,U) &= \left\{ \begin{pmatrix} x \\ 0\,y \end{pmatrix} \in X \times Y : \begin{pmatrix} x \\ y \end{pmatrix} \in G(U) \right\} \\ &= \left\{ \begin{pmatrix} x \\ 0 \end{pmatrix} \in X \times Y : x \in \mathcal{D}(U) \right\} \\ &= G(0_{\mathcal{D}(U)}). \end{aligned}$$

Hence, $0\,U$ is a linear operator.
(iii) We signal one of the differences between the theory of linear relations and the theory of the linear operators. If U is an operator, then we have $U - U = 0_{\mathcal{D}(U)}$, otherwise $U - U \neq 0_{\mathcal{D}(U)}$. ◇

Proposition 2.7.1 *Let X, Y be two vector spaces and $U \in L\mathcal{R}(X, Y)$. Then, the relation $U - U$ is given by*

$$G(U - U) = \mathcal{D}(U) \times U(0). \qquad \diamond$$

Proof Let $\begin{pmatrix} x \\ t \end{pmatrix} \in G(U - U)$. It derives from Remark 2.7.1 (ii) that there exists $y, z \in Y$ such that $\begin{pmatrix} x \\ y \end{pmatrix} \in G(U)$, $\begin{pmatrix} x \\ z \end{pmatrix} \in G(-U)$ and $t = y + z$. Since $\begin{pmatrix} x \\ z \end{pmatrix} \in G(-U)$, then by using Remark 2.7.1 (iv), we obtain $\begin{pmatrix} x \\ -z \end{pmatrix} \in G(U)$. Because $U \in L\mathcal{R}(X, Y)$ acquired from Proposition 2.6.2 that

$$\begin{pmatrix} 0 \\ y + z \end{pmatrix} = \begin{pmatrix} x \\ y \end{pmatrix} - \begin{pmatrix} x \\ -z \end{pmatrix} \in G(U)$$

which yields $t = y + z \in U(0)$. Hence,

$$\begin{pmatrix} x \\ t \end{pmatrix} \in \mathcal{D}(U) \times U(0).$$

Conversely, let $\begin{pmatrix} x \\ t \end{pmatrix} \in \mathcal{D}(U) \times U(0)$. Since $U \in L\mathcal{R}(X, Y)$ and $x \in \mathcal{D}(U)$, then

$$U(x) + U(-x) = U(0).$$

This implies that there exists $y \in U(x)$ and $z \in U(-x)$ such that $t = y + z$. Consequently, $\begin{pmatrix} x \\ y \end{pmatrix} \in G(U)$ and $\begin{pmatrix} -x \\ z \end{pmatrix} \in G(U)$. As a result of the fact $U \in L\mathcal{R}(X, Y)$

implies from Proposition 2.6.2 that $\begin{pmatrix} x \\ -z \end{pmatrix} \in G(U)$. The use of Remark 2.7.1 (iv) we mean that $\begin{pmatrix} x \\ z \end{pmatrix} \in G(-U)$. This enables us presume that $\begin{pmatrix} x \\ t \end{pmatrix} = \begin{pmatrix} x \\ y+z \end{pmatrix} \in G(U - U)$. Q.E.D.

Proposition 2.7.2 *Let $V, U \in L\mathscr{R}(X, Y)$. Then,*
(i) *If $V(0) \subset U(0)$ and $\mathscr{D}(U) \subset \mathscr{D}(V)$, then $U + V - V = U$.*
(ii) *If $G(U) \subset G(V)$, then $G(U) \subset G(U + V)$.* ◇

Proof (i) Let $\begin{pmatrix} x \\ y \end{pmatrix} \in G(U + V - V)$. Then, $x \in \mathscr{D}(U - V + V) = \mathscr{D}(U)$ and $y \in (U - V + V)x$, in order that, $y \in Ux + V(0)$. Since $V(0) \subset U(0)$, in that case

$$Ux + V(0) \subset Ux + U(0)$$

which harvests $y \in Ux$ and $x \in \mathscr{D}(U)$, that is $\begin{pmatrix} x \\ y \end{pmatrix} \in G(U)$. On that account

$$G(U + V - V) \subset G(U).$$

Conversely, let $\begin{pmatrix} x \\ y \end{pmatrix} \in G(U)$, in order that $x \in \mathscr{D}(U)$ and $Ux = y + U(0)$. On the other hand, out of $(U + V - V)(0) = U(0)$ and $\mathscr{D}(U - V + V) = \mathscr{D}(U)$, we can infer $x \in \mathscr{D}(U - V + V)$ and

$$(U + V - V)x = y + (U + V - V)(0).$$

Then,

$$G(U) \subset G(U + V - V).$$

(ii) We have

$$\begin{aligned} G(U) &= \left\{ \begin{pmatrix} x \\ y \end{pmatrix} \in X \times Y \text{ such that } x \in \mathscr{D}(U) \subset \mathscr{D}(V) \text{ and } y \in Ux \subset Vx \right\}, \\ &\subset \left\{ \begin{pmatrix} x \\ y \end{pmatrix} \in X \times Y \text{ such that } x \in \mathscr{D}(U) \bigcap \mathscr{D}(V) \textit{ and } y \in Ux + Vx \right\}, \\ &= \left\{ \begin{pmatrix} x \\ y \end{pmatrix} \in X \times Y \text{ such that } x \in \mathscr{D}(U + V) \textit{ and } y \in (U + V)x \right\}, \\ &= G(U + V). \end{aligned}$$

This achieves the proof. Q.E.D.

2.7.2 Extensions and Restrictions of Linear Relation

Definition 2.7.2 (i) Let M be a subspace of X and $U \in L\mathcal{R}(X, Y)$ be such that $M \bigcap \mathcal{D}(U) \neq \emptyset$. The restriction of U to M, denoted by $U_{|M}$, is defined by

$$U_{|M} \in L\mathcal{R}(X, Y)$$

with

$$\begin{aligned} G(U_{|M}) &= \left\{ \begin{pmatrix} x \\ y \end{pmatrix} \in G(U) \text{ such that } x \in M \right\} \\ &= G(U) \bigcup (M \times Y), \end{aligned}$$

$$\mathcal{D}(U_{|M}) = \mathcal{D}(U) \bigcap M$$

and

$$(U_{|M})m = Um \ \text{ for all } \ m \in M.$$

(ii) Let $V \in L\mathcal{R}(X, Y)$. We claim that V is said an extension of U, if $V \mid_{\mathcal{D}(U)} = U$. We write $U \subset V$. It is obvious that

$$U_{|M} = U_{|\mathcal{D}(U) \bigcap M}.$$

◇

Definition 2.7.3 Let X be a Banach space, M be a subspace of X and $U \in L\mathcal{R}(X)$. We say that M is a core of U if

$$G(U) \subset \overline{G(U_{|M})}.$$

◇

It becomes apparent that if V is an extension of U, then

$$G(U) \subset G(V).$$

Lemma 2.7.1 *Let U, $V \in L\mathcal{R}(X, Y)$ with $G(U) \subset G(V)$. Then,*

(i) $\alpha(U) \leq \alpha(V)$.
(ii) $\beta(V) \leq \beta(U)$.
(iii) $i(U) \leq i(V)$.

◇

Proof (i) It becomes clear that

$$\begin{aligned} N(U) &= \left\{ x \in \mathscr{D}(U) \text{ such that } \begin{pmatrix} x \\ 0 \end{pmatrix} \in G(U) \right\}, \\ &\subset \left\{ x \in \mathscr{D}(U) \text{ such that } \begin{pmatrix} x \\ 0 \end{pmatrix} \in G(U) \right\}, \\ &= N(V_{|\mathscr{D}(U)}), \\ &\subset N(V). \end{aligned}$$

So,

$$\alpha(U) \leq \alpha(V).$$

(ii) Let $y \in R(U)$. Then, $y \in Ux$ for all $x \in \mathscr{D}(U)$. We get by $G(U) \subset G(V)$, $y \in Vx$ for all $x \in \mathscr{D}(U)$. So, $y \in R(V_{|\mathscr{D}(U)})$. Thus, $y \in R(V)$. We infer that

$$Y/_{R(V)} \subset Y/_{R(U)}.$$

Hence,

$$\beta(V) \leq \beta(U).$$

(iii), we have

$$\begin{aligned} i(U) &= \alpha(U) - \beta(U) \\ &\leq \alpha(V) - \beta(V) \\ &= i(V). \end{aligned}$$

This achieves the proof. Q.E.D.

Proposition 2.7.3 *(i) If $G(V) \subset G(U)$, then U is an extension of V if, and only if, $V(0) = U(0)$.*
(ii) If $G(V) \subset G(U)$, $N(U) \subset N(V)$, and $R(U) \subset R(V)$, then $U = V$.
(iii) If $U(0) = V(0)$ and $\mathscr{D}(U) = \mathscr{D}(V)$, then $U = V$ or the graphs of U and V are incomparable. ◇

Proof Exercise. Q.E.D.

Example 2.7.1 Let V be single-valued operator such that $V \neq 0$ and let $W = -V$, and $U(0) = Y$. Then, $(W + V)U(0) = \{0\}$ but $(WU + VU)(0) = VU(0) - VU(0) = VU(0) = V(Y) \neq \{0\}$. So, by Proposition 2.7.3 (i), $WU + VU$ is not an extension of $(W + V)U$.

Example 2.7.2 Let $X = Y = \mathbb{R}^2$ and consider the matrix operator

$$U = \begin{pmatrix} 1 & 0 \\ 0 & 0 \end{pmatrix}.$$

Writing $u = \begin{pmatrix} u_1 \\ u_2 \end{pmatrix}$, $y = \begin{pmatrix} v_1 \\ y_2 \end{pmatrix}$ for arbitrary vectors in $\mathbb{R}^2$, we have

$$G(U) = \left\{ \left(\begin{pmatrix} u_1 \\ u_2 \end{pmatrix}, \begin{pmatrix} v_1 \\ v_2 \end{pmatrix} \right) \ : \ \begin{pmatrix} u_1 \\ u_2 \end{pmatrix} \in \mathbb{R}^2, \begin{pmatrix} v_1 \\ v_2 \end{pmatrix} = \begin{pmatrix} u_1 \\ 0 \end{pmatrix} \right\}$$

and

$$U^{-1} \begin{pmatrix} u_1 \\ 0 \end{pmatrix} = \begin{pmatrix} u_1 \\ 0 \end{pmatrix} + U^{-1}(0),$$

where $U^{-1}(0) = \left\{ \begin{pmatrix} 0 \\ u_2 \end{pmatrix}, \text{ for } u_2 \in \mathbb{R} \right\}$ is a linear subspace of $\mathbb{R}^2$. It is simple to verify that U^{-1} is a linear relation with domain

$$\mathscr{D}(U^{-1}) = \left\{ \begin{pmatrix} u_1 \\ 0 \end{pmatrix} : u_1 \in \mathbb{R} \right\}.$$

Consider the relation $V \in L\mathscr{R}(X)$ defined by

$$V \begin{pmatrix} u_1 \\ u_2 \end{pmatrix} = \begin{pmatrix} u_1 \\ 0 \end{pmatrix} + U^{-1}(0).$$

Then, V is an extension of U^{-1}. Moreover, writing $J = J_{R(U)}$, and according to Example 2.5.2, we have

$$\begin{aligned} V I_{R(T)} &= V J J^{-1} \\ &= U^{-1}. \end{aligned}$$

Thus, $U = J J^{-1} V^{-1}$ and $U \neq V^{-1}$.

In the next proposition we give conditions for an extension V of a linear relation U to be equal to U.

Proposition 2.7.4 *Let W, U, and $V \in L\mathscr{R}(X, Y)$ be such that $N(V) = N(U)$ and $R(V) = R(U)$.*

(i) If V is an extension of U, then $V = U$.
(ii) If $W(0) \subset U(0)$, $\mathscr{D}(U) \subset \mathscr{D}(W)$ and $V \subset U - W$, then $V + W \subset U$.
(iii) If V be an extension of U and $i(U) = i(V)$, then $U = V$. ◇

Proof (i) Let $\begin{pmatrix} x \\ y \end{pmatrix} \in G(V)$. Then, $y \in Vx \subset R(V) = R(U)$. For this reason, there exists $x' \in \mathscr{D}(U)$ such that $y \in Ux' = Vx'$. Since $Vx = Vx'$ implies $x - x' \in N(V) = N(U) \in \mathscr{D}(U)$. So,

$$x = (x - x') + x' \in \mathscr{D}(U).$$

Thus,

$$\begin{pmatrix} x \\ y \end{pmatrix} \in G(U).$$

(ii) Let $\begin{pmatrix} x \\ y \end{pmatrix} \in G(V+W)$, then there exist u, v such that $y = u + v$, $\begin{pmatrix} x \\ u \end{pmatrix} \in G(V)$ and $\begin{pmatrix} x \\ v \end{pmatrix} \in G(W)$. On the other hand, $G(V) \subset G(U-W)$, then $\begin{pmatrix} x \\ u \end{pmatrix} \in G(U - W)$. Hence, $\begin{pmatrix} x \\ y \end{pmatrix} \in G(U - W + W)$. However, $W(0) \subset U(0)$ and $\mathscr{D}(U) \subset \mathscr{D}(W)$. Then, on the basis of Proposition 2.7.2 (i), we have

$$G(U - W + W) = G(U).$$

This implies that

$$G(V + W) \subset G(U).$$

(iii) To establishe that $U = V$ it suffices to prove that $V \subset U$. It is simple to verify that

$$\alpha(U) \leq \alpha(V)$$

and

$$\beta(V) \leq \beta(U).$$

Therefore, $i(U) = i(V)$, we have $N(U) = N(V)$ and $R(U) = R(V)$ and by using (i), we have $U = V$. This achieves the proof. Q.E.D.

2.7.3 Addition of 2 × 2 Matrices Linear Relations

Lemma 2.7.2 *Let $A_1, A_2 \in L\mathscr{R}(X)$, $B_1, B_2 \in L\mathscr{R}(Y, X)$, $C_1, C_2 \in L\mathscr{R}(X, Y)$ and $D_1, D_2 \in L\mathscr{R}(Y)$. Then,*

$$\begin{pmatrix} A_1 & B_1 \\ C_1 & D_1 \end{pmatrix} + \begin{pmatrix} A_2 & B_2 \\ C_2 & D_2 \end{pmatrix} = \begin{pmatrix} A_1 + A_2 & B_1 + B_2 \\ C_1 + C_2 & D_1 + D_2 \end{pmatrix}. \tag{2.7.1}$$

◇

Proof It is obvious that

$$\left[\begin{pmatrix} A_1 & B_1 \\ C_1 & D_1 \end{pmatrix} + \begin{pmatrix} A_2 & B_2 \\ C_2 & D_2 \end{pmatrix}\right]\begin{pmatrix} 0 \\ 0 \end{pmatrix} = \begin{pmatrix} A_1 & B_1 \\ C_1 & D_1 \end{pmatrix}\begin{pmatrix} 0 \\ 0 \end{pmatrix} + \begin{pmatrix} A_2 & B_2 \\ C_2 & D_2 \end{pmatrix}\begin{pmatrix} 0 \\ 0 \end{pmatrix},$$
$$= \begin{pmatrix} A_1(0) + B_1(0) \\ C_1(0) + D_1(0) \end{pmatrix} + \begin{pmatrix} A_2(0) + B_2(0) \\ C_2(0) + D_2(0) \end{pmatrix},$$
$$= \begin{pmatrix} A_1(0) + A_2(0) + B_1(0) + B_2(0) \\ C_1(0) + C_2(0) + D_1(0) + D_2(0) \end{pmatrix},$$
$$= \begin{pmatrix} (A_1 + A_2)(0) + (B_1 + B_2)(0) \\ (C_1 + C_2)(0) + (D_1 + D_2)(0) \end{pmatrix}.$$

We infer that

$$\left[\begin{pmatrix} A_1 & B_1 \\ C_1 & D_1 \end{pmatrix} + \begin{pmatrix} A_2 & B_2 \\ C_2 & D_2 \end{pmatrix}\right]\begin{pmatrix} 0 \\ 0 \end{pmatrix} = \begin{pmatrix} A_1 + A_2 & B_1 + B_2 \\ C_1 + C_2 & D_1 + D_2 \end{pmatrix}\begin{pmatrix} 0 \\ 0 \end{pmatrix}. \tag{2.7.2}$$

Furthermore,

$$\mathscr{D}\left[\begin{pmatrix} A_1 & B_1 \\ C_1 & D_1 \end{pmatrix} + \begin{pmatrix} A_2 & B_2 \\ C_2 & D_2 \end{pmatrix}\right]$$
$$= \mathscr{D}\left[\begin{pmatrix} A_1 & B_1 \\ C_1 & D_1 \end{pmatrix}\right]\bigcap \mathscr{D}\left[\begin{pmatrix} A_2 & B_2 \\ C_2 & D_2 \end{pmatrix}\right]$$
$$= \left(\mathscr{D}(A_1)\bigcap \mathscr{D}(C_1)\right) \times \left(\mathscr{D}(B_1)\bigcap \mathscr{D}(D_1)\right)\bigcap\left(\mathscr{D}(A_2)\bigcap \mathscr{D}(C_2)\right) \times \left(\mathscr{D}(B_2)\bigcap \mathscr{D}(D_2)\right)$$
$$= \left(\mathscr{D}(A_1)\bigcap \mathscr{D}(A_2)\right)\bigcap\left(\mathscr{D}(C_1)\bigcap \mathscr{D}(C_2)\right) \times \left(\mathscr{D}(B_1)\bigcap \mathscr{D}(B_2)\right)\bigcap\left(\mathscr{D}(D_1)\bigcap \mathscr{D}(D_2)\right)$$
$$= \left(\mathscr{D}(A_1 + A_2)\bigcap \mathscr{D}(C_1 + C_2)\right) \times \left(\mathscr{D}(B_1 + B_2)\bigcap \mathscr{D}(D_1 + D_2)\right).$$

We obtain

$$\mathscr{D}\left[\begin{pmatrix} A_1 & B_1 \\ C_1 & D_1 \end{pmatrix} + \begin{pmatrix} A_2 & B_2 \\ C_2 & D_2 \end{pmatrix}\right] = \mathscr{D}\left[\begin{pmatrix} A_1 + A_2 & B_1 + B_2 \\ C_1 + C_2 & D_1 + D_2 \end{pmatrix}\right]. \tag{2.7.3}$$

Let $\left[\begin{pmatrix} x_1 \\ y_1 \end{pmatrix}, \begin{pmatrix} x_2 \\ y_2 \end{pmatrix}\right] \in G\left[\begin{pmatrix} A_1 & B_1 \\ C_1 & D_1 \end{pmatrix} + \begin{pmatrix} A_2 & B_2 \\ C_2 & D_2 \end{pmatrix}\right]$. Then, there exist $\begin{pmatrix} x_3 \\ y_3 \end{pmatrix}$ and $\begin{pmatrix} x_4 \\ y_4 \end{pmatrix}$ such that

$$\begin{pmatrix} x_2 \\ y_2 \end{pmatrix} = \begin{pmatrix} x_3 \\ y_3 \end{pmatrix} + \begin{pmatrix} x_4 \\ y_4 \end{pmatrix},$$

with

$$\begin{pmatrix} x_3 \\ y_3 \end{pmatrix} \in \begin{pmatrix} A_1 & B_1 \\ C_1 & D_1 \end{pmatrix}\begin{pmatrix} x_1 \\ y_1 \end{pmatrix}$$

and

$$\begin{pmatrix} x_4 \\ y_4 \end{pmatrix} \in \begin{pmatrix} A_2 & B_2 \\ C_2 & D_2 \end{pmatrix} \begin{pmatrix} x_1 \\ y_1 \end{pmatrix}.$$

We get $x_3 \in A_1x_1 + B_1y_1$, $y_3 \in C_1x_1 + D_1y_1$, $x_4 \in A_2x_1 + B_2y_1$ and $y_4 \in C_2x_1 + D_2y_1$. It means that

$$\begin{aligned} \begin{pmatrix} x_2 \\ y_2 \end{pmatrix} &= \begin{pmatrix} x_3 + x_4 \\ y_3 + y_4 \end{pmatrix} \\ &\in \begin{pmatrix} A_1x_1 + B_1y_1 + A_2x_1 + B_2y_1 \\ C_1x_1 + D_1y_1 + C_2x_1 + D_2y_1 \end{pmatrix}. \end{aligned}$$

In addition

$$\begin{pmatrix} A_1x_1 + B_1y_1 + A_2x_1 + B_2y_1 \\ C_1x_1 + D_1y_1 + C_2x_1 + D_2y_1 \end{pmatrix} = \begin{pmatrix} A_1 + A_2 & B_1 + B_2 \\ C_1 + C_2 & D_1 + D_2 \end{pmatrix} \begin{pmatrix} x_1 \\ y_1 \end{pmatrix}.$$

Consequently,

$$G\left[\begin{pmatrix} A_1 & B_1 \\ C_1 & D_1 \end{pmatrix} + \begin{pmatrix} A_2 & B_2 \\ C_2 & D_2 \end{pmatrix}\right] \subset G\left[\begin{pmatrix} A_1 + A_2 & B_1 + B_2 \\ C_1 + C_2 & D_1 + D_2 \end{pmatrix}\right]. \tag{2.7.4}$$

Using estimations (2.7.2), (2.7.3), (2.7.4) and Proposition 2.7.3 (iii), we have Eq. (2.7.1). This achieves the proof. Q.E.D.

2.8 Product of Linear Relations

2.8.1 Product of two Linear Relations

Definition 2.8.1 Given $U \in L\mathscr{R}(X, Y)$ and $V \in L\mathscr{R}(Y, Z)$. The product VU is a linear relation defined as

$$VUx = V(Ux), \quad x \in \mathscr{D}(VU).$$

If $\alpha \in \mathbb{K}$, we define the relation αU by

$$(\alpha U)x = \alpha(Ux), \quad x \in \mathscr{D}(U). \qquad \diamond$$

Clearly

$$G(VU) = \left\{\begin{pmatrix} x \\ z \end{pmatrix} \in X \times Z \text{ such that } \begin{pmatrix} x \\ y \end{pmatrix} \in G(U) \text{ and } \begin{pmatrix} y \\ z \end{pmatrix} \in G(V) \text{ for some } y \in Y\right\}, \tag{2.8.1}$$

and

$$\begin{aligned}\mathscr{D}(VU) &= \{x \in X \text{ such that } VUx \neq \emptyset\}, \\ &= \left\{x \in X \text{ such that } Ux \bigcap \mathscr{D}(V) \neq \emptyset\right\}, \\ &= U^{-1}(\mathscr{D}(V)), \qquad (2.8.2) \\ &= U^{-1}V^{-1}(R(V)). \qquad (2.8.3)\end{aligned}$$

Lemma 2.8.1 [2, Corollary I.3.2] *Let X, Y, Z be vector spaces and let $U \in L\mathscr{R}(X, Y)$ and $V \in L\mathscr{R}(Y, Z)$. Then, $VU \in L\mathscr{R}(X, Z)$.* ◇

Lemma 2.8.2 (Index of product) [2, Corollary I.6.12] *Let X, Y, Z be three vector spaces, $U \in L\mathscr{R}(X, Y)$, $V \in L\mathscr{R}(Y, Z)$ and $\mathscr{D}(V) = Y$. If we believe that $i(U) < \infty$ and $i(V) < \infty$, then $i(VU) < \infty$ and in this case we have*

$$i(VU) = i(V) - \dim\left(U(0) \bigcap V^{-1}(0)\right) + i(U).$$

◇

Proposition 2.8.1 [2, Proposition I.3.1] *Let $U \in L\mathscr{R}(X)$, $\alpha \in \mathbb{K}^*$, and let $M, N \subset X$. Then,*

(i) $U(\alpha M) = \alpha U(M)$,
(ii) $U(M) + U(N) \subset U(M + N)$,
(iii) if $M \subset \mathscr{D}(U)$ or $N \subset \mathscr{D}(U)$, then $U(M + N) = U(M) + U(N)$. Moreover, if $M \bigcap N = \{0\}$, then $U(M \bigcap N) = U(0) = U(M) \bigcap U(N)$,
(iv) $UU^{-1}(M) = M \bigcap R(U) + U(0)$, for all $M \subset X$, where $R(U)$ is the range of U,
(v) $U^{-1}U(M) = M \bigcap \mathscr{D}(U) + N(U)$, for all $M \subset X$, where $N(U)$ is the kernel of the linear relation U. ◇

Theorem 2.8.1 [2, Proposition I.2.8 and Corollary I.2.11] *Let $U \in L\mathscr{R}(X, Y)$. Then,*

(i) for $x \in \mathscr{D}(U)$, we have

$$y \in Ux \text{ if, and only if, } Ux = y + U(0).$$

Specially,

$$0 \in Ux \text{ if, and only if, } Ux = U(0).$$

(ii) For $x_1, x_2 \in \mathscr{D}(U)$, we have

$$Ux_1 \bigcap Ux_2 \neq \emptyset \text{ if, and only if, } Ux_1 = Ux_2.$$

(iii) $U^{-1}Ux = x + U^{-1}(0)$, for all $x \in \mathscr{D}(U)$.
(iv) $UU^{-1}y = y + U(0)$, for all $y \in R(U)$. ◇

Remark 2.8.1 (*i*) The statement (*i*) of Theorem 2.8.1 is equivalent to say that

$$N(U) = \{x \in \mathscr{D}(U) \text{ such that } Ux = U(0)\}.$$

(ii) Obvious U is one-to-one if, and only if, $U^{-1}U = I_{\mathscr{D}(U)}$.
(iii) $U(0) = \{0\}$ if, and only if, $UU^{-1} = I_{R(U)}$. ◇

Proposition 2.8.2 *Let $U \in L\mathscr{R}(X, Y)$ and $R, V \in L\mathscr{R}(Z, X)$. Then,*

(i) $\alpha(VU) = (\alpha V)U = V(\alpha U)$ for all $\alpha \in \mathbb{K} \setminus \{0\}$.
(ii) If $G(V) \subset G(U)$, then $G(VR) \subset G(UR)$.
(iii) $(R + V)U \subset RU + VU$ with equality if $U(0) = \{0\}$.
(iv) $UR + UV = U(R + V)$ if $\mathscr{D}(U)$ contains both the range of R and V. ◇

Proof (i) Let $\alpha \in \mathbb{K} \setminus \{0\}$. Then,

$$\begin{aligned}
\begin{pmatrix} x \\ z \end{pmatrix} \in G((\alpha V)U) &\Longleftrightarrow \exists y \in \mathscr{D}(V) \text{ such that } \begin{pmatrix} x \\ y \end{pmatrix} \in G(U) \text{ and } (y, z) \in G(\alpha V) \\
&\Longleftrightarrow \exists y \in \mathscr{D}(V) : \begin{pmatrix} x \\ y \end{pmatrix} \in G(U) \text{ and } \begin{pmatrix} y \\ \alpha^{-1} z \end{pmatrix} \in G(V) \\
&\Longleftrightarrow \begin{pmatrix} x \\ z \end{pmatrix} \in G(\alpha(VU)) \\
&\Longleftrightarrow \begin{pmatrix} x \\ \alpha^{-1} z \end{pmatrix} \in G(VU) \\
&\Longleftrightarrow \exists y \in \mathscr{D}(V) : \begin{pmatrix} x \\ \alpha^{-1} y \end{pmatrix} \in G(U) \text{ and } \begin{pmatrix} \alpha^{-1} y \\ \alpha^{-1} z \end{pmatrix} \in G(V) \\
&\Longleftrightarrow \exists y \in \mathscr{D}(V) : \begin{pmatrix} x \\ y \end{pmatrix} \in G(\alpha T) \text{ and } \begin{pmatrix} y \\ z \end{pmatrix} \in G(V) \\
&\Longleftrightarrow \begin{pmatrix} x \\ z \end{pmatrix} \in G(V(\alpha U)).
\end{aligned}$$

(ii) Let $\begin{pmatrix} z \\ y \end{pmatrix} \in G(VR)$. Then, there exists $x \in \mathscr{D}(V)$ such that $\begin{pmatrix} z \\ x \end{pmatrix} \in G(R)$ and $\begin{pmatrix} x \\ y \end{pmatrix} \in G(V) \subset G(U)$. So, $\begin{pmatrix} z \\ y \end{pmatrix} \in G(UR)$.
(iii) Let $R, V \in L\mathscr{R}(Y, Z)$ and $U \in L\mathscr{R}(X, Y)$. Then,

$$\begin{aligned}
G((R+V)U) &= \left\{ \begin{pmatrix} x \\ z \end{pmatrix} \text{ such that there is } y \in \mathscr{D}(R+V),\ \begin{pmatrix} x \\ y \end{pmatrix} \in G(U) \text{ and } \begin{pmatrix} y \\ z \end{pmatrix} \in G(R+V) \right\} \\
&= \left\{ \begin{pmatrix} x \\ z \end{pmatrix} : \begin{pmatrix} x \\ y \end{pmatrix} \in G(U) \text{ and } z = z_1 + z_2, \text{ where } \begin{pmatrix} y \\ z_1 \end{pmatrix} \in G(R) \text{ and } \begin{pmatrix} y \\ z_2 \end{pmatrix} \in G(V) \right\} \\
&\subset \left\{ \begin{pmatrix} x \\ z = z_1 + z_2 \end{pmatrix} : \begin{pmatrix} x \\ z_1 \end{pmatrix} \in G(RU) \text{ and } \begin{pmatrix} x \\ z_2 \end{pmatrix} \in G(VU) \right\} \\
&= G(RU + VU).
\end{aligned}$$

Now, we suppose that U is single-valued. Seeing that Proposition 2.7.3 (iii),

$$\begin{cases} \mathscr{D}(RU + VU) = \mathscr{D}((R+V)U), \\ (RU + VU)(0) = (R+V)U(0). \end{cases}$$

On the one hand,

$$\begin{aligned} x \in \mathscr{D}(RU+VU) &\Longleftrightarrow x \in \mathscr{D}(RU) \bigcap \mathscr{D}(VU) \\ &\Longleftrightarrow x \in U^{-1}(\mathscr{D}(R)) \bigcap U^{-1}(\mathscr{D}(V)) \\ &\Longleftrightarrow x \in U^{-1}y, \quad y \in \mathscr{D}(R) \text{ and } x \in U^{-1}z, \quad z \in \mathscr{D}(V) \\ &\Longleftrightarrow Ux \in UU^{-1}y = y + U(0) = y \text{ and } Ux \in UU^{-1}z = z + U(0) = z \\ &\Longleftrightarrow Ux \in \mathscr{D}(R) \text{ and } Ux \in \mathscr{D}(V) \\ &\Longleftrightarrow Ux \in \mathscr{D}(R) \bigcap \mathscr{D}(V) = \mathscr{D}(R+V) \\ &\Longleftrightarrow x \in U^{-1}(\mathscr{D}(R+V)) = \mathscr{D}\left((R+V)U\right). \end{aligned}$$

So, $\mathscr{D}(RU+VU) = \mathscr{D}\left((R+V)U\right)$. On the other hand,

$$\begin{aligned} (RU+VU)(0) &= RU(0) + UV(0) \\ &= R(0) + V(0) \\ &= (R+V)(0) \\ &= (R+V)U(0). \end{aligned}$$

(iv) Let $U \in L\mathscr{R}(X, Y)$ and $R, V \in L\mathscr{R}(Z, X)$. We have

$$\begin{aligned} G(UR+UV) &= \left\{ \begin{pmatrix} z \\ y \end{pmatrix} \text{ such that } y = r + s \text{ where } (z, r) \in G(UR) \text{ and } (z, s) \in G(UV) \right\} \\ &= \{(z, r+s) : (z, x_1) \in G(R), (x_1, r) \in G(U), (z, x_2) \in G(V) \text{ and } (x_2, s) \in G(U)\} \\ &\subset \{(z, r+s) : (z, x_1 + x_2) \in G(R+V) \text{ and } (x_1 + x_2, r+s) \in G(U)\} \\ &= G\left(U(R+S)\right). \end{aligned}$$

Since $U(R+V)(0) = (UR+UV)(0)$, then $U(R+V)$ is an extension of $UR+UV$. Now, let $z \in \mathscr{D}(TR+UV)$, then according to the foregoing, $z \in \mathscr{D}\left(U(R+V)\right)$. Let $y \in U(R+V)z$, then $y \in Ux$, where $x \in (R+V)z = Rz + Vz$. Hence, $x = x_1 + x_2$, where $x_1 \in Rz$ and $x_2 \in Vz$. Thus, $y \in U(x_1 + x_2) = Ux_1 + Ux_2 \subset URz + UVz = (TR+UV)(z)$. So,

$$U(R+V)z \subset (UR+UV)(z).$$

Therefore,

$$U(R+V)z = (UR+UV)(z).$$

If $\mathscr{D}(U)$ contains both the range of R and V, then

$$\mathscr{D}\left(U(R+V)\right) = \mathscr{D}(R+V).$$

Indeed,

$$\begin{aligned} x \in \mathscr{D}(R+V) &\Longleftrightarrow (R+V)x \neq \emptyset \\ &\Longleftrightarrow \exists y \in (R+V)x \subset \mathscr{D}(U) \\ &\Longleftrightarrow \exists y \text{ such that } Uy \neq \emptyset \\ &\Longleftrightarrow U(R+V)x \neq \emptyset \\ &\Longleftrightarrow x \in \mathscr{D}\left(U(R+V)\right). \end{aligned}$$

So,

$$\begin{aligned} \mathscr{D}\left(U(R+V)\right) &= \mathscr{D}(R) \bigcap \mathscr{D}(V) \\ &= \mathscr{D}(UR) \bigcap \mathscr{D}(UV) \\ &= \mathscr{D}(UR+UV). \end{aligned}$$

This gives the desired result. Q.E.D.

Example 2.8.1 Let $U = I_{\mathbb{R}^2}$ and V be the inverse of the single valued map $\begin{pmatrix} 0 & 1 \\ 0 & 0 \end{pmatrix}$. Then,

$$G(V) = \left\{ \left(\begin{pmatrix} x \\ 0 \end{pmatrix}, \begin{pmatrix} w \\ x \end{pmatrix} \right) \text{ for } x,\ w \in \mathbb{R} \right\}.$$

$$T = I_{|M}, \text{ where } M = \left\{ \begin{pmatrix} 0 \\ y \end{pmatrix} \text{ for } y \in \mathbb{R} \right\}.$$

Hence, $\mathscr{D}(TU) = \mathscr{D}(T)$ and $\mathscr{D}(TV) = \left\{ \begin{pmatrix} x \\ 0 \end{pmatrix} \text{ for } x \in \mathbb{R} \right\}$.
So, $\mathscr{D}(TU + TV) = \{0\}$.
Now,

$$G(U+V) = \left\{ \left(\begin{pmatrix} x \\ 0 \end{pmatrix}, \begin{pmatrix} z \\ x \end{pmatrix} \right) \text{ for } x,\ z \in \mathbb{R} \right\}.$$

Hence,

$$\begin{aligned} \mathscr{D}(T(U+V)) &= (U+V)^{-1}(\mathscr{D}(T)) \\ &= \left\{ \begin{pmatrix} x \\ 0 \end{pmatrix} \ for\ x \in \mathbb{R} \right\}. \end{aligned}$$

Therefore, $TU + TV \neq T(U+V)$.

Lemma 2.8.3 [20, Lemmas 3.1 and 3.3] *Let X, Y, and Z be Hilbert spaces.*

(i) Let $U, V \in L\mathscr{R}(X,Y)$. If V is one-to-one, then $U + V = (I + UV^{-1})V$.
(ii) Let $U \in L\mathscr{R}(X,Y)$ and $V, W \in L\mathscr{R}(Y,Z)$. Then, $(V+W)U \subset VU + WU$. Further, if U is an single-valued operator, then $(V+W)U = VU + WU$. ◇

Theorem 2.8.2 *Let X, Y, and Z be linear spaces and let $U \in L\mathscr{R}(X, Y)$, $V \in L\mathscr{R}(Y, Z)$. Then,*

(i) $N(U) \subset N(VU) \subset \mathscr{D}(VU) \subset \mathscr{D}(U)$, and
(ii) $V(0) \subset VU(0) \subset R(VU) \subset R(V)$. ◇

Proof (i) Let $x \in N(U)$, so that by using Theorem 2.8.1 (i) and Remark 2.8.1 (i), we have $\begin{pmatrix} x \\ 0 \end{pmatrix} \in G(U)$. Before $\begin{pmatrix} 0 \\ 0 \end{pmatrix} \in G(V)$, we have $\begin{pmatrix} x \\ 0 \end{pmatrix} \in G(VU)$. Then, $x \in N(VU)$. So,

$$N(U) \subset N(VU).$$

With clarity,

$$N(VU) \subset \mathscr{D}(VU).$$

Let $x \in \mathscr{D}(VU)$, so that $\begin{pmatrix} x \\ z \end{pmatrix} \in G(VU)$ for some $z \in Z$. Then, $\begin{pmatrix} x \\ y \end{pmatrix} \in G(U)$ and $\begin{pmatrix} y \\ z \end{pmatrix} \in G(V)$ for some $y \in Y$, which indicates that $x \in \mathscr{D}(U)$. Thus,

$$\mathscr{D}(VU) \subset \mathscr{D}(U).$$

(ii) Let $z \in V(0)$, so that $\begin{pmatrix} 0 \\ z \end{pmatrix} \in G(V)$. Since $\begin{pmatrix} 0 \\ 0 \end{pmatrix} \in G(U)$, accordingly $\begin{pmatrix} 0 \\ z \end{pmatrix} \in G(VU)$ also $z \in VU(0)$ and $V(0) \subset VU(0)$. It is clear that

$$VU(0) \subset R(VU).$$

Now, we take $z \in R(VU)$, so that $\begin{pmatrix} x \\ z \end{pmatrix} \in G(VU)$ for some $x \in X$. Afterwards, $\begin{pmatrix} x \\ y \end{pmatrix} \in G(U)$ and $\begin{pmatrix} y \\ z \end{pmatrix} \in G(V)$ for some $y \in Y$. Thus, $z \in R(V)$. So, $R(VU) \subset R(V)$. This completes the proof. Q.E.D.

2.8.2 *Ascent and Descent of a Linear Relation*

The composition of linear relations is clearly associative and hence for all $n \in \mathbb{N}$, U^n is defined as usual with $U^0 = I$ and $U^1 = U$. It is easy to see that

$$(U^{-1})^n = (U^n)^{-1}, \quad \text{for all } n \in \mathbb{N}.$$

The kernels and the range of the iterates U^n, $n \in \mathbb{N}$, form two increasing and decreasing chains, respectively, i.e., the chain of kernels

$$\{0\} = N(U^0) \subset N(U) \subset N(U^2) \subset \cdots$$

and the chain of range

$$X = R(U^0) \supset R(U) \supset R(U^2) \supset \cdots .$$

Furthermore, if $N(U^k) = N(U^{k+1})$ for some $k \in \mathbb{N}$, then $N(U^n) = N(U^k)$ for all nonnegative integers $n \geq k$ and if $R(U^k) = R(U^{k+1})$ for some $k \in \mathbb{N}$, then $R(U^n) = R(U^k)$ for all nonnegative integers $n \geq k$. Giving $U \in L\mathscr{R}(X)$, $n \in \mathbb{N}$ and $\beta_n(U) = c_n(U) = \dim R(U^n)/R(U^{n+1})$. The descent of U is defined by

$$\begin{aligned} d(U) &= \inf\{n \in \mathbb{N} : c_n(U) = 0\} \\ &= \inf\{n : R(U^n) = R(U^{n+1})\} \end{aligned}$$

and the essential descent of U is defined by

$$d_e(U) = \inf\{n \in \mathbb{N} : c_n(U) < \infty\}.$$

Let $U \in L\mathscr{R}(X)$, $n \in \mathbb{N}$ and let $\alpha_n(U) = c'_n(U) = \dim N(U^{n+1})/N(U^n)$. The ascent of U is defined by

$$\begin{aligned} a(U) &= \inf\{n \in \mathbb{N} : c'_n(U) = 0\} \\ &= \inf\{n : N(U^n) = N(U^{n+1})\} \end{aligned}$$

and the essential ascent of U is defined by

$$a_e(U) = \inf\{n \in \mathbb{N} : c'_n(U) < \infty\}.$$

Obviously

$$a(U) = \inf\{n \in \mathbb{N} : \alpha_n(U) = 0\}$$

If $a_e(U) < \infty$, then we denote

$$p(U) := \inf\{p \in \mathbb{N} :\ \alpha_n(U) = \alpha_p(U),\ \forall n \geq p\}.$$

Clearly

$$d(U) = \inf\{n \in \mathbb{N} : \beta_n(U) = 0\}.$$

If $d_e(U) < \infty$, then we denote

$$q(U) := \inf\{q \in \mathbb{N} :\ \beta_n(U) = \beta_q(U),\ \forall n \geq q\}.$$

The singular chain manifold of U, denoted by $R_c(U)$, is defined by

$$R_c(U) = R_0(U) \bigcap R_\infty(U),$$

where $R_0(U) = \bigcup_{i=1}^{\infty} N(U_i)$ (*or* $N_\infty(U)$) and $R_\infty(U) = \bigcup_{i=1}^{\infty} U_i(0)$. The linear space $R_c(U)$ is non-trivial if, and only if, there exists a number $s \in \mathbb{N}$ and elements $x_i \in X$, $1 \leq i \leq s$, not all equal to zero, such that

$$\begin{pmatrix} 0 \\ x_1 \end{pmatrix}, \begin{pmatrix} x_1 \\ x_2 \end{pmatrix}, \ldots, \begin{pmatrix} x_{s-1} \\ x_s \end{pmatrix}, \begin{pmatrix} x_s \\ 0 \end{pmatrix} \in G(U).$$

2.8.3 *Product of 2 × 2 Matrices of Linear Relations*

Lemma 2.8.4 *Let X and Y be two Banach spaces, A_1, $A_2 \in L\mathscr{R}(X)$, B_1, $B_2 \in L\mathscr{R}(Y, X)$, C_1, $C_2 \in L\mathscr{R}(X, Y)$ and D_1, $D_2 \in L\mathscr{R}(Y)$. Then,*
(i)

$$\begin{pmatrix} A_1 & B_1 \\ C_1 & D_1 \end{pmatrix} \begin{pmatrix} A_2 & B_2 \\ C_2 & D_2 \end{pmatrix} \subset \begin{pmatrix} A_1A_2 + B_1C_2 & A_1B_2 + B_1D_2 \\ C_1A_2 + D_1C_2 & C_1B_2 + D_1D_2 \end{pmatrix}. \tag{2.8.4}$$

Moreover, if $A_2(0) + B_2(0) \subset N(A_1) \bigcap N(C_1)$ and $C_2(0) + D_2(0) \subset N(B_1) \bigcap N(D_1)$ such that $\mathscr{D}(A_1)$ and $\mathscr{D}(C_1)$ contain the ranges of both A_2 and B_2 with $\mathscr{D}(B_1)$ and $\mathscr{D}(D_1)$ containing the ranges of both C_2 and D_2, then

$$\begin{pmatrix} A_1 & B_1 \\ C_1 & D_1 \end{pmatrix} \begin{pmatrix} A_2 & B_2 \\ C_2 & D_2 \end{pmatrix} = \begin{pmatrix} A_1A_2 + B_1C_2 & A_1B_2 + B_1D_2 \\ C_1A_2 + D_1C_2 & C_1B_2 + D_1D_2 \end{pmatrix}. \tag{2.8.5}$$

(ii) If A_2 and B_2 are single-valued, then

$$\begin{pmatrix} A_1 & 0 \\ C_1 & D_1 \end{pmatrix} \begin{pmatrix} A_2 & B_2 \\ C_2 & D_2 \end{pmatrix} = \begin{pmatrix} A_1A_2 & A_1B_2 \\ C_1A_2 + D_1C_2 & C_1B_2 + D_1D_2 \end{pmatrix}. \tag{2.8.6}$$

(iii) If C_2 and D_2 are single-valued, then

$$\begin{pmatrix} A_1 & B_1 \\ 0 & D_1 \end{pmatrix} \begin{pmatrix} A_2 & B_2 \\ C_2 & D_2 \end{pmatrix} = \begin{pmatrix} A_1A_2 + B_1C_2 & A_1B_2 + B_1D_2 \\ D_1C_2 & D_1D_2 \end{pmatrix}.$$

(iv) If C_2 and D_2 are single-valued, then

$$\begin{pmatrix} 0 & B_1 \\ C_1 & D_1 \end{pmatrix} \begin{pmatrix} A_2 & B_2 \\ C_2 & D_2 \end{pmatrix} = \begin{pmatrix} B_1C_2 & B_1D_2 \\ C_1A_2 + D_1C_2 & C_1B_2 + D_1D_2 \end{pmatrix}.$$

(v) If A_2 and B_2 are single-valued, then

$$\begin{pmatrix} A_1 & B_1 \\ C_1 & 0 \end{pmatrix} \begin{pmatrix} A_2 & B_2 \\ C_2 & D_2 \end{pmatrix} = \begin{pmatrix} A_1A_2 + B_1C_2 & A_1B_2 + B_1D_2 \\ C_1A_2 & C_1B_2 \end{pmatrix}. \qquad \diamond$$

Proof (*i*) Let $\left(\begin{pmatrix} x_1 \\ y_1 \end{pmatrix}, \begin{pmatrix} x_2 \\ y_2 \end{pmatrix}\right) \in G\left(\begin{pmatrix} A_1 & B_1 \\ C_1 & D_1 \end{pmatrix}\begin{pmatrix} A_2 & B_2 \\ C_2 & D_2 \end{pmatrix}\right)$. Then, there exists $\begin{pmatrix} x_3 \\ y_3 \end{pmatrix} \in X \times Y$, such that

$$\left(\begin{pmatrix} x_1 \\ y_1 \end{pmatrix}, \begin{pmatrix} x_3 \\ y_3 \end{pmatrix}\right) \in G\left(\begin{pmatrix} A_2 & B_2 \\ C_2 & D_2 \end{pmatrix}\right)$$

and

$$\left(\begin{pmatrix} x_3 \\ y_3 \end{pmatrix}, \begin{pmatrix} x_2 \\ y_2 \end{pmatrix}\right) \in G\left(\begin{pmatrix} A_1 & B_1 \\ C_1 & D_1 \end{pmatrix}\right).$$

We obtain

$$\begin{pmatrix} x_3 \\ y_3 \end{pmatrix} \in \begin{pmatrix} A_2 & B_2 \\ C_2 & D_2 \end{pmatrix}\begin{pmatrix} x_1 \\ y_1 \end{pmatrix}$$

and

$$\begin{pmatrix} x_2 \\ y_2 \end{pmatrix} \in \begin{pmatrix} A_1 & B_1 \\ C_1 & D_1 \end{pmatrix}\begin{pmatrix} x_3 \\ y_3 \end{pmatrix}.$$

This implies that $x_3 \in A_2x_1 + B_2y_1$, $y_3 \in C_2x_1 + D_2y_1$, $x_2 \in A_1x_3 + B_1y_3$, and $y_2 \in C_1x_3 + D_1y_3$. Thus,

$$\begin{cases} x_2 \in A_1(A_2x_1 + B_2y_1) + B_1(C_2x_1 + D_2y_1) \subset A_1A_2x_1 + A_1B_2y_1 + B_1C_2x_1 + B_1D_2y_1, \\ y_2 \in C_1(A_2x_1 + B_2y_1) + D_1(C_2x_1 + D_2y_1) \subset C_1A_2x_1 + C_1B_2y_1 + D_1C_2x_1 + D_1D_2y_1. \end{cases}$$

We can get

$$\begin{pmatrix} x_2 \\ y_2 \end{pmatrix} \in \begin{pmatrix} (A_1A_2 + B_1C_2)x_1 + (A_1B_2 + B_1D_2)y_1 \\ (C_1A_2 + D_1C_2)x_1 + (D_1C_2 + D_1D_2)y_1 \end{pmatrix}.$$

It follows that

$$\left(\begin{pmatrix} x_1 \\ y_1 \end{pmatrix}, \begin{pmatrix} x_2 \\ y_2 \end{pmatrix}\right) \in G\left(\begin{pmatrix} A_1A_2 + B_1C_2 & A_1B_2 + B_1D_2 \\ C_1A_2 + D_1C_2 & C_1B_2 + D_1D_2 \end{pmatrix}\right).$$

We finish with the below inclusion

$$G\left(\begin{pmatrix} A_1 & B_1 \\ C_1 & D_1 \end{pmatrix}\begin{pmatrix} A_2 & B_2 \\ C_2 & D_2 \end{pmatrix}\right) \subset G\left(\begin{pmatrix} A_1A_2 + B_1C_2 & A_1B_2 + B_1D_2 \\ C_1A_2 + D_1C_2 & C_1B_2 + D_1D_2 \end{pmatrix}\right). \quad (2.8.7)$$

Then, Eq. (2.8.4). In other words,

$$\begin{aligned}\begin{pmatrix} A_1 & B_1 \\ C_1 & D_1 \end{pmatrix}\begin{pmatrix} A_2 & B_2 \\ C_2 & D_2 \end{pmatrix}\begin{pmatrix} 0 \\ 0 \end{pmatrix} &\subset \begin{pmatrix} A_1A_2 + B_1C_2 & A_1B_2 + B_1D_2 \\ C_1A_2 + D_1C_2 & C_1B_2 + D_1D_2 \end{pmatrix}\begin{pmatrix} 0 \\ 0 \end{pmatrix} \\ &= \begin{pmatrix} A_1A_2(0) + B_1C_2(0) + A_1B_2(0) + B_1D_2(0) \\ C_1A_2(0) + D_1C_2(0) + C_1B_2(0) + D_1D_2(0) \end{pmatrix} \\ &= \begin{pmatrix} A_1(A_2(0) + B_2(0)) + B_1(C_2(0) + D_2(0)) \\ C_1(A_2(0) + B_2(0) + D_1(C_2(0) + D_2(0)) \end{pmatrix} \\ &\subset \begin{pmatrix} A_1A_1^{-1}(0) + B_1B_1^{-1}(0) \\ C_1C_1^{-1}(0) + D_1D_1^{-1}(0) \end{pmatrix}.\end{aligned}$$

Moreover, by applying Lemma 2.6.3, we notice that

$$\begin{pmatrix} A_1A_1^{-1}(0) + B_1B_1^{-1}(0) \\ C_1C_1^{-1}(0) + D_1D_1^{-1}(0) \end{pmatrix} = \begin{pmatrix} A_1(0) + B_1(0) \\ C_1(0) + D_1(0) \end{pmatrix}.$$

Hence,

$$\begin{pmatrix} A_1 & B_1 \\ C_1 & D_1 \end{pmatrix}\begin{pmatrix} A_2 & B_2 \\ C_2 & D_2 \end{pmatrix}\begin{pmatrix} 0 \\ 0 \end{pmatrix} \subset \begin{pmatrix} A_1 & B_1 \\ C_1 & D_1 \end{pmatrix}\begin{pmatrix} 0 \\ 0 \end{pmatrix}.$$

Therefore, we obtain

$$\begin{aligned}\begin{pmatrix} A_1 & B_1 \\ C_1 & D_1 \end{pmatrix}\begin{pmatrix} A_2 & B_2 \\ C_2 & D_2 \end{pmatrix}\begin{pmatrix} 0 \\ 0 \end{pmatrix} &\subset \begin{pmatrix} A_1A_2 + B_1C_2 & A_1B_2 + B_1D_2 \\ C_1A_2 + D_1C_2 & C_1B_2 + D_1D_2 \end{pmatrix}\begin{pmatrix} 0 \\ 0 \end{pmatrix} \\ &\subset \begin{pmatrix} A_1 & B_1 \\ C_1 & D_1 \end{pmatrix}\begin{pmatrix} A_2 & B_2 \\ C_2 & D_2 \end{pmatrix}\begin{pmatrix} 0 \\ 0 \end{pmatrix}.\end{aligned}$$

From the above inequality, we have

$$\begin{pmatrix} A_1 & B_1 \\ C_1 & D_1 \end{pmatrix}\begin{pmatrix} A_2 & B_2 \\ C_2 & D_2 \end{pmatrix}\begin{pmatrix} 0 \\ 0 \end{pmatrix} = \begin{pmatrix} A_1A_2 + B_1C_2 & A_1B_2 + B_1D_2 \\ C_1A_2 + D_1C_2 & C_1B_2 + D_1D_2 \end{pmatrix}\begin{pmatrix} 0 \\ 0 \end{pmatrix}. \quad (2.8.8)$$

Moreover,

$$\mathscr{D}\left[\begin{pmatrix} A_1 & B_1 \\ C_1 & D_1 \end{pmatrix}\begin{pmatrix} A_2 & B_2 \\ C_2 & D_2 \end{pmatrix}\right] = \left\{\begin{pmatrix} x \\ y \end{pmatrix} \in X \times Y : \begin{pmatrix} A_1 & B_1 \\ C_1 & D_1 \end{pmatrix}\begin{pmatrix} A_2 & B_2 \\ C_2 & D_2 \end{pmatrix}\begin{pmatrix} x \\ y \end{pmatrix} \neq \emptyset\right\}.$$

Let $\begin{pmatrix} x \\ y \end{pmatrix} \in \mathscr{D}\left[\begin{pmatrix} A_1 & B_1 \\ C_1 & D_1 \end{pmatrix}\begin{pmatrix} A_2 & B_2 \\ C_2 & D_2 \end{pmatrix}\right]$ if, and only if,

$$\begin{pmatrix} A_1 & B_1 \\ C_1 & D_1 \end{pmatrix} \begin{pmatrix} A_2 & B_2 \\ C_2 & D_2 \end{pmatrix} \begin{pmatrix} x \\ y \end{pmatrix} \neq \emptyset$$

if, and only if,

$$\begin{pmatrix} A_1 & B_1 \\ C_1 & D_1 \end{pmatrix} \begin{pmatrix} A_2 x + B_2 y \\ C_2 x + D_2 y \end{pmatrix} \neq \emptyset$$

if, and only if,

$$\begin{pmatrix} A_1(A_2 x + B_2 y) + B_1(C_2 x + D_2 y) \\ C_1(A_2 x + B_2 y) + D_1(C_2 x + D_2 y) \end{pmatrix} \neq \emptyset. \tag{2.8.9}$$

Since $\mathscr{D}(A_1)$ and $\mathscr{D}(C_1)$ contain the ranges of both A_2 and B_2, with $\mathscr{D}(B_1)$ and $\mathscr{D}(D_1)$ containing the ranges of both C_2 and D_2, then we have the equivalence: Eq. (2.8.9) is satisfied if, and only if,

$$\begin{pmatrix} A_1 A_2 x + A_1 B_2 y + B_1 C_2 x + B_1 D_2 y \\ C_1 A_2 x + C_1 B_2 y + D_1 C_2 x + D_1 D_2 y \end{pmatrix} \neq \emptyset$$

if, and only if,

$$\begin{pmatrix} (A_1 A_2 + B_1 C_2)x + (A_1 B_2 + B_1 D_2)y \\ (C_1 A_2 + D_1 C_2)x + (C_1 B_2 + D_1 D_2)y \end{pmatrix} \neq \emptyset$$

if, and only if,

$$\begin{pmatrix} A_1 A_2 + B_1 C_2 & A_1 B_2 + B_1 D_2 \\ C_1 A_2 + D_1 C_2 & C_1 B_2 + D_1 D_2 \end{pmatrix} \begin{pmatrix} x \\ y \end{pmatrix} \neq \emptyset$$

if, and only if,

$$\begin{pmatrix} x \\ y \end{pmatrix} \in \mathscr{D}\left[\begin{pmatrix} A_1 A_2 + B_1 C_2 & A_1 B_2 + B_1 D_2 \\ C_1 A_2 + D_1 C_2 & C_1 B_2 + D_1 D_2 \end{pmatrix}\right].$$

Consequently,

$$\mathscr{D}\left[\begin{pmatrix} A_1 A_2 + B_1 C_2 & A_1 B_2 + B_1 D_2 \\ C_1 A_2 + D_1 C_2 & C_1 B_2 + D_1 D_2 \end{pmatrix}\right] = \mathscr{D}\left[\begin{pmatrix} A_1 & B_1 \\ C_1 & D_1 \end{pmatrix} \begin{pmatrix} A_2 & B_2 \\ C_2 & D_2 \end{pmatrix}\right]. \tag{2.8.10}$$

Using the inclusion (2.8.7), Eqs. (2.8.8), (2.8.10) and Proposition 2.7.3 (iii), we find Eq. (2.8.5).

(ii) Let $\left[\begin{pmatrix} x_1 \\ y_1 \end{pmatrix}, \begin{pmatrix} x_2 \\ y_2 \end{pmatrix}\right] \in G\left[\begin{pmatrix} A_1 A_2 & A_1 B_2 \\ C_1 A_2 + D_1 C_2 & C_1 B_2 + D_1 D_2 \end{pmatrix}\right]$. This implies that $x_2 \in A_1 A_2 x_1 + A_1 B_2 y_1 \subset A_1(A_2 x_1 + B_2 y_1)$ and $y_2 \in C_1 A_2 x_1 + D_1 C_2 x_1 + C_1 B_2 y_1 + D_1 D_2 y_1 \subset C_1(A_2 x_1 + B_2 y_1) + D_1(C_2 x_1 + D_2 y_1)$. It follows that there exist $x_3 \in A_2 x_1 + B_2 y_1$ and $y_3 \in C_2 x_1 + D_2 y_1$ such that $x_2 \in A_1 x_3$ and $y_2 \in C_1 x_3 + D_1 y_3$. Thus,

$$\begin{pmatrix} x_2 \\ y_2 \end{pmatrix} \in \begin{pmatrix} A_1 & 0 \\ C_1 & D_1 \end{pmatrix} \begin{pmatrix} x_3 \\ y_3 \end{pmatrix} \subset \begin{pmatrix} A_1 & 0 \\ C_1 & D_1 \end{pmatrix} \begin{pmatrix} A_2 & B_2 \\ C_2 & D_2 \end{pmatrix} \begin{pmatrix} x_1 \\ y_1 \end{pmatrix}.$$

This is equivalent to

$$\left[\begin{pmatrix} x_1 \\ y_1 \end{pmatrix}, \begin{pmatrix} x_2 \\ y_2 \end{pmatrix} \right] \in G \left[\begin{pmatrix} A_1 & 0 \\ C_1 & D_1 \end{pmatrix} \begin{pmatrix} A_2 & B_2 \\ C_2 & D_2 \end{pmatrix} \right].$$

Hence,

$$\begin{pmatrix} A_1 A_2 & A_1 B_2 \\ C_1 A_2 + D_1 C_2 & C_1 B_2 + D_1 D_2 \end{pmatrix} \subset \begin{pmatrix} A_1 & 0 \\ C_1 & D_1 \end{pmatrix} \begin{pmatrix} A_2 & B_2 \\ C_2 & D_2 \end{pmatrix}. \tag{2.8.11}$$

From estimation (2.8.4) and (2.8.11), we get the equality (2.8.6).
(iii), (iv), and (v) can be checked in the same way as (ii). Q.E.D.

Definition 2.8.2 Let X and Y be two Banach spaces. In the product space $X \times Y$, we consider an multivalued linear operator formally defined by a matrix

$$\mathcal{L}_0 = \begin{pmatrix} A & B \\ C & D \end{pmatrix}, \tag{2.8.12}$$

where $A \in L\mathcal{R}(X)$, $B \in L\mathcal{R}(Y, X)$, $C \in L\mathcal{R}(X, Y)$, and $D \in L\mathcal{R}(Y)$ with its natural domain

$$\mathcal{D}(\mathcal{L}_0) = \left(\mathcal{D}(A) \bigcap \mathcal{D}(C)\right) \times \left(\mathcal{D}(B) \bigcap \mathcal{D}(D)\right). \tag{2.8.13}$$

The graph of $\mathcal{L}_0$ is defined by

$$G(\mathcal{L}_0) = \{((x_1, x_2), (y_1, y_2)) : (x_1, x_2) \in \mathcal{D}(\mathcal{L}_0),\ y_1 \in Ax_1 + Bx_2 \text{ and } y_2 \in Cx_1 + Dx_2\}.$$

◇

Lemma 2.8.5 *Let $A \in L\mathcal{R}(X)$, $B \in L\mathcal{R}(X)$, $C \in L\mathcal{R}(X)$ and $D \in L\mathcal{R}(X)$. Assume that $\mathcal{D}(A) \subset N(C)$, $\mathcal{D}(D) \subset N(B)$, $B(0) \subset A(0)$, and $C(0) \subset D(0)$, then*

(i) $N(\mathcal{L}_0) = N(A) \times N(D)$.
(ii) $R(\mathcal{L}_0) = R(A) \times R(D)$. ◇

Proof (i) Let us assume that $\begin{pmatrix} x \\ y \end{pmatrix} \in N(\mathcal{L}_0)$. Then, by using Remark 2.8.1 (i), we have

$$\mathcal{L}_0 \begin{pmatrix} x \\ y \end{pmatrix} = \mathcal{L}_0 \begin{pmatrix} 0 \\ 0 \end{pmatrix}.$$

This implies that

$$Ax + By = A(0) + B(0) \text{ and } Dy + Cx = D(0) + C(0). \tag{2.8.14}$$

From the hypothesis $B(0) \subset A(0)$ and $C(0) \subset D(0)$, we infer that

$$A(0) = A(0) + B(0), \tag{2.8.15}$$

and

$$D(0) = C(0) + D(0). \tag{2.8.16}$$

Since $\mathscr{D}(A) \subset N(C)$ and $\mathscr{D}(D) \subset N(B)$, then by using (2.8.13), we have

$$\begin{aligned} \mathscr{D}(\mathscr{L}_0) =& \mathscr{D}(A) \times \mathscr{D}(D) \\ &\subset N(C) \times N(B). \end{aligned} \tag{2.8.17}$$

Since $N(\mathscr{L}_0) \subset \mathscr{D}(\mathscr{L}_0)$ implies

$$N(\mathscr{L}_0) \subset N(C) \times N(B).$$

Thus, $x \in N(C)$ and $y \in N(B)$. So, by using Remark 2.8.1 (i), we have

$$Cx = C(0) \text{ and } By = B(0). \tag{2.8.18}$$

Hence, by referring to (2.8.14), we infer that

$$Ax + B(0) = A(0) + B(0), \tag{2.8.19}$$

and

$$Dy + C(0) = D(0) + C(0). \tag{2.8.20}$$

On the other hand, by the linearity of A, we obtain

$$\begin{aligned} Ax + B(0) &= Ax + A(0) + B(0) \\ &= Ax + A(0) \quad \text{(from (2.8.15))} \\ &= Ax. \end{aligned} \tag{2.8.21}$$

Similarly as (2.8.21), we infer that

$$Dy + C(0) = Dy. \tag{2.8.22}$$

The use of (2.8.15), (2.8.19), and (2.8.21) makes us to deduce that $Ax = A(0)$, and from (2.8.16), (2.8.20), and (2.8.22), we infer that $Dy = D(0)$. Again by using Remark 2.8.1 (i), we deduce that

$$\begin{pmatrix} x \\ y \end{pmatrix} \in N(A) \times N(D).$$

Conversely, let $\begin{pmatrix} x \\ y \end{pmatrix} \in N(A) \times N(D)$. Then, by using Remark 2.8.1 (i), we have $Ax = A(0)$ and $Dy = D(0)$. This implies from (2.8.15) and (2.8.16) that

$$Ax = A(0) + B(0)$$

and

$$Dy = D(0) + C(0).$$

By referring to (2.8.21), (2.8.22), and (2.8.18), we get

$$Ax + By = A(0) + B(0)$$

and

$$Dy + Cx = D(0) + C(0).$$

Hence,

$$\mathscr{L}_0 \begin{pmatrix} x \\ y \end{pmatrix} = \mathscr{L}_0 \begin{pmatrix} 0 \\ 0 \end{pmatrix}.$$

In view of Remark 2.8.1 (i) implies that $\begin{pmatrix} x \\ y \end{pmatrix} \in N(\mathscr{L}_0)$. As a result, $N(\mathscr{L}_0) = N(A) \times N(D)$, as desired.

(ii) We have

$$R(\mathscr{L}_0) = \left\{ \begin{pmatrix} z \\ t \end{pmatrix} \in X \times X : \exists \begin{pmatrix} x \\ y \end{pmatrix} \in \mathscr{D}(\mathscr{L}_0) \text{ such that } \begin{pmatrix} z \\ t \end{pmatrix} \in \mathscr{L}_0 \begin{pmatrix} x \\ y \end{pmatrix} \right\}$$

$$= \left\{ \begin{pmatrix} z \\ t \end{pmatrix} \in X \times X : \begin{matrix} \exists\, x \in \mathscr{D}(A) \\ \exists\, y \in \mathscr{D}(D) \end{matrix} \text{ such that } \begin{matrix} z \in Ax + By \\ t \in Dy + Cx \end{matrix} \right\} \quad \text{(by (1.8.17))}.$$

Based on the hypotheses $\mathscr{D}(A) \subset N(C)$ and $\mathscr{D}(D) \subset N(B)$, we infer from Theorem 2.8.1 (i) that

$$R(\mathscr{L}_0) = \left\{ \begin{pmatrix} z \\ t \end{pmatrix} \in X \times X : \begin{matrix} \exists\, x \in \mathscr{D}(A) \\ \exists\, y \in \mathscr{D}(D) \end{matrix} \text{ such that } \begin{matrix} z \in Ax + B(0) \\ t \in Dy + C(0) \end{matrix} \right\}.$$

The use of (2.8.21) and (2.8.22) makes us to conclude that

$$R(\mathscr{L}_0) = \left\{ \begin{pmatrix} z \\ t \end{pmatrix} \in X \times X : \begin{matrix} \exists\, x \in \mathscr{D}(A) \\ \exists\, y \in \mathscr{D}(D) \end{matrix} \text{ such that } \begin{matrix} z \in Ax \\ t \in Dy \end{matrix} \right\}$$
$$= R(A) \times R(D).$$

This completes the proof. Q.E.D.

Lemma 2.8.6 *Suppose that $A \in L\mathscr{R}(X)$, $B \in L\mathscr{R}(Y, X)$, $C \in L\mathscr{R}(X, Y)$, and $D \in L\mathscr{R}(Y)$ satisfies $\mathscr{D}(A) \subset \mathscr{D}(C)$. Let $\eta \in \mathbb{C}$ be such that $(A - \eta)^{-1}$ is single-valued and everywhere defined and $C(A - \eta I)^{-1}B(0) \subset C(0)$, then*

$$\mathscr{L}_0 - \eta I = \begin{pmatrix} I & 0 \\ C(A - \eta I)^{-1} & I \end{pmatrix} \begin{pmatrix} A - \eta I & 0 \\ 0 & M(\eta) - \eta I \end{pmatrix} \begin{pmatrix} I & (A - \eta I)^{-1}B \\ 0 & I \end{pmatrix},$$

where $\mathscr{L}_0$ is given in (2.8.12) and $M(\eta) = D - C(A - \eta I)^{-1}B$. ◇

Proof Let us assume that

$$H = \begin{pmatrix} I & 0 \\ F(\eta) & I \end{pmatrix},$$

$$K = \begin{pmatrix} A - \eta I & 0 \\ 0 & M(\eta) - \eta I \end{pmatrix},$$

and

$$J = \begin{pmatrix} I & E(\eta) \\ 0 & I \end{pmatrix},$$

where $F(\eta) = C(A - \eta I)^{-1}$ and $E(\eta) = (A - \eta I)^{-1}B$. Let $\begin{pmatrix} x_1 \\ x_2 \end{pmatrix} \in \mathscr{D}(HKJ)$, then

$$HKJ \begin{pmatrix} x_1 \\ x_2 \end{pmatrix} = \begin{pmatrix} I & 0 \\ F(\eta) & I \end{pmatrix} \begin{pmatrix} A - \eta I & 0 \\ 0 & M(\eta) - \eta I \end{pmatrix} \begin{pmatrix} I & E(\eta) \\ 0 & I \end{pmatrix} \begin{pmatrix} x_1 \\ x_2 \end{pmatrix},$$
$$= \begin{pmatrix} I & 0 \\ F(\eta) & I \end{pmatrix} \begin{pmatrix} (A - \eta I)x_1 + (A - \eta I)E(\eta)x_2 \\ (M(\eta) - \eta I)x_2 \end{pmatrix}.$$

Since

$$(A - \eta I)E(\eta)x_2 = (A - \eta I)(A - \eta I)^{-1}Bx_2,$$

then by using Theorem 2.8.1 (iv), we have

$$(A - \eta I)E(\eta)x_2 = Bx_2 + (A - \eta I)(0)$$
$$= Bx_2 + A(0).$$

Thus, we get

$$
\begin{aligned}
HKJ\begin{pmatrix} x_1 \\ x_2 \end{pmatrix} &= \begin{pmatrix} I & 0 \\ F(\eta) & I \end{pmatrix}\begin{pmatrix} (A-\eta I)x_1 + (A-\eta I)(0) + Bx_2 \\ (M(\eta)-\eta I)x_2 \end{pmatrix}, \\
&= \begin{pmatrix} (A-\eta I)x_1 + Bx_2 \\ F(\eta)\left[(A-\eta I)x_1 + Bx_2\right] + M(\eta)x_2 - \eta x_2 \end{pmatrix}.
\end{aligned}
$$

Since $\mathscr{D}(A) \subset \mathscr{D}(C)$ implies that

$$
\begin{aligned}
(A-\eta)(\mathscr{D}(A)) &\subset (A-\eta)(\mathscr{D}(C)) \\
&= \mathscr{D}(C(A-\eta)^{-1}) \\
&= \mathscr{D}(F(\eta)).
\end{aligned}
$$

Hence, $(A-\eta I)x_1 \subset \mathscr{D}(F(\eta))$. Therefore, by using the fact that $A-\eta$ is one-to-one, we can write

$$
\begin{aligned}
F(\eta)\left[(A-\eta I)x_1 + Bx_2\right] &= F(\eta)(A-\eta I)x_1 + F(\eta)Bx_2 \\
&= C(A-\eta I)^{-1}(A-\eta I)x_1 + C(A-\eta I)^{-1}Bx_2 \\
&= Cx_1 + C(A-\eta I)^{-1}Bx_2.
\end{aligned}
$$

Therefore, we have

$$
\begin{aligned}
HKJ\begin{pmatrix} x_1 \\ x_2 \end{pmatrix} &= \begin{pmatrix} (A-\eta I)x_1 + Bx_2 \\ Cx_1 + C(A-\eta I)^{-1}Bx_2 + Dx_2 - C(A-\eta I)^{-1}Bx_2 - \eta x_2 \end{pmatrix} \\
&= \begin{pmatrix} (A-\eta I)x_1 + Bx_2 \\ Cx_1 + C(A-\eta I)^{-1}B(0) + (D-\eta I)x_2 \end{pmatrix}.
\end{aligned}
$$

At this level, since $C(A-\eta I)^{-1}B(0) \subset C(0)$, then we obtain

$$
Cx_1 + C(A-\eta I)^{-1}B(0) \subset Cx_1.
$$

This implies that

$$
\begin{aligned}
(HKJ)\begin{pmatrix} x_1 \\ x_2 \end{pmatrix} &= \begin{pmatrix} (A-\eta I)x_1 + Bx_2 \\ Cx_1 + (D-\eta I)x_2 \end{pmatrix} \\
&= \left[\begin{pmatrix} A & B \\ C & D \end{pmatrix} - \eta\begin{pmatrix} I & 0 \\ 0 & I \end{pmatrix}\right]\begin{pmatrix} x_1 \\ x_2 \end{pmatrix} \\
&= (\mathscr{L}_0 - \eta I)\begin{pmatrix} x_1 \\ x_2 \end{pmatrix}.
\end{aligned}
$$

Moreover, we have

$$\begin{aligned}\mathscr{D}(HKJ) &= \left\{\begin{pmatrix} x_1 \\ x_2 \end{pmatrix} \in X \times Y : x_1 + E(\eta)x_2 \in \mathscr{D}(A) \text{ and } x_2 \in \mathscr{D}(B) \bigcap \mathscr{D}(D)\right\} \\ &= \mathscr{D}(A) \times \left(\mathscr{D}(B) \bigcap \mathscr{D}(D)\right) \\ &= \left[\mathscr{D}(A) \bigcap \mathscr{D}(C)\right] \times \left[\mathscr{D}(B) \bigcap \mathscr{D}(D)\right] \\ &= \mathscr{D}(\mathscr{L}_0).\end{aligned}$$

We conclude that $\mathscr{L}_0 - \eta I = HKJ$. Q.E.D.

Now, giving X and Y be two Banach spaces. Let $A \in L\mathscr{R}(X)$ and $B \in L\mathscr{R}(Y)$. For a given $C \in L\mathscr{R}(Y, X)$, we denote by

$$\mathscr{M}_C = \begin{pmatrix} A & C \\ 0 & B \end{pmatrix}$$

the linear relation acting on $X \bigoplus Y$ defined by

$$G(\mathscr{M}_C) = \left\{\left(\begin{pmatrix} u_1 \\ u_2 \end{pmatrix}, \begin{pmatrix} v_1 \\ v_2 \end{pmatrix}\right) \in (X \times Y)^2 \text{ such that } v_1 \in Au_1 + Cu_2 \text{ and } v_2 \in Bu_2\right\}.$$

Corollary 2.8.1 *If $C \in L\mathscr{R}(Y, X)$, then the matrix linear relation $\mathscr{M}_C$ admits the following decomposition*

$$\mathscr{M}_C = \begin{pmatrix} I & 0 \\ 0 & B \end{pmatrix}\begin{pmatrix} I & C \\ 0 & I \end{pmatrix}\begin{pmatrix} A & 0 \\ 0 & I \end{pmatrix}. \tag{2.8.23}$$

◇

Theorem 2.8.3 *Let X and Y be two Banach spaces, $\mathscr{D}(CA^{-1})$ contain the ranges of both A and B, $\mathscr{D}(C)$ contain the range of A and let's put*

$$\mathscr{A} = \begin{pmatrix} A & B \\ C + CA^{-1} - CA^{-1} & D + CA^{-1}B - CA^{-1}B \end{pmatrix}.$$

If B is single-valued and $D(0) \subset C(0)$, then

$$\mathscr{A} = \begin{pmatrix} I & 0 \\ CA^{-1} & I \end{pmatrix}\begin{pmatrix} A & 0 \\ 0 & D - CA^{-1}B \end{pmatrix}\begin{pmatrix} I & A^{-1}B \\ 0 & I \end{pmatrix}.$$

◇

Proof On one hand, let

$$\left[\begin{pmatrix} x_1 \\ x_2 \end{pmatrix}, \begin{pmatrix} y_1 \\ y_2 \end{pmatrix}\right] \in G\left[\begin{pmatrix} I & 0 \\ CA^{-1} & I \end{pmatrix}\begin{pmatrix} A & 0 \\ 0 & D - CA^{-1}B \end{pmatrix}\begin{pmatrix} I & A^{-1}B \\ 0 & I \end{pmatrix}\right].$$

This is equivalent to

$$\begin{pmatrix} y_1 \\ y_2 \end{pmatrix} \in \begin{pmatrix} I & 0 \\ CA^{-1} & I \end{pmatrix}\begin{pmatrix} A & 0 \\ 0 & D - CA^{-1}B \end{pmatrix}\begin{pmatrix} I & A^{-1}B \\ 0 & I \end{pmatrix}\begin{pmatrix} x_1 \\ x_2 \end{pmatrix}.$$

We can get

$$\begin{pmatrix} y_1 \\ y_2 \end{pmatrix} \in \begin{pmatrix} I & 0 \\ CA^{-1} & I \end{pmatrix}\begin{pmatrix} A & 0 \\ 0 & D - CA^{-1}B \end{pmatrix}\begin{pmatrix} x_1 + A^{-1}Bx_2 \\ x_2 \end{pmatrix}$$

if, and only if,

$$\begin{pmatrix} y_1 \\ y_2 \end{pmatrix} \in \begin{pmatrix} I & 0 \\ CA^{-1} & I \end{pmatrix}\begin{pmatrix} A(x_1 + A^{-1}Bx_2) \\ Dx_2 - CA^{-1}Bx_2 \end{pmatrix}.$$

Since

$$\begin{aligned} R(A^{-1}B) &= \mathscr{D}(B^{-1}A) \\ &= A^{-1}(R(B)) \\ &= \left\{x \in \mathscr{D}(A) \text{ such that } Ax \bigcap R(B) \neq \emptyset\right\}, \end{aligned}$$

then $\mathscr{D}(A)$ contain the range of $A^{-1}B$. Hence,

$$\begin{pmatrix} y_1 \\ y_2 \end{pmatrix} \in \begin{pmatrix} I & 0 \\ CA^{-1} & I \end{pmatrix}\begin{pmatrix} Ax_1 + AA^{-1}Bx_2 \\ Dx_2 - CA^{-1}Bx_2 \end{pmatrix}.$$

Therefore,

$$\begin{pmatrix} y_1 \\ y_2 \end{pmatrix} \in \begin{pmatrix} I & 0 \\ CA^{-1} & I \end{pmatrix}\begin{pmatrix} Ax_1 + Bx_2 + A(0) \\ Dx_2 - CA^{-1}Bx_2 \end{pmatrix}.$$

This is equivalent to

$$\begin{pmatrix} y_1 \\ y_2 \end{pmatrix} \in \begin{pmatrix} I & 0 \\ CA^{-1} & I \end{pmatrix}\begin{pmatrix} Ax_1 + Bx_2 \\ Dx_2 - CA^{-1}Bx_2 \end{pmatrix}$$

if, and only if,

$$\begin{pmatrix} y_1 \\ y_2 \end{pmatrix} \in \begin{pmatrix} Ax_1 + Bx_2 \\ CA^{-1}(Ax_1 + Bx_2) + Dx_2 - CA^{-1}Bx_2 \end{pmatrix}.$$

Hence, $\mathscr{D}(CA^{-1})$ contain the ranges of both A and B. Thus,

$$\begin{pmatrix} y_1 \\ y_2 \end{pmatrix} \in \begin{pmatrix} Ax_1 + Bx_2 \\ CA^{-1}Ax_1 + CA^{-1}Bx_2 + Dx_2 - CA^{-1}Bx_2 \end{pmatrix}.$$

Therefore,

$$\begin{pmatrix} y_1 \\ y_2 \end{pmatrix} \in \begin{pmatrix} Ax_1 + Bx_2 \\ C(x_1 + A^{-1}(0)) + Dx_2 + CA^{-1}Bx_2 - CA^{-1}Bx_2 \end{pmatrix}.$$

Since $\mathscr{D}(C)$ contain the range of A, then

$$\begin{pmatrix} y_1 \\ y_2 \end{pmatrix} \in \begin{pmatrix} Ax_1 + Bx_2 \\ Cx_1 + CA^{-1}(0) + Dx_2 + CA^{-1}Bx_2 - CA^{-1}Bx_2 \end{pmatrix}.$$

We conclude that

$$G\left[\begin{pmatrix} I & 0 \\ CA^{-1} & I \end{pmatrix}\begin{pmatrix} A & 0 \\ 0 & D - CA^{-1}B \end{pmatrix}\begin{pmatrix} I & A^{-1}B \\ 0 & I \end{pmatrix}\right]$$
$$\subset G\left[\begin{pmatrix} A & B \\ C + CA^{-1} - CA^{-1} & D + CA^{-1}B - CA^{-1}B \end{pmatrix}\right]. \tag{2.8.24}$$

On the other hand

$$\begin{aligned}
&\begin{pmatrix} I & 0 \\ CA^{-1} & I \end{pmatrix}\begin{pmatrix} A & 0 \\ 0 & D - CA^{-1}B \end{pmatrix}\begin{pmatrix} I & A^{-1}B \\ 0 & I \end{pmatrix}\begin{pmatrix} 0 \\ 0 \end{pmatrix} \\
&= \begin{pmatrix} I & 0 \\ CA^{-1} & I \end{pmatrix}\begin{pmatrix} A & 0 \\ 0 & D - CA^{-1}B \end{pmatrix}\begin{pmatrix} A^{-1}(0) \\ 0 \end{pmatrix} \\
&= \begin{pmatrix} I & 0 \\ CA^{-1} & I \end{pmatrix}\begin{pmatrix} AA^{-1}(0) \\ CA^{-1}(0) \end{pmatrix} \\
&= \begin{pmatrix} I & 0 \\ CA^{-1} & I \end{pmatrix}\begin{pmatrix} AA^{-1}(0) \\ CA^{-1}(0) \end{pmatrix} \\
&= \begin{pmatrix} A(0) \\ CA^{-1}A(0) + CA^{-1}(0) \end{pmatrix} \\
&= \begin{pmatrix} A(0) \\ CA^{-1}(0) \end{pmatrix}.
\end{aligned}$$

Since $D(0) \subset C(0) \subset CA^{-1}(0)$, then

$$\begin{aligned}
&\begin{pmatrix} I & 0 \\ CA^{-1} & I \end{pmatrix}\begin{pmatrix} A & 0 \\ 0 & D - CA^{-1}B \end{pmatrix}\begin{pmatrix} I & A^{-1}B \\ 0 & I \end{pmatrix}\begin{pmatrix} 0 \\ 0 \end{pmatrix} \\
&= \begin{pmatrix} A(0) \\ C(0) + CA^{-1}(0) + D(0) + CA^{-1}(0) \end{pmatrix} \\
&= \begin{pmatrix} A & B \\ C + CA^{-1} - CA^{-1} & D + CA^{-1}B - CA^{-1}B \end{pmatrix}\begin{pmatrix} 0 \\ 0 \end{pmatrix}.
\end{aligned} \tag{2.8.25}$$

Finally, let $\begin{pmatrix} x \\ y \end{pmatrix} \in \mathscr{D}\left[\begin{pmatrix} I & 0 \\ CA^{-1} & I \end{pmatrix}\begin{pmatrix} A & 0 \\ 0 & D - CA^{-1}B \end{pmatrix}\begin{pmatrix} I & A^{-1}B \\ 0 & I \end{pmatrix}\right]$. Then,

$$\begin{pmatrix} I & 0 \\ CA^{-1} & I \end{pmatrix}\begin{pmatrix} A & 0 \\ 0 & D - CA^{-1}B \end{pmatrix}\begin{pmatrix} I & A^{-1}B \\ 0 & I \end{pmatrix}\begin{pmatrix} x \\ y \end{pmatrix} \neq \emptyset$$

if, and only if,

$$\begin{pmatrix} I & 0 \\ CA^{-1} & I \end{pmatrix}\begin{pmatrix} A & 0 \\ 0 & D - CA^{-1}B \end{pmatrix}\begin{pmatrix} x + A^{-1}By \\ y \end{pmatrix} \neq \emptyset$$

if, and only if,

$$\begin{pmatrix} I & 0 \\ CA^{-1} & I \end{pmatrix}\begin{pmatrix} Ax + By \\ Dy - CA^{-1}By \end{pmatrix} \neq \emptyset$$

if, and only if,

$$\begin{pmatrix} Ax + By \\ Cx + CA^{-1}(0) + CA^{-1}By + Dy - CA^{-1}By \end{pmatrix} \neq \emptyset$$

if, and only if,

$$\begin{pmatrix} A & B \\ C + CA^{-1} - CA^{-1} & D + CA^{-1}B - CA^{-1}B \end{pmatrix}\begin{pmatrix} x \\ y \end{pmatrix} \neq \emptyset.$$

Therefore,

$$\mathscr{D}\left[\begin{pmatrix} A & B \\ C + CA^{-1} - CA^{-1} & D + CA^{-1}B - CA^{-1}B \end{pmatrix}\right]$$
$$= \mathscr{D}\left[\begin{pmatrix} I & 0 \\ CA^{-1} & I \end{pmatrix}\begin{pmatrix} A & 0 \\ 0 & D - CA^{-1}B \end{pmatrix}\begin{pmatrix} I & A^{-1}B \\ 0 & I \end{pmatrix}\right]. \tag{2.8.26}$$

By using Eqs. (2.8.24), (2.8.25), (2.8.26), and Proposition 2.7.3 (iii), we conclude that

$$\begin{pmatrix} A & B \\ C + CA^{-1} - CA^{-1} & D + CA^{-1}B - CA^{-1}B \end{pmatrix}$$
$$= \begin{pmatrix} I & 0 \\ CA^{-1} & I \end{pmatrix}\begin{pmatrix} A & 0 \\ 0 & D - CA^{-1}B \end{pmatrix}\begin{pmatrix} I & A^{-1}B \\ 0 & I \end{pmatrix}.$$

This achieves the proof. Q.E.D.

2.8.4 Product of 3 × 3 Matrices of Linear Relations

Lemma 2.8.7 *Let X, Y, and Z be three Banach spaces. If A, $A_1 \in L\mathscr{R}(X)$, B, $B_1 \in L\mathscr{R}(Y)$ and C, $C_1 \in L\mathscr{R}(Z)$, then*

$$\begin{pmatrix} A & 0 & 0 \\ 0 & B & 0 \\ 0 & 0 & C \end{pmatrix}\begin{pmatrix} A_1 & 0 & 0 \\ 0 & B_1 & 0 \\ 0 & 0 & C_1 \end{pmatrix} = \begin{pmatrix} AA_1 & 0 & 0 \\ 0 & BB_1 & 0 \\ 0 & 0 & CC_1 \end{pmatrix}.$$

◇

Proof Let

$$\left[\begin{pmatrix} x_1 \\ x_2 \\ x_3 \end{pmatrix}, \begin{pmatrix} y_1 \\ y_2 \\ y_3 \end{pmatrix}\right] \in G\left[\begin{pmatrix} A & 0 & 0 \\ 0 & B & 0 \\ 0 & 0 & C \end{pmatrix}\begin{pmatrix} A_1 & 0 & 0 \\ 0 & B_1 & 0 \\ 0 & 0 & C_1 \end{pmatrix}\right].$$

Then, by using Definition 2.8.1 there exists $\begin{pmatrix} z_1 \\ z_2 \\ z_3 \end{pmatrix} \in X \times Y \times Z$ such that

$$\left[\begin{pmatrix} x_1 \\ x_2 \\ x_3 \end{pmatrix}, \begin{pmatrix} z_1 \\ z_2 \\ z_3 \end{pmatrix}\right] \in G\left[\begin{pmatrix} A_1 & 0 & 0 \\ 0 & B_1 & 0 \\ 0 & 0 & C_1 \end{pmatrix}\right],$$

and

$$\left[\begin{pmatrix} z_1 \\ z_2 \\ z_3 \end{pmatrix}, \begin{pmatrix} y_1 \\ y_2 \\ y_3 \end{pmatrix}\right] \in G\left[\begin{pmatrix} A & 0 & 0 \\ 0 & B & 0 \\ 0 & 0 & C \end{pmatrix}\right].$$

We get $z_1 \in A_1x_1$, $z_2 \in B_1x_2$, $z_3 \in C_1x_3$, $y_1 \in Az_1$, $y_2 \in Bz_2$, and $y_3 \in Cz_3$. Thus, $y_1 \in AA_1x_1$, $y_2 \in BB_1x_2$, and $y_3 \in CC_1x_3$. This is equivalent to

$$\left[\begin{pmatrix} x_1 \\ x_2 \\ x_3 \end{pmatrix}, \begin{pmatrix} y_1 \\ y_2 \\ y_3 \end{pmatrix}\right] \in G\left[\begin{pmatrix} AA_1 & 0 & 0 \\ 0 & BB_1 & 0 \\ 0 & 0 & CC_1 \end{pmatrix}\right].$$

The same technique is used for the converse inclusion. Q.E.D.

Proposition 2.8.3 *Let X, Y, and Z be three Banach spaces. If A, $A_1 \in L\mathscr{R}(X)$, B, $B_1 \in L\mathscr{R}(Y, X)$, C, $C_1 \in L\mathscr{R}(Z, Y)$, D, $D_1 \in L\mathscr{R}(X, Y)$, E, $E_1 \in L\mathscr{R}(Y)$, F, $F_1 \in L\mathscr{R}(Z, Y)$, G, $G_1 \in L\mathscr{R}(Z, Y)$, H, $H_1 \in L\mathscr{R}(Y, Z)$ and K, $K_1 \in L\mathscr{R}(Z)$, then*

(i)

$$\left[\begin{pmatrix} A & B & C \\ D & E & F \\ G & H & K \end{pmatrix}\begin{pmatrix} A_1 & B_1 & C_1 \\ D_1 & E_1 & F_1 \\ G_1 & H_1 & K_1 \end{pmatrix}\right] \subset$$

$$\begin{pmatrix} AA_1+BD_1+CG_1 & AB_1+BE_1+CH_1 & AC_1+BF_1+CK_1 \\ DA_1+ED_1+FG_1 & DB_1+EE_1+FH_1 & DC_1+EF_1+FK_1 \\ GA_1+HD_1+KG_1 & GB_1+HE_1+KH_1 & GC_1+HF_1+KK_1 \end{pmatrix}.$$

(ii) Moreover, if A_1, B_1, C_1, D_1, E_1, F_1, G_1, H_1, and K_1 are single-valued (or $D = G = H = D_1 = G_1 = H_1 = 0$ and K_1 is single-valued), then we have the equality. ◇

Proof (i) Let

$$\left[\begin{pmatrix} x_1 \\ x_2 \\ x_3 \end{pmatrix}, \begin{pmatrix} y_1 \\ y_2 \\ y_3 \end{pmatrix}\right] \in G\left[\begin{pmatrix} A & B & C \\ D & E & F \\ G & H & K \end{pmatrix}\begin{pmatrix} A_1 & B_1 & C_1 \\ D_1 & E_1 & F_1 \\ G_1 & H_1 & K_1 \end{pmatrix}\right].$$

Then, there exists $\begin{pmatrix} z_1 \\ z_2 \\ z_3 \end{pmatrix} \in X \times Y \times Z$ such that

$$\left[\begin{pmatrix} x_1 \\ x_2 \\ x_3 \end{pmatrix}, \begin{pmatrix} z_1 \\ z_2 \\ z_3 \end{pmatrix}\right] \in G\left[\begin{pmatrix} A_1 & B_1 & C_1 \\ D_1 & E_1 & F_1 \\ G_1 & H_1 & K_1 \end{pmatrix}\right]$$

and

$$\left[\begin{pmatrix} z_1 \\ z_2 \\ z_3 \end{pmatrix}, \begin{pmatrix} y_1 \\ y_2 \\ y_3 \end{pmatrix}\right] \in G\left[\begin{pmatrix} A & B & C \\ D & E & F \\ G & H & K \end{pmatrix}\right].$$

Then,

$$\begin{cases} z_1 \in A_1x_1 + B_1x_2 + C_1x_3 \\ z_2 \in D_1x_1 + E_1x_2 + F_1x_3 \\ z_3 \in G_1x_1 + H_1x_2 + K_1x_3 \end{cases}$$

and

$$\begin{cases} y_1 \in Az_1 + Bz_2 + Cz_3 \\ y_2 \in Dz_1 + Ez_2 + Fz_3 \\ y_3 \in Gz_1 + Hz_2 + Kz_3. \end{cases}$$

Thus,

$$\begin{aligned} y_1 &\in A(A_1x_1 + B_1x_2 + C_1x_3) + B(D_1x_1 + E_1x_2 + F_1x_3) \\ &\quad + C(G_1x_1 + H_1x_2 + K_1x_3), \\ &= (AA_1 + BD_1 + CG_1)x_1 + (AB_1 + BE_1 + CH_1)x_2 \\ &\quad + (AC_1 + BF_1 + CK_1)x_3, \end{aligned}$$

$$\begin{aligned} y_2 &\in D(A_1x_1 + B_1x_2 + C_1x_3) + E(D_1x_1 + E_1x_2 + F_1x_3) \\ &\quad + F(G_1x_1 + H_1x_2 + K_1x_3), \\ &= (DA_1 + ED_1 + FG_1)x_1 + (DB_1 + EE_1 + FH_1)x_2 \\ &\quad + (DC_1 + EF_1 + FK_1)x_3, \end{aligned}$$

and

$$\begin{aligned} y_3 &\in G(A_1x_1 + B_1x_2 + C_1x_3) + H(D_1x_1 + E_1x_2 + F_1x_3) \\ &\quad + K(G_1x_1 + H_1x_2 + K_1x_3, \\ &= (GA_1 + HD_1 + KG_1)x_1 + (GB_1 + HE_1 + KH_1)x_2 \\ &\quad + (GC_1 + HF_1 + KK_1)x_3. \end{aligned}$$

Hence,

$$\left[\begin{pmatrix} x_1 \\ x_2 \\ x_3 \end{pmatrix}, \begin{pmatrix} y_1 \\ y_2 \\ y_3 \end{pmatrix}\right] \in$$

$$G\left[\begin{pmatrix} AA_1 + BD_1 + CG_1 & AB_1 + BE_1 + CH_1 & AC_1 + BF_1 + CK_1 \\ DA_1 + ED_1 + FG_1 & DB_1 + EE_1 + FH_1 & DC_1 + EF_1 + FK_1 \\ GA_1 + HD_1 + KG_1 & GB_1 + HE_1 + KH_1 & GC_1 + HF_1 + KK_1 \end{pmatrix}\right].$$

(ii) Assume that $A_1,\ B_1,\ C_1,\ D_1,\ E_1,\ F_1,\ G_1,\ H_1$, and K_1 are single-valued. Let

$$\left[\begin{pmatrix} x_1 \\ x_2 \\ x_3 \end{pmatrix}, \begin{pmatrix} y_1 \\ y_2 \\ y_3 \end{pmatrix}\right] \in$$

$$G\left[\begin{pmatrix} AA_1 + BD_1 + CG_1 & AB_1 + BE_1 + CH_1 & AC_1 + BF_1 + CK_1 \\ DA_1 + ED_1 + FG_1 & DB_1 + EE_1 + FH_1 & DC_1 + EF_1 + FK_1 \\ GA_1 + HD_1 + KG_1 & GB_1 + HE_1 + KH_1 & GC_1 + HF_1 + KK_1 \end{pmatrix}\right].$$

Then,

$$\begin{aligned} y_1 &\in (AA_1 + BD_1 + CG_1)x_1 + (AB_1 + BE_1 + CH_1)x_2 + (AC_1 + BF_1 + CK_1)x_3, \\ &= A(A_1x_1 + B_1x_2 + C_1x_3) + B(D_1x_1 + E_1x_2 + F_1x_3) + C(G_1x_1 + H_1x_2 + K_1x_3), \end{aligned}$$

$$\begin{aligned} y_2 &\in (DA_1 + ED_1 + FG_1)x_1 + (DB_1 + EE_1 + FH_1)x_2 + (DC_1 + EF_1 + FK_1)x_3, \\ &= D(A_1x_1 + B_1x_2 + C_1x_3) + E(D_1x_1 + E_1x_2 + F_1x_3) + F(G_1x_1 + H_1x_2 + K_1x_3), \end{aligned}$$

and

$$\begin{aligned} y_3 &\in (GA_1+HD_1+KG_1)x_1+(GB_1+HE_1+KH_1)x_2+(GC_1+HF_1+KK_1)x_3,\\ &= G(A_1x_1+B_1x_2+C_1x_3)+H(D_1x_1+E_1x_2+F_1x_3)+K(G_1x_1+H_1x_2+K_1x_3). \end{aligned}$$

Let

$$\begin{cases} z_1 \in A_1x_1+B_1x_2+C_1x_3\\ z_2 \in D_1x_1+E_1x_2+F_1x_3\\ z_3 \in G_1x_1+H_1x_2+K_1x_3. \end{cases}$$

Then,

$$\begin{cases} y_1 \in Az_1+Bz_2+Cz_3\\ y_2 \in Dz_1+Ez_2+Fz_3\\ y_3 \in Gz_1+Hz_2+Kz_3. \end{cases}$$

Hence,

$$\left[\begin{pmatrix} x_1\\ x_2\\ x_3 \end{pmatrix}, \begin{pmatrix} y_1\\ y_2\\ y_3 \end{pmatrix}\right] \in G\left[\begin{pmatrix} A & B & C\\ D & E & F\\ G & H & K \end{pmatrix}\begin{pmatrix} A_1 & B_1 & C_1\\ D_1 & E_1 & F_1\\ G_1 & H_1 & K_1 \end{pmatrix}\right].$$

Thus,

$$\left[\begin{pmatrix} A & B & C\\ D & E & F\\ G & H & K \end{pmatrix}\begin{pmatrix} A_1 & B_1 & C_1\\ D_1 & E_1 & F_1\\ G_1 & H_1 & K_1 \end{pmatrix}\right] =$$

$$\begin{pmatrix} AA_1+BD_1+CG_1 & AB_1+BE_1+CH_1 & AC_1+BF_1+CK_1\\ DA_1+ED_1+FG_1 & DB_1+EE_1+FH_1 & DC_1+EF_1+FK_1\\ GA_1+HD_1+KG_1 & GB_1+HE_1+KH_1 & GC_1+HF_1+KK_1 \end{pmatrix}.$$

Now, we assume that $D = G = H = D_1 = G_1 = H_1 = 0$ and K_1 is single-valued. Let

$$\left[\begin{pmatrix} x_1\\ x_2\\ x_3 \end{pmatrix}, \begin{pmatrix} y_1\\ y_2\\ y_3 \end{pmatrix}\right] \in G\left[\begin{pmatrix} AA_1 & AB_1+BE_1 & AC_1+BF_1+CK_1\\ 0 & EE_1 & EF_1+FK_1\\ 0 & 0 & KK_1 \end{pmatrix}\right].$$

Then,

$$\begin{cases} y_1 \in A(A_1x_1+B_1x_2+C_1x_3)+B(E_1x_2+F_1x_3)+CK_1x_3\\ y_2 \in E(E_1x_2+F_1x_3)+FK_1x_3\\ y_3 \in KK_1x_3. \end{cases}$$

So, if we take

$$\begin{cases} z_1 \in A_1x_1+B_1x_2+C_1x_3\\ z_2 \in E_1x_2+F_1x_3,\\ z_3 \in K_1x_3 \end{cases}$$

such that

$$\begin{cases} y_1 \in Az_1 + Bz_2 + CK_1x_3 \\ y_2 \in Ez_2 + FK_1x_3 \\ y_3 \in Kz_3. \end{cases}$$

We deduce that

$$\left[\begin{pmatrix} x_1 \\ x_2 \\ x_3 \end{pmatrix}, \begin{pmatrix} y_1 \\ y_2 \\ y_3 \end{pmatrix}\right] \in G\left[\begin{pmatrix} A & B & C \\ 0 & E & F \\ 0 & 0 & K \end{pmatrix}\begin{pmatrix} A_1 & B_1 & C_1 \\ 0 & E_1 & F_1 \\ 0 & 0 & K_1 \end{pmatrix}\right].$$

This achieves the proof. Q.E.D.

2.9 Quotient Map of Linear Relation

Let X be a normed space and F be a closed vector subspace of X. The operator

$$Q_F^X : X \longrightarrow X/_F$$

is the natural quotient map with $\mathscr{D}\left(Q_F^X\right) = X$ and $N\left(Q_F^X\right) = F$. We must denote $Q^Y_{\overline{U(0)}}$ by Q_U, or simply Q when U is understood. Then, $Q^Y_{\overline{U(0)}}$ (or simply, Q_U) is a bounded operator with $N(Q_U) = \overline{U(0)}$.

Lemma 2.9.1 *Let D be a bounded closed linear subspace of a space X. For every sequence $(x_n)_n$ in X such that $(Q_D x_n)_n$ is bounded, $(x_n)_n$ is bounded.* ◇

Proof Let D be a bounded closed linear subspace of X and let $(x_n)_n$ be a sequence in X such that $(Q_D x_n)_n$ is bounded, then for all $n \in \mathbb{N}$, there exists a real number $M_0 \geq 0$ such that $\|Q_D x_n\| \leq M_0$, i.e., $\operatorname{dist}(x_n, D) \leq M_0$. But $\operatorname{dist}(x_n, D) = \inf\{\|x_n - d\|;\ d \in D\}$, then for all $\varepsilon > 0$, there exists $d_\varepsilon \in D$ such that

$$\|x_n - d_\varepsilon\| < \operatorname{dist}(x_n, D) + \varepsilon.$$

Moreover,

$$\begin{aligned} \|x_n\| &\leq \|x_n - d_\varepsilon\| + \|d_\varepsilon\| \\ &< \operatorname{dist}(x_n, D) + \varepsilon + \|d_\varepsilon\|. \end{aligned}$$

Since D is bounded, there exists $M_1 \geq 0$ such that $\|d_\varepsilon\| \leq M_1$. Hence, for all $\varepsilon > 0$,

$$\|x_n\| < M_0 + M_1 + \varepsilon.$$

So,

$$\|x_n\| \leq M_0 + M_1.$$

Therefore, $(x_n)_n$ is bounded. Q.E.D.

Proposition 2.9.1 *Let $U \in L\mathscr{R}(X, Y)$. Then,*

$$Q_U U : X \longrightarrow Y/\overline{U(0)}$$

is a single-valued operator. ◇

Proof Let $Q_U y_1, Q_U y_2 \in Q_U Ux$, then, $Q_U y_1 - Q_U y_2 \in Q_U Ux - Q_U Ux$. Since $Q_U Ux - Q_U Ux = Q_U U(0) \subset Q_U \overline{U(0)} = \{0\}$, then $y_1 = y_2$. So, $Q_U U$ is an single-valued operator. Q.E.D.

Remark 2.9.1 We note that for $x \in \mathscr{D}(U)$, $Q_U Ux = Q_U v$ for all $v \in Ux$. Indeed, since $v \in U(x)$ if, and only if, $U(x) = v + U(0)$, then $Q_U Ux = Q_U v$. ◇

Lemma 2.9.1 ([2, Corollary I.6.10]) *Let $V \in L\mathscr{R}(X, Y)$. Then,*

(i) $\alpha(Q_U V) = \alpha(V)$.
(ii) $\beta(Q_U V) = \beta(V)$. ◇

2.9.1 Quotient Map of 2 × 2-Block Matrix Linear Relations

Unfortunately, there is no uniqueness of the expression of $Q_{\mathscr{L}_0}\mathscr{L}_0$.

Lemma 2.9.2 *(i) If $A(0) \subset B(0)$ and $D(0) \subset C(0)$, then*

$$Q_{\mathscr{L}_0}\mathscr{L}_0 = \begin{pmatrix} Q_B A & Q_B B \\ Q_C C & Q_C D \end{pmatrix}.$$

(ii) If $B(0) \subset A(0)$ and $C(0) \subset D(0)$, then

$$Q_{\mathscr{L}_0}\mathscr{L}_0 = \begin{pmatrix} Q_A A & Q_A B \\ Q_D C & Q_D D \end{pmatrix}.$$

(iii) If $B(0) \subset A(0)$ and $D(0) \subset C(0)$, then

$$Q_{\mathscr{L}_0}\mathscr{L}_0 = \begin{pmatrix} Q_A A & Q_A B \\ Q_C C & Q_C D \end{pmatrix}.$$

(iv) If $A(0) \subset B(0)$ and $C(0) \subset D(0)$, then

$$Q_{\mathscr{L}_0}\mathscr{L}_0 = \begin{pmatrix} Q_B A & Q_B B \\ Q_D C & Q_D D \end{pmatrix}.$$ ◇

Proof (*i*) Let $\begin{pmatrix} x \\ y \end{pmatrix} \in (\mathscr{D}(A) \bigcap \mathscr{D}(C)) \times (\mathscr{D}(B) \bigcap \mathscr{D}(D))$ and

$$\begin{pmatrix} a \\ b \end{pmatrix} \in \mathscr{L}_0 \begin{pmatrix} x \\ y \end{pmatrix}.$$

Then,

$$\begin{pmatrix} a \\ b \end{pmatrix} \in \begin{pmatrix} A & B \\ C & D \end{pmatrix} \begin{pmatrix} x \\ y \end{pmatrix} = \begin{pmatrix} Ax + By \\ Cx + Dy \end{pmatrix}.$$

So, $a \in Ax + By$ and $b \in Cx + Dy$. Thus, there exist $a_1 \in Ax$, $a_2 \in By$, $b_1 \in Cx$, and $b_2 \in Dy$ such that $a = a_1 + a_2$ and $b = b_1 + b_2$. Hence

$$Q_{\mathscr{L}_0} \mathscr{L}_0 \begin{pmatrix} x \\ y \end{pmatrix} = Q_{\mathscr{L}_0} \begin{pmatrix} a \\ b \end{pmatrix}.$$

Now, let us find the expression of $Q_{\mathscr{L}_0} \begin{pmatrix} a \\ b \end{pmatrix}$. Let $\begin{pmatrix} u \\ v \end{pmatrix} \in Q_{\mathscr{L}_0} \begin{pmatrix} a \\ b \end{pmatrix}$ if, and only if,

$$\begin{pmatrix} u \\ v \end{pmatrix} - \begin{pmatrix} a \\ b \end{pmatrix} \in \overline{\mathscr{L}_0(0)}.$$

Notice that $u - a \in \overline{A(0) + B(0)} = \overline{B(0)}$ and $v - b \in \overline{C(0) + D(0)} = \overline{C(0)}$. Therefore,

$$\begin{cases} u \in Q_B a = Q_B(a_1 + a_2) \\ v \in Q_C b = Q_C(b_1 + b_2). \end{cases}$$

So,

$$\begin{cases} u \in Q_B(Ax + By) = Q_B Ax + Q_B By \\ v \in Q_C(Cx + Dy) = Q_C Cx + Q_C Dy. \end{cases}$$

Finally,

$$\begin{pmatrix} u \\ v \end{pmatrix} \in \begin{pmatrix} Q_B A & Q_B B \\ Q_C C & Q_C D \end{pmatrix} \begin{pmatrix} x \\ y \end{pmatrix}$$

i.e.,

$$Q_{\mathscr{L}_0} \mathscr{L}_0 = \begin{pmatrix} Q_B A & Q_B B \\ Q_C C & Q_C D \end{pmatrix}.$$

The proofs of (ii), (iii), and (iv) may be achieved in a similar way as (i) Q.E.D.

Now, we may deduce the following proposition.

Proposition 2.9.2 *Let* $A \in L\mathscr{R}(H)$, $B \in L\mathscr{R}(K)$, *and* $C \in L\mathscr{R}(K, H)$ *be such that* $C(0) \subseteq A(0)$. *Then,*

$$Q_{\mathcal{M}_C}\mathcal{M}_C = \begin{pmatrix} Q_A A & Q_A C \\ 0 & Q_B B \end{pmatrix},$$

with

$$\mathcal{M}_C = \begin{pmatrix} A & C \\ 0 & B \end{pmatrix}. \tag{2.9.1}$$

◇

Proof Let $\begin{pmatrix} x \\ y \end{pmatrix} \in H \times K$ and $\begin{pmatrix} u \\ v \end{pmatrix} \in \mathcal{M}_C \begin{pmatrix} x \\ y \end{pmatrix}$ i.e., there exist $u_1 \in Ax$ and $u_2 \in Cy$ such that $u = u_1 + u_2 \in Ax + Cy$ and $v \in By$. This implies that

$$Q_{\mathcal{M}_C}\mathcal{M}_C \begin{pmatrix} x \\ y \end{pmatrix} = Q_{\mathcal{M}_C} \begin{pmatrix} u \\ v \end{pmatrix}.$$

Now, we will search the expression of $Q_{\mathcal{M}_C} \begin{pmatrix} u \\ v \end{pmatrix}$. We remark that $\begin{pmatrix} \widetilde{u} \\ \widetilde{v} \end{pmatrix} \in Q_{\mathcal{M}_C} \begin{pmatrix} u \\ v \end{pmatrix}$ if, and only if, $\begin{pmatrix} \widetilde{u} \\ \widetilde{v} \end{pmatrix} - \begin{pmatrix} u \\ v \end{pmatrix} \in \overline{\mathcal{M}_C(0)}$ i.e.,

$$\begin{cases} \widetilde{u} - u \in \overline{A(0) + C(0)} = \overline{A(0)}, \\ \widetilde{v} - v \in \overline{B(0)}. \end{cases}$$

That's equivalent to

$$\begin{cases} \widetilde{u} \in Q_A(u_1 + u_2), \\ \widetilde{v} \in Q_B(v). \end{cases}$$

This shows that

$$Q_{\mathcal{M}_C} \begin{pmatrix} u \\ v \end{pmatrix} = \begin{pmatrix} Q_A u_1 + Q_A u_2 \\ Q_B v \end{pmatrix}.$$

Since $u_1 \in A(x), u_2 \in Cy$, and $v \in By$, it follows that $Q_A u_1 = Q_A Ax$, $Q_A u_2 = Q_A Cy$ and $Q_B v = Q_B By$. As a conclusion

$$\begin{aligned} Q_{\mathcal{M}_C}\mathcal{M}_C \begin{pmatrix} x \\ y \end{pmatrix} &= \begin{pmatrix} Q_A Ax + Q_A Cy \\ Q_B By \end{pmatrix} \\ &= \begin{pmatrix} Q_A A & Q_A C \\ 0 & Q_B B \end{pmatrix} \begin{pmatrix} x \\ y \end{pmatrix}. \end{aligned}$$

This achieves the proof. Q.E.D.

Lemma 2.9.3 *Let $\mathcal{L}_1$ and $\mathcal{L}_2$ be the block matrix linear relations defined by*

$$\mathcal{L}_1 = \begin{pmatrix} A_1 & B_1 \\ C_1 & D_1 \end{pmatrix}$$

and

$$\mathscr{L}_2 = \begin{pmatrix} A_2 & B_2 \\ C_2 & D_2 \end{pmatrix},$$

where $A_1, A_2 \in L\mathscr{R}(X)$, $B_1, B_2 \in L\mathscr{R}(Y, X)$, $C_1, C_2 \in L\mathscr{R}(X, Y)$, *and* $D_1, D_2 \in L\mathscr{R}(Y)$. *Then,*

(i) *If* $A_1(0) \subset B_1(0)$ *and* $D_1(0) \subset C_1(0)$, *then*

$$Q_{\mathscr{L}_1}\mathscr{L}_2 = \begin{pmatrix} Q_{B_1}A_2 & Q_{B_1}B_2 \\ Q_{C_1}C_2 & Q_{C_1}D_2 \end{pmatrix}.$$

(ii) *If* $A_1(0) \subset B_1(0)$ *and* $C_1(0) \subset D_1(0)$, *then*

$$Q_{\mathscr{L}_1}\mathscr{L}_2 = \begin{pmatrix} Q_{B_1}A_2 & Q_{B_1}B_2 \\ Q_{D_1}C_2 & Q_{D_1}D_2 \end{pmatrix}.$$

(iii) *If* $B_1(0) \subset A_1(0)$ *and* $C_1(0) \subset D_1(0)$, *then*

$$Q_{\mathscr{L}_1}\mathscr{L}_2 = \begin{pmatrix} Q_{A_1}A_2 & Q_{A_1}B_2 \\ Q_{D_1}C_2 & Q_{D_1}D_2 \end{pmatrix}.$$

(iv) *If* $B_1(0) \subset A_1(0)$ *and* $D_1(0) \subset C_1(0)$, *then*

$$Q_{\mathscr{L}_1}\mathscr{L}_2 = \begin{pmatrix} Q_{A_1}A_2 & Q_{A_1}B_2 \\ Q_{C_1}C_2 & Q_{C_1}D_2 \end{pmatrix}.$$

◇

Proof (i) Let $\begin{pmatrix} x \\ y \end{pmatrix} \in (\mathscr{D}(A_2)\bigcap\mathscr{D}(C_2)) \times (\mathscr{D}(B_2)\bigcap\mathscr{D}(D_2))$ and $\begin{pmatrix} a \\ b \end{pmatrix} \in \mathscr{L}_2\begin{pmatrix} x \\ y \end{pmatrix}$. Then,

$$\begin{pmatrix} a \\ b \end{pmatrix} \in \begin{pmatrix} A_2 & B_2 \\ C_2 & D_2 \end{pmatrix}\begin{pmatrix} x \\ y \end{pmatrix} = \begin{pmatrix} A_2x + B_2y \\ C_2x + D_2y \end{pmatrix}.$$

Hence, $a \in A_2x + B_2y$ and $b \in C_2x + D_2y$, i.e., there exist $a_1 \in A_2x$, $a_2 \in B_2y$, $b_1 \in C_2x$ and $b_2 \in D_2y$ such that $a = a_1 + a_2$ and $b = b_1 + b_2$. On the other hand,

$$Q_{\mathscr{L}_1}\mathscr{L}_2\begin{pmatrix} x \\ y \end{pmatrix} = Q_{\mathscr{L}_1}\begin{pmatrix} a \\ b \end{pmatrix}.$$

Now, let us examine the expression $Q_{\mathscr{L}_1}\begin{pmatrix} a \\ b \end{pmatrix}$. In fact, let $\begin{pmatrix} u \\ v \end{pmatrix} \in Q_{\mathscr{L}_1}\begin{pmatrix} a \\ b \end{pmatrix}$ if, and only if,

$$\begin{pmatrix} u \\ v \end{pmatrix} - \begin{pmatrix} a \\ b \end{pmatrix} \in \overline{\mathscr{L}(0)}.$$

Notice that $u - a \in \overline{A_1(0) + B_1(0)} = \overline{B_1(0)}$ and $v - b \in \overline{C_1(0) + D_1(0)} = \overline{C_1(0)}$. Hence,

$$\begin{cases} u - a \in \overline{B_1(0)} \\ v - b \in \overline{C_1(0)}. \end{cases}$$

Therefore,

$$\begin{cases} u \in Q_{B_1} a = Q_{B_1}(a_1 + a_2) \\ v \in Q_{C_1} b = Q_{C_1}(b_1 + b_2). \end{cases}$$

So,

$$\begin{cases} u \in Q_{B_1}(A_2 x + B_2 y) = Q_{B_1} A_2 x + Q_{B_1} B_2 y \\ v \in Q_{C_1}(C_2 x + D_2 y) = Q_{C_1} C_2 x + Q_{C_1} D_2 y. \end{cases}$$

Finally,

$$\begin{pmatrix} u \\ v \end{pmatrix} \in \begin{pmatrix} Q_{B_1} A_2 & Q_{B_1} B_2 \\ Q_{C_1} C_2 & Q_{C_1} D_2 \end{pmatrix} \begin{pmatrix} x \\ y \end{pmatrix}$$

i.e.,

$$Q_{\mathcal{L}_1} \mathcal{L}_2 = \begin{pmatrix} Q_{B_1} A_2 & Q_{B_1} B_2 \\ Q_{C_1} C_2 & Q_{C_1} D_2 \end{pmatrix}.$$

The proofs of (ii), (iii), and (iv) may be achieved in a similar way as (i). This achieves the proof. Q.E.D.

2.9.2 *Quotient Map of* 3 × 3*-block matrix*

Let X_1, X_2, and X_3 be three Banach spaces. In the product space $X_1 \times X_2 \times X_3$, we consider the multivalued linear operators defined by a matrix

$$\mathcal{T} = \begin{pmatrix} A & 0 & 0 \\ 0 & E & 0 \\ 0 & 0 & K \end{pmatrix}, \tag{2.9.2}$$

$$\mathcal{S} = \begin{pmatrix} 0 & B & 0 \\ 0 & 0 & F \\ G & 0 & 0 \end{pmatrix} \tag{2.9.3}$$

and

$$\mathcal{K} = \begin{pmatrix} 0 & 0 & C \\ D & 0 & 0 \\ 0 & H & 0 \end{pmatrix}, \tag{2.9.4}$$

where the relation A acts on X_1 with domain $\mathscr{D}(A)$, the multivalued linear operator E acts on X_2 with domain $\mathscr{D}(E)$, and the multivalued linear operator K acts on X_3 with domain $\mathscr{D}(K)$. The intertwining relation linear B is defined on the domain $\mathscr{D}(B) \subset X_2$ into X_1, multivalued linear operator H is defined on the domain $\mathscr{D}(H) \subset X_2$ into X_3, the relation C is defined on the domain $\mathscr{D}(C) \subset X_3$ into X_1, the relation F is defined on the domain $\mathscr{D}(F) \subset X_3$ into X_2, the relation linear D is defined on the domain $\mathscr{D}(D) \subset X_1$ into X_2, and the multivalued linear operator G is defined on the domain $\mathscr{D}(G) \subset X_1$ into X_3.

Proposition 2.9.3 *Let $\mathscr{T}$, $\mathscr{S}$, $\mathscr{K}$ be given, respectively, in (2.9.2), (2.9.3), and (2.9.4). Then,*

(i) $Q_{\mathscr{T}}\mathscr{T} = \begin{pmatrix} Q_A A & 0 & 0 \\ 0 & Q_E E & 0 \\ 0 & 0 & Q_K K \end{pmatrix}$,

(ii) $Q_{\mathscr{S}}\mathscr{S} = \begin{pmatrix} 0 & Q_B B & 0 \\ 0 & 0 & Q_F F \\ Q_G G & 0 & 0 \end{pmatrix}$, *and*

(iii) $Q_{\mathscr{K}}\mathscr{K} = \begin{pmatrix} 0 & 0 & Q_C C \\ Q_D D & 0 & 0 \\ 0 & Q_H H & 0 \end{pmatrix}$. ◇

Proof (*i*) Giving $\begin{pmatrix} x \\ y \\ z \end{pmatrix} \in \mathscr{D}(A) \times \mathscr{D}(E) \times \mathscr{D}(K)$ and $\begin{pmatrix} u \\ v \\ w \end{pmatrix} \in \mathscr{T}\begin{pmatrix} x \\ y \\ z \end{pmatrix}$, then $u \in Ax$, $v \in Ey$ and $w \in Kz$. This yields

$$Q_{\mathscr{T}}\mathscr{T}\begin{pmatrix} x \\ y \\ z \end{pmatrix} = Q_{\mathscr{T}}\begin{pmatrix} u \\ v \\ w \end{pmatrix}.$$

Now, let us find the expression of $Q_{\mathscr{T}}\begin{pmatrix} u \\ v \\ w \end{pmatrix}$. Notice that $\begin{pmatrix} \widetilde{u} \\ \widetilde{v} \\ \widetilde{w} \end{pmatrix} \in Q_{\mathscr{T}}\begin{pmatrix} u \\ v \\ w \end{pmatrix}$ if, and only if,

$$\begin{pmatrix} \widetilde{u} \\ \widetilde{v} \\ \widetilde{w} \end{pmatrix} - \begin{pmatrix} u \\ v \\ w \end{pmatrix} \in \begin{pmatrix} \overline{A(0)} \\ \overline{E(0)} \\ \overline{K(0)} \end{pmatrix}.$$

Then,

$$\begin{cases} \widetilde{u} - u \in \overline{A(0)} \\ \widetilde{v} - v \in \overline{E(0)} \\ \widetilde{w} - w \in \overline{K(0)}. \end{cases}$$

This is equivalent to

$$\begin{cases} \widetilde{u} \in Q_A u, \\ \widetilde{v} \in Q_E v, \\ \widetilde{w} \in Q_K w. \end{cases}$$

This shows that

$$Q_{\mathscr{T}} \begin{pmatrix} u \\ v \\ w \end{pmatrix} = \begin{pmatrix} Q_A u \\ Q_E v \\ Q_K w \end{pmatrix} = \begin{pmatrix} Q_A A x \\ Q_E E y \\ Q_K K z \end{pmatrix}.$$

Then, for all $\begin{pmatrix} x \\ y \\ z \end{pmatrix} \in \mathscr{D}(A) \times \mathscr{D}(E) \times \mathscr{D}(K)$, we have

$$Q_{\mathscr{T}} \mathscr{T} \begin{pmatrix} x \\ y \\ z \end{pmatrix} = \begin{pmatrix} Q_A A & 0 & 0 \\ 0 & Q_E E & 0 \\ 0 & 0 & Q_K K \end{pmatrix} \begin{pmatrix} x \\ y \\ z \end{pmatrix}.$$

The proof of (ii) and (iii) may be checked in a similar fashion as that of in (i). Q.E.D.

Proposition 2.9.4 *Let X_1, X_2, and X_3 be Banach spaces. In the product space $X_1 \times X_2 \times X_3$, we define the multivalued linear operator by a matrix*

$$\mathscr{M}_C = \begin{pmatrix} A_1 & B & C \\ 0 & A_2 & D \\ 0 & 0 & A_3 \end{pmatrix},$$

where $A_1 \in L\mathscr{R}(X_1)$, $A_2 \in L\mathscr{R}(X_2)$, $A_3 \in L\mathscr{R}(X_3)$, $B \in L\mathscr{R}(X_2, X_1)$, $C \in L\mathscr{R}(X_3, X_1)$ and $D \in L\mathscr{R}(X_3, X_2)$. If $C(0) \subset B(0) \subset A_1(0)$ and $D(0) \subset A_2(0)$, then

$$Q_{\mathscr{M}_C} \mathscr{M}_C = \begin{pmatrix} Q_{A_1} A_1 & Q_{A_1} B & Q_{A_1} C \\ 0 & Q_{A_2} A_2 & Q_{A_2} D \\ 0 & 0 & Q_{A_3} A_3 \end{pmatrix}. \qquad \diamond$$

Proof Let $\begin{pmatrix} x_1 \\ x_2 \\ x_3 \end{pmatrix} \in X_1 \times X_2 \times X_3$ and $\begin{pmatrix} u \\ v \\ w \end{pmatrix} \in \mathscr{M}_C \begin{pmatrix} x_1 \\ x_2 \\ x_3 \end{pmatrix}$. Then, there exist $u_1 \in Ax_1$, $u_2 \in Cx_2$, $u_3 \in Bx_3$, $v_1 \in A_2x_2$, and $v_2 \in Dx_3$ such that $u = u_1 + u_2 + u_3 \in A_1x_1 + Bx_2 + Cx_3$, $v = v_1 + v_2 \in A_2x_2 + Dx_3$, and $w \in A_3x_3$. This implies that

$$Q_{\mathcal{M}_C}\mathcal{M}_C\begin{pmatrix} x_1 \\ x_2 \\ x_3 \end{pmatrix} = Q_{\mathcal{M}_C}\begin{pmatrix} u \\ v \\ w \end{pmatrix}.$$

Now, let us find the expression of

$$Q_{\mathcal{M}_C}\begin{pmatrix} u \\ v \\ w \end{pmatrix}.$$

Notice that

$$\begin{pmatrix} \bar{u} \\ \bar{v} \\ \bar{w} \end{pmatrix} \in Q_{\mathcal{M}_C}\begin{pmatrix} u \\ v \\ w \end{pmatrix}$$

if, and only if,

$$\begin{pmatrix} \bar{u} \\ \bar{v} \\ \bar{w} \end{pmatrix} - \begin{pmatrix} u \\ v \\ w \end{pmatrix} \in \overline{\mathcal{M}_C\begin{pmatrix} 0 \\ 0 \\ 0 \end{pmatrix}}$$

if, and only if,

$$\begin{cases} \bar{u} - u \in A_1(0) + B(0) + C(0) \\ \bar{v} - v \in A_2(0) + D(0) \\ \bar{w} - w \in A_3(0). \end{cases}$$

This is equivalent to say that

$$\begin{cases} \bar{u} \in Q_{A_1}(u_1 + u_2 + u_3) \\ \bar{v} \in Q_{A_2}(v_1 + v_2) \\ \bar{w} \in Q_{A_3}(w). \end{cases}$$

This shows that

$$Q_{\mathcal{M}_C}\begin{pmatrix} u \\ v \\ w \end{pmatrix} = \begin{pmatrix} Q_{A_1}u_1 + Q_{A_1}u_2 + Q_{A_1}u_3 \\ Q_{A_2}v_1 + Q_{A_2}v_2 \\ Q_{A_3}w \end{pmatrix}.$$

Since $u_1 \in Ax_1$, $u_2 \in Cx_2$, $u_3 \in Bx_3$, $v_1 \in A_2x_2$, $v_2 \in Dx_3$, and $w \in A_3x_3$, it follows that $Q_{A_1}u_1 = Q_{A_1}A_1x_1$, $Q_{A_1}u_2 = Q_{A_1}Bx_2$, $Q_{A_1}u_3 = Q_{A_1}Cx_3$, $Q_{A_2}v_1 = Q_{A_2}A_2x_2$, $Q_{A_2}v_2 = Q_{A_2}Dx_3$, and $Q_{A_3}w = Q_{A_3}A_3x_3$. As a conclusion,

$$Q_{\mathcal{M}_C}\mathcal{M}_C\begin{pmatrix}x_1\\x_2\\x_3\end{pmatrix}=\begin{pmatrix}Q_{A_1}u_1+Q_{A_1}u_2+Q_{A_1}u_3\\Q_{A_2}v_1+Q_{A_2}v_2\\Q_{A_3}w\end{pmatrix},$$
$$=\begin{pmatrix}Q_{A_1}A_1 & Q_{A_1}B & Q_{A_1}C\\0 & Q_{A_2}A_2 & Q_{A_2}D\\0 & 0 & Q_{A_3}A_3\end{pmatrix}\begin{pmatrix}x_1\\x_2\\x_3\end{pmatrix}.$$

This achieves the proof. Q.E.D.

2.10 Norm of Linear Relations

Let X and Y be two normed linear spaces over a number field $\mathbb{K}$ ($\mathbb{R}$ or $\mathbb{C}$). The cartesian product $X \times Y$ is a normed space with the usual definition of addition and multiplication by scalars, and its norm is defined as

$$\left\|\begin{pmatrix}u\\v\end{pmatrix}\right\|_p=\left(\|u\|^p+\|v\|^p\right)^{\frac{1}{p}},\ u\in X,\ v\in Y \text{ and } 1\le p<\infty. \tag{2.10.1}$$

Lemma 2.10.1 *Let us supply the space $X \times Y$ with the norm*

$$\left\|\begin{pmatrix}u\\v\end{pmatrix}\right\|_1=\|u\|+\|v\|,\ u\in X \text{ and } v\in Y. \tag{2.10.2}$$

Then, the norms defined in (2.10.1), with $p = 2$, and (2.10.2) are equivalent. ◇

Proof Follows immediately from the inequalities

$$x+y\le\sqrt{2(x^2+y^2)}\le\sqrt{2}(x+y)\quad x,\ y\ge 0.$$ Q.E.D.

2.10.1 Norm of Multivalued Linear Operator

Let X, Y be two normed spaces and $U \in L\mathcal{R}(X, Y)$. We define the norm of Ux and U by

$$\|Ux\|=\|Q_U Ux\|,\quad \text{for all } x\in\mathcal{D}(U)$$

and

$$\|U\|_{L\mathcal{R}(X,Y)}=\|Q_U U\|$$

with

$$\|U\|:=\|U\|_{L\mathcal{R}(X,Y)}$$

$$= \sup_{x \in B_{\mathscr{D}(U)}} \|Ux\|.$$

Lemma 2.10.2 *Let $U \in L\mathscr{R}(X, Y)$. Then,*

(i) For $u \in \mathscr{D}(U)$,

$$\begin{aligned} \|Uu\| &= dist(v, U(0)) \text{ for all } v \in Uu \\ &= \inf_{v \in Ux} \|v\| \\ &= dist(v + U(0), 0) \text{ for all } v \in Uu \\ &= dist(Uu, 0) \\ &= dist(Uu, U(0)). \end{aligned}$$

(ii) $\|U\| = \sup_{u \in B_{\mathscr{D}(U)}} \|Uu\|$, *where* $B_{\mathscr{D}(U)} := \{x \in \mathscr{D}(T) : \|u\| \leq 1\}$. ◇

Proof (i) The primary equality takes from definition of $\|Uu\|$ and Theorem 2.8.1. The second equality is preserved from the definition and properties of the norm on $Y/\overline{U(0)}$. The rest are obvious.
(ii) They consist of

$$\|U\| = \|Q_U U\|.$$

Since

$$\begin{aligned} \|Q_U U\| &= \sup_{u \in B_{\mathscr{D}(U)}} \|Q_U Uu\| \\ &= \sup_{x \in B_{\mathscr{D}(U)}} \|Uu\|. \end{aligned}$$

Then,

$$\|U\| = \sup_{u \in B_{\mathscr{D}(U)}} \|Uu\|.$$

This completes the proof. Q.E.D.

$\|.\|$ is not a true "norm" function seeing that $\|U\| = 0$ does not imply that $U = 0$.

Example 2.10.1 Let X be a normed space and let M be a nonzero vector subspace of X. Define $U \in L\mathscr{R}(X)$ by

$$Ux = M \text{ for all } x \in X.$$

It is obvious that U is nonzero linear relation. Moreover, we have $Ux = U(0)$ for all $x \in X$. Then, $\|Q_U Ux\| = \|Q_U U(0)\| = 0$ for all $x \in X$. This implies that $\|Ux\| = 0$ for all $x \in X$. Hence, by using Lemma 2.10.2, we have $\|U\| = 0$.

Lemma 2.10.3 *For U and $V \in L\mathscr{R}(X, Y)$, we have*

(i) $\|Uu + Vu\| \leq \|Uu\| + \|Vu\|$, *for* $u \in \mathscr{D}(U + V)$.
(ii) $\|\lambda Uu\| = |\lambda| \|Uu\|$, *for* $\lambda \in \mathbb{K}$ *and* $u \in \mathscr{D}(U)$. ◇

Proof *(i)* Let $v \in Vu$ and $t \in Uu$, then $v + t \in Vu + Uu = (V + U)u$. As a result, by means of Lemma 2.10.2 *(i)*,

$$\|Uu + Vu\| = \text{dist}\,(t + v, (U + V)(0))\,.$$

Since

$$\begin{aligned}\text{dist}\,(t + v, (U + V)(0)) &\leq \text{dist}\,(t, U(0) + V(0)) + \text{dist}\,(v, U(0) + V(0)) \\ &\leq \text{dist}\,(t, U(0)) + \text{dist}\,(v, V(0)) \\ &= \|Uu\| + \|Vu\|.\end{aligned}$$

Consequently,

$$\|Uu + Vu\| \leq \|Uu\| + \|Vu\|, \quad \text{for } u \in \mathscr{D}(U + V).$$

(ii) The case $\lambda = 0$ is obviously. If $\lambda \neq 0$, then

$$\begin{aligned}\|\lambda Uu\| &= \|Q_U(\alpha U)(x)\| \text{ (since } \lambda U(0) = U(0) \text{ for } \lambda \neq 0) \\ &= \|\lambda Q_U U(u)\| \\ &= |\lambda| \|Q_U U(u)\|.\end{aligned}$$

This achieves the proof. Q.E.D.

Lemma 2.10.4 *If* $V(0) \subset \overline{U(0)}$ *and* $\mathscr{D}(U) \subset \mathscr{D}(V)$, *then* $Q_U V(0) = \{0\}$ *and* $\|Q_U V\| \leq \|Q_V V\|$. ◇

Proof Since $V(0) \subset \overline{U(0)}$, then

$$Q_U(V(0)) \subset Q_U(\overline{U(0)}) = \{0\}.$$

So, $Q_U(V)$ is single-valued operator, and

$$\|Q_U V\| = \text{dist}(Vu, \overline{U(0)}).$$

On the other hand

$$\begin{aligned}\text{dist}(Vu, \overline{U(0)}) &\leq \text{dist}(Vu, V(0)) \\ &= \|Vu\| \\ &= \|Q_V V(u)\|.\end{aligned}$$

Then,

$$\|Q_U V\| \leq \|Q_V V\|.$$

This achieves the proof. Q.E.D.

Corollary 2.10.1 *Let $U \in L\mathcal{R}(X, Y)$ and $V \in L\mathcal{R}(X, Y)$. Then,*

$$\|VU\| \leq \|V\|\|I_{\mathcal{D}(V)}U\| \quad (\infty.0 \text{ excluded})$$

with $\|VU\| = 0$ whenever $\|V\| = 0$ even if $\|I_{\mathcal{D}(V)}U\| = \infty$. Furthermore, we have the implication

$$U(0) \subset \mathcal{D}(V) \Longrightarrow \|VU\| \leq \|V\|\|U\|.$$ ◇

Remark 2.10.1 By considering the restriction of U to a one-dimensional subspace of $\mathcal{D}(VU)$ with $U(0) \subset \mathcal{D}(V)$, then by applying Corollary 2.10.1, we deduce that

$$\|VUx\| \leq \|V\|\|Ux\|$$

holds for each $x \in \mathcal{D}(VU)$. ◇

2.10.2 *Continuity and Openness of a Multivalued Linear Operator*

Definition 2.10.1 (i) Let U be an arbitrary multivalued linear operator from one topological space to another. The multivalued linear operator U is said continuous if for each neighborhood $\mathcal{O}$ in $R(U)$ the range $U^{-1}(\mathcal{O})$ is a neighborhood in $\mathcal{D}(U)$.
(ii) U is called open if whenever $\mathcal{O}$ is a neighborhood in $\mathcal{D}(U)$, the range $U(\mathcal{O})$ is a neighborhood of $R(U)$.
(iii) If $\mathcal{D}(U) = X$ and U is a continuous, then we shall say that U is bounded multivalued linear operator.
(iv) Let $U, V \in L\mathcal{R}(X, Y)$. The multivalued linear operator U is said to be V-co-continuous if there exist constants $\alpha, \beta > 0$ such that

$$U(B_X) \subset \alpha V(B_X) + \beta B_Y + U(0),$$

where B_X and B_Y designate, respectively, the closed unit ball in X and Y. ◇

The class of bounded multivalued linear operator as defined above will be denoted by $B\mathcal{R}(X, Y)$ and as useful we write $B\mathcal{R}(X) := B\mathcal{R}(X, X)$ and the class of bounded single valued-linear operators as defined above will be denoted by $\mathcal{L}(X, Y)$ and as useful we write $\mathcal{L}(X) := \mathcal{L}(X, X)$.

Remark 2.10.2 (i) We mean that U is continuous multivalued linear operator if, and only if, U^{-1} is open.

(ii) Let $\mathcal{O}$ be a subspace of a normed vector space X, then the injection operator $J_{\mathcal{O}}^{X}$ is continuous. ◇

Lemma 2.10.5 *Let $U \in L\mathcal{R}(X, Y)$. If U is open multivalued linear operator, then $Q_U U$ is open linear single-valued operator.* ◇

Proof Exercise. Q.E.D.

Lemma 2.10.6 *Let X and Y be two normed spaces, $A \in L\mathcal{R}(X)$, $B \in L\mathcal{R}(Y)$, $C \in L\mathcal{R}(Y, X)$ and $D \in L\mathcal{R}(X, Y)$. Then,*

(i) For all $\begin{pmatrix} v_1 \\ v_2 \end{pmatrix} \in \mathcal{D}(\mathcal{L}_0)$, we have

$$\left\| \mathcal{L}_0 \begin{pmatrix} v_1 \\ v_2 \end{pmatrix} \right\| \leq \|Av_1\| + \|Bv_2\| + \|Cv_1\| + \|Dv_2\|.$$

(ii) If $A = D = 0$, then

$$\left\| \mathcal{L}_0 \begin{pmatrix} v_1 \\ v_2 \end{pmatrix} \right\| = \sqrt{\|Cv_1\|^2 + \|Bv_2\|^2}.$$

(iii) If $C = B = 0$, and A, D are continuous, then

$$\|\mathcal{L}_0\| = \max(\|A\|, \|D\|) \leq \sqrt{\|A\|^2 + \|D\|^2}.$$ ◇

Proof (i) By using Lemma 2.10.2 (i), we have

$$\left\| \mathcal{L}_0 \begin{pmatrix} v_1 \\ v_2 \end{pmatrix} \right\| = \inf \left\{ \left\| \begin{pmatrix} y_1 \\ y_2 \end{pmatrix} \right\|_{X \oplus Y} \text{ such that } \begin{pmatrix} y_1 \\ y_2 \end{pmatrix} \in \mathcal{L}_0 \begin{pmatrix} v_1 \\ v_2 \end{pmatrix} \right\}.$$

Let $\begin{pmatrix} y_1 \\ y_2 \end{pmatrix} \in \mathcal{L}_0 \begin{pmatrix} v_1 \\ v_2 \end{pmatrix}$. Then, there exist $y_{1,1} \in Av_1$, $y_{1,2} \in Bv_2$, $y_{2,1} \in Cv_1$ and $y_{2,2} \in Dv_2$ such that $y_1 = y_{1,1} + y_{1,2}$ and $y_2 = y_{2,1} + y_{2,2}$. We have

$$\begin{aligned} \sqrt{\|y_1\|^2 + \|y_2\|^2} &\leq \left((\|y_{1,1}\| + \|y_{1,2}\|)^2 + (\|y_{2,1}\| + \|y_{2,2}\|)^2 \right)^{\frac{1}{2}} \\ &\leq \|y_{1,1}\| + \|y_{1,2}\| + \|y_{2,1}\| + \|y_{2,2}\|. \end{aligned}$$

Thus,

$$\begin{aligned} \left\| \mathcal{L}_0 \begin{pmatrix} v_1 \\ v_2 \end{pmatrix} \right\| &\leq \|y_{1,1}\| + \|y_{1,2}\| + \|y_{2,1}\| + \|y_{2,2}\| \\ &\leq \|Av_1\| + \|Bv_2\| + \|Cv_1\| + \|Dv_2\|. \end{aligned}$$

(ii) If $A = D = 0$, then

$$\left\| \mathscr{L}_0 \begin{pmatrix} v_1 \\ v_2 \end{pmatrix} \right\| = \inf \left\{ \left\| \begin{pmatrix} y_1 \\ y_2 \end{pmatrix} \right\|_{X \bigoplus Y} \text{ such that } y_1 \in Bv_2 \text{ and } y_2 \in Cv_1 \right\}.$$

On the other hand, $\|Bv_2\| \leq \|y_1\|$, for all $y_1 \in Bv_2$ and $\|Cv_1\| \leq \|y_2\|$, for all $y_2 \in Cv_1$. Thus,

$$\sqrt{\|Cv_1\|^2 + \|Bv_2\|^2} \leq \sqrt{\|y_1\|^2 + \|y_2\|^2},$$

for all $y_1 \in Bv_2$ and$y_2 \in Cv_1$. This gives the required result.
(iii) If $C = B = 0$, then by the same way we get

$$\begin{aligned} \left\| \mathscr{L}_0 \begin{pmatrix} v_1 \\ v_2 \end{pmatrix} \right\| =& \sqrt{\|Av_1\|^2 + \|Dv_2\|^2} \\ \leq& \sqrt{\|A\|^2 + \|D\|^2} \sqrt{\|v_1\|^2 + \|v_2\|^2}. \end{aligned}$$

Hence,

$$\|\mathscr{L}_0\| \leq \sqrt{\|A\|^2 + \|D\|^2}.$$

We have $\|\mathscr{L}_0\| = \|Q_{\mathscr{L}_0}\mathscr{L}_0\|$. By using Lemma 2.9.2, we have

$$Q_{\mathscr{L}_0}\mathscr{L}_0 = \begin{pmatrix} Q_A A & 0 \\ 0 & Q_D D \end{pmatrix}.$$

Thus,

$$\begin{aligned} \|Q_{\mathscr{L}_0}\mathscr{L}_0\| &= \max\left\{\|Q_A A\|, \|Q_D D\|\right\} \\ &= \max\left\{\|A\|, \|D\|\right\} \\ &= \|\mathscr{L}_0\|. \end{aligned}$$

This completes the proof. Q.E.D.

Proposition 2.10.1 *Let* $U \in L\mathscr{R}(X, Y)$. *Then,*

(i) $\|U\| < \infty$ *if, and only if, there exists* $\eta > 0$ *such that*

$$U(B_{\mathscr{D}(U)}) \subset \eta B_{R(U)} + U(0). \tag{2.10.3}$$

(ii) *If* $\|U\| < \infty$, *then*

$$\|U\| = \inf_{\eta > 0} \left\{ \eta \text{ such that } U(B_{\mathscr{D}(U)}) \subset \eta B_{R(U)} + U(0) \right\}. \tag{2.10.4}$$

◇

Proof (i) Assume $\|U\| < \infty$. Hence, if we take $u \in B_{\mathcal{D}(U)}$ and $v \in Uu$, then by using Lemma 2.10.2, there exists $q \in U(0)$ such that for all $\varepsilon > 0$,

$$\|v - q\| < \|U\| + \varepsilon.$$

Hence, $v - q \in (\|U\| + \varepsilon)B_{R(U)}$. Consequently,

$$v \in (\|u\| + \varepsilon)B_{R(U)} + U(0). \tag{2.10.5}$$

Looking at it the other way, if we suppose that Eq. (2.10.3) holds. Now, giving $u \in B_{\mathcal{D}(U)}$ and choose $v \in Uu$. Then,

$$v = \lambda v_1 + q,$$

with $\|v_1\| \leq 1$ and $q \in U(0)$. Thus,

$$\|v - q\| \leq \eta.$$

Hence,

$$\text{dist}(v, U(0)) \leq \eta.$$

By applying Lemma 2.10.2, we deduce that $\|U\| \leq \eta < \infty$.

(ii) Assume $\|U\| < \infty$. If $\|U\| = 0$, then

$$U(B_{\mathcal{D}(U)}) \subset \overline{U(0)}$$

and Eq. (2.10.4) holds. Suppose $\|U\| > 0$. It follows from Eq. (2.10.5) that

$$\|U\| \geq \inf_{\eta > 0} \left\{\eta \text{ such that } U(B_{\mathcal{D}(U)}) \subset \eta B_{R(U)} + U(0)\right\}.$$

Let $\delta \in]0, \|U\|[$ and choose $u \in B_{\mathcal{D}(U)}$ and $v \in Uu$ such that

$$\alpha < \text{dist}(v, U(0)). \tag{2.10.6}$$

If $U(B_{\mathcal{D}(U)}) \subset \delta B_{R(U)} + U(0)$, hence there exists $v_1 \in B_{R(U)}$ and $q \in U(0)$ such that

$$v = \delta v_1 + q.$$

So,

$$\|v - q\| \leq \alpha,$$

which contradicts Eq. (2.10.6). Thus,

$$\alpha < \inf_{\eta>0} \left\{\eta \text{ such that } U(B_{\mathscr{D}(U)}) \subset \eta B_{R(U)} + U(0)\right\}$$

and the result (2.10.4) follows. Q.E.D.

Corollary 2.10.2 *If $U \in L\mathscr{R}(X, Y)$ is continuous, then U is V-co-continuous for all $V \in L\mathscr{R}(X, Y)$.* ◇

Lemma 2.10.7 *Let $U \in B\mathscr{R}(X, Y)$ be such that $U(0)$ is closed and $(x_n)_n$ be a sequence of X such that $x_n \in Ux_{n+1}$ for all $n \in \mathbb{N}$. Let R denotes the convergence radius of the entire series*

$$\sum_{n\geq 1} \lambda^{n-1} x_n.$$

Then, for all scalar λ such that $|\lambda| < R$, we have

$$U\left(\sum_{n\geq 1} \lambda^{n-1} x_n\right) = \sum_{n\geq 1} \lambda^{n-1} x_{n-1} + U(0).$$ ◇

Proof Let λ be such that $|\lambda| < R$. We have

$$\begin{aligned} U\left(\sum_{n\geq 1} \lambda^{n-1} x_n\right) &= \sum_{n=1}^{N} \lambda^{n-1} U x_n + U\left(\sum_{n\geq N+1} \lambda^{n-1} x_n\right) \\ &= \sum_{n\geq 1} \lambda^{n-1} x_{n-1} - \sum_{n\geq N+1} \lambda^{n-1} x_{n-1} + U\left(\sum_{n\geq N+1} \lambda^{n-1} x_n\right). \end{aligned}$$

Now, since the operators Q_U and $Q_U U$ are bounded, then

$$\begin{aligned} Q_U\left(U\left(\sum_{n\geq 1} \lambda^{n-1} x_n\right)\right) &= Q_U\left[\sum_{n\geq 1} \lambda^{n-1} x_{n-1} - \sum_{n\geq N+1} \lambda^{n-1} x_{n-1} + U\left(\sum_{n\geq N+1} \lambda^{n-1} x_n\right)\right] \\ &= Q_U\left(\sum_{n\geq 1} \lambda^{n-1} x_{n-1}\right). \end{aligned}$$

Hence,

$$Q_U\left(U\left(\sum_{n\geq 1} \lambda^{n-1} x_n\right) - \sum_{n\geq 1} \lambda^{n-1} x_{n-1}\right) = 0,$$

which implies that

$$U\left(\sum_{n\geq 1} \lambda^{n-1} x_n\right) - \sum_{n\geq 1} \lambda^{n-1} x_{n-1} \subseteq U(0).$$

Thus,

$$U\left(\sum_{n\geq 1}\lambda^{n-1}x_n\right)=\sum_{n\geq 1}\lambda^{n-1}x_{n-1}+U(0).$$

This completes the proof. Q.E.D.

Lemma 2.10.8 *Let Q be bounded linear and f be an analytic function on $\mathbb{D}(\lambda, r)$. Then, Qf is an analytic function on $\mathbb{D}(\overline{\lambda}, r)$.* ◇

Proof Since f is an analytic function on $\mathbb{D}(\lambda, r)$. Then,

$$f(\mu)=\sum_{k=0}^{\infty}a_k\,(\lambda-\mu)^k,\quad \text{for all } \mu\in\mathbb{D}(\lambda, r).$$

This implies that

$$\begin{aligned}
Qf(\mu) &= Q\left(\sum_{k=0}^{\infty}a_k\,(\lambda-\mu)^k\right)\\
&= Q\left(\lim_{n\to\infty}\sum_{k=0}^{n}a_k\,(\lambda-\mu)^k\right)\\
&= \lim_{n\to\infty}Q\left(\sum_{k=0}^{n}a_k\,(\lambda-\mu)^k\right)\quad (\text{as } Q \text{ is bounded})\\
&= \lim_{n\to\infty}\sum_{k=0}^{n}\underbrace{Q(a_k)}_{=b_k}\,(\lambda-\mu)^k\quad (\text{as } Q \text{ is linear})\\
&= \sum_{k=0}^{\infty}b_k\,(\lambda-\mu)^k,\quad \text{for all } \mu\in\mathbb{D}(\lambda, r).
\end{aligned}$$

Hence,

$$Qf(\overline{\mu})=\sum_{k=0}^{\infty}b_k\,\left(\overline{\lambda}-\overline{\mu}\right)^k,\quad \text{for all } \overline{\mu}\in\mathbb{D}(\overline{\lambda}, r).$$

This is equivalent to saying that Qf is an analytic function on $\mathbb{D}(\overline{\lambda}, r)$. Q.E.D.

Proposition 2.10.2 *Let X, Y, Z, and W be Banach spaces. Let U be a bounded operator and $R\in\mathscr{L}(Z, W)$ be bijective and $V\in L\mathscr{R}(Y, Z)$. Then, V is a onto relation if, and only if, RVU is a onto relation.* ◇

Proof Let us assume that $V\in L\mathscr{R}(Y, Z)$ be a onto relation, we propose to show that RVU is also onto. Since U is bounded single-valued and bijective, then for all $y\in Y$, there exists $x\in X$ such that

$$Ux = y.$$

Now, by using the fact that V is onto, we infer that for all $z \in Z$, there exists $x \in X$ such that $z \in VUx$. Then, by Theorem 2.8.1 (i), we deduce that

$$VUx = z + VU(0).$$

Hence, by applying (i) of Theorem 2.8.1, we get

$$\begin{aligned} RVUx &= R(z + VU(0)) \\ &= Rz + RVU(0). \end{aligned} \tag{2.10.7}$$

In view of the fact that R is a bijective single-valued operator, then for all $w \in R(R) = W$, we can write $Rz = w$. Thus, by using (2.10.7), together with Theorem 2.8.1, we get $w \in RVUx$. Therefore, RVU is onto. Conversely, we suppose that RVU is a onto relation. Then, for all $w \in W$, there exists $x \in \mathscr{D}(RVU)$ such that $w \in RVUx$. Hence, by referring to Theorem 2.8.1, we have

$$RVUx = w + RVU(0). \tag{2.10.8}$$

Since U is a bijective operator, then for all $y \in R(U) = Y$, we infer from (2.10.8) that $RSy = w + RV(0)$. Now, the fact that $R \in B\mathscr{R}(Z, W)$ is a single-valued bijective implies that

$$R^{-1}RVy = R^{-1}w + R^{-1}RV(0).$$

Thus, again by using Theorem 2.8.1, we deduce that $R^{-1}w \in Vy$. Therefore, for all $z = R^{-1}w \in \mathscr{D}(R) = Z$, we have $z \in R(V)$. As a result, V is onto, as desired. Q.E.D.

2.10.3 Minimum Modulus of Multivalued Operator

Definition 2.10.2 Let $U \in L\mathscr{R}(X, Y)$. The minimum modulus of U is the quantity

$$\begin{aligned} \gamma(U) &= \begin{cases} \infty & \text{if } \mathscr{D}(U) \subset \overline{N(U)} \\ \inf\left\{ \dfrac{\|Ux\|}{\operatorname{dist}(x, N(U))} \text{ such that } x \in \mathscr{D}(U)\backslash\overline{N(U)} \right\} & \text{otherwise,} \end{cases} \\ &= \sup\left\{ \lambda \text{ such that } \|Ux\| \geq \lambda \operatorname{dist}(x, N(U)) \text{ for } x \in \mathscr{D}(U) \right\}. \end{aligned}$$

◇

Lemma 2.10.9 *Let X, Y be two normed spaces and $U \in L\mathscr{R}(X, Y)$. Then,*

$$\gamma(U) = \|U^{-1}\|^{-1}.$$

◇

Proof Exercise. Q.E.D.

Proposition 2.10.3 [2, Proposition II.3.10] *Let X, Y be normed spaces, $U \in L\mathcal{R}(X, Y)$ be with minimum modulus of U is fine and let M be a nonempty subset of $R(U)$. Then, for $N \subset \mathcal{D}(U)$, we have*

$$dist(UM, N) \geq \gamma(U) dist(N, U^{-1}M). \qquad \diamondsuit$$

Proposition 2.10.4 *Let $U \in L\mathcal{R}(X, Y)$. Then,*

(i) U is continuous if, and only if, $\|U\| < \infty$.
(ii) If $\dim \mathcal{D}(U) < \infty$, then U is continuous.
(iii) U is open if, and only if, $\gamma(U) > 0$.
(iv) If $\dim(R(U)) < \infty$, then U is open. $\diamondsuit$

Proof (i) First we suppose U is continuous multivalued operator. The fact

$$U(0) + B_Y$$

is a neighborhood of $U(0)$, implies

$$U^{-1}(U(0) + B_Y) = U^{-1}(B_{R(U)})$$

is a neighborhood of 0. Hence, there is $\lambda > 0$ such that

$$\lambda B_{\mathcal{D}(U)} \subset U^{-1}(B_{R(U)}).$$

Therefore,

$$\begin{aligned}\lambda U(B_{\mathcal{D}(U)}) \subset & B_{R(U)} + U(0) \\ = & UU^{-1}(B_{R(U)}).\end{aligned}$$

Now, in look of Proposition 2.10.1, we have $\|U\| < \infty$. Conversely, we assume that $\|U\| < \infty$. Hence, if we take $x \in \mathcal{D}(U)$, $y \in Ux$ and giving V be a nontrivial closed ball in $R(U)$ with center y. Hence, $\mathcal{V}_0 = \mathcal{V} \backslash \{y\} \subset \alpha B_{R(U)}$ for some $\alpha > 0$. So, by using Proposition 2.10.1, there exists $\lambda > 0$ such that

$$U(B_{\mathcal{D}(U)}) \subset \lambda B_{R(U)} + U(0).$$

So,

$$\begin{aligned}B_{\mathcal{D}(U)} + U^{-1}(0) \subset & \lambda U^{-1}(B_{R(U)}) \\ = & \alpha^{-1}\lambda U^{-1}(\mathcal{V}_0),\end{aligned}$$

and

$$\lambda^{-1}\alpha B_{\mathscr{D}(U)} + U^{-1}(0) \subset U^{-1}(\mathscr{V}_0)$$
$$= U^{-1}(\mathscr{V}\backslash\{y\}).$$

Hence,

$$\lambda^{-1}\alpha B_{\mathscr{D}(U)} + U^{-1}y \subset U^{-1}(V\backslash\{y\}) + U^{-1}y$$
$$\subset U^{-1}(\mathscr{V}).$$

So, $\lambda^{-1}\alpha B_{\mathscr{D}(U)} + U^{-1}y$ is a neighborhood of x in $\mathscr{D}(U)$. We take $\mathscr{W}$ a neighborhood of Ux, let $O \subset \mathscr{W}$ be an open set containing $y \in Ux$ and let $\mathscr{V} \subset \mathscr{O}$ be a non-trivial closed ball with center y. Then, $U^{-1}(\mathscr{W})$ is neighborhood of x. This implies that U is continuous multivalued operator.
(ii) If we suppose $\dim \mathscr{D}(U) < \infty$. Then, we have $Q_U U$ is a continuous linear operator, that is, $\|Q_U U\| < \infty$. Since

$$\|U\| = \|Q_U U\|,$$

the result follows from (i).
(iii) See [2, Proposition 2.3.2].
(iv) Since $\dim(D(U^{-1}) = \dim(R(U)) < \infty$, then by (ii), we have U^{-1} is continuous. Hence, by using Remark 2.10.2, we deduce that U is open. Q.E.D.

Corollary 2.10.3 *Let X and Y be normed spaces, and $U \in L\mathscr{R}(X, Y)$. If U is continuous linear relation, then*

$$\|Uu\| \leq \|U\|\|u\|$$

for all $u \in \mathscr{D}(U)$. ◇

Proposition 2.10.5 *Let U, $V \in L\mathscr{R}(X, Y)$, where X and Y be normed spaces . If $\overline{V(0)} \subset \overline{U(0)}$ and $\mathscr{D}(U) \subset \mathscr{D}(V)$. Then,*

$$\|(U - V)u\| \geq \|Uu\| - \|Vu\|, \quad u \in \mathscr{D}(U). \tag{2.10.9}$$

Moreover, if U and V are bounded multivalued operators. Then, $U - V \in B\mathscr{R}(X, Y)$, and

$$\|U - V\| \geq \|U\| - \|V\|. \tag{2.10.10}$$

Additionally, if $V \in B\mathscr{R}(X, Y)$ and U is unbounded multivalued operator or U is bounded multivalued operator, $V_{|\mathscr{D}(T)}$ is unbounded, and $\overline{V(0)} = \overline{U(0)}$, then $U - V$ is unbounded multivalued operator. ◇

Proof First we calmly that

$$\overline{U(0)} = \overline{(U - V)(0)}$$

since $\overline{V(0)} \subset \overline{U(0)}$. Via (i) of Lemma 2.10.2, believe that for any $u \in \mathscr{D}(U)$, $v_1 \in Uu$ and $v_2 \in Vu$,

$$\begin{aligned}\|(U - V)u\| &= \text{dist}(v_1 - v_2, \overline{(U - V)(0)})\\ &= \text{dist}(v_1 - v_2, \overline{U(0)})\\ &\geq \text{dist}(v_1, \overline{U(0)}) - \text{dist}(v_2, \overline{U(0)})\\ &\geq \text{dist}(v_1, \overline{U(0)}) - \text{dist}(v_2, \overline{V(0)})\\ &= \|Uu\| - \|Vu\|, \end{aligned} \tag{2.10.11}$$

which yields that (2.10.9) holds. Also, we presume that U and V are bounded multivalued operators. Hence, by using both Lemma 2.10.3 (i) and Theorem 2.10.4, we have $U - V$ is bounded multivalued operator. It follows from (2.10.11) that

$$\|Uu\| \leq \|(U - V)u\| + \|Vu\|,$$

for all $u \in \mathscr{D}(U)$ with $\|u\| \leq 1$. Hence, by using (ii) of Lemma 2.10.2, we deduce that

$$\|U\| \leq \|(U - V)\| + \|V\|.$$

So, (2.10.10) holds. Likewise, if we assume that V is bounded multivalued operator and U is unbounded multivalued operator it might be ready to verify that $U - V$ is unbounded by both (2.10.11), Lemma 2.10.2 and Corollary 2.10.3. Finally, assume that U is bounded multivalued operator, $V_{|\mathscr{D}(U)}$ is unbounded, and $\overline{V(0)} = \overline{U(0)}$. Hence, by using (2.10.11) that

$$\begin{aligned}\|(U - V)u\| \geq& \text{dist}(v_2, \overline{V(0)}) - \text{dist}(v_1, \overline{U(0)})\\ =&\|Vu\| - \|Uu\|,\end{aligned}$$

which, collectively with Lemma 2.10.2, that $U - V$ is unbounded multivalued operator. Q.E.D.

Theorem 2.10.1 [2, Theorem III.7.4] *Let X, Y be two normed spaces and let U, $V \in L\mathscr{R}(X, Y)$ satisfy $\gamma(U) > 0$, $V(0) \subset \overline{U(0)}$, $\mathscr{D}(U) \subset \mathscr{D}(V)$, and $\|V\|_{L\mathscr{R}(X,Y)} < \gamma(U)$. Then,*

$$\alpha(U + V) \leq \alpha(U)$$

and

$$\overline{\beta}(U + V) \leq \overline{\beta}(U). \qquad \diamond$$

Proposition 2.10.6 [2, Proposition II.6.3] *Let $U \in L\mathscr{R}(X, Y)$ be such that $N(U)$ is topologically complemented in $\mathscr{D}(U)$ (or in X) and let P be a continuous projection defined on $\mathscr{D}(U)$ (or on X) with $N(U)$. Then,*

$$0 \leq \|P\|^{-1}\gamma(U) \leq \gamma(U_{|R(P)}) \leq \gamma(U). \qquad \diamond$$

Lemma 2.10.10 *Let $U, V \in L\mathscr{R}(X)$ be such that U is one-to-one, open, $\overline{V(0)} \subset \overline{U(0)}$ and $\mathscr{D}(U) \subset \mathscr{D}(V)$. Then, we have*

$$\gamma(U - V) \geq \gamma(U) - \|V\|.$$

Moreover, if $\|V\| < \gamma(U)$, then $U - V$ is open and one-to-one. $\diamond$

Proof If we take $x \in \mathscr{D}(U - V)$, then by applying Proposition 2.10.5, we embody

$$\begin{aligned} \|(U - V)x\| &\geq \|Ux\| - \|Vx\| \\ &\geq \gamma(U)\|x\| - \|V\|\|x\| \\ &\geq (\gamma(U) - \|V\|)\|x\|. \end{aligned}$$

This implies that

$$\gamma(U - V) \geq \gamma(U) - \|V\|.$$

As $\|V\| < \gamma(U)$, we implies that $\gamma(U - V) > 0$. So, as part of Proposition 2.10.4 (iii), we deduce $U - V$ is open. Now, in view of Theorem 2.10.1 (i), we have $U - V$ is one-to-one. Q.E.D.

Remark 2.10.3 Let $U \in L\mathscr{R}(X, Y)$. If U is open with $N(U)$ is topologically complemented in $\mathscr{D}(U)$ (or in X), then

$$0 < \|P\|^{-1}_{L\mathscr{R}(X)} \leq 1,$$

where P is any continuous projection defined on $\mathscr{D}(U)$ (or on X) with $N(U)$. Indeed, we have $\|P\|^{-1} > 0$. It follows from Proposition 2.10.6 that

$$0 \leq \|P\|^{-1}_{L\mathscr{R}(X)}\gamma(U) \leq \gamma(U). \tag{2.10.12}$$

Since U is open, then by using Proposition 2.10.4 (ii), we have $\gamma(U) > 0$. Hence, multiplying (2.10.12) by $\gamma(U)^{-1}$, we infer that $\|P\|^{-1}_{L\mathscr{R}(X)} \leq 1$. $\diamond$

Proposition 2.10.7 ([2, Proposition II.3.4]) *Let $U \in L\mathscr{R}(X, Y)$. Then,*

(i) $N(U) \subset N(Q_U U)$ with equality if $U(0)$ is relatively closed in $R(U)$.
(ii) $\gamma(U) \leq \gamma(Q_U U)$ with equality if $U(0)$ is relatively closed in $R(U)$. $\diamond$

Proposition 2.10.8 [2, Corollary II.3.9] *Let U be an open linear relation such that the kernel $N(U) = \overline{N(U)}$. Then,*

(i) $N(QU) = N(U)$.
(ii) $\gamma(QU) = \gamma(U)$. $\diamond$

Lemma 2.10.11 [2, Corollary III.7.7] *Let X, Y be two normed spaces and $U \in L\mathscr{R}(X)$ be open and one-to-one with dense range. Then, for any relation V such that $V(0) \subset \overline{U(0)}$, $\mathscr{D}(V) \supset \mathscr{D}(U)$ and $\|V\| < \gamma(U)$, we have $U + V$ is open one-to-one with dense range.* ◇

Lemma 2.10.12 *Let $A \in L\mathscr{R}(H)$ and $B \in L\mathscr{R}(K)$. Let $(y_n)_n$ be an orthonormal sequence of $N(B)^{\perp}$ such that $\|Q_B B y_n\| \to 0$. Suppose that there exists $C \in L\mathscr{R}(K, H)$ such that $C(0) \subseteq A(0)$ for which $R(\mathscr{M}_C)$ is closed. Then, there exist $\epsilon_0 > 0$ and a subsequence $(y_{n_k})_k$ of $(y_n)_n$ such that dist$(R(Q_A A), Q_A C y_{n_k}) \geq \epsilon_0$, for all $k \in \mathbb{N}$.* ◇

Proof Since $R(\mathscr{M}_C)$ is closed, we have

$$\gamma(\mathscr{M}_C) := \inf \left\{ \frac{\left\| Q_{\mathscr{M}_C} \mathscr{M}_C \begin{pmatrix} u \\ v \end{pmatrix} \right\|}{\text{dist}\left(\begin{pmatrix} u \\ v \end{pmatrix} \text{ such that } N(\mathscr{M}_C) \right)} \text{ such that } \begin{pmatrix} u \\ v \end{pmatrix} \notin N(\mathscr{M}_C) \right\} \neq 0.$$

Suppose that for all $\epsilon > 0$ and $(y_{n_k})_k$ subsequence of $(y_n)_n$, we have

$$\text{dist}(R(Q_A A), Q_A C y_{n_k}) \leq \epsilon.$$

Hence,

$$\text{dist}(R(Q_A A), Q_A C y_n) \to 0.$$

Then, there exists $(x_n)_n \in H$ such that

$$\|Q_A A x_n - Q_A C y_n\| \to 0.$$

For each $n \in \mathbb{N}$, we can prove that

$$\text{dist}\left(\begin{pmatrix} x_n \\ -y_n \end{pmatrix}, N(\mathscr{M}_C) \right) \geq 1.$$

To see this, observe that

$$\begin{aligned}
\text{dist}\left(\begin{pmatrix} x_n \\ -y_n \end{pmatrix}, N(\mathscr{M}_C) \right) &= \inf \left\{ \left\| \begin{pmatrix} x_n \\ -y_n \end{pmatrix} - \begin{pmatrix} u \\ v \end{pmatrix} \right\| : \begin{pmatrix} u \\ v \end{pmatrix} \in N(\mathscr{M}_C) \right\} \\
&= \inf \left\{ \sqrt{\|x_n - u\|^2 + \|y_n + v\|^2} : \begin{pmatrix} u \\ v \end{pmatrix} \in N(\mathscr{M}_C) \right\} \\
&= \inf \left\{ \sqrt{\|x_n - u\|^2 + \|y_n\|^2 + \|v\|^2} \text{ such that } \begin{pmatrix} u \\ v \end{pmatrix} \in N(\mathscr{M}_C) \right\} \\
&\geq \|y_n\| \\
&= 1.
\end{aligned}$$

Then,

$$\frac{\left\| Q_{\mathscr{M}_C}\mathscr{M}_C\begin{pmatrix} x_n \\ -y_n \end{pmatrix}\right\|}{\operatorname{dist}\left(\begin{pmatrix} x_n \\ -y_n \end{pmatrix}, N(\mathscr{M}_C)\right)} \leq \left\| Q_{\mathscr{M}_C}\mathscr{M}_C\begin{pmatrix} x_n \\ -y_n \end{pmatrix}\right\|$$

$$= \sqrt{\|Q_A A x_n - Q_A C y_n\|^2 + \|Q_B B y_n\|^2} \to 0$$

which is a contradiction with the fact that $\gamma(\mathscr{M}_C) \neq 0$. Q.E.D.

2.11 Closed Linear Relation

2.11.1 Closure and Closed Linear Relation

Let $U \in L\mathscr{R}(X, Y)$. U is called closed if its graph $G(U)$ is closed in $X \times Y$. The closure of U, denoted $\overline{U}$, is defined in terms of their corresponding graphs in $X \times Y$ as follows

$$G(\overline{U}) = \overline{G(U)} \subset X \times Y.$$

We denote the collection of all closed multivalued linear operators from X into Y by $C\mathscr{R}(X, Y)$ and as useful we write $C\mathscr{R}(X) := C\mathscr{R}(X, X)$.

Proposition 2.11.1 [2, Proposition II.5.3], [21, Lemma 5.3] (i) *Let* $U \in L\mathscr{R}(X, Y)$. *The following properties are equivalent*

(i_1) $U \in C\mathscr{R}(X, Y)$,

(i_2) $U^{-1} \in C\mathscr{R}(Y, X)$, *and*

(i_3) $Q_U U$ *is closed linear operator and* $U(0)$ *is closed space.*

Especially, $N(U)$ *is closed, if* $U \in C\mathscr{R}(Y, X)$.

(ii) *Suppose that* $U \in C\mathscr{R}(X, Y)$. *Then,* $R(U)$ *is closed if, and only if,* $R(Q_U U)$ *is closed.* ◇

Remark 2.11.1 $U \in C\mathscr{R}(X, Y)$, if, and only if, $U(0)$ is closed and $x_n \to x$ in X for every sequence $(x_n)_n$ in $\mathscr{D}(U)$ and $Q_U U x_n \to y$ in $Y/\overline{U(0)}$ implies that $x \in \mathscr{D}(U)$ and $Q_U U x = y$. ◇

Remark 2.11.2 In the beliefs of linear relations, we do not have that every bounded multivalued linear operator is closed as in the operator theory. ◇

Example 2.11.1 Let X be a normed closed vector space and let Q be a not closed vector subspace of X. Define $U \in L\mathscr{R}(X, Y)$ by

$$Ux = Q \text{ for all } x \in X.$$

Hence, U is bounded multivalued linear operator but not closed. Indeed, we have $\mathscr{D}(U) = X$ which harvests it is closed. Also, $Ux = U(0)$ for all $x \in X$. By ethics of Lemma 2.10.2 that $\|Ux\| = 0$ for all $x \in X$. This ascertained from Lemma 2.10.2 that $\|U\|_{L\mathscr{R}(X,Y)} = 0 < \infty$. Then, U is bounded. Since $U(0) = Q$ and Q is not closed it implies from Theorem 2.11.1 (iv) that U is not closed multivalued linear operator.

We denote the collection of all bounded and closed multivalued linear operators from X into Y by $BC\mathscr{R}(X, Y)$ and as useful we write $BC\mathscr{R}(X) := BC\mathscr{R}(X, X)$.

Theorem 2.11.1 [2, Theorem III.5.3 and Corollary III.5.4]

(i) If $U \in C\mathscr{R}(X, Y)$ and $\mathscr{D}(U) = X$, then $U \in B\mathscr{R}(X, Y)$.
(ii) If U is open multivalued linear operator, then $R(U) = \overline{R(U)}$.
(iii) If $R(U) = \overline{R(U)}$, then U is open multivalued linear operator.
(iv) If U is continuous multivalued linear operator and $\mathscr{D}(U)$ and $U(0)$ are closed, then $U \in C\mathscr{R}(X, Y)$. ◇

Let $U \in L\mathscr{R}(X, Y)$. The U graph norm $\|\cdot\|_U$ is defined on $\mathscr{D}(U)$ by

$$\|x\|_U := \|x\| + \|Ux\|, \quad for\ x \in \mathscr{D}(U).$$

We designate by X_U the vector space $(\mathscr{D}(U), \|\cdot\|_U)$ and $i_U \in L\mathscr{R}(X_U, X)$ the identity injection (or the graph operator) of X_U into X, i.e.,

$$\mathscr{D}(i_U) = X_U,$$

and

$$i_U(x) = G_U(x) = x \text{ for all } x \in X_U.$$

G_U is called the graph operator of U.

Definition 2.11.1 Let $\widetilde{M}$ and $\widetilde{N}$ be the completion of the normed space M and N, respectively and $U \in L\mathscr{R}(M, N)$. The completion of U, denoted $\widetilde{U}$, is defined in terms of her corresponding graph

$$G(\widetilde{U}) \subset \widetilde{M} \times \widetilde{N}. \tag{2.11.1}$$

Remark 2.11.3 The ramification of Theorem 2.11.1 rest supportable if we replaced U by $\widetilde{U}$, where $\widetilde{U}$ is provided in (2.11.1). ◇

Remark 2.11.4 (i) Let U be a multivalued linear operator such that $G(U)$ is complete, then U is said to be completely closed (or complete).
(ii) Giving X, Y be two normed spaces and $U \in L\mathscr{R}(X, Y)$. Then, $\widetilde{U} \in C\mathscr{R}(\widetilde{X}, \widetilde{Y})$. Indeed, we have $G(\widetilde{U})$ is complete which yields it is closed. Then, $\widetilde{U}$ is closed.
(iii) If X and Y are two Banach spaces and U is a closed multivalued linear operator. Then, U is a completely closed and $\widetilde{U} = U$. Indeed, we have $G(U)$ is closed. The fact that $X \times Y$ is Banach spaces implies $G(U)$ is complete in $X \times Y$. Consequently, $G(\widetilde{U}) = G(U)$. ◇

Lemma 2.11.1 *Let U, F, $V \in L\mathscr{R}(X)$ and let $\eta \in \mathbb{C}$. Then,*

(i) $R((\eta V - U)G_F) = R(\eta V - U)$.
(ii) $N((\eta V - U)G_F) = N(\eta V - U)$. ◇

Proof (i) Using the fact that $G_F x = (G_F)^{-1}x = x$, we have

$$R(U) = U(\mathscr{D}(U))$$

and

$$\mathscr{D}(UF) = F^{-1}(\mathscr{D}(U)).$$

So,

$$R((\eta V - U)G_F) = (\eta V - U)G_F(\mathscr{D}((\eta V - U)G_F)).$$

On the other hand

$$\begin{aligned}(\eta V - U)G_F(\mathscr{D}((\eta V - U)G_F)) &= (\eta V - U)(\mathscr{D}((\eta V - U)G_F)),\\ &= (\eta V - U)G_F(\mathscr{D}(\eta V - U)),\\ &= (\eta V - U)(\mathscr{D}(\eta V - U)),\\ &= R(\eta V - U).\end{aligned}$$

Then,

$$R((\eta V - U)G_F) = R(\eta V - U).$$

(ii) We have

$$\begin{aligned}N((\eta V - U)G_F) &= \{x \in \mathscr{D}((\eta V - U)G_F) \text{ such that } (\eta V - U)G_F(x) = (\eta V - U)G_F(0)\}\\ &= \{x \in \mathscr{D}(\eta V - U) \text{ such that } (\eta V - U)(x) = (\eta V - U)(0)\}.\end{aligned}$$

Then,

$$N((\eta V - U)G_F) = N(\eta V - U).$$ Q.E.D.

Lemma 2.11.2 *Let U in $BC\mathscr{R}(X)$ and $(x_n)_n \subset X$ be such that $x_n \in Ux_{n-1}$ for all $n \in \mathbb{N}^*$. Let R be the convergence radius of the power series*

$$\sum_{n\geq 1} \eta^n x_{n-1}.$$

Then, for all scalar η such that $|\eta| > \frac{1}{R}$, we have

$$U\left(\sum_{n\geq 1} \eta^{-n} x_{n-1}\right) = \sum_{n\geq 1} \eta^{-n} x_n + U(0).$$ ◇

Proof For a scalar η such that $|\eta| > \frac{1}{R}$, we have

$$U\left(\sum_{n\geq 1}\eta^{-n}x_{n-1}\right) = \sum_{n=1}^{N}\eta^{-n}Ux_{n-1} + U\left(\sum_{n\geq N+1}\eta^{-n}x_{n-1}\right)$$
$$= \sum_{n\geq 1}\eta^{-n}x_n - \sum_{n\geq N+1}\eta^{-n}x_n + U\left(\sum_{n\geq N+1}\eta^{-n}x_{n-1}\right).$$

Now, since the operators Q_U and $Q_U U$ are bounded, then we can easily see that

$$Q_U\left(U\left(\sum_{n\geq 1}\eta^{-n}x_{n-1}\right)\right) = Q_U\left(\sum_{n\geq 1}\eta^{-n}x_n\right).$$

Then,

$$U\left(\sum_{n\geq 1}\eta^{-n}x_{n-1}\right) - \sum_{n\geq 1}\eta^{-n}x_n \subseteq U(0).$$

Thus,

$$U\left(\sum_{n\geq 1}\eta^{-n}x_{n-1}\right) = \sum_{n\geq 1}\eta^{-n}x_n + U(0).$$ Q.E.D.

Lemma 2.11.3 *Let U be in $BC\mathscr{R}(X)$, $\varepsilon > 0$ and $(x_n)_n \subset X$ be such that*

$$\limsup_{n\to\infty}\|x_n\|^{\frac{1}{n}} \leq \varepsilon.$$

Then, for all $n \in \mathbb{N}$, there exists $\alpha_{n+1} \in Ux_n$ such that for all $|\eta| < \frac{1}{\varepsilon}$, we have

$$U\left(\sum_{n\geq 0}\eta^n x_n\right) = \sum_{n\geq 0}\eta^n\alpha_{n+1} + U(0).$$ ◇

Proof Let $\alpha'_{n+1} \in Ux_n$. Then,

$$\begin{aligned}\operatorname{dist}(\alpha'_{n+1}, U(0)) &= \|Ux_n\|\\ &\leq \|U\|\|x_n\|.\end{aligned}$$

Hence, for a fixed $\gamma > 0$, there exists $\beta_{n+1} \in U(0)$ such that

$$\|\alpha'_{n+1} - \beta_{n+1}\| \leq (\|U\| + \gamma)\|x_n\|.$$

Take $\alpha_{n+1} = \alpha'_{n+1} - \beta_{n+1}$. Then, $\alpha_{n+1} \in Ux_n$ and the series

$$\sum_{n\geq 0} \eta^n \alpha_{n+1}$$

is absolutely convergent for all η such that $|\eta| < \dfrac{1}{\varepsilon}$. Thus, we obtain

$$Q_U U \left(\sum_{n\geq 0} \eta^n x_n \right) = Q_U \left(\sum_{n\geq 0} \eta^n \alpha_{n+1} \right).$$

Hence,

$$U \left(\sum_{n\geq 0} \eta^n x_n \right) = \sum_{n\geq 0} \eta^n \alpha_{n+1} + U(0). \qquad \text{Q.E.D.}$$

2.11.2 *Sum of Closed Linear Relations in Banach Space*

Proposition 2.11.2 *Let $V, U \in L\mathscr{R}(X, Y)$ be such that $V(0) \subseteq U(0)$ and $\overline{\mathscr{D}(U)} \subset \mathscr{D}(V)$. If U is closed and V is continuous, then $U + V$ is closed. Specifically, $\eta - U$ is closed for all $\eta \in \mathbb{C}$.* ◇

Proof First we assume that $V(0) = U(0) = \{0\}$. If we take $(x_n)_n \subset \mathscr{D}(U + V) = \mathscr{D}(U)$ such that x_n converge to x and $(U + V)x_n$ converge to y. It follows that

$$Ux_n = (U + V)x_n - Vx_n.$$

Since the single-valued operator V is continuous, then Vx_n converge to Vx. Accordingly, Ux_n converge to $y - Vx$. As U is closed, we imply that $Ux = y - Vx$ and $x \in \mathscr{D}(U)$. Hence, $y = (V + U)x$ and $x \in \mathscr{D}(V + U)$. For the general case, i.e., V, $U \in L\mathscr{R}(X, Y)$. As $V(0) \subseteq U(0)$, we deduce that

$$\begin{aligned}(U + V)(0) &= U(0) + V(0)\\ &= U(0)\end{aligned}$$

and $(U + V)(0)$ is closed space. Since

$$\begin{aligned}Q_{U+V}(U + V) &= Q_U(U + V)\\ &= Q_U U + Q_U V,\end{aligned} \qquad (2.11.2)$$

then $\overline{V(0)} \subset \overline{U(0)} = U(0)$. Consequently, by using Lemma 2.1.1, we infer that

$$Y/\overline{V(0)}\Big/R \equiv Y/U(0)$$

and

$$Q_U = Q_R Q_V,$$

where $R = U(0)/\overline{V(0)}$ wherefore,

$$Q_U V = Q_R Q_V V.$$

Since V is continuous, then $Q_U V$ is a continuous single-valued operator. Using Eq. (2.11.2), we obtain that $Q_{U+V}(U+V)$ is closed. Again, it follows from Proposition 2.11.1 (i), that $U+V$ is closed. Q.E.D.

Theorem 2.11.2 *Let* $U,\ V \in L\mathscr{R}(X,Y)$ *be such that* $U \in C\mathscr{R}(X,Y)$ *and* V *is continuous linear relation such that* $\dim V(0) < +\infty$ *and* $\mathscr{D}(V) \supset \overline{\mathscr{D}(U)}$. *Then,* $U+V \in C\mathscr{R}(X,Y)$. ◇

Proof As $\mathscr{D}(Q_{U+V}) = Y$ and by applying Proposition 2.7.3, we deduce

$$Q_{U+V}(U+V) = Q_{U+V}U + Q_{U+V}V.$$

In addition, since U is closed, then, using Proposition 2.11.1, $U(0)$ is closed and $Q_U U$ is closed. Hence,

$$\dim\left(\frac{U(0)+V(0)}{U(0)}\right) < \infty.$$

Then, we obtain

$$Q_{U+V}U = Q^{Y/U(0)}_{\frac{U(0)+V(0)}{U(0)}} Q^Y_U U$$

is closed. On that account $Q_{U+V}V$ is a continuous single-valued relation, in fact, $Q_{U+V}V(0) = 0$. On the other hand, $\mathscr{D}(Q_{U+V}) = Y$,

$$\|Q_{U+V}V\| \leq \|Q_{U+V}\|\|V\|.$$

As $\mathscr{D}(Q_{U+V}V) = \mathscr{D}(V) \supset \overline{\mathscr{D}(Q_{U+V}V)} = \overline{\mathscr{D}(U)}$ and on the basis of Proposition 2.11.2, we obtain that

$$Q_{U+V}(U+V) = Q_{U+V}U + Q_{U+V}V$$

is closed linear operator. It is absolutely clear that $(U+V)(0) = U(0)+V(0)$ is closed space. Accordingly, in view of Proposition 2.11.1, we have $U+V \in C\mathscr{R}(X,Y)$. Q.E.D.

Theorem 2.11.3 *Let* $V, U \in L\mathscr{R}(X,Y)$ *be such that* $\mathscr{D}(V) \supset \mathscr{D}(U)$, $V(0) \subset \overline{U(0)}$ *and*

$$\|Vu\| \leq a\|u\| + b\|Uu\|, \quad u \in \mathscr{D}(U), \tag{2.11.3}$$

for some constants a, b *with* $b < 1$. *Then,* $U \in C\mathscr{R}(X,Y)$ *if, and only if,* $U+V \in C\mathscr{R}(X,Y)$. ◇

Proof First we assume that $U \in C\mathscr{R}(X, Y)$. Let us examine two cases for V and U.
First case. If V and U are two linear operators. Hence,

$$\begin{aligned}\|Uu\| =& \|(U+V-V)u\| \\ \leq& \|(U+V)u\| + \|Vu\|, \quad u \in \mathscr{D}(U),\end{aligned}$$

hence by using (2.11.3), we obtain

$$(1-b)\|Uu\| \leq \|(U+V)u\| + a\|x\|, \quad u \in \mathscr{D}(U).$$

if we take $(u_n)_n$ is sequence in $\mathscr{D}(U+V) = \mathscr{D}(U)$ such that $u_n \to u$ and $(U+V)u_n \to \eta$, then there exists $N_0 \in \mathbb{N}$ such that for all $n,\ m \geq N_0$, we have

$$(1-b)\|Uu_n - Uu_m\| \leq a\|u_n - u_m\| + \|(U+V)\|\|u_n - u_n\|.$$

Consequently, $(Uu_n)_n$ is a Cauchy sequence in the Banach space Y. Thus, $Uu_n \to u_0$. Since U is closed linear operator, $u \in \mathscr{D}(U)$ and $Uu = u_0$, then $Vu_n = (U+V)u_n - Uu_n \to \eta - x_0$ as $n \to \infty$. But

$$\|V(u_n - u)\| \leq (a\|u_n - u\| + b\|Uu_n - u_0\|) \to 0 \text{ as } n \to \infty,$$

which shows that $Vu = \eta - U_0$. Hence, $(U+V)u = \eta$ and $x \in \mathscr{D}(U+V)$.

Second case. U and V are two linear operators. As $V(0) \subset \overline{U(0)} = U(0)$, it is obvious that $Q_U = Q_{U+V}$. Then,

$$Q_{U+V}(U+V) = Q_U(U) + Q_U(V).$$

By applying Lemma 2.10.4, we conclude that $Q_U(V)$ two linear operators and

$$\begin{aligned}\|Q_U(V)u\| \leq& \|Q_V(V)u\| \\ =& \|Vu\| \\ \leq& a\|u\| + b\|Uu\|, \ u \in \mathscr{D}(U).\end{aligned}$$

Hence,

$$\|Q_U(V)u\| \leq a\|u\| + b\|Q_U(U)u\|, \quad u \in \mathscr{D}(U).$$

Since $Q_U(U)$ is closed, then $Q_U(U) + Q_U(V)$ is closed linear operator. The fact that $(U+V)(0) = U(0)$ is closed, that $U+V$ is closed. Conversely, assume that $U+V$ is closed. From Eq.(2.11.3), there exists $a > 0$, and $0 < b < 1$, such that for all $u \in \mathscr{D}(U)$,

$$\begin{aligned}\|-Vu\| &= \|Vu\| \\ &\leq a\|u\| + b\|Uu\| \\ &= a\|u\| + b\|Uu - Vu - Vu\| \text{ (as Proposition 1.7.2 } (i)) \\ &\leq \left(\frac{a}{1-b}\right)\|u\| + \left(\frac{b}{1-b}\right)\|(U+V)u\| \text{ (as Lemma 1.10.3 } (i)).\end{aligned}$$

This implies that $U + V - V$ is closed and making use of Proposition 2.7.2, we have $U \in C\mathscr{R}(X, Y)$. Q.E.D.

Theorem 2.11.4 *Let $S, U, V \in L\mathscr{R}(X, Y)$ be such that $\mathscr{D}(U) \subset \mathscr{D}(S) \subset \mathscr{D}(V)$, $V(0) \subset S(0) \subset \overline{U(0)}$. Assume that*

(i) there exist two constants $a_1, b_1 > 0$ such that

$$\|Sx\| \leq a_1\|x\| + b_1\|Ux\|, \quad x \in \mathscr{D}(U),$$

(ii) there exist two constants $a_2, b_2 > 0$ such that $b_1(1 + b_2) < 1$ and

$$\|Vx\| \leq a_2\|x\| + b_2\|Sx\|, \quad x \in \mathscr{D}(S).$$

Then, $U \in C\mathscr{R}(X, Y)$ if, and only if, $U + S + V \in C\mathscr{R}(X, Y)$. ◇

Proof Let us consider two cases for S, U, and V.

First case. If S, U, and V are operators, then for all $x \in \mathscr{D}(U)$, we have

$$\begin{aligned}\|(U+S+V)x\| &\leq \|Ux\| + \|Sx\| + \|Vx\| \\ &\leq \|Ux\| + a_1\|x\| + b_1\|Ux\| + a_2\|x\| + b_2\|Sx\| \\ &= (1+b_1)\|Ux\| + (a_1+a_2)\|x\| + b_2(a_1\|x\| + b_1\|Ux\|).\end{aligned}$$

Hence,

$$\|(U+S+V)x\| \leq (a_1 + a_2 + a_1b_2)\|x\| + (1 + b_1 + b_1b_2)\|Ux\|. \tag{2.11.4}$$

Similarly, for all $x \in \mathscr{D}(U)$, we have

$$\begin{aligned}\|(S+V)x\| &\leq \|Sx\| + \|Vx\| \\ &\leq a_1\|x\| + b_1\|Ux\| + a_2\|x\| + b_2\|Sx\| \\ &= b_1\|Ux\| + (a_1+a_2)\|x\| + b_2(a_1\|x\| + b_1\|Ux\|).\end{aligned}$$

Hence,

$$\|(S+V)x\| \leq (a_1 + a_2 + a_1b_2)\|x\| + b_1(1 + b_2)\|Ux\|. \tag{2.11.5}$$

Combining (2.11.4) and (2.11.5), it follows that for all $x \in \mathscr{D}(U)$, we have

$$\begin{aligned}\|(U+S+V)x\| &\geq \|Ux\| - \|(S+V)x\| \\ &\geq \|Ux\| - (a_1+a_2+a_1b_2)\|x\| - b_1(1+b_2)\|Ux\| \\ &= -(a_1+a_2+a_1b_2)\|x\| + [1-b_1(1+b_2)]\|Ux\|\end{aligned}$$

which yields

$$[1-b_1(1+b_2)]\|Ux\| \leq \|(U+S+V)x\| + (a_1+a_2+a_1b_2)\|x\|.$$

Let $\omega = 1-b_1(1+b_2)$ $(0<\omega<1)$ and $\sigma = a_1+a_2+a_1b_2$ $(\sigma>0)$. Then,

$$\|Ux\| \leq \omega^{-1}\left(\|(U+S+V)x\| + \sigma\|x\|\right). \tag{2.11.6}$$

Let $(x_n)_n$ be a sequence in $\mathscr{D}(U+S+V) = \mathscr{D}(U)$ such that $x_n \to x$ and

$$(U+S+V)x_n \to \sigma.$$

Then, by using Eq. (2.11.6), there exists $N_1 \in \mathbb{N}$ such that for all $n,\ m \geq N_1$, we have

$$\|Ux_n - Ux_m\| \leq \omega^{-1}\left(\|(U+S+V)(x_n-x_m)\| + \sigma\|x_n-x_m\|\right).$$

Consequently, $(Ux_n)_n$ is a Cauchy sequence in the Banach space Y and therefore, there exists $\theta_1 \in Y$ such that Ux_n converge to θ_1. On the other hand T is closed, this implies that $x \in \mathscr{D}(U)$ and

$$Ux = \theta_1.$$

From Eq. (2.11.4), it follows that

$$\|(U+S+V)(x_n-x)\| \leq (a_1+a_2+a_1b_2)\|x_n-x\| + (1+b_1+b_1b_2)\|U(x_n-x)\| \to 0.$$

which, by letting $n \to +\infty$, we have $(U+S+V)x_n$ converge to $\psi = (U+S+V)x$. Accordingly, $U+S+V$ is a closed linear operator.

Second case. U, S, and V are multivalued linear operators. It follows from $V(0) \subset S(0) \subset U(0)$, it is clear that $Q_{U+S+V} = Q_U$. Then,

$$\begin{aligned}Q_{U+S+K}(U+S+V) &= Q_U(U+S+V) \\ &= Q_U(U) + Q_U(S) + Q_U(V).\end{aligned}$$

As $Q_U(U)$ is a closed linear operator, hence by using Lemma 2.10.4, we have $Q_U S$, $Q_U(V)$ are single-valued operators. Then, by using Lemma 2.10.3 (iii), we have

$$\begin{aligned}\|Q_U(S)\| &\leq \|Q_S(S)\| \\ &= \|Sx\|\end{aligned}$$

$$\leq a_1\|x\| + b_1\|Ux\|, \quad x \in \mathscr{D}(U),$$

and so,

$$\|Q_U(S)\| \leq a_1\|x\| + b_1\|Q_U(U)x\|, \quad x \in \mathscr{D}(U).$$

Similarly, by using Lemma 2.10.4, we have

$$\begin{aligned}\|Q_U(V)\| \leq& \|Q_V(V)\| \\ =& a_2\|x\| + b_2\|Q_S(S)x\|, \quad x \in \mathscr{D}(S),\end{aligned}$$

where $b_1(1+a_2) < 1$. So, $Q_U(U) + Q_U(S) + Q_U(V)$ is a closed operator. On the other hand $U(0) + S(0) + V(0) = U(0)$ is closed space, hence $U + S + V$ is a closed. Conversely, we suppose that $U + V + S$ is closed relation. Then,

$$\| - Sx\| \leq (a_1 + b_1\sigma)\,\|x\| + b_1\omega^{-1}\|(U + V + S)x\|, \quad x \in \mathscr{D}(U),$$

$$\| - Vx\| \leq a_2\|x\| + b_2\| - Sx\|, \quad x \in \mathscr{D}(S)$$

and

$$b_1\omega^{-1}(1 + a_2) < 1.$$

By referring to the relations $U + S + V$, $-V$ and $-S$, we conclude that $U + S + V - V - S$ is closed relation. On the other hand $(V + S)(0) \subset U(0)$ implies that

$$\begin{aligned}U &= U + (V + S) - (V + S) \\ &= (U + V + S) - V - S.\end{aligned}$$

Consequently, U is closed. Q.E.D.

Theorem 2.11.5 *Let*

$$\mathscr{T} := \begin{pmatrix} A & 0 & 0 \\ 0 & E & 0 \\ 0 & 0 & K \end{pmatrix}, \quad \mathscr{S} := \begin{pmatrix} 0 & B & 0 \\ 0 & 0 & F \\ G & 0 & 0 \end{pmatrix} \quad \text{and} \quad \mathscr{K} := \begin{pmatrix} 0 & 0 & C \\ D & 0 & 0 \\ 0 & H & 0 \end{pmatrix}. \tag{2.11.7}$$

such that $\mathscr{D}(E) \subset \mathscr{D}(B)$, $\mathscr{D}(K) \subset \mathscr{D}(F)$, $\mathscr{D}(A) \subset \mathscr{D}(G)$, $\mathscr{D}(F) \subset \mathscr{D}(C)$, $\mathscr{D}(G) \subset \mathscr{D}(D)$, $\mathscr{D}(B) \subset \mathscr{D}(H)$, $\mathscr{K}(0) \subset \mathscr{S}(0) \subset \overline{\mathscr{T}(0)}$,

$$\begin{aligned}\|B\varphi\| &\leq a_1\|\varphi\| + b_1\|E\varphi\| \text{ for all } \varphi \in \mathscr{D}(E),\\ \|F\varphi\| &\leq a_2\|\varphi\| + b_2\|K\varphi\| \text{ for all } \varphi \in \mathscr{D}(K),\\ \|G\varphi\| &\leq a_3\|\varphi\| + b_3\|A\varphi\| \text{ for all } \varphi \in \mathscr{D}(A),\\ \|C\varphi\| &\leq a_4\|\varphi\| + b_4\|F\varphi\| \text{ for all } \varphi \in \mathscr{D}(F),\\ \|D\varphi\| &\leq a_5\|\varphi\| + b_5\|G\varphi\| \text{ for all } \varphi \in \mathscr{D}(G),\end{aligned}$$

and

$$\|H\varphi\| \leq a_6\|\varphi\| + b_6\|B\varphi\| \text{ for all } \varphi \in \mathscr{D}(B),$$

with

$$\max\{b_1,\ b_2,\ b_3\}\,(1+\max\{a_4,\ a_5,\ a_6\}) < 1,$$

where $\mathscr{T}$, $\mathscr{S}$, and $\mathscr{K}$ are given in (2.11.7). Then,

$$\mathscr{A} := \begin{pmatrix} A & B & C \\ D & E & F \\ G & H & K \end{pmatrix} \tag{2.11.8}$$

is closed multivalued block matrix if, and only if, $\mathscr{T}$ is closed multivalued diagonal matrix if, and only if, A, E, K are closed linear relations. ◇

Proof First as

$$\begin{cases} \mathscr{D}(\mathscr{T}) = \mathscr{D}(A) \times \mathscr{D}(E) \times \mathscr{D}(K), \\ \mathscr{D}(\mathscr{S}) = \mathscr{D}(G) \times \mathscr{D}(B) \times \mathscr{D}(F), \\ \mathscr{D}(\mathscr{K}) = \mathscr{D}(D) \times \mathscr{D}(H) \times \mathscr{D}(C). \end{cases}$$

Then,

$$\mathscr{D}(\mathscr{T}) \subset \mathscr{D}(\mathscr{S}) \subset \mathscr{D}(\mathscr{K}).$$

On the other hand,

$$\|\mathscr{S}\varphi\|^2 \leq \eta_1^2\,\|\varphi\|^2 + \chi_1^2\,\|\mathscr{T}\varphi\|^2,\ \ \varphi \in \mathscr{D}(\mathscr{T})$$

$$\|\mathscr{K}\varphi\|^2 \leq \eta_2^2\,\|\varphi\|^2 + \chi_2^2\,\|\mathscr{S}\varphi\|^2,\ \ \varphi \in \mathscr{D}(S),$$

where

$$\eta_1 = \max\left\{\sqrt{a_1^2 + a_1 b_1},\ \sqrt{a_2^2 + a_2 b_2},\ \sqrt{a_3^2 + a_3 b_3}\right\},$$

$$\chi_1 = \max\left\{\sqrt{b_1^2 + a_1 b_1},\ \sqrt{b_2^2 + a_2 b_2},\ \sqrt{b_3^2 + a_3 b_3}\right\},$$

$$\eta_2 = \max\left\{\sqrt{a_4^2 + a_4 b_4},\ \sqrt{a_5^2 + a_5 b_5},\ \sqrt{a_6^2 + a_6 b_6}\right\},$$

$$\chi_2 = \max\left\{\sqrt{b_4^2 + a_4 b_4},\ \sqrt{b_5^2 + a_5 b_5},\ \sqrt{b_6^2 + a_6 b_6}\right\}.$$

From Remark 2.12.1, it follows that

$$\|\mathscr{S}\varphi\| \leq \max\{a_1, a_2, a_3\}\,\|\varphi\| + \max\{b_1, b_2, b_3\}\,\|\mathscr{T}\varphi\|,\ \ \varphi \in \mathscr{D}(\mathscr{T})$$

$$\|\mathcal{K}\varphi\| \leq \max\{a_4, a_5, a_6\} \|\varphi\| + \max\{b_4, b_5, b_6\} \|\mathcal{S}\varphi\|, \quad \varphi \in \mathcal{D}(\mathcal{S}),$$

where

$$\max\{b_1, b_2, b_3\} (1 + \max\{a_4, a_5, a_6\}) < 1.$$

Using Theorem 2.11.4, we deduce that

$$\begin{pmatrix} A & B & C \\ D & E & F \\ G & H & K \end{pmatrix}$$

is closed if, and only if,

$$\begin{pmatrix} A & 0 & 0 \\ 0 & E & 0 \\ 0 & 0 & K \end{pmatrix}$$

is closed, which in turn is equivalent to the fact A, E, and K are closed. Q.E.D.

2.11.3 Composition of Closed Linear Relations

Lemma 2.11.4 [2, Proposition II.5.17] *Let $V \in C\mathcal{R}(Z, X)$ and $U \in L\mathcal{R}(X, Y)$ be have a closed range and satisfy $\alpha(U) < \infty$ and $\gamma(U) > 0$. Then, UV is closed.* ◇

Proposition 2.11.3 [2, Proposition II.5.17] *Let $V \in C\mathcal{R}(X, Y)$ and $U \in L\mathcal{R}(Y, Z)$ be have a closed range and satisfy $\alpha(U) < \infty$ and $\gamma(U) > 0$, then $UV \in C\mathcal{R}(X, Z)$.* ◇

Lemma 2.11.5 *If $U \in L\mathcal{R}(X, Y)$ and $V \in L\mathcal{R}(Y, Z)$ are closed with $\alpha(V) < \infty$ and $R(V)$ is closed, then VU is closed.* ◇

Proof By Theorem 2.11.1 (iii), V is open and thus the result follows immediately from Lemma 2.11.4. Q.E.D.

Lemma 2.11.6 *Let $V \in C\mathcal{R}(Y, Z)$ and $U \in \mathcal{L}(X, Y)$. Then, VU is closed.* ◇

Proof Let $\begin{pmatrix} x_n \\ y_n \end{pmatrix} \in G(VU)$ be such that $\begin{pmatrix} x_n \\ y_n \end{pmatrix}$ converges to $\begin{pmatrix} x \\ y \end{pmatrix}$. Then, $y_n \in VUx_n = V(Ux_n)$. Hence, $\begin{pmatrix} Ux_n \\ y_n \end{pmatrix} \in G(V)$. Since $U \in \mathcal{L}(X, Y)$, and $V \in C\mathcal{R}(Y, Z)$, then $(Ux, y) \in G(V)$. Consequently, $y \in V(Ux) = VUx$.Q.E.D.

Remark 2.11.5 Let $U \in C\mathcal{R}(H, K)$ and $A \in \mathcal{L}(H, K)$. Then, $U^{-1}A \in C\mathcal{R}(H)$ and $UA^{-1} \in C\mathcal{R}(K)$. Indeed, since $U \in C\mathcal{R}(H, K)$, then $U^{-1} \in C\mathcal{R}(K, H)$. Based on the assumption $A \in \mathcal{L}(H, K)$, we infer from Lemma 2.11.6 that $U^{-1}A \in C\mathcal{R}(H)$. This implies that $UA^{-1} \in C\mathcal{R}(K)$. ◇

Remark 2.11.6 (i) If X and Y are complete spaces and let $U : X \longrightarrow Y$ be a closed multivalued linear operator. Then, the space X_U is also a Banach space and $U(0)$ is complete.
(ii) Let X be a complete space. If we denote by j_U the inverse of i_U, then j_U and i_U are bijective and everywhere defined. Hence,

$$\begin{aligned} R(Ui_U) &= Ui_U(\mathscr{D}(Ui_U)) \\ &= Ui_U j_U(\mathscr{D}(U)) \\ &= R(U) \end{aligned}$$

and $\alpha(Ui_U) = \alpha(U)$.
(iii) Let $U \in C\mathscr{R}(X, Y)$. Since i_U is a bounded operator, then it follows from Lemma 2.11.6 that $Ui_U \in C\mathscr{R}(X_U, Y)$.
(iv) V is U-bounded if, and only if, $\mathscr{D}(U) \subset \mathscr{D}(V)$, and VG_U is bounded. ◇

Lemma 2.11.7 *Let $V \in C\mathscr{R}(Y, Z)$ and $U \in B\mathscr{R}(X, Y)$ be with* $\dim(U(0)) < \infty$. *Then, $VU \in C\mathscr{R}(X, Z)$.* ◇

Proof Since $U \in B\mathscr{R}(X, Y)$, then

$$\|Q_U U\| = \|U\| < \infty$$

and

$$\mathscr{D}(Q_U(U)) = \mathscr{D}(U) = X.$$

This implies that $Q_U(U) \in \mathscr{L}(X, Y/U(0))$. Moreover, we have $S \in C\mathscr{R}(Y, Z)$. This implies that $V^{-1} \in C\mathscr{R}(Z, Y)$. Since $\dim(U(0)) < \infty$, we get $Q_U V^{-1}$ is closed which implies that

$$VQ_U^{-1} \in C\mathscr{R}(Y/U(0), Z).$$

Since

$$VU = VQ_U^{-1}Q_U U, \tag{2.11.9}$$

then by using both Lemma 2.11.6 and Eq. (2.11.9), we conclude that $VU \in C\mathscr{R}(X, Z)$. This achieves the proof. Q.E.D.

2.11.4 Closed Linear Relations in Hilbert Spaces

In this subsection, we recall an important decomposition for closed linear relations on Hilbert spaces and some consequences of it. Let H be a Hilbert space and let U, $V \in C\mathscr{R}(H)$ such that

$$G(U)\bigcap G(V)=\left\{\begin{pmatrix}0\\0\end{pmatrix}\right\},$$

denote

$$G(U)\dot{+}G(V)=\{(f+h,g+k):\ (f,g)\in G(U)\text{ and }(h,k)\in G(V)\}.$$

Further, if U and V are orthogonal, i.e., $G(U)\perp G(V)$ that is

$$\langle (f,g),(h,k)\rangle=\langle (f,g)\rangle+\langle (h,k)\rangle=0$$

for any $(f,g)\in G(U)$ and $(h,k)\in G(V)$, then we denote

$$G(U)\bigoplus G(V)=G(U)\dot{+}G(V).$$

In addition, we introduce the following notation for convenience

$$G(U)\ominus G(V)=\{(f,g)\in G(U):\langle (f,g)\rangle+\langle (h,k)\rangle=0\ \text{ for all }(h,k)\in G(V)\}.$$

In [22], R. Arens introduced the following decomposition for a closed linear relation U on Hilbert space H

$$G(U)=G(U_s)\bigoplus G(U_\infty), \tag{2.11.10}$$

where

$$G(U_\infty)=G(U)\bigcap(\{0\}\times H)\text{ and }G(U_s)=G(U)\ominus G(U_\infty). \tag{2.11.11}$$

Remark 2.11.7 Let H be a Hilbert space and let $U\in C\mathcal{R}(H)$. Then, $G(U_s)$ and $G(U_\infty)$ are orthogonal closed vector subspaces, where U_s and U_∞ are given in (2.11.11). ◇

Lemma 2.11.8 *Let H be a Hilbert space and let $U\in C\mathcal{R}(H)$. Then,*

(i) $G(U_\infty)=\{0\}\times U(0)$, where U_∞ is given in (2.11.11).
(ii) U_s is an operator, where U_s is given in (2.11.11). ◇

Proof (i) Let $\begin{pmatrix}x\\y\end{pmatrix}\in G(U_\infty)$. Then, by using (2.11.11), we have $\begin{pmatrix}x\\y\end{pmatrix}\in G(U)$ and $\begin{pmatrix}x\\y\end{pmatrix}\in\{0\}\times X$. This implies that $x=0$ and $y\in U(0)$. Hence, we infer that $G(U_\infty)\subset\{0\}\times U(0)$. Conversely, let $(0,y)\in\{0\}\times U(0)$. Then, $(0,y)\in\{0\}\times H$ and $y\in U(0)$ which yields $(0,y)\in G(U)\bigcap(\{0\}\times H)$. It follows from (2.11.11) that $(0,y)\in G(U_\infty)$.

(ii) Let us assume that $y\in U_s(0)$. It follows from (2.11.11) that $(0,y)\in G(U)\bigoplus G(U_\infty)$. This implies that $(x,y_1)\in G(U)$ and $(x,y_1-y)\in G(U_\infty)$. By referring to (i), we have $x = 0$ and $y_1 - y \in U(0)$. Consequently, $y_1 \in U(0)$. Then, by virtue of Lemma 2.6.1 (i), we obtain

$$y = y_1 - (y_1 - y) \in U(0).$$

Hence, $(0, y) \in G(U_\infty) \bigcap G(U_s)$. The use of (2.11.10) allows us to conclude that $y = 0$. In view of Lemma 2.6.2 (i) implies that U_s is an operator. Q.E.D.

Remark 2.11.8 Let X be a Hilbert space and let $U \in C\mathscr{R}(X)$. Then,

(i) U is an operator if, and only if, $U = U_s$.
(ii) $\mathscr{D}(U_s) = \mathscr{D}(U)$, $\mathscr{D}(U_\infty) = \{0\}$, $R(U_s) \subset U(0)^\perp$ and $R(U_\infty) = U(0)$. ◇

Lemma 2.11.9 [23, Theorem 2.4] *Let H be a Hilbert space and let $U \in C\mathscr{R}(H)$. Then,*

(i) $\|U_s x\| = \|Ux\|$, *for $x \in \mathscr{D}(U)$, where U_s is given in (2.11.11).*
(ii) $\|U_s\|_{L\mathscr{R}(H)} = \|U\|_{L\mathscr{R}(H)}$. ◇

Proposition 2.11.4 *Let X be a Hilbert space, $U \in C\mathscr{R}(H)$ and $\eta \in \mathbb{C}$. Then,*

(i) $G(\eta - U) = G(\eta - U_s) \bigoplus (\{0\} \times U(0))$, *where U_s is given in (2.11.11).*
(ii) *For $x \in \mathscr{D}(\eta - U)$, we have*

$$(\eta - U)x = (\eta - U_s)x + U(0). \qquad ◇$$

Proof (i) Let $\begin{pmatrix} x \\ y \end{pmatrix} \in G(\eta - U)$. Then, it follows from Remark 2.7.1 (iii) and (iv) that $(x, \eta\, x) \in G(\eta I_{\mathscr{D}(U)})$ and $(x, \eta x - y) \in G(U)$. By referring to (2.11.10) and Lemma 2.11.8 (i), we obtain

$$(x, \eta x - y + z) \in G(U_s)$$

and

$$\begin{pmatrix} 0 \\ z \end{pmatrix} \in \{0\} \times U(0).$$

Again, by using Lemma 2.11.8 (ii), we get that $U_s(x) = \eta x - y + z$ which yields $\begin{pmatrix} x \\ y - z \end{pmatrix} \in G(\eta - U_s)$. Hence, we deduce that

$$\begin{pmatrix} x \\ y \end{pmatrix} = \begin{pmatrix} x \\ y - z \end{pmatrix} + \begin{pmatrix} 0 \\ z \end{pmatrix} \in G(\eta - U_s) + (\{0\} \times U(0)).$$

Conversely, let $\begin{pmatrix} x \\ y \end{pmatrix} \in G(\eta - U_s) + (\{0\} \times U(0))$. This leads to $(0, y_1) \in \{0\} \times U(0)$ and $(x, y - y_1) \in G(\eta - U_s)$. By referring to Lemma 2.11.8 (i), we have

$$\begin{pmatrix} x \\ \eta x - y + y_1 \end{pmatrix} \in G(U_s).$$

The fact that $(0, y_1) \in \{0\} \times U(0)$ implies from Lemma 2.6.1 (i) that $(0, -y_1) \in \{0\} \times U(0)$. Hence, by virtue of (2.11.10) and Lemma 2.11.8 (i), we obtain

$$\begin{pmatrix} x \\ \eta x - y \end{pmatrix} \in G(U) = G((U - \eta) + \eta).$$

The use of Remark 2.7.1 (iii) and (iv) allows us to conclude that $\begin{pmatrix} x \\ y \end{pmatrix} \in G(\eta - U)$. It remains to prove that

$$G(\eta - U_s) \bigcap (\{0\} \times U(0)) = \{0\}.$$

Let us assume that $\begin{pmatrix} x \\ y \end{pmatrix} \in G(\eta - U_s) \bigcap (\{0\} \times U(0))$. Then, by using (ii) of Lemma 2.11.8, we have $y = (\eta - U_s)x$, $y \in U(0)$ and $x = 0$. Consequently, $y = 0$.

(ii) Let $x \in \mathscr{D}(\eta - U)$. Let us assume that $y \in (\eta - U)x$. This implies from (i) that

$$(x, y_1) \in G(\eta - U_s)$$

and

$$y - y_1 \in U(0).$$

Consequently, $y \in (\eta - U_s)x + U(0)$. This enables us to conclude that

$$(\eta - U)x \subset (\eta - U_s)x + U(0).$$

Conversely, suppose that $y_1 + y_2 \in (\eta - U_s)x + U(0)$. Then, $y_1 \in (\eta - U_s)x$ and $y_2 \in T(0)$ which yields

$$\begin{pmatrix} x \\ y_1 \end{pmatrix} \in G(\eta - U_s)$$

and

$$(0, y_2) \in \{0\} \times U(0).$$

This implies from (i) that $\begin{pmatrix} x \\ y_1 + y_2 \end{pmatrix} \in G(\eta - U)$. This is equivalent to saying that $y_1 + y_2 \in (\eta - U)x$. This shows that $(\eta - U_s)x + U(0) \subset (\eta - U)x$. Q.E.D.

Remark 2.11.9 Let X, Y, Z be normed spaces, $U \in C\mathscr{R}(Y, Z)$ and $V \in \mathscr{L}(X, Y)$. Then,

$$UV \in C\mathscr{R}(X, Z).$$

In particular, $\eta U \in C\mathscr{R}(Y, Z)$ for all nonzero scalar η. Indeed, let $\begin{pmatrix} x_n \\ z_n \end{pmatrix}_n \in G(UV)$ such that $\begin{pmatrix} x_n \\ z_n \end{pmatrix} \to \begin{pmatrix} x \\ z \end{pmatrix}$ as $n \to +\infty$. This implies from Lemma 2.5.2 (ii) that there exists $(y_n)_n \subset Y$ with $\begin{pmatrix} x_n \\ y_n \end{pmatrix}_n \in G(V)$ and $\begin{pmatrix} x_n \\ z_n \end{pmatrix} \in G(U)$. The fact that S is a bounded operator and $x_n \to x$ as $n \to +\infty$ implies that

$$y_n = Vx_n \to Vx \text{ as } n \to +\infty.$$

Hence, $(y_n, z_n) \to (Vx, z)$ as $n \to +\infty$. Since U is closed, we infer that $\begin{pmatrix} Vx \\ z \end{pmatrix} \in G(U)$. Since $\begin{pmatrix} x \\ Vx \end{pmatrix} \in G(V)$, then $\begin{pmatrix} x \\ z \end{pmatrix} \in G(UV)$. This is equivalent to saying that on the other hand $\eta U = U\eta = U\eta I_{\mathscr{D}(U)}$ and $\eta I_{\mathscr{D}(U)}$ is a bounded operator on $\mathscr{D}(U)$, then $\eta U \in C\mathscr{R}(Y, Z)$. ◇

Remark 2.11.10 Let $U \in C\mathscr{R}(H)$ and $Q \in \mathscr{L}(H)$ be such that $Q^2 = I$. Let us define the operator

$$U^Q = QUQ. \tag{2.11.12}$$

Then, it is easy to seen that

(i) $(U^Q)^Q = QT^QQ = \underbrace{QQ}_{=I}U\underbrace{QQ}_{=I} = U.$

(ii) U^Q is a closed linear relation on H. Since $H = R(I) = R\left(Q^2\right) \subseteq R(Q)$, then Q is surjective. In addition, we have $N(Q) \subseteq N\left(Q^2\right) = N(I) = \{0\}$. Hence, Q is invertible. Since Q is onto, then there exists a constant $c > 0$ such that

$$Q(B_H(0, 1)) \supset B_H(0, c).$$

Now, by assumption that Q is one-to-one imply that if $g \in H$ is chosen so that $\|Qg\| < c$, then $\|g\| < 1$. By homogeneity, we find that

$$\|g\| < \frac{1}{c}\|Qg\| \ \forall g \in H$$

and therefore Q^{-1} is a bounded single-valued operator. Then, $U^{-1} \in C\mathscr{R}(H)$ and $Q^{-1} \in \mathscr{L}(H)$. Hence, by using the relation $N(Q^{-1}) = Q(0) = \{0\}$, we have $\gamma(Q^{-1}) > 0$ and $R(Q^{-1}) = \mathscr{D}(Q) = H$. So, we infer from Lemma 2.11.4 that $Q^{-1}U^{-1} \in C\mathscr{R}(H)$. Thus, by referring to Lemma 2.11.7, we deduce that $Q^{-1}U^{-1}Q^{-1} \in C\mathscr{R}(H)$. This implies that $QUQ = \left(Q^{-1}U^{-1}Q^{-1}\right)^{-1} \in C\mathscr{R}(H)$. ◇

2.12 Closable Linear Relation

2.12.1 Definition and Some Auxiliary Results

The linear relation U is said to be closable if $\overline{U}$ is an extension of U (i.e., $Ux = \overline{U}x$ for all $x \in \mathscr{D}(U)$).

Proposition 2.12.1 [2, Proposition II .5.1] *Let $U \in L\mathscr{R}(X, Y)$, then $Q_U\overline{U} = \overline{Q_U U}$.* ◇

Lemma 2.12.1 [2, Exercise .5.19] *Let $V \in L\mathscr{R}(X, Y)$. Then, $\overline{V(0)} \subset \overline{V}(0)$.* ◇

Example 2.12.1 Let U be any linear relation with $U(0) \neq \overline{U(0)}$. Let W be any finite-dimensional subspace of $\mathscr{D}(U)$ and let $V = U|_W$, then V is continuous. However,

$$\begin{aligned}\overline{V(0)} &= \overline{U(0)}\\ &\neq U(0)\\ &= V(0).\end{aligned}$$

Therefore, V is not closable. Hence, V is continuous but does not imply V is closable.

Lemma 2.12.2 [2, Definition II.5.7] *Let $U \in L\mathscr{R}(X, Y)$. Then, U is closable if, and only if, $U(0) = \overline{U}(0)$ if, and only if, $Q_U U$ is closable and $U(0)$ is closed. In particular, if U is continuous and $U(0)$ is closed, then U is closable.* ◇

Proposition 2.12.2 *Let $V, U \in L\mathscr{R}(X, Y)$ be such that $V(0) \subset U(0)$ and $\overline{\mathscr{D}(U)} \subset \mathscr{D}(V)$. If U is closable and V is continuous, then $U + V$ is closable and $\overline{V + U} = V + \overline{U}$.* ◇

Proof (a) $\overline{U} + V$ is closed. Indeed, U is closable, then $U(0) = \overline{U}(0)$. It follows from Lemma 2.12.1, we infer that $\overline{U(0)} \subset \overline{U}(0)$. Hence,

$$\begin{aligned}U(0) &= \overline{U}(0)\\ &= U(0).\end{aligned}$$

So,

$$U(0) \subset \overline{U}(0).$$

Moreover, $\mathscr{D}(\overline{U}) \subset \overline{\mathscr{D}(U)}$ and since

$$\mathscr{D}(U) \times Y = G(U) + \{0\} \times Y,$$

then

$$\overline{G(U)} + \{0\} \times Y \subset \overline{G(U) + \{0\} \times Y} = \overline{\mathscr{D}(U)} \times Y \tag{2.12.1}$$

and clearly,

$$\overline{G(U)} + \{0\} \times Y = G(\overline{U}) + \{0\} \times Y = \mathscr{D}(\overline{U}) \times Y. \tag{2.12.2}$$

Using Eqs. (2.12.1), (2.12.2), and Proposition 2.11.2, we infer that $\overline{U} + V$ is closed, as desired.

(*b*) $\overline{U + V} \subset \overline{U} + V$. Clearly $U \subset \overline{U}$, so that $U + V \subset \overline{U} + V$ and thus

$$\begin{aligned}\overline{U + V} &\subset \overline{\overline{U} + V}\\ &= \overline{U} + V\\ &\subset \overline{U} + V \text{ (by } (a)).\end{aligned}$$

(*c*) $U + V$ is closable. It follows from Lemma 2.12.1 that $(U + V)(0) = (\overline{U + V})(0)$. Further,

$$\begin{aligned}(U + V)(0) &= U(0) + S(0)\\ &= U(0)\\ &= \overline{U}(0)\\ &= \overline{U(0)}\\ &= \overline{(U + V)(0)}. \end{aligned} \tag{2.12.3}$$

Moreover, by using both (*b*) and (2.12.3), we have

$$\overline{(U + V)(0)} \subset (\overline{U + V})(0) \subset (U + V)(0). \tag{2.12.4}$$

So, in view of both Eqs. (2.12.3) and (2.12.4), we infer

$$(U + V)(0) = (\overline{U + V})(0).$$

Applying Lemma 2.12.2 (*iii*) that $U + V$ is closable.

(*d*) $\overline{U + V} = \overline{U} + V$. By using (*b*), it only remains to show that

$$\overline{U} + V \subset \overline{U + V}.$$

Hence, it follows from Proposition 2.7.2 that

$$\begin{aligned}U &= U + V - V\\ &\subset \overline{U + V} - V.\end{aligned}$$

In addition, by the hypotheses, we can use the part (*i*) to conclude that $\overline{U + V} - V$ is closed. Consequently,

$$\begin{aligned}\overline{U} &\subset \overline{\overline{U+V}-V}\\ &= \overline{U+V}-V\end{aligned}$$

and by appling Proposition 2.7.2, we infer that $\overline{U}+V \subset \overline{U+V}$. Q.E.D.

Corollary 2.12.1 *Let $V, U \in L\mathscr{R}(X,Y)$ be such that $V(0) \subset U(0)$, and $\overline{\mathscr{D}(U)} \subset \mathscr{D}(V)$. If $U \in C\mathscr{R}(X,Y)$ and V is continuous, then $\eta V - U \in C\mathscr{R}(X,Y)$ for all $\eta \in \mathbb{C}$.*

(i) If $U \in C\mathscr{R}(X,Y)$ and V is continuous, then $\eta V - U \in C\mathscr{R}(X,Y)$ for all $\eta \in \mathbb{C}$.
(ii) If U is closable and V is continuous, then $\eta V - U$ is closable and $\overline{\eta V - U} = \eta V - \overline{U}$. ◇

Let X and Y be two normed spaces and $U, V \in L\mathscr{R}(X,Y)$. The linear relation V is called relatively bounded with respect to U (or U-bounded) if $\mathscr{D}(V) \supset \mathscr{D}(U)$ and there exist constants a, b such that

$$\|Vx\| \le a\|x\| + b\|Ux\| \tag{2.12.5}$$

holds for all $x \in \mathscr{D}(U)$. The infimum δ of all b such that the inequality (2.12.5) holds for some $a \ge 0$ is called the relative bound of S with respect to T (or U-bound of V).

Remark 2.12.1 Let $V, U \in L\mathscr{R}(X,Y)$. Then,

(i) If V is 0-bounded, then V is bounded.
(ii) The inequality (2.12.5) is equivalent to

$$\|Vx\|^2 \le a_1^2\|x\|^2 + b_1^2\|Ux\|^2\,,\quad x \in \mathscr{D}(U), \tag{2.12.6}$$

where $a_1 = \sqrt{a^2+ab}$, and $b_1 = \sqrt{b^2+ab}$
(iii) V is U-bounded if, and only if, S is $(\eta - U)$-bounded for some $\eta \in \mathbb{C}$.
(iv) V is U-bounded with U-bound δ if, and only if, $Q_V V$ is $Q_U U$-bounded with $Q_U U$-bound δ. In fact,

$$\begin{aligned}\|Q_U Vx\| &= \operatorname{dist}(\overline{U(0)}, Vx)\\ &\le \operatorname{dist}(V(0), Vx)\\ &= \|Vx\|\\ &\le a\|x\| + b\|Ux\|\\ &= a\|x\| + b\|Q_U Ux\|.\end{aligned}$$

◇

Proposition 2.12.3 *If V is U-bounded with U-bound $\delta < 1$ and $V(0) \subset U(0)$, then U is $(U+V)$-bounded with $(U+V)$-bound $\le \frac{\delta}{1-\delta}$.* ◇

Proof Since $\mathcal{D}(U+V) = \mathcal{D}(V) \bigcap \mathcal{D}(U) = \mathcal{D}(U)$ with $\mathcal{D}(U+V) \subset \mathcal{D}(V)$. As V is U-bounded, it implies that there exist $a > 0$, $0 < b < 1$, such that for all $x \in \mathcal{D}(U)$,

$$\begin{aligned} \|Vx\| &\leq a\|x\| + b\|Ux\| \\ &= a\|x\| + b\|Ux + Vx - Vx\| \text{ (as Proposition 1.7.2 } (i)) \\ &\leq a\|x\| + b\|Ux + Vx\| + b\|Vx\| \text{ (as Lemma 1.10.3 } (i)). \end{aligned}$$

Since $0 < b < 1$, it follows that

$$\|Vx\| \leq \frac{a}{1-b}\|x\| + \frac{b}{1-b}\|(U+V)x\|, \quad x \in \mathcal{D}(U).$$

This achieves the proof. Q.E.D.

Theorem 2.12.1 *Let $U \in C\mathcal{R}(X, Y)$, and V, $W \in L\mathcal{R}(X, Y)$ be such that $W(0) \subset V(0) \subset U(0)$. If V is U-bounded with U-bound $\delta_1 < 1$ and W is V-bounded with V-bound δ_2 such that $\frac{\delta_1\delta_2}{1-\delta_1} < 1$. Then, $U + V + W \in C\mathcal{R}(X, Y)$.* $\diamond$

Proof As U is closed and V is U-bounded with U-bound $\delta_1 < 1$, then by applying Theorem 2.11.3, we have $U + V \in C\mathcal{R}(X, Y)$. On the one hand, V is U-bounded with U-bound $\delta_1 < 1$, then by applying Preposition 2.12.3, we have S is $(U+V)$-bounded with $(U+V)$-bound $\leq \frac{\delta_1}{1-\delta_1}$. On the other hand, W is V-bounded with V-bound δ_2. Hence, by using Lemma 2.12.3, we obtain W is $(U+V)$-bounded with $(U+V)$-bound $\leq \frac{\delta_1\delta_2}{1-\delta_1}$. As K is $(U+V)$-bounded with $(T+S)$-bound < 1, and $U + V$ is closed, then by applying Theorem 2.11.3, we obtain $U + V + W \in C\mathcal{R}(X, Y)$. Q.E.D.

Lemma 2.12.3 *Let V, U, A and $B \in L\mathcal{R}(X, Y)$.*

(i) If V and U are closable, and $\mathcal{D}(U) \subset \mathcal{D}(V)$, then V is U-bounded with U-bound α if, and only if, $\overline{V}$ is $\overline{U}$-bounded with $\overline{U}$-bound α.

(ii) If $V = E + F$ is U-bounded with U-bound β, F is E-bounded with E-bound $\delta < 1$, and $F(0) \subset E(0)$, then E is U-bounded with T-bound $\gamma \leq \frac{\beta}{1-\delta}$. $\diamond$

Proof (i) Let V be U-bounded with U-bound α. Then, $Q_V V$ is $Q_U U$-bounded with $Q_U U$-bound α. This is equivalent to $\overline{Q_V V}$ is $\overline{Q_U U}$-bounded with $\overline{Q_U V}$-bound α. On the other hand $\mathcal{D}(U) = \mathcal{D}(Q_U U)$, and $\overline{Q_U U} = Q_{\overline{U}}\overline{U}$, the latter holds if, and only if, $\overline{V}$ is $\overline{U}$-bounded with $\overline{U}$-bound α.

(ii) As F is E-bounded with E-bound $\delta < 1$, and $F(0) \subset E(0)$, then F is V-bounded with V-bound $\alpha \leq \frac{\delta}{1-\delta}$. Consequently, V is U-bounded with U-bound β. This implies that F is U-bounded with U-bound $\alpha\beta \leq \frac{\delta\beta}{1-\delta}$. So, F is U-bounded with U-bound $\alpha\beta \leq \frac{\delta\beta}{1-\delta}$, and heedful that V is U-bounded with U-bound β and $A = V - B$. So, A is U-bounded with U-bound $\gamma \leq \frac{\beta}{1-\delta}$. Q.E.D.

Lemma 2.12.4 *Let $S, U \in L\mathcal{R}(X, Y)$ be such that $S(0) \subset U(0)$. We assume that S is U-bounded with U-bound δ.*

(i) If U is closable and $\delta < 1$, then $U + S$ is closable.
(ii) If $U + S$ is closable, and $\delta < \frac{1}{2}$, then U is closable.
If, in addition, S is closable, then $\overline{S+U} = \overline{S} + \overline{U}$ and $\mathcal{D}(\overline{U+S}) = \mathcal{D}(\overline{U})$. ◇

Proof (i) First, it is clear that $(U + S)(0) = U(0)$ and U is closable. Then, $(U + S)(0)$ is closed, and $Q_U U$ is closable. The fact that S is U-bounded with U-bound $\delta < 1$, then $Q_U S$ is $Q_U U$-bounded with $Q_U U$-bound < 1. So,

$$
\begin{aligned}
Q_{S+U}(U+S) &= Q_U(U+S) \\
&= Q_U U + Q_U S
\end{aligned}
$$

is closable. It follows from Lemma 2.12.2, we have $U + S$ is closable.

(ii) Since $U + S$ is closable, S is U-bounded with U-bound $\delta < \frac{1}{2}$, and by using Proposition 2.7.2, we get

$$U = U + S - S.$$

Consequently, S is $(U + S)$-bounded with $(U + S)$-bound < 1. So, by using (i), it follows that U is closable. As U and S are closable, and S is U-bounded with U-bound < 1, then, by applying Lemma 2.12.3 (i), we obtain $\overline{S}$ is $\overline{U}$-bounded with $\overline{U}$-bound < 1. So, $\overline{S} + \overline{U}$ is closed. Since $U \subset \overline{U}$, and $S + U \subset \overline{S} + \overline{U}$, then

$$
\begin{aligned}
\overline{U+S} &\subset \overline{\overline{U} + \overline{S}} \\
&= \overline{U} + \overline{S}
\end{aligned}
$$

($\overline{S} + \overline{U}$ is closed). Since by applying Proposition 2.7.2, we get

$$U = U + S - S.$$

Then,

$$
\begin{aligned}
U &= U + S - S \\
&\subset \overline{U+S} - \overline{S}.
\end{aligned}
$$

Since S is U-bounded with U-bound $\delta < \frac{1}{2}$, then S is $(U + S)$-bounded with $(U + S)$-bound < 1. By using Lemma 2.12.3 (i), we have $\overline{S}$ is $(\overline{U+S})$-bounded with $(\overline{U+S})$-bound < 1. Consequently, $\overline{U+S} - \overline{S}$ is closed. Then,

$$
\begin{aligned}
\overline{U} &\subset \overline{\overline{U+S} - \overline{S}} \\
&= \overline{U+S} - \overline{S},
\end{aligned}
$$

and, we have

$$\overline{U} + \overline{S} \subset \overline{U + S}.$$

As consequence

$$\overline{U} + \overline{S} = \overline{U + S},$$

and

$$\begin{aligned}\mathscr{D}(\overline{U + S}) &= \mathscr{D}(\overline{U} + \overline{S}) \\ &= \mathscr{D}(\overline{U}).\end{aligned}$$

This achieves the proof.

2.12.2 *Product of Closable Linear Relations*

Lemma 2.12.5 *Let $S \in L\mathscr{R}(Y, Z)$ be closable and $U \in \mathscr{L}(X, Y)$. Then,*

(i) SU is closable and $\overline{SU} \subset \overline{S}U$.
(ii) If U is bijective, then $\overline{SU} = \overline{S}U$. ◇

Proof By using Lemma 2.11.6, we have $\overline{S}U$ is closed. Moreover, $S \subset \overline{S}$, so that $SU \subset \overline{S}U$. Hence,

$$\begin{aligned}\overline{SU} &\subset \overline{\overline{S}U} \\ &= \overline{S}U.\end{aligned}$$

Now, we show that SU is closable. In fact, SU is closable if, and only if, $\overline{SU}(0) = SU(0)$. So, by applying Lemma 2.12.1, we obtain

$$\overline{SU(0)} \subset \overline{SU}(0). \tag{2.12.7}$$

We deduce from (2.12.7) that

$$\overline{SU(0)} \subset \overline{S}U(0). \tag{2.12.8}$$

On the other hand,

$$\begin{aligned}\overline{S}U(0) &= \overline{S}(0) \text{ (as } U \text{ is single-valued)} \\ &= S(0) \text{ (as } S \text{ is closable)} \\ &= SU(0).\end{aligned} \tag{2.12.9}$$

Now, contribute both Eqs. (2.12.8) and (2.12.9), we obtain

$$\begin{aligned}\overline{SU}(0) &= \overline{S}U(0)\\ &= S(0)\\ &= SU(0).\end{aligned}$$

So, SU is closable.
(ii) By (i), it only remains to proof $\overline{S}U \subset \overline{SU}$. Since $U^{-1}(0) = N(U) = \{0\}$, then U^{-1} is single-valued, $R(U) = Y$ and $\mathscr{D}(U^{-1}) = R(U)$. Thus, $\mathscr{D}(U^{-1}) = Y$. As U is closed, then by applying Lemma 2.12.2, we obtain U^{-1} is closed. On the one hand, by using Theorem 2.11.1 (i), U^{-1} is bounded single-valued. On the other hand, SU is closable by (i), we can claim (i), to SU instead S and U^{-1} instead of U, we have SUU^{-1} is closable. Hence,

$$\overline{SUU^{-1}} \subset \overline{SU}U^{-1},$$

but $U^{-1}U = I_{\mathscr{D}(U)}$ and $UU^{-1} = I_{R(U)} = I_Y$. As consequence,

$$S \subset SUU^{-1}.$$

This implies that

$$\begin{aligned}\overline{S} &\subset \overline{SUU^{-1}}\\ &\subset \overline{SU}U^{-1}\end{aligned}$$

and

$$\begin{aligned}\overline{S}U &\subset \overline{SU}U^{-1}U\\ &= \overline{SU}.\end{aligned}$$

This achieves the proof. Q.E.D.

Theorem 2.12.2 *Let $S \in L\mathscr{R}(Y, Z)$ be closable and $U \in L\mathscr{R}(X, Y)$ be a bounded linear relation with* $\dim(U(0)) < \infty$ *and* $\overline{S}U(0) \subset SU(0)$. *Then,*

(i) SU is closable and $\overline{SU} \subset \overline{S}U$.
(ii) If U is bijective, then $\overline{SU} = \overline{S}U$. ◇

Proof (i) Let us study the following two cases for U.
First case. If U is single-valued, then the result follows immediately from Lemma 2.12.5.

Second case. If U is a linear relation, when applying Lemma 2.11.7, we infer that $\overline{S}U \in C\mathscr{R}(X, Z)$. Hence,

$$\begin{aligned}\overline{SU} &\subset \overline{\overline{S}U}\\ &= \overline{S}U.\end{aligned}$$

Now, let's assume that SU is closable. Indeed, since $U(0)$ is closed, we can write

$$SU = SQ_U^{-1}Q_U.$$

Since $\dim(Q_U^{-1})(0) = \dim(U(0)) < \infty$, by applying Lemma 2.11.7, we deduce that $\overline{S}Q_U^{-1}$ is closed. Furthermore, $S \subset \overline{S}$. Thus,

$$SQ_U^{-1} \subset \overline{S}Q_U^{-1},$$

which implies that

$$\overline{SQ_U^{-1}} \subset \overline{\overline{S}Q_U^{-1}} = \overline{S}Q_U^{-1}. \tag{2.12.10}$$

Grounded on Lemma 2.12.1, we have

$$\overline{SQ_U^{-1}(0)} \subset \overline{SQ_U^{-1}}(0). \tag{2.12.11}$$

Combining Eqs. (2.12.10) and (2.12.11), we obtain

$$\begin{aligned} \overline{SQ_U^{-1}(0)} &\subset \overline{SQ_U^{-1}}(0) \\ &\subset \overline{S}Q_U^{-1}(0). \end{aligned}$$

On the other side, we have

$$\overline{S}Q_U^{-1}(0) = \overline{S}U(0) \subset SU(0) = SQ_U^{-1}(0). \tag{2.12.12}$$

As a matter of fact, it follows from Eqs. (2.12.11) and (2.12.12) that

$$\begin{aligned} SQ_U^{-1}(0) &\subset \overline{SQ_U^{-1}(0)} \\ &\subset \overline{SQ_U^{-1}}(0) \\ &\subset SQ_T^{-1}(0). \end{aligned}$$

As a result, this last inclusion allows us to conclude that

$$\overline{SQ_U^{-1}}(0) = SQ_U^{-1}(0).$$

Consequently, by using Theorem 2.11.2, we infer that SQ_U^{-1} is closable. Now, since $U \in B\mathscr{R}(X, Y)$, we get Q_U is a bounded single-valued. Therefore, investing SQ_U^{-1} and $Q_U U$ in the first case, we conclude that

$$SU = SQ_U^{-1}Q_U U$$

is closable.

(ii) Departing from the fact that $U \in B\mathscr{R}(X, Y)$ and $U(0)$ is closed and using Theorem 2.11.2 (ii), we have $U \in C\mathscr{R}(X, Y)$, which is equivalent to $U^{-1} \in C\mathscr{R}(Y, X)$. Since U is bijective, we infer that

$$U^{-1}(0) = N(U) = \{0\}$$

and

$$\begin{aligned}\mathscr{D}(U^{-1}) &= R(U)\\ &= Y.\end{aligned}$$

When applying Theorem 2.11.1, we deduce that U^{-1} is bounded single-valued. Now, referring to the fact that (i), SU is closable and $\overline{SU} \subset \overline{S}U$, we obtain

$$\overline{SU}U^{-1} \subset \overline{S}UU^{-1}.$$

Then,

$$\begin{aligned}\overline{SU}U^{-1}(0) &\subset \overline{S}UU^{-1}(0)\\ &\subset \overline{S}U(0) \text{ (as } U^{-1} \text{ is single-valued)}\\ &\subset SU(0)\\ &\subset SUU^{-1}(0).\end{aligned}$$

So, by applying (i), we obtain SUU^{-1} which is closable and

$$\overline{SUU^{-1}} \subset \overline{SU}U^{-1}.$$

But $UU^{-1} = I_{R(T)} = I_Y$, then

$$\begin{aligned}\overline{S} &\subset \overline{SUU^{-1}}\\ &\subset \overline{SU}U^{-1}.\end{aligned}$$

Thus, $\overline{S}U \subset \overline{SU}$, which completes the proof. Q.E.D.

Corollary 2.12.2 *Let $U \in \mathscr{L}(X, Y)$ be bijective and let $V \in L\mathscr{R}(Z, Y)$ be closable. Then, $U^{-1}S$ is closable and $\overline{U^{-1}V} = U^{-1}\overline{V}$.* ◇

Proof As $U \in \mathscr{L}(X, Y)$, then U is closed. So, by applying Proposition 2.11.1 (i), we have U^{-1} is closed and U^{-1} is bounded single-valued. In fact, $U^{-1}(0) = U(0) = \{0\}$, and $R(U^{-1}) = \mathscr{D}(U) = X$. Hence, $U^{-1}\overline{V}$ is closed. Moreover,

$$\begin{aligned}\overline{U^{-1}\overline{S}}(0) &= U^{-1}\overline{V}(0)\\ &= U^{-1}V(0) \text{ (as } V \text{ is closable).}\end{aligned} \tag{2.12.13}$$

What's more, $U^{-1}V \subset U^{-1}\overline{V}$, then

$$\begin{aligned}\overline{U^{-1}V} &\subset \overline{U^{-1}\overline{V}}\\ &= U^{-1}\overline{V}.\end{aligned}$$

Hence,

$$\overline{U^{-1}V} \subset U^{-1}\overline{V}.$$

A combination of both Eq. (2.12.13) and Lemma 2.12.1 leads to

$$\begin{aligned}\overline{U^{-1}V(0)} &\subset \overline{U^{-1}V}(0)\\ &\subset \overline{U^{-1}\overline{V}}(0)\\ &= U^{-1}V(0).\end{aligned}$$

Hence,

$$\begin{aligned}U^{-1}V(0) &\subset \overline{U^{-1}V(0)}\\ &\subset \overline{U^{-1}V}(0)\\ &\subset U^{-1}V(0).\end{aligned}$$

Consequently,

$$\begin{aligned}\overline{U^{-1}V(0)} &= \overline{U^{-1}V}(0)\\ &= U^{-1}V(0).\end{aligned}$$

Therefore, $U^{-1}V$ is closable. The proof of $U^{-1}\overline{V} \subset U^{-1}V$ may be checked in the same way as the proof of Lemma 2.12.5 (ii). Q.E.D.

Lemma 2.12.6 *Let $A \in L\mathscr{R}(X)$, $B \in L\mathscr{R}(Y, X)$, $C \in L\mathscr{R}(X, Y)$ and $D \in L\mathscr{R}(Y)$ be such that $\mathscr{D}(A) \subset \mathscr{D}(C)$. Then,*

(i) If $\overline{\mathscr{D}(B)} \subset \mathscr{D}(D)$, $\rho(A) \neq \emptyset$ and D is continuous such that for some $\mu \in \rho(A)$, we have $D(A-\mu)^{-1}B$ is closable and $D(0) \subset \left(C(A-\mu)^{-1}B\right)(0)$. Then, $D - \mu - C(A-\mu)^{-1}B$ is closable and

$$\overline{D - \mu - C(A-\mu)^{-1}B} = D - \mu - \overline{C(A-\mu)^{-1}B}.$$

(ii) If C is closable, then for all $\mu \in \rho(A)$, we have $G(\mu) = C(A-\mu)^{-1}$ is closable and $\overline{G(\mu)}$ is a bounded linear relation. ◇

Proof (i) Follows straight away from Proposition 2.12.2 and Lemma 2.12.5.
(ii) Using Lemma 2.12.5, we have $C(A-\mu)^{-1}$ is closable. Since $\overline{G(\mu)}$ is closed and $\mathscr{D}(\overline{G(\mu)}) = X$, then by using Theorem 2.11.1 (i), we have $\overline{G(\mu)}$ is a bounded linear relation. Q.E.D.

Theorem 2.12.3 *Let $U \in L\mathcal{R}(X, Y)$ be single-valued bounded bijective, $S \in L\mathcal{R}(Y, Z)$ be closable and assume that $R \in L\mathcal{R}(Z, W)$ is bounded and bijective with $R(0)$ is closed. Then, RSU is closable and $\overline{RSU} = R\overline{S}U$.* ◇

Proof Let us study two cases for R.
First case. R is single-valued. Then, by using Lemma 2.12.5 (ii), we obtain SU is closable and $\overline{SU} = \overline{S}U$. Since R^{-1} is bounded single-valued bijective. In fact, $R^{-1}(0) = N(R) = \{0\}$ (as R is one-to-one) implies R^{-1} is single-valued. Further, $\mathcal{D}(R^{-1}) = R(R) = W$ (as R is onto) is closed, then R is open. On the other hand R^{-1} is continuous, hence R^{-1} is bounded single-valued and

$$N(R^{-1}) = R(0) = \{0\}$$

(since R is single-valued) and $R(R^{-1}) = \mathcal{D}(R) = Z$. Thus,

$$R = (R^{-1})^{-1}.$$

So, by using Lemma 2.12.5 (ii)

$$RSU = (R^{-1})^{-1}SU$$

is closable with

$$\overline{RSU} = R\overline{SU}.$$

Since U is bounded single-valued, then by using Lemma 2.12.5, we have

$$\overline{SU} = \overline{S}U.$$

Hence,

$$\overline{RSU} = R\overline{S}U.$$

Second case. R is linear relation. Then,
(a) $R\overline{S}$ is closed. Indeed, S is closable, then $\overline{S}$ is closed. Since R is bounded with $R(0)$ is closed, then by virtue of Lemma 2.11.4, $R\overline{S}$ is closed.
(b) We show that $R\overline{S}U$ is closed. This is equivalently, by using Proposition 2.11.1 (i), to $(R\overline{S}U)^{-1}$ is closed. Indeed,

$$(R\overline{S}U)^{-1} = U^{-1}(R\overline{S})^{-1}$$

and, by using (a), $R\overline{S}$ is closed. By using again Proposition 2.11.1, we infer that $(R\overline{S})^{-1}$ is closed. This implies $(R\overline{S})^{-1}$ is closable. On the other hand T is bounded single-valued bijective. So, by applying Lemma 2.12.5 (ii), we deduce $T^{-1}(R\overline{S})^{-1}$ is closable and

$$\begin{aligned}\overline{U^{-1}(R\overline{S})^{-1}} &= U^{-1}\overline{(R\overline{S})^{-1}} \\ &= U^{-1}(R\overline{S})^{-1} \\ &= (R\overline{S}U)^{-1}.\end{aligned}$$

So, we obtain (b).
(c) We show that $RSU(0)$ is closed. Indeed,

$$\begin{aligned}RSU(0) &= RS(0) \text{ (again } U \text{ is single-valued so that } U(0) = \{0\}) \\ &= R\overline{S}(0) \text{ (as } S \text{ is closable)} \\ &= R\overline{S}U(0).\end{aligned}$$

Therefore, $RSU(0)$ is closed by virtue of both (b) and Proposition 2.11.1 (i).

(d) $Q_{RSU}RSU$ is a closable operator. Indeed, by using Lemma 2.1.1 (ii),

$$W\Big/RS(0) \equiv (W/R(0))\Big/(RS(0)/R(0)),$$

and $Q_{RS} = Q_A Q_R$ with $A := RS(0)/R(0)$. Since $RS(0) = RSU(0)$ is closed (by (c)), then

$$\begin{aligned}Q_{RS} &= Q_{RSU} \\ &= Q_{R\overline{S}} \\ &= Q_{R\overline{S}U}.\end{aligned}$$

Clearly

$$\begin{aligned}\mathscr{D}(Q_{RSU}RSU) &= \mathscr{D}(Q_A Q_R RSU) \\ &= \mathscr{D}(Q_R RSU) \\ &= \mathscr{D}(RSU)\end{aligned}$$

and

$$Q_{R\overline{S}U}R\overline{S}U = Q_{RSU}R\overline{S}U.$$

This implies that

$$\begin{aligned}Q_R RSU &= Q_R R\overline{S}U|_{\mathscr{D}(Q_R RSU)} \\ &= Q_R R\overline{S}U|_{\mathscr{D}(RSU)}.\end{aligned}$$

So, $Q_{R\overline{S}U}R\overline{S}U$ is an extension of $Q_{RSU}RSU$ and as $R\overline{S}U$ is closed (by (b)). By applying Lemma 2.12.2, we have $Q_{RSU}R\overline{S}U$ is closed. Therefore, $Q_{RSU}R\overline{S}U$ is a closed extension single-valued of the single-valued $Q_{RSU}RSU$. Hence, (d) is true.
(e) RSU is closable. It is simple consequence form of the properties (c), (d) and Lemma 2.12.2. It only remains to see that

$$\overline{RSU} = R\overline{S}U.$$

(f) $\overline{RSU} \subset R\overline{S}U$. Indeed, $RSU \subset R\overline{S}U$, then

$$\begin{aligned}\overline{RSU} &\subset \overline{R\overline{S}U}\\ &= R\overline{S}U.\end{aligned}$$

(g) $R\overline{S}U \subset \overline{RSU}$. Indeed, $\overline{Q_R RSU} = Q_R R\overline{S}T$ implies

$$Q_A\overline{Q_R RSU} = Q_A Q_R R\overline{S}U,$$

where $Q_A Q_R = Q_{RSU}$. But

$$Q_A\overline{Q_R RSU} \subset \overline{Q_A Q_R RSU}$$

so that

$$Q_{RSU} R\overline{S}U \subset \overline{Q_{RSU} RSU}$$

with $\overline{Q_{RSU} RSU} = Q_{RSU}\overline{RSU}$. Hence,

$$Q_{RSU} R\overline{S}U \subset Q_{RSU}\overline{RSU}$$

and by using (f), we have the other inclusion. Hence,

$$Q_{RSU} R\overline{S}U = Q_{RSU}\overline{RSU}.$$

In particular

$$\begin{aligned}\mathscr{D}(R\overline{S}U) &= \mathscr{D}(Q_{RSU} R\overline{S}U)\\ &= \mathscr{D}(Q_{RSU}\overline{RSU}\\ &= \mathscr{D}(\overline{RSU}).\end{aligned}$$

Hence,

$$\mathscr{D}(\overline{RSU}) = \mathscr{D}(R\overline{S}U).$$

Since $R\overline{S}U(0) = RSU(0)$ by (c), we have

$$\begin{aligned}\overline{RSU(0)} &= R\overline{S}U(0)\\ &\subset \overline{RSU}(0)\end{aligned}$$

and

$$\overline{RSU}(0) \subset R\overline{S}U(0) \ \text{ by } (f),$$

we deduce that

$$\mathscr{D}(\overline{RSU}) = \mathscr{D}(R\overline{S}U),$$

$$\overline{RSU(0)} = R\overline{S}U(0)$$

and

$$\overline{RSU} \subset R\overline{S}U.$$

Thus, by virtue of Proposition 2.7.3 (iii), we have $\overline{RSU} = R\overline{S}U$. Q.E.D.

Theorem 2.12.4 *Let* $U \in B\mathscr{R}(X, Y)$ *and* $R \in B\mathscr{R}(Z, W)$ *be bijective with* $\dim(U(0)) < \infty$ *and* $R(0)$ *is closed. If* $V \in L\mathscr{R}(Y, Z)$ *is closable such that* $\overline{V}U(0) \subset VU(0)$, *then* RVU *is closable and* $\overline{RVU} = R\overline{V}U$. ◇

Proof Resting on Theorem 2.12.2 (ii), we obtain that VU is closable and $\overline{VU} = \overline{V}U$. Now, taking into account the fact that $R \in B\mathscr{R}(X, Y)$ and $R(0)$ is closed, and using Theorem 2.11.1 (iv), we infer that $R \in C\mathscr{R}(X, Y)$ which is equivalent to $R^{-1} \in C\mathscr{R}(Y, X)$. Furthermore, we have R is bijective. Then, $R^{-1}(0) = N(R) = \{0\}$ and $\mathscr{D}(R^{-1}) = W$. Relying up on the closed graph theorem (see Theorem 2.11.1 (i)), we deduce that R^{-1} is bounded single-valued. The use of Lemma 2.12.5 (ii) leads to conclude that $(R^{-1})^{-1}VU = RVU$ is closable and

$$\begin{aligned}\overline{RVU} &= R\overline{VU}\\ &= R\overline{V}U.\end{aligned}$$

This achieves the proof. Q.E.D.

2.13 Diagonally and Off-Diagonally Dominant Block Matrix Linear Relation

2.13.1 Definitions and Some Results

Definition 2.13.1 (i) The block matrix linear relations $\mathscr{L}_0$, given in (2.8.12), is called diagonally dominant if C is A-bounded and B is D-bounded.
(ii) The block matrix linear relations $\mathscr{L}_0$, given in (2.8.12), is called off-diagonally dominant if A is C-bounded and D is B-bounded. ◇

Definition 2.13.2 (i) The block matrix of the linear relation $\mathscr{L}_0$, given in (2.8.12), is called diagonally dominant of order δ if C is A-bounded with A-bound δ_C, B is D-bounded with D-bound δ_B, and $\delta = \max(\delta_C, \delta_B)$.
(ii) The block matrix of the linear relation $\mathscr{L}_0$, given in (2.8.12), is called off-diagonally dominant of order δ if A is C-bounded with C-bound δ_A, D is B-bounded with B-bound δ_D, and $\delta = \max(\delta_A, \delta_D)$. ◇

Proposition 2.13.1 *Let us define the block matrix linear relations*

$$\mathscr{T} := \begin{pmatrix} A & 0 \\ 0 & D \end{pmatrix}$$

and

$$\mathscr{S} := \begin{pmatrix} 0 & B \\ C & 0 \end{pmatrix}.$$

Then, the following implications hold

(i) If $\mathscr{L}_0$, given in (2.8.12), is diagonally dominant of the order δ, then $\mathscr{S}$ is $\mathscr{T}$-bounded with $\mathscr{T}$-bound δ.
(ii) If $\mathscr{L}_0$, given in (2.8.12), is off-diagonally dominant of the order δ, then $\mathscr{T}$ is $\mathscr{S}$-bounded with $\mathscr{S}$-bound δ. ◇

Proof We prove (i), the proof of (ii) is similar. Let $\varepsilon > 0$ be arbitrary. By the assumptions and using inequality (2.12.6), there exist constants $a, b, c, d \geq 0$ such that $\delta_B \leq b < \delta_B + \varepsilon$, $\delta_C \leq d < \delta_C + \varepsilon$ and

$$\begin{array}{ll} \|Bg\|^2 \leq a^2\|g\|^2 + b^2\|Dg\|^2, & g \in \mathscr{D}(D) \\ \|Cf\|^2 \leq c^2\|f\|^2 + d^2\|Af\|^2, & f \in \mathscr{D}(A). \end{array}$$

Hence, we obtain for $z^t = (f, g)^t \in \mathscr{D}(A) \times \mathscr{D}(D)$, we have

$$\begin{aligned}
\|\mathscr{S}z\|^2 =& \|Q_{\mathscr{S}}\mathscr{S}z\|^2 \\
=& \left\| \begin{pmatrix} 0 & Q_B B \\ Q_C C & 0 \end{pmatrix} \begin{pmatrix} f \\ g \end{pmatrix} \right\|^2 \\
=& \|Q_B Bg\|^2 + \|Q_C Cf\|^2 \\
\leq& \|Bg\|^2 + \|Cf\|^2 \text{ see Lemma 1.10.6 (iii)} \\
\leq& a^2\|g\|^2 + b^2\|Dg\|^2 + c^2\|f\|^2 + d^2\|Af\|^2 \\
\leq& (\max\{a, c\})^2 \left\| \begin{pmatrix} f \\ g \end{pmatrix} \right\|^2 + (\max\{b, d\})^2(\|Dg\|^2 + \|Af\|^2) \\
\leq& (\max\{a, c\})^2 \left\| \begin{pmatrix} f \\ g \end{pmatrix} \right\|^2 + (\max\{b, d\})^2(\|Q_D Dg\|^2 + \|Q_A Af\|^2) \\
\leq& (\max\{c, a\})^2 \left\| \begin{pmatrix} f \\ g \end{pmatrix} \right\|^2 + (\max\{b, d\})^2 \left\| \begin{pmatrix} Q_A A & 0 \\ 0 & Q_D D \end{pmatrix} \begin{pmatrix} f \\ g \end{pmatrix} \right\|^2 \\
=& (\max\{c, a\})^2\|z\|^2 + (\max\{b, d\})^2\|Q_{\mathscr{T}}\mathscr{T}z\|^2 \text{ see Lemma 1.10.6 (ii)} \\
=& (\max\{c, a\})^2\|z\|^2 + (\max\{b, d\})^2\|\mathscr{T}z\|^2.
\end{aligned}$$

Since $\max\{b, d\} < \max(\delta_B + \varepsilon, \delta_C + \varepsilon) = \delta + \varepsilon$, then $\mathcal{S}$ is $\mathcal{T}$-bounded with $\mathcal{T}$-bound $< \delta$, a contradiction follows if we take $f = 0$ and $g = 0$ in the above inequality. This achieves the proof. Q.E.D.

As a directly consequence of Proposition 2.13.1.

Corollary 2.13.1 *(i) If, in Proposition 2.13.1 (i), the order of dominance δ is < 1, $B(0) \subset A(0)$ and $C(0) \subset D(0)$, then $\mathcal{S}$ is $\mathcal{L}_0$-bounded with $\mathcal{L}_0$-bound $\leq \frac{\delta}{1-\delta}$.*
(ii) If, in Proposition 2.13.1 (ii), the order of dominance δ is < 1, $A(0) \subset B(0)$ and $D(0) \subset C(0)$, then $\mathcal{T}$ is $\mathcal{L}_0$-bounded with $\mathcal{L}_0$-bound $\leq \frac{\delta}{1-\delta}$. ◇

Lemma 2.13.1 *Define the block matrix of linear relation*

$$\widetilde{\mathcal{T}} = \begin{pmatrix} A & B \\ 0 & D \end{pmatrix}$$

and

$$\widetilde{\mathcal{S}} = \begin{pmatrix} 0 & 0 \\ C & 0 \end{pmatrix}.$$

(i) If $\mathcal{L}_0$, given in (2.8.12), is diagonally dominant, for the relative bounds δ_C of C, δ_B of B with $\delta_C^2(1+\delta_B^2) < 1$ or $\delta_B^2(1+\delta_C^2) < 1$, then $\widetilde{\mathcal{S}}$ is $\widetilde{\mathcal{T}}$-bounded with $\widetilde{\mathcal{T}}$-bound $\delta < 1$.
(ii) If $\mathcal{L}_0$, given in (2.8.12), is off-diagonally dominant, for the relative bounds δ_A of A, δ_D of D with $\delta_A^2(1+\delta_D^2) < 1$ or $\delta_D^2(1+\delta_A^2) < 1$, then $\widetilde{\mathcal{T}}$ is $\widetilde{\mathcal{S}}$-bounded with $\widetilde{\mathcal{T}}$-bound $\delta < 1$. ◇

Proof We show (i) and the proof of (ii) is similar. Choose $\varepsilon > 0$ such that $(\delta_C + \varepsilon)^2(1 + (\delta_B + \varepsilon)^2) < 1$ and by assumption, $\mathcal{L}_0$ is diagonally dominant, using the inequality (2.12.6), there exist constants $a \geq 0, b \geq 0, c \geq 0, d \geq 0$ with $\delta_B \leq b \leq \delta_B + \varepsilon$, $\delta_C \leq d \leq \delta_C + \varepsilon$ and

$$\begin{aligned} \|Bg\|^2 &\leq a^2\|g\|^2 + b^2\|Dg\|^2, \quad g \in \mathcal{D}(D) \\ \|Cf\|^2 &\leq c^2\|f\|^2 + d^2\|Af\|^2, \quad f \in \mathcal{D}(A). \end{aligned}$$

For $z^t = (f, g)^t \in \mathcal{D}(A) \times \mathcal{D}(D)$, we obtain

$$\begin{aligned} \|\widetilde{\mathcal{S}}z\|^2 &= \|Q_{\widetilde{\mathcal{S}}}\widetilde{\mathcal{S}}z\|^2 \\ &\leq \left\| \begin{pmatrix} 0 & 0 \\ Q_C C & 0 \end{pmatrix} \begin{pmatrix} f \\ g \end{pmatrix} \right\|^2 \\ &\leq \|Q_C C f\|^2 \\ &= \|Cf\|^2 \\ &\leq c^2\|f\|^2 + d^2\|Af\|^2 \\ &\leq c^2\|f\|^2 + d^2\|Q_A Af\|^2 \\ &\leq c^2\|f\|^2 + d^2(\|Q_A Af + Q_A Bg\| + \|Q_A Bg\|)^2 \end{aligned}$$

$$
\begin{aligned}
&\leq c^2\|f\|^2 + d^2[(1+\varepsilon^{-1})\|Q_A Af + Q_A Bg\|^2 + (1+\varepsilon)\|Q_A Bg\|^2] \\
&\leq c^2\|f\|^2 + d^2[(1+\varepsilon^{-1})\|Q_A Af + Q_A Bg\|^2 + (1+\varepsilon)\|Bg\|^2] \\
&\leq c^2\|f\|^2 + d^2[(1+\varepsilon^{-1})\|Q_A Af + Q_A Bg\|^2 + (1+\varepsilon)(a^2\|g\|^2 + b^2\|Dg\|^2)] \\
&\leq c^2\|f\|^2 + d^2[(1+\varepsilon^{-1})\|Q_A Af + Q_A Bg\|^2 + (1+\varepsilon)(a^2\|g\|^2 + b^2\|Q_D Dg\|^2)] \\
&\leq c^2\|f\|^2 + d^2[b'(\|Q_A Af + Q_A Bg\|^2 + \|Q_D Dg\|^2) + a'\|g\|^2] \\
&\quad \textit{with } a' = (1+\varepsilon)a^2 \textit{ et } b' = \max\{(1+\varepsilon^{-1}), (1+\varepsilon)b^2\} \\
&\leq c'(\|f\|^2 + \|g\|^2) + d^2b'(\|Q_A Af + Q_A Bg\|^2 + \|Q_D Dg\|^2) \\
&\quad \textit{with } c' = \max\{c^2, a'\} \\
&\leq c'\left\| \begin{pmatrix} f \\ g \end{pmatrix} \right\|^2 + d^2b'\left\| \begin{pmatrix} Q_A A & Q_A B \\ 0 & Q_D D \end{pmatrix} \begin{pmatrix} f \\ g \end{pmatrix} \right\|^2 \\
&\leq c'\|z\|^2 + d^2b'\|Q_{\widetilde{\mathscr{T}}}\widetilde{\mathscr{T}}z\|^2 \\
&= c'\|z\|^2 + d^2b'\|\widetilde{\mathscr{T}}z\|^2.
\end{aligned}
$$

For $\varepsilon = \dfrac{1}{b^2}$, we have $d^2b' < 1$. Then, $\widetilde{\mathscr{S}}$ is $\widetilde{\mathscr{T}}$-bounded with $\widetilde{\mathscr{T}}$-bound $\delta < 1$. This achieves the proof. Q.E.D.

2.13.2 Closability of Diagonally and Off-Diagonally Dominant Block Matrix Linear Relation

Theorem 2.13.1 (i) *If the block matrix linear relations $\mathscr{L}_0$, given in (2.8.12), is diagonally dominant of the order < 1 such that $B(0) \subset A(0)$ and $C(0) \subset D(0)$, then $\mathscr{L}_0$ is closable if, and only if, A and D are closable. In this case, the closure of $\mathscr{L}_0$ is given by*

$$\overline{\mathscr{L}_0} = \begin{pmatrix} \overline{A} & \overline{B} \\ \overline{C} & \overline{D} \end{pmatrix}.$$

(ii) *If the block matrix linear relations $\mathscr{L}_0$, given in (2.8.12), is off-diagonally dominant of the order < 1 such that $A(0) \subset B(0)$ and $D(0) \subset C(0)$, then $\mathscr{L}_0$ is closable if, and only if, B and C are closable. In this case, the closure of $\mathscr{L}_0$ is given by*

$$\overline{\mathscr{L}_0} = \begin{pmatrix} \overline{A} & \overline{B} \\ \overline{C} & \overline{D} \end{pmatrix}.$$ ◇

Proof We prove (i), the proof of (ii) is similar. Let us define the block matrix linear relations $\mathscr{L}_0 = \mathscr{T} + \mathscr{S}$ with

$$\mathscr{T} = \begin{pmatrix} A & 0 \\ 0 & D \end{pmatrix}$$

and

$$\mathscr{S} = \begin{pmatrix} 0 & B \\ C & 0 \end{pmatrix}.$$

Suppose that $\mathscr{L}_0 = \mathscr{T} + \mathscr{S}$ is closable. We have $\mathscr{L}_0$ is diagonally dominant of the order < 1 and by Proposition 2.13.1 (i), we obtain $\mathscr{S}$ is $\mathscr{T}$-bounded with $\mathscr{T}$-bound < 1. Then, $\mathscr{T}$ is closable. So, A and D are closable. Conversely,

$$\begin{aligned}
\mathscr{T}(0) &= \begin{pmatrix} A(0) \\ D(0) \end{pmatrix} \\
&= \begin{pmatrix} \overline{A}(0) \\ \overline{D}(0) \end{pmatrix} \text{ as } A \text{ and } D \text{ are closable} \\
&= \begin{pmatrix} \overline{A} & 0 \\ 0 & \overline{D} \end{pmatrix} \begin{pmatrix} 0 \\ 0 \end{pmatrix} \\
&= \overline{\mathscr{T}}(0).
\end{aligned}$$

So, $\mathscr{T}(0) = \overline{\mathscr{T}}(0)$, we obtain $\mathscr{T}$ is closable. On the other hand, $\mathscr{L}_0$ is diagonally dominant of the order < 1, then, by Proposition 2.13.1 (i), we obtain $\mathscr{S}$ is $\mathscr{T}$-bounded with $\mathscr{T}$-bound < 1. Finally, applying Lemma 2.12.4, we get $\mathscr{L}_0 = \mathscr{T} + \mathscr{S}$ is closable and $\overline{\mathscr{L}_0} = \overline{\mathscr{T} + \mathscr{S}} = \overline{\mathscr{T}} + \overline{\mathscr{S}}$. Q.E.D.

Corollary 2.13.2 *The block matrix linear relations $\mathscr{L}_0$, given in (2.8.12), is closed if one of the following conditions holds*

(i) $\mathscr{L}_0$ is diagonally dominant of the order < 1, A and D are closed, $B(0) \subset A(0)$ and $C(0) \subset D(0)$.
(ii) $\mathscr{L}_0$ is off-diagonally dominant of order < 1, B and C are closed, $A(0) \subset B(0)$ and $D(0) \subset C(0)$. ◇

Proof We prove (i), the proof of (ii) is similar. Let us define the block matrix of linear relation $\mathscr{L}_0 = \mathscr{T} + \mathscr{S}$ with

$$\mathscr{T} = \begin{pmatrix} A & 0 \\ 0 & D \end{pmatrix}$$

and

$$\mathscr{S} = \begin{pmatrix} 0 & B \\ C & 0 \end{pmatrix}.$$

If A and D are closed, then $\mathscr{T}$ is closed and, we have $\mathscr{L}_0$ is diagonally dominant of the order < 1. By using Proposition 2.13.1 (i), we obtain $\mathscr{S}$ is $\mathscr{T}$-bounded with $\mathscr{T}$-bound < 1. Hence, $\mathscr{L}_0 = \mathscr{T} + \mathscr{S}$ is closed. Q.E.D.

Remark 2.13.1 (i) The block matrix linear relations $\mathscr{L}_0$, given in (2.8.12), is diagonally dominant if A, D are closed, $\mathscr{D}(A) \subset \mathscr{D}(C)$ and $\mathscr{D}(D) \subset \mathscr{D}(B)$.

(ii) The block matrix linear relation $\mathscr{L}_0$, given in (2.8.12), is off-diagonally dominant if B, C are closed, $\mathscr{D}(C) \subset \mathscr{D}(A)$ and $\mathscr{D}(B) \subset \mathscr{D}(D)$.

In both cases, the block matrix relation $\mathscr{L}_0$, given in (2.8.12), is defined with domain $\mathscr{D}(\mathscr{L}_0) := \mathscr{D}(A) \times \mathscr{D}(D)$ if $\mathscr{L}_0$ is diagonally dominant and $\mathscr{D}(\mathscr{L}_0) := \mathscr{D}(C) \times \mathscr{D}(B)$ if $\mathscr{L}_0$ is off-diagonally dominant. ◇

Remark 2.13.2 In the Hilbert space $H \times K$, we consider the linear relations $\mathscr{L}_0$ and $\mathscr{M}_B$ given by the block matrix of linear relations

$$\mathscr{L}_0 = \begin{pmatrix} A & B \\ C & D \end{pmatrix} \tag{2.13.1}$$

and

$$\mathscr{M}_B = \begin{pmatrix} A & B \\ 0 & D \end{pmatrix}, \tag{2.13.2}$$

where $A \in L\mathscr{R}(H)$, $B \in L\mathscr{R}(K, H)$, $C \in L\mathscr{R}(H, K)$ and $D \in L\mathscr{R}(K)$ with

$$\mathscr{D}(\mathscr{L}_0) = \left(\mathscr{D}(A) \bigcap \mathscr{D}(C)\right) \times \left(\mathscr{D}(B) \bigcap \mathscr{D}(D)\right), \tag{2.13.3}$$

and

$$\mathscr{D}(\mathscr{M}_B) = \mathscr{D}(A) \times \left(\mathscr{D}(B) \bigcap \mathscr{D}(D)\right).$$

Let $A \in C\mathscr{R}(H)$, $D \in C\mathscr{R}(K)$ and let B, C be continuous linear relations from K into H and from H into K, respectively. Suppose that $\mathscr{D}(A) \subset \mathscr{D}(C)$, $\mathscr{D}(D) \subset \mathscr{D}(B)$, $B(0) \subset A(0)$ and $C(0) \subset D(0)$. Then, $\mathscr{M}_B$ and $\mathscr{L}_0$, given in (2.13.1) and (2.13.2), respectively, are closed and

$$\begin{aligned} \mathscr{D}(\mathscr{M}_B) &= \mathscr{D}(\mathscr{L}_0) \\ &= \mathscr{D}(A) \times \mathscr{D}(D). \end{aligned}$$

Indeed, since $\mathscr{D}(A) \subset \mathscr{D}(C)$, $\mathscr{D}(D) \subset \mathscr{D}(B)$, then by using (2.13.3), we have $\mathscr{D}(\mathscr{L}_0) = \mathscr{D}(A) \times \mathscr{D}(D)$. The use of Lemma 2.7.2 makes us to decompose $\mathscr{L}_0$ in the following form $\mathscr{L}_0 = \mathscr{T} + \mathscr{S}$, where

$$\mathscr{T} = \begin{pmatrix} A & 0 \\ 0 & D \end{pmatrix}$$

and

$$\mathscr{S} = \begin{pmatrix} 0 & B \\ C & 0 \end{pmatrix}.$$

It follows from that $\mathscr{T}$ is closed. Moreover, we have

$$\begin{aligned}\|\mathscr{S}\| &= \left\|\begin{pmatrix} 0 & B \\ C & 0 \end{pmatrix}\right\| \\ &= \left\|\begin{pmatrix} 0 & Q_B B \\ Q_C C & 0 \end{pmatrix}\right\| \quad \text{(by [108, Lemma 2.3])} \\ &\leq \left\|\begin{pmatrix} 0 & Q_B B \\ 0 & 0 \end{pmatrix}\right\| + \left\|\begin{pmatrix} 0 & 0 \\ Q_C C & 0 \end{pmatrix}\right\| \\ &\leq \|Q_B B\| + \|Q_C C\| \\ &\leq \|B\| + \|C\|, \end{aligned} \tag{2.13.4}$$

which yields that $\mathscr{S}$ is continuous. This implies from Proposition 2.11.2 that $\mathscr{L}_0$ is closed. Substituting C for the null relation in (2.13.1) yields that $\mathscr{M}_B$ is also closed and $\mathscr{D}(\mathscr{M}_B) = \mathscr{D}(A) \times \mathscr{D}(D)$. ◇

2.14 Selection of Linear Relations

Definition 2.14.1 Let $U \in L\mathscr{R}(X, Y)$. A linear operator V is called a selection of U, if

$$U = \underbrace{V}_{\textbf{single-valued part}} + U - U \quad \text{and} \quad \mathscr{D}(U) = \mathscr{D}(V). \qquad ◇$$

If V is a selection of U, then for all $x \in \mathscr{D}(U)$

$$\begin{aligned} Ux &= Vx + Ux - Ux \\ &= Vx + U(0). \end{aligned} \tag{2.14.1}$$

Lemma 2.14.1 [25, Lemma 7.3.1] *Let X, Y be two vector spaces and let $U \in L\mathscr{R}(X, Y)$. A is a linear selection of U if, and only if, $\eta - A$ is a linear selection of $\eta - U$ for all $\eta \in \mathbb{C}$.* ◇

Proposition 2.14.1 [2, Proposition I.5.2] *Let X, Y be two vector spaces and $U \in L\mathscr{R}(X, Y)$. If P is a linear projection with $\mathscr{D}(P) = R(U)$ and $N(P) = U(0)$, then PU is a selection of U.* ◇

Proposition 2.14.2 [2, Proposition I.5.3] *Let X, Y be two vector spaces, $U \in L\mathscr{R}(X, Y)$ and let A be a selection of U. Then,*

(i) $G(A) \bigcap (\{0\} \times U(0)) = \{0\} \times \{0\}$.
(ii) $G(U) = G(A) + (\{0\} \times U(0))$. ◇

Proposition 2.14.3 [2, Propositions II.4.1] *Let* $U \in L\mathscr{R}(X, Y)$.

If U *has a continuous selection* A*, then* U *is continuous. Further,*

$$\|U\| \leq \|A\|.$$

(ii) *If* U *is continuous and if* $U(0)$ *is the kernel of continuous projection* P *defined on* $R(U)$*, then* PU *is a continuous selection of* U. ◇

Proposition 2.14.4 *Let* $U \in C\mathscr{R}(H)$ *and* $\eta \in \mathbb{C}$*. Then,* $\eta - U_s$ *is a selection of* $\eta - U$*, where* U_s *is given in (2.11.11).* ◇

Proof By referring to (i) and Remark 2.11.8 (ii), we obtain

$$\mathscr{D}(\eta - U) = \mathscr{D}(\eta - U_s).$$

This implies from (ii) that $\eta - U_s$ is a selection of $\eta - U$, for all $\eta \in \mathbb{C}$. Q.E.D.

We define the linear relation U_G associated with U by

$$\begin{cases} U_G \in L\mathscr{R}(X, X \times Y), \\ \mathscr{D}(U_G) = \mathscr{D}(U), \text{ and} \\ U_G x = \left\{ \begin{pmatrix} x \\ y \end{pmatrix} \in X \times Y \text{ such that } y \in Ux \right\}. \end{cases} \tag{2.14.2}$$

Remark 2.14.1 If T is continuous, then U_G, given in (2.14.2), is continuous. In fact,

$$\begin{aligned} \|U_G x\| &= \inf\{\|x\| + \|y\| \text{ such that } y \in Ux\} \\ &= \|x\| + \|Ux\| \\ &\leq \|x\| + \|U\|\|x\| \text{ as Corollary 1.10.3} \\ &= (1 + \|U\|)\|x\|. \end{aligned}$$

◇

Theorem 2.14.1 *Let an analytic function of the linear relation* $T : \mathscr{U} \longrightarrow L\mathscr{R}(X, Y)$ *on an open set* $\mathscr{U} \subset \mathbb{C}$ *for which the mapping* $T(\eta) : X \longrightarrow Y$ *is onto for all* $\eta \in \mathscr{U}$*. If* T *has a selection* E *satisfying* $T(0) \subset R(E)$*, then for every analytic function of the linear relation* $\mathscr{K} : U \longrightarrow Y$*, there exists an analytic function* $f : \mathscr{U} \longrightarrow X$ *such that*

$$T(\eta) f(\eta) = \mathscr{K}(\eta), \quad \textit{for all } \eta \in \mathscr{U}.$$ ◇

Proof Let $T(\eta) : X \longrightarrow Y$ be onto for all $\eta \in \mathscr{U}$ and E be a selection of T satisfying $T(0) \subset R(E)$. Thus,

$$\begin{aligned} R(T) &= R(E) + T(0) \\ &= R(E) \end{aligned}$$

and

$$\mathscr{D}(T) = \mathscr{D}(E).$$

This implies that $E(\eta)$ is onto for all $\eta \in U$, where $T(\eta) = E(\eta) + T(0)$. Grounded on Leiterer's theorem, we state that for every analytic function $g : \mathscr{U} \longrightarrow Y$, there exists an analytic operator function $f : \mathscr{U} \longrightarrow X$ such that

$$E(\eta)f(\eta) = g(\eta), \quad for\ all\ \eta \in \mathscr{U}.$$

Therefore, for all $\eta \in \mathscr{U}$, we have

$$\begin{aligned} T(\eta)f(\eta) &= (E + T - T)(\eta)f(\eta) \\ &= E(\eta)f(\eta) + T(0) \\ &= g(\eta) + T(0) \\ &= \mathscr{K}(\eta). \end{aligned}$$

This achieves the proof. Q.E.D.

Proposition 2.14.5 (i) *Let P_1, P_2, P_3, and P_4 denote the linear projections with domains $R(A)$, $R(B)$, $R(C)$, and $R(D)$, with kernel $A(0)$, $B(0)$, $C(0)$, and $D(0)$, respectively. Then,*

$$\begin{pmatrix} P_1A & P_2B \\ P_3C & P_4D \end{pmatrix}$$

is a selection of $\mathscr{L}_0$, where $\mathscr{L}_0$ is given in (2.8.12).
(ii) *If A_1, B_1, C_1, and D_1 are selections of A, B, C, and D, respectively such that $R(A_1)\bigcap A(0) = \{0\}$, $R(B_1)\bigcap B(0) = \{0\}$, $R(C_1)\bigcap C(0) = \{0\}$ and $R(D_1)\bigcap D(0) = \{0\}$. Then,*

$$\begin{pmatrix} A_1 & B_1 \\ C_1 & D_1 \end{pmatrix}$$

is selection of $\mathscr{L}_0$.
(iii) *If A, B, C, and D have continuous selections, then $\mathscr{L}_0$ is continuous.*
(iv) *If $\mathscr{L}_0$ is continuous and*

$$\begin{pmatrix} 0 \\ 0 \end{pmatrix} \times \begin{pmatrix} A(0) + B(0) \\ C(0) + D(0) \end{pmatrix}$$

is topologically complemented in $G(\mathscr{L}_0)$, then $\mathscr{L}_0$ has a continuous selection. ◇

Proof (i) It is clear that P_1A is single-valued. Let $y \in Ax$, $x \in \mathscr{D}(A)$, then

$$\begin{aligned} Ax &= y + A(0) \\ &= P_1y + (I - P_1)y + A(0). \end{aligned}$$

Since $P_1(I - P_1)y = P_1y - P_1y = 0$, then $(I - P_1)y \in N(P_1) = A(0)$. Because $Ax = y + A(0)$, then

$$\begin{aligned} P_1Ax &= P_1y + P_1A(0) \\ &= P_1y. \end{aligned}$$

So,

$$\begin{aligned} Ax &= y + A(0) \\ &= P_1y + A(0) \\ &= P_1Ax + A(0), \end{aligned}$$

as follows P_1A is a selection of A. Likewise we have P_2B, P_3C, and P_4D are selections of B, C, and D, respectively. Then,

$$\begin{aligned} \begin{pmatrix} A & B \\ C & D \end{pmatrix} &= \begin{pmatrix} P_1A + A - A & P_2B + B - B \\ P_3C + C - C & P_4D + D - D \end{pmatrix}, \\ &= \begin{pmatrix} P_1A & P_2B \\ P_3C & P_4D \end{pmatrix} + \begin{pmatrix} A & B \\ C & D \end{pmatrix} - \begin{pmatrix} A & B \\ C & D \end{pmatrix}, \end{aligned}$$

and

$$\begin{aligned} \mathscr{D}\left[\begin{pmatrix} A & B \\ C & D \end{pmatrix}\right] &= \left(\mathscr{D}(A) \bigcap \mathscr{D}(C)\right) \bigotimes \left(\mathscr{D}(B) \bigcap \mathscr{D}(D)\right) \\ &= \mathscr{D}\left[\begin{pmatrix} P_1A & P_2B \\ P_3C & P_4D \end{pmatrix}\right]. \end{aligned}$$

(ii) For $x \in \mathscr{D}(A)$, we have

$$Ax = A_1x + A(0).$$

Hence,

$$\begin{aligned} P_1Ax &= P_1(A_1x + A(0)), \\ &= P_1A_1x + P_1(A(0)), \\ &= P_1A_1x \text{ (since } P_1(A(0)) = P_1P_1^{-1}(0) = \{0\}), \\ &= P_1A_1x + A_1x - A_1x, \\ &= A_1x + (P_1 - I)A_1x. \end{aligned}$$

Since $P_1(P_1 - I)A_1x = P_1A_1x - P_1A_1x = 0$, then

$$(P_1 - I)A_1x \in N(P_1) = A(0).$$

We conclude that

$$(P_1 - I)A_1x \in R(A_1) \bigcap A(0) = \{0\}.$$

Therefore,

$$P_1Ax = A_1x$$

for all $x \in \mathscr{D}(A)$. In the same way, we shall show that $B_1 = P_2B$, $C_1 = P_3C$, and $D_1 = P_4D$. Then,

$$\begin{pmatrix} A_1 & B_1 \\ C_1 & D_1 \end{pmatrix} = \begin{pmatrix} P_1A & P_2B \\ P_3C & P_4D \end{pmatrix}.$$

(iii) Let A_1, B_1, C_1, and D_1 be continuous selections of A, B, C, and D, respectively. Then,

$$\begin{pmatrix} A & B \\ C & D \end{pmatrix} = \begin{pmatrix} A_1 & B_1 \\ C_1 & D_1 \end{pmatrix} + \begin{pmatrix} A & B \\ C & D \end{pmatrix} - \begin{pmatrix} A & B \\ C & D \end{pmatrix}.$$

Hence, for all $\begin{pmatrix} x \\ y \end{pmatrix} \in \mathscr{D}\left[\begin{pmatrix} A & B \\ C & D \end{pmatrix}\right]$, we have

$$\begin{aligned}
\left\| \begin{pmatrix} A & B \\ C & D \end{pmatrix} \begin{pmatrix} x \\ y \end{pmatrix} \right\| &= \left\| \begin{pmatrix} A_1 & B_1 \\ C_1 & D_1 \end{pmatrix} \begin{pmatrix} x \\ y \end{pmatrix} + \begin{pmatrix} A & B \\ C & D \end{pmatrix} \begin{pmatrix} 0 \\ 0 \end{pmatrix} \right\| \\
&\le \left\| \begin{pmatrix} A_1 & B_1 \\ C_1 & D_1 \end{pmatrix} \begin{pmatrix} x \\ y \end{pmatrix} \right\| + \left\| \begin{pmatrix} A & B \\ C & D \end{pmatrix} \begin{pmatrix} 0 \\ 0 \end{pmatrix} \right\| \\
&= \left\| \begin{pmatrix} A_1 & B_1 \\ C_1 & D_1 \end{pmatrix} \begin{pmatrix} x \\ y \end{pmatrix} \right\| \\
&= \left\| \begin{pmatrix} A_1x + B_1y \\ C_1x + D_1y \end{pmatrix} \right\| \\
&= \|A_1x + B_1y\| + \|C_1x + D_1y\| \\
&\le \|A_1\|\|x\| + \|B_1\|\|y\| + \|C_1\|\|x\| + \|D_1\|\|y\| \\
&= (\|A_1\| + \|B_1\| + \|C_1\| + \|D_1\|)\, \|(x, y)\| < \infty.
\end{aligned}$$

(iv) Since $\mathscr{L}_0$ is continuous, then $\mathscr{L}_{0G(x_1,x_2)}$ is continuous, where $\mathscr{L}_{0G(x_1,x_2)}$ is the linear relation associated with $\mathscr{L}_0$ defined in (2.14.2). Indeed,

$$\begin{aligned}
\left\| \mathscr{L}_{0G(x_1,x_2)} \right\| &= \inf\left\{ \|(x_1, x_2)\| + \|(y_1, y_2)\| \text{ such that } \begin{pmatrix} y_1 \\ y_2 \end{pmatrix} \in \mathscr{L}_0 \begin{pmatrix} x_1 \\ x_2 \end{pmatrix} \right\} \\
&\le \|(x_1, x_2)\| + \left\| \mathscr{L}_0 \begin{pmatrix} x_1 \\ x_2 \end{pmatrix} \right\| \\
&\le (1 + \|\mathscr{L}_0\|)\, \|(x_1, x_2)\|.
\end{aligned}$$

Let P be a bounded projection defined on $G(\mathscr{L}_0)$ with kernel

$$\begin{pmatrix} 0 \\ 0 \end{pmatrix} \times \begin{pmatrix} A(0) + B(0) \\ C(0) + D(0) \end{pmatrix}.$$

Then, $P\mathscr{L}_{0G(x_1,x_2)}$ is single-valued, since

$$P\mathscr{L}_{0G(0,0)} = P\left(\begin{pmatrix} 0 \\ 0 \end{pmatrix} \times \begin{pmatrix} A(0) + B(0) \\ C(0) + D(0) \end{pmatrix}\right) = \begin{pmatrix} 0 \\ 0 \end{pmatrix}.$$

Since

$$\begin{aligned} N(P) &= \mathscr{L}_{0G(0,0)} \\ &= \begin{pmatrix} 0 \\ 0 \end{pmatrix} \times \begin{pmatrix} A(0) + B(0) \\ C(0) + D(0) \end{pmatrix}, \end{aligned}$$

and

$$\begin{aligned} \mathscr{D}(P) &= R(\mathscr{L}_{0G(x_1,x_2)}) \\ &= G(\mathscr{L}_0), \end{aligned}$$

then $P\mathscr{L}_{0G(x_1,x_2)}$ is a selection of $\mathscr{L}_{0G(x_1,x_2)}$. Hence,

$$\|P\mathscr{L}_{0G(x_1,x_2)}\| \leq \|P\|\|\mathscr{L}_{0G(x_1,x_2)}\| < \infty.$$

Therefore, $\mathscr{L}_0$ has a continuous selection. Q.E.D.

Theorem 2.14.2 *Let A, $B \in B\mathscr{R}(X)$, $C \in B\mathscr{R}(X)$, $T_1 \in B\mathscr{R}(X)$ and $T_2 \in B\mathscr{R}(X)$. We consider the multivalued linear operators matrix*

$$\mathscr{M}_C = \begin{pmatrix} A & C \\ 0 & B \end{pmatrix}$$

and

$$\mathscr{T} = \begin{pmatrix} T_1 & 0 \\ 0 & T_2 \end{pmatrix}.$$

Then, for all $\lambda \in \mathbb{C}$, we have

(*i*) $\lambda - \mathscr{M}_C - \mathscr{T} = \begin{pmatrix} I & 0 \\ 0 & \lambda - B - T_2 \end{pmatrix} \begin{pmatrix} I & -C \\ 0 & I \end{pmatrix} \begin{pmatrix} \lambda - A - T_1 & 0 \\ 0 & I \end{pmatrix}.$

(ii) Let C_1 be a selection of C such that $(\lambda - A - T_1)^{-1}$ is a single-valued, then

$$\lambda - \mathcal{M}_C - \mathcal{T} = \begin{pmatrix} I & 0 \\ 0 & \lambda - B - T_2 \end{pmatrix} \begin{pmatrix} I & -C_1 \\ 0 & I \end{pmatrix} \begin{pmatrix} \lambda - A - T_1 & 0 \\ 0 & I \end{pmatrix} + \begin{pmatrix} 0 & C - C \\ 0 & (\lambda - B - T_2)(C - C) \end{pmatrix}.$$

◇

Proof (*i*) Let $\lambda \in \mathbb{C}$. By using Corollary 2.8.1, we have

$$\begin{pmatrix} I & 0 \\ 0 & \lambda - B - T_2 \end{pmatrix} \begin{pmatrix} I & -C \\ 0 & I \end{pmatrix} \begin{pmatrix} \lambda - A - T_1 & 0 \\ 0 & I \end{pmatrix} = \begin{pmatrix} I & -C \\ 0 & \lambda - B - T_2 \end{pmatrix} \begin{pmatrix} (\lambda - A - T_1)^{-1} & 0 \\ 0 & I \end{pmatrix}.$$

The use of Lemma 2.8.4 (*iv*) leads to

$$\begin{pmatrix} I & -C \\ 0 & \lambda - B - T_2 \end{pmatrix} \begin{pmatrix} \lambda - A - T_1 & 0 \\ 0 & I \end{pmatrix} = \lambda - \mathcal{M}_C - \mathcal{T}.$$

(*ii*) Let $\lambda \in \mathbb{C}$ and C_1 be a selection of C, then by using the assertion (*i*), we get

$$\lambda - \mathcal{M}_C - \mathcal{T} = \begin{pmatrix} I & 0 \\ 0 & \lambda - B - T_2 \end{pmatrix} \begin{pmatrix} I & -C_1 - C + C \\ 0 & I \end{pmatrix} \begin{pmatrix} \lambda - A - T_1 & 0 \\ 0 & I \end{pmatrix}.$$

According to Lemma 2.8.4 (*i*), the matrix $\lambda - \mathcal{M}_C - \mathcal{T}$ can be written as follows

$$\lambda - \mathcal{M}_C - \mathcal{T} = \begin{pmatrix} I & 0 \\ 0 & \lambda - B - T_2 \end{pmatrix} \left[\begin{pmatrix} I & -C_1 \\ 0 & I \end{pmatrix} + \begin{pmatrix} 0 & C - C \\ 0 & 0 \end{pmatrix} \right] \begin{pmatrix} \lambda - A - T_1 & 0 \\ 0 & I \end{pmatrix}.$$

We claim that

$$\begin{pmatrix} \lambda - A - T_1 & 0 \\ 0 & I \end{pmatrix} \begin{pmatrix} 0 \\ 0 \end{pmatrix} \subset N\left[\begin{pmatrix} 0 & C - C \\ 0 & 0 \end{pmatrix} \right].$$

Indeed, by using Proposition 2.7.3, we have

$$\lambda - \mathcal{M}_C - \mathcal{T} = \begin{pmatrix} I & 0 \\ 0 & \lambda - B - T_2 \end{pmatrix}$$
$$\left[\begin{pmatrix} I & -C_1 \\ 0 & I \end{pmatrix} \begin{pmatrix} \lambda - A - T_1 & 0 \\ 0 & I \end{pmatrix} + \begin{pmatrix} 0 & C - C \\ 0 & 0 \end{pmatrix} \begin{pmatrix} \lambda - A - T_1 & 0 \\ 0 & I \end{pmatrix} \right].$$

From Proposition 2.7.3, we have

$$\lambda - \mathcal{M}_C - \mathcal{T} = \begin{pmatrix} I & 0 \\ 0 & \lambda - B - T_2 \end{pmatrix} \left[\begin{pmatrix} \lambda - A - T_1 & -C_1 \\ 0 & I \end{pmatrix} + \begin{pmatrix} 0 & C - C \\ 0 & 0 \end{pmatrix} \right].$$

Furthermore, since $\mathcal{D}\left[\begin{pmatrix} I & 0 \\ 0 & \lambda - B - T_2 \end{pmatrix} \right]$ containing the ranges of both

$$\begin{pmatrix} \lambda - A - T_1 & -C_1 \\ 0 & I \end{pmatrix}$$

and

$$\begin{pmatrix} 0 & C - C \\ 0 & 0 \end{pmatrix},$$

then

$$\lambda - \mathcal{M}_C - \mathcal{T} = \begin{pmatrix} I & 0 \\ 0 & \lambda - B - T_2 \end{pmatrix} \begin{pmatrix} I & -C_1 \\ 0 & I \end{pmatrix} \begin{pmatrix} \lambda - A - T_1 & 0 \\ 0 & I \end{pmatrix} + \begin{pmatrix} I & 0 \\ 0 & \lambda - B - T_2 \end{pmatrix} \begin{pmatrix} 0 & C - C \\ 0 & 0 \end{pmatrix}.$$

Therefore, by using Lemma 2.8.4 (iv), we conclude that

$$\lambda - \mathcal{M}_C - \mathcal{T} = \begin{pmatrix} I & 0 \\ 0 & \lambda - B - T_2 \end{pmatrix} \begin{pmatrix} I & -C_1 \\ 0 & I \end{pmatrix} \begin{pmatrix} \lambda - A - T_1 & 0 \\ 0 & I \end{pmatrix} + \begin{pmatrix} 0 & C - C \\ 0 & (\lambda - B - T_2)(C - C) \end{pmatrix}.$$

This achieves the proof. Q.E.D.

Theorem 2.14.3 *(i) If $(A + B + T_1)(0) \subset N(C)$, then for all $\lambda \in \mathbb{C}\backslash\{0\}$, we have*

$$\lambda - \mathcal{L}_0 - \mathcal{T} = -\frac{1}{\lambda} \begin{pmatrix} 0 & 0 \\ C(A + T_1) & CB \end{pmatrix} + \begin{pmatrix} I & 0 \\ -\frac{1}{\lambda} C & I \end{pmatrix} \begin{pmatrix} \lambda - A - T_1 & -B \\ 0 & \lambda - D - T_2 \end{pmatrix}.$$

(ii) Let C_1 be a selection of C. If $(A + B + T_1)(0) \subset N(C)$, then for all $\lambda \in \mathbb{C}\backslash\{0\}$, we have

$$\lambda - \mathcal{L}_0 - \mathcal{T} = -\frac{1}{\lambda} \begin{pmatrix} 0 & 0 \\ C(A + T_1) & CB \end{pmatrix} + \begin{pmatrix} I & 0 \\ -\frac{1}{\lambda} C_1 & I \end{pmatrix} \begin{pmatrix} \lambda - A - T_1 & -B \\ 0 & \lambda - D - T_2 \end{pmatrix}.$$

(iii) Let C_1 (resp. B_1) be a selection of C (resp. B). If $(A + B + T_1)(0) \subset N(C)$, then for all $\lambda \in \mathbb{C}\backslash\{0\}$, we have

$$\lambda - \mathcal{L}_0 - \mathcal{T} = \begin{pmatrix} I & 0 \\ -\frac{1}{\lambda} C_1 & I \end{pmatrix} \begin{pmatrix} \lambda - A - T_1 & -B_1 \\ 0 & \lambda - D - T_2 \end{pmatrix} - \frac{1}{\lambda} \begin{pmatrix} 0 & \lambda(B - B) \\ C(A + T_1) & CB + C_1(B - B) \end{pmatrix}.$$

◇

Proof (*i*) Let $\lambda \in \mathbb{C}\backslash\{0\}$ and

$$\left[\begin{pmatrix} x_1 \\ x_2 \end{pmatrix}, \begin{pmatrix} y_1 \\ y_2 \end{pmatrix}\right] \in G\left[\begin{pmatrix} I & 0 \\ -\frac{1}{\lambda}C & I \end{pmatrix}\begin{pmatrix} \lambda - A - T_1 & -B \\ 0 & \lambda - D - T_2 \end{pmatrix}\right].$$

Therefore,

$$\begin{pmatrix} y_1 \\ y_2 \end{pmatrix} \in \begin{pmatrix} I & 0 \\ -\frac{1}{\lambda}C & I \end{pmatrix}\begin{pmatrix} \lambda - A - T_1 & -B \\ 0 & \lambda - D - T_2 \end{pmatrix}\begin{pmatrix} x_1 \\ x_2 \end{pmatrix}$$

equivalent to

$$\begin{pmatrix} y_1 \\ y_2 \end{pmatrix} \in \begin{pmatrix} I & 0 \\ -\frac{1}{\lambda}C & I \end{pmatrix}\begin{pmatrix} (\lambda - A - T_1)x_1 - Bx_2 \\ (\lambda - D - T_2)x_2 \end{pmatrix}$$

equivalent to

$$\begin{pmatrix} y_1 \\ y_2 \end{pmatrix} \in \begin{pmatrix} (\lambda - A - T_1)x_1 - Bx_2 \\ -\frac{1}{\lambda}C[(\lambda - A - T_1)x_1 - Bx_2] + (\lambda - D - T_2)x_2 \end{pmatrix}.$$

Since $\mathscr{D}(C) = X$, then

$$\begin{cases} y_1 \in (\lambda - A - T_1)x_1 - Bx_2 \\ y_2 \in -\frac{1}{\lambda}C(\lambda - A - T_1)x_1 + \frac{1}{\lambda}CBx_2 + (\lambda - D - T_2)x_2. \end{cases}$$

We can get

$$\begin{pmatrix} y_1 \\ y_2 \end{pmatrix} \in \begin{pmatrix} \lambda - A - T_1 & -B \\ -\frac{1}{\lambda}C(\lambda - A - T_1) & \frac{1}{\lambda}CB + \lambda - D - T_2 \end{pmatrix}\begin{pmatrix} x_1 \\ x_2 \end{pmatrix}.$$

It follows that

$$\left[\begin{pmatrix} x_1 \\ x_2 \end{pmatrix}, \begin{pmatrix} y_1 \\ y_2 \end{pmatrix}\right] \in G\left[\begin{pmatrix} \lambda - A - T_1 & -B \\ -\frac{1}{\lambda}C(\lambda - A - T_1) & \frac{1}{\lambda}CB + \lambda - D - T_2 \end{pmatrix}\right].$$

We end up with the following inclusion

$$G\left[\begin{pmatrix} I & 0 \\ -\frac{1}{\lambda}C & I \end{pmatrix}\begin{pmatrix} \lambda - A - T_1 & -B \\ 0 & \lambda - D - T_2 \end{pmatrix}\right] \subset G\left[\begin{pmatrix} \lambda - A - T_1 & -B \\ -\frac{1}{\lambda}C(\lambda - A - T_1) & \frac{1}{\lambda}CB + \lambda - D - T_2 \end{pmatrix}\right].$$

Hence,

$$\begin{aligned}
&\begin{pmatrix} I & 0 \\ -\frac{1}{\lambda}C & I \end{pmatrix}\begin{pmatrix} \lambda - A - T_1 & -B \\ 0 & \lambda - D - T_2 \end{pmatrix}\begin{pmatrix} 0 \\ 0 \end{pmatrix} \\
&= \begin{pmatrix} I & 0 \\ -\frac{1}{\lambda}C & I \end{pmatrix}\begin{pmatrix} A(0) + T_1(0) + B(0) \\ D(0) + T_2(0) \end{pmatrix} \\
&= \begin{pmatrix} A(0) + T_1(0) + B(0) \\ \frac{1}{\lambda}C(A(0) + T_1(0) + B(0)) + D(0) + T_2(0) \end{pmatrix} \\
&= \begin{pmatrix} \lambda - A - T_1 & -B \\ -\frac{1}{\lambda}C(\lambda - A - T_1) & \frac{1}{\lambda}CB + \lambda - D - T_2 \end{pmatrix}\begin{pmatrix} 0 \\ 0 \end{pmatrix}.
\end{aligned}$$

In addition, $\mathscr{D}(C) = X$, then

$$\begin{aligned}
\mathscr{D}\left[\begin{pmatrix} I & 0 \\ -\frac{1}{\lambda}C & I \end{pmatrix}\begin{pmatrix} \lambda - A - T_1 & -B \\ 0 & \lambda - D - T_2 \end{pmatrix}\right] &= \mathscr{D}(\lambda - A - T_1) \times \left(\mathscr{D}(B)\bigcap \mathscr{D}(\lambda - D - T_2)\right) \\
&= \mathscr{D}\left[\begin{pmatrix} \lambda - A - T_1 & -B \\ -\frac{1}{\lambda}C(\lambda - A - T_1) & \frac{1}{\lambda}CB + \lambda - D - T_2 \end{pmatrix}\right].
\end{aligned}$$

A consequence of Proposition 2.7.3 (iii), we have

$$\begin{pmatrix} I & 0 \\ -\frac{1}{\lambda}C & I \end{pmatrix}\begin{pmatrix} \lambda - A - T_1 & -B \\ 0 & \lambda - D - T_2 \end{pmatrix} = \begin{pmatrix} \lambda - A - T_1 & -B \\ -\frac{1}{\lambda}C(\lambda - A - T_1) & \frac{1}{\lambda}CB + \lambda - D - T_2 \end{pmatrix}.$$

By using Lemma 2.7.2, it is clear that

$$\begin{aligned}
&\frac{1}{\lambda}\begin{pmatrix} 0 & 0 \\ -C(A + T_1) & -CB \end{pmatrix} + \begin{pmatrix} \lambda - A - T_1 & -B \\ -\frac{1}{\lambda}C(\lambda - A - T_1) & \frac{1}{\lambda}CB + \lambda - D - T_2 \end{pmatrix} \\
&= \begin{pmatrix} \lambda - A - T_1 & -B \\ -\frac{1}{\lambda}C(A + T_1 + \lambda - A - T_1) & \frac{1}{\lambda}(CB - CB) + \lambda - D - T_2 \end{pmatrix}.
\end{aligned}$$

Let

$$\left[\begin{pmatrix} x_1 \\ x_2 \end{pmatrix}, \begin{pmatrix} y_1 \\ y_2 \end{pmatrix}\right] \in G\left[\begin{pmatrix} \lambda - A - T_1 & -B \\ -\frac{1}{\lambda}C(A + T_1 + \lambda - A - T_1) & \frac{1}{\lambda}C(B - B) + \lambda - D - T_2 \end{pmatrix}\right].$$

It follows that

$$\begin{pmatrix} y_1 \\ y_2 \end{pmatrix} \in \begin{pmatrix} \lambda - A - T_1 & -B \\ -\frac{1}{\lambda}C(A + T_1 + \lambda - A - T_1) & \frac{1}{\lambda}C(B - B) + \lambda - D - T_2 \end{pmatrix} \begin{pmatrix} x_1 \\ x_2 \end{pmatrix}$$

if, and only if,

$$\begin{pmatrix} y_1 \\ y_2 \end{pmatrix} \in \begin{pmatrix} (\lambda - A - T_1)x_1 - Bx_2 \\ -\frac{1}{\lambda}C(A + T_1 + \lambda - A - T_1)x_1 + (\frac{1}{\lambda}C(B - B) + \lambda - D - T_2)x_2 \end{pmatrix}.$$

Since $C \in B\mathscr{R}(X)$, then

$$\begin{pmatrix} y_1 \\ y_2 \end{pmatrix} \in \begin{pmatrix} (\lambda - A - T_1)x_1 - Bx_2 \\ -\frac{CA + CT_1}{\lambda}x_1 - Cx_1 + \frac{1}{\lambda}CAx_1 + \frac{1}{\lambda}CT_1x_1 - \frac{1}{\lambda}CBx_2 + \frac{1}{\lambda}CBx_2 + (\lambda - D - T_2)x_2 \end{pmatrix}.$$

If, and only if,

$$\begin{pmatrix} y_1 \\ y_2 \end{pmatrix} \in \begin{pmatrix} (\lambda - A - T_1)x_1 - Bx_2 \\ CA(0) + CT_1(0) - Cx_1 + CB(0) + (\lambda - D - T_2)x_2 \end{pmatrix}.$$

Since $CA(0) + CB(0) + CT_1(0) = C(A + B + T_1)(0) \subset CN(C) = CC^{-1}(0) = C(0)$, then

$$CA(0) + CB(0) + CT_1(0) - Cx_1 = -Cx_1.$$

Therefore,

$$\begin{pmatrix} y_1 \\ y_2 \end{pmatrix} \in \begin{pmatrix} (\lambda - A - T_1)x_1 - Bx_2 \\ -Cx_1 + (\lambda - D - T_2)x_2 \end{pmatrix}.$$

This proves that

$$\left[\begin{pmatrix} x_1 \\ x_2 \end{pmatrix}, \begin{pmatrix} y_1 \\ y_2 \end{pmatrix}\right] \in G(\lambda - \mathscr{L}_0 - \mathscr{T}).$$

So,

$$G\left[\begin{pmatrix} \lambda - A - T_1 & -B \\ -\frac{1}{\lambda}C(A + T_1 + \lambda - A - T_1) & \frac{1}{\lambda}(CB - CB) + \lambda - D - T_2 \end{pmatrix}\right] \subset G(\lambda - \mathscr{L}_0 - \mathscr{T}). \tag{2.14.3}$$

Hence, it is easy to verify that

$$\begin{pmatrix} \lambda - A - T_1 & -B \\ -\frac{1}{\lambda}C(A + T_1 + \lambda - A - T_1) & \frac{1}{\lambda}(CB - CB) + \lambda - D - T_2 \end{pmatrix} \begin{pmatrix} 0 \\ 0 \end{pmatrix}$$

$$= \begin{pmatrix} (\lambda - A - T_1)(0) - B(0) \\ -\frac{1}{\lambda}C(A + T_1 + \lambda - A - T_1)(0) + \frac{1}{\lambda}CB(0) + (\lambda - D - T_2)(0) \end{pmatrix},$$

$$= \begin{pmatrix} A(0) + T_1(0) + B(0) \\ C(A + T_1)(0) + CB(0) + (D + T_2)(0) \end{pmatrix},$$

$$= (\lambda - \mathscr{L}_0 - \mathscr{T}) \begin{pmatrix} 0 \\ 0 \end{pmatrix}.$$

We infer that

$$\begin{pmatrix} \lambda - A - T_1 & -B \\ -\frac{1}{\lambda}C(A + T_1 + \lambda - A - T_1) & \frac{1}{\lambda}C(B - B) + \lambda - D - T_2 \end{pmatrix} \begin{pmatrix} 0 \\ 0 \end{pmatrix}$$
$$= (\lambda - \mathscr{L}_0 - \mathscr{T}) \begin{pmatrix} 0 \\ 0 \end{pmatrix}. \tag{2.14.4}$$

Now, let us define the domain as follows

$$\mathscr{D}\left[\begin{pmatrix} \lambda - A - T_1 & -B \\ -\frac{1}{\lambda}C(A + T_1 + \lambda - A - T_1) & \frac{1}{\lambda}C(B - B) + \lambda - D - T_2 \end{pmatrix}\right]$$

$$= \left[\mathscr{D}(\lambda - A - T_1) \bigcap \mathscr{D}\left(-\frac{1}{\lambda}C(A + T_1 + \lambda - A - T_1)\right)\right]$$
$$\times \left[\mathscr{D}(B) \bigcap \mathscr{D}\left(\frac{1}{\lambda}(CB - CB) + \lambda - D - T_2\right)\right]$$
$$= \left[\mathscr{D}(A) \bigcap \mathscr{D}(T_1)\right] \times \left[\mathscr{D}(B) \bigcap \mathscr{D}(D) \bigcap \mathscr{D}(T_2)\right]$$
$$= \mathscr{D}(\lambda - \mathscr{L}_0 - \mathscr{T}).$$

We deduce that

$$\mathscr{D}\left[\begin{pmatrix} \lambda - A - T_1 & -B \\ -\frac{1}{\lambda}C(A + T_1 + \lambda - A - T_1) & \frac{1}{\lambda}(CB - CB) + \lambda - D - T_2 \end{pmatrix}\right]$$
$$= \mathscr{D}(\lambda - \mathscr{L}_0 - \mathscr{T}). \tag{2.14.5}$$

By using Eqs. (2.14.3), (2.14.4), (2.14.5) and Proposition 2.7.3 (iii), we conclude that

$$\lambda - \mathcal{L}_0 - \mathcal{T} = -\frac{1}{\lambda}\begin{pmatrix} 0 & 0 \\ C(A+T_1) & CB \end{pmatrix} + \begin{pmatrix} I & 0 \\ -\frac{1}{\lambda}C & I \end{pmatrix}\begin{pmatrix} \lambda - A - T_1 & -B \\ 0 & \lambda - D - T_2 \end{pmatrix}.$$

(ii) Let $\lambda \in \mathbb{C}\backslash\{0\}$. Let C_1 be a selection of C, then by using the assertion (i), we get

$$\lambda - \mathcal{L}_0 - \mathcal{T} = \frac{1}{\lambda}\begin{pmatrix} 0 & 0 \\ -C(A+T_1) & -CB \end{pmatrix}$$
$$+ \begin{pmatrix} I & 0 \\ -\frac{1}{\lambda}(C_1 + C - C) & I \end{pmatrix}\begin{pmatrix} \lambda - A - T_1 & -B \\ 0 & \lambda - D - T_2 \end{pmatrix}.$$

Then, by Lemma 2.8.4 (i), we conclude that

$$\lambda - \mathcal{L}_0 - \mathcal{T} = \frac{1}{\lambda}\begin{pmatrix} 0 & 0 \\ -C(A+T_1) & -CB \end{pmatrix} +$$
$$\left[\begin{pmatrix} I & 0 \\ -\frac{1}{\lambda}C_1 & I \end{pmatrix} + \begin{pmatrix} 0 & 0 \\ \frac{1}{\lambda}(C - C) & 0 \end{pmatrix}\right]\begin{pmatrix} \lambda - A - T_1 & -B \\ 0 & \lambda - D - T_2 \end{pmatrix}.$$

By using Proposition 2.7.3, we can write

$$\lambda - \mathcal{L}_0 - \mathcal{T} = \frac{1}{\lambda}\begin{pmatrix} 0 & 0 \\ -C(A+T_1) & -CB \end{pmatrix} + \begin{pmatrix} I & 0 \\ -\frac{1}{\lambda}C_1 & I \end{pmatrix}\begin{pmatrix} \lambda - A - T_1 & -B \\ 0 & \lambda - D - T_2 \end{pmatrix}$$
$$+ \begin{pmatrix} 0 & 0 \\ \frac{1}{\lambda}(C - C) & 0 \end{pmatrix}\begin{pmatrix} \lambda - A - T_1 & -B \\ 0 & \lambda - D - T_2 \end{pmatrix}.$$

Conversely, by using the same reasoning as above, we find

$$\begin{pmatrix} 0 & 0 \\ \frac{1}{\lambda}(C - C) & 0 \end{pmatrix}\begin{pmatrix} \lambda - A - T_1 & -B \\ 0 & \lambda - D - T_2 \end{pmatrix} = \begin{pmatrix} 0 & 0 \\ \frac{1}{\lambda}(C - C)(\lambda - A - T_1) & \frac{1}{\lambda}(C - C)B \end{pmatrix}.$$

Lemma 2.8.4 (i) implies that

$$\lambda - \mathcal{L}_0 - \mathcal{T} = \begin{pmatrix} I & 0 \\ -\frac{1}{\lambda}C_1 & I \end{pmatrix}\begin{pmatrix} \lambda - A - T_1 & -B \\ 0 & \lambda - D - T_2 \end{pmatrix}$$
$$+\frac{1}{\lambda}\begin{pmatrix} 0 & 0 \\ (C - C)(\lambda - A - T_1) - C(A+T_1) & (C - C)B - CB \end{pmatrix}.$$

Let $\left[\begin{pmatrix} x_1 \\ x_2 \end{pmatrix}, \begin{pmatrix} y_1 \\ y_2 \end{pmatrix}\right] \in G\left[\begin{pmatrix} 0 & 0 \\ (C - C)(\lambda - A - T_1) - C(A+T_1) & (C - C)B - CB \end{pmatrix}\right].$

In other words,

$$\begin{pmatrix} y_1 \\ y_2 \end{pmatrix} \in \begin{pmatrix} 0 & 0 \\ (C - C)(\lambda - A - T_1) - C(A + T_1) & (C - C)B - CB \end{pmatrix} \begin{pmatrix} x_1 \\ x_2 \end{pmatrix}$$

if, and only if,

$$\begin{pmatrix} y_1 \\ y_2 \end{pmatrix} \in \begin{pmatrix} 0 \\ (C - C)(\lambda - A - T_1)x_1 - C(A + T_1)x_1 + (C - C)Bx_2 - CBx_2 \end{pmatrix}$$

if, and only if,

$$\begin{pmatrix} y_1 \\ y_2 \end{pmatrix} \in \begin{pmatrix} 0 \\ C(0) - C(A + T_1)x_1 + C(A + T_1)(0) + CBx_2 + CB(0) \end{pmatrix}$$

if, and only if,

$$\begin{pmatrix} y_1 \\ y_2 \end{pmatrix} \in \begin{pmatrix} 0 \\ -C(A + T_1)x_1 - CBx_2 \end{pmatrix}$$

if, and only if,

$$\left[\begin{pmatrix} x_1 \\ x_2 \end{pmatrix}, \begin{pmatrix} y_1 \\ y_2 \end{pmatrix} \right] \in G\left[\begin{pmatrix} 0 & 0 \\ -C(A + T_1) & -CB \end{pmatrix} \right].$$

Therefore,

$$\lambda - \mathscr{L}_0 - \mathscr{T} = \begin{pmatrix} I & 0 \\ -\frac{1}{\lambda}C_1 & I \end{pmatrix} \begin{pmatrix} \lambda - A - T_1 & -B \\ 0 & \lambda - D - T_2 \end{pmatrix} - \frac{1}{\lambda} \begin{pmatrix} 0 & 0 \\ C(A + T_1) & CB \end{pmatrix}.$$

(iii) Let $\lambda \in \mathbb{C} \backslash \{0\}$. By using the assertion (ii), we have

$$\begin{aligned} \lambda - \mathscr{L}_0 - \mathscr{T} &= \begin{pmatrix} I & 0 \\ -\frac{1}{\lambda}C_1 & I \end{pmatrix} \begin{pmatrix} \lambda - A - T_1 & -B_1 + B - B \\ 0 & \lambda - D - T_2 \end{pmatrix} - \frac{1}{\lambda} \begin{pmatrix} 0 & 0 \\ C(A + T_1) & CB \end{pmatrix} \\ &= \begin{pmatrix} I & 0 \\ -\frac{1}{\lambda}C_1 & I \end{pmatrix} \left[\begin{pmatrix} \lambda - A - T_1 & -B_1 \\ 0 & \lambda - D - T_2 \end{pmatrix} + \begin{pmatrix} 0 & B - B \\ 0 & 0 \end{pmatrix} \right] \\ &\quad - \frac{1}{\lambda} \begin{pmatrix} 0 & 0 \\ C(A + T_1) & CB \end{pmatrix} \\ &= \begin{pmatrix} I & 0 \\ -\frac{1}{\lambda}C_1 & I \end{pmatrix} \begin{pmatrix} \lambda - A - T_1 & -B_1 \\ 0 & \lambda - D - T_2 \end{pmatrix} \\ &\quad + \begin{pmatrix} I & 0 \\ -\frac{1}{\lambda}C_1 & I \end{pmatrix} \begin{pmatrix} 0 & B - B \\ 0 & 0 \end{pmatrix} - \frac{1}{\lambda} \begin{pmatrix} 0 & 0 \\ C(A + T_1) & CB \end{pmatrix}. \end{aligned}$$

Then, by using Lemma 2.8.4 (i) and (ii), we conclude that

$$\lambda - \mathscr{L}_0 - \mathscr{T} = \begin{pmatrix} I & 0 \\ -\frac{1}{\lambda}C_1 & I \end{pmatrix} \begin{pmatrix} \lambda - A - T_1 & -B_1 \\ 0 & \lambda - D - T_2 \end{pmatrix}$$
$$- \frac{1}{\lambda} \begin{pmatrix} 0 & \lambda(B - B) \\ C(A + T_1) & CB + C_1(B - B) \end{pmatrix}.$$

This achieves the proof. Q.E.D.

2.15 Adjoint of Linear Relation

2.15.1 *Definitions and Properties*

Let $U \in L\mathscr{R}(X, Y)$, the adjoint U^* of U is defined by

$$\begin{aligned} G(U^*) =& G(-U^{-1})^{\perp} \\ & \subset X^* \times Y^*, \end{aligned}$$

where

$$\begin{aligned} [(y, x) + (y', x')] &= [x, x'] + [y, y'] \\ &= x'(x) + y'(y). \end{aligned}$$

Thus, means that $(y', x') \in G(U^*)$ if, and only if, $y'(y) - x'(x) = 0$ for all $\begin{pmatrix} x \\ y \end{pmatrix} \in G(U)$. Thus,

$$y'(y) = x'(x)$$

for all $y \in Ux$, $x \in \mathscr{D}(U)$. Then, $x' \in U^*y'$ if, and only if, $y'(Ux) = x'(x)$ for all $x \in \mathscr{D}(U)$. We can identify the adjoint as follows:

$$G(U^*) := \left\{(y', x') \in Y^* \times X^* : x' \text{ is an extension of } y'U\right\}.$$

It is clear that this definition coincides with the classical one when U is a densely defined operator and it permits one to define the conjugate of a non-densely defined operator.

In the next we recall some basis properties of the adjoint of a linear relation.

Proposition 2.15.1 [2, Chapter III] *Let $U \in L\mathscr{R}(X)$. Then,*

(i) U^* is a closed linear relation in $L\mathscr{R}(X^*)$.
(ii) $G(U^* + V^*) \subset G((U + V)^*)$.
(iii) If U is continuous with $\mathscr{D}(U) \subset \mathscr{D}(V)$, then $(U + V)^* = U^* + V^*$.

(iv) $N(U^*) = R(U)^{\perp}$, $U^*(0) = \mathscr{D}(U)^{\perp}$, $N(\overline{U}) = R(U^*)^{\top}$, *and* $\overline{U}(0) = \mathscr{D}(U)^{\top}$.
(v) If U *is closed, then* $R(U)$ *is closed if, and only if,* $R(U^*)$ *is closed if, and only if,* U *is open if, and only if,* U^* *is open.*
(vi) If U *is continuous, then* $\|U\| = \|U^*\|$.
(vii) $\mathscr{D}(U^*)^{\top\perp} = \overline{\mathscr{D}(U^*)}$.
(viii) T *is continuous if, and only if,* $\mathscr{D}(U^*) = U(0)^{\perp}$. ◇

Proposition 2.15.2 [2, Proposition III.7.2] *Let* X, Y *be two normed spaces and* $U \in L\mathscr{R}(X, Y)$. *Then,* $\alpha(U^*) = \overline{\beta}(U)$. ◇

Proposition 2.15.3 [2, Proposition, II.3.2] *Let* $V \in L\mathscr{R}(X, Y)$. *Then,*

(i) V *is open if, and only if,* $R(V^*) = N(V)^{\perp}$.
(ii) If V *is open, then* $\gamma(V) = \gamma(V^*)$. ◇

Proposition 2.15.4 [2, Proposition III.1.3] *Let* X, Y *be two normed spaces and* $V \in L\mathscr{R}(X, Y)$. *Then,*

(i) $(\overline{V})^* = V^*$.
(ii) $(V^*)^{-1} = (V^{-1})^*$.
(iii) $(\lambda V)^* = \lambda V^*$ $(\lambda \neq 0)$. ◇

Lemma 2.15.1 [2, Proposition III.1.2] *Let* X, Y *be two normed spaces and* $U \in L\mathscr{R}(X, Y)$. *Then,* $U^* \in C\mathscr{R}(Y^*, X^*)$. ◇

Remark 2.15.1 Let X, Y be two normed spaces and $V \in C\mathscr{R}(X, Y)$. Then,

$$V^{**} = V.$$

Indeed, we have $G((V^*)^*) = G(-(V^*)^{-1})^{\top}$. Then, it follows from Proposition 2.15.4 *(ii)* and *(iii)* that

$$\begin{aligned} G((V^*)^*) &= G((-V^{-1})^*)^{\top} \\ &= (G(-(-V^{-1})^{-1})^{\perp})^{\top} \\ &= (G(V)^{\perp})^{\top}. \end{aligned}$$

This implies Remark 2.1.1 *(ii)* that $G((V^*)^*) = \overline{G(V)}$. Since $V \in C\mathscr{R}(X, Y)$, then $G((V^*)^*) = G(V)$. ◇

Theorem 2.15.1 *Let* $V \in L\mathscr{R}(X, Y)$. *Then, the following properties are equivalent*

(i) $R(\widetilde{V})$ *is closed, where* $\widetilde{V}$ *is the completion of* V,
(ii) $R(V^*)$ *is closed,*
(iii) $R(\widetilde{V}^*)$ *is weak*-closed.* ◇

As a direct consequence of Theorem 2.15.1 and Remark 2.11.4, we infer the following result.

Corollary 2.15.1 *Let X, Y be two complete spaces and $U \in C\mathscr{R}(X, Y)$. Then, $R(U)$ is closed if, and only if, $R(U^*)$ is closed.* ◇

Lemma 2.15.2 *Let $U \in L\mathscr{R}(X)$ and S be a continuous linear relation such that $V(0) \subset \overline{U(0)}$ and $\mathscr{D}(V) \supset \mathscr{D}(U)$. Then, S^* is continuous, $V^*(0) \subset V^*(0)$ and $\mathscr{D}(V^*) \supset \mathscr{D}(U^*)$.* ◇

Proof By Proposition 2.15.1 (vi), we have $\|V^*\| = \|V\|$. Thus, V^* is continuous. Consequently,

$$\begin{aligned}
\mathscr{D}(V^*) &= V(0)^{\perp} \\
&\supset \overline{V(0)}^{\perp} \\
&\supset \overline{V}(0)^{\perp} \\
&= \mathscr{D}(V^*)^{\top\perp} \\
&= \overline{\mathscr{D}(V^*)},
\end{aligned}$$

and by applying Proposition 2.15.1 (iv), we have

$$\begin{aligned}
V^*(0) &= \mathscr{D}(V)^{\perp} \\
&\subset \mathscr{D}(U)^{\perp} \\
&= U^*(0).
\end{aligned}$$

This completes the proof. Q.E.D.

Lemma 2.15.3 [26, Theorem 3.1] *Let X, Y be two Banach spaces, U and $V \in L\mathscr{R}(X, Y)$. Suppose that $\mathscr{D}(VU)$ is a core of V. Then, $(VU)^* = U^*V^*$ if, and only if, U^* is $(UV)^*$-bounded.* ◇

Lemma 2.15.4 *Let X, Y be two Banach spaces, U and $V \in L\mathscr{R}(X, Y)$. If U is V-co-continuous, then there are $a, b > 0$ such that*

$$\|y'U\| \leq a\|y'V\| + b\|y'\|, \text{ for all } y' \in \mathscr{D}(V^*). \quad ◇$$

Proposition 2.15.5 [2, Proposition III.1.5] *Let X, Y be two normed spaces and $U, V \in L\mathscr{R}(X, Y)$. Then,*

(i) $G(V^* + U^*) \subset G((V + U)^*)$.
(ii) If $\mathscr{D}(U) \subset \mathscr{D}(V)$, and V is continuous, then $V^* + U^* = (V + U)^*$. ◇

Theorem 2.15.2 *1. Let $U \in L\mathscr{R}(X, Y)$ and $V \in L\mathscr{R}(Y, Z)$. Then, $G(U^*V^*) \subset G((VU)^*)$. Furthermore, if either*

(i) $R(U^*) = X^*$ and $\mathscr{D}(V) \subset R(U)$, or
(ii) $\mathscr{D}(V^*) = Z^*$ and $R(U) \subset \mathscr{D}(V)$,

then

$$(VU)^* = U^*V^*.$$

2. Let M be a finite codimensional subspace of $\mathscr{D}(U)$, closed in $\mathscr{D}(U)$. Then,

$$(UJ_M^X)^* = (J_M^X)^*U^*. \qquad \diamond$$

Proof Exercise Q.E.D.

Additionally, by using Proposition 2.15.1 (iv) that the adjoint of a linear operator is single-valued if, and only if, it is densely defined.

Example 2.15.1 Let $X = L^2([-a, a])$, we consider the following single-valued operator defined by

$$\begin{aligned} A : \mathscr{D}(A) \subset X &\longrightarrow X \\ \psi &\longrightarrow A(\psi) \end{aligned}$$

with if $\psi \in \mathscr{D}(A)$, we have

$$\begin{cases} \psi(x) = 0 & \forall x \in [-a, 0] \\ x\psi(x) \in L^2([-a, a]) & \forall x \in [0, a]. \end{cases}$$

It is clear that $\overline{\mathscr{D}(A)} \neq X$. Further,

$$\begin{aligned} G(A^*) &:= \{(f, g) \in X^2 : \langle g, h\rangle = \langle f, k\rangle \ \forall\, (h, k) \in G(A)\} \\ &= \{(f, g) \in X^2 : \langle g, h\rangle = \langle f, k\rangle \ h \in \mathscr{D}(A);\ k = A(h)\} \\ &= \{(f, g) \in X^2 : \langle g, h\rangle = \langle f, A(h)\rangle \ \forall h \in \mathscr{D}(A)\}. \end{aligned}$$

On the other hand,

$$\begin{aligned} \langle f, A(h)\rangle &:= \int_{-a}^{a} f(x)(Ah)(x)dx \\ &= \int_{-a}^{0} f(x)(Ah)(x)dx + \int_{0}^{a} f(x)(Ah)(x)dx \\ &= 0 + \int_{0}^{a} f(x)xh(x)dx \\ &= \int_{0}^{a} xf(x)h(x)dx. \end{aligned}$$

Therefore, $(A^* f)(x) = xf(x)\ \forall x \in [0, a]$ and it is arbitrary in $[-a, 0]$. Let

$$F := \{\psi \in X \ : \ \psi(x) = 0 \text{ if } x \in [0, a]\}.$$

Then, A^* is the linear relation defined by

$$(A^*h)(x) = xh(x) + F \ \ \forall h \in \mathscr{D}(A^*),$$

with

$$\begin{aligned}\mathscr{D}(A^*) &:= \{\psi \in X : (A^*)(\psi) \neq \emptyset\} \\ &= \{\psi \in X : x\psi(x) + F \neq \emptyset \, \forall \psi \in \mathscr{D}(A), \; x \in [-a, a]\} \\ &= X \bigcap \mathscr{D}(A).\end{aligned}$$

Thus,

$$\mathscr{D}(A^*) := \{\psi \in X : \exists c \geq 0 \, \forall h \in \mathscr{D}(A), \; |\langle \psi, Ah \rangle| \leq c\|h\|_X\}.$$

2.15.2 Hermitian Linear Relation

Let H be a Hilbert space. U is said to be an Hermitian relation if $U \subset U^*$, and U is said to be a self-adjoint relation if $U = U^*$.

Definition 2.15.1 Let H be a vector space. A linear involution is a linear operator U acting from H into H, such that U^2 is the identity operator on H. ◇

Remark 2.15.2 Let $Q : H \longrightarrow H$ be a bounded involutive single-valued linear operator such that $QQ^* = Q^*Q = I$ and $U \in L\mathscr{R}(H)$. If U is a self-adjoint and $QU = UQ$, then $U = U^* = U^Q$, where U^Q is given in (2.11.12). ◇

Proposition 2.15.6 *U is an Hermitian relation if, and only if, the quadratic form of U, which is defined by*

$$\langle Ux, x \rangle = \left\{ \langle y, x \rangle \text{ such that } \begin{pmatrix} x \\ y \end{pmatrix} \in G(U) \right\},$$

takes real values. ◇

Proof We suppose that U is an Hermitian relation. Hence, for all $x \in \mathscr{D}(U)$, we have $Ux = U^*x$. It follows from the definition of the adjoint relation on Hilbert space that

$$\langle U^*x, x \rangle = \langle x, y \rangle, \text{ for all } y \in Ux.$$

Hence, for all $y \in Ux$, we have

$$\begin{aligned}\langle Ux, x \rangle &= \langle U^*x, x \rangle \\ &= \langle x, y \rangle \\ &= \overline{\langle Ux, x \rangle}.\end{aligned}$$

Conversely, suppose that $\begin{pmatrix} x \\ y \end{pmatrix} \in G(U)$. Since the quadratic form of U takes real values, then for all $y \in Ux$, we have

$$\begin{aligned}\langle x, y\rangle &= \overline{\langle x, y\rangle}\\ &= \langle y, x\rangle.\end{aligned}$$

Then, $\begin{pmatrix} x \\ y \end{pmatrix} \in G(U^*)$. As a result, $G(U) \subset G(U^*)$, as desired. Q.E.D.

Lemma 2.15.5 *Let $U \in L\mathscr{R}(X)$. Then, U is Hermitian if, and only if, for all u, $v \in \mathscr{D}(U)$, and for all $x \in Uu$, $y \in Uv$, we have $\langle x, v\rangle = \langle u, y\rangle$.* ◇

Proof Let u, $v \in \mathscr{D}(U)$, $x \in Uu$, and $y \in Uv$. We have

$$\begin{aligned}(u, x) &\in G(U) \subset G(U^*),\\ (v, y) &\in G(U) \subset G(U^*).\end{aligned}$$

Since $(u, x) \in G(U^*)$ and $(v, y) \in G(U)$, then by the definition of the adjoint, we get $\langle x, v\rangle = \langle u, y\rangle$ as desired. For the reverse implication, let $\begin{pmatrix} x \\ y \end{pmatrix} \in G(U)$ and let $\begin{pmatrix} z \\ t \end{pmatrix} \in G(U)$. Thus, $y \in Ux$ and $t \in Uz$. So, by hypothesis, we have

$$\langle y, z\rangle = \langle x, t\rangle.$$

Hence, $\begin{pmatrix} x \\ y \end{pmatrix} \in G(U^*)$. Q.E.D.

Lemma 2.15.6 *Let $U \in L\mathscr{R}(X)$. Then, U is Hermitian if, and only if, for all $x \in \mathscr{D}(U)$ and $y \in Ux$, we have $\langle y, x\rangle$ is real.* ◇

Proof Let $x \in \mathscr{D}(U)$ and $y \in Ux$. We claim that $\langle y, x\rangle \in \mathbb{R}$. Indeed, as $\begin{pmatrix} x \\ y \end{pmatrix} \in G(U)$ and U is Hermitian, then $\begin{pmatrix} x \\ y \end{pmatrix} \in G(U^*)$. Hence,

$$\langle y, x\rangle = \langle x, y\rangle.$$

So,

$$\langle y, x\rangle = \overline{\langle y, x\rangle}.$$

Therefore, $\langle y, x\rangle$ is real. For the reverse implication, let u, $v \in \mathscr{D}(U)$ and x, y be such that $x \in Uu$ and $y \in Uv$. We shows that $\langle x, v\rangle = \langle u, y\rangle$. To do this, we make the following calculation.

$$\begin{aligned}\langle x, v\rangle - \langle u, y\rangle =& \langle x, v - u\rangle + \langle x, u\rangle - \langle u - v, y\rangle - \langle v, y\rangle\\ =& -\langle x - y, u - v\rangle + \overline{\langle v - u, y\rangle} + \langle v - u, y\rangle + \langle x, u\rangle - \langle v, y\rangle.\end{aligned}$$

On the other hand, since $x \in Uu$, $y \in Uv$ and $x - y \in t(u - v)$, then by hypothesis, we get that $\langle x, v\rangle - \langle u, y\rangle$ is real. On the other hand, we note that the previous calculation remains true if we replace (x, u) by (ix, iu). This shows that

$$\langle ix, v\rangle - \langle iu, y\rangle = i(\langle x, v\rangle - \langle u, y\rangle)$$

is also real. Hence, we get that $\langle x, v\rangle - \langle u, y\rangle = 0$ as claimed. Now, using Lemma 2.15.5, we get that U is Hermitian. Q.E.D.

Lemma 2.15.7 *Let H be an Hilbert space and $U \in L\mathscr{R}(H)$. If U is bounded, then for all $n \in \mathbb{N}^*$,*

$$(U^n)^* = (U^*)^n. \qquad \diamond$$

Proof As $\mathscr{D}(U) = H$, then $\mathscr{D}(U^2) = H$. Also, $(U^2)^*$ is a single-valued relation. Let $y \in H$, then

$$\langle (U^2)^* x, y\rangle = \langle x, t\rangle$$

for all $t \in U^2(y)$. Let $t \in U^2(y)$, then there exists z such that $(y, z) \in G(U)$ and $(z, t) \in G(U)$. Hence,

$$\begin{aligned}\langle x, t\rangle &= \langle U^* x, z\rangle \\ &= \langle U^*(U^* x), y\rangle.\end{aligned}$$

Consequently, $(U^2)^* = (U^*)^2$. Proceeding by induction, we can prove the result for each $n \geq 3$. Q.E.D.

Lemma 2.15.8 *Let $Q : H \longrightarrow H$ be a bounded involutive single-valued linear operator such that $QQ^* = Q^*Q = I$. Then,*

(i) $\langle Qx, Qy\rangle = \langle y, x\rangle$, for all $x, y \in H$.

(ii) Q is a self-adjoint and Q^* is a involutive.

(iii) Q is invertible and $Q^{-1} \in \mathscr{L}(H)$. $\diamond$

Proof (i) Supported on the hypothesis $Q^*Q = I$, we infer that

$$\begin{aligned}\langle Qx, Qy\rangle &= \langle x, Q^*Qy\rangle \\ &= \langle y, x\rangle, \quad \text{for all } x, \; y \in H.\end{aligned}$$

(ii) The fact that $Q^2 = I$ implies that

$$Q^* = Q^2 Q^*$$

$$= Q\underbrace{QQ^*}_{=I}$$
$$= Q.$$

This implies that

$$\begin{aligned}(Q^*)^2 &= Q^*Q^* \\ &= Q^*Q \quad (\text{as } Q^* = Q) \\ &= I.\end{aligned}$$

(iii) Since $H = R(I) = R\left(Q^2\right) \subseteq R(Q)$, then Q is surjective. In addition, we have $N(Q) \subseteq N\left(Q^2\right) = N(I) = \{0\}$. Hence, Q is invertible. Since Q is onto, then there exists a constant $c > 0$ such that

$$Q(B_H(0, 1)) \supset B_H(0, c).$$

Now, by assumption that Q is one-to-one imply that if $h \in H$ is chosen so that $\|Qh\| < c$, then $\|h\| < 1$. By homogeneity, we find that

$$\|h\| < \frac{1}{c}\|Qh\| \;\; \forall h \in H$$

and therefore Q^{-1} is a bounded single-valued operator. Q.E.D.

Lemma 2.15.9 *Let $Q : H \longrightarrow H$ be a bounded involutive single-valued linear operator such that $QQ^* = Q^*Q = I$ and $U \in L\mathscr{R}(H)$. If $U^Q = QUQ$, then*

(i) $QU^QQ = U$.
(ii) $U = U^Q$ if, and only if, $QU = UQ$. ◇

Proof (i) We have

$$\begin{aligned}QU^QQ &= QQUQQ \\ &= U \;(\text{since } Q^2 = I).\end{aligned}$$

(ii) We assume that $U = U^Q$. Let $(x, y) \in G(QU)$. Then, there exists $z \in R(U)$ such that $(x, z) \in G(U)$ and $(z, y) \in G(Q)$. The fact that $U = U^Q = QUQ$ implies that there exists $u \in R(UQ)$ such that $(x, u) \in G(UQ)$ and $(u, z) \in G(Q)$. Moreover, we have

$$y = Qz = Q(Qu) = u \;(\text{since } Q^2 = I).$$

Hence, $(x, y) \in G(UQ)$. In order to prove the opposite inclusion, it is sufficient to proceed by the same reasoning as $G(QU) \subset G(UQ)$. Conversely, assume that $QU = UQ$. Then,

$$\begin{aligned} U^Q &= QUQ \\ &= QQU \\ &= U \text{ (since } Q^2 = I). \end{aligned}$$

This completes the proof. Q.E.D.

2.15.3 *Self-adjointness of* 2 × 2 *Block Matrix of Linear Relations*

Theorem 2.15.3 *Let* $A \in C\mathscr{R}(H)$, $D \in C\mathscr{R}(K)$, $B \in L\mathscr{R}(K, H)$ *and* $C \in L\mathscr{R}(H, K)$ *be four continuous linear relations such that* $B(0) \subset A(0)$ *and* $C(0) \subset D(0)$. *Then, the adjoint of* $\mathscr{L}_0$ *is a closed linear relation from* $A(0)^{\perp} \times D(0)^{\perp}$ *into* $H \times K$ *and defined by*

$$\mathscr{L}_0^* = \begin{pmatrix} A^* & C^* \\ B^* & D^* \end{pmatrix}. \tag{2.15.1}$$

◇

Proof We carve this proof into two steps.

Step 1. We show that $\mathscr{D}(\mathscr{L}_0^*) = A(0)^{\perp} \times D(0)^{\perp}$. Let $\begin{pmatrix} x \\ y \end{pmatrix} \in \mathscr{D}(\mathscr{L}_0)$, then

$$\begin{aligned} \left\| \mathscr{L}_0 \begin{pmatrix} x \\ y \end{pmatrix} \right\| &= \left\| \begin{pmatrix} Ax + By \\ Cx + Dy \end{pmatrix} \right\| \\ &= \|Ax + By\| + \|Cx + Dy\| \quad \text{(by Remark 2.8.4)} \\ &\leq \|Ax\| + \|By\| + \|Cx\| + \|Dy\| \\ &\leq \max\{\|A\|, \|B\|, \|C\|, \|D\|\} \left\| \begin{pmatrix} x \\ y \end{pmatrix} \right\|. \end{aligned}$$

This yields that $\mathscr{L}_0$ is continuous. Hence, from Proposition 2.15.1, we have

$$\begin{aligned} \mathscr{D}(\mathscr{L}_0^*) &= \mathscr{L}_0 \begin{pmatrix} 0 \\ 0 \end{pmatrix}^{\perp} \\ &= \begin{pmatrix} A(0) + B(0) \\ D(0) + C(0) \end{pmatrix}^{\perp} \\ &= \begin{pmatrix} A(0) \\ D(0) \end{pmatrix}^{\perp} \quad \text{(by (1.8.15) and (1.8.16))} \\ &= A(0)^{\perp} \times D(0)^{\perp} \quad \text{(by Lemma 1.1.2 } (ii)). \end{aligned} \tag{2.15.2}$$

The use of both (2.15.2) and Proposition 2.15.1, we deduce that

$$\mathscr{D}(\mathscr{L}_0^*) = \mathscr{D}(A^*) \times \mathscr{D}(D^*). \tag{2.15.3}$$

The fact that $B(0) \subset A(0)$ and $C(0) \subset D(0)$ implies that $\mathscr{D}(A^*) \subset \mathscr{D}(B^*)$ and $\mathscr{D}(D^*) \subset \mathscr{D}(C^*)$. Thus,

$$\mathscr{D}(\mathscr{L}_0^*) = \mathscr{D}\left[\begin{pmatrix} A^* & C^* \\ B^* & D^* \end{pmatrix}\right].$$

Step 2. We are going to prove that (2.15.1) holds. Let $\left[\begin{pmatrix} x^* \\ y^* \end{pmatrix}, \begin{pmatrix} z^* \\ t^* \end{pmatrix}\right] \in G(\mathscr{L}_0^*)$. By using the definition of the adjoint relation on Hilbert space, we get

$$\begin{aligned}
&\left\{\left\langle \begin{pmatrix} z^* \\ t^* \end{pmatrix}, \begin{pmatrix} x \\ y \end{pmatrix}\right\rangle = \left\langle \begin{pmatrix} x^* \\ y^* \end{pmatrix}, \begin{pmatrix} z \\ t \end{pmatrix}\right\rangle \text{ such that } \begin{pmatrix} z \\ t \end{pmatrix} \in \mathscr{L}_0 \begin{pmatrix} x \\ y \end{pmatrix}\right\} \\
&= \left\{\left\langle \begin{pmatrix} z^* \\ t^* \end{pmatrix}, \begin{pmatrix} x \\ y \end{pmatrix}\right\rangle = \left\langle \begin{pmatrix} x^* \\ y^* \end{pmatrix}, \begin{pmatrix} z \\ t \end{pmatrix}\right\rangle \text{ such that } \begin{matrix} z \in Ax + By \\ t \in Cx + Dy \end{matrix}\right\} \\
&= \left\{\left\langle \begin{pmatrix} z^* \\ t^* \end{pmatrix}, \begin{pmatrix} x \\ y \end{pmatrix}\right\rangle = \left\langle \begin{pmatrix} x^* \\ y^* \end{pmatrix}, \begin{pmatrix} z_1 + z_2 \\ t_1 + t_2 \end{pmatrix}\right\rangle \text{ such that } \begin{matrix} (x, z_1) \in G(A) \\ (y, z_2) \in G(B) \\ (x, t_1) \in G(C) \\ (y, t_2) \in G(D) \end{matrix}\right\}.
\end{aligned}$$

On the other hand, for $z_1 \in Ax,\ z_2 \in By,\ t_1 \in Cx$, and $t_2 \in Dy$, we have

$$\begin{aligned}
\left\langle \begin{pmatrix} x^* \\ y^* \end{pmatrix}, \begin{pmatrix} z_1 + z_2 \\ t_1 + t_2 \end{pmatrix}\right\rangle &= \langle x^*, z_1 + z_2\rangle + \langle y^*, t_1 + t_2\rangle \\
&= \langle x^*, z_1\rangle + \langle x^*, z_2\rangle + \langle y^*, t_1\rangle + \langle y^*, t_2\rangle.
\end{aligned}$$

In addition, for $\begin{pmatrix} x^* \\ y^* \end{pmatrix} \in \mathscr{D}(\mathscr{L}_0^*) = \mathscr{D}\left[\begin{pmatrix} A^* & C^* \\ B^* & D^* \end{pmatrix}\right]$, we obtain from the definition of $G(A^*)$, $G(B^*)$, $G(C^*)$ and $G(D^*)$ on a Hilbert space that

$$\begin{aligned}
\left\langle \begin{pmatrix} z^* \\ t^* \end{pmatrix}, \begin{pmatrix} x \\ y \end{pmatrix}\right\rangle &= \langle A^*x^*, x\rangle + \langle B^*x^*, y\rangle + \langle C^*y^*, x\rangle + \langle D^*y^*, y\rangle \\
&= \langle A^*x^* + C^*y^*, x\rangle + \langle B^*x^* + D^*y^*, y\rangle \\
&= \left\langle \begin{pmatrix} A^*x^* + C^*y^* \\ B^*x^* + D^*y^* \end{pmatrix}, \begin{pmatrix} x \\ y \end{pmatrix}\right\rangle \\
&= \left\langle \begin{pmatrix} A^* & C^* \\ B^* & D^* \end{pmatrix}\begin{pmatrix} x^* \\ y^* \end{pmatrix}, \begin{pmatrix} x \\ y \end{pmatrix}\right\rangle.
\end{aligned}$$

This leads us to deduce that

$$\left\{\left\langle \begin{pmatrix} z^* \\ t^* \end{pmatrix}, \begin{pmatrix} x \\ y \end{pmatrix} \right\rangle \text{ such that } \begin{pmatrix} x^* \\ y^* \end{pmatrix} \in \mathscr{D}(\mathscr{L}_0^*) \text{ and } \begin{pmatrix} z^* \\ t^* \end{pmatrix} \in \mathscr{L}_0^* \begin{pmatrix} x^* \\ y^* \end{pmatrix}\right\}$$
$$= \left\{\left\langle \begin{pmatrix} A^* & C^* \\ B^* & D^* \end{pmatrix} \begin{pmatrix} x^* \\ y^* \end{pmatrix}, \begin{pmatrix} x \\ y \end{pmatrix} \right\rangle \text{ such that } \begin{pmatrix} x^* \\ y^* \end{pmatrix} \in \mathscr{D}\left[\begin{pmatrix} A^* & C^* \\ B^* & D^* \end{pmatrix}\right]\right\}.$$

Consequently, from Proposition 2.7.3 (ii), we infer that (2.15.1) holds. Q.E.D.

Proposition 2.15.7 *Let $A \in L\mathscr{R}(H)$, $B \in L\mathscr{R}(K)$ and $C \in L\mathscr{R}(K, H)$. Then, the adjoint of $\mathscr{M}_C$ is the single-valued relation acting on $H \bigoplus K$ of the form*

$$\mathscr{M}_C^* = \begin{pmatrix} A^* & 0 \\ C^* & B^* \end{pmatrix}. \qquad \diamond$$

Proof Let $\begin{pmatrix} u \\ v \end{pmatrix}$ and $\begin{pmatrix} x \\ y \end{pmatrix} \in H \times K$. By the definition of the adjoint relation on Hilbert spaces, we have

$$\left\langle \mathscr{M}_C^* \begin{pmatrix} u \\ v \end{pmatrix}, \begin{pmatrix} x \\ y \end{pmatrix} \right\rangle = \left\langle \begin{pmatrix} u \\ v \end{pmatrix}, \begin{pmatrix} x' \\ y' \end{pmatrix} \right\rangle$$
$$= \langle u, x' \rangle + \langle v, y' \rangle.$$

Since $x' \in Ax + Cy$, then there exist $x_1 \in Ax$ and $x_2 \in Cy$ such that $x' = x_1 + x_2$. Hence,

$$\begin{aligned} \langle u, x' \rangle + \langle v, y' \rangle &= \langle u, x_1 \rangle + \langle u, x_2 \rangle + \langle v, y' \rangle \\ &= \langle A^*u, x \rangle + \langle C^*u, y \rangle + \langle B^*v, y \rangle \\ &= \langle A^*u, x \rangle + \langle C^*u + B^*v, y \rangle \\ &= \left\langle \begin{pmatrix} A^* & 0 \\ C^* & B^* \end{pmatrix} \begin{pmatrix} u \\ v \end{pmatrix}, \begin{pmatrix} x \\ y \end{pmatrix} \right\rangle. \end{aligned}$$

This achieves the proof. Q.E.D.

Let X and Y be two Hilbert spaces with inner product $\langle \cdot, \cdot \rangle$. Then, the product space $X \times Y$ is still a Hilbert space with the following induced inner product, still denoted by $\langle \cdot, \cdot \rangle$ without any confusion

$$\left\langle \begin{pmatrix} x \\ y \end{pmatrix}, \begin{pmatrix} z \\ t \end{pmatrix} \right\rangle = \langle x, z \rangle + \langle y, t \rangle,$$

for all $\begin{pmatrix} x \\ y \end{pmatrix}, \begin{pmatrix} z \\ t \end{pmatrix} \in X \times Y$.

Remark 2.15.3 Let $A \in L\mathcal{R}(H)$. Then, $H/\overline{A(0)}$ is a Hilbert space when equipped which the scalar product $\langle \cdot, \cdot \rangle_1$ defined by

$$\langle \overline{x}, \overline{y} \rangle_1 := \langle P(x), P(y) \rangle$$

for all $x, y \in H$, when P is the orthogonal projection with direction $\overline{A(0)}$. ◇

Lemma 2.15.10 *Let X and Y be two Hilbert space. Suppose that $A \in L\mathcal{R}(X)$, $C \in L\mathcal{R}(X, Y)$ and $D \in L\mathcal{R}(Y)$. If A, D are self-adjoint relations and $C = B^*$, then the quadratic form of the block matrix of the linear relation $\mathcal{L}_0$, given in (2.8.12), takes real values.* ◇

Proof Let us assume that $\left[\begin{pmatrix} x \\ y \end{pmatrix}, \begin{pmatrix} z \\ t \end{pmatrix}\right] \in G(\mathcal{L}_0)$. We propose to show that

$$\left\langle \mathcal{L}_0 \begin{pmatrix} x \\ y \end{pmatrix}, \begin{pmatrix} x \\ y \end{pmatrix} \right\rangle = \left\{ \left\langle \begin{pmatrix} z \\ t \end{pmatrix}, \begin{pmatrix} x \\ y \end{pmatrix} \right\rangle \text{ such that } \begin{pmatrix} z \\ t \end{pmatrix} \in \mathcal{L}_0 \begin{pmatrix} x \\ y \end{pmatrix} \right\},$$

takes real values. For this purpose, let $\begin{pmatrix} z \\ t \end{pmatrix} \in \begin{pmatrix} Ax + By \\ Cx + Dy \end{pmatrix}$, then there exists $z_1 \in Ax$, $z_2 \in By$, $t_1 \in Cx$ and $t_2 \in Dy$ such that $z = z_1 + z_2$ and $t = t_1 + t_2$. Hence,

$$\begin{aligned} \left\langle \begin{pmatrix} z \\ t \end{pmatrix}, \begin{pmatrix} x \\ y \end{pmatrix} \right\rangle &= \langle z, x \rangle + \langle t, y \rangle \\ &= \langle z_1 + z_2, x \rangle + \langle t_1 + t_2 y \rangle \\ &= \langle z_1, x \rangle + \langle z_2, x \rangle + \langle t_1, y \rangle + \langle t_2, y \rangle. \end{aligned}$$

Since $G(C) = G(B^*)$, then $(x, t_1) \in G(B^*)$. So, by the definition of the adjoint relation on Hilbert spaces, we obtain

$$\langle t_1, y \rangle = \langle x, z_2 \rangle,$$

for all $(y, z_2) \in G(B)$. Thus,

$$\left\langle \begin{pmatrix} x \\ y \end{pmatrix}, \begin{pmatrix} z \\ t \end{pmatrix} \right\rangle = \langle z_1, x \rangle + \overline{\langle t_1, y \rangle} + \langle t_1, y \rangle + \langle t_2, y \rangle. \tag{2.15.4}$$

Since A and D are self-adjoint relations, then by using (2.15.4) and Proposition 2.15.6, we conclude that the quadratic form of $\mathcal{L}_0$ takes real values. Q.E.D.

Remark 2.15.4 For a closed linear relation D in a Hilbert space Y, we have the following decomposition:

$$G(D) = G(D_S) \bigoplus G(D_\infty), \tag{2.15.5}$$

where $G(D_\infty) = \{(0, y) \in Y^2 : (0, y) \in G(D)\}$ and $G(D_S) = G(D) \bigoplus G(D_\infty)$ (see [22]). Since D is self-adjoint in Y, then by using [27, p. 26], we infer that D_S is also self-adjoint in $D(0)^\perp$.

We assume that

$$\mathscr{L}_0 = \begin{pmatrix} A & B \\ B^* & D \end{pmatrix}, \tag{2.15.6}$$

where $B \in C\mathscr{R}(Y, X)$, A, and D are self-adjoint relation in X and Y, respectively. Therefore, by applying Lemma 2.15.10 and Proposition 2.15.6, we have $\mathscr{L}_0$ is an Hermitian relation. However, it is not necessarily self-adjoint. For this reason, the following result is meaningful.

Theorem 2.15.4 *Let $A \in B\mathscr{R}(X)$ and $D \in B\mathscr{R}(X)$ be self-adjoint relations such that* $\dim(A(0)) < \infty$. *Let us assume that $B \in C\mathscr{R}(X)$ with B^* and B are relatively bounded with respect to A and D with A-bound and D-bound δ_{B^*} and δ_B, respectively. If $B(0) \subset A(0)$ and $\delta_{B^*}\delta_B < 1$, then the linear relation $\mathscr{L}_0$, given in (2.15.6), is self-adjoint.* ◇

Proof Since A and D are self-adjoint, then $\rho(A)$ and $\rho(D)$ are not empty sets and contain the set $\mathbb{C}\backslash\mathbb{R}$. In what follows, we will assume that $\lambda \in \rho(A) \bigcap \rho(D)$. On the one hand, since B and $(A - \lambda)^{-1}$ are closed with $\mathscr{D}(A)$ is closed and $\dim(A(0)) < \infty$, then $(A - \lambda)^{-1}B$ is closed. On the other hand,

$$R(B) \subset X = \mathscr{D}((A - \lambda)^{-1})$$

and

$$\begin{aligned} \mathscr{D}(((A - \lambda)^{-1})^*) &= \mathscr{D}(((A - \lambda)^*)^{-1}) \\ &= \mathscr{D}((A^* - \lambda)^{-1}) \\ &= X^*. \end{aligned}$$

Hence,

$$((A - \lambda)^{-1}B)^* = B^*(A - \lambda)^{-1}. \tag{2.15.7}$$

The fact that B^* is A-bounded implies that there exist two constants $a_{B^*}, b_{B^*} \geq 0$ such that

$$\|B^*(A - \lambda)^{-1}x\| \leq a_{B^*}\|(A - \lambda)^{-1}x\| + b_{B^*}\|A(A - \lambda)^{-1}x\|, \text{ for all } x \in X. \tag{2.15.8}$$

Since $\lambda \in \rho(A)$, then it follows from both Eq. (2.15.7) and (2.15.8) that

$$\begin{aligned} \|B^*(A - \lambda)^{-1}x\| &\leq a_{B^*}\|(A - \lambda)^{-1}x\| + b_{B^*}\|A\|\|(A - \lambda)^{-1}x\| \\ &\leq \left[(a_{B^*} + b_{B^*}\|A\|)\,\|(A - \lambda)^{-1}\|\right]\|x\|. \end{aligned}$$

Since $A \in B\mathscr{R}(X)$ and self-adjoint, then by using Eq. (2.15.7) that $B^*(A-\lambda)^{-1}$ is continuous. Therefore, $(A-\lambda)^{-1}B$ is continuous. At this level, since $B(0) \subset A(0)$, we infer that

$$\begin{aligned}(A-\lambda)^{-1}B(0) &\subset (A-\lambda)^{-1}(A-\lambda)(0)\\ &= (A-\lambda)^{-1}(0)\\ &= \{0\}.\end{aligned}$$

This implies that $(A-\lambda)^{-1}B$ is a bounded single-valued operator. Hence, by using Eq. (2.15.7), we have $B^*(A-\lambda)^{-1}$ is a bounded single-valued operator. Now, by applying Lemma 2.8.6, we get

$$\mathscr{L}_0 - \lambda I = \begin{pmatrix} I & 0 \\ B^*(A-\lambda I)^{-1} & I \end{pmatrix}\begin{pmatrix} A-\lambda I & 0 \\ 0 & M(\lambda) \end{pmatrix}\begin{pmatrix} I & (A-\lambda I)^{-1}B \\ 0 & I \end{pmatrix}, \tag{2.15.9}$$

where $M(\lambda) = D - \lambda I - B^*(A-\lambda I)^{-1}B$. Consequently, the first and the third factors in the right-hand side of (2.15.9) are bounded and boundedly invertible operators. Now, since $D - \lambda I$ is one-to-one, then by referring to Lemma 2.8.3, we infer that

$$M(\lambda) = \left[I + B^*(A-\lambda I)^{-1}B(D-\lambda I)^{-1}\right](D-\lambda I).$$

Now, we propose to show that $I + B^*(A-\lambda I)^{-1}B(D-\lambda I)^{-1}$ is a bijective relation. Hence, it is sufficient to prove that

$$\|B^*(A \pm it)^{-1}B(D \pm it)^{-1}\| < 1, \tag{2.15.10}$$

provided that $|t| > t_0$ and t_0 is sufficiently large. Since D is self-adjoint in Y, then D_S is also self-adjoint in $D(0)^{\perp}$. According to (2.15.5), we can write

$$G(D \pm it) = G(D_S \pm it) \bigoplus G(D_\infty).$$

Hence,

$$\begin{aligned}\|(D \pm it)y\|^2 &= \|(D_S \pm it)y\|^2\\ &= \|D_S y\|^2 + t^2\|y\|^2\\ &= \|Dy\|^2 + t^2\|y\|^2.\end{aligned}$$

This implies that

$$\begin{aligned}\|D(D \pm it)^{-1}y\| &\le \|(D \pm it)(D \pm it)^{-1}y\|\\ &\le \|y + (D \pm it)(0)\|\\ &\le \|y\|,\end{aligned}$$

and

$$\begin{aligned}\|(D \pm it)^{-1}y\| &\leq \frac{1}{|t|}\|(D \pm it)(D \pm it)^{-1}y\| \\ &\leq \frac{1}{|t|}\|y + (D \pm it)(0)\| \\ &\leq \frac{1}{|t|}\|y\|.\end{aligned}$$

Now, by using the fact that B is D-bounded, and the above estimates, we deduce that there exist $a_B, b_B \geq 0$ such that

$$\begin{aligned}\|B(D \pm it)^{-1}y\| &\leq a_B\|(D \pm it)^{-1}y\| + b_B\|D(D \pm it)^{-1}y\| \\ &\leq \frac{a_B}{|t|}\|y\| + b_B\|y\|.\end{aligned}$$

Therefore, we obtain

$$\|B(D \pm it)^{-1}y\| \leq \left(\frac{a_B}{|t|} + b_B\right)\|y\|. \tag{2.15.11}$$

Analogously, the fact that B^* is A-bounded entails the following estimate

$$\|B^*(A \pm it)^{-1}x\| \leq \left(\frac{a_{B^*}}{|t|} + b_{B^*}\right)\|x\|. \tag{2.15.12}$$

One can take numbers b_{B^*} and b_B close to the numbers δ_{B^*} and δ_B, so that inequality $b_{B^*}b_B < 1$ preserved. Thus, the use of (2.15.11) and (2.15.12) makes us conclude that

$$\begin{aligned}\|B^*(A \pm it)^{-1}B(D \pm it)^{-1}\| &\leq \|B^*(A \pm it)^{-1}\|\|B(D \pm it)^{-1}\| \\ &< 1.\end{aligned}$$

Hence, (2.15.10) holds. This shows that $M(\lambda)$ is onto for some (and then for all) $\lambda \in \mathbb{C}\backslash\mathbb{R}$. At this level, since A is self-adjoint, then $A \pm \lambda I$ is onto for some (and then for all) $\lambda \in \mathbb{C}\backslash\mathbb{R}$. This implies that

$$\begin{pmatrix} A \pm \lambda I & 0 \\ 0 & M(\lambda) \end{pmatrix}$$

is onto for some (and then for all) $\lambda \in \mathbb{C}\backslash\mathbb{R}$. This means from Proposition 2.10.2 that $\mathscr{L}_0 \pm \lambda I$ is onto for some (and then for all) $\lambda \in \mathbb{C}\backslash\mathbb{R}$. This leads to deduce that the linear relation $\mathscr{L}_0$, given in (2.15.6), is self-adjoint. Q.E.D.

Theorem 2.15.5 *Let A and D be self-adjoint relations in X and Y, respectively. Suppose that $B \in L\mathscr{R}(Y, X)$ such that B^* and B are relatively bounded with respect to A and D with A-bound and D-bound δ_{B^*} and δ_B, respectively. If* $\max\{\delta_{B^*}, \delta_B\} < 1$, *then the linear relation $\mathscr{L}_0$, given in (2.15.6), is self-adjoint.* ◇

Proof Let us decompose the linear relation $\mathscr{L}_0$ in the following form

$$\mathscr{L}_0 = \mathscr{T} + \mathscr{S},$$

where

$$\mathscr{S} = \begin{pmatrix} 0 & B \\ B^* & 0 \end{pmatrix}$$

and

$$\mathscr{T} = \begin{pmatrix} A & 0 \\ 0 & D \end{pmatrix}$$

with $\mathscr{D}(\mathscr{T}) = \mathscr{D}(A) \times \mathscr{D}(D)$. First, we can prove that $\mathscr{T}$ is self-adjoint. Indeed, since A and D are self-adjoint, then

$$R(A - \lambda I) = R(A - \overline{\lambda} I) = X$$

and

$$R(D - \lambda I) = R(D - \overline{\lambda} I) = Y,$$

for $\lambda \in \mathbb{C}$. This implies that

$$\begin{aligned} R(\mathscr{T} - \lambda I) &= R(\mathscr{T} - \overline{\lambda} I) \\ &= R(A - \overline{\lambda} I) \times R(D - \overline{\lambda} I) \\ &= X \times Y. \end{aligned}$$

Then, $\mathscr{T}$ is self-adjoint. Second, we have to prove that $\mathscr{S}$ is a Hermitian relation. Let $\begin{pmatrix} x \\ y \end{pmatrix} \in \mathscr{D}(\mathscr{S})$ such that

$$\begin{pmatrix} z \\ t \end{pmatrix} \in \begin{pmatrix} 0 & B \\ B^* & 0 \end{pmatrix} \begin{pmatrix} x \\ y \end{pmatrix} = \begin{pmatrix} By \\ B^* x \end{pmatrix}.$$

This implies that $(y, z) \in G(B)$ and $(x, t) \in G(B^*)$. So, by the definition of the adjoint relation on Hilbert spaces, we obtain

$$\langle t, y \rangle = \langle x, z \rangle. \tag{2.15.13}$$

It follows from (2.15.13) that

$$\left\langle \begin{pmatrix} z \\ t \end{pmatrix}, \begin{pmatrix} x \\ y \end{pmatrix} \right\rangle = \langle x, z\rangle + \langle y, t\rangle$$
$$= \overline{\langle y, t\rangle} + \langle y, t\rangle.$$

Hence, the quadratic form of $\mathscr{S}$, which is defined by

$$\left\langle \mathscr{S}\begin{pmatrix} x \\ y \end{pmatrix}, \begin{pmatrix} x \\ y \end{pmatrix} \right\rangle = \left\{ \left\langle \begin{pmatrix} z \\ t \end{pmatrix}, \begin{pmatrix} x \\ y \end{pmatrix} \right\rangle \text{ such that } \begin{pmatrix} z \\ t \end{pmatrix} \in \mathscr{S}\begin{pmatrix} x \\ y \end{pmatrix} \right\},$$

takes real values. Thus, by applying Proposition 2.15.6, we deduce that $\mathscr{S}$ is Hermitian. At this level, let us show that the linear relation $\mathscr{S}$ is $\mathscr{T}$-bounded in the norm defined in (2.10.2). According to the assumptions, there exist $a_{B^*}, b_{B^*}, a_B, b_B \geq 0$ such that

$$\|B^*x\| \leq a_{B^*}\|x\| + b_{B^*}\|Ax\|, \text{ for all } x \in \mathscr{D}(A), \tag{2.15.14}$$

and

$$\|By\| \leq a_B\|y\| + b_B\|Dy\|, \text{ for all } y \in \mathscr{D}(D). \tag{2.15.15}$$

Let $\begin{pmatrix} x \\ y \end{pmatrix} \in \mathscr{D}(A) \times \mathscr{D}(D)$. Then, by using (2.15.14) and (2.15.15), we deduce that

$$\begin{aligned}
\left\| \mathscr{S}\begin{pmatrix} x \\ y \end{pmatrix} \right\| &= \left\| Q_{\mathscr{S}}\mathscr{S}\begin{pmatrix} x \\ y \end{pmatrix} \right\| \\
&= \left\| \begin{pmatrix} Q_B By \\ Q_{B^*}B^*x \end{pmatrix} \right\| \\
&= \|Q_B By\| + \|Q_{B^*}B^*x\| \\
&\leq a_B\|y\| + b_B\|Q_D Dy\| + a_{B^*}\|x\| + b_{B^*}\|Q_A Ax\| \\
&\leq \max\{a_B, a_{B^*}\}\left(\|x\| + \|y\|\right) + \max\{b_B, b_{B^*}\}\left(\|Q_A Ax\| + \|Q_D Dy\|\right) \\
&\leq \max\{a_B, a_{B^*}\}\left\| \begin{pmatrix} x \\ y \end{pmatrix} \right\| + \max\{b_B, b_{B^*}\}\left\| Q_{\mathscr{T}}\mathscr{T}\begin{pmatrix} x \\ y \end{pmatrix} \right\| \\
&\leq \max\{a_B, a_{B^*}\}\left\| \begin{pmatrix} x \\ y \end{pmatrix} \right\| + \max\{b_B, b_{B^*}\}\left\| \mathscr{T}\begin{pmatrix} x \\ y \end{pmatrix} \right\|.
\end{aligned}$$

Thus, the $\mathscr{T}$-bound of the relation $\mathscr{S}$ is equal to $\max\{b_B, b_{B^*}\} < 1$. Since $\mathscr{T}$ is self-adjoint, then the relation $\mathscr{L}_0$ is self-adjoint as well. Q.E.D.

2.15.4 The Closeness of 3 × 3 Diagonal Matrices Linear Relations

Lemma 2.15.11 *The linear relation*

$$\mathcal{T} = \begin{pmatrix} A & 0 & 0 \\ 0 & E & 0 \\ 0 & 0 & K \end{pmatrix} \tag{2.15.16}$$

is closed if, and only if, A, E, and K are closed. ◇

Proof Suppose that

$$\begin{pmatrix} A & 0 & 0 \\ 0 & E & 0 \\ 0 & 0 & K \end{pmatrix}$$

is closed. By Lemma 2.12.2 and Proposition 2.9.3, it follows that

$$\begin{pmatrix} Q_A(A) & 0 & 0 \\ 0 & Q_E(E) & 0 \\ 0 & 0 & Q_K(K) \end{pmatrix}$$

is a closed linear operator and that

$$\begin{pmatrix} \overline{A(0)} \\ \overline{E(0)} \\ \overline{K(0)} \end{pmatrix} = \begin{pmatrix} A(0) \\ E(0) \\ K(0) \end{pmatrix}.$$

Thus, $Q_A(A)$, $Q_E(E)$, and $Q_K(K)$ are closed. Further, $\overline{A(0)} = A(0)$, $E(0) = \overline{E(0)}$ and $\overline{K(0)} = K(0)$. Consequently, A, E, and K are closed linear relations. Conversely, if we suppose that A, E, and K are closed linear relations, then by using Lemma 2.12.2, it follows that $Q_A(A)$, $Q_E(E)$, and $Q_K(K)$ are closed operators and $\overline{A(0)} = A(0)$, $E(0) = \overline{E(0)}$ and $\overline{K(0)} = K(0)$. Thus, $Q_{\mathcal{T}}\mathcal{T}$ is a closed linear operator and $\mathcal{T}(0)$ is closed. Hence, by applying Lemma 2.12.2, we infer that $\mathcal{T}$ is closed. This completes the proof. Q.E.D.

2.15.5 Adjoint of 3 × 3 Matrices of Linear Relations

Let $A_1 \in BC\mathcal{R}(X_1)$, $A_2 \in BC\mathcal{R}(X_2)$, $A_3 \in BC\mathcal{R}(X_3)$, $B \in B\mathcal{R}(X_2)$, $C \in B\mathcal{R}(X_3)$, and $D \in B\mathcal{R}(X_3)$. Then,

$$\begin{pmatrix} A_1 & B & C \\ 0 & A_2 & D \\ 0 & 0 & A_3 \end{pmatrix} = \begin{pmatrix} I & 0 & 0 \\ 0 & I & 0 \\ 0 & 0 & A_3 \end{pmatrix} \begin{pmatrix} I & 0 & C \\ 0 & I & D \\ 0 & 0 & I \end{pmatrix} \begin{pmatrix} I & 0 & 0 \\ 0 & A_2 & 0 \\ 0 & 0 & I \end{pmatrix} \begin{pmatrix} I & B & 0 \\ 0 & I & 0 \\ 0 & 0 & I \end{pmatrix} \begin{pmatrix} A_1 & 0 & 0 \\ 0 & I & 0 \\ 0 & 0 & I \end{pmatrix}. \tag{2.15.17}$$

Theorem 2.15.6 *Let* $A_1 \in BC\mathscr{R}(X_1)$, $A_2 \in BC\mathscr{R}(X_2)$, *and* $A_3 \in BC\mathscr{R}(X_3)$. *Then,*

$$\begin{pmatrix} A_1 & B & C \\ 0 & A_2 & D \\ 0 & 0 & A_3 \end{pmatrix}^* = \begin{pmatrix} A_1^* & 0 & 0 \\ B^* & A_2^* & 0 \\ C^* & D^* & A_3^* \end{pmatrix},$$

where B, C, *and* D *are bounded linear relations.* ◇

Proof Let $\mathscr{M}_C = \left[\begin{pmatrix} A_1 & B & C \\ 0 & A_2 & D \\ 0 & 0 & A_3 \end{pmatrix} \right]$. Then, in view of Eq. (2.15.17), we have

$$\mathscr{M}_C^* = \left[\begin{pmatrix} I & 0 & 0 \\ 0 & I & 0 \\ 0 & 0 & A_3 \end{pmatrix} \begin{pmatrix} I & 0 & C \\ 0 & I & D \\ 0 & 0 & I \end{pmatrix} \begin{pmatrix} I & 0 & 0 \\ 0 & A_2 & 0 \\ 0 & 0 & I \end{pmatrix} \begin{pmatrix} I & B & 0 \\ 0 & I & 0 \\ 0 & 0 & I \end{pmatrix} \begin{pmatrix} A_1 & 0 & 0 \\ 0 & I & 0 \\ 0 & 0 & I \end{pmatrix} \right]^*.$$

Since

$$\begin{pmatrix} I & B & C \\ 0 & A_2 & D \\ 0 & 0 & A_3 \end{pmatrix}$$

is bounded, so it is continuous. According to Corollary 2.10.2, we get

$$\begin{pmatrix} I & B & C \\ 0 & A_2 & D \\ 0 & 0 & A_3 \end{pmatrix}$$

is $\begin{pmatrix} I & B & C \\ 0 & A_2 & D \\ 0 & 0 & A_3 \end{pmatrix} \begin{pmatrix} A_1 & 0 & 0 \\ 0 & I & 0 \\ 0 & 0 & I \end{pmatrix}$-co-continuous. Hence,

$$\begin{pmatrix} I & B & C \\ 0 & A_2 & D \\ 0 & 0 & A_3 \end{pmatrix} \begin{pmatrix} 0 \\ 0 \\ 0 \end{pmatrix} \subset \begin{pmatrix} I & B & C \\ 0 & A_2 & D \\ 0 & 0 & A_3 \end{pmatrix} \begin{pmatrix} A_1 & 0 & 0 \\ 0 & I & 0 \\ 0 & 0 & I \end{pmatrix} \begin{pmatrix} 0 \\ 0 \\ 0 \end{pmatrix}.$$

By using Lemma 2.15.4, we have

$$\begin{pmatrix} I & B & C \\ 0 & A_2 & D \\ 0 & 0 & A_3 \end{pmatrix}^*$$

is $\left[\begin{pmatrix} I & B & C \\ 0 & A_2 & D \\ 0 & 0 & A_3 \end{pmatrix}\begin{pmatrix} A_1 & 0 & 0 \\ 0 & I & 0 \\ 0 & 0 & I \end{pmatrix}\right]^*$-bounded. Now, we show that

$$\mathscr{D}\left[\begin{pmatrix} I & B & C \\ 0 & A_2 & D \\ 0 & 0 & A_3 \end{pmatrix}\begin{pmatrix} A_1 & 0 & 0 \\ 0 & I & 0 \\ 0 & 0 & I \end{pmatrix}\right]$$

is a core of

$$\begin{pmatrix} A_1 & 0 & 0 \\ 0 & I & 0 \\ 0 & 0 & I \end{pmatrix}.$$

Indeed, since

$$\begin{aligned}\mathscr{D}\left[\begin{pmatrix} I & B & C \\ 0 & A_2 & D \\ 0 & 0 & A_3 \end{pmatrix}\begin{pmatrix} A_1 & 0 & 0 \\ 0 & I & 0 \\ 0 & 0 & I \end{pmatrix}\right] &= \mathscr{D}\left[\begin{pmatrix} A_1 & B & C \\ 0 & A_2 & D \\ 0 & 0 & A_3 \end{pmatrix}\right] \\ &= X_1 \times X_2 \times X_3,\end{aligned}$$

and A_1, A_2, A_3, B, C, and D are bounded, in addition

$$\mathscr{D}\left[\begin{pmatrix} A_1 & 0 & 0 \\ 0 & I & 0 \\ 0 & 0 & I \end{pmatrix}\right] = X \times X_2 \times X_3,$$

as A_1 is bounded, so using Definition 2.7.3, we get the result. According to Lemma 2.15.3, we get

$$\mathscr{M}_C^* = \left[\begin{pmatrix} A_1 & 0 & 0 \\ 0 & I & 0 \\ 0 & 0 & I \end{pmatrix}\begin{pmatrix} I & B & C \\ 0 & A_2 & D \\ 0 & 0 & A_3 \end{pmatrix}\right]^*.$$

Now, applying the same technique to $\begin{pmatrix} I & B & C \\ 0 & A_2 & D \\ 0 & 0 & A_3 \end{pmatrix}^*$, we infer that

$$\begin{aligned}\mathscr{M}_C^* &= \begin{pmatrix} A_1 & 0 & 0 \\ 0 & I & 0 \\ 0 & 0 & I \end{pmatrix}^*\begin{pmatrix} I & B & 0 \\ 0 & I & 0 \\ 0 & 0 & I \end{pmatrix}^*\begin{pmatrix} I & 0 & 0 \\ 0 & A_2 & 0 \\ 0 & 0 & I \end{pmatrix}^*\begin{pmatrix} I & 0 & C \\ 0 & I & D \\ 0 & 0 & I \end{pmatrix}^*\begin{pmatrix} I & 0 & 0 \\ 0 & I & 0 \\ 0 & 0 & A_3 \end{pmatrix}^* \\ &= \begin{pmatrix} A_1 & 0 & 0 \\ 0 & I & 0 \\ 0 & 0 & I \end{pmatrix}^*\begin{pmatrix} I & 0 & 0 \\ B^* & I & 0 \\ 0 & 0 & I \end{pmatrix}\begin{pmatrix} I & 0 & 0 \\ 0 & A_2 & 0 \\ 0 & 0 & I \end{pmatrix}^*\begin{pmatrix} I & 0 & 0 \\ 0 & I & 0 \\ C^* & D^* & I \end{pmatrix}\begin{pmatrix} I & 0 & 0 \\ 0 & I & 0 \\ 0 & 0 & A_3 \end{pmatrix}^*.\end{aligned}$$

It is enough to prove that

$$\begin{pmatrix} I & 0 & 0 \\ 0 & I & 0 \\ 0 & 0 & A_3 \end{pmatrix}^* = \begin{pmatrix} I & 0 & 0 \\ 0 & I & 0 \\ 0 & 0 & A_3^* \end{pmatrix}.$$

Let

$$\left[\begin{pmatrix} y_1' \\ y_2' \\ y_3' \end{pmatrix}, \begin{pmatrix} x_1' \\ x_2' \\ x_3' \end{pmatrix} \right] \in G \left[\begin{pmatrix} I & 0 & 0 \\ 0 & I & 0 \\ 0 & 0 & A_3 \end{pmatrix}^* \right].$$

Then, for all

$$\left[\begin{pmatrix} x_1 \\ x_2 \\ x_3 \end{pmatrix}, \begin{pmatrix} y_1 \\ y_2 \\ y_3 \end{pmatrix} \right] \in G \left[\begin{pmatrix} I & 0 & 0 \\ 0 & I & 0 \\ 0 & 0 & A_3 \end{pmatrix} \right],$$

we have $y_1' y_1 = x_1' x_1,\ y_2' y_2 = x_2' x_2$, and $y_3' y_3 = x_3' x_3$. Since $y_1 = x_1,\ y_2 = x_2$ and $y_3 \in A_3 x_3$ and for all $(x_3, y_3) \in G(A_3),\ y_3' y_3 = x_3' x_3$. So, we get

$$\left[\begin{pmatrix} y_1' \\ y_2' \\ y_3' \end{pmatrix}, \begin{pmatrix} x_1' \\ x_2' \\ x_3' \end{pmatrix} \right] \in G \left[\begin{pmatrix} I & 0 & 0 \\ 0 & I & 0 \\ 0 & 0 & A_3^* \end{pmatrix} \right].$$

We conclude that

$$\begin{pmatrix} I & 0 & 0 \\ 0 & I & 0 \\ 0 & 0 & A_3 \end{pmatrix}^* = \begin{pmatrix} I & 0 & 0 \\ 0 & I & 0 \\ 0 & 0 & A_3^* \end{pmatrix}.$$

By analogy on

$$\begin{pmatrix} I & 0 & 0 \\ 0 & A_2 & 0 \\ 0 & 0 & I \end{pmatrix}^*$$

and

$$\begin{pmatrix} A_1 & 0 & 0 \\ 0 & I & 0 \\ 0 & 0 & I \end{pmatrix}^*,$$

we get the proof of Theorem 2.15.6. Q.E.D.

2.16 Gap Between Two 2 × 2 Closed Multivalued Linear Operator Matrices

In this section we investigate the notion of gap of 2 × 2-block matrices of multivalued linear operators.

2.16.1 *Gap Between Two Linear Relations*

Let $V, U \in C\mathscr{R}(X, Y)$. The gap between V and U is defined by

$$\delta(V, U) := \begin{cases} \sup\limits_{\varphi \in G(V)\ \|\varphi\|=1} \operatorname{dist}(\varphi, G(U)) \text{ if } G(V) \neq \{0\} \\ 0 \qquad\qquad\qquad\qquad\qquad \text{otherwise} \end{cases}$$

and

$$\widehat{\delta}(V, U) := \max\left\{\delta(V, U), \delta(U, V)\right\}.$$

Theorem 2.16.1 *Let* $V, U \in C\mathscr{R}(X, Y)$, *then*

(i) $\delta(V, U) = \delta(V^{-1}, U^{-1})$, *and*
(ii) $\widehat{\delta}(V, U) = \widehat{\delta}(V^{-1}, U^{-1})$. ◇

Proof (*i*) Fix any $\varphi = \begin{pmatrix} x \\ y \end{pmatrix} \in G(V)$ with $\|\varphi\| = 1$, then $\varphi^{-1} = \begin{pmatrix} y \\ x \end{pmatrix} \in G(V^{-1})$ and $\|\varphi^{-1}\| = \|\varphi\| = 1$. Then, we have

$$\begin{aligned} \delta(V, U) &= \sup_{\varphi \in G(V)\ \|\varphi\|=1} \operatorname{dist}(\varphi, G(S)) \\ &= \sup_{\varphi^{-1} \in G(V^{-1})\ \|\varphi^{-1}\|=1} \operatorname{dist}(\varphi, G(S)). \end{aligned}$$

Let $\psi = \begin{pmatrix} u \\ v \end{pmatrix} \in G(U)$, so $\psi^{-1} = (v, u) \in G(U^{-1})$. For this reason

$$\begin{aligned} \|\varphi - \psi\|^2 &= \| \begin{pmatrix} x \\ y \end{pmatrix} - \begin{pmatrix} u \\ v \end{pmatrix} \|^2 \\ &= \|x - u\|^2 + \|y - v\|^2 \\ &= \|(y, x) - (v, u)\|^2 \\ &= \|\varphi^{-1} - \psi^{-1}\|^2. \end{aligned}$$

It means if

$$\operatorname{dist}(\varphi, G(U)) = \operatorname{dist}(\varphi^{-1}, G(U^{-1})).$$

As a result,

$$\delta(V, U) = \delta(V^{-1}, U^{-1}).$$

(ii) First it is clear that

$$\widehat{\delta}(V, U) = \max\{\delta(V, U), \delta(V, U)\}.$$

On the other hand, by using (i), we have

$$\max\{\delta(V, U), \delta(V, U)\} = \max\left\{\delta(V^{-1}, U^{-1}), \delta(U^{-1}, V^{-1})\right\}$$

Then,

$$\widehat{\delta}(V, U) = \widehat{\delta}(V^{-1}, U^{-1}).$$

This achieves the proof. Q.E.D.

Remark 2.16.1 By using Theorem 2.16.1, we deduce that the gap is invariant with respect to inversion. ◇

Theorem 2.16.2 *Let $V \in C\mathscr{R}(X, Y)$. If V is everywhere defined, then*

$$\delta(V, 0_X) \leq \frac{\|V\|}{(1 + \|V\|^p)^{\frac{1}{p}}}, \quad p \geq 1.$$ ◇

Proof As $V \in B\mathscr{R}(X, Y)$, then $\|Q_V V\| < \infty$, which implies that $Q_V V \in \mathscr{L}(X, Y/V(0))$. This implies that

$$\delta(Q_V V, 0_X) = \frac{\|Q_V V\|}{(1 + \|Q_V V\|^p)^{\frac{1}{p}}}.$$

Let $\varphi = \begin{pmatrix} x \\ y \end{pmatrix} \in G(V)$ be such that

$$\|\varphi\| = \|(x, y)\|^p = \|x\|^p + \|y\|^p = 1. \tag{2.16.1}$$

In that case

$$\delta(V, 0_X) = \sup_{\varphi \in G(V)\ \|\varphi\|=1} \text{dist}(\varphi, G(0_X)).$$

Since

$$\sup_{\varphi \in G(V)\ \|\varphi\|=1} \text{dist}(\varphi, G(0_X)) = \sup_{x \in X\ \|x\|^p + \|y\|^p = 1} \left[\inf_{z \in X} \|(x, y) - (z, 0)\|\right]$$

and

$$\sup_{x \in X\ \|x\|^p + \|y\|^p = 1} \left[\inf_{z \in X} \|(x, y) - (z, 0)\|\right] = \sup_{x \in X\ \|x\|^p + \|y\|^p = 1} \left[\inf_{z \in X} (\|x - z\|^p + \|y\|^p)^{\frac{1}{p}}\right],$$

we conclude that

$$\sup_{\varphi\in G(V)\ \|\varphi\|=1} \operatorname{dist}(\varphi, G(0_X)) = \sup_{x\in X\ \|x\|^p+\|y\|^p=1} \|y\|.$$

As $x \in \mathscr{D}(V) = X$ and $y \in Vx$, then for any given $\varepsilon > 0$, we have

$$\|y\| \leq \|Vx\| + \varepsilon. \tag{2.16.2}$$

This implies that

$$\|x\|^p + \|y\|^p \leq \|x\|^p + (\|Vx\| + \varepsilon)^p.$$

On the other hand

$$(\|Vx\| + \varepsilon)^p = \|Vx\|^p + \sum_{k=0}^{p-1} C_p^k \|Vx\|^k \varepsilon^{p-k}.$$

Since ε is arbitrary, then we obtain

$$\begin{aligned}\|x\|^p + \|y\|^p &\leq \|x\|^p + \|Vx\|^p \\ &= \|x\|^p + \|Q_V Vx\|^p.\end{aligned}$$

Hence, according to the condition (2.16.1), we deduce that

$$\|x\|^p + \|Q_V Vx\|^p = 1. \tag{2.16.3}$$

Let $x = cw$, where $w \in X$ and $c = \frac{1}{(1+\|Q_V Vw\|^p)^{\frac{1}{p}}}$. So, by using (2.16.3), we have

$$\sup_{x\in X\ \|x\|^p+\|Q_V Vx\|^p=1} \|Q_V Vx\| = \sup_{w\in X\ \|w\|=1} \frac{\|Q_V Vw\|}{(1+\|Q_V Vw\|^p)^{\frac{1}{p}}}.$$

Since

$$\sup_{w\in X\ \|w\|=1} \frac{\|Q_V Vw\|}{(1+\|Q_V Vw\|^p)^{\frac{1}{p}}} = \frac{\|Q_V V\|}{(1+\|Q_V V\|^p)^{\frac{1}{p}}},$$

then

$$\sup_{x\in X\ \|x\|^p+\|Q_V Vx\|^p=1} \|Q_V Vx\| = \delta(Q_V V, 0_X).$$

Thus, for any given $\varepsilon > 0$, and by using (2.16.2), we get

$$\delta(V, 0_X) = \sup_{x\in X\ \|x\|^p+\|y\|^p=1} \|y\|.$$

Since

$$\sup_{x\in X\ \|x\|^p+\|y\|^p=1} \|y\| \leq \sup_{x\in X\ \|x\|^p+\|Q_V Vx\|^p=1} \|Q_V Vx\| + \varepsilon,$$

then

$$\delta(V, 0_X) \leq \sup_{x\in X\ \|x\|^p+\|Q_V Vx\|^p=1} \|Q_V Vx\| + \varepsilon.$$

By the arbitrariness of ε, we conclude that

$$\delta(V, 0_X) \leq \delta(Q_V V, 0_X).$$

As

$$\begin{aligned}\delta(Q_V V, 0_X) &= \frac{\|Q_V V\|}{(1+\|Q_V V\|^p)^{\frac{1}{p}}}\\ &= \frac{\|V\|}{(1+\|V\|^p)^{\frac{1}{p}}},\end{aligned}$$

we have

$$\delta(V, 0_X) \leq \frac{\|V\|}{(1+\|V\|^p)^{\frac{1}{p}}}.$$

This achieves the proof. Q.E.D.

Remark 2.16.2 If V is a closed linear relation everywhere defined, then Theorem 2.16.2 proves the relationship between $\delta(V, 0_X)$ and $\delta(Q_V V, 0_X)$, that is

$$\delta(V, 0_X) \leq \delta(Q_V V, 0_X). \qquad \diamond$$

2.16.2 Relationship Between the Gap and Norm of Multivalued Linear Operator

Theorem 2.16.3 *Let $U \in L\mathcal{R}(X, Y)$. If U is a closed linear relation everywhere defined and $p = 2$, then*

$$\widehat{\delta}(U, 0_X) \leq \frac{\|U\|}{\sqrt{1+\|U\|^2}}. \qquad \diamond$$

Proof By using Theorem 2.16.2 it is sufficient to prove that

$$\delta(0_X, U) \leq \frac{\|U\|}{\sqrt{1+\|U\|^2}}.$$

Let us observe that

$$\begin{aligned}\delta(0_X, U) &= \sup_{x\in X\ \|x\|=1} \operatorname{dist}((x,0), G(U)) \\ &= \sup_{x\in X\ \|x\|=1} \left[\inf_{(y,z)\in G(U)} \|(x,0)-(y,z)\| \right] \\ &= \sup_{x\in X\ \|x\|=1} \left[\inf_{(y,z)\in G(U)} (\|x-y\|^2 + \|z\|^2)^{\frac{1}{2}} \right].\end{aligned}$$

Since $y \in \mathscr{D}(U) = X$ and $z \in Uy$, then for any given $\varepsilon > 0$, we have

$$\|z\| \leq \|Uy\| + \varepsilon.$$

So, for any $x, y \in X$, we obtain that

$$\begin{aligned}\|x-y\|^2 + \|z\|^2 &\leq \|x-y\|^2 + (\|Uy\| + \varepsilon)^2 \\ &= \|x-y\|^2 + \|Uy\|^2 + 2\varepsilon\|Uy\| + \varepsilon^2.\end{aligned}$$

Since ε is arbitrary, we get

$$(\|x-y\|^2 + \|z\|^2)^{\frac{1}{2}} \leq (\|x-y\|^2 + \|Uy\|^2)^{\frac{1}{2}}. \tag{2.16.4}$$

Consider the function

$$f(x, y) = \|x-y\|^2 + \|Uy\|^2.$$

Let $x \in X$ be such that $\|x\| = 1$ and $y = \frac{x}{1+\|Ux\|^2}$. Then,

$$\begin{aligned}f(x,y) &= f\left(x, \frac{x}{1+\|Ux\|^2}\right) \\ &= \left\| x - \frac{x}{1+\|Ux\|^2} \right\|^2 + \frac{\|Ux\|^2}{(1+\|Ux\|^2)^2} \\ &= \frac{\|x\|^2\|Ux\|^4 + \|Ux\|^2}{(1+\|Ux\|^2)^2} \\ &= \frac{\|Ux\|^2(\|x\|^2\|Ux\|^2 + 1)}{(1+\|Ux\|^2)^2} \\ &= \frac{\|Ux\|^2(\|Ux\|^2 + 1)}{(1+\|Ux\|^2)^2} \quad \text{as } \|x\| = 1 \\ &= \frac{\|Ux\|^2}{1+\|Ux\|^2}.\end{aligned}$$

Hence, by using (2.16.4), we deduce that

$$\begin{aligned}\operatorname{dist}((x,0),G(U)) &\le (\|x-y\|^2+\|z\|^2)^{\frac{1}{2}}\\ &\le \frac{\|Ux\|}{\sqrt{1+\|Ux\|^2}}.\end{aligned}$$

Also, the function $g(u)=\dfrac{u}{1+u}$ is increasing, it follows that

$$\begin{aligned}\delta(0_X,U) &= \sup_{x\in X\ \|x\|=1}\operatorname{dist}((x,0),G(U))\\ &\le \sup_{x\in X\ \|x\|=1}\frac{\|Ux\|}{\sqrt{1+\|Ux\|^2}}\\ &\le \frac{\|U\|}{\sqrt{1+\|U\|^2}}.\end{aligned}$$

Finally, we have proved the following result

$$\begin{aligned}\widehat{\delta}(U,0_X) &= \max\{\delta(U,0_X),\delta(0_X,U)\}\\ &\le \frac{\|U\|}{\sqrt{1+\|U\|^2}}.\end{aligned}$$

This achieves the proof. Q.E.D.

Next, we give a relationship between $\delta(V+E,U+E)$ and $\delta(V,U)$.

Theorem 2.16.4 *Let $U,V\in C\mathcal{R}(X,Y)$ and let $E:X\longrightarrow Y$ be a continuous linear relation satisfy that $\mathcal{D}(E)\supset\mathcal{D}(U)\bigcup\mathcal{D}(V)$ and $A(0)\subset U(0)\bigcap V(0)$. Then,*

$$\delta(V+E,U+E)\le(2+\|E\|^2)\delta(V,U), \tag{2.16.5}$$

and

$$\widehat{\delta}(V+E,U+E)\le(2+\|E\|^2)\widehat{\delta}(V,U). \tag{2.16.6}$$

◇

Proof By using Proposition 2.11.2, we have $V+E$ and $U+E$ are closed. Let $\varphi=\begin{pmatrix}x\\y\end{pmatrix}\in G(V+E)$ be with $\|\varphi\|=1$. Then, there exist $(x,y_1)\in G(V)$ and $(x,y_2)\in G(E)$ such that $y=y_1+y_2$. Hence,

$$\|\varphi\|^2=\|x\|^2+\|y_1+y_2\|^2=1. \tag{2.16.7}$$

Since $y_2\in Ax$, then for any given $\varepsilon>0$, we have

$$\begin{aligned}\|y_2\| &< \|Ex\| + \varepsilon \\ &< \|E\|\|x\| + \varepsilon.\end{aligned} \tag{2.16.8}$$

Hence, by using (2.16.8), we have

$$\begin{aligned}\|y_1\| &= \|y_1 + y_2 - y_2\| \\ &\leq \|y_1 + y_2\| + \|y_2\| \\ &\leq \|y_1 + y_2\| + \|E\|\|x\| + \varepsilon.\end{aligned}$$

Thus, we get

$$\begin{aligned}\|y_1\|^2 &\leq ((\|y_1 + y_2\| + \|E\|\|x\|) + \varepsilon)^2 \\ &\leq (\|y_1 + y_2\| + \|E\|\|x\|)^2 + 2\varepsilon\,(\|y_1 + y_2\| + \|E\|\|x\|) + \varepsilon^2.\end{aligned}$$

In addition, for any given $\varepsilon > 0$ and by Cauchy–Schwarz inequality, we obtain

$$\|y_1\|^2 \leq \left(1 + \|E\|^2\right)\left(\|x\|^2 + \|y_1 + y_2\|^2\right) + 2\varepsilon\,(\|y_1 + y_2\| + \|E\|\|x\|) + \varepsilon^2.$$

By using (2.16.7), we have

$$\|y_1\|^2 \leq \left(1 + \|A\|^2\right) + 2\varepsilon\,(\|y_1 + y_2\| + \|E\|\|x\|) + \varepsilon^2,$$

and by arbitrariness of ε, we deduce that

$$\left(\|x\|^2 + \|y_1\|^2\right)^{\frac{1}{2}} \leq \left(2 + \|E\|^2\right)^{\frac{1}{2}}. \tag{2.16.9}$$

Therefore , for any $u \in \mathscr{D}(U + E)$, there is $v \in (U + E)u$. Then, there exists $\begin{pmatrix} u \\ v_1 \end{pmatrix} \in G(U)$ and $\begin{pmatrix} u \\ v_2 \end{pmatrix} \in G(E)$ such that $v = v_1 + v_2$. Since $y_2 \in Ex$ and $v_2 \in Au$, then we infer that $y_2 - v_2 \in E(x - u)$. Hence, for any given $\varepsilon_0 > 0$, we have

$$\|y_2 - v_2\| \leq \|E\|\|x - u\| + \varepsilon_0. \tag{2.16.10}$$

It follows from (2.16.10), we can get

$$\begin{aligned}\|y_1 + y_2 - v_1 - v_2\| &\leq \|y_1 - v_1\| + \|y_2 - v_2\| \\ &\leq \|y_1 - v_1\| + \|E\|\|x - u\| + \varepsilon_0.\end{aligned}$$

Similarly, for all $u \in \mathscr{D}(U + E) = \mathscr{D}(U)$ and by arbitrariness of ε_0, we have

$$\left(\|x - u\|^2 + \|y_1 + y_2 - v_1 - v_2\|^2\right)^{\frac{1}{2}} \leq \left(2 + \|A\|^2\right)^{\frac{1}{2}}\left(\|x - u\|^2 + \|y_1 - v_1\|^2\right)^{\frac{1}{2}}.$$

We not that $\psi = \begin{pmatrix} u \\ v \end{pmatrix} \in G(U + A)$, $\varphi_1 = (x, y_1) \in G(S)$ and $\psi_1 = \begin{pmatrix} u \\ v_1 \end{pmatrix} \in G(U)$, we get

$$\operatorname{dist}(\varphi, G(U + E)) \leq \left(2 + \|E\|^2\right)^{\frac{1}{2}} \operatorname{dist}(\varphi_1, G(U)). \tag{2.16.11}$$

Therefore, by using (2.2.2), (2.16.9), and (2.16.11) that

$$\begin{aligned}\delta(U + E, V + E) = & \sup_{\substack{\varphi \in G(V+E) \\ \|x\|^2 + \|y_1 + y_2\|^2 = 1}} \operatorname{dist}(\varphi, G(U + E)) \\ & \leq \left(2 + \|E\|^2\right)^{\frac{1}{2}} \sup_{\substack{\varphi_1 \in G(S) \\ (\|x\|^2 + \|y_1\|^2)^{\frac{1}{2}} \leq (2 + \|E\|^2)^{\frac{1}{2}}}} \operatorname{dist}(\varphi_1, G(U)).\end{aligned}$$

Hence, by using Eq. (2.2.2), we have

$$\delta(U + E, V + E) \leq (2 + \|E\|^2)\delta(V, U).$$

Thus, (2.16.5) holds and it also implies (2.16.6) as a direct consequence. Q.E.D.

Theorem 2.16.5 *Let U, $V \in L\mathcal{R}(X, Y)$. If U and $V \in BC\mathcal{R}(X, Y)$ satisfies $U(0) = V(0)$, then*

$$\widehat{\delta}(U, V) \leq (2 + \min\{\|U\|, \|V\|\}^2) \frac{\|U - V\|}{\sqrt{1 + \|U - V\|^2}}. \qquad \diamond$$

Proof It follows from Proposition 2.11.2, we have $U - V$ and $V - U$ are closed. On the other hand $U = U - V + V$ and $V = V - U + U$ based on the assumption that $U(0) = V(0)$ (see Proposition 2.7.2). Since $U \in B\mathcal{R}(X, Y)$, then by using Theorem 2.16.4, we have

$$\begin{aligned}\widehat{\delta}(U, V) =& \widehat{\delta}((U - V) + V, 0_X + V) \\ \leq& (2 + \|V\|^2)\widehat{\delta}(U - V, 0_X),\end{aligned}$$

and

$$\begin{aligned}\widehat{\delta}(U, V) =& \widehat{\delta}(0_X + U, (V - U) + U) \\ \leq& (2 + \|U\|^2)\widehat{\delta}(V - U, 0_X).\end{aligned}$$

Which, together with Theorem 2.16.3, yields that

$$\widehat{\delta}(U, V) \leq (2 + \min\{\|U\|, \|V\|\}^2) \frac{\|U - V\|}{\sqrt{1 + \|U - V\|^2}}.$$

Q.E.D.

Theorem 2.16.6 *Let $V, U \in C\mathcal{R}(X,Y)$ satisfy that $\|U\| < \infty$, $U(0) \subset V(0)$, and $\mathcal{D}(U) \supset \mathcal{D}(V)$. If $\widehat{\delta}(V,U) < (1+\|U\|^2)^{-\frac{1}{2}}$, then $\|V\| < \infty$, $\mathcal{D}(V) = \mathcal{D}(U)$ and*

$$\|V - U\| \leq \frac{1+\|U\|^2}{1-\sqrt{1+\|U\|^2}\delta(V,U)}\delta(V,U). \qquad \diamond$$

Proof We carve this proof into two steps.

Step 1. First we show that $V - U$ is continuous. Indeed, if we take any $x \in \mathcal{D}(V - U) \subset \mathcal{D}(U)$ such that $x \neq 0$, there exist $y \in (V-U)x$ and $z \in Ux$. Hence, for any given $\varepsilon > 0$, we have

$$\|y\| < \|(V-U)x\| + \frac{\varepsilon}{2} \text{ and } \|z\| < \|U\|\|x\| + \frac{\varepsilon}{2}. \tag{2.16.12}$$

By using assumption that $U(0) \subset V(0)$ and $\mathcal{D}(V) \subset \mathcal{D}(U)$, we can write $V = V - U + U$. This implies that $(x, y+z) \in G(V)$. In addition, for any $\begin{pmatrix} u \\ v \end{pmatrix} \in G(U)$, we obtain $(x-u, z-v) \in G(U)$ and for any $\varepsilon_0 > 0$, there is $w \in U(x-u)$ such that

$$\|w\| < \|U\|\|x-u\| + \varepsilon_0. \tag{2.16.13}$$

Hence, we deduce that $(0, z-v-w) \in G(U)$. The fact that $U(0) \subset (V-U)(0)$ implies that $(0, z-v-w) \in G(V-U)$. Thus, we have $(x, y+z-v-w) \in G(V-U)$. Consequently, for any $\begin{pmatrix} u \\ v \end{pmatrix} \in G(U)$, we deduce from (2.16.13) that

$$\begin{aligned} \|y+z-v-w\| &\leq \|y+z-v\| + \|w\| \\ &< \|y+z-v\| + \|U\|\|x-u\| + \varepsilon_0. \end{aligned}$$

By using Cauchy–Schwarz inequality, we get

$$\|y+z-v-w\| < \left(1+\|U\|^2\right)^{\frac{1}{2}} \left(\|x-u\|^2 + \|y+z-v\|^2\right)^{\frac{1}{2}} + \varepsilon_0.$$

Hence, we have

$$\|y+z-v-w\| < \left(1+\|U\|^2\right)^{\frac{1}{2}} \inf_{(u,v)\in G(U)} \left(\|x-u\|^2 + \|y+z-v\|^2\right)^{\frac{1}{2}} + \varepsilon_0.$$

Thus, by arbitrariness of ε_0, we have

$$\|(V-U)x\| \leq \delta(V,U)(1+\|U\|^2)^{\frac{1}{2}}. \tag{2.16.14}$$

Now, let us assume that

$$\|x\|^2 + \|y+z\|^2 = 1. \tag{2.16.15}$$

Then, by using (2.16.12), we have

$$\begin{aligned}1 &\le \|x\|^2 + (\|y\| + \|z\|)^2 \\ &< \|x\|^2 + (\|(V-U)x\| + \|U\|\|x\| + \varepsilon)^2 \\ &< (1 + \|U\|^2)\|x\|^2 + 2\|U\|\|x\|\|(V-U)x\| + \|(V-U)x\|^2 \\ &+ 2\varepsilon(\|(V-U)x\| + \|U\|\|x\|) + \varepsilon^2.\end{aligned}$$

Then, by arbitrariness of ε, we get

$$\begin{aligned}1 &\le \left((1 + \|U\|^2)\|x\|^2 + 2\|U\|\|x\|\|(V-U)x\| + \|(V-U)x\|^2\right)^{\frac{1}{2}} \\ &\le \left((1 + \|U\|^2)\|x\|^2 + 2(1 + \|U\|^2)^{\frac{1}{2}}\|x\|\|(V-U)x\| + \|(V-U)x\|^2\right)^{\frac{1}{2}} \\ &\le \left(\left((1 + \|U\|^2)^{\frac{1}{2}}\|x\| + \|(V-U)x\|\right)^2\right)^{\frac{1}{2}} \\ &\le (1 + \|U\|^2)^{\frac{1}{2}}\|x\| + \|(V-U)x\|. \end{aligned} \tag{2.16.16}$$

Therefore, it follows from both(2.16.14) and (2.16.16) that

$$\|(V - Ux\| \le (1 + \|U\|^2)\delta(V, U)\|x\| + (1 + \|U\|^2)^{\frac{1}{2}}\delta(V, U)\|(V-U)x\|.$$

Then, for all $x \in \mathscr{D}(V-U)$ satisfying (2.16.15), we deduce that

$$\|(V-U)x\| \le \frac{1 + \|U\|^2}{1 - (1 + \|U\|^2)^{\frac{1}{2}}\delta(V, U)}\delta(V, U)\|x\|. \tag{2.16.17}$$

Since (2.16.17) is homogeneous in x, hence it is true for any $x \in \mathscr{D}(V-U) = \mathscr{D}(V)$ without normalization (2.16.15). So, $V - U$ is continuous. We conclude that

$$\begin{aligned}\|V\| &= \|V - U + U\| \\ &\le \|V - U\| + \|U\| < \infty.\end{aligned}$$

As a result, makes us conclude that $\mathscr{D}(V)$ is closed.

Step 2. We show that $\mathscr{D}(V)$ is dense in $\mathscr{D}(U)$.

we take u be any given nonzero vector of $\mathscr{D}(U)$. Hence, for any $\varepsilon_1 > 0$, there is $v \in Uu$ such that

$$\begin{aligned}\|v\| &< \|Uu\| + \varepsilon_1 \\ &< \|U\|\|u\| + \varepsilon_1.\end{aligned} \tag{2.16.18}$$

We suppose that $u \in \mathscr{D}(U)$ so normalized that $\begin{pmatrix} u \\ v \end{pmatrix}$ is in unit sphere of $G(U)$,

$$\|u\|^2 + \|v\|^2 = 1. \tag{2.16.19}$$

Let δ be such that

$$\delta(V, U) < \delta < (1 + \|U\|^2)^{-\frac{1}{2}}.$$

Then, there exists $\begin{pmatrix} x \\ y \end{pmatrix} \in G(V)$ satisfying

$$\|x - u\|^2 + \|y - v\|^2 < \delta^2.$$

So,

$$\|x - u\| < \delta \ \text{ for all } x \in \mathscr{D}(V). \tag{2.16.20}$$

Let M be the closure of $\mathscr{D}(V)$. Since $\operatorname{dist}(u, \mathscr{D}(V)) = \operatorname{dist}(u, M)$, then by using (2.16.20), we have

$$\operatorname{dist}(u, M) < \delta.$$

Moreover, by using (2.16.18) and (2.16.19) and the Cauchy–Schwarz inequality that

$$\begin{aligned} 1 &< \|u\|^2 + (\|U\|\|u\| + \varepsilon_1)^2 \\ &< (1 + \|U\|^2)\|u\|^2 + 2\varepsilon_1\|T\|\|u\| + \varepsilon_1^2. \end{aligned}$$

So, by arbitrariness of ε_1, we obtain

$$1 \leq (1 + \|U\|^2)^{\frac{1}{2}} \|u\|.$$

As consequently,

$$\operatorname{dist}(u, M) < (1 + \|U\|^2)^{\frac{1}{2}} \delta \|u\|. \tag{2.16.21}$$

Since (2.16.21) is homogeneous in u, then it is true for any $u \in \mathscr{D}(U)$. The fact that $(1 + \|U\|^2)^{\frac{1}{2}} \delta < 1$ implies that $M = \overline{\mathscr{D}(V)} = \mathscr{D}(U)$. Since $\mathscr{D}(V)$ is closed (by Step 1), then $\mathscr{D}(V) = \mathscr{D}(U)$, as desired. Q.E.D.

The next result can be derived from Theorem 2.16.6.

Theorem 2.16.7 *Let $U, V \in C\mathscr{R}(X, Y)$ and $U^{-1} \in \mathscr{L}(Y, X)$. If*

$$\widehat{\delta}(V, U) < (1 + \|U^{-1}\|^2)^{-\frac{1}{2}},$$

then $V^{-1} \in \mathscr{L}(Y, X)$. ◇

Proof Keeping in mind that $\delta(V^{-1}, U^{-1}) = \delta(V, U)$. Then, the result of this theorem would be direct consequence of Theorem 2.16.6. Thus, it remains to show that V is one-to-one, which yields that V^{-1} is an operator. Let S be such that $\delta(V, U) < (1 + \|U^{-1}\|^2)^{-\frac{1}{2}}$, and we assume, by contradiction, that $\begin{pmatrix} x \\ 0 \end{pmatrix} \in G(U)$ such that $\|x\| = 1$ can be found. Hence, $(x, 0)$ is in the unit sphere of $G(V)$. Then, there is $(y, z) \in G(U)$ such that

$$\|x - y\|^2 + \|z\|^2 < \delta'^2, \tag{2.16.22}$$

for some number δ' satisfy

$$\delta(V, U) < \delta' < (1 + \|U^{-1}\|^2)^{-\frac{1}{2}}.$$

Hence,

$$\begin{aligned} 1 &= \|x\|^2 \\ &\leq (\|x - y\| + \|y\|)^2 \\ &\leq (\|x - y\| + \|U^{-1}\| \|z\|)^2. \end{aligned}$$

By using Cauchy–Schwarz inequality and from Eq. (2.16.22), we obtain

$$\begin{aligned} 1 &\leq (1 + \|U^{-1}\|^2)(\|x - y\|^2 + \|z\|^2) \\ &< \delta'^2(1 + \|U^{-1}\|^2) \\ &< 1, \end{aligned}$$

which is a contradiction. Q.E.D.

Theorem 2.16.8 *Let $U, V \in C\mathscr{R}(X, Y)$. If U_1 and V_1 are selections of U and V, respectively, then*

$$\delta(U, V) \leq 2\delta(U_1, V_1).$$ ◇

Proof Since U and V are closed, then U_1 and V_1 are closed with $\mathscr{D}(U_1) = \mathscr{D}(U)$ and $\mathscr{D}(V_1) = \mathscr{D}(V)$. Fix any δ' such that

$$\begin{aligned} \delta(U_1, V_1) &= \delta(G(U_1), G(V_1)) \\ &< \delta'. \end{aligned} \tag{2.16.23}$$

We shall show that

$$\delta(\varphi, V) \leq 2\,\delta',$$

for all $\varphi \in G(U)$ with $\|\varphi\| = 1$. For this purpose, let $\varphi = \begin{pmatrix} x \\ y \end{pmatrix} \in G(U)$ be with $\|\varphi\| = 1$. Then, there exists $(x, U_1 x) \in G(U_1)$ and $(0, \alpha_1) \in \{0\} \times U(0)$ such that

$y = U_1x + \alpha_1$. Hence,

$$\|\varphi\| = \|(x, U_1x + \alpha_1)\| = \|x\|^2 + \|U_1x + \alpha_1\|^2 = 1. \tag{2.16.24}$$

Put $\gamma^2 = \|x\|^2 + \|U_1x\|^2$. It is clear that $\gamma > 0$, then we can suppose that

$$\varphi_1 = \gamma^{-1}(x, U_1x) \in G(U_1)$$

with $\|\varphi_1\| = 1$. Moreover, by (2.16.23), we get $\text{dist}(\varphi_1, G(V_1)) < \delta'$. Hence, there is a $z \in \mathscr{D}(V_1)$ such that

$$\|x - z\|^2 + \|U_1x - V_1z\|^2 < \gamma^2\delta'^2. \tag{2.16.25}$$

suppose that $\psi = \begin{pmatrix} z \\ t \end{pmatrix} \in G(V)$ such that $\begin{pmatrix} z \\ t \end{pmatrix} = (z, V_1z) + (0, \alpha_2)$, where $\alpha_2 \in V(0)$. Since by (2.16.25), we obtain that

$$\begin{aligned}
\|\varphi - \psi\|^2 &= \|\, ((x, U_1x) + (0, \alpha_1)) - ((z, V_1z) + (0, \alpha_2))\, \|^2 \\
&= \|(x - z, U_1x - V_1z) + (0, \alpha_1 - \alpha_2)\|^2 \\
&= \|x - z\|^2 + \|(U_1x - V_1z) + (\alpha_1 - \alpha_2)\|^2 \\
&\le \|x - z\|^2 + 2\|U_1x - V_1z\|^2 + 2\|\alpha_1 - \alpha_2\|^2 \\
&< 2\gamma^2\delta'^2 + 2\|\alpha_1 - \alpha_2\|^2.
\end{aligned}$$

Since, by using (2.16.24), we get

$$\begin{aligned}
\gamma^2 &= \|x\|^2 + \|U_1x\|^2 \\
&= \|x\|^2 + \|U_1x + \alpha_1 - \alpha_1\|^2 \\
&\le \|x\|^2 + 2\|U_1x + \alpha_1\|^2 + 2\|\alpha_1\|^2 \\
&\le 2(1 + \|\alpha_1\|^2).
\end{aligned}$$

Hence,

$$\|\varphi - \psi\|^2 \le 4(1 + \|\alpha_1\|^2)\delta'^2 + 2\|\alpha_1 - \alpha_2\|^2.$$

Since $\alpha_1 \in U(0)$, then for any given $\varepsilon > 0$, we infer that

$$\|\alpha_1\| < \|U(0)\| + \varepsilon.$$

Since $\|\alpha_1\| < \varepsilon$, then for any given $\varepsilon_0 > 0$, we have

$$\|\alpha_1 - \alpha_2\| \le \varepsilon_0.$$

Thus,

$$\|\varphi - \psi\|^2 \leq 4(1+\varepsilon^2)\delta'^2 + 2\varepsilon_0^2.$$

It follows from $\psi \in G(V)$ and the arbitrariness of ε and ε_0, we get

$$\text{dist}(\varphi, G(V)) \leq 2\delta'.$$

Hence, by the hypothesis, $\varphi \in G(U)$ and $\|\varphi\| = 1$, we deduce that

$$\delta(U, V) \leq 2\delta'.$$

As δ' may be arbitrarily close to $\delta(U_1, V_1)$, we conclude that

$$\delta(T, S) \leq 2\,\delta(U_1, V_1). \qquad \text{Q.E.D.}$$

The next result can be easily derived from Theorem 2.16.8.

Corollary 2.16.1 *Let U, $V \in C\mathcal{R}(X, Y)$. If U_1 and V_1 are selections of U and V respectively, then*

$$\widehat{\delta}(U, V) \leq 2\widehat{\delta}(U_1, V_1). \qquad \diamond$$

Theorem 2.16.9 *(i) Let U_1, $V_1 \in C\mathcal{R}(X)$ and U_2, $V_2 \in C\mathcal{R}(Y)$. Consider the multivalued linear operators matrices*

$$\mathscr{U} = \begin{pmatrix} U_1 & 0 \\ 0 & U_2 \end{pmatrix}$$

and

$$\mathscr{V} = \begin{pmatrix} V_1 & 0 \\ 0 & V_2 \end{pmatrix}.$$

Then,

$$\widehat{\delta}\,(\mathscr{U}, \mathscr{V}) \leq \widehat{\delta}\,(U_1, V_1) + \widehat{\delta}\,(U_2, V_2)\,.$$

(ii) Let A, $B \in C\mathcal{R}(X, Y)$ and C, $D \in C\mathcal{R}(Y, X)$. We take the multivalued linear operator matrices

$$\mathscr{C} = \begin{pmatrix} 0 & C \\ 0 & 0 \end{pmatrix},$$

$$\mathscr{D} = \begin{pmatrix} 0 & D \\ 0 & 0 \end{pmatrix},$$

$$\mathscr{A} = \begin{pmatrix} 0 & 0 \\ A & 0 \end{pmatrix},$$

and

$$\mathscr{B} = \begin{pmatrix} 0 & 0 \\ B & 0 \end{pmatrix}.$$

Then, $\widehat{\delta}(\mathscr{C}, \mathscr{D}) \leq \widehat{\delta}(C, D)$ *and* $\widehat{\delta}(\mathscr{A}, \mathscr{B}) \leq \widehat{\delta}(A, B)$. ◇

Proof (i) We suppose that $\varphi = \left[\begin{pmatrix} x_1 \\ x_2 \end{pmatrix}, \begin{pmatrix} y_1 \\ y_2 \end{pmatrix}\right] \in G\left[\begin{pmatrix} U_1 & 0 \\ 0 & U_2 \end{pmatrix}\right]$. This is equivalent to saying that $(x_1, y_1) \in G(U_1)$ and $(x_2, y_2) \in G(U_2)$. Let $\begin{pmatrix} x_1 \\ x_2 \end{pmatrix} \in \mathscr{D}(\mathscr{T})$ and

$$\begin{aligned} \left\|\left[\begin{pmatrix} x_1 \\ x_2 \end{pmatrix}, \begin{pmatrix} y_1 \\ y_2 \end{pmatrix}\right]\right\|^2 &= \left\|\begin{pmatrix} x_1 \\ x_2 \end{pmatrix}\right\|^2 + \left\|\begin{pmatrix} y_1 \\ y_2 \end{pmatrix}\right\|^2 \\ &= \|(x_1, y_1)\|^2 + \|(x_2, y_2)\|^2 \\ &= 1. \end{aligned}$$

Hence, by using the above equality, we deduce that

$$\|(x_1, y_1)\| \leq 1 \text{ and } \|(x_2, y_2)\| \leq 1.$$

Let $\psi = \left[\begin{pmatrix} z_1 \\ z_2 \end{pmatrix}, \begin{pmatrix} t_1 \\ t_2 \end{pmatrix}\right] \in G\left[\begin{pmatrix} V_1 & 0 \\ 0 & V_2 \end{pmatrix}\right]$. It follows that $\begin{pmatrix} t_1 \\ t_2 \end{pmatrix} \in \begin{pmatrix} V_1 z_1 \\ V_2 z_2 \end{pmatrix}$. In addition , for any $\begin{pmatrix} z_1 \\ z_2 \end{pmatrix} \in \mathscr{D}\left[\begin{pmatrix} V_1 & 0 \\ 0 & V_2 \end{pmatrix}\right]$, we get

$$\begin{aligned} \|\varphi - \psi\| &= \left\|\left[\begin{pmatrix} x_1 - z_1 \\ x_2 - z_2 \end{pmatrix}, \begin{pmatrix} y_1 - t_1 \\ y_2 - t_2 \end{pmatrix}\right]\right\| \\ &= \left(\left\|\begin{pmatrix} x_1 - z_1 \\ x_2 - z_2 \end{pmatrix}\right\|^2 + \left\|\begin{pmatrix} y_1 - t_1 \\ y_2 - t_2 \end{pmatrix}\right\|^2\right)^{\frac{1}{2}} \\ &= \left(\|x_1 - z_1\|^2 + \|x_2 - z_2\|^2 + \|y_1 - t_1\|^2 + \|y_2 - t_2\|^2\right)^{\frac{1}{2}} \\ &\leq \left(\|x_1 - z_1\|^2 + \|y_1 - t_1\|^2\right)^{\frac{1}{2}} + \left(\|x_2 - z_2\|^2 + \|y_2 - t_2\|^2\right)^{\frac{1}{2}} \\ &= \|(x_1, y_1) - (z_1, t_1)\| + \|(x_2, y_2) - (z_2, t_2)\|. \end{aligned}$$

This implies that

$$\operatorname{dist}(\varphi, G(\mathscr{S})) \leq \operatorname{dist}((x_1, y_1), G(V_1)) + \operatorname{dist}((x_2, y_2), G(V_2)). \tag{2.16.26}$$

Therefore, by using (2.16.26), we infer that

$$
\begin{aligned}
\delta(\mathscr{U}, \mathscr{V}) &= \sup_{\substack{\begin{pmatrix} y_1 \\ y_2 \end{pmatrix} \in \begin{pmatrix} U_1 x_1 \\ U_2 x_2 \end{pmatrix} \\ \left\| \left[\begin{pmatrix} x_1 \\ x_2 \end{pmatrix}, \begin{pmatrix} y_1 \\ y_2 \end{pmatrix} \right] \right\| = 1}} \operatorname{dist}\left(\left[\begin{pmatrix} x_1 \\ x_2 \end{pmatrix}, \begin{pmatrix} y_1 \\ y_2 \end{pmatrix} \right], G\left[\begin{pmatrix} V_1 & 0 \\ 0 & V_2 \end{pmatrix} \right] \right) \\
&\leq \sup_{\substack{y_1 \in U_1 x_1 \\ y_2 \in U_2 x_2 \ \|(x_1, y_1)\| \leq 1, \ \|(x_2, y_2)\| \leq 1}} \{\operatorname{dist}((x_1, y_1), G(V_1)) + \operatorname{dist}((x_2, y_2), G(V_2))\} \\
&= \sup_{y_1 \in U_1 x_1 \ \|(x_1, y_1)\| \leq 1} \operatorname{dist}((x_1, y_1), G(V_1)) + \sup_{y_2 \in U_2 x_2 \ \|(x_2, y_2)\| \leq 1} \operatorname{dist}((x_2, y_2), G(V_2)) \\
&= \delta(U_1, V_1) + \delta(U_2, V_2).
\end{aligned}
$$

Since $\mathscr{U}$ and $\mathscr{V}$ may be exchanged in the above argument, then we obtain

$$\widehat{\delta}(\mathscr{U}, \mathscr{V}) \leq \widehat{\delta}(U_1, V_1) + \widehat{\delta}(U_2, V_2).$$

The proof of (ii) may be achieved by using the same reasoning as for (i). Q.E.D.

Theorem 2.16.10 *Let $U_1, V_1 \in C\mathscr{R}(X)$ and $U_2, V_2 \in C\mathscr{R}(Y)$. If $E \in B\mathscr{R}(Y, X)$ and $F \in B\mathscr{R}(X, Y)$ satisfies $E(0) \subset U_1(0) \bigcap V_1(0)$ and $F(0) \subset U_2(0) \bigcap V_2(0)$, then*

$$\widehat{\delta}\left(\begin{pmatrix} U_1 & E \\ F & U_2 \end{pmatrix}, \begin{pmatrix} V_1 & E \\ F & V_2 \end{pmatrix} \right) \leq \left(2 + (\|E\| + \|F\|)^2\right) \left(\widehat{\delta}(U_1, V_1) + \widehat{\delta}(U_2, V_2)\right). \quad \diamond$$

Proof Let $E \in B\mathscr{R}(Y, X)$ and $F \in B\mathscr{R}(X, Y)$. Then, for all $\begin{pmatrix} x \\ y \end{pmatrix} \in \mathscr{D}\left[\begin{pmatrix} 0 & E \\ F & 0 \end{pmatrix} \right]$, we have

$$
\begin{aligned}
\left\| \begin{pmatrix} 0 & E \\ F & 0 \end{pmatrix} \begin{pmatrix} x \\ y \end{pmatrix} \right\|^2 &= \left\| \begin{pmatrix} Ey \\ Fx \end{pmatrix} \right\|^2 \\
&\leq \|E\|^2 \|y\|^2 + \|F\|^2 \|x\|^2 \\
&\leq \left(\|A\|^2 + \|B\|^2\right) \left(\|x\|^2 + \|y\|^2\right) \\
&= (\|E\|^2 + \|F\|^2) \left\| \begin{pmatrix} x \\ y \end{pmatrix} \right\|^2 < \infty.
\end{aligned}
$$

We know that

$$
\begin{aligned}
\mathscr{D}\left[\begin{pmatrix} 0 & E \\ F & 0 \end{pmatrix} \right] &= \mathscr{D}(F) \times \mathscr{D}(E) \\
&= X \times Y.
\end{aligned}
$$

On the other hand,

$$\left\| \begin{pmatrix} 0 & E \\ F & 0 \end{pmatrix} \right\| = \left\| \begin{pmatrix} 0 & E \\ 0 & 0 \end{pmatrix} + \begin{pmatrix} 0 & 0 \\ F & 0 \end{pmatrix} \right\|$$

$$\begin{aligned}
&\leq \left\| \begin{pmatrix} 0 & E \\ 0 & 0 \end{pmatrix} \right\| + \left\| \begin{pmatrix} 0 & 0 \\ F & 0 \end{pmatrix} \right\| \\
&= \left\| \begin{pmatrix} 0 & Q_E E \\ 0 & 0 \end{pmatrix} \right\| + \left\| \begin{pmatrix} 0 & 0 \\ Q_F F & 0 \end{pmatrix} \right\| \\
&= \|Q_E E\| + \|Q_F F\|.
\end{aligned}$$

This implies that

$$\left\| \begin{pmatrix} 0 & E \\ F & 0 \end{pmatrix} \right\| \leq \|E\| + \|F\| < \infty. \tag{2.16.27}$$

Then, $\begin{pmatrix} 0 & E \\ F & 0 \end{pmatrix}$ is bounded. Since $U_1, V_1 \in C\mathcal{R}(X)$ and $U_2, V_2 \in C\mathcal{R}(Y)$, then by using Proposition 2.11.2, we infer that $\begin{pmatrix} U_1 & E \\ F & U_2 \end{pmatrix}$ and $\begin{pmatrix} V_1 & E \\ F & V_2 \end{pmatrix}$ are closed. Consider $\mathcal{U} = \begin{pmatrix} U_1 & 0 \\ 0 & U_2 \end{pmatrix}$, $\mathcal{V} = \begin{pmatrix} V_1 & 0 \\ 0 & V_2 \end{pmatrix}$ and $\mathcal{A} = \begin{pmatrix} 0 & E \\ F & 0 \end{pmatrix}$. On the one hand, using the fact that

$$\begin{cases} \mathcal{D}(\mathcal{U}) = \mathcal{D}(U_1) \times \mathcal{D}(U_2), \\ \mathcal{D}(\mathcal{V}) = \mathcal{D}(V_1) \times \mathcal{D}(V_2), \\ \mathcal{D}(\mathcal{A}) = \mathcal{D}(F) \times \mathcal{D}(E), \end{cases}$$

we infer that $\mathcal{D}(\mathcal{U}) \bigcup \mathcal{D}(\mathcal{V}) \subset \mathcal{D}(\mathcal{A})$. On the other hand, it is obvious that

$$\begin{aligned}
\begin{pmatrix} 0 & E \\ F & 0 \end{pmatrix} \begin{pmatrix} 0 \\ 0 \end{pmatrix} &= \begin{pmatrix} E(0) \\ F(0) \end{pmatrix} \\
&\subset \begin{pmatrix} U_1(0) \bigcap V_1(0) \\ U_2(0) \bigcap V_2(0) \end{pmatrix}.
\end{aligned}$$

This implies that $\mathcal{A}(0) \subset \mathcal{U}(0) \bigcap \mathcal{S}(0)$. Hence, by using Theorem 2.16.4, we conclude that

$$\widehat{\delta}(\mathcal{U} + \mathcal{E}, \mathcal{V} + \mathcal{E}) \leq (2 + \|\mathcal{E}\|^2) \widehat{\delta}(\mathcal{U}, \mathcal{V}).$$

Consequently, by applying Theorem 2.16.9 (i) and Eq. (2.16.27), we conclude that

$$\widehat{\delta}\left(\begin{pmatrix} U_1 & E \\ A & U_2 \end{pmatrix}, \begin{pmatrix} V_1 & E \\ F & V_2 \end{pmatrix} \right) \leq \left(2 + (\|E\| + \|F\|)^2\right) \left(\widehat{\delta}(U_1, V_1) + \widehat{\delta}(U_2, V_2)\right).$$

This achieves the proof. Q.E.D.

Remark 2.16.3 Let $U_1, V_1 \in C\mathcal{R}(X)$ and $U_2, V_2 \in C\mathcal{R}(Y)$. If E, F, G, and D are selections of U_1, U_2, V_1, and V_2, respectively, such that $R(E) \bigcap U_1(0) = \{0\}$, $R(F) \bigcap U_2(0) = \{0\}$, $R(G) \bigcap V_1(0) = \{0\}$, and $R(D) \bigcap V_2(0) = \{0\}$. Then,

$$\widehat{\delta}\left(\begin{pmatrix} U_1 & 0 \\ 0 & U_2 \end{pmatrix}, \begin{pmatrix} V_1 & 0 \\ 0 & V_2 \end{pmatrix} \right) \leq 2\, \widehat{\delta}\left(\begin{pmatrix} E & 0 \\ 0 & F \end{pmatrix}, \begin{pmatrix} G & 0 \\ 0 & D \end{pmatrix} \right). \qquad \diamond$$

References

1. Ammar, A., Jeribi, A.: Spectral Theory of Multivalued Linear Operators, p. 314 (English). CRC Press, Boca Raton, FL (2021). ISBN 9781771889667
2. Cross, R.: Multivalued linear operators. In: Monographs and Textbooks in Pure and Applied Mathematics, vol. 213, pp. x+335. Marcel Dekker, Inc., New York (1998). ISBN: 0-8247-0219-0
3. Jeribi, A.: Spectral Theory and Applications of Linear Operators and Block Operator Matrices. Springer, New York (2015)
4. Schechter, M.: Spectra of Partial Differential Operators. 2nd edn., North-Holland Series in Applied Mathematics and Mechanics, vol. 14. North-Holland Publishing Co., Amsterdam (1986)
5. Schechter, M.: Principles of Functional Analysis, 2nd edn., Graduate Studies in Mathematics, vol. 36. American Mathematical Society, Providence, RI (2002)
6. Knapp, A.W.: Basic algebra. Along with a companion volume Advanced algebra. Cornerstones. Birkhäuser Boston, Inc., Boston, MA (2006)
7. Müller, V.: Spectral Theory of Linear Operator and Spectral Systems in Banach Algebras. Birkhäuser Verlag (2003)
8. Labuschagne, L.E.: Certain norm related quantities of unbounded linear operators. Math. Nachr. **157**, 137–162 (1992)
9. Petryshyn, W.V.: Remarks on condensing and k-set-contractive mappings. J. Math. Anal. Appl. **39**, 717–741 (1972)
10. Krein, M.G., Krasnoselski, M.A.: Theoremes fundamentaux sur l'extension d'operateurs Hermitiens. Uspekhi Matematicheskikh Nauk **2**(3(19)), 60–106 (1947)
11. Ammar, A.: Some properties of the Wolf and Weyl essential spectra of a sequence of linear operators ν-convergent. Indag. Math. (N.S.) **28**(2), 424–435 (2017)
12. Ammar, A., Jeribi, A.: The essential pseudo-spectra of a sequence of linear operators. Complex Anal. Oper. Theory **12**(3), 835–848 (2018)
13. Ammar, A., Jeribi, A.: The Weyl essential spectrum of a sequence of linear operators in Banach spaces. Indag. Math. (N.S.) **27**(1), 282–295 (2016)
14. Ammar, A., Jeribi, A., Lazrag, N.: Sequence of linear operators in non-Archimedean Banach spaces. Mediterr. J. Math. **16**(5), Paper No. 130, 20 (2019)
15. Ammar, A., Jeribi, A., Lazrag, N.: Sequence of multivalued linear operators converging in the generalized sense. Bull. Iranian Math. Soc. **46**(6), 1697–1729 (2020)
16. Ammar, A., Jeribi, A., Lazrag, N.: Sequence of linear operators converging in the generalized sense and its essential pseudospectra. (English) Zbl 07361676; Jeribi, A. (ed.).: Operator theory. In: Proceedings of the International Conference on Operator Theory, pp. 147–155, Hammamet, Tunisia, 2018. Berlin: De Gruyter. De Gruyter Proc. Math (2021)
17. Cvetković, D.: On gaps between bounded operators. Publ. Inst. Math. (Beograd) (N.S.) **72**(86), 49–54 (2002)
18. Labrousse, J.P., Sandovici, A., de Snoo, H., Winkler, H.: The Kato decomposition of quasi-Fredholm relations. Oper. Matrices **4**, 1–51 (2010)
19. Laursen, K.B., Neumann, M.M.: An introduction to local spectral theory. In: London Mathematical Society Monographs. New Series, vol. 20. The Clarendon Press, Oxford University Press, New York (2000)
20. Shi, Y., Shao, C., Liu, Y.: Resolvent convergence and spectral approximations of sequences of self-adjoint subspaces. J. Math. Anal. Appl. **409**(2), 1005–1020 (2014)
21. Álvarez, T., Cross, R.W., Wilcox, D.: Perturbation theorems for upper and lower semi-Fredholm linear relations. Publ. Math. Debrecen **65**, 179–191 (2004)
22. Arens, R.: Operational calculus of linear relations. Pac. J. Math. **11**, 9–23 (1961)
23. Shi, Y.: The Glazman-Krein-Naimark theory for Hermitian subspaces. J. Operator Theor. **68**, 241–256 (2012)

24. Ammar, A., Fakhfakh, S., Jeribi, A.: Stability of the Essential Spectrum of the Diagonally and Off-diagonally dominant Block Matrix Linear Relations. J. Pseudo-Differ. Oper. Appl. **7**(4), 493–509 (2016)
25. Wilcox, D.:Multivalued SemiFredholm operators in normed linear spaces, Ph.D. Thesis, Univ. Cape Town (2001)
26. Jaftha, J.: The conjugate of a product of linear relations. Comme. Math. Univ. Carol. **47**, 265–273 (2006)
27. Dijksma, A., de Snoo, H.S.V.: Eigenfunction expansions associated with pairs of ordinary differential expressions. J. Differ. Equ. **60**(1), 21–56 (1985)

Chapter 3
Spectra and Local Spectral Theory for Block Multivalued Operator Matrices

One of the most useful tools in the local spectral theory of bounded or unbounded closed operators on Banach spaces is called the single-valued extension property (SVEP) (see [1, 2]). This notion was firstly introduced by Dunford [3] and studied by a variety of mathematicians. We can cite Finch [4], Laursen and Neumann [2] and Aiena and her co-authors [5–8]. We refer the reader to Finch's article [4] for the definition of the single-valued extension property of the closed linear operator U on Banach space. Motivated by Aiena et al.'s work [9], we introduce some concepts of the local spectral theory and the single-valued extension property abbreviated SVEP, of the closed linear relations on a Banach space. After that, we analyze basic properties of these notions and establish a relationship between the analytic spectral subspace and the analytic core.

3.1 Resolvent Set of Linear Relation

Let $U \in L\mathscr{R}(X)$ and $\lambda \in \mathbb{C}$. The quantities

$$R(\lambda, U) = (\lambda I_X - U)^{-1} = (\lambda - U)^{-1}$$

is called the resolvent of U (corresponding to λ) and

$$U_\lambda = (\lambda - \widetilde{U})^{-1}$$

is called the complete resolvent of U, where $\widetilde{U}$ is given in (2.11.1).

A. Ammar and A. Jeribi, *Spectral Theory of Block Multivalued Operator Matrices*, Infosys Science Foundation Series, https://doi.org/10.1007/978-981-95-0397-1_3

Definition 3.1.1 Let $U \in L\mathscr{R}(X)$. The resolvent set of U is defined as

$$\rho(U) = \{\lambda \in \mathbb{C} \text{ such that } U_\lambda \text{ is everywhere defined and single-valued}\}. \quad (3.1.1)$$

◇

Lemma 3.1.1 *Let $U \in L\mathscr{R}(X)$. Then,*

$$\rho(U) = \{\lambda \in \mathbb{C} \textit{ such that } \lambda - U \textit{ is one-to-one, open with dense range on } X\}. \quad ◇$$

Proof Exercise. Q.E.D.

Remark 3.1.1 Let $\mu \in \rho(U)$. Then, for all $\mu \neq \lambda \in \mathbb{C}$ and $V = (\mu - \lambda)((\mu - \lambda)^{-1} - U_\mu)$, we have

$$\lambda - U = V(\mu - U). \quad ◇$$

Proposition 3.1.1 [10, Proposition VI.1.3] *Let $U \in L\mathscr{R}(X)$. Then,*

(i) $\rho(U)$ is an open set.
(ii) [Resolvent Equation] Let $\lambda,\ \mu \in \rho(U)$. Then, $U_\mu - U_\lambda = (\lambda - \mu)U_\mu U_\lambda$. ◇

Lemma 3.1.2 *Let $U \in C\mathscr{R}(H)$ be such that $\|U\| < \infty$ and $\lambda \in \rho(U)$. Then, for every $x \in H$, we have*

$$\lambda(\lambda - U)^{-1}x \to x \quad \textit{as } |\lambda| \to \infty. \quad ◇$$

Proof Fix $x \in \mathscr{D}\left((\lambda - U)^{-1}\right) = H$. Then,

$$\begin{aligned}\lambda(\lambda - U)^{-1}x - (x + (\lambda - U)(0)) &= \lambda(\lambda - U)^{-1}x - (\lambda - U)(\lambda - U)^{-1}x \\ &= (\lambda - \lambda + U)(\lambda - U)^{-1}x \\ &= U(\lambda - U)^{-1}x.\end{aligned}$$

This implies that

$$(\lambda(\lambda - U)^{-1}x - x) + U(0) = U(\lambda - U)^{-1}x.$$

Hence, $\lambda(\lambda - U)^{-1}x - x \in U(\lambda - U)^{-1}x$, which yields that

$$\|U(\lambda - U)^{-1}x\| = \operatorname{dist}(\lambda(\lambda - U)^{-1}x - x, U(0)).$$

The fact that $\{0\} = (\lambda - U)^{-1}(0) \in \mathscr{D}(U)$ implies that

$$\|U(\lambda - U)^{-1}x\| \leq \|U\|\|(\lambda - U)^{-1}x\|.$$

Therefore,

$$\operatorname{dist}(\lambda(\lambda - U)^{-1}x - x, U(0)) \leq \|U\|\|(\lambda - U)^{-1}x\|.$$

Using the definition of the upper bound, for any given $\eta > 0$, we obtain that

$$\begin{aligned}\|\lambda(\lambda - U)^{-1}x - x\| &\le \operatorname{dist}(\lambda(\lambda - U)^{-1}x - x, U(0)) + \eta \\ &\le \|U\|\|(\lambda - U)^{-1}x\| + \eta.\end{aligned} \tag{3.1.2}$$

For sufficiently large $|\lambda|$, we can get that

$$\begin{aligned}(\lambda - U)^{-1} &= \left(\lambda(I - \lambda^{-1}U)\right)^{-1} \\ &= (I - \lambda^{-1}U)^{-1}\lambda^{-1}.\end{aligned}$$

This yields that

$$\begin{aligned}\left\|(\lambda - U)^{-1}\right\| &= \left\|(I - \lambda^{-1}U)^{-1}\lambda^{-1}\right\| \\ &= \left|\lambda^{-1}\right| \left\|(I - \lambda^{-1}U)^{-1}\right\| \\ &= |\lambda|^{-1} \left\|(I - \lambda^{-1}U)^{-1}\right\|.\end{aligned}$$

The fact that $(\lambda - U)^{-1}$ is a bounded linear operator implies that $(I - \lambda^{-1}U)^{-1}$ so is. This leads us to saying that

$$\|(\lambda - U)^{-1}x\| \to 0 \text{ when } |\lambda| \to +\infty.$$

Thus, by arbitrariness of η in (3.1.2) and by the fact that $\|U\| \le \infty$, we deduce that

$$\|\lambda(\lambda - U)^{-1}x\| \to x \text{ as } |\lambda| \to +\infty.$$

Q.E.D.

Proposition 3.1.2 *Let $\lambda \in \rho(U)$. If $|\lambda - \mu| \le \|U_\lambda\|^{-1}$, then*

$$U_\mu = \sum_{n=0}^{\infty} (\lambda - \mu)^n U_\lambda^{n+1}.$$

◇

Proof Exercise. Q.E.D.

Remark 3.1.2 (i) Let H and K be two Hilbert spaces and let $U \in L\mathscr{R}(H)$. It follows from Proposition 2.15.1 (iii) that the conjugate of $\lambda - U$ is

$$(\lambda - U)^* = \overline{\lambda} - U^*, \text{ for all } \lambda \in \mathbb{C}. \tag{3.1.3}$$

(ii) For $U \in L\mathscr{R}(H, K)$, we have

$$N(U^*) = R(U)^\perp,\ U^*(0) = \mathscr{D}(U)^\perp,\ N(\overline{U}) = R(U^*)^\perp \text{ and } \overline{U}(0) = \mathscr{D}(U^*)^\perp. \tag{3.1.4}$$

(iii) Let $U \in L\mathscr{R}(H)$. It follows immediately from (3.1.3) and (3.1.4) that the resolvent set of U^* is defined as

$$\rho(U^*) = \{\overline{\lambda} \text{ such that } \lambda \in \rho(U)\}. \tag{3.1.5}$$

Example 3.1.1 Let $X = L^2([-1,1])$, we consider the following single-valued operator defined by

$$\begin{aligned} U : \mathscr{D}(U) \subset X &\longrightarrow X \\ \psi &\longrightarrow U(\psi) \end{aligned}$$

with if $\psi \in \mathscr{D}(U)$, we have

$$\begin{cases} \psi(x) = 0 & \forall x \in [-1,0] \\ x\psi(x) \in L^2([-a,a]) & \forall x \in [0,1]. \end{cases}$$

It is clear that $\overline{\mathscr{D}(U)} \neq X$.

$$\begin{aligned} G(U^*) := & \ \{(f,g) \in X^2 : \langle g, h\rangle = \langle f, k\rangle \ \forall\, (h,k) \in G(U)\} \\ = & \ \{(f,g) \in X^2 : \langle g, h\rangle = \langle f, k\rangle \ h \in \mathscr{D}(U);\ k = U(h)\} \\ = & \ \{(f,g) \in X^2 : \langle g, h\rangle = \langle f, U(h)\rangle \ \forall h \in \mathscr{D}(U)\}. \end{aligned}$$

On the other hand,

$$\begin{aligned} \langle f, U(h)\rangle &= \int_{-1}^{1} f(x)(Uh)(x)dx \\ &= \int_{-1}^{0} f(x)(Uh)(x)dx + \int_{0}^{1} f(x)(Uh)(x)dx \\ &= 0 + \int_{0}^{1} f(x)xh(x)dx \\ &= \int_{0}^{1} xf(x)h(x)dx. \end{aligned}$$

Therefore,

$$(U^* f)(x) = xf(x) \ \forall x \in [0,1]$$

and it is arbitrary in $[-1,0]$. Let Ξ be the set

$$\Xi = \{\psi \in X \ : \ \psi(x) = 0 \text{ if } x \in [0,1]\}.$$

Hence, U^* is the linear relation defined by

$$(U^* h)(x) = xh(x) + \Xi \ \ \forall h \in \mathscr{D}(U^*).$$

With

$$\begin{aligned} \mathscr{D}(U^*) = & \ \{\psi \in X : \ (U^*)(\psi) \neq \emptyset\} \\ = & \ \{\psi \in X : \ x\psi(x) + \Xi \neq \emptyset \ \forall \psi \in \mathscr{D}(U), \ x \in [-1,1]\} \\ = & \ X \bigcap \mathscr{D}(U). \end{aligned}$$

Thus,

$$\mathscr{D}(U^*) = \{\psi \in X : \exists c \geq 0\, \forall h \in \mathscr{D}(U),\ |\langle \psi, Ah\rangle| \leq c\|h\|_X\}.$$

Now, we will study the resolvent of U^*. Let $\lambda \in \mathbb{C}$. Then,

$$\begin{aligned} G((U^* - \lambda I)^{-1}) := &\ \left\{(g, f) \in X^2 \ :\ (f, g) \in G((A^* - \lambda I)\right\} \\ = &\ \left\{(g, f) \in X^2 \ : f \in \mathscr{D}(A^*)\,, g \in (U^* - \lambda I) f\right\} \\ = &\ \left\{(g, f) \in X^2 : f \in \mathscr{D}(A^*), g(x) \in (x - \lambda) f(x) + \varXi\right\}. \end{aligned}$$

Hence,

$$\begin{aligned} N(U^* - \lambda I) = &\ ((U^* - \lambda I)^{-1})(0) \\ = &\ \varXi \\ \neq &\ \{0\}. \end{aligned}$$

So, $\rho(U^*) = \emptyset$.

The following result shows that the resolvent of a closed linear relation can not have constant norm on any open set.

Theorem 3.1.1 *Let $U \in L\mathscr{R}(H)$ be such that $U^2 \in C\mathscr{R}(H)$ and $\mathscr{D}(U^2)$ is closed. Suppose that $(\lambda - U)^{-1} \in \mathscr{L}(H)$ for all λ in some open subset O of $\mathbb{C}$. Then, $\left\|(\lambda - U)^{-1}\right\|$ is not constant for all $\lambda \in O$.* ◇

Proof Without loss of generality we may assume that O is an open subset of $\mathbb{C}$ containing 0. Since $(\lambda - U)^{-1} \in \mathscr{L}(H)$, then there exists $M > 0$ such that

$$\left\|(\lambda - U)^{-1}\right\| \leq M, \quad \text{for all } \lambda \in O. \tag{3.1.6}$$

We shall prove that

$$\left\|(\lambda - U)^{-1}\right\| < M, \quad \text{for all } \lambda \in O. \tag{3.1.7}$$

We argue by contradiction that is $\left\|(\lambda - U)^{-1}\right\| = M$, for all $\lambda \in O$. Based on the assumption $0 \in O$, we can get $\lambda = 0$. Hence, $U^{-1} \in \mathscr{L}(H)$, which yields from (3.1.6) that

$$\left\|U^{-1}\right\| \leq M. \tag{3.1.8}$$

It follows from Proposition 3.1.2 that for $|\lambda| < \|U^{-1}\|$

$$\begin{aligned} (\lambda - U)^{-1} = &\ \sum_{n=0}^{\infty} \left(U^{-1}\lambda\right)^n U^{-1} \\ = &\ \sum_{n=0}^{\infty} \lambda^n \left(U^{-1}\right)^{n+1}. \end{aligned}$$

Hence, for $x \in H$ such that $\|x\| = 1$, we get that

$$\begin{aligned}\left\|(\lambda - U)^{-1}x\right\|^2 &= \left\langle(\lambda - U)^{-1}x, (\lambda - U)^{-1}x\right\rangle \\ &= \sum_{n,m=0}^{\infty} \lambda^n \,(\overline{\lambda})^m \left\langle(U^{-1})^{n+1}x, (U^{-1})^{m+1}x\right\rangle. \end{aligned} \tag{3.1.9}$$

Let $0 < r < \|U^{-1}\|$. Integration (3.1.9) along the circle centered at 0 and with radius r. Then,

$$\begin{aligned}\int_0^1 \left\|(re^{2i\pi t} - U)^{-1}x\right\|^2 dt &= \sum_{n,m=0}^{\infty} r^{n+m} e^{2i\pi t(n-m)} \left\langle(U^{-1})^{n+1}x, (U^{-1})^{m+1}x\right\rangle \\ &\geq \left\|U^{-1}x\right\|^2 + \left\|(U^{-1})^2x\right\|^2 r^2.\end{aligned}$$

Consequently,

$$\begin{aligned}\left\|U^{-1}x\right\|^2 + \left\|(U^{-1})^2x\right\|^2 r^2 &\leq \int_0^1 \left\|(re^{2i\pi t} - U)^{-1}\right\|^2 dt \\ &\leq M^2 \ \ \text{(from (3.1.6))}. \end{aligned} \tag{3.1.10}$$

Using the definition of the upper bound, for any given $\varepsilon_1 > 0$, we obtain from (3.1.8) that

$$M^2 - \varepsilon_1^2 \leq \left\|U^{-1}x\right\|^2. \tag{3.1.11}$$

Hence, from (3.1.10) and (3.1.11), we infer that

$$M^2 - \varepsilon_1^2 + \left\|(U^{-1})^2x\right\|^2 r^2 \leq M^2.$$

This leads to

$$\left\|(U^{-1})^2x\right\| \leq \frac{\varepsilon_1}{r}.$$

On the other hand, by using the fact that $((U^{-1})^2)^{-1} = U^2$, we infer from Theorem 2.8.1 that

$$U^2(U^{-1})^2x = x + U^2(0).$$

This implies from Theorem 2.8.1 (i) that $x \in U^2(U^{-1})^2x$. Hence, for all $\varepsilon_2 > 0$, we get

$$\begin{aligned}1 &= \|x\| \\ &\leq \left\|U^2(U^{-1})^2x\right\| + \varepsilon_2.\end{aligned}$$

Since $(U^{-1})^2(0) = 0$, then we infer from Remark 2.10.1 that

$$\begin{aligned} 1 &\leq \left\|U^2\right\| \left\|(U^{-1})^2 x\right\| + \varepsilon_2 \\ &\leq \left\|U^2\right\| \frac{\varepsilon_1}{r} + \varepsilon_2. \end{aligned} \tag{3.1.12}$$

The fact that $U^2 \in C\mathscr{R}(H)$ and $\mathscr{D}(U^2)$ is closed, then by referring to Theorem 2.11.1 (i), we obtain $\left\|U^2\right\| < \infty$. Therefore, by arbitrariness of $\varepsilon_1, \varepsilon_2$, we deduce from (3.1.12) that $1 \leq 0$. This contradiction shows that (3.1.7) holds. Q.E.D.

Lemma 3.1.3 *Let A, $B \in L\mathscr{R}(X)$ be with $\lambda \in \rho(A) \bigcap \rho(B)$. Suppose that C, $D \in B\mathscr{R}(X)$ satisfies $D(0) \subset B(0)$, $C(0) \subset A(0)$. If*

$$\|C\| < \frac{1}{1 + \|D\|} \text{ and } 1 + \|D\| < \min\{\gamma(\lambda - A), \gamma(\lambda - B)\}. \tag{3.1.13}$$

Then, $\lambda \in \rho\left(\begin{pmatrix} A & C \\ D & B \end{pmatrix}\right)$. ◇

Proof Without loss of generality, we suppose that $\lambda = 0$. Let us assume that A and B are open and one-to-one with dense range. Consider the multivalued linear operator matrix

$$\mathscr{T} = \begin{pmatrix} A & 0 \\ 0 & B \end{pmatrix}.$$

First, we have to prove that $\mathscr{T}$ is one-to-one. Let $\begin{pmatrix} x \\ y \end{pmatrix} \in N\begin{pmatrix} A & 0 \\ 0 & B \end{pmatrix} = N(A) \times N(B)$, then

$$\begin{pmatrix} 0 \\ 0 \end{pmatrix} \in \begin{pmatrix} Ax \\ By \end{pmatrix}.$$

This implies that $x \in N(A)$ and $y \in N(B)$. Hence, by using the fact that $N(A) = N(B) = \{0\}$, we conclude that

$$N\left(\begin{pmatrix} A & 0 \\ 0 & B \end{pmatrix}\right) = \left\{\begin{pmatrix} 0 \\ 0 \end{pmatrix}\right\}.$$

This is equivalent to saying that $\mathscr{T}$ is one-to-one. Second, we have to show that $\mathscr{T}$ is open. Since A and B are open and one-to-one, then

$$\|Ax\| \geq \gamma(A)\|x\|$$

and

$$\|By\| \geq \gamma(B)\|y\|.$$

Hence, for any $\begin{pmatrix} x \\ y \end{pmatrix} \in \mathscr{D}\left[\begin{pmatrix} A & 0 \\ 0 & B \end{pmatrix}\right]$ such that $\begin{pmatrix} x \\ y \end{pmatrix} \notin N\left[\begin{pmatrix} A & 0 \\ 0 & B \end{pmatrix}\right]$, we infer that

$$\begin{aligned}\left\|\begin{pmatrix} A & 0 \\ 0 & B \end{pmatrix}\begin{pmatrix} x \\ y \end{pmatrix}\right\|^2 &= \left\|\begin{pmatrix} Ax \\ By \end{pmatrix}\right\|^2 \\ &= \|Ax\|^2 + \|By\|^2 \\ &\geq (\gamma(A))^2\|x\|^2 + (\gamma(B))^2\|y\|^2 \\ &\geq \min\{(\gamma(A))^2, (\gamma(B))^2\}\left(\|x\|^2 + \|y\|^2\right) \\ &= \min(\gamma(A), \gamma(B))^2\|(x, y)\|^2.\end{aligned}$$

The fact that $\mathscr{T}$ is one-to-one implies that

$$\operatorname{dist}\left(\begin{pmatrix} x \\ y \end{pmatrix}, N(\mathscr{T})\right) = \|(x, y)\|.$$

Thus,

$$\left\|\begin{pmatrix} A & 0 \\ 0 & B \end{pmatrix}\begin{pmatrix} x \\ y \end{pmatrix}\right\| \geq \min(\gamma(A), \gamma(B)) \operatorname{dist}\left(\begin{pmatrix} x \\ y \end{pmatrix}, N\left[\begin{pmatrix} A & 0 \\ 0 & B \end{pmatrix}\right]\right).$$

This implies that

$$\gamma(\mathscr{T}) \geq \min(\gamma(A), \gamma(B)) > 0. \tag{3.1.14}$$

We can easily deduce that $\mathscr{T}$ is open. Finally, we have to prove that $R(\mathscr{T})$ is dense in $X \times X$. Since $R(A)$ and $R(B)$ are dense in X, we deduce that

$$\begin{aligned}\overline{R(\mathscr{T})} &= \overline{R(A) \times R(B)} \\ &= \overline{R(A)} \times \overline{R(B)} \\ &= X \times X.\end{aligned}$$

As a result,

$$\mathscr{T} = \begin{pmatrix} A & 0 \\ 0 & B \end{pmatrix}$$

is open and one-to-one with a dense range. Now, consider the multivalued linear operator matrix

$$\mathscr{S} = \begin{pmatrix} 0 & C \\ D & 0 \end{pmatrix}.$$

On the one hand, the fact that $D(0) \subset B(0),\ C(0) \subset A(0)$ and

$$\begin{cases} \mathscr{D}(\mathscr{T}) = \mathscr{D}(A) \times \mathscr{D}(B), \\ \mathscr{D}(\mathscr{S}) = \mathscr{D}(D) \times \mathscr{D}(C), \end{cases}$$

implies that $\mathscr{D}(\mathscr{T}) \subset \mathscr{D}(\mathscr{S})$ and $\mathscr{S}(0) \subset \mathscr{T}(0)$. On the other hand,

$$\begin{aligned}
\|\mathscr{S}\| &= \left\| \begin{pmatrix} 0 & C \\ 0 & 0 \end{pmatrix} + \begin{pmatrix} 0 & 0 \\ D & 0 \end{pmatrix} \right\| \\
&\le \left\| \begin{pmatrix} 0 & C \\ 0 & 0 \end{pmatrix} \right\| + \left\| \begin{pmatrix} 0 & 0 \\ D & 0 \end{pmatrix} \right\| \\
&= \left\| \begin{pmatrix} 0 & Q_C C \\ 0 & 0 \end{pmatrix} \right\| + \left\| \begin{pmatrix} 0 & 0 \\ Q_D D & 0 \end{pmatrix} \right\| \\
&\le \|Q_C C\| + \|Q_D D\|.
\end{aligned}$$

This implies that

$$\|\mathscr{S}\| \le \|C\| + \|D\| < \infty. \tag{3.1.15}$$

It follows from (3.1.15), (3.1.13) and (3.1.14) that

$$\begin{aligned}
\|\mathscr{S}\| &\le \|C\| + \|D\| \\
&< \min(\gamma(A), \gamma(B)) \\
&< \gamma(\mathscr{T}).
\end{aligned}$$

As a result, Lemma 2.10.11 makes us conclude that $\mathscr{T} + \mathscr{S}$ is open and one-to-one with dense range. Hence, by using Lemma 3.1.1, we have $0 \in \rho \begin{pmatrix} A & C \\ D & B \end{pmatrix}$. Q.E.D.

3.2 Spectrum of Linear Relation

Let X and Y be two normed spaces and $U \in C\mathscr{R}(X, Y)$. We say that U is bounded below if it is one-to-one and open, U is left invertible if U is one-to-one and $R(U)$ is topologically complemented in Y, and U is right invertible if U is onto and $N(U)$ is topologically complemented in X. The sets of all left invertible linear relations, right invertible relations, and invertible linear relations on X are defined, respectively, by

$$\begin{aligned}
\mathscr{G}\mathscr{R}_l(X, Y) &= \{U \in C\mathscr{R}(X, Y) \text{ such that } U \text{ is a left invertible linear relation}\}, \\
\mathscr{G}\mathscr{R}_r(X, Y) &= \{U \in C\mathscr{R}(X, Y) \text{ such that } U \text{ is a right invertible linear relation}\}, \\
\mathscr{G}\mathscr{R}(X, Y) &= \mathscr{G}\mathscr{R}_l(X, Y) \bigcap \mathscr{G}\mathscr{R}_r(X, Y).
\end{aligned}$$

If $X = Y$, then the sets $\mathscr{G}\mathscr{R}_l(X, Y)$, $\mathscr{G}\mathscr{R}_r(X, Y)$, and $\mathscr{G}\mathscr{R}(X, Y)$ are replaced, respectively, by $\mathscr{G}\mathscr{R}_l(X)$, $\mathscr{G}\mathscr{R}_r(X)$, and $\mathscr{G}\mathscr{R}(X)$.

3.2.1 Spectrum

Let $U \in L\mathscr{R}(X)$. The spectrum of U is defined as

$$\sigma(U) = \mathbb{C} \backslash \rho(U).$$

The left spectrum is defined by

$$\sigma_l(U) := \{\lambda \in \mathbb{C} \text{ such that } \lambda - U \notin \mathscr{G}\mathscr{R}_l(X)\},$$

and the right spectrum is defined by

$$\sigma_r(U) := \{\lambda \in \mathbb{C} \text{ such that } \lambda - U \notin \mathscr{G}\mathscr{R}_r(X)\}.$$

The defect spectrum of U is the set defined by

$$\sigma_\delta(U) := \{\lambda \in \mathbb{C} \text{ such that } \lambda - U \text{ is not onto}\}.$$

The point spectrum, continuous, and the residual spectrum are defined as

$$\begin{aligned} P\sigma(U) &= \{\lambda \in \mathbb{C} \text{ such that } N(\lambda - U) \neq 0\}, \\ R\sigma(U) &= \left\{\lambda \in \mathbb{C} \text{ such that } \lambda - U \text{ is one-to-one and } \overline{R(\lambda - U)} \neq X\right\}, \\ C\sigma(U) &= \left\{\lambda \in \mathbb{C} \text{ such that } \lambda - U \text{ is one-to-one and } \overline{R(\lambda - U)} = X \text{ but is not open}\right\}. \end{aligned}$$

Remark 3.2.1 Let $V \in L\mathscr{R}(X)$. Then,

(i) The spectrum $\sigma(V)$ of a closed linear relation V is closed.
(ii) We observe that

$$\sigma(V) = \sigma(\widetilde{V}) = \sigma(\overline{V})$$

and

$$\rho(V) = \rho(\overline{V}),$$

where $\widetilde{V}$ is given in (2.11.1) and $\overline{V}$ is the closure of V. ◇

Theorem 3.2.1 [10, Theorem VI.3.3] *Let X be a non zero normed space and let U be a continuous and densely defined linear relation on X. Then, $\sigma(U) \neq \emptyset$.* ◇

Proposition 3.2.1 [10, Proposition VI.1.11] *Let $U \in C\mathscr{R}(X)$, then*

$$\sigma(U) = \sigma(U^*).$$ ◇

Theorem 3.2.2 *Let $U \in C\mathscr{R}(X)$. If $\lambda \notin \sigma(U)$, then*

$$\|(\lambda - U)^{-1}\| \geq \frac{1}{dist(\lambda, \sigma(U))}.$$ ◇

Proof Let $\lambda \notin \sigma(U)$, then $\lambda \in \rho(U)$. Since $\text{dist}(\lambda, \sigma(U)) = \inf\{|z-\lambda| \text{ such that } z \in \sigma(U)\}$, so for all $\eta > 0$, there exists $z_\eta \in \sigma(U)$ such that

$$|\lambda - z_\eta| < \text{dist}(\lambda, \sigma(U)) + \eta.$$

Let us suppose $|\lambda - z_\eta| < \gamma(\lambda - U)$. As, $\lambda - U$ is one-to-one open with dense range, then

$$(z_\eta - \lambda)(0) = 0 \subset (\lambda - U)(0),$$

$$\begin{aligned} \mathcal{D}((z_\eta - \lambda)) &= X \\ &\supset \mathcal{D}((\lambda - U)) \\ &= \mathcal{D}(U) \end{aligned}$$

and

$$|\lambda - z_\eta| < \gamma(\lambda - U).$$

Hence, by Lemma 2.10.11, we get

$$\lambda - U + z_\eta - \lambda = z_\eta - U$$

is one-to-one open with dense range. Then, $z_\eta \in \rho(U)$, which is a contradiction. Consequently,

$$|\lambda - z_\eta| \geq \gamma(\lambda - U) \text{ for all } \eta > 0.$$

Therefore,

$$\begin{aligned} \gamma(\lambda - U) &\leq |\lambda - z_\eta| \\ &< \text{dist}(\lambda, \sigma(U)) + \eta \quad \text{for all } \eta > 0. \end{aligned}$$

Thus,

$$\gamma(\lambda - U) \leq \text{dist}(\lambda, \sigma(U)).$$

So,

$$\|(\lambda - U)^{-1}\| \geq \frac{1}{\text{dist}(\lambda, \sigma(U))}.$$

This completes the proof. Q.E.D.

Lemma 3.2.1 *Let $U \in BC\mathcal{R}(X)$. Then, $\sigma_\delta(U)$ is a closed subset in $\mathbb{C}$.* ◇

Proof Exercise. Q.E.D.

Lemma 3.2.2 *Let X, Y be two normed space and let $U \in L\mathcal{R}(X, Y)$. If A is a linear selection of U, then*

$$P\sigma(U) = P\sigma(A).$$ ◇

Proof This follows from the fact that $N(A-\lambda) = N(U-\lambda)$. Q.E.D.

Remark 3.2.2 Let X be a normed spaces and $U \in L\mathcal{R}(X)$. Then,

$$\sigma(U) = \sigma(\widetilde{U}) = \sigma(\overline{U}). \qquad \diamond$$

3.2.2 *Approximate Point Spectrum of Linear Relation*

The approximate point spectrum of U is the set defined by

$$\sigma_{ap}(U) = \left\{\lambda \in \mathbb{C} \text{ such that } \exists (x_n) \in \mathcal{D}(U) \text{ with } \|x_n\| = 1 \text{ and } \lim_{n\to+\infty} \|(\lambda - U)x_n\| = 0\right\}.$$

Proposition 3.2.2 *Let $V \in C\mathcal{R}(H)$ and $\lambda \in \mathbb{C}$. Then, the following statements are equivalent*

(i) $\lambda \notin \sigma_{ap}(V)$.
(ii) *There exists a constant $c > 0$ such that $\|(\lambda - V)x\| \geq c\|x\|$ for all $x \in \mathcal{D}(V)$.*
(iii) $\lambda - V$ *is one-to-one and open.* $\diamond$

Proof $(i) \Longrightarrow (ii)$ Suppose that there exists $x \in \mathcal{D}(V)$ such that

$$\|(\lambda - V)x\| < c\|x\|, \quad \text{for all } c > 0.$$

This implies that there is a non zero sequence $(x_n) \subset \mathcal{D}(V)$ such that

$$\|(\lambda - V)x_n\| < \frac{1}{n}\|x_n\|, \quad \text{for any } n \geq 1.$$

Put $y_n = \frac{x_n}{\|x_n\|}$. Then, $(y_n) \subset \mathcal{D}(V)$, $\|y_n\| = 1$ and

$$\|(\lambda - V)y_n\| < \frac{1}{n}, \quad \text{for any } n \geq 1.$$

Letting $n \to \infty$, we infer that $\lambda \in \sigma_{ap}(V)$.

$(ii) \Longrightarrow (iii)$ It follows from Sect. 2.8.1 that $(\lambda - V)^{-1}$ is a continuous operator. Then, by using Remark 2.10.2, we deduce that $\lambda - V$ is one-to-one and open.

$(iii) \Longrightarrow (i)$ Since $\lambda - V$ is one-to-one and open, then by using Remark 2.10.2, we obtain $(\lambda - V)^{-1}$ is a continuous operator. In view of Remark 2.6.1 (v) implies that

$$\|x\| = \|(\lambda - V)^{-1}(\lambda - V)x\|, \quad \text{for all } x \in \mathcal{D}(V). \tag{3.2.1}$$

For $x \in \mathcal{D}(V)\backslash\{0\}$. By setting $y = \frac{x}{\|x\|}$, we infer that $y \in \mathcal{D}(V)$, $\|y\| = 1$. This leads from (3.2.1) us to

$$\|(\lambda - V)^{-1}(\lambda - V)y\| = 1. \tag{3.2.2}$$

Let $z \in (\lambda - V)y$. Then, by using Theorem 2.8.1 (i), we have

$$(\lambda - V)y = z + V(0). \tag{3.2.3}$$

This implies from (3.2.2) and (3.2.3) that

$$\begin{aligned} 1 = &\ \|(\lambda - V)^{-1}(z + V(0))\| \\ = &\ \|(\lambda - V)^{-1}z + (\lambda - V)^{-1}(\lambda - V)(0))\| \ \ \text{(as Proposition 2.8.1 } (iii)) \\ \leq &\ \|(\lambda - V)^{-1}z\| + \|(\lambda - V)^{-1}(\lambda - V)(0))\| \\ \leq &\ \|(\lambda - V)^{-1}z\| \ \ \text{(by (3.2.1))} \\ \leq &\ \|(\lambda - V)^{-1}\|\|z\| \ \ \text{(since } (\lambda - V)^{-1} \text{ is continuous).} \end{aligned} \tag{3.2.4}$$

By referring to Lemma 2.10.2, we have

$$\begin{aligned} \|(\lambda - V)y\| = &\ \text{dist}((\lambda - V)y, 0) \\ = &\ \inf_{z \in (\lambda - V)y} \|z\|. \end{aligned}$$

Then, for any given $\varepsilon > 0$, we obtain that

$$\|z\| \leq \|(\lambda - V)y\| + \varepsilon.$$

Hence, we deduce from (3.2.4) that

$$1 \leq \|(\lambda - V)^{-1}\|(\|(\lambda - V)y\| + \varepsilon).$$

By the arbitrariness of ε, we infer that

$$1 \leq \|(\lambda - V)^{-1}\|\|(\lambda - V)y\|.$$

For each $n \in \mathbb{N}$, we have

$$1 \leq \|(\lambda - V)^{-1}\|\|(\lambda - V)y_n\|, \ (y_n) \subset \mathscr{D}(V) \text{ with } \|y_n\| = 1.$$

Based on the assumption $\lambda - V$ is open, we infer from Lemma 2.10.9 and Proposition 2.15.3 (i) that

$$0 < \|(\lambda - V)^{-1}\|^{-1} \leq \|(\lambda - V)y_n\|.$$

Thus, there is no sequence $(y_n) \subset \mathscr{D}(V)$ with $\|y_n\| = 1$ such that

$$\|(\lambda - T)y_n\| \to 0 \text{ as } n \to \infty.$$

This is equivalent to saying that $\lambda \notin \sigma_{ap}(V)$. Q.E.D.

Remark 3.2.3 Let $V \in C\mathscr{R}(H)$. Then,

$$\sigma_{ap}(V) \subset \sigma(V).$$

Indeed, let $\lambda \notin \sigma(V)$. Then, $\lambda \in \rho(V)$ which yields from Lemma 3.1.1 that $\lambda - V$ is one-to-one, open and has dense range. This implies from Proposition 3.2.2 that $\lambda \notin \sigma_{ap}(V)$. ◇

Lemma 3.2.3 *Let $V \in C\mathscr{R}(H)$. Then,*

$$\sigma_{\delta}(V^*) = \{\overline{\lambda} \in \mathbb{C} \text{ such that } \lambda \in \sigma_{ap}(V)\}. \quad \diamondsuit$$

Proof We are going to prove that $\sigma_{\delta}(V^*) \subset \{\overline{\lambda} \in \mathbb{C}$ such that $\lambda \in \sigma_{ap}(V)\}$. Suppose that $\lambda \notin \sigma_{ap}(V)$. Then, by using Proposition 3.2.2, we infer that $\lambda - V$ is one-to-one and open. It follows from the closedness, openness of $\lambda - V$ and Theorem 2.11.1 (ii) that $R(\lambda - V)$ is closed. Hence, by using (3.1.3) and Theorem 2.15.1, we infer that $R(\overline{\lambda} - V^*)$ is closed. In addition, $\lambda - V \in C\mathscr{R}(H)$, we infer from (3.1.3) and (3.1.4) that

$$R(\overline{\lambda} - V^*)^{\perp} = N(\lambda - V) = \{0\}.$$

This implies that $R(\overline{\lambda} - V^*) = H$. This is equivalent to saying that $\overline{\lambda} \notin \sigma_{\delta}(V^*)$. Conversely, assume that $\overline{\lambda} \notin \sigma_{\delta}(V^*)$. Then, by using (3.1.3), we have $(\lambda - V)^*$ is onto. Hence, the fact that $\lambda - T$ is closed implies from (3.1.4) that

$$N(\lambda - V) = \{0\}.$$

Since $R((\lambda - V)^*)$ is closed, then by applying Theorem 2.15.1, we infer that $R(\lambda - V)$ is closed. This implies from Theorem 2.11.1 that $\lambda - V$ is open. This yields to deduce that $\lambda - V$ is one-to-one and open. The use of Proposition 3.2.2 makes us conclude that $\lambda \notin \sigma_{ap}(V)$. Q.E.D.

3.2.3 Riesz Projection and Isolated Point of the Spectrum of a Linear Relation

Let $V \in BC\mathscr{R}(X)$ be such that $\rho(V) \neq \emptyset$. It is well known that the resolvent $R_{\lambda}(V)$ is holomorphic in $\rho(V)$. Therefore, if λ_0 is an isolated point of $\sigma(V)$, and $\epsilon > 0$ such that $\sigma(V) \setminus \{\lambda_0\}$ lies on or outside the circle $|\lambda - \lambda_0| = \epsilon$, the resolvent $R_{\lambda}(V)$ admits a Laurent expansion in powers of $\lambda - \lambda_0$ in the form

$$R_{\lambda}(V) = \sum_{n=0}^{\infty}(\lambda - \lambda_0)^n A_n + \sum_{n=1}^{\infty}(\lambda - \lambda_0)^{-n} B_n,$$

with

$$A_n = \frac{1}{2\pi i}\int_C (\lambda - \lambda_0)^{-n-1} R_\lambda(V)d\lambda$$

and

$$B_n = \frac{1}{2\pi i}\int_C (\lambda - \lambda_0)^{n-1} R_\lambda(V)d\lambda,$$

where C is any counterclockwise circle $|\lambda - \lambda_0| = \mu$ with $0 < \mu < \epsilon$. We note that A_n and B_n are bounded operators and satisfy the following statements (see Lemma 4.1 in [11]):

$$\begin{aligned}(V - \lambda_0)A_{n+1} &= A_n + V(0) \quad \text{for all } n \geq 0,\\ (V - \lambda_0)B_n &= B_{n+1} + V(0) \quad \text{for all } n \geq 1,\\ (V - \lambda_0)^n B_1 &= B_{n+1} + V^n(0) \quad \text{for all } n \geq 0.\end{aligned}$$

These last equalities allow us to conclude that if we have $R(B_{n+1}) \subset V^n(0)$ for some $n \geq 0$, then $R(B_{m+1}) \subset V^m(0)$ for all $m \geq n$. It is also known that $P_{\lambda_0} := -B_1$ is a projection. The operator P_{λ_0} is called the Riesz integral projection of V at λ_0. Let $\lambda_0 \in \sigma(V)$ be an isolated point of the spectrum, and let Γ_{λ_0} be a simple closed contour around λ_0 such that the all of $\sigma(V) \setminus \{\lambda_0\}$ lies on or outside of Γ_{λ_0}

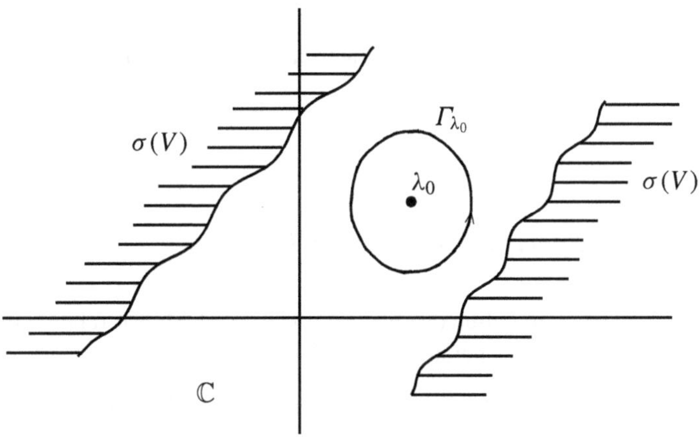

An admissible contour Γ_{λ_0} for a Riesz projection.

Definition 3.2.1 Let V be a closed multivalued linear operator on X and λ_0 be an isolated point of $\sigma(V)$. For an admissible contour Γ_{λ_0} as described above

$$P_{\lambda_0} = \frac{-1}{2i\pi}\int_{\Gamma_{\lambda_0}} (V - \mu)^{-1} d\mu$$

is called the Riesz integral for V and λ_0.

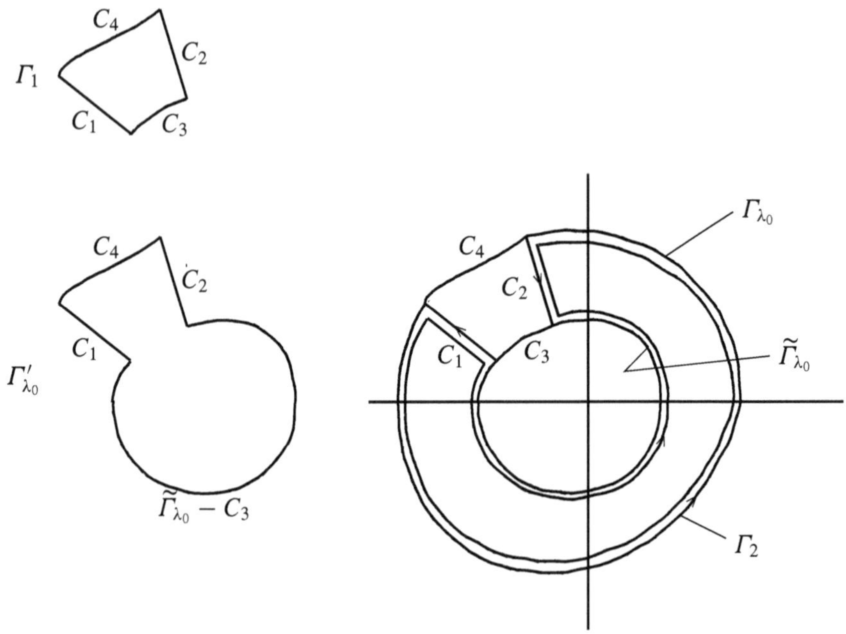

Principal contours Γ_{λ_0} and $\widetilde{\Gamma}_{\lambda_0}$, and the intermediate contours Γ_1, Γ_2, and Γ'_{λ_0}.

◇

3.2.4 Discrete Spectrum of a Linear Relation

We are now in position to define the discrete spectrum of a linear relation in light of the definition in the case of operators (see [12]).

Definition 3.2.2 Let $V \in BC\mathscr{R}(X)$. We say that $\lambda_0 \in \sigma(V)$ is discrete if λ_0 is an isolated point in $\sigma(V)$ and the Laurent expansion $R_\lambda(V)$ at λ_0 satisfies $R(B_1)$ is finite dimensional. In the case where $R(B_1)$ is of dimension one we say that λ_0 is a nondegenerate eigenvalue. We denote by $\sigma_{dis}(V)$ the set of all discrete points of the spectrum of T. ◇

We begin our discussion of the discrete spectrum by the following theorem.

Theorem 3.2.3 *Let $V \in BC\mathscr{R}(X)$ be such that for some $\lambda \in \mathbb{C}$, we have $C_\lambda = \{\mu \in \mathbb{C}$ such that $|\mu - \lambda| = r\} \subset \rho(V)$. Define P by*

$$P = \frac{-1}{2i\pi} \int_{|\mu-\lambda|=r} (V - \mu)^{-1} d\mu.$$

Then,

(i) *P is a bounded operator satisfying $P^2 = P$.*
(ii) *If* $\dim(R(P)) = n < +\infty$, *then V has at most n points of its spectrum in $\{\mu$ such that $|\mu - \lambda| < r\}$ and each is discrete.*
(iii) *If* $\dim(R(P)) = 1$, *then V has exactly one spectral point in C_λ and it is nondegenerate.* ◇

Proof (i) The proof of Lemma 8 in [13] carries without change to prove that P is a bounded operator satisfying $P^2 = P$.
(ii) If $M = R(P)$ and $N = N(P)$, then M and N are complementary closed subspaces, that is $M \bigoplus N = X$. In addition, we have $V(0) \subset N$, for all $\mu \in \rho(V)$, $R_\mu(T)P = PR_\mu(V)$, $VP = PV + V(0), V(N) \subset N$ and for all $x \in M, Vx \bigcap M \neq \emptyset$. Indeed we have for all $\mu \in \rho(V)$,

$$\begin{aligned} R_\mu(V)V(0) = &\ (V-\mu)^{-1}(V-\mu)(0) \\ = &\ N(V-\mu) \\ = &\ \{0\}. \end{aligned}$$

So, $PV(0) = \{0\}$ and hence $V(0) \subset N(P) = N$. Let $\mu \in \rho(V)$, then

$$\begin{aligned} R_\mu(V)P &= \frac{-1}{2i\pi}\int_{C_\mu} R_\mu(V)R_\alpha(V)d\alpha \\ &= \frac{-1}{2i\pi}\int_{C_\mu} R_\mu(V)R_\alpha(V)d\alpha \\ &= PR_\mu(V), \end{aligned}$$

where $C_\mu = \{\lambda \in \mathbb{C}$ such that $|\lambda - \mu| = r\}$. Thus,

$$(V-\mu)R_\mu(V)P = (V-\mu)PR_\mu(V).$$

So,

$$P + V(0) = (V-\mu)PR_\mu(V)$$

and hence,

$$\begin{aligned} (P + V(0))(V-\mu) = &\ (V-\mu)P \\ = &\ VP - \mu P. \end{aligned}$$

Whence

$$VP = PV + V(0).$$

Now, let $x \in N$. We have

$$\begin{aligned} PVx + V(0) &= VPx \\ &= V(0). \end{aligned}$$

Then,

$$PVx \subset V(0).$$

Therefore,

$$PVx \subset N \bigcap M = \{0\}.$$

Then, $Vx \subset N$. So, $V(N) \subset N$. If $x \in M$, then

$$\begin{aligned} Vx &= VPx \\ &= PVx + V(0). \end{aligned}$$

So,

$$Vx \bigcap M \neq \emptyset.$$

We denote by V_M the relation defined by $G(V_M) = G(V) \bigcap (M \times M)$ and by V_N the relation defined by $G(V_N) = G(V) \bigcap (N \times N)$. Then, we claim that V_M is a bounded operator and $V = V_M \bigoplus V_N$. Indeed, let $y \in V_M(0)$. Then, $y \in V(0) \subset N$ and $y \in M$. Hence, $y = 0$. On another hand, $G(V_M)$ is closed and $\mathscr{D}(V_M) = M$. So, V_M is bounded. Now, let $x \in X$, then there exist $x_1 \in M$ and $x_2 \in N$ such that $x = x_1 + x_2$. Take $y \in V_M x_1$. Then, $y \in V x_1$ and,

$$Vx_1 = y + V(0).$$

Therefore,

$$\begin{aligned} V_M x_1 + V_N x_2 &= V x_2 + y + V(0) \\ &= V x_2 + V x_1 \\ &= Vx. \end{aligned}$$

Then,

$$V = V_M \bigoplus V_N.$$

Now, let ν be such that $|\nu - \lambda| < r$. We claim that $\nu \notin \sigma(V_N)$. Indeed, if we put

$$B = -\frac{1}{2i\pi} \int_{|\mu-\lambda|=R} (\nu - \mu)^{-1} (V - \mu)^{-1} d\mu$$

where R is such that $0 < r < R$ and

$$\sigma(V)\bigcap\{\mu \text{ such that } |\mu-\lambda|<r\}=\sigma(V)\bigcap\{\mu \text{ such that } |\mu-\lambda|<R\}.$$

We claim that

$$BP = PB = 0.$$

Indeed,

$$\begin{aligned}
BP &= \frac{1}{(2i\pi)^2}\int_{|\mu-\lambda|=R} d\mu \int_{|s-\lambda|=r} (\nu-\mu)^{-1}(V-\mu)^{-1}(V-s)^{-1}ds \\
&= \frac{1}{(2i\pi)^2}\int_{|\mu-\lambda|=R} d\mu \int_{|s-\lambda|=r} (\nu-\mu)^{-1}(s-\mu)^{-1}\left[(V-\mu)^{-1}-(V-s)^{-1}\right]ds \\
&= \frac{1}{(2i\pi)^2}\int_{|\mu-\lambda|=R} d\mu \int_{|s-\lambda|=r} (\nu-\mu)^{-1}(V-\mu)^{-1}(s-\mu)^{-1}ds \\
&\quad \frac{1}{(2i\pi)^2}\int_{|\mu-\lambda|=R} d\mu \int_{|s-\lambda|=r} (\nu-\mu)^{-1}(V-\mu)^{-1}(s-\mu)^{-1}ds.
\end{aligned}$$

Let

$$I=\int_{|\mu-\lambda|=R} d\mu \int_{|s-\lambda|=r} (\nu-\mu)^{-1}(V-\mu)^{-1}(s-\mu)^{-1}ds$$

and

$$J=\int_{|\mu-\lambda|=R} d\mu \int_{|s-\lambda|=r} (\nu-\mu)^{-1}(V-\mu)^{-1}(s-\mu)^{-1}ds.$$

On another hand,

$$\begin{aligned}
I &= \int_{|\mu-\lambda|=R} d\mu \int_{|s-\lambda|=r} (\nu-\mu)^{-1}(V-\mu)^{-1}(s-\mu)^{-1}ds \\
&= \int_{|\mu-\lambda|=R} d\mu \left[\int_{|s-\lambda|=r} (s-\mu)^{-1}ds\right](\nu-\mu)^{-1}(V-\mu)^{-1} \\
&= 0
\end{aligned}$$

and

$$\begin{aligned}
J &= \int_{|\mu-\lambda|=R} d\mu \int_{|s-\lambda|=r} (\nu-\mu)^{-1}(V-s)^{-1}(s-\mu)^{-1}ds \\
&= \int_{|s-\lambda|=r} ds \left[\int_{|\mu-\lambda|=R} (s-\mu)^{-1}(\nu-\mu)^{-1}d\mu\right](V-s)^{-1} \\
&= \int_{|s-\lambda|=r} ds \left[\int_{|\mu-\lambda|=R} \left(\frac{1}{s-\nu}\,\frac{1}{\mu-\nu}\frac{1}{s-\nu}\,\frac{1}{s-\mu}\right)d\mu\right](V-s)^{-1} \\
&= 0.
\end{aligned}$$

Hence,

$$BP = 0.$$

Furthermore,

$$
\begin{aligned}
PB &= \frac{1}{(2i\pi)^2} \int_{|s-\lambda|=r} ds \int_{|\mu-\lambda|=R} d\mu(\nu-\mu)^{-1}(V-s)^{-1}(s-\mu)^{-1} \\
&= \frac{1}{(2i\pi)^2} \int_{|s-\lambda|=r} ds \int_{|\mu-\lambda|=R} d\mu(\nu-\mu)^{-1}(s-\mu)^{-1}[(V-\mu)^{-1}-(T-s)^{-1}] \\
&= \frac{1}{(2i\pi)^2} \int_{|s-\lambda|=r} ds \int_{|\mu-\lambda|=R} d\mu(\nu-\mu)^{-1}(s-\mu)^{-1}(V-\mu)^{-1} \\
&\quad \frac{1}{(2i\pi)^2} \int_{|s-\lambda|=r} ds \int_{|\mu-\lambda|=R} d\mu(\nu-\mu)^{-1}(s-\mu)^{-1}(T-s)^{-1}.
\end{aligned}
$$

Let

$$
I' = \int_{|s-\lambda|=r} ds \int_{|\mu-\lambda|=R} d\mu(\nu-\mu)^{-1}(s-\mu)^{-1}(V-\mu)^{-1}
$$

and

$$
J' = \int_{|s-\lambda|=r} ds \int_{|\mu-\lambda|=R} d\mu(\nu-\mu)^{-1}(s-\mu)^{-1}(V-s)^{-1}.
$$

Then,

$$
\begin{aligned}
I' &= \int_{|s-\lambda|=r} ds \int_{|\mu-\lambda|=R} d\mu(\nu-\mu)^{-1}(s-\mu)^{-1}(V-\mu)^{-1} \\
&= \int_{|\mu-\lambda|=R} d\mu \left[\int_{|s-\lambda|=r} (s-\mu)^{-1} ds\right] (V-\mu)^{-1}(\nu-\mu)^{-1} \\
&= 0 \left(\text{because} \int_{|s-\lambda|=r} (s-\mu)^{-1} ds = 0\right)
\end{aligned}
$$

and

$$
\begin{aligned}
J' &= \int_{|s-\lambda|=r} ds \int_{|\mu-\lambda|=R} d\mu(\nu-\mu)^{-1}(s-\mu)^{-1}(V-s)^{-1} \\
&= \int_{|s-\lambda|=r} ds \left[\int_{|\mu-\lambda|=R} d\mu(\nu-\mu)^{-1}(s-\mu)^{-1}\right] (V-s)^{-1} \\
&= \int_{|s-\lambda|=r} ds \left[\int_{|\mu-\lambda|=R} d\mu \left(\frac{1}{s-\nu} \frac{1}{\nu-\nu} \frac{1}{s-\nu} \frac{1}{s-\mu}\right)\right] (V-s)^{-1} \\
&= 0.
\end{aligned}
$$

Therefore,

$$
BP = PB = 0.
$$

This implies in particular that $B(M) \subset M$ and $B(N) \subset N$. Furthermore, we have

$$
(V-\nu)B = B(V-\nu)
$$

and

$$\begin{aligned}(V-\nu)B &= -\frac{1}{2i\pi}\int_{|\mu-\lambda|=r}(\nu-\mu)^{-1}(V-\nu)(V-\mu)^{-1}d\mu\\ &= -\frac{1}{2i\pi}\int_{|\mu-\lambda|=r}(\nu-\mu)^{-1}(V-\mu+\mu-\nu)(V-\mu)^{-1}d\mu\\ &= -\frac{1}{2i\pi}\int_{|\mu-\lambda|=r}(\nu-\mu)^{-1}(I+V-V)+(\mu-\nu)^{-1}(\nu-\mu)(V-\mu)^{-1}d\mu\\ &= -\frac{1}{2i\pi}(-2i\pi)I+\frac{1}{2i\pi}\int_{|\mu-\lambda|=r}(V-\mu)^{-1}d\mu\\ &= I-P.\end{aligned}$$

Thus, if we consider the restriction on N, we get

$$\begin{aligned}(V_N-\nu)B &= B(V_N-\nu)\\ &= I_N.\end{aligned}$$

This implies that $\nu \notin \sigma(V_N)$. As $V_M \bigoplus V_N = V$, we conclude that for $\nu \in \{\mu \text{ such that } |\mu-\lambda|<r\}$, we have $\nu \in \rho(V)$ if, and only if, $\nu \in \rho(V_M)$. But V_M is an operator and $\dim(M)=n<+\infty$, then

$$\sigma(V_M)=\{\lambda_1,\ldots,\lambda_k\},$$

where $k \le n$. Therefore, $\sigma(V)\bigcap\{\mu \text{ such that } |\lambda-\mu|<r\}$ is a finite set. It remain to see that if $\nu \in \sigma(V)\bigcap\{\mu \text{ such that } |\mu-\lambda|<r\}$, then ν is discrete. To do this, we note that ν is isolated in $\sigma(V)$. Let P_ν be the Riesz integral projection of V at ν. Then,

$$\begin{aligned}P_\nu P &= PP_\nu\\ &= -P_\nu.\end{aligned}$$

Thus, $R(P_\nu) \subset R(P)$. Hence,

$$\dim(R(P_\nu)) \simeq \dim(R(B_1)) < +\infty.$$

So, ν is discrete.
(iii) Follows immediately from (ii). Q.E.D.

To continue our discussion of the discrete spectrum, let us give the following definition of a pole of the resolvent of a linear relation.

Definition 3.2.3 Let $U \in BC\mathscr{R}(X)$. We say that λ_0 is a pole of the resolvent $R_\lambda(U)$ of order $m \in \mathbb{N}^*$ if λ_0 is an isolated point of the spectrum $\sigma(U)$ of U, and the Laurent expansion of $R_\lambda(U)$ at λ_0 satisfies

$$R(B_m) \not\subset U^{m-1}(0)$$

and

$$R(B_{m+1}) \subset U^m(0),$$

where

$$B_m = \frac{1}{2\pi i}\int_C (\lambda - \lambda_0)^{m-1} R_\lambda(U) d\lambda$$

with C is any counterclockwise $|\lambda - \lambda_0| = \eta$ such that

$$[\sigma(U) \setminus \{\lambda_0\}] \bigcap \{\lambda \in \mathbb{C} \text{ such that } |\lambda - \lambda_0| < \eta\} = \emptyset. \qquad \diamond$$

The following theorem is proved by Lajnef and Mnif in [11] and plays an important role in our considerations.

Proposition 3.2.3 [11, Theorem 4.1] *Let $U \in BC\mathcal{R}(X)$ and η_0 be an isolated point of the spectrum of U. If η_0 is a pole of the resolvent $R_\eta(U)$ with order $m > 0$ and P_{η_0} be the Riesz integral projection of U at η_0, then the following assertions hold*

(i) $R(P_{\eta_0}) = N(\eta_0 I - U)^m$ *and* $a(\eta_0 I - U) = m$.
(ii) η_0 *is an eigenvalue of* U.
(iii) $N(P_{\eta_0}) = R(\eta_0 I - U)^n$ *for all* $n \geq m$.
(iv) $d(\eta_0 I - U) = m$.
(v) $X = N(\eta_0 I - U)^m \bigoplus R(\eta_0 I - U)^m$. $\diamond$

3.3 Pencil Resolvent Set of Linear Relation

Definition 3.3.1 Let X be a normed vector space, $U \in C\mathcal{R}(X)$, V be a continuous linear relation such that $V \neq 0$, $V(0) \subset \overline{U(0)}$, and $\mathcal{D}(V) \supset \mathcal{D}(U)$. Then, we define the pencil resolvent set of U by

$$\rho_V(U) := \{\lambda \in \mathbb{C} \text{ such that } \lambda V - U \text{ is one-to-one, open with dense range on } X\},$$

and the pencil spectrum of U by

$$\sigma_V(U) = \mathbb{C} \backslash \rho_V(U).$$

We define the pencil approximate spectrum of U by

$$\sigma_{ap,V}(U) = \left\{\lambda \in \mathbb{C} \text{ such that } \exists (x_n) \in \mathcal{D}(\lambda V - U) \text{ with } \|x_n\| = 1 \text{ and } \lim_{n\to+\infty} \|(\lambda V - U)x_n\| = 0\right\},$$

and the pencil defect spectrum of U by

$$\sigma_{\delta,V}(U) := \{\lambda \in \mathbb{C} \text{ such that } \lambda V - U \text{ is not onto}\}.$$

Let $U, V \in L\mathscr{R}(H, K)$. The compression spectrum $\sigma_{V,0}(U)$ is defined as

$$\sigma_{V,0}(U) = \{\lambda \in \mathbb{C} \text{ such that } \overline{R(\lambda V - U)} \subsetneq K\}. \qquad \diamondsuit$$

Proposition 3.3.1 *Let $U, V \in L\mathscr{R}(H, K)$ and $\lambda \in \mathbb{C}$ be such that $\mathscr{D}(U) \subset \mathscr{D}(V)$ and $V(0) \subset U(0)$. Then, the following statements are equivalent*

(i) $\lambda \notin \sigma_{ap,V}(U)$.
(ii) There is $c > 0$ such that $\|(\lambda V - U)x\| \geq c\|x\|$ for all $x \in \mathscr{D}(U)$.
(iii) $\lambda V - U$ is one-to-one and open. $\diamondsuit$

Proof This proof is similar to proof of Proposition 3.2.2. Q.E.D.

Remark 3.3.1 It is obvious that if $U \in C\mathscr{R}(X)$ and X is complete, we will return to the pencil spectrum definition in a Banach space with closed linear relation. In such a case

$$\begin{aligned} \rho_V(U) &= \{\lambda \in \mathbb{C} \; : \; \lambda V - U \text{ is bijective}\} \\ &= \left\{\lambda \in \mathbb{C} : (\lambda V - U)^{-1} \text{ is a bounded linear operator on } X\right\}. \end{aligned} \qquad \diamondsuit$$

We recall the pencil resolvent of U at λ the operator defined by

$$R_V(\lambda, U) := (\lambda V - U)^{-1}.$$

The following results may be found in [14]:

Theorem 3.3.1 *Let $U \in C\mathscr{R}(X)$ be one-to-one with dense range and V be a continuous linear relation such that $V(0) \subset \overline{U(0)}$, $\mathscr{D}(V) \supset \mathscr{D}(U)$ and $\|V\| \neq 0$. Then,*

$$\sigma_V(U) \subset \left\{\lambda \in \mathbb{C} \text{ such that } |\lambda| \geq \frac{\gamma(U)}{\|V\|}\right\}. \qquad \diamondsuit$$

Theorem 3.3.2 *Let $U \in L\mathscr{R}(X)$ and V be a continuous linear relation such that $V(0) \subset \overline{U(0)}$ and $\mathscr{D}(V) \supset \mathscr{D}(U)$. Then, $\rho_V(U)$ is an open set.* $\diamondsuit$

Theorem 3.3.3 *Let X be a Banach space. Let $U \in C\mathscr{R}(X)$ and $V \in B\mathscr{R}(X)$ be such that $\mathscr{D}(U) = X$, $V(0) = U(0)$ and $\dim V(0) < \infty$. If $0 \in \rho(V)$, then*

$$\rho_V(U) = \rho(V^{-1}U) \bigcap \rho(UV^{-1}). \qquad \diamondsuit$$

Theorem 3.3.4 (The pencil resolvent identity) *Let $U \in C\mathscr{R}(X)$ and let $V \in B\mathscr{R}(X)$ be such that $V(0) \subset U(0)$. Then, for all λ, $\mu \in \rho_V(U)$, we have*

$$R_V(\mu, U) - R_V(\lambda, U) = (\lambda - \mu) R_V(\mu, U) V R_V(\lambda, U). \qquad \diamondsuit$$

3.4 The Local Resolvent Set of Linear Relation

It is well known that the resolvent function $(\lambda - U)^{-1}$ is a holomorphic operator defined on the resolvent set $\rho(U)$ and everywhere defined and single-valued. By referring to Theorem 2.8.1 (iv), we infer that the vector-valued analytic function $f_x : \rho(U) \longrightarrow X$ defined as

$$f_x(\lambda) = (\lambda - U)^{-1}x, \text{ for any } x \in X,$$

is an analytic function which satisfies

$$(\lambda - U)f_x(\lambda) = x + U(0), \text{ for all } \lambda \in \rho(U).$$

Definition 3.4.1 Let X be a Banach space, $U \in C\mathscr{R}(X)$ and $x \in X$. Let $\rho_U(x)$ be denote the set of all $\lambda \in \mathbb{C}$ for which there exist an open neighborhood $\mathscr{U}_\lambda$ and an analytic function $f_{\lambda,x} : \mathscr{U}_\lambda \longrightarrow X$ such that the equation

$$(\mu I - U)f_{\lambda,x}(\mu) = x + U(0),$$

holds for all $\mu \in \mathscr{U}_\lambda$. ◇

$\rho_U(x)$ is called the local resolvent set of U at the point x.

Lemma 1 *Let X be a Banach space, $x, y \in X$ and let $U \in BC\mathscr{R}(X)$. Then,*

$$\rho(U) \subset \rho_U(x).$$

Proof If $\rho(U) = \emptyset$, then

$$\rho(U) \subset \rho_U(x).$$

If $\rho(U) \neq \emptyset$, then there exists $\lambda \in \rho(U)$. Hence, the fact that $U \in C\mathscr{R}(X)$ implies that $(\lambda - U)^{-1}x$ is a vector-valued analytic function on $\rho(U)$. Moreover, it follows from Theorem 2.8.1, that

$$\begin{aligned}(\lambda - U)(\lambda - U)^{-1}x &= x + (\lambda - U)(0),\\ &= x + U(0).\end{aligned}$$

This enables us to conclude that $\rho(U) \subset \rho_U(x)$. Q.E.D.

Remark 3.4.1 Let X be a non zero Banach space and let $U \in BC\mathscr{R}(X)$. Then,

$$\rho_U(x) \neq \emptyset.$$

Indeed, by using Lemma 1, we have $\rho(U) \subset \rho_U(x)$, then by using Theorem 3.2.1, we have $\rho_U(x) \neq \emptyset$. ◇

Proposition 3.4.1 *Let X be a Banach space, $x, y \in X$ and let $U \in BC\mathscr{R}(X)$. Then,*

(i) $\rho_U(x)$ *is open.*
(ii) $\rho_U(x) = \mathbb{C}$, *for all* $x \in U(0)$.
(iii) $\rho_U(\alpha x) = \rho_U(x)$, *for all* $\alpha \in \mathbb{C}\backslash\{0\}$.
(iv) $\rho_U(x) \bigcap \rho_U(y) \subset \rho_U(\alpha x + \beta y)$, *for all* $\alpha, \beta \in \mathbb{C}$.
(v) Let $F \subset \mathbb{C}$ *and* $\lambda \in \mathbb{C}$. *Then,* $\mathbb{C}\backslash(F\backslash\{\lambda\}) \subset \rho_U(x)$ *if, and only if,* $\mathbb{C}\backslash F \subset \rho_{\lambda+U}(x)$. *In particular,* $\mathbb{C}\backslash\{\lambda\} \subset \rho_U(x)$ *if, and only if,* $\mathbb{C}\backslash\{0\} \subset \rho_{U-\lambda}(x)$. ◇

Proof (i) By using Remark 3.4.1, we have $\rho_U(x) \neq \emptyset$. Then, there exists $\lambda_0 \in \rho_U(x)$. It sufficient to find an open ball $B(\lambda_0, r)$ such that

$$B(\lambda_0, r) \subset \rho_U(x), \tag{3.4.1}$$

where $B(\lambda_0, r)$ designates the open ball centered at λ_0 with radius $r > 0$. Using the fact that $\lambda_0 \in \rho_U(x)$, then there exists an open neighborhood $\mathscr{U}$ of λ_0 in $\mathbb{C}$ and an analytic function $f : \mathscr{U} \longrightarrow \mathscr{D}(U)$ which satisfies

$$(\lambda - U)f(\lambda) = x + U(0), \text{ for all } \lambda \in \mathscr{U}. \tag{3.4.2}$$

Since $\mathscr{U}$ is open neighborhood of λ_0, then there exists $r > 0$ such that

$$B(\lambda_0, r) \subset \mathscr{U}.$$

Let us assume that $\mu \in B(\lambda_0, r)$ and g be the restriction of f in $B(\lambda_0, r)$. It is clear that g is analytic function and $B(\lambda_0, r)$ is neighborhood of μ. Hence, by referring to (3.4.2), we infer that

$$(\lambda - U)g(\lambda) = x + U(0), \text{ for all } \lambda \in B(\lambda_0, r).$$

This implies that $\mu \in \rho_U(x)$. Therefore, (3.4.1) holds. This enables us to conclude that $\rho_U(x)$ is open.
(ii) Let $x \in U(0)$. It is clear that $\rho_U(x) \subset \mathbb{C}$. Then, it is sufficient to show that

$$\mathbb{C} \subset \rho_U(x).$$

Let $\lambda \in \mathbb{C}$ and let $f : \mathbb{C} \longrightarrow \mathscr{D}(U)$ be an analytic function. Then, there exists $y \in R(\lambda - U)$ such that $y \in (\lambda - U)f(\lambda)$. It follows, from Theorem 2.8.1, that

$$(\lambda - U)f(\lambda) = y + U(0). \tag{3.4.3}$$

Then, $f(\lambda)$ coalesces with $(\lambda - U)^{-1}y$ on $\rho(U)$, so $f(\lambda) \to 0$ as $|\lambda| \to \infty$. By the vector-valued version of Liouville's theorem, we obtain $f \equiv 0$. By referring to (3.4.3), we infer that $y \in U(0)$. Hence, the fact $x \in U(0)$ and $y \in U(0)$ implies from Theorem 2.8.1 (i) that

$$x + U(0) = y + U(0).$$

Again, by using (3.4.3), we conclude that

$$(\lambda - U)f(\lambda) = x + U(0).$$

This is equivalent to saying that $\lambda \in \rho_U(x)$. Thus,

$$\rho_U(x) = \mathbb{C},$$

for all $x \in U(0)$.
(iii) Let us assume that $\lambda \in \rho_U(\alpha x)$. Then, there exists an open neighborhood $\mathscr{U}$ of λ in $\mathbb{C}$ and an analytic function $f : \mathscr{U} \longrightarrow \mathscr{D}(U)$ such that

$$(\lambda - U)f(\lambda) = \alpha x + U(0),$$

for all $\lambda \in \mathscr{U}$. Using the fact that $\alpha \neq 0$, we infer that $g : U \longrightarrow \mathscr{D}(U)$ which defined by

$$g(\lambda) = \frac{1}{\alpha} f(\lambda),$$

is an analytic function. Hence, by the linearity of $\lambda - U$, we obtain

$$\begin{aligned}(\lambda - U)g(\lambda) &= \frac{1}{\alpha}((\lambda - U)f(\lambda)) \text{ for all } \lambda \in \mathscr{U} \\ &= \frac{1}{\alpha}(\alpha x + U(0)) \\ &= x + \frac{1}{\alpha}U(0) \\ &= x + U(0).\end{aligned}$$

This shows that

$$\rho_U(\alpha x) \subset \rho_U(x).$$

Conversely, a same reasoning as before leads to the result.
(iv) If $\alpha = \beta = 0$, then it follows from (iii) that

$$\rho_U(0\,x + 0\,y) = \rho_U(0) = \mathbb{C}.$$

Hence,

$$\rho_U(x) \bigcap \rho_U(y) \subset \rho_U(0).$$

It remains to prove that

$$\rho_U(x) \bigcap \rho_U(y) \subset \rho_U(\alpha\,x + \beta\,y)$$

for all $(\alpha, \beta) \in \mathbb{C}^2 \backslash \left\{ \begin{pmatrix} 0 \\ 0 \end{pmatrix} \right\}$. Let $\lambda \in \rho_U(x) \bigcap \rho_U(y)$. Then, there exist two open neighborhoods $\mathcal{U}_1$, $\mathcal{U}_2$ of λ in $\mathbb{C}$ and two analytic functions $f : \mathcal{U}_1 \longrightarrow \mathcal{D}(U)$, $g : \mathcal{U}_2 \longrightarrow \mathcal{D}(U)$ such that

$$(\xi - U)f(\xi) = x + U(0),$$

for all $\xi \in \mathcal{U}_1$ and

$$(\zeta - U)g(\zeta) = x + U(0),$$

for all $\zeta \in \mathcal{U}_2$. Consequently, for all $\mu \in \mathcal{U}_1 \bigcap \mathcal{U}_2$, we have

$$(\mu - U)f(\mu) = x + U(0) \text{ and } (\mu - U)g(\mu) = y + U(0). \tag{3.4.4}$$

Let $h : \mathcal{U}_1 \bigcap \mathcal{U}_2 \longrightarrow \mathcal{D}(U)$ be given by

$$h(\mu) = \alpha f(\mu) + \beta g(\mu), \tag{3.4.5}$$

for all $(\alpha, \beta) \in \mathbb{C}^2$ and $\mu \in \mathcal{U}_1 \bigcap \mathcal{U}_2$. Since $\mathcal{U}_1 \bigcap \mathcal{U}_2$ is an open neighborhood of λ in $\mathbb{C}$, f and g are analytic functions on $\mathcal{U}_1 \bigcap \mathcal{U}_2$, then h is an analytic function. At this level, we propose to prove that

$$(\mu - U)h(\mu) = (\alpha x + \beta y) + U(0), \tag{3.4.6}$$

for all $\alpha, \beta \in \mathbb{C}$ and $\mu \in \mathcal{U}_1 \bigcap \mathcal{U}_2$.

<u>First case</u>. If $\alpha, \beta \in \mathbb{C}\backslash\{0\}$, then by referring to (3.4.5), for all $\mu \in \mathcal{U}_1 \bigcap U_2$, we obtain

$$\begin{aligned} (\mu - U)h(\mu) &= (\mu - U)(\alpha f(\mu) + \beta g(\mu)), \\ &= \alpha\left((\mu - U)f(\mu)\right) + \beta\left((\mu - U)g(\mu)\right). \end{aligned}$$

This implies from (3.4.4), for all $\mu \in \mathcal{U}_1 \bigcap \mathcal{U}_2$, that

$$\begin{aligned} (\mu - U)h(\mu) &= \alpha(x + U(0)) + \beta\,(y + U(0)), \\ &= \alpha x + U(0) + \beta\, y + U(0), \\ &= \alpha x + \beta\, y + U(0). \end{aligned}$$

<u>Second case</u>. If $\alpha = 0$ and $\beta \in \mathbb{C}\backslash\{0\}$, then it follows from (3.4.4) and (3.4.5), for all $\mu \in \mathcal{U}_1 \bigcap \mathcal{U}_2$ that

$$\begin{aligned} (\mu - U)h(\mu) &= (\mu - U)(\beta g(\mu)), \text{ for all } \mu \in \mathcal{U}_1 \bigcap \mathcal{U}_2 \\ &= \beta y + U(0). \end{aligned}$$

Third case: If $\beta = 0$ and $\alpha \in \mathbb{C}\backslash\{0\}$, then by (3.4.4) and (3.4.5), for all $\mu \in \mathscr{U}_1 \bigcap \mathscr{U}_2$, we infer that

$$(\mu - U)h(\mu) = \alpha x + U(0),$$

for all $\mu \in \mathscr{U}_1 \bigcap \mathscr{U}_2$. Hence, (3.4.6) holds. This means that $\lambda \in \rho_U(\alpha x + \beta y)$, for all $(\alpha, \beta) \in \mathbb{C}^2\backslash\left\{\begin{pmatrix} 0 \\ 0 \end{pmatrix}\right\}$. So,

$$\rho_U(x) \bigcap \rho_U(y) \subset \rho_U(\alpha x + \beta y),$$

for all $\alpha, \beta \in \mathbb{C}$.

(vi) Assume that $\mathbb{C}\backslash(F\backslash\{\lambda\}) \subset \rho_U(x)$ for all $\lambda \in \mathbb{C}$. Let $\mu \in \mathbb{C}\backslash F$. Then, $\mu - \lambda \notin F\backslash\{\lambda\}$. So, $\mu - \lambda \in \mathbb{C}\backslash(F\backslash\{\lambda\})$ which yields $\mu - \lambda \in \rho_U(x)$. This implies that there exist open neighborhood U of $\mu - \lambda$ in $\mathbb{C}$ and an analytic function $f : \mathscr{U} \longrightarrow \mathscr{D}(U)$ which satisfies

$$(\zeta - U)f(\zeta) = x + U(0),$$

for all $\zeta \in \mathscr{U}$. This leads to

$$(\zeta + \lambda - (U + \lambda))f(\zeta + \lambda - \lambda) = x + U(0),$$

for all $\zeta + \lambda \in U + \{\lambda\}$. Consequently,

$$(\xi - (U + \lambda))f(\xi - \lambda) = x + U(0),$$

for all $\xi \in \mathscr{U} + \{\lambda\}$. Consider $g : \mathscr{U} + \{\lambda\} \longrightarrow \mathscr{D}(U)$ defined by

$$g(\xi) = f(\xi - \lambda).$$

Then, g is an analytic function. This leads to

$$(\xi - (U + \lambda))\, g(\xi) = x + U(0),$$

for all $\xi \in \mathscr{U} + \{\lambda\}$. Hence, $\mu \in \rho_{U+\lambda}(x)$. This enables us to conclude that if

$$\mathbb{C}\backslash(F\backslash\{\lambda\}) \subset \rho_U(x), \text{ then } \mathbb{C}\backslash F \subset \rho_{U+\lambda}(x). \tag{3.4.7}$$

The converse implication holds by substituting λ, F and U, respectively, for $-\lambda$, $F\backslash\{\lambda\}$ and $U + \lambda$ in (3.4.7). Q.E.D.

Proposition 3.4.2 *Let $U \in C\mathscr{R}(H)$ and Q be a involutive single-valued such that $U^Q = UQU$, where U^Q is given in (2.11.12). Then, $\rho_{U^Q}(Qx) = \rho_U(x)^c$.* ◇

Proof Let $\lambda \in \rho_U(x)$. Then, there exist an open neighborhood $\mathbb{D}(\lambda, r)$ of λ in $\mathbb{C}$ and an analytic function $f : \mathbb{D}(\lambda, r) \longrightarrow \mathscr{D}(U)$ which satisfies

$$(\mu - U)f(\mu) = x + U(0) \text{ for all } \mu \in \mathbb{D}(\lambda, r).$$

This implies that

$$\begin{aligned} Q(\mu - U)f(\mu) &= Q\,(x + U(0)) \\ &= Q(x) + QU(0) \\ &= Q(x) + QUQ(0) \text{ (as } Q \text{ is an operator).} \end{aligned}$$

Based on the assumption $Q^2 = I$, we infer that

$$Q(\mu - U)Q^2 f(\mu) = Q(x) + QUQ(0).$$

The fact that $Q \in \mathscr{L}(H)$ implies

$$(\mu - QUQ)Qf(\mu) = Q(x) + QUQ(0). \tag{3.4.8}$$

Let us define $g : \mathbb{D}(\overline{\lambda}, r) \longrightarrow \mathscr{D}(U)$ by $g(\overline{\mu}) = \underline{Qf(\mu)}$ for $\overline{\mu} \in \mathbb{D}(\overline{\lambda}, r)$. It follows from Lemma 2.10.8 that g is an analytic function on $\mathbb{D}(\overline{\lambda}, r)$. By referring to (3.4.8), we have

$$(\overline{\mu} - QUQ)g(\overline{\mu}) = Q(x) + QUQ(0) \text{ for all } \overline{\mu} \in \mathbb{D}(\overline{\lambda}, r).$$

Hence, $\overline{\lambda} \in \rho_{U^Q}(Qx)$. This shows that

$$\rho_U(x)^c \subset \rho_{U^Q}(Qx). \tag{3.4.9}$$

By substituting U by U^Q and x by Qx, respectively, in (3.4.9), we obtain

$$\begin{aligned} \rho_{U^Q}(Qx) &\subset \rho_{(U^Q)^Q}(x)^c. \\ &\subset \rho_U(x)^c \text{ (by using Remark 2.11.10).} \end{aligned}$$

This completes the proof. Q.E.D.

3.4.1 Local Spectrum Set of Linear Relation

Let $U \in L\mathscr{R}(X)$ and $x \in X$. The local spectrum of U at the point x is then defined by

$$\sigma_U(x) := \mathbb{C} \setminus \rho_U(x).$$

Evidently, by Definition 3.4.1, we have

$$\rho_U(x) := \bigcup_{\lambda \in \rho_U(x)} \mathscr{U}_\lambda,$$

where $\mathscr{U}_\lambda$ is an open neighborhoods of λ. Hence, it is an open subset of $\mathbb{C}$. Moreover, for all $x \in X$,

$$\rho(U) \subseteq \rho_U(x)$$

and

$$\sigma_U(x) \subseteq \sigma(U).$$

At this level, we assume that $P\sigma(U) \neq \emptyset$ and $U(0)$ is the kernel of continuous projection P defined on $R(U)$. Since $A = PU + U$ is continuous selection of U, then $\mathscr{D}(U) = X$ that A is a bounded selection of U. Hence, the fact that $P\sigma(U) \neq \emptyset$ implies, from Lemma 3.2.2, that $\sigma(A) \neq \emptyset$. It follows from [15] that it is possible to find an analytic function g_x such that

$$(\lambda - A)g_x(\lambda) = x, \tag{3.4.10}$$

for some, and sometimes even for all, values of λ that are in the spectrum of A. Let $\mu \in P\sigma(U)$. If (3.4.10) holds for some $\mu \in P\sigma(A)$, then again by Lemma 3.2.2, for some $\mu \in P\sigma(U)$, we conclude that

$$\begin{aligned}(\mu - U)g_x(\mu) &= (\mu - A)g_x(\mu)) + U(0),\\ &= x + U(0).\end{aligned}$$

This property suggests the following concepts. We can find the following result in [16].

Proposition 3.4.3 *Let X be a Banach space and let $U \in BC\mathscr{R}(X)$. Then,*

$$\sigma_\delta(U) = \bigcup_{x \in X} \sigma_U(x)$$ ◇

Let us investigate some auxiliary results.

Proposition 3.4.4 *Let $A \in L\mathscr{R}(X)$, $B \in L\mathscr{R}(Y)$ and $C \in L\mathscr{R}(Y, X)$. If A has a selection E which satisfies $A(0) \subset R(E)$ and $A(0) \subset C(0)$, then*

$$\sigma_{\mathscr{M}_C}(x) \subset \sigma_\delta(A) \bigcup \sigma_B(z),$$

for all $x = y \bigoplus z \in X \bigoplus Y$, where $\mathscr{M}_C$ is given in (2.9.1). ◇

Proof Let $\lambda \notin \sigma_\delta(A) \bigcup \sigma_B(z)$. Then, there exist two open neighborhoods $\mathscr{V}_1$ and $\mathscr{V}_2$ of λ in $\mathbb{C}$ and an analytic operator function $f_1 : \mathscr{V}_1 \longrightarrow Y$ such that

$$(B-\mu)f_1(\mu) = z + B(0), \quad \text{for all } \mu \in \mathscr{V}_1$$

and

$$V_2 \bigcap \sigma_\delta(A) = \emptyset.$$

Let us assume that $\mathscr{V} = \mathscr{V}_1 \bigcap \mathscr{V}_2$. Hence, we get

$$(B-\mu)f_1(\mu) = z + B(0), \quad \text{for all } \mu \in \mathscr{V}.$$

Furthermore, by applying Theorem 2.14.1, there exists an analytic operator function $f_2 : V \longrightarrow X$ satisfying

$$(A-\mu)f_2(\mu) = y - Cf_1(\mu) + A(0), \quad \text{for all } \mu \in V.$$

Moreover, the nonzero analytic relation linear function $f_2 \bigoplus f_1 : \mathscr{V} \longrightarrow X \bigoplus Y$ defined by

$$(f_2 \bigoplus f_1)(\mu) = f_2(\mu) \bigoplus f_1(\mu)$$

satisfies

$$\begin{aligned}(\mathscr{M}_C - \mu)(f_2(\mu) \bigoplus f_1(\mu)) &= (y + A(0)) \bigoplus (z + B(0)) \\ &= x + \mathscr{M}_C(0),\end{aligned}$$

for all $\mu \in \mathscr{V}$. Thus, $\lambda \notin \sigma_{\mathscr{M}_C}(x)$. Q.E.D.

Corollary 3.4.1 *Let $A \in L\mathscr{R}(X)$, $B \in L\mathscr{R}(Y)$ and $C \in L\mathscr{R}(Y, X)$. If A has a selection E satisfying $A(0) \subset R(E)$ and $A(0) \subset C(0)$, then*

$$\sigma_\delta(\mathscr{M}_C) \subset \sigma_\delta(A) \bigcup \sigma_\delta(B),$$

where $\mathscr{M}_C$ is given in (2.9.1). ◇

3.4.2 Analytic Residuum of Multivalued Linear Operators

Let us now exhibit the definition of the analytic residuum of multivalued linear operators. Let $U \in B\mathscr{R}(X)$. The analytic residuum S_U is the set of $\lambda_0 \in \mathbb{C}$ for which there exist a neighborhood $\mathscr{V}_{\lambda_0}$ and $f : \mathscr{V}_{\lambda_0} \longrightarrow \mathscr{D}(U)$, a nonzero analytic function such that

$$(U-\lambda)f(\lambda) = U(0), \quad \text{for all } \lambda \in \mathscr{V}_{\lambda_0}.$$

Remark 3.4.2 Let $A \in L\mathcal{R}(X)$, $B \in L\mathcal{R}(Y)$ and $C \in L\mathcal{R}(Y, X)$.

(i) If $C = 0$, then $\sigma_{\mathcal{M}_0}(y \bigoplus z) = \sigma_A(y) \bigcup \sigma_B(z)$.
(ii) $\sigma_B(z) \subset \sigma_{\mathcal{M}_C}(y \bigoplus z)$ for all $y \in X$. Indeed, let $y \in X$ and $\lambda \notin \sigma_{\mathcal{M}_C}(y \bigoplus z)$. As a matter of fact, there exist an open neighborhood V_λ and a nonzero analytic operator function $f : V_\lambda \longrightarrow X \bigoplus Y$ such that

$$(\mathcal{M}_C - \mu) f(\mu) = (y \bigoplus z) + \mathcal{M}_C(0), \quad \text{for all } \mu \in V_\lambda.$$

Hence, for $f = f_1 \bigoplus f_2$ with $f_1 : V_\lambda \longrightarrow X$ and $f_2 : V_\lambda \longrightarrow Y$ which are analytic operators functions, we obtain

$$(A - \mu) f_1(\mu) + C f_2(\mu) = y + A(0)$$

and

$$(B - \mu) f_2(\mu) = z + B(0), \quad \text{for all } \mu \in V_\lambda.$$

This implies that $\lambda \notin \sigma_B(z)$. Thus, we conclude that $\sigma_B(z) \subset \sigma_{\mathcal{M}_C}(y \bigoplus z)$.
(iii) It is easy to notice that $S_A \subset S_{\mathcal{M}_C} \subset S_A \bigcup S_B$. As a consequence, by referring to Corollary 3.4.1, we infer that $\sigma(\mathcal{M}_C) \subset \sigma(A) \bigcup \sigma(B)$. ◇

Proposition 3.4.5 *Let $A \in L\mathcal{R}(X)$, $B \in L\mathcal{R}(Y)$ and $C \in L\mathcal{R}(Y, X)$. If A has a selection E satisfying $A(0) \subset R(E)$ and $A(0) \subset C(0)$, then*

$$S_B \subset \sigma_\delta(A) \bigcup S_{\mathcal{M}_C},$$

where $\mathcal{M}_C$ is given in (2.9.1). ◇

Proof Let us suppose that $\lambda_0 \in S_B \setminus \sigma_\delta(A)$. Then, there exist a neighborhood V_{λ_0} and a nonzero analytic operator function $f_1 : V_{\lambda_0} \longrightarrow Y$ which satisfies

$$(B - \mu) f_1(\mu) = B(0),$$

for all $\mu \in V_{\lambda_0}$ and $V_{\lambda_0} \bigcap \sigma_\delta(A) = \emptyset$. Applying Theorem 2.14.1, there exists an analytic operator function $f_2 : V_{\lambda_0} \longrightarrow X$ satisfying

$$(A - \mu) f_2(\mu) = -C f_1(\mu) + A(0), \quad \text{for all } \mu \in V_{\lambda_0}.$$

Moreover, the nonzero analytic function of the linear relation $f_2 \bigoplus f_1 : V_{\lambda_0} \longrightarrow X \bigoplus Y$ defined by

$$(f_2 \bigoplus f_1)(\mu) = f_2(\mu) \bigoplus f_1(\mu)$$

satisfies

$$(\mathcal{M}_C - \mu)(f_2(\mu) \bigoplus f_1(\mu)) = \mathcal{M}_C(0), \quad \text{for all } \mu \in V_{\lambda_0}.$$

Therefore, $\lambda_0 \in S_{\mathcal{M}_C}$, which proves our claim. Q.E.D.

Now, departing from Remark 3.4.2 (ii), (iii) and Proposition 3.4.5, we build up the following result.

Corollary 3.4.2 *Let $A \in L\mathcal{R}(X)$, $B \in L\mathcal{R}(Y)$ and $C \in L\mathcal{R}(Y, X)$. If A has a selection E which satisfies $A(0) \subset R(E)$ and $A(0) \subset C(0)$, then*

$$\sigma(B) \subset \sigma_\delta(A) \bigcup \sigma(\mathcal{M}_C),$$

where $\mathcal{M}_C$ is given in (2.9.1). ◇

We can define the upper triangular 3×3-block matrix of linear relations

$$\mathcal{U} = \begin{pmatrix} U_{11} & U_{12} & U_{13} \\ 0 & U_{22} & U_{23} \\ 0 & 0 & U_{33} \end{pmatrix} \tag{3.4.11}$$

on $X_1 \bigoplus X_2 \bigoplus X_3$, where $U_{ii} \in L\mathcal{R}(X_i)$, and $U_{ij} \in L\mathcal{R}(X_j, X_i)$ for all $i \in \{1, 2, 3\}$, $j \in \{2, 3\}$ and X_1, X_2, X_3 are complex Banach spaces.

Proposition 3.4.6 *If $U_{ii} \in B\mathcal{R}(X_i)$, and U_{ij} are bounded operators from X_j into X_i for all $i \in \{1, 2, 3\}$ and $j \in \{2, 3\}$. Then,*

$$P\sigma(U_{11}) \subset P\sigma(\mathcal{U}) \subset P\sigma(U_{11}) \bigcup P\sigma(U_{22}) \bigcup P\sigma(U_{33}),$$

where $\mathcal{U}$ is given in (3.4.11). ◇

Proof To prove that $P\sigma(\mathcal{U}) \subset P\sigma(U_{11}) \bigcup P\sigma(U_{22}) \bigcup P\sigma(U_{33})$, we assume that $\lambda \notin P\sigma(U_{11}) \bigcup P\sigma(U_{22}) \bigcup P\sigma(T_{33})$. Then, $\lambda - U_{11}$, $\lambda - U_{22}$ and $\lambda - U_{33}$ are one-to-one. Let $(x, y, z) \in N(\lambda - \mathcal{U})$. Hence,

$$\begin{pmatrix} 0 \\ 0 \\ 0 \end{pmatrix} \in (\lambda - \mathcal{U}) \begin{pmatrix} x \\ y \\ z \end{pmatrix}.$$

Also, since $\lambda - U_{33}$ is one-to-one, then $z = 0$. Further, based on the hypotheses $\lambda - T_{22}$ is one-to-one and U_{23} is an operator, we conclude that $y = 0$. Again, using the fact that $\lambda - U_{11}$ is one-to-one, U_{12} and U_{13} are operators, we infer that $x = 0$. Thus,

$$N(\lambda - \mathcal{T}) = \{(0, 0, 0)\},$$

which leads to conclude that $\lambda - \mathcal{U}$ is one-to-one. Conversely, let $\lambda \notin P\sigma(\mathcal{U})$, hence $\lambda - \mathcal{U}$ is one-to-one. Therefore, $0 \in (\lambda - U_{11})x$ and so

$$\begin{cases} 0 \in (\lambda - U_{11})x - U_{12}(0) - U_{13}(0) \\ 0 \in (\lambda - U_{22})(0) - U_{23}(0) \\ 0 \in (\lambda - U_{33})(0). \end{cases}$$

As a result, $(x, 0, 0) \in N(\lambda - \mathscr{U})$, which implies that $x = 0$. As a consequence, $\lambda \notin P\sigma(U_{11})$. Q.E.D.

3.5 Compact Linear Relations

Throughout this section, we give some fundamental results about compact and demicompact linear relations see [17–24, 24–28, 28–36]. Let X and Y be two normed spaces and $T \in L\mathscr{R}(X, Y)$. We say that T is precompact (resp. compact), if the analogous property holds for $Q_T T$, i.e., T is precompact (resp. compact) if $Q_T T(B_X)$ (resp. $\overline{Q_T T(B_X)}$) is totally bounded (resp. a compact subset) in Y. A relation $S \in L\mathscr{R}(X, Y)$ is said to be T-compact if $\mathscr{D}(T) \subset \mathscr{D}(S)$, and SG_T is compact. The set of precompact (resp. precompact and everywhere defined, compact) is denoted $\mathscr{P}\mathscr{R}\mathscr{K}(X, Y)$ (resp. $\mathscr{E}\mathscr{P}\mathscr{R}\mathscr{K}(X, Y)$, $K\mathscr{R}(X, Y)$). If $X = Y$, we may write $\mathscr{P}\mathscr{R}\mathscr{K}(X)$, $\mathscr{E}\mathscr{P}\mathscr{R}\mathscr{K}(X)$ and $K\mathscr{R}(X)$ instead of $\mathscr{P}\mathscr{R}\mathscr{K}(X, Y)$, $\mathscr{E}\mathscr{P}\mathscr{R}\mathscr{K}(X, Y)$ and $K\mathscr{R}(X, Y)$, respectively.

Proposition 3.5.1 *Let X be complete, A, B and $S \in L\mathscr{R}(X)$ be such that $B(0) \subset A(0)$ and let $\lambda \in \mathbb{C}$. If B is A-precompact, then $i(\lambda S - A) = i(\lambda S - (A + B))$.* ◇

Proof Since BG_A is precompact, and X is complete, then BG_A is compact and by using Lemma 2.11.1, we have

$$\begin{aligned} i(\lambda S - A) &= i((\lambda S - A)G_A), \\ &= i((\lambda S - A)G_A + BG_A), \\ &= i((\lambda S - (A + B))G_A), \\ &= i(\lambda S - (A + B)). \end{aligned}$$

This achieves the proof. Q.E.D.

Proposition 3.5.2 [10, Proposition V.5.3] *Let T be a continuous linear relation. Then, T is precompact if, and only if, T^* is compact.* ◇

Let us first recall some facts about bounded linear operators in Banach spaces. Suppose that X and Y are Banach spaces and $T : X \longrightarrow Y$ is a bounded linear operator. Then, T is said to be weakly compact if $\overline{T(B_X)}$ is weakly compact (equivalently, if T transforms each bounded sequence into a sequence having a weakly convergent subsequence).

Definition 3.5.1 Let $T \in L\mathscr{R}(X, Y)$. Then, T is said to be weakly compact if $T(B_X)$ is relatively $\sigma(Y, \mathscr{D}(T^*))$-compact, where $\sigma(\cdot, \cdot)$ is the weak topology. ◇

We denote by $W\mathscr{R}(X, Y)$ the set of weakly compact linear relation from X into Y. If $X = Y$, we may write $W\mathscr{R}(X)$ instead of $W\mathscr{R}(X, Y)$. Notice that

(i) T is weakly compact if, and only if, the single-valued relation QT is weakly compact, since the domains of T^* and $(QT)^*$ coincide, and $\mathcal{D}(T^*) \subset T(0)^\perp$.
(ii) If T is continuous, then

$$\begin{aligned}(Q(Y))^* &= T(0)^\perp \\ &= \mathcal{D}(T^*),\end{aligned}$$

and so the definition is the standard one for the continuous linear operator QT.

Definition 3.5.2 Let $T \in L\mathcal{R}(X, Y)$. The linear relation T is called strictly singular if there is no infinite-dimensional subspace M of $\mathcal{D}(T)$ for which $T_{|M}$ is one-to-one and open. ◇

The set of strictly singular linear relations is denoted by $\mathcal{S}\mathcal{S}\mathcal{R}(X, Y)$. If $X = Y$, we may write $\mathcal{S}\mathcal{S}\mathcal{R}(X)$ instead of $\mathcal{S}\mathcal{S}\mathcal{R}(X, Y)$.

Theorem 3.5.1 [10, Proposition V.2.12] *Let $T \in L\mathcal{R}(X, Y)$ be single-valued and continuous, and let $S \in L\mathcal{R}(Y, Z)$ be strictly singular. Then, ST is strictly singular.* ◇

Lemma 3.5.1 [10, Corollary V.2.3 and V.2.8] *Let $K \in L\mathcal{R}(X, Y)$ be precompact, then K is a continuous strictly singular linear relation.* ◇

Theorem 3.5.2 [10, Proposition V.2.10] *Let $T \in L\mathcal{R}(X, Y)$ be strictly singular and let $S \in L\mathcal{R}(Y, Z)$ be continuous. If $\overline{T(0)} \subset \mathcal{D}(S)$, then ST is strictly singular.* ◇

Definition 3.5.3 Let X and Y be two normed spaces, and $\mathcal{I}(X)$, $\mathcal{C}(X)$, and $\mathcal{P}(X)$ be denote, respectively, the infinite dimensional, finite codimensional, and closed finite codimensional subspaces of a normed linear space X. We define the quantities (also called measures of compactness) $\Gamma(T)$, $\Gamma_0(T)$, $\overline{\Gamma}_0(T)$, and $\widetilde{\Delta}(T)$ as follows:

First case. If $\dim(\mathcal{D}(T)) < \infty$ and $\dim(Y) < \infty$, then all quantities are zero. Thus,

$$\begin{aligned}\Gamma(T) &= \widetilde{\Delta}(T) \\ &= \Gamma_0(T) \\ &= \overline{\Gamma}_0(T) \\ &= 0.\end{aligned}$$

Second case. If $\dim(\mathcal{D}(T)) = \infty$ and $\dim(Y) = \infty$, then

$$\begin{aligned}\Gamma(T) &= \inf_{M \in \mathcal{I}(\mathcal{D}(T))} \|T_{|M}\| \\ \Gamma_0(T) &= \inf_{M \in \mathcal{C}(\mathcal{D}(T))} \|T_{|M}\| \\ \overline{\Gamma}_0(T) &= \inf_{M \in \mathcal{P}(\mathcal{D}(T))} \|T_{|M}\| \\ \Delta(T) &= \sup_{M \in \mathcal{I}(\mathcal{D}(T))} \Gamma(T_{|M}).\end{aligned}$$

◇

The following inequalities hold

$$\Gamma(T) \leq \Delta(T) \leq \Gamma_0(T) \leq \overline{\Gamma}_0(T).$$

Let $\mathscr{E}(X)$ be denote the collection of closed subspaces of X of infinite dimension. If $T \in L\mathscr{R}(X, Y)$, we define the quantities

$$\Gamma'(T) := \inf_{M \in \mathscr{E}(X)} ||Q_M T||$$

and

$$\Delta'(T) := \sup_{M \in \mathscr{E}(X)} \Gamma'(Q_M T).$$

The linear relation T is called strictly cosingular if $\Delta'(T) = 0$. We denote the class of strictly cosingular linear relations from X into Y by $\mathscr{SRC}(X, Y)$. If $X = Y$, we may write $\mathscr{SRC}(X)$ instead of $\mathscr{SRC}(X, Y)$. The following proposition gives a characterization of the strictly singular multivalued linear operators by means of the compactness measure.

Proposition 3.5.3 [10, Theorem V. 2.6] *Let X and Y be two normed spaces and let $T \in L\mathscr{R}(X, Y)$. Then, T is strictly singular if, and only if, $\Delta(T) = 0$.* ◇

Let M be a subset of X and let J_M^X (abbreviated J_M when X is understood) be denote the natural injection map of M into X, i.e., $J_M \in \mathscr{R}(M, X)$, $\mathscr{D}(J_M) = M$ and $J_M(m) = m$ for all $m \in M$. Then,

$$T_{|M} = T J_M J_M^{-1}.$$

Remark 3.5.1 Let M be a subspace of a normed vector space X, then the injection operator J_M^X is continuous. ◇

Lemma 3.5.2 [10, Proposition IV.5.11] *(i) Let $T, S \in L\mathscr{R}(X, Y)$. Then,*

$$\Gamma'(T + S) \leq \Delta'(J_Y T) + \Gamma'(S).$$

(ii) Let $T \in L\mathscr{R}(X, Y)$. Then,

$$\Gamma'(J_Y T) = \Gamma'(T).$$

(iii) Let $M \subset Y$ be such that $\dim(M) < \infty$ *and* $\dim(Y) = \infty$. *Then,*

$$\Gamma'(Q_M T) = \Gamma'(T)$$

and

$$\Delta'(Q_M T) = \Delta'(T).$$ ◇

3.5.1 Demicompact Linear Relations

A linear relation $U : \mathscr{D}(U) \subset X \to Y$ is called demicompact if for every bounded sequence $(x_n)_n$ in $\mathscr{D}(U)$ such that

$$Q_{I-U}(I-U)x_n \to y \in Y/\overline{(I-U)(0)},$$

there exists a subsequence of $(Q_U x_n)_n$ that converges.
The collection of all demicompact linear relations between X and Y is denoted by $\mathscr{DC}(X, Y)$. When $X = Y$, this set is abbreviated as $\mathscr{DC}(X) := \mathscr{DC}(X, X)$.

Remark 3.5.2 A linear relation $R : \mathscr{D}(R) \subset X \to X$ is demicompact precisely when, for every bounded sequence $(x_n)_n$ in $\mathscr{D}(R)$ for which

$$Q_R x_n - Q_R R x_n \to y \in X/\overline{R(0)},$$

there exists a convergent subsequence of $(Q_R x_n)_n$.

Since $R(0)$ is a linear subspace of X, it follows that

$$(I - R)(0) = 0 - R(0) = -R(0) = R(0).$$

Therefore, the quotient operators satisfy

$$Q_{I-R} = Q_R,$$

and since the domain of Q_R is the whole space X, we have

$$Q_{I-R}(I - R) = Q_R(I - R) = Q_R - Q_R R. \qquad \diamondsuit$$

Theorem 3.5.3 *Let $\Lambda \in L\mathscr{R}(X)$. If Λ is a compact linear relation, then Λ is also a demicompact linear relation.* $\diamondsuit$

Proof Consider a bounded sequence $(x_n)_n \subset \mathscr{D}(\Lambda)$ for which

$$Q_\Lambda(x_n - \Lambda x_n) = Q_\Lambda x_n - Q_\Lambda \Lambda x_n \to y_0.$$

Because the sequence $(x_n)_n$ is bounded, there exists a constant $M > 0$ such that $\|x_n\| \leq M$ holds for every index n. Dividing the terms $Q_\Lambda \Lambda x_n$ by M, we observe that

$$\frac{1}{M} Q_\Lambda \Lambda x_n \in Q_\Lambda \Lambda B_X,$$

where B_X denotes the closed unit ball in X. Since Λ is a compact linear relation, the closure $\overline{Q_\Lambda \Lambda B_X}$ is compact. This compactness ensures the existence of a subsequence $\left(\frac{1}{M} Q_\Lambda \Lambda x_{\psi(n)}\right)_n$ which converges to some element $y_1 \in X$. Combining these facts, the subsequence $(Q_\Lambda x_{\psi(n)})_n$ converges to the limit $y_0 + M y_1$. Hence, from the boundedness and compactness assumptions, the desired convergence of the subsequence of $(Q_A x_n)_n$ is established. Q.E.D.

Proposition 3.5.4 *Let $\mathscr{R} \in L\mathscr{R}(X)$. Assume that $I - Q_{\mathscr{R}}$ is compact. Then, $\mathscr{R}$ is a demicompact linear relation if and only if the operator $Q_{\mathscr{R}}\mathscr{R}$ is demicompact.*

Proof Assume that $\mathscr{R}$ is demicompact. Let $(x_n)_n$ be a bounded sequence in $\mathscr{D}(\mathscr{R})$ such that

$$x_n - Q_{\mathscr{R}}\mathscr{R}x_n \to y.$$

We can write

$$x_n - Q_{\mathscr{R}}\mathscr{R}x_n = (I - Q_{\mathscr{R}})x_n + \big(Q_{\mathscr{R}}x_n - Q_{\mathscr{R}}\mathscr{R}x_n\big).$$

Since $I - Q_{\mathscr{R}}$ is compact and $(x_n - Q_{\mathscr{R}}\mathscr{R}x_n)_n$ converges, it follows that the sequence $(Q_{\mathscr{R}}x_n - Q_{\mathscr{R}}\mathscr{R}x_n)_n$ admits a convergent subsequence. Then, using the demicompactness of $\mathscr{R}$, we conclude that $(Q_{\mathscr{R}}x_n)_n$ also has a convergent subsequence. Moreover,

$$x_n = (I - Q_{\mathscr{R}})x_n + Q_{\mathscr{R}}x_n.$$

Since $I - Q_{\mathscr{R}}$ is compact and $(Q_{\mathscr{R}}x_n)_n$ has a convergent subsequence, the sequence $(x_n)_n$ itself admits a convergent subsequence. Hence, $Q_{\mathscr{R}}\mathscr{R}$ is demicompact.

Conversely, suppose $Q_{\mathscr{R}}\mathscr{R}$ is demicompact. Let $(x_n)_n$ be a bounded sequence in $\mathscr{D}(\mathscr{R})$ such that

$$Q_{\mathscr{R}}x_n - Q_{\mathscr{R}}\mathscr{R}x_n \to y.$$

Observe that

$$Q_{\mathscr{R}}x_n - Q_{\mathscr{R}}\mathscr{R}x_n = -(I - Q_{\mathscr{R}})x_n + \big(x_n - Q_{\mathscr{R}}\mathscr{R}x_n\big).$$

Since $I - Q_{\mathscr{R}}$ is compact and $(Q_{\mathscr{R}}x_n - Q_{\mathscr{R}}\mathscr{R}x_n)_n$ converges, the sequence $(x_n - Q_{\mathscr{R}}\mathscr{R}x_n)_n$ admits a convergent subsequence. By the demicompactness of $Q_{\mathscr{R}}\mathscr{R}$, it follows that $(x_n)_n$ has a convergent subsequence.
Finally,

$$Q_{\mathscr{R}}x_n = -(I - Q_{\mathscr{R}})x_n + x_n.$$

Since $I - Q_{\mathscr{R}}$ is compact and $(x_n)_n$ has a convergent subsequence, $(Q_{\mathscr{R}}x_n)_n$ also admits a convergent subsequence. Q.E.D.

Lemma 3.5.3 *Let A and B be closed linear relations. Then the following statements hold:*
(i) The block-diagonal linear relation

$$\mathscr{U} = \begin{pmatrix} A & 0 \\ 0 & B \end{pmatrix}$$

is demicompact if and only if both A and B are demicompact.

(ii) Suppose C is a linear relation such that $C(0) \subset A(0)$, and the quotient map Q_A is bounded. If both A and B are demicompact, then the upper block-triangular relation

$$\mathcal{M}_C = \begin{pmatrix} A & C \\ 0 & B \end{pmatrix}$$

is also demicompact.

◇

Proof (i) Let $((x_n, y_n))_n$ be a bounded sequence in $\mathcal{D}(\mathcal{U}) = \mathcal{D}(A) \times \mathcal{D}(B)$ such that

$$Q_{I-\mathcal{U}}(I - \mathcal{U})\begin{pmatrix} x_n \\ y_n \end{pmatrix} \to \begin{pmatrix} x \\ y \end{pmatrix}.$$

i.e.,

$$\begin{pmatrix} Q_{I-A}(I-A) & 0 \\ 0 & Q_{I-B}(I-B) \end{pmatrix}\begin{pmatrix} x_n \\ y_n \end{pmatrix} \to \begin{pmatrix} x \\ y \end{pmatrix}.$$

The latter is equivalent to

$$\begin{pmatrix} Q_A(I-A) & 0 \\ 0 & Q_B(I-B) \end{pmatrix}\begin{pmatrix} x_n \\ y_n \end{pmatrix} \to \begin{pmatrix} x \\ y \end{pmatrix}.$$

It follows that

$$\begin{pmatrix} Q_A(I-A)x_n \\ Q_B(I-B)y_n \end{pmatrix} \to \begin{pmatrix} x \\ y \end{pmatrix}.$$

Since $I - A$ is demicompact, then $Q_A x_n$ has a convergent subsequence. Denote by $Q_A x_{\varrho(n)}$. However, $Q_B(I - B)y_n \to y$, then

$$Q_B(I-B)y_{\varrho(n)} \to y.$$

As a result, $I - B$ is demicompact, then $Q_B x_{\varrho(n)}$ has a convergent subsequence that we denote by $Q_B x_{\psi(\varrho(n))}$. Thus, $Q_{\mathcal{U}}\begin{pmatrix} x_n \\ y_n \end{pmatrix}$ has a convergent subsequence. The converse implication holds trivially.

(ii) Let $((x_n, y_n))_n$ be a bounded sequence in $\mathcal{D}(\mathcal{M}_C)$ such that

$$Q_{I-\mathcal{M}_C}(I - \mathcal{M}_C)\begin{pmatrix} x_n \\ y_n \end{pmatrix} \to \begin{pmatrix} x \\ y \end{pmatrix}$$

i.e.,

$$\begin{pmatrix} Q_{I-A}(I-A) & -Q_{I-A}C \\ 0 & Q_{I-B}(I-B) \end{pmatrix}\begin{pmatrix} x_n \\ y_n \end{pmatrix} \to \begin{pmatrix} x \\ y \end{pmatrix}.$$

Since $\mathscr{M}_C \begin{pmatrix} 0 \\ 0 \end{pmatrix} = \begin{pmatrix} A(0) \\ B(0) \end{pmatrix}$, then

$$\begin{pmatrix} Q_A(I-A)x_n - Q_A C y_n \\ Q_B(I-B)y_n \end{pmatrix} \to \begin{pmatrix} x \\ y \end{pmatrix}.$$

We establish that that $Q_B(I-B)y_n$ is convergent. In conjunction with the demicompactness of B, this leads to show that $(y_n)_n$ has a convergent subsequence. Since $Q_A C$ is bounded operator and A is demicompact, we infer that $(x_n)_n$ has a convergent subsequence, which demonstrates the demicompactness of $\mathscr{M}_C$. Q.E.D.

Proposition 3.5.5 *Let $V \in C\mathscr{R}(X)$. If V is 1-set-contraction, then μV is demicompact for each $\mu \in [0, 1[$.* ◇

Proof Let $(x_n)_n$ be a bounded sequence of $\mathscr{D}(V)$ such that

$$y_n = Q_{\mu V}(I - \mu V)x_n \to y.$$

If $\delta(Q_{\mu V}(x_n)) \neq 0$, then since $Q_{\mu V}(x_n) \subset y_n + Q_{\mu V}(Vx_n)$ and using the fact that μV is 1-set-contraction, we get

$$\begin{aligned} \delta(\{Q_{\mu V}x_n\}) &\leq \delta(\{y_n) + \delta(Q_{\mu V}\mu V x_n\}) \\ &\leq \delta(\{Q_{\mu V}\mu V x_n)\}) \\ &\leq \mu\delta(\{Q_{\mu V}x_n\}) \\ &< \delta(\{Q_{\mu V}x_n\}), \end{aligned}$$

which is impossible. Hence, $\delta(\{Q_{\mu V}x_n\}) = 0$. So, $(Q_{\mu V}x_n)_n$ is relatively compact, and we deduce that μV is demicompact Q.E.D.

Definition 3.5.4 Let $V \in L\mathscr{R}(X)$, and let δ_1 and δ_2 denote two Kuratowski measures of non-compactness. The following definitions apply:
(i) The multivalued linear operator V is said to be *D-condensing* if, for every bounded subset $B \subset \mathscr{D}(V)$, the image $Q_V V(B)$ is a bounded subset of the quotient space $Y/\overline{V(0)}$, and satisfies

$$\delta_2(Q_V V(B)) < \delta_1(Q_D B),$$

whenever $\delta_1(Q_D B) > 0$.
(ii) If $D = \{0\}$, the operator V is called a *condensing linear relation*, meaning it is $\{0\}$-condensing in the sense described above. ◇

Theorem 3.5.4 *Let $R \in L\mathscr{R}(X)$ be such that $\overline{R(0)} \subset N(R)$. If R is $\overline{R(0)}$-condensing, then R is demicompact.* ◇

Proof Let $(x_n)_n$ be a bounded sequence in $\mathscr{D}(R)$ such that

$$y_n = Q_R(I - R)x_n \to y.$$

Suppose that $\delta(\{x_n\}) \neq 0$. Since $\{Q_R x_n\} \subset \{y_n\} + \{Q_R R x_n\}$, it follows, using Proposition 2.2.1 (iv), that

$$\begin{aligned}\delta(\{Q_R x_n\}) &\leq \delta(\{y_n\}) + \delta(\{Q_R R x_n\}) \\ &\leq \delta(\{Q_R R x_n\}) \\ &< \delta(\{Q_R x_n\}),\end{aligned}$$

which is impossible. Hence, $\delta(\{Q_R x_n\}) = 0$. So, $(Q_R x_n)_n$ is relatively compact. Therefore, $(Q_R x_n)_n$ has a convergent subsequence. Q.E.D.

3.5.2 *Relatively Demicompact Linear Relation*

Let X and Y be Banach spaces. Consider two linear relations $U : \mathscr{D}(U) \subset X \longrightarrow Y$ and $V : \mathscr{D}(V) \subset X \longrightarrow Y$. Suppose that $V(0) \subset U(0)$ and $\mathscr{D}(V) \subset \mathscr{D}(U)$. We say that U is V-*demicompact* (or demicompact relative to V) if, for every bounded sequence $(x_n)_n$ in $\mathscr{D}(U)$ such that

$$Q_{V-U}(V-U)x_n = Q_U(V-U)x_n \to y \in Y/\overline{U(0)},$$

there exists a convergent subsequence of $(Q_U V x_n)_n$.
We denote:

$$\mathscr{DC}_V(X,Y) = \{U \in C\mathscr{R}(X,Y) \text{ such that } U \text{ is } V\text{-demicompact}\}.$$

In the case where $X = Y$, we write:

$$\mathscr{DC}_V(X,X) := \mathscr{DC}_V(X).$$

Lemma 3.5.4 *Let $U : \mathscr{D}(U) \subseteq X \longrightarrow Y$ and $V : \mathscr{D}(V) \subseteq X \longrightarrow Y$ be two linear relations such that $\mathscr{D}(U) \subseteq \mathscr{D}(V)$ and $V(0) \subseteq U(0)$. Assume further that the operator $Q_U(V-I)$ is compact. Under these conditions, U is considered V-demicompact if and only if the operator UG_V possesses the demicompactness property.*

Proof Assume that U is V-demicompact. Consider a bounded sequence $(x_n)_n$ in $\mathscr{D}(U)$, which is contained in $\mathscr{D}(V)$, such that the sequence

$$Q_{UG_V}(I - UG_V)x_n = Q_U(I-U)x_n$$

converges in the quotient space $Y/\overline{U(0)}$. Expanding this, we have

$$Q_U(V-U)x_n = Q_U(I-U)x_n + Q_U(V-I)x_n.$$

Hence, the sequence $\{Q_U(V-U)x_n\}$ converges. Since U is V-demicompact, it follows that there exists a convergent subsequence of $\{Q_U V x_n\}$. Moreover, the relation

$$Q_U x_n = Q_U(I - V)x_n + Q_U V x_n$$

allows us to conclude that $\{Q_U x_n\}$ also admits a convergent subsequence.

Conversely, suppose the operator UG_V is demicompact. Let $(x_n)_n$ be a bounded sequence in $\mathscr{D}(U) \subseteq \mathscr{D}(V)$ such that $\{Q_U(V - U)x_n\}$ converges. Observe that

$$\begin{aligned} Q_U(I - UG_V)x_n &= Q_U(I - U)x_n \\ &= Q_U(V - U)x_n - Q_U(V - I)x_n. \end{aligned}$$

Because the sequences $\{Q_U(V - U)x_n\}$ and $\{Q_U(V - I)x_n\}$ converge, the sequence $\{Q_U(I - UG_V)x_n\}$ also converges. Since UG_V is demicompact, it follows that the sequence $\{Q_{UG_V} x_n\} = \{Q_U x_n\}$ contains a convergent subsequence. Finally, by the identity

$$Q_U V x_n = Q_U x_n - Q_U(I - V)x_n,$$

we conclude that $\{Q_U V x_n\}$ admits a convergent subsequence as well, proving that U is V-demicompact. Q.E.D.

Proposition 3.5.6 *Let $\mu \in \mathbb{C}^*$, and consider the linear relations*

$$U : \mathscr{D}(U) \subseteq X \to Y, \quad V : \mathscr{D}(V) \subseteq X \to Y,$$

such that the operator $\mu V N(\mu V - U)$ is compact, the range of $\mu V - U$ is closed, and $\mu V - U$ is an open mapping. Suppose also that the composed map $Q_{\mu V N(\mu V - U)} \circ \mu V$ is continuous. Under these assumptions, U is μV-demicompact. Moreover, if the domains coincide, i.e., $\mathscr{D}(U) = \mathscr{D}(V)$, then μV is compact relative to U.

Proof Since the operator $\mu V - U$ is open, its range $R(\mu V - U)$ is closed, and the mapping

$$Q_{\mu V N(\mu V - U)} \circ \mu V$$

is continuous, it follows that $Q_U(\mu V - U)$ is also an open operator and that $Q_{\mu V N(\mu V - U)} \mu V$ remains continuous. Consequently, by applying Remark 2.10.2, the composition

$$Q_{\mu V N(\mu V - U)} \mu V \big(Q_U(\mu V - U)\big)^{-1}$$

is continuous as well. Now, let $(x_n)_n$ be a bounded sequence in $\mathscr{D}(U)$ for which the sequence $(Q_U(\mu V - U)x_n)_n$ converges to some $Q_U x$. Using the continuity established above, the sequence

$$\big(Q_{\mu V N(\mu V - U)} \mu V (Q_U(\mu V - U))^{-1} Q_U(\mu V - U)x_n\big)_n$$

converges to

$$Q_{\mu V N(\mu V - U)} \mu V (Q_U(\mu V - U))^{-1} Q_U x.$$

Since $(Q_{\mu VN(\mu V-U)}\mu V x_n)_n$ converges to

$$Q_{\mu VN(\mu V-U)}\mu V(\mu V-U)^{-1}x,$$

and $\mu VN(\mu V-U)$ is compact, it follows that $(\mu V x_n)_n$ admits a convergent subsequence. This shows that U is μV-demicompact. Finally, if the domains coincide, that is $\mathscr{D}(U)=\mathscr{D}(V)$, then μV is U-compact. Q.E.D.

Lemma 3.5.5 *Let $U:\mathscr{D}(U)\subset X\longrightarrow Y$ and $V:\mathscr{D}(V)\subset X\longrightarrow Y$ be two linear relations such that $V(0)\subset U(0)$, $\mathscr{D}(U)\subset\mathscr{D}(V)$, and the operator $Q_U(I-V)$ is compact. Then, U is V-demicompact if and only if the composition UG_V is demicompact.* ◇

Proof Suppose that U is V-demicompact. Let $(x_n)_n$ be a bounded sequence in $\mathscr{D}(U)\subset\mathscr{D}(V)$ with $V(0)\subset U(0)$. Then,

$$Q_U(V-U)x_n=Q_U(I-U)x_n+Q_U(V-I)x_n. \tag{3.5.1}$$

If the sequence $\left(Q_{UG_V}(I-UG_V)x_n\right)_n:=\left(Q_U(I-U)x_n\right)_n$ converges, then by applying (3.5.1) and the compactness of $Q_U(I-V)$, the sequence $\left(Q_U(V-U)x_n\right)_n$ admits a convergent subsequence. Since U is V-demicompact, it follows that $\left(Q_U V x_n\right)_n$ has a convergent subsequence. Furthermore, using

$$Q_U x_n=Q_U(I-V)x_n+Q_U V x_n,$$

we conclude that $\left(Q_U x_n\right)_n$ also contains a convergent subsequence.
Conversely, assume that UG_V is demicompact and let $(x_n)_n$ be a bounded sequence in $\mathscr{D}(U)\subset\mathscr{D}(V)$ with $V(0)\subset U(0)$. Then,

$$\begin{aligned}Q_U(I-UG_V)x_n&=Q_U(I-U)x_n\\&=Q_U(V-U)x_n-Q_U(V-I)x_n.\end{aligned} \tag{3.5.2}$$

If $\left(Q_U(V-U)x_n\right)_n$ converges, then from (3.5.2) the sequence $\left(Q_U(I-UG_V)x_n\right)_n$ converges as well. Since UG_V is demicompact, the sequence $\left(Q_{UG_V}x_n\right)_n:=\left(Q_U x_n\right)_n$ has a convergent subsequence. Finally, from

$$Q_U V x_n=Q_U x_n-Q_U(I-V)x_n,$$

it follows that $\left(Q_U V x_n\right)_n$ has a convergent subsequence.

Q.E.D.

Consider the set

$$\mathscr{DC}_\mu(X,Y)=\{V\in\mathscr{R}(X,Y)\ :\ \|Vx\|\le|\mu|\|x\|\text{ for all }x\in\mathscr{D}(V)\text{ and }|\mu|<1\}.$$

Proposition 3.5.7 *Let $\mu \in \mathbb{C}$, and let $U : \mathcal{D}(U) \subset X \longrightarrow X$. If $U \in \mathcal{DC}_\mu(X) := \mathcal{DC}_\mu(X, X)$, then $U \in \mathcal{DC}(X)$.* ◇

Proof Assume that $U \in \mathcal{DC}_\mu(X)$. Then we have:

$$\|Ux\| \leq |\mu|\|x\| \quad \text{for all } x \in \mathcal{D}(U).$$

It follows that:

$$\|x\| - \|Ux\| \geq (1 - |\mu|)\|x\|,$$

and therefore,

$$\begin{aligned} \|x\| &\leq \frac{\|(I-U)x\|}{1-|\mu|} \\ &= \frac{\|Q_U(I-U)x\|}{1-|\mu|}. \end{aligned} \tag{3.5.3}$$

Let $(x_n)_n$ be a bounded sequence in $\mathcal{D}(U)$ such that:

$$Q_U(I-U)x_n \to y.$$

Applying inequality (3.5.3), we deduce that:

$$\|x_n - y\| \to 0.$$

Thus, the sequence $(x_n)_n$ converges, and so does $(Q_U x_n)_n$. Hence, $U \in \mathcal{DC}(X)$. Q.E.D.

Proposition 3.5.8 *Let $\mu \in \mathbb{C} \setminus \{0\}$, and let $U : \mathcal{D}(U) \subset X \to X$ be a linear relation satisfying:*

$$\|Ux\| \leq |\mu|\|Q_U x\| \quad \textit{for all } x \in \mathcal{D}(U).$$

If $|\mu| < 1$, then U is demicompact.

Proof From the assumption, we obtain:

$$\|Q_U x\| - \|Ux\| \geq (1 - |\mu|)\|Q_U x\|,$$

which implies:

$$\|Q_U x\| \leq \frac{\|Q_U(I-U)x\|}{1-|\mu|}. \tag{3.5.4}$$

Let $(x_n)_n$ be a bounded sequence in $\mathcal{D}(U)$ such that:

$$Q_U(I-U)x_n \to y.$$

Then, from (3.5.4), we conclude that $\|Q_U(x_n - y)\| \to 0$, and thus $(Q_U x_n)_n$ has a convergent subsequence. Q.E.D.

Proposition 3.5.9 *Let $U, U_0 : \mathscr{D}(U) = \mathscr{D}(U_0) \subset X \to Y$ be linear relations, and let $V : \mathscr{D}(V) \subset X \to Y$ be a closed linear relation with:*

$$\mathscr{D}(U) \subset \mathscr{D}(V), \quad V(0) \subset U_0(0) \subset U(0).$$

Assume that U_0 is V-demicompact and there exist scalars $a, b \in \mathbb{C}$ such that $|b| < 1$ and:

$$\|Ux - U_0x\| \leq |a|\|Vx - Ux\| + |b|\|Vx - U_0x\| \quad \text{for all } x \in \mathscr{D}(U).$$

Then U is also V-demicompact.

Proof From the inequality:

$$\begin{aligned}\|Vx - U_0x\| - \|Vx - Ux\| &\leq \|Ux - U_0x\| \\ &\leq |a|\|Vx - Ux\| + |b|\|Vx - U_0x\|,\end{aligned}$$

we derive:

$$(1 - |b|)\|Vx - U_0x\| \leq (1 + |a|)\|Vx - Ux\|,$$

which implies:

$$\|Vx - U_0x\| \leq \left(\frac{1 + |a|}{1 - |b|}\right)\|Vx - Ux\|. \tag{3.5.5}$$

Let $(x_n)_n$ be a bounded sequence in $\mathscr{D}(U)$ such that:

$$Q_U(V - U)x_n \to y.$$

Using (3.5.5), we obtain:

$$\|Q_{U_0}(V - U_0)(x_n - y)\| \to 0.$$

Since U_0 is V-demicompact, the sequence $(Q_{U_0}Vx_n)_n$ has a convergent subsequence. Furthermore, since:

$$\begin{aligned}\|Q_U Vx_n\| &= \operatorname{dist}(U(0), Vx_n) \\ &\leq \operatorname{dist}(U_0(0), Vx_n) \\ &= \|Q_{U_0}Vx_n\|,\end{aligned}$$

the sequence $(Q_U Vx_n)_n$ also has a convergent subsequence. Q.E.D.

Proposition 3.5.10 *Let U, $U_0 : \mathscr{D}(U) = \mathscr{D}(U_0) \subset X \longrightarrow Y$ be linear relations, and let $V : \mathscr{D}(V) \subset X \longrightarrow Y$ be a continuous and closed linear relation such that*

$$\mathscr{D}(U) \subset \mathscr{D}(V) \quad and \quad V(0) \subset U_0(0) \subset U(0).$$

If U is continuous and U_0 is V-demicompact, and there exist $\alpha, \beta \in \mathbb{C}$ such that $|\alpha| > 1$, and for all $x \in \mathscr{D}(U)$,

$$\|Ux - U_0x\| \geq |\alpha| \cdot \|Vx - U_0x\| - |\beta| \cdot \|Vx - Ux\|,$$

then U is a V-demicompact linear relation. ◇

Proof Observe that

$$\begin{aligned}\|Ux - U_0x\| &= \|Ux - U_0x + Vx - Vx\| \\ &\leq \|Vx - U_0x\| + \|Vx - Ux\|,\end{aligned}$$

therefore,

$$|\alpha| \cdot \|Vx - U_0x\| - |\beta| \cdot \|Vx - U_0x\| \leq \|Vx - Ux\| + \|Vx - Ux\|.$$

Consequently,

$$(|\alpha| - 1) \cdot \|Vx - U_0x\| \leq (1 + |\beta|) \cdot \|Vx - Ux\|.$$

Thus, we deduce that

$$\|Vx - U_0x\| \leq \left(\frac{1 + |\beta|}{|\alpha| - 1}\right) \cdot \|Vx - Ux\|. \tag{3.5.6}$$

Now, let $(x_n)_n$ be a bounded sequence in $\mathscr{D}(U)$ satisfying

$$Q_U(V - U)x_n \to y.$$

Making use of inequality (3.5.6), it follows that

$$\|Q_{U_0}(V - U_0)(x_n - y)\| \to 0.$$

Hence, by the assumption that U_0 is V-demicompact, we infer that

$$Q_{U_0}(V - U_0)x_n \to y.$$

This implies that the sequence $(Q_{U_0}Vx_n)_n$ possesses a convergent subsequence. On the other hand, we note

$$\begin{aligned}\|Q_U Vx_n\| &= \operatorname{dist}(U(0), Vx_n) \\ &\leq \operatorname{dist}(U_0(0), Vx_n) \\ &= \|Q_{U_0}Vx_n\|.\end{aligned}$$

In conclusion, the sequence $(Q_U V x_n)_n$ admits a convergent subsequence. Q.E.D.

Proposition 3.5.11 *Let α, μ, $k \in \mathbb{C}$ be such that $\mu \neq 0$. Let $U : \mathcal{D}(U) \subset X \longrightarrow Y$ be a continuous linear relation and $V : \mathcal{D}(V) \subset X \longrightarrow Y$. If U is a $(|k| - \overline{U(0)})$-set-contraction, then $|\alpha|U$ is μV-demicompact for each $|\alpha k| < 1$.* ◇

Proof Let $(x_n)_n$ be a bounded sequence in $\mathcal{D}(U)$ such that

$$Q_{|\alpha|U}\mu V x_n - Q_{|\alpha|U}|\alpha|U x_n \to y.$$

Then,

$$Q_{|\alpha|U}\mu V x_n = Q_{|\alpha|U}\big(\mu V x_n - |\alpha|U x_n\big) + Q_{|\alpha|U}|\alpha|U x_n. \tag{3.5.7}$$

If $\gamma(\{Q_{|\alpha|U}\mu V x_n\}) \neq 0$, then by applying (3.5.7), we obtain

$$\begin{aligned}\gamma(\{Q_{|\alpha|U}\mu V x_n\}) &\leq \gamma\big(\{Q_{|\alpha|U}(\mu V x_n - |\alpha|U x_n)\}\big) + \gamma(\{Q_{|\alpha|U}|\alpha|U x_n\}),\\ &\leq |\alpha k|\,\gamma(\{Q_{|\alpha|U} x_n\}),\\ &< \gamma(\{Q_{|\alpha|U} x_n\}).\end{aligned}$$

Therefore, this leads to a contradiction. Consequently,

$$\gamma(\{Q_{|\alpha|U}\mu V x_n\}) = 0.$$

Hence, the sequence $(Q_{|\alpha|U}\mu V x_n)_n$ is relatively compact. Q.E.D.

A direct implication of Proposition 3.5.11 is the corollary stated below:

Corollary 3.5.1 *Let $a, b \in \mathbb{C}$ with $a \neq 0$ and $b \neq -1$. Let $U : \mathcal{D}(U) \subset X \to Y$ be a continuous linear relation, and let $V : \mathcal{D}(V) \subset X \to Y$ be another linear relation. If U is a $(|b| - \overline{U(0)})$-set-contractive operator, then the scaled relation $\frac{1}{|1+b|}U$ is aV-demicompact.* ◇

Proposition 3.5.12 *Let α, μ, $k \in \mathbb{C}$ with $\mu \neq 0$. Consider a continuous linear relation $U : \mathcal{D}(U) \subset X \to Y$ and another linear relation $V : \mathcal{D}(V) \subset X \to Y$. Suppose that U is a $(|k| - \overline{U(0)})$-set-contractive operator. Then, for every scalar α satisfying $|\alpha k| < 1$, the operator $|\alpha|U$ is μV-demicompact.*

Proof Let $(x_n)_n$ be a bounded sequence in $\mathcal{D}(U)$ such that

$$Q_{|\alpha|U}\mu V x_n - Q_{|\alpha|U}|\alpha|U x_n \to y.$$

We then have the decomposition

$$Q_{|\alpha|U}\mu V x_n = Q_{|\alpha|U}(\mu V x_n - |\alpha|U x_n) + Q_{|\alpha|U}|\alpha|U x_n.$$

Assume that the measure of non-compactness satisfies $\gamma(\{Q_{|\alpha|U}\mu V x_n\}) \neq 0$. By the triangle inequality for γ, we get

$$\gamma(\{Q_{|\alpha|U}\mu V x_n\}) \leq \gamma(\{Q_{|\alpha|U}(\mu V x_n - |\alpha|U x_n)\}) + \gamma(\{Q_{|\alpha|U}|\alpha|U x_n\}).$$

Since the sequence $Q_{|\alpha|U}(\mu V x_n - |\alpha|U x_n)$ converges to y, it is relatively compact, so

$$\gamma(\{Q_{|\alpha|U}(\mu V x_n - |\alpha|U x_n)\}) = 0.$$

On the other hand, there exists a constant $k > 0$ such that

$$\gamma(\{Q_{|\alpha|U}|\alpha|U x_n\}) \leq |\alpha|k\, \gamma(\{Q_{|\alpha|U} x_n\}).$$

Combining these results, we obtain

$$\gamma(\{Q_{|\alpha|U}\mu V x_n\}) \leq |\alpha|k\, \gamma(\{Q_{|\alpha|U} x_n\}).$$

If $|\alpha|k < 1$, this inequality implies that

$$\gamma(\{Q_{|\alpha|U}\mu V x_n\}) < \gamma(\{Q_{|\alpha|U} x_n\}),$$

which contradicts the initial assumption. Therefore, to avoid this contradiction, we must have

$$\gamma(\{Q_{|\alpha|U}\mu V x_n\}) = 0,$$

meaning that the sequence $(Q_{|\alpha|U}\mu V x_n)_n$ is relatively compact. Q.E.D.

Theorem 3.5.5 *Let $\nu \in \mathbb{C}^*$, $\alpha \in (0, 1)$, and let $U : \mathcal{D}(U) \subset X \longrightarrow X$ be a linear relation, and $V : \mathcal{D}(V) \subset X \longrightarrow X$ be a continuous linear relation. If there exists $m \in \mathbb{N}^*$ such that $(Q_U \alpha U)^m$ is $Q_U \nu V$-compact and $Q_V - Q_U$ is compact, then αU is νV-demicompact.* ◇

Proof Let us consider the various cases for m:

First case. For $m = 1$. Using Lemma 3.5.5, we noticed that αU is νV-demicompact for each $\alpha \in (0, 1)$.

Second case. For $m \in \mathbb{N}^* \setminus \{1\}$. Let $(x_n)_n$ be a bounded sequence of $\mathcal{D}(U)$ such that

$$y_n = Q_U \nu V x_n - Q_U \alpha U x_n \to y.$$

Then,

$$\begin{aligned}&\sum_{k=0}^{m-1}(Q_U\alpha U)^k Q_U \nu V x_n\\ &= \sum_{k=0}^{m-1}(Q_U\alpha U)^k y_n + \sum_{k=0}^{m-1}(Q_U\alpha U)^{k+1} x_n,\end{aligned}$$

$$= \sum_{k=0}^{m-1}(Q_U\alpha U)^k y_n + \sum_{k=0}^{m-2}(Q_U\alpha U)^{k+1}x_n + (Q_U\alpha U)^m x_n,$$

$$= \sum_{k=0}^{m-1}(Q_U\alpha U)^k y_n + \sum_{k=0}^{m-2}(Q_U\alpha U)^{k+1}x_n + (Q_U\alpha U)^m x_n. \tag{3.5.8}$$

On the other hand,

$$\sum_{k=0}^{m-1}(Q_U\alpha U)^k Q_U\nu V x_n = Q_U\nu V x_n + \sum_{k=0}^{m-2}(Q_U\alpha U)^{k+1} Q_U\nu V x_n. \tag{3.5.9}$$

Then, by using both (3.5.8) and (3.5.9), we have

$$Q_U\nu V x_n = \sum_{k=0}^{m-1}(Q_U\alpha U)^k y_n + \sum_{k=0}^{m-2}(Q_U\alpha U)^{k+1}(Q_V - Q_U)x_n + (Q_U\alpha U)^m x_n. \tag{3.5.10}$$

Since $Q_U U Q_U$ and $(Q_U U)^n Q_U\nu V$ are single-valued linear operators for all $n \geq 1$, we get $\sum_{k=0}^{m-2}(Q_U\alpha U)^{k+1}Q_{\alpha U}\nu V x_n$ is single-valued. Then, by using (3.5.10), we have

$$\begin{aligned}\gamma(Q_U\nu V x_n) \leq\ & \sum_{k=0}^{m-1}\overline{\gamma}((Q_U\alpha U)^k)\gamma(\{y_n\}) + \overline{\gamma}((Q_U\alpha U)^m)\gamma(\{x_n\}) + \\ & \sum_{k=0}^{m-2}\overline{\gamma}((Q_U\alpha U)^{k+1})\overline{\gamma}(Q_V - Q_U)\gamma(\{x_n\}) \\ =\ & 0.\end{aligned}$$

We conclude that $\gamma(\{Q_U\nu V x_n\}) = 0$, hence $(Q_U\nu V x_n)_n$ is relatively compact. Then there exists a convergent subsequence of $(Q_U\nu V x_n)_n$. Q.E.D.

An immediate consequence of Theorem 3.5.5.

Corollary 3.5.2 *Let $\mu, \nu, k \in \mathbb{C}$ with $\nu \neq 0$. Let $L : \mathcal{D}(L) \subset X \to X$ and $M : \mathcal{D}(M) \subset X \to X$ be continuous linear relations. Then:*

(i) If $m \in \mathbb{N}^*$, $k \neq -1$, $(Q_L L)^m$ is νM-compact, and $Q_M - Q_L$ is compact, then $\frac{1}{|1+k|}L$ is $Q_L\nu M$-demicompact.

(ii) If $(Q_L|\mu|L)^m$ is $Q_L\nu M$-compact for every $\mu \in (0, 1)$ and some $m > 0$, then $|\mu|L$ is νM-demicompact for each $\mu \in (0, 1)$.

(iii) If $m \in \mathbb{N}^*$, $k \geq 0$, $\overline{\gamma}((Q_L|\mu|L)^m) \leq k$, and $Q_M - Q_L$ is compact, then $|\mu|L$ is νM-demicompact for each μ satisfying $0 < \mu^m k < 1$.

(iv) If $m \in \mathbb{N}^*$, $k \geq 0$, $\overline{\gamma}(L^m) \leq k$, and $Q_M - Q_L$ is compact, then $\frac{1}{1+k}L$ is νM-demicompact.

◇

3.6 Linear Relations with Fredholm Property

3.6.1 Linear Relations That are Upper and Lower Semi-Fredholm

In this section, we introduce foundational concepts and findings that underpin this book, specifically, the established characteristics of Fredholm linear relations, along with several results concerning semi-Fredholm perturbations in Banach spaces.

Definition 3.6.1 Let X and Y be two normed spaces. A linear relation $R \in L\mathcal{R}(X, Y)$ is said to be upper semi-Fredholm, denoted by $R \in \mathcal{F}_+(X, Y)$, if there exists a closed subspace $M \subset X$ of finite codimension such that the restriction

$$R|_M : M \to Y$$

is single-valued and admits a continuous inverse.
A linear relation R is said to be **lower semi-Fredholm**, denoted by $R \in \mathcal{F}_-(X, Y)$, if its adjoint R^* is upper semi-Fredholm. R is said to be Fredholm, denoted by $R \in \mathcal{F}(X, Y)$, if it is both upper and lower semi-Fredholm. Finally, R is said to be semi-Fredholm, denoted by $R \in \mathcal{F}_\pm(X, Y)$, if it is either upper or lower semi-Fredholm. ◇

The following result may be found in [10]:

Theorem 3.6.1 *Let X and Y be two normed spaces. Let $T \in L\mathcal{R}(X, Y)$ be such that* $\dim(\mathcal{D}(T)) < \infty$. *Then, $T \in \mathcal{F}_+(X, Y)$ if, and only if, $\Gamma(T) > 0$.* ◇

Corollary 3.6.1 [10, Corollary V.1.7]*Let X, Y be two Banach spaces and let $T \in C\mathcal{R}(X, Y)$. Then, $T \in \mathcal{F}_+(X, Y)$ if, and only if, $R(T)$ is closed and $\alpha(T)$ is finite.* ◇

Theorem 3.6.2 *Let X and Y be normed spaces. The following statements are equivalent: (i) V is not an upper semi-Fredholm operator, i.e., $V \notin \mathcal{F}_+(X, Y)$.*
(ii) There exists a bounded subset $M \subset \mathcal{D}(V)$ which is not precompact, but whose image $Q_V(M)$ is precompact.
(iii) V admits a singular sequence; that is, a sequence $(x_n)_n \subset \mathcal{D}(V)$ of norm-one vectors $\|x_n\| = 1$ such that $(x_n)_n$ has no Cauchy subsequence and $\lim_{n\to\infty} \|Vx_n\| = 0$. ◇

Proposition 3.6.1 *Let $R \in \mathcal{F}_+(X, Y)$. Then any bounded sequence $(x_n)_n$ in $\mathcal{D}(R)$ such that $(Q_R x_n)_n$ is a Cauchy sequence has a Cauchy subsequence.* ◇

Theorem 3.6.3 *Let X be a Banach space, Y a normed vector space, and $V \in L\mathcal{R}(X, Y)$. Then the following properties are equivalent:*

(i) V is upper semi-Fredholm.
(ii) There exist a bounded linear operator U and a bounded finite rank projection operator P such that

$$UV = I_{\mathcal{D}(V)} - P.$$ ◇

Proposition 3.6.2 *Let $V \in L\mathscr{R}(X, Y)$. Consider the following conditions:*

(i) V is upper semi-Fredholm and $\overline{\beta}(V) < \infty$.
(ii) V is lower semi-Fredholm and $\alpha(V) < \infty$.

Then, condition (i) implies (ii). Moreover, if X is complete and V is closed, the converse also holds, that is, (ii) implies (i).

◇

Lemma 3.6.1 *Let $V \in L\mathscr{R}(X, Y)$ and let $W \subset Y$ be a finite-dimensional subspace. Then, the following hold:*

(i) V is upper semi-Fredholm, i.e., $V \in \mathscr{F}_+(X, Y)$, if and only if the induced operator

$$Q_W V : X \to Y/W$$

belongs to $\mathscr{F}_+(X, Y/W)$.
(ii) V is lower semi-Fredholm, i.e., $V \in \mathscr{F}_-(X, Y)$, if and only if $Q_W V \in \mathscr{F}_-(X, Y/W)$.

◇

Lemma 3.6.2 *Let $R \in L\mathscr{R}(X, Y)$ and $G \in L\mathscr{R}(Y, Z)$. Then:*

(i) If R is single-valued, $R \in \mathscr{F}_+(X, Y)$, and $G \in \mathscr{F}_+(Y, Z)$, then $GR \in \mathscr{F}_+(X, Z)$.
(ii) If $G \in B\mathscr{R}(Y, Z)$ and $GR \in \mathscr{F}_+(X, Z)$, then $R \in \mathscr{F}_+(X, Y)$. ◇

Lemma 3.6.3 *Let $Q, V \in L\mathscr{R}(X, Y)$. If* $\dim Q(0) < \infty$, *then*

$$V + Q - Q \in \mathscr{F}_+(X, Y) \quad \text{if and only if} \quad V \in \mathscr{F}_+(X, Y). \qquad \diamond$$

Lemma 3.6.4 [10, Proposition V.5.2] *Let X, Y be two normed spaces and let $T \in L\mathscr{R}(X, Y)$. Then, $T \in \mathscr{F}_-(X, Y)$ if, and only if, $R(\widetilde{T})$ is closed and $\beta(\widetilde{T})$ is finite.* ◇

As a direct consequence of Remark 2.11.4 and Lemma 3.6.4, we infer the following result:

Corollary 3.6.2 *Let X, Y be two Banach spaces and let $T \in C\mathscr{R}(X, Y)$. Then, $T \in \mathscr{F}_-(X, Y)$ if, and only if, $R(T)$ is closed and $\beta(T)$ is finite.* ◇

Proposition 3.6.3 *Let $S \in L\mathscr{R}(X, Y)$, $T \in \mathscr{F}_+(X, Y)$ be with $G(S) \subset G(T)$ and* $\dim \mathscr{D}(S) = \infty$, *then $S \in \mathscr{F}_+(X, Y)$.* ◇

3.6.2 Closed Linear Relations That are Upper and Lower Semi-Fredholm

Let X and Y be two Banach spaces. We aim to generalize the classical concept of Fredholm operators by extending the classes of closed, single-valued Fredholm-type operators to include closed multivalued linear relations.

In this generalized setting, the definitions of the classes $\mathscr{F}_+(X, Y)$ and $\mathscr{F}_-(X, Y)$ correspond, respectively, to the following sets:

- The class $\Phi_+(X, Y)$, defined as

$$\Phi_+(X, Y) := \{U \in C\mathscr{R}(X, Y) : \ \alpha(U) < \infty \text{ and } R(U) \text{ is closed in } Y\},$$

where $\alpha(U)$ denotes the dimension of the null space of U;
- The class $\Phi_-(X, Y)$, defined as

$$\Phi_-(X, Y) := \{U \in C\mathscr{R}(X, Y) : \ \ \beta(U) < \infty \text{ and } R(U) \text{ is closed in } Y\},$$

where $\beta(U)$ denotes the codimension of the range of U.

We also define:

$$\Phi_\pm(X, Y) := \Phi_+(X, Y) \cup \Phi_-(X, Y), \quad \text{and} \quad \Phi(X, Y) := \Phi_+(X, Y) \cap \Phi_-(X, Y),$$

so that $\Phi(X, Y)$ represents the class of Fredholm linear relations from X into Y. In the special case where $X = Y$, the notations simplify to $\Phi_+(X)$, $\Phi_-(X)$, $\Phi_\pm(X)$, and $\Phi(X)$, respectively. In addition, if $U \in C\mathscr{R}(X)$, a complex number λ is believed to be part of Φ_{+U} (respectively, Φ_U) if $\lambda - U \in \Phi_+(X)$ (respectively, $\Phi(X)$). Likewise, for $U \in C\mathscr{R}(X, Y)$ and $V \in L\mathscr{R}(X, Y)$, if $\lambda V - U$ is closed, then λ belongs to $\Phi_{+U,V}$ (respectively, $\Phi_{U,V}$) if $\lambda V - U \in \Phi_+(X, Y)$ (respectively, $\Phi(X, Y)$). For a detailed exposition of the Fredholm theory in the context of linear relations and multivalued operators, the reader is referred to the relevant literature [17, 19, 23, 25, 30, 37–72, 72–75, 75–81, 81–103, 103–109].

Lemma 3.6.5 *Let $U \in L\mathscr{R}(X, Y)$. If $U \in \Phi(X, Y)$, then there exist B bounded operator from Y into X and A a finite rank operator such that*

$$UB = (I - A) + UB - UB,$$

(that is, $I - A$ is a selection of UB). ◇

Proof Given that U is closed, its inverse U^{-1} is also closed. Consequently, the set $U^{-1}(0) = N(U)$ is closed as well. Thus, the assertion follows directly from [110, Corollary 3.25].

Q.E.D.

Theorem 3.6.4 [110, Theorem 6.1] *Let X be a Banach space, V, $U \in C\mathscr{R}(X)$. Then,*

(i) Assuming that V, $U \in \Phi_+(X)$, and $VU \in C\mathscr{R}(X)$, then $VU \in \Phi_+(X)$ and $UV \in \Phi_+(X)$.
(ii) Assuming that V, $U \in \Phi_-(X)$ with UV (resp. VU) is closed, then $UV \in \Phi_-(X)$ (resp. $VU \in \Phi_-(X)$).
(iii) Assuming that V, $U \in \Phi(X)$, and $VU \in C\mathscr{R}(X)$, then $UV \in \Phi(X)$ and

$$i(UV) + \dim(V(0) \bigcap N(U)) = i(U) + i(V) + \dim\left(\frac{X}{R(V) + \mathscr{D}(U)}\right).$$

(iv) Assuming that V and U are everywhere defined and $UV \in \Phi_+(X)$*, then* $V \in \Phi_+(X)$.
(v) Assuming that V and U are everywhere defined and closed such that $UV \in \Phi(X)$ *and* $VU \in \Phi(X)$*, then* $V \in \Phi(X)$ *and* $U \in \Phi(X)$. ◇

Proposition 3.6.4 *Let X be a Banach space.*

(i) Let $U, V \in L\mathscr{R}(X)$ *be such that* $\gamma(U) > 0$, $V(0) \subset U(0)$ *and* $\alpha(U) < \infty$. *Then,*

$$\gamma(U - V) \geq \|P\|^{-1}_{L\mathscr{R}(X)} \gamma(U) - \|V\|,$$

where P is any continuous projection defined on $\mathscr{D}(U)$ *(or on X) with kernel* $N(U)$.

(ii) Let $U \in C\mathscr{R}(X)$ *and* $V \in L\mathscr{R}(X)$ *be such that* $\gamma(U) > 0$, $V(0) \subset U(0)$, $\overline{\mathscr{D}(U)} \subset \mathscr{D}(V)$, $\|V\| < \gamma(U)$ *and* $\beta(U) < \infty$. *Then,*

$$\gamma(U - V) \geq \|P\|^{-1}_{L\mathscr{R}(X^*)} \gamma(U) - \|V\|,$$

where P is any continuous projection defined on $\mathscr{D}(U^*)$ *(or on* X^**) with kernel* $N(U^*)$. ◇

Proof *(i)* Since $\alpha(U) < \infty$, then by Remark 2.1.2 *(ii)*, we infer that $N(U)$ is topologically complemented in $\mathscr{D}(U)$. Using the fact that $\gamma(U) > 0$ and Proposition 2.10.6, we obtain

$$0 < \|P\|^{-1}_{L\mathscr{R}(X)} \gamma(U) \leq \gamma(U_{|R(P)}) \leq \gamma(U), \tag{3.6.1}$$

where P is any continuous projection defined on $\mathscr{D}(U)$ (or on X) with kernel $N(U)$. Let $\widehat{U} : R(P) \longrightarrow R(U)$ be the bijection associated with U. It follows from (3.6.1) that

$$0 < \|P\|^{-1}_{L\mathscr{R}(X)} \gamma(U) \leq \gamma(\widehat{U}) \leq \gamma(U). \tag{3.6.2}$$

Consequently, $\gamma(\widehat{U}) > 0$ which implies that $\widehat{U}$ is open. Using the fact that $\widehat{U}$ is one-to-one, $\widehat{U}(0) = U(0) \supset V(0)$ and also Lemma 2.10.10, we infer that

$$\gamma(\widehat{U} - V) \geq \gamma(\widehat{U}) - \|V\|_{L\mathscr{R}(X)}. \tag{3.6.3}$$

By virtue of (3.6.2) and (3.6.3), we infer that

$$\gamma(\widehat{U} - V) \geq \|P\|^{-1}_{L\mathscr{R}(X)} \gamma(U) - \|V\|_{L\mathscr{R}(X)}. \tag{3.6.4}$$

Now, we propose to show that

$$\gamma(U - V) \geq \gamma(\widehat{U} - V).$$

It is clear that $\widehat{U} - V$ is a restriction of $U - V$ to $R(P)$. This implies that $N(\widehat{U} - V) \subset N(U - V)$. Consequently,

$$\operatorname{dist}(x, N(U-V)) \le \operatorname{dist}(x, N(\widehat{U}-V)), \text{ for all } x \in X. \tag{3.6.5}$$

For $x \in \mathscr{D}(\widehat{U}-V)$, by applying (3.6.5), we have

$$\begin{aligned}\gamma(\widehat{U}-V)\operatorname{dist}(x, N(U-S)) &\le \gamma(\widehat{U}-V)\operatorname{dist}(x, N(\widehat{U}-V))\\ &\le \|(\widehat{U}-V)x\|\\ &\le \|(U-V)x\|.\end{aligned}$$

This leads to

$$\gamma(U-V) \ge \gamma(\widehat{U}-V). \tag{3.6.6}$$

It follows from (3.6.4) and (3.6.6) that

$$\gamma(U-V) \ge \|P\|^{-1}_{L\mathscr{R}(X)}\gamma(U) - \|V\|_{L\mathscr{R}(X)}.$$

(ii) The fact that $\gamma(U) > 0$ implies from Proposition 2.10.4 (ii) that U is open. Since U is closed, then by using Theorem 2.11.1 (ii), we infer that $R(U)$ is closed. Accordingly, $\beta(U) < \infty$, we obtain $U \in \Phi_-(X)$. By virtue of (i) of Lemma 3.6.7, we get that $R(U^*)$ is closed and

$$\alpha(U^*) < \infty.$$

This implies from Propositions 2.11.1 (iii) and 2.10.4 (ii) that

$$\gamma(U^*) > 0.$$

Combining the fact that U is open and Proposition 2.15.3, we have

$$\begin{aligned}\|S\|_{L\mathscr{R}(X)} &< \gamma(U)\\ &= \gamma(U^*).\end{aligned}$$

According to $V(0) \subset U(0)$, $\overline{\mathscr{D}(U)} \subset \mathscr{D}(V)$, Lemma 3.6.8, we obtain $U - V \in \Phi_-(X)$. As a consequence of the Lemma 3.6.7 (i) we have $(U-V)^* \in \Phi_+(X)$. By referring to (iii) of Theorem 2.11.1, we infer that $(U-V)^*$ is open. Making use of the fact that V is continuous and $\mathscr{D}(U) \subset \mathscr{D}(S)$, we infer from Proposition 2.15.5 (ii) that

$$(U-V)^* = U^* - V^*. \tag{3.6.7}$$

Our goal is to show that $V^*(0) \subset U^*(0)$. Since $\mathscr{D}(U) \subset \mathscr{D}(V)$. It follows by using Proposition 2.15.1 (ii), we have

$$\begin{aligned}V^*(0) &= \mathscr{D}(V)^\perp\\ &\subset \mathscr{D}(U)^\perp\\ &= U^*(0).\end{aligned}$$

As stated in (i) and (3.6.7), we obtain

$$\begin{aligned}\gamma((U-V)^*) = &\ \gamma(U^*-V^*) \\ \geq &\ \|P\|^{-1}_{L\mathscr{R}(X^*)}\gamma(U^*) - \|V^*\|_{L\mathscr{R}(X^*)}\end{aligned}$$

where P is any continuous projection defined on $\mathscr{D}(U^*)$ with kernel $N(U^*)$. The fact that $(U-V)^*$ is open implies from Proposition 2.15.3 (iv) that

$$\gamma(U-V) \geq \|P\|^{-1}_{L\mathscr{R}(X^*)}\gamma(U) - \|V\|_{L\mathscr{R}(X^*)}.$$ Q.E.D.

Lemma 3.6.6 *Let* $U \in L\mathscr{R}(X)$ *and* $\lambda \neq 0$, *then* $\lambda - U \in \Phi(X)$ *if, and only if,* $\lambda^{-1} - U^{-1} \in \Phi(X)$ *and*

$$i(\lambda - U) = i(U^{-1} - \lambda^{-1}).$$ ◇

Proof (a) For all $\lambda \neq 0$, we have

$$\lambda - U = -\lambda(\lambda^{-1} - U^{-1})U.$$

Indeed, by using Proposition 3.1.1 (with $\mu = 0$), we have

$$\lambda - U = (0-\lambda)[(0-\lambda)^{-1} - (0-U)^{-1}](0-U).$$

(b) $R(\lambda - U) = R(\lambda^{-1} - U^{-1})$. Indeed, by using (a), we have

$$\begin{aligned}R(\lambda - U) = &\ R\left(\lambda^{-1} - U^{-1})U\right) \\ = &\ (\lambda^{-1} - U^{-1})R(U) \\ = &\ R(\lambda^{-1} - U^{-1}).\end{aligned}$$

These equalities combined with the fact that $R(U) = X$ ensure that

$$R(\lambda - U) = R(\lambda^{-1} - U^{-1}). \tag{3.6.8}$$

Now, using Proposition 3.1.1 (with $\mu = 0$), we infer that

$$N(\lambda - U) = N(\lambda^{-1} - U^{-1}). \tag{3.6.9}$$

Applying Eqs. (3.6.8) and (3.6.9), we have $\lambda - U \in \Phi(X)$ if, and only if, $\lambda^{-1} - U^{-1} \in \Phi(X)$ and

$$\begin{aligned}i(\lambda - U) = &\ \alpha(\lambda - U) - \beta(\lambda - U) \\ = &\ \alpha(\lambda^{-1} - U^{-1}) - \beta(\lambda^{-1} - U^{-1}) \\ = &\ i(\lambda^{-1} - U^{-1}).\end{aligned}$$

This achieves the proof. Q.E.D.

Lemma 3.6.7 [10, Chapter V] *Let X, Y be two Banach spaces and let $U \in C\mathscr{R}(X, Y)$. Then,*

(i) $U \in \Phi_-(X, Y)$ if, and only if, $U^* \in \Phi_+(Y^*, X^*)$.
(ii) $U \in \Phi_+(X, Y)$ if, and only if, $U^* \in \Phi_-(Y^*, X^*)$. ◇

Lemma 3.6.8 *[110, Corollary 2.2] Let X, Y be two complete spaces and let V, $U \in L\mathscr{R}(X, Y)$ be such that $V(0) \subset U(0)$ and $\overline{\mathscr{D}(U)} \subset \mathscr{D}(V)$.*

(i) If $\|V\|_{L\mathscr{R}(X,Y)} < \gamma(U^*)$ and $U \in \Phi_-(X, Y)$, then $U \pm V \in \Phi_-(X, Y)$.
(ii) If $\|V\|_{L\mathscr{R}(X,Y)} < \gamma(U)$ and $U \in \Phi_+(X, Y)$, then $U \pm V \in \Phi_+(X, Y)$. ◇

Theorem 3.6.5 [111, Theorem 6.1] *Let X be a Banach space and let $U \in C\mathscr{R}(X)$. Then,*

(i) $U \in \Phi_+(X)$ if, and only if, $U^* \in \Phi_-(X^*)$.
(ii) $U \in \Phi_-(X)$ if, and only if, $U^* \in \Phi_+(X^*)$. Moreover, if U is semi-Fredholm then $\alpha(U^*) = \beta(U)$ and $\alpha(U) = \beta(U^*)$.
(iii) Let $U \in \Phi_+(X)$. Then, there exists $\varepsilon > 0$ such that for every $\lambda \in \mathbb{K}$ with $0 < |\lambda| < \varepsilon$, we have $\lambda - U \in \Phi_+(X)$, $\alpha(\lambda - U) \leq \alpha(U)$ and $i(\lambda - U) = i(U)$.
(iv) Let $U \in \Phi_-(X)$. Then, there exists $\varepsilon > 0$ such that for every $\lambda \in \mathbb{K}$ with $0 < |\lambda| < \varepsilon$, we have $\lambda - U \in \Phi_-(X)$, $\beta(\lambda - U) \leq \beta(U)$ and $i(\lambda - U) = i(U)$. ◇

Theorem 3.6.6 [111, Theorem 25] *(i) If $U \in \Phi_+(X)$, then there exists $\varepsilon > 0$ such that $\lambda - U \in \Phi_+(X)$ and $\alpha(\lambda - U)$ is constant on the annulus with $0 < |\lambda| < \varepsilon$. Moreover, $\alpha(\lambda - U) \leq \alpha(U)$ for all $|\lambda| < \varepsilon$.*
(ii) If $U \in \Phi_-(X)$, then there exists $\varepsilon > 0$ such that $\lambda - U \in \Phi_-(X)$ and $\beta(\lambda - U)$ is constant on the annulus with $0 < |\lambda| < \varepsilon$. Moreover, $\beta(\lambda - U) \leq \beta(U)$ for all $|\lambda| < \varepsilon$. ◇

Lemma 3.6.9 *Let X be a Banach space and let $U \in C\mathscr{R}(X)$. Then,*

(i) $U \in \Phi_+(X)$ if, and only if, $Q_U U \in \Phi_+(X, X/U(0))$. In such case $i(U) = i(Q_U U)$.
(ii) $U \in \Phi_-(X)$ if, and only if, $Q_U U \in \Phi_-(X, X/U(0))$. In such case $i(U) = i(Q_U U)$.
(iii) $U \in \Phi(X, X/U(0))$ if, and only if, $Q_U(U) \in \Phi(X, X/U(0))$. In such case $i(U) = i(Q_U U)$. ◇

Proof Since U is closed, then by Proposition 2.11.1, we have $Q_U(U)$ is closed and $U(0)$ is closed. Second $R(U)$ is closed if, and only if, $R(Q_U U)$ is closed. In such case $\beta(U) = \beta(Q_U U)$. Indeed, $U(0)$ is closed implies

$$\begin{aligned} R(Q_U(U)) &= \left(R(U) + \overline{U(0)}\right) / \overline{U(0)} \\ &= R(U)/U(0), \end{aligned}$$

and so by Lemma 2.1.1 (i), we have $R(U)$ is closed if, and only if, $R(Q_U U)$ is closed. In such case, we have

$$\beta(U) = \beta(Q_U U).$$

Furthermore, $N(U) \subset N(Q_U(U))$ with equality if $U(0)$ is closed in $R(U)$. Q.E.D.

Lemma 3.6.10 *If $U \in L\mathscr{R}(X)$ and V is single-valued on X. If $UV \in \Phi(X)$, then $U \in \Phi(X)$.* ◇

Proof Assuming that $UV \in \Phi(X)$, then by Lemma 3.6.9

$$Q_{UV}UV \in \Phi(X).$$

Observe that, $UV(0) = V(0)$ which means that

$$\begin{aligned} Q_{UV}UV &= Q_U UV \\ &= Q_U(U)V. \end{aligned}$$

Hence, $Q_U UV \in \Phi(X)$ and since $Q_U U \in \Phi(X)$ and V are operators, we have $U \in \Phi(X)$.

Q.E.D.

Lemma 3.6.11 *Let $U \in B\mathscr{R}(X) \bigcap K\mathscr{R}(X)$ and $\lambda \in \mathbb{K}\backslash\{0\}$. If $\dim(U(0)) < \infty$, then $\lambda - U \in \Phi(X)$ and $i(\lambda - U) = \dim(U(0))$.* ◇

Proof Inasmuch as $\dim U(0) < \infty$ and $U \in B\mathscr{R}(X)$, then U and $\lambda - U$ are bounded and closed. Also,

$$\begin{aligned} Q_{\lambda - U}(\lambda - U) &= Q_U(\lambda - U) \\ &= Q_U(\lambda) - Q_U(U). \end{aligned}$$

Since $Q_U(U)$ is bounded compact operator, $N(Q_U(\lambda)) = U(0)$ and $R(Q_U(\lambda)) = X\big/U(0)$, then $Q_U(\lambda)$ is a Fredholm operator. For this reason, $Q_{\lambda-U}(\lambda - U)$ is a Fredholm operator and

$$i(Q_{\lambda-U}(\lambda - U)) = \dim U(0).$$

So, by using Lemma 3.6.9 (iii), we have $\lambda - U$ is a Fredholm relation and $i(\lambda - U) = \dim U(0)$. Q.E.D.

3.6.3 Stability and Perturbation Results Concerning Linear Relations

Theorem 3.6.7 [10, Theorem V.5.12] *Let X and Y be two normed spaces and let $V, U \in L\mathscr{R}(X, Y)$ be such that $\mathscr{D}(V) \supset \mathscr{D}(U)$. Then,*

(i) Assuming that $V \in \mathcal{PRK}(X, Y)$ *and* $U \in \mathcal{F}_{-}(X, Y)$*, then* $U + V \in \mathcal{F}_{-}(X, Y)$.
(ii) Assuming that $\|V\| < \gamma(U^*)$ *and* $U \in \mathcal{F}_{-}(X, Y)$*, then* $U + V \in \mathcal{F}_{-}(X, Y)$.
(iii) Assuming that $\dim(R(V)) < \infty$ *and* $U \in \mathcal{F}_{-}(X, Y)$*, then* $U + V \in \mathcal{F}_{-}(X, Y)$. ◇

An immediate consequence of Theorem 3.6.7 is the following result.

Corollary 3.6.3 *Let X and Y be two complete spaces and let* $V, U \in L\mathcal{R}(X, Y)$ *be such that* $V(0) \subset U(0)$ *and* $\overline{\mathcal{D}(U)} \subset \mathcal{D}(V)$*. Then,*

(i) If $V \in \mathcal{PRK}(X, Y)$ *and* $U \in \Phi_{-}(X, Y)$*, then* $U \pm V \in \Phi_{-}(X, Y)$.
(ii) Assuming $U \in \Phi_{-}(X, Y)$ *and* $\|V\| < \gamma(U^*)$*, then* $U \pm V \in \Phi_{-}(X, Y)$.
(iii) Assuming $U \in \Phi_{-}(X, Y)$ *and* $\dim(R(V)) < \infty$*, then* $U + V \in \Phi_{-}(X, Y)$.
(iv) In the event that $U \in \Phi_{+}(X, Y)$ *and* $V \in \mathcal{SSR}(X, Y)$*, then* $U \pm V \in \Phi_{+}(X, Y)$.
(v) If $U + V \in \Phi_{+}(X, Y)$ *or* $U - V \in \Phi_{+}(X, Y)$ *and* $V \in \mathcal{SSR}(X, Y)$*, then* $U \in \Phi_{+}(X, Y)$.
(vi) If $\|V\| < \gamma(U)$ *and* $U \in \Phi_{+}(X, Y)$*, then* $U \pm V \in \Phi_{+}(X, Y)$. ◇

Theorem 3.6.8 *Let* $U, V, W \in L\mathcal{R}(X, Y)$ *satisfy*

$$\mathcal{D}(V) \subset \mathcal{D}(U) \subset \mathcal{D}(W), \quad W(0) \subset U(0) \subset \overline{V(0)},$$

and let $\widehat{V}$ *be the bijection associated with* V*. Suppose the following hold:*
(a) There exists a constant α_1 *such that*

$$\|U\varphi\| \le \alpha_1\big(\|\varphi\| + \|V\varphi\|\big), \quad \varphi \in \mathcal{D}(V),$$

(b) There exists a constant β_1 *such that*

$$\alpha_1(1 + \beta_1) < 1, \quad (1 + \beta_1) < \gamma(\widehat{V}),$$

and

$$\|W\varphi\| \le \beta_1\big(\|\varphi\| + \|U\varphi\|\big), \quad \varphi \in \mathcal{D}(V).$$

If $V \in \Phi(X, Y)$*, then the sum* $V + U + W$ *belongs to* $\Phi(X, Y)$ *and satisfies*
(i) $\alpha(V + U + W) \le \alpha(V)$,
(ii) $\overline{\beta}(V + U + W) \le \overline{\beta}(V)$.

Proof From Theorem 2.11.4, we deduce that the sum $V + U + W$ defines a closed linear relation. Let V_1, U_1, and W_1 denote the respective restrictions of V, U, and W to the subspace X_V. It is clear that V is a Fredholm linear relation, and both U_1 and W_1 are bounded linear relations. Moreover, it is straightforward to verify the estimate

$$\begin{aligned}\|U_1 + W_1\| &\le \beta_1 + \alpha_1(1 + \beta_1)\\ &\le \gamma(\widehat{V})\\ &= \gamma(\widehat{V^*}),\end{aligned}$$

which, in conjunction with Lemma 2.10.11 and Corollary 3.6.3, implies that the sum $U_1 + W_1 + V_1$ is a Fredholm linear relation.

Q.E.D.

Corollary 3.6.4 *Let U, V, and W be three operators such that*

$$\mathscr{D}(V) \subset \mathscr{D}(U) \subset \mathscr{D}(W),$$

and let $\widehat{V}$ denote the bijection associated with V. Suppose the following conditions are satisfied:

(a) There exists a constant α_1 such that

$$\|U\varphi\| \leq \alpha_1 \left(\|\varphi\| + \|V\varphi\|\right), \quad \varphi \in \mathscr{D}(V),$$

(b) There exists a constant β_1 such that

$$\alpha_1(1 + \beta_1) < 1, \quad (1 + \beta_1) < \gamma(\widehat{V}),$$

and

$$\|W\varphi\| \leq \beta_1 \left(\|\varphi\| + \|U\varphi\|\right), \quad \varphi \in \mathscr{D}(V).$$

If $V \in \Phi(X, Y)$, then the sum $V + U + W$ also belongs to $\Phi(X, Y)$ and satisfies the following properties:
(i) $\alpha(V + U + W) \leq \alpha(V)$,
(ii) $\beta(V + U + W) \leq \beta(V)$,
(iii) $i(V + U + W) = i(V)$. ◇

Theorem 3.6.9 *Let $\widehat{\mathscr{T}}$ be the bijection associated with $\mathscr{T}$, where $\mathscr{T}$ is given in (2.11.7). If we suppose that the conditions of Theorem 2.11.5 are satisfied and we suppose that $\alpha_1(1 + \beta_1) < 1$ and $(1 + \beta_1) < \gamma(\widehat{\mathscr{T}})$, where*

$$\alpha_1 = \max\left\{\max\left\{a_1, a_2, a_3\right\}, \max\left\{b_1, b_2, b_3\right\}\right\}$$

and

$$\beta_1 = \max\left\{\max\{a_4, a_5, a_6\}, \max\{b_4, b_5, b_6\}\right\}.$$

Then, if A, E and K are Fredholm linear relations, then $\mathscr{A}$, given in (2.11.8), is a Fredholm linear relation, which satisfies the following properties

(i) $\alpha(\mathscr{A}) \leq \alpha(\mathscr{T})$
(ii) $\beta(\mathscr{A}) \leq \beta(\mathscr{T})$

and if $\mathscr{A}$ is single-valued, then $i(\mathscr{A}) = i(A) + i(E) + i(K)$. ◇

Proof The proofs follows from Theorems 3.6.8 and 2.11.5. Now, if $\mathscr{A}$ is single-valued, then by Corollary 3.6.4, we have $i(\mathscr{A}) = i(\mathscr{T})$. On the other hand,

$$\begin{pmatrix} A & 0 & 0 \\ 0 & E & 0 \\ 0 & 0 & K \end{pmatrix} = \begin{pmatrix} A & 0 & 0 \\ 0 & I & 0 \\ 0 & 0 & I \end{pmatrix} \begin{pmatrix} I & 0 & 0 \\ 0 & E & 0 \\ 0 & 0 & I \end{pmatrix} \begin{pmatrix} I & 0 & 0 \\ 0 & I & 0 \\ 0 & 0 & K \end{pmatrix}$$

and $i(\mathscr{A}) = i(\mathscr{T}) = i(A) + i(E) + i(K)$. This completes the proof. Q.E.D.

3.6.4 Fredholm of an Upper Triangular Matrix of Linear Relations

In this subsection we will generalize the results obtained in the last section, in the case of $n \times n$ upper triangular matrix of linear relations with $n \geq 2$.

Lemma 3.6.12 *Let $A \in L\mathscr{R}(X_1)$, $B \in L\mathscr{R}(X_2)$ be closed and everywhere defined. We consider the 2×2 matrix of linear relation*

$$\mathscr{M}_C := \begin{pmatrix} A & C \\ 0 & B \end{pmatrix},$$

where C is a bounded single-valued. Then,

(i) If $\mathscr{M}_C \in \Phi_+(X_1 \times X_2)$, then $A \in \Phi_+(X_1)$.
(ii) If $\mathscr{M}_C \in \Phi_-(X_1 \times X_2)$, then $B \in \Phi_-(X_2)$.
(iii) If $\mathscr{M}_C \in \Phi(X_1 \times X_2)$, then $A \in \Phi_+(X_1)$ and $B \in \Phi_-(X_2)$. ◇

Proof The result follows immediately from Eq. (2.8.23) and Theorem 3.6.4 (iv), (v). Q.E.D.

Remark 3.6.1 Using the same reasoning as the proof of Lemma 3.6.12, we can show that if linear relation

$$\mathscr{M}_D := \begin{pmatrix} A & 0 \\ D & B \end{pmatrix}$$

is in $\Phi(X_1 \times X_2)$ for some bounded single-valued D, then $A \in \Phi_-(X_1)$ and $B \in \Phi_+(X_2)$. ◇

In a similar manner, the case of a 2×2 matrix of linear relations can also be proved:

Proposition 3.6.5 *Consider n infinite-dimensional separable Hilbert spaces $H_1, H_2, \ldots, H_n$ with $n \geq 2$. For each $1 \leq i \leq n$, let $A_i \in L\mathscr{R}(H_i)$ be a linear relation on H_i, and for each $1 \leq i < j \leq n$, let $C_{ij} \in L\mathscr{R}(H_j, H_i)$ be a linear relation from H_j to H_i. Consider the linear operator $M_n = \big((M_n)_{ij}\big)_{1 \leq i,j \leq n}$ defined on the direct sum $\bigoplus_{i=1}^n H_i$, with entries given by*

$$(M_n)_{ij} = \begin{cases} C_{ij}, & \text{if } i < j, \\ A_i, & \text{if } i = j, \\ 0, & \text{if } i > j, \end{cases}$$

and represented by the block matrix

$$M_n = \begin{pmatrix} A_1 & C_{12} & C_{13} & \cdots & C_{1n} \\ 0 & A_2 & C_{23} & \cdots & C_{2n} \\ 0 & 0 & A_3 & \cdots & C_{3n} \\ \vdots & \vdots & \vdots & \ddots & \vdots \\ 0 & 0 & 0 & \cdots & A_n \end{pmatrix}.$$

Furthermore, we assume that for every pair $i < j$, the following inclusion holds:

$$C_{ij}(0) \subseteq A_i(0),$$

where $C_{ij}(0)$ and $A_i(0)$ denote the multivalued parts (kernels at zero) of the respective linear relations. Then, the operator part $Q_{M_n} M_n$ of M_n acts on the space $\bigoplus_{i=1}^n H_i$ and admits the block matrix representation

$$Q_{M_n} M_n = \left((Q_{M_n} M_n)_{ij}\right)_{1 \leq i, j \leq n},$$

where the blocks are given by

$$(Q_{M_n} M_n)_{ij} = \begin{cases} Q_{A_i} C_{ij}, & i < j, \\ Q_{A_i} A_i, & i = j, \\ 0, & i > j, \end{cases}$$

and thus

$$Q_{M_n} M_n = \begin{pmatrix} Q_{A_1} A_1 & Q_{A_1} C_{12} & Q_{A_1} C_{13} & \cdots & Q_{A_1} C_{1n} \\ 0 & Q_{A_2} A_2 & Q_{A_2} C_{23} & \cdots & Q_{A_2} C_{2n} \\ 0 & 0 & Q_{A_3} A_3 & \cdots & Q_{A_3} C_{3n} \\ \vdots & \vdots & \vdots & \ddots & \vdots \\ 0 & 0 & 0 & \cdots & Q_{A_n} A_n \end{pmatrix}.$$

Here, Q_{A_i} denotes the operator part (i.e., the single-valued component) of the linear relation A_i.

Theorem 3.6.10 *Let A_1 be a bounded linear relation such that $A_1 \in \Phi_+(H_1)$. For each $2 \leq i \leq n$, assume that $A_i \in CB\mathscr{R}(H_i)$, and suppose that the range $R(A_n)$ is not closed. Then, the following equivalence holds:*

$$\beta(A_1) = +\infty \iff \exists\,(C_{ij})_{1\leq i<j\leq n} \subset B\mathscr{R}(H_j, H_i)$$

such that the following conditions hold:

- *For all* $j = 2, \ldots, n$, *we have* $C_{1j}(0) \subseteq A_1(0)$,
- *For all* $i = 2, \ldots, n-1$, *we have* $C_{ij} = 0$.

Under these assumptions, the operator matrix M_n, *defined on the Hilbert space direct sum* $\bigoplus_{i=1}^{n} H_i$, *is given by*

$$M_n = \begin{pmatrix} A_1 & C_{12} & C_{13} & \cdots & C_{1,n-1} & C_{1n} \\ 0 & A_2 & 0 & \cdots & 0 & 0 \\ 0 & 0 & A_3 & \cdots & 0 & 0 \\ \vdots & \vdots & \vdots & \ddots & \vdots & \vdots \\ 0 & 0 & 0 & \cdots & A_{n-1} & 0 \\ 0 & 0 & 0 & \cdots & 0 & A_n \end{pmatrix},$$

and belongs to the class $\Phi_+\left(\bigoplus_{i=1}^{n} H_i\right)$. ◇

Proof Given that $\beta(A_1) = +\infty$, we conclude that

$$\dim R(A_1)^{\perp} = +\infty.$$

It follows that there exists an isometric mapping

$$U : H_2 \times H_3 \times \cdots \times H_n \longrightarrow R(A_1)^{\perp}.$$

We introduce, for each index $2 \leq j \leq n$,

$$\begin{aligned} C_{1j} : H_j &\longrightarrow \begin{pmatrix} R(A_1)^{\perp} \\ R(A_1) \end{pmatrix} \\ x_j &\longmapsto C_{1j}x_j = \begin{pmatrix} U(0, \cdots, 0, x_j, 0, \ldots, 0) \\ 0 \end{pmatrix}, \end{aligned}$$

which ensures that for any $x = (x_2, \ldots, x_n) \in \prod_{2\leq i\leq n} H_i$, the equality

$$\sum_{2\leq j\leq n} C_{1j}(x_j) = U(x)$$

holds true.

We assert that with C_{1j} defined above, and for arbitrary operators $C_{ij} \in L\mathscr{R}(H_j, H_i)$ for all $2 \leq i < j \leq n$, the operator matrix M_n defines an upper semi-Fredholm relation.

Step 1: Finiteness of $\alpha(M_n)$. Take any vector $x = (x_1, \dots, x_n) \in N(M_n)$. Then,

$$0 \in A_1 x_1 + \sum_{2 \le j \le n} C_{1j}(x_j),$$

which implies

$$-\sum_{2 \le j \le n} C_{1j}(x_j) \in A_1 x_1.$$

Thus,

$$-U(x_2, \dots, x_n) \in A_1 x_1,$$

leading to

$$U(x_2, \dots, x_n) \in R(A_1) \cap R(A_1)^{\perp} = \{0\}.$$

Hence, $U(x_2, \dots, x_n) = 0$ and $0 \in A_1 x_1$, from which it follows that

$$N(M_n) \subseteq N(A_1) \oplus \{0\} \oplus \cdots \oplus \{0\},$$

implying

$$\alpha(M_n) \le \alpha(A_1) < +\infty.$$

Step 2: Closedness of $R(M_n)$. Let $X_m = (x_{1m}, x_{2m}, \dots, x_{nm})$ and suppose that

$$Q_{M_n} M_n X_m \to \widetilde{X} = (\widetilde{x_1}, \widetilde{x_2}, \cdots, \widetilde{x_n}).$$

Then,

$$Q_{A_1} A_1 x_{1m} + \sum_{2 \le j \le n} Q_{A_1} C_{1j} x_{jm} \to \widetilde{x_1}.$$

Since $Q_{A_1} A_1 x_{1m} \perp \sum_{2 \le j \le n} Q_{A_1} C_{1j} x_{jm}$, both sequences

$$(Q_{A_1} A_1 x_{1m})_m \quad \text{and} \quad \left(\sum_{2 \le j \le n} Q_{A_1} C_{1j} x_{jm} \right)_m$$

are Cauchy sequences. Utilizing the form of C_{1j},

$$\sum_{2 \le j \le n} C_{1j} x_{jm} = U(x_{2m}, x_{3m}, \dots, x_{nm}),$$

we derive

$$\begin{aligned}\left\| Q_{A_1}\left(\sum_{2\le j\le n} C_{1j}x_{jm}\right)\right\| &= \|Q_{A_1}U(x_{2m}, x_{3m}, \dots, x_{nm})\| \\ &= \operatorname{dist}(U(x_{2m}, \dots, x_{nm}), A_1(0)) \\ &= \inf_{u\in A_1(0)} \|U(x_{2m}, \dots, x_{nm}) - u\| \\ &= \inf_{u\in A_1(0)} \sqrt{\|U(x_{2m}, \dots, x_{nm})\|^2 + \|u\|^2} \\ &= \|U(x_{2m}, \dots, x_{nm})\| \\ &= \|(x_{2m}, x_{3m}, \dots, x_{nm})\|.\end{aligned}$$

This confirms that each sequence $(x_{jm})_m$ is Cauchy for $2 \le j \le n$. Assume

$$Q_{A_1}A_1x_{1m} \to Q_{A_1}A_1a_1 \quad \text{and} \quad x_{jm} \to a_j \quad \text{for all } 2 \le j \le n.$$

It follows that

$$\begin{aligned}\widetilde{X} &= \left(Q_{A_i}A_ia_i + \sum_{i<j\le n} Q_{A_i}C_{ij}(a_j), Q_{A_n}A_na_n \right) \\ &= Q_{M_n}M_n \begin{pmatrix} a_1 \\ a_2 \\ \vdots \\ a_n \end{pmatrix}.\end{aligned}$$

Therefore, $R(M_n)$ is a closed subspace.

We have established that for any operators C_{ij} with indices satisfying $2 \le i < j \le n$, the operator M_n qualifies as an upper semi-Fredholm operator. This conclusion remains valid even in the special case where $C_{ij} = 0$ for every $2 \le i \le n-1$.

For the reverse implication. Suppose that $\beta(A_1)$ is finite, say $\beta(A_1) = N < +\infty$. Since the range $R(A_n)$ fails to be closed, there exists a sequence $(t_m)_m$ in $N(A_n)^{\perp}$ consisting of orthonormal vectors such that each $\|t_m\| = 1$ and

$$\|Q_{A_n}A_nt_m\| \to 0 \quad \text{as } m \to \infty.$$

This fact follows from the non-closedness of $R(A_n)$, which implies

$$\gamma(A_n) = \inf\left\{ \frac{\|Q_{A_n}A_nt\|}{\operatorname{dist}(t, N(A_n))} : t \notin N(A_n) \right\} = 0.$$

Consequently,

$$\inf\left\{ \|Q_{A_n}A_nt\| : \|t\| = 1, t \in N(A_n)^{\perp} \right\} = 0,$$

confirming the existence of such an orthonormal sequence $(t_m)_m$ with the stated properties. We now assert the existence of a positive constant $\varepsilon_0 > 0$ and a subsequence $(t_{m_k})_k$ of $(t_m)_m$ such that the distance

$$\operatorname{dist}\left(R(Q_{A_1}A_1), Q_{A_1}C_{1n}t_{m_k}\right) \geq \varepsilon_0$$

holds for every natural number k. Assuming the contrary, suppose

$$\operatorname{dist}\left(R(Q_{A_1}A_1), Q_{A_1}C_{1n}t_m\right) \to 0 \quad \text{as } m \to +\infty.$$

Then, there is a sequence $(x_m)_m \subseteq H_1$ such that

$$\|Q_{A_1}A_1x_m - Q_{A_1}C_{1n}t_m\| \to 0.$$

Observe the vectors

$$X_m = (x_m, 0, \ldots, 0, -t_m) \in \bigoplus_{1\leq i\leq n} H_i.$$

One can easily verify that

$$\operatorname{dist}(X_m, N(M_n)) \geq 1.$$

Hence,

$$\frac{\|Q_{M_n}M_nX_m\|}{\operatorname{dist}(X_m, N(M_n))} \leq \|Q_{M_n}M_nX_m\|$$

On the other hand

$$= \sqrt{\|Q_{A_1}A_1x_m - Q_{A_1}C_{1n}t_m\|^2 + \|Q_{A_n}A_nt_m\|^2} \to 0,$$

which contradicts the fact that the gap $\gamma(M_n)$ is strictly positive. Therefore, there must exist an $\varepsilon_0 > 0$ and a subsequence $(t_{m_k})_k$ such that

$$\operatorname{dist}\left(R(Q_{A_1}A_1), Q_{A_1}C_{1n}t_{m_k}\right) \geq \varepsilon_0$$

for all k. This implies the existence of a sequence of angles $(\theta_m)_m$ in the interval $[0, 2\pi]$ satisfying

$$\left\| Q_{A_1}C_{1n}\left(\sum_{m\geq 1}\frac{1}{m}t_me^{i\theta_m}\right)\right\| = +\infty.$$

However, since

$$\left\|\sum_{m\geq 1}\frac{1}{m}t_me^{i\theta_m}\right\|^2 = \sum_{m\geq 1}\frac{1}{m^2} < +\infty,$$

we deduce that

$$t=\sum_{m\geq 1}\frac{1}{m}t_m e^{i\theta_m}\in H_n,$$

which leads to a contradiction. Therefore, the only possibility is

$$\beta(A_1)=+\infty.$$

Theorem 3.6.11 *Let A_1 be a bounded linear relation, such that $A_1\in\Phi_+(H_1)$, and for each index i with $2\leq i\leq n$, suppose $A_i\in CB\mathcal{R}(H_i)$ with the range $R(A_i)$ being closed. Assume further that for every $i\in\{2,3,\ldots,n\}$, if $\alpha(A_i)=+\infty$,and $\beta(A_{i-1})=+\infty$. Then, for all pairs of indices $1\leq i<j\leq n$, there exists an operator $C_{ij}\in B\mathcal{R}(H_j,H_i)$ such that the block operator matrix*

$$M_n\in\Phi_+\left(\bigoplus_{1\leq i\leq n}H_i\right).$$

◇

Proof The proof is carried out via induction. The base case $n=2$ has been demonstrated in the previous section. Assume the statement holds for an arbitrary integer n, and suppose that for every $i\in\{2,\ldots,n+1\}$, the range $R(A_i)$ is closed. Our objective is to establish that if $\beta(A_{i-1})=+\infty$ for all $i\in\{2,\ldots,n+1\}$ with $\alpha(A_i)=+\infty$, then there exist operators $C_{ij}\in B\mathcal{R}(H_j,H_i)$ such that

$$M_{n+1}\in\Phi_+\left(\bigoplus_{1\leq i\leq n+1}H_i\right).$$

<u>Definition of $(C_{ij})_{1\leq i<j\leq n}$</u> : since $A_1\in\Phi_+(H_1)$ and A_1 is bounded linear relation, each $R(A_i)$ is closed for all $i\in\{2,3,\ldots,n\}$, and we assume $\beta(A_{i-1})=+\infty$ for every $i\in\{2,3,\ldots,n\}$ such that $\alpha(A_i)=+\infty$, then by the inductive hypothesis, there exist operators $C_{ij}\in B\mathcal{R}(H_j,H_i)$ ensuring that

$$M_n\in\Phi_+\left(\bigoplus_{1\leq i\leq n}H_i\right).$$

<u>Definition of $(C_{i(n+1)})_{1\leq i\leq n}$</u> : we know that $M_n\in\Phi_+\left(\bigoplus_{1\leq i\leq n}H_i\right)$, and that $R(A_{n+1})$ is closed. Therefore, if $\alpha(A_{n+1})<+\infty$, then for any operator $C=(C_{i(n+1)})_{1\leq i\leq n}$, we obtain

$$M_{n+1}=\begin{pmatrix}M_n & C\\ 0 & A_{n+1}\end{pmatrix}\in\Phi_+\left(\bigoplus_{1\leq i\leq n+1}H_i\right).$$

On the other hand, if $\alpha(A_{n+1}) = +\infty$, then by the induction assumption we have

$$\beta(A_n) = +\infty.$$

Since $\beta(A_n) \leq \beta(M_n)$, it follows that

$$\beta(M_n) = +\infty.$$

Then, by Theorem 3.6.10, there exists an operator $C = (C_{i(n+1)})_{1\leq i\leq n}$ such that

$$M_{n+1} \in \Phi_+\left(\bigoplus_{1\leq i\leq n+1} H_i\right).$$

Q.E.D.

Theorem 3.6.12 *Let A_1 be a bounded linear relation such that $A_1 \in \Phi_+(H_1)$, and for all $2 \leq i \leq n$, let $A_i \in C B\mathscr{R}(H_i)$ be bounded linear relations with $R(A_n)$ closed. Suppose there exists some $i \in \{2, 3, \ldots, n-1\}$ such that $R(A_i)$ is not closed. Define*

$$p = \inf\{i \in \{2, 3, \ldots, n-1\} \ : R(A_i) \text{ is not closed}\}.$$

If $\beta(A_1) = +\infty$, and $\beta(A_{i-1}) = +\infty$ for all $i \in \{p+1, \ldots, n\}$ such that $\alpha(A_i) = +\infty$, then for every $1 \leq i < j \leq n$, there exists $C_{ij} \in B\mathscr{R}(H_j, H_i)$ such that

$$M_n \in \Phi_+\left(\bigoplus_{1\leq i\leq n} H_i\right).$$

◇

Proof In the same manner, during the proof of Theorem 3.6.11. Presume the result holds for order n. Suppose that $R(A_{n+1})$ is closed and that there exists some $i \in \{2, 3, \ldots, n\}$ such that $R(A_i)$ is not closed. Let

$$\widetilde{p} = \inf\{i = 2, 3, \cdots, n \ \text{ such that } \ R(A_i) \ \text{ is not closed}\}.$$

<u>Construction of C_{ij}, $1 \leq i < j \leq n$:</u>

• If $\widetilde{p} \leq n-1$. In this case, we consider two scenarios:

<u>First case:</u> If $R(A_n)$ is closed, then the conditions $\beta(A_1) = +\infty$ and $\beta(A_{i-1}) = +\infty$ for all

$$i \in \{\widetilde{p}+1, \ldots, n\} \text{ such that } \alpha(A_i) = +\infty$$

imply, by the inductive hypothesis, the existence of operators C_{ij} for $1 \leq i < j \leq n$ such that

$$M_n \in \Phi_+\left(\bigoplus_{1\leq i\leq n} H_i\right).$$

Second case: If $R(A_n)$ is not closed, then since $\beta(A_1) = +\infty$, it follows from Theorem 3.4.2 that there exist operators C_{ij}, $1 \leq i < j \leq n$, such that

$$M_n \in \Phi_+\left(\bigoplus_{1\leq i\leq n} H_i\right).$$

• If $\widetilde{p} = n$, the situation reduces to the second case.
Combining these two cases, we conclude: if $\beta(A_1) = +\infty$ and $\beta(A_{i-1}) = +\infty$ for all

$$i \in \{\widetilde{p}+1, \ldots, n\} \text{ such that } \alpha(A_i) = +\infty,$$

then there exist operators C_{ij}, $1 \leq i < j \leq n$, such that

$$M_n \in \Phi_+\left(\bigoplus_{1\leq i\leq n} H_i\right).$$

Construction of $C = C_{i(n+1)},\ 1 \leq i \leq n$:

Since $M_n \in B\Phi_+\left(\bigoplus_{1\leq i\leq n} H_i\right)$ and $R(A_{n+1})$ is closed, if either

$$\alpha(A_{n+1}) < +\infty \quad \text{or} \quad \alpha(A_{n+1}) = \beta(A_n) = +\infty,$$

then, by Theorem 3.6.10, there exists

$$C = C_{i(n+1)}, \quad 1 \leq i \leq n,$$

such that

$$M_{n+1} \in \Phi_+\left(\bigoplus_{1\leq i\leq n+1} H_i\right).$$

Hence, the proof is finished. Q.E.D.

Proposition 3.6.6 *Let A be a bounded linear relation such that $A \in \Phi_+(H_1)$, and let $B \in CB\mathscr{R}(K)$ be a linear relation with non-closed range, i.e., $R(B)$ is not closed. Then, $\beta(A) = +\infty$ if, and only if, $\exists\, C \in \mathscr{L}(K, H)$ such that $\mathscr{M}_C \in \Phi_\pm(H \oplus K)$.* ◇

Proof Assume that $\beta(A) = +\infty$. Then, it follows that $R(A)^\perp$ is an infinite-dimensional separable Hilbert space. Consequently, there exists an isometric mapping $U : K \longrightarrow R(A)^\perp$. We define an operator $C : K \longrightarrow H$ by

$$C = \begin{pmatrix} U \\ 0 \end{pmatrix} : K \longrightarrow \begin{pmatrix} R(A)^\perp \\ R(A) \end{pmatrix}.$$

We claim that $\mathcal{M}_C$ is an upper semi-Fredholm relation with negative index.

Initially, we show that $\mathcal{M}_C$ is an upper semi-Fredholm relation. Let $\begin{pmatrix} u \\ v \end{pmatrix} \in N(\mathcal{M}_C)$, then $0 \in Au + Cv$ and $0 \in Bv$. Thus, $-Cv \in Au$ which implies that

$$-Cv \in R(A) \bigcap R(A)^{\perp} = \{0\}.$$

Since $Cv = Uv$, then $v = 0$ and $u \in N(A)$. As a consequence,

$$N(\mathcal{M}_C) \subseteq N(A) \bigoplus \{0\}.$$

Then,

$$\alpha(\mathcal{M}_C) \leq \alpha(A) < +\infty.$$

We bring the argument to a close by demonstrating that

$$R(\mathcal{M}_C) \text{ is closed.}$$

By invoking Lemma 2.10.5 and Theorem 2.11.1, it is sufficient to prove the closedness of

$$R(Q_{\mathcal{M}_C}\mathcal{M}_C).$$

By applying Proposition 2.9.2, we obtain

$$Q_{\mathcal{M}_C}\mathcal{M}_C = \begin{pmatrix} Q_A A & Q_A C \\ 0 & Q_B B \end{pmatrix}.$$

Let $(x_n, y_n) \in H \bigoplus K$ be such that

$$Q_{\mathcal{M}_C}\mathcal{M}_C \begin{pmatrix} x_n \\ y_n \end{pmatrix} \to \begin{pmatrix} \widetilde{u} \\ \widetilde{v} \end{pmatrix}.$$

Consequently,

$$Q_A A x_n + Q_A C y_n \to \widetilde{u}$$

and

$$Q_B B y_n \to \widetilde{v}.$$

Let $u_n \in A x_n$. Then,

$$Q_A A x_n = Q_A u_n.$$

Owing to

$$\begin{aligned}\langle Q_A u_n, Q_A C y_n\rangle_1 &= \langle P_{A(0)^\perp}(u_n), P_{A(0)^\perp}(C y_n)\rangle \\ &= \langle u_n, C y_n\rangle - \langle P_{A(0)}(u_n), P_{A(0)}(C y_n)\rangle \\ &= -\langle P_{A(0)}(u_n), P_{A(0)}(C y_n)\rangle \\ &= -\langle P_{A(0)}(u_n), C y_n\rangle \\ &= 0.\end{aligned}$$

From this, we deduce that $Q_A A x_n$ is orthogonal to $Q_A C y_n$. Consequently, the sequences $(Q_A A x_n)_n$ and $(Q_A C y_n)_n$ are both Cauchy. By the definition of C, we have

$$C y_n = U y_n,$$

and thus

$$\begin{aligned}\|Q_A C y_n\| &= \|Q_A U y_n\| \\ &= \operatorname{dist}(U y_n, A(0)) \\ &= \inf_{u \in A(0)} \sqrt{\|U y_n\|^2 + \|u\|^2} \\ &= \|U y_n\| \\ &= \|y_n\|.\end{aligned}$$

For this reason $(y_n)_n$ is a Cauchy sequence. Let $y_n \to \widetilde{y}$ and $Q_A A x_n \to Q_A A \widetilde{x}$. Then, $\widetilde{u} = Q_A A \widetilde{x} + Q_A C \widetilde{y}$ and $\widetilde{v} = Q_B B \widetilde{y}$. That is to say

$$\begin{pmatrix} \widetilde{u} \\ \widetilde{v} \end{pmatrix} = Q_{\mathcal{M}_C} \mathcal{M}_C \begin{pmatrix} \widetilde{x} \\ \widetilde{y} \end{pmatrix},$$

and hence $R(\mathcal{M}_C)$ is closed. The prior proof demonstrates that $\mathcal{M}_C$ is an upper semi-Fredholm relation. Our next step is to prove that $i(\mathcal{M}_C) \leq 0$. It can be shown that $R(\mathcal{M}_C) \subseteq H \bigoplus R(B)$. Thus,

$$\beta(\mathcal{M}_C) \geq \beta(B) = +\infty,$$

due to the fact that $R(B)$ is not closed. Which means that

$$i(\mathcal{M}_C) = \alpha(\mathcal{M}_C) - \beta(\mathcal{M}_C) \leq 0,$$

since $\alpha(\mathcal{M}_C)$ is finite.

To establish the reverse implication, suppose that $\beta(A) = N < +\infty$. Since $R(B)$ is not closed, there exists an orthonormal sequence $(y_n)_n$ in $N(B)^\perp$ such that $\|Q_B B y_n\| \to 0$. In fact, since $R(B)$ is not closed, it follows that

$$\gamma(B) = \inf\left\{\frac{\|Q_B By\|}{\text{dist}(y, N(B))},\quad y \notin N(B)\right\} = 0.$$

Let $y \in K\backslash N(B)$. Note down $y = y_1 + y_2$, where $y_1 \in N(B)$ and $y_2 \in N(B)^{\perp}$. Then,

$$\begin{aligned}
\frac{\|Q_B By\|}{\text{dist}(y, N(B))} &= \frac{\|Q_B By\|}{\inf\{\|y_1 + y_2 - z\|, z \in N(B)\}} \\
&= \frac{\|Q_B By_2\|}{\inf\{\|y_2 - z\|,\ z \in N(B)\}} \\
&= \frac{\|Q_B By_2\|}{\inf\left\{\sqrt{\|y_2\|^2 + \|z\|^2},\ z \in N(B)\right\}} \\
&= \frac{\|Q_B By_2\|}{\|y_2\|} \\
&= \left\| B\frac{y_2}{\|y_2\|}\right\|.
\end{aligned}$$

Thus,

$$\inf\left\{\|Q_B By\|\text{such that } \|y\| = 1 \text{ and } y \in N(B)^{\perp}\right\} = 0.$$

Thus, there exists an orthonormal sequence $(y_n)_n \subseteq N(B)^{\perp}$ for which

$$\|Q_B By_n\| \to 0.$$

Because $R(\mathcal{M}_C)$ is closed (due to $\mathcal{M}_C$ being an upper semi-Fredholm relation), Lemma 2.10.12 guarantees the existence of a constant

$$\varepsilon_0 > 0.$$

and a subsequence $(y_{n_k})_k$ of $(y_n)_n$ for which

$$\text{dist}(R(Q_A A), Q_A C y_{n_k}) \geq \varepsilon_0,$$

for all $k \in \mathbb{N}$. Next, considering a suitable subsequence, we assume that

$$\text{dist}(R(Q_A A), Q_A C y_n) \geq \varepsilon_0 \quad \text{for all} \quad n \in \mathbb{N}.$$

Hence, by [112, (2.4) (p. 765)], there exists a sequence $(\theta_n)_n \in [0, 2\pi]$ such that

$$\left\| C\left(\sum_{n\geq 0}\frac{1}{n}y_n e^{i\theta_n}\right)\right\| = +\infty$$

but the fact

$$\left\|\sum_{n\geq 1}\frac{1}{n}y_n e^{i\theta_n}\right\|^2 = \sum_{n\geq 1}\frac{1}{n^2} < +\infty,$$

implies that

$$y = \sum_{n\geq 1} \frac{1}{n} y_n e^{i\theta_n} \in K$$

which is a contradiction. As a result, we must have $\beta(A) = +\infty$. Q.E.D.

Proposition 3.6.7 *Let $A \in L\mathscr{R}(H)$, $B \in L\mathscr{R}(K)$ and $C \in L\mathscr{R}(K, H)$ be such that $R(A)$ is closed and $\mathscr{M}_C \in \Phi_+(H \bigoplus K)$. If $\alpha(B) = +\infty$, then $\beta(A) = +\infty$.* ◇

Proof Assuming the opposite, we have that $\beta(A) < +\infty$. There are two cases to consider.
First case. Suppose that $C(N(B))$ is finite dimensional. Then,

$$\dim(N(C) \bigcap N(B)) = +\infty.$$

Thus, there exists an orthonormal sequence $\{y_n\} \in N(C) \bigcap N(B)$ such that

$$\begin{pmatrix} 0 \\ 0 \end{pmatrix} \in \mathscr{M}_C \begin{pmatrix} 0 \\ y_n \end{pmatrix}$$

for each n, which implies that $N(\mathscr{M}_C)$ is infinite dimensional. This is a contradiction.
Second case. Presume tha $C(N(B))$ is infinite dimensional. Since $R(A)$ is closed, we know, by our assumption, that

$$\dim R(A)^{\perp} = \beta(A) < +\infty.$$

If $\dim(C(N(B)) \bigcap R(A)) < +\infty$, then

$$\dim C(N(B)) < +\infty,$$

which is a contradiction. For this reason $C(N(B)) \bigcap R(A)$ is infinite dimensional. Thus, we can find an orthonormal sequence $(y_n)_n \in N(B)$ for which there exists a sequence $(x_n)_n \in H$ such that $Cy_n \in Ax_n$ for each n. Then,

$$\begin{pmatrix} 0 \\ 0 \end{pmatrix} \in \mathscr{M}_C \begin{pmatrix} x_n \\ -y_n \end{pmatrix}$$

for each n, which implies that $N(\mathscr{M}_C)$ is infinite dimensional. It is a contradiction again. Therefore, $\beta(A) = +\infty$. Q.E.D.

Proposition 3.6.8 *Let H and K be Hilbert spaces. Consider the following:*

- *A is a bounded linear relation in $\Phi_+(H)$,*
- *$B \in CB\mathscr{R}(K)$ is a closed, bounded linear relation with closed range $R(B)$.*

Assume that the defect numbers satisfy:

$$\alpha(B) = \dim \ker B = +\infty \quad \text{and} \quad \beta(A) = \operatorname{codim} R(A) = +\infty.$$

Then, there exists a bounded linear operator $C \in \mathscr{L}(K, H)$ such that the operator matrix

$$\mathscr{M}_C = \begin{pmatrix} A & C \\ 0 & B \end{pmatrix}$$

is semi-Fredholm; that is, $\mathscr{M}_C \in \Phi_\pm(H \oplus K)$. ◇

Proof There are two possible situations to examine
First case. Consider the case where $\beta(B) = +\infty$. Given that $N(B)$ and $R(A)^\perp$ are infinite-dimensional separable Hilbert spaces, there exists an isometric isomorphism $U : N(B) \longrightarrow R(A)^\perp$. Define the operator $C : K \to H$ by

$$C = \begin{pmatrix} U & 0 \\ 0 & 0 \end{pmatrix} : \begin{pmatrix} N(B) \\ N(B)^\perp \end{pmatrix} \longrightarrow \begin{pmatrix} R(A)^\perp \\ R(A) \end{pmatrix}.$$

We now claim that $\mathscr{M}_C \in \Phi_\pm(H \oplus K)$. Indeed, suppose

$$\begin{pmatrix} u \\ v \end{pmatrix} \in N(\mathscr{M}_C).$$

Then $0 \in Au + Cv$ and $0 \in Bv$, so $-Cv \in Au$, which implies $Cv \in R(A) \cap R(A)^\perp$. Hence, $Cv = 0$, and consequently $0 \in Au$. Since $v \in N(B)$, we deduce that

$$Cv = Uv = 0 \quad \Rightarrow \quad v = 0.$$

Thus, we have shown that

$$N(\mathscr{M}_C) \subseteq N(A) \oplus \{0\},$$

which yields

$$\alpha(\mathscr{M}_C) \leq \alpha(A) < +\infty.$$

Now, assume that

$$Q_{\mathscr{M}_C}\mathscr{M}_C \begin{pmatrix} u_n \\ v_n \end{pmatrix} \to \begin{pmatrix} \widetilde{u} \\ \widetilde{v} \end{pmatrix}.$$

This implies that

$$Q_A A u_n + Q_A C v_n \to \widetilde{u}, \quad \text{and} \quad Q_B B v_n \to \widetilde{v}.$$

Write

$$v_n = \alpha_n + \beta_n, \quad \text{with } \alpha_n \in N(B),\ \beta_n \in N(B)^\perp.$$

We claim that $(\beta_n)_n$ is a Cauchy sequence. Indeed, since

$$Q_B B v_n = Q_B B \beta_n \to \widetilde{v}$$

and $R(Q_B B)$ is closed, there exists $v \in N(B)^{\perp}$ such that

$$Q_B B \beta_n \to Q_B B v = \widetilde{v},$$

so

$$Q_B B(\beta_n - v) \to 0.$$

As $\beta_n - v \in N(B)^{\perp}$ and $\gamma(Q_B B) > 0$, it follows that

$$\beta_n - v \to 0.$$

Thus, $(\beta_n)_n$ is a Cauchy sequence. Also, since $(Q_A A u_n + Q_A C v_n)_n$ is a Cauchy sequence and $Q_A A u_n \perp Q_A C v_n$, both sequences $(Q_A A u_n)_n$ and $(Q_A C v_n)_n$ are Cauchy. By using the definition of C, we have:

$$C v_n = C(\alpha_n + \beta_n) = U \alpha_n,$$

which implies that $(\alpha_n)_n$ is a Cauchy sequence. Hence, $(v_n)_n$ is a Cauchy sequence. Let $v_n \to \widetilde{y}$ and $Q_A A u_n \to Q_A A \widetilde{x}$, then

$$\widetilde{u} = Q_A A \widetilde{x} + Q_A C \widetilde{y}, \quad \widetilde{v} = Q_B B \widetilde{y}.$$

Therefore,

$$\begin{pmatrix} \widetilde{u} \\ \widetilde{v} \end{pmatrix} = Q_{\mathscr{M}_C} \mathscr{M}_C \begin{pmatrix} \widetilde{x} \\ \widetilde{y} \end{pmatrix}.$$

By means of both Lemma 2.10.5 and Theorem 2.11.1, it follows that $R(\mathscr{M}_C)$ is closed.
Due to the fact that $R(\mathscr{M}_C) \subseteq H \oplus R(B)$, it follows that

$$\beta(\mathscr{M}_C) \geq \beta(B),$$

and thus

$$\beta(\mathscr{M}_C) = +\infty.$$

As a result, we conclude that $\mathscr{M}_C \in \Phi_{\pm}(H \oplus K)$.
Second case. Suppose $\beta(B) < +\infty$. Then,

$$\beta(Q_B B) < +\infty.$$

Hence,

$$\dim R((Q_B B)^*) = +\infty.$$

Indeed, if we suppose that $\dim R((Q_B B)^*) < +\infty$, then $(Q_B B)^*$ is a compact operator. By using Schauder theorem, we have $(Q_B B)^{**}$ is also a compact operator. As $R((Q_B B)^{**})$ is a closed subspace of $K/B(0)$ (because $R((Q_B B)^{**}) = N((Q_B B)^*)^{\perp})$, then

$$\dim R((Q_B B)^{**}) < +\infty.$$

It follows that

$$\dim N((Q_B B)^*)^{\perp} < +\infty.$$

Like this $\dim N((Q_B B)^*) = \beta(Q_B B) = +\infty$. A contradiction. As a result, by using Lemma 2.9.1, therefore

$$\dim R(B^*) = +\infty.$$

Hence,

$$\dim N(B)^{\perp} = \dim R(B^*) = +\infty.$$

Consider $\{e_1, e_2, \ldots, e_{\alpha(A)}\}$ be an orthonormal set in $N(B)^{\perp}$ and take

$$M = \text{span}\{e_1, e_2, \ldots, e_{\alpha(A)}\}.$$

As both $N(B) + M$ and $R(A)^{\perp}$ are infinite-dimensional separable Hilbert spaces, an isometric map $J : N(B) + M \to R(A)^{\perp}$ exists. Assign an operator $C : K \longrightarrow H$ by

$$C = \begin{pmatrix} J & 0 \\ 0 & 0 \end{pmatrix} : \begin{pmatrix} N(B) + M \\ (N(B) + M)^{\perp} \end{pmatrix} \longrightarrow \begin{pmatrix} R(A)^{\perp} \\ R(A) \end{pmatrix}.$$

Consequently, we claim that $\mathscr{M}_C \in \Phi_{\pm}(H \bigoplus K)$. As a matter of fact, if $\begin{pmatrix} u \\ v \end{pmatrix} \in N(\mathscr{M}_C)$, then $0 \in Au + Cv$ and $0 \in Bv$. Thus, $-Cv \in Au$ and so $Cv \in R(A) \bigcap R(A)^{\perp}$. Accordingly $Cv = 0$ and $0 \in Au$. Since $v \in N(B) + M$, we find that $Cv = Jv = 0$. It follows that $v = 0$ because J is injective. This establishes that

$$N(\mathscr{M}_C) = N(A) \bigoplus \{0\},$$

so

$$\alpha(\mathscr{M}_C) = \alpha(A).$$

For the moment, assume that

$$Q_{\mathscr{M}_C} \mathscr{M}_C \begin{pmatrix} u_n \\ v_n \end{pmatrix} \to \begin{pmatrix} \widetilde{u} \\ \widetilde{v} \end{pmatrix}.$$

Thus,

$$Q_A A u_n + Q_A C v_n \to \widetilde{u}$$

and

$$Q_B B v_n \to \widetilde{v}.$$

Thus, $(Q_A A u_n)_n$ and $(Q_A C v_n)_n$ are all Cauchy sequences. Write $v_n = \alpha_n + \beta_n$, where $\alpha_n \in N(B) + M$ and $\beta_n \in (N(B) + M)^{\perp}$. In light of the definition of C, $C v_n = J(\alpha_n)$, and hence $(\alpha_n)_n$ is a Cauchy sequence. We claim that $(\beta_n)_n$ is a Cauchy sequence. Clearly, since $Q_B B \beta_n = Q_B B v_n - Q_B B \alpha_n$, it follows that $(Q_B B \beta_n)_n$ is a Cauchy sequence. Since $(N(B) + M)^{\perp} \subseteq N(B)^{\perp}$ and $R(Q_B B)$ is closed, we can get that $(\beta_n)_n$ is a Cauchy sequence. For this reason, $(v_n)_n$ is a Cauchy sequence. Let $v_n \to \widetilde{y}$ and $Q_A A u_n \to Q_A A \widetilde{x}$, it means that

$$\begin{pmatrix} \widetilde{u} \\ \widetilde{v} \end{pmatrix} = Q_{\mathcal{M}_C} \mathcal{M}_C \begin{pmatrix} \widetilde{x} \\ \widetilde{y} \end{pmatrix}$$

and hence $R(\mathcal{M}_C)$ is closed, by Lemma 2.10.5 and Theorem 2.11.1. Moreover, we can demonstrate that $\beta(\mathcal{M}_C) = \alpha(A)$. To see this, we need to prove that $\alpha(\mathcal{M}_C^*) = \alpha(A)$. Let

$$\left\{ \begin{pmatrix} x_n \\ y_n \end{pmatrix} : 1 \leq n \leq k \right\} \subseteq N(\mathcal{M}_C^*)$$

be an orthonormal set. Then, $A^* x_n = 0$ and hence

$$x_n \in N(A^*) = R(A)^{\perp}.$$

By definition of C,

$$\begin{aligned} C^* x_n &= J^* x_n \\ &= -B^* y_n \in (N(B) + M) \bigcap N(B)^{\perp} = M. \end{aligned}$$

It should be observed that J^* is one-to-one, then

$$k \leq \alpha(A)$$

which implies that

$$\alpha(\mathcal{M}_C^*) \leq \alpha(A).$$

Since $e_n \in N(B)^{\perp} \bigcap M$ for each $n \in \{1, 2, \ldots, \alpha(A)\}$, then there exist $u_n \in R(A)^{\perp} = N(A^*)$ and $v_n \in R(B^*)$ such that $e_n = J^* u_n = B^* v_n$. Hence, $\{u_1, u_2, \ldots, u_{\alpha(A)}\}$ is linear independent and

$$\left\{ \begin{pmatrix} u_n \\ -v_n \end{pmatrix} : \ n \leq \alpha(A) \right\} \subset N(\mathcal{M}_C^*).$$

It implies that $\alpha(\mathscr{M}_C^*) \geq \alpha(A)$. For this reason

$$\begin{aligned}\alpha(\mathscr{M}_C^*) &= \alpha(A)\\ &= \beta(\mathscr{M}_C).\end{aligned}$$

This indicates that $\mathscr{M}_C$ is Weyl. Then, $\mathscr{M}_C \in \Phi_{\pm}(H \bigoplus K)$. Q.E.D.

Proposition 3.6.9 *Let $A \in \Phi_+(H) \cap B\mathscr{R}(H)$ and $B \in \Phi_+(K) \cap B\mathscr{R}(K)$, where $B\Phi_+(H)$ denotes the set of bounded upper semi-Fredholm operators on the Hilbert space H. Then,*

$$\alpha(A) + \alpha(B) \leq \beta(A) + \beta(B) \quad \textit{if and only if} \quad \mathscr{M}_C \in \Phi_{\pm}(H \oplus K)$$

for some operator $C \in \mathscr{L}(K, H)$. ◇

Proof Suppose that $\alpha(A) + \alpha(B) \leq \beta(A) + \beta(B)$. We aim to show that for every operator $C \in \mathscr{LR}(K, H)$ satisfying $C(0) \subseteq A(0)$, the operator $\mathscr{M}_C$ belongs to $\Phi_+(H \oplus K)$.

To see this, let $C \in \mathscr{LR}(K, H)$ such that $C(0) \subseteq A(0)$. Then, $\mathscr{M}_C$ can be written as a product of two bounded operators:

$$\mathscr{M}_C = \begin{pmatrix} I & 0 \\ 0 & B \end{pmatrix} \begin{pmatrix} A & C \\ 0 & I \end{pmatrix}.$$

Use the notation by

$$D_1 = \begin{pmatrix} I & 0 \\ 0 & B \end{pmatrix}$$

and

$$D_2 = \begin{pmatrix} A & C \\ 0 & I \end{pmatrix}.$$

Let $\begin{pmatrix} x \\ y \end{pmatrix} \in N(D_1)$, then $x = 0$ and $0 \in By$. Hence,

$$N(D_1) \subseteq \{0\} \bigoplus N(B)$$

so

$$\alpha(D_1) \leq \alpha(B) < +\infty.$$

Now, to prove that $R(D_1)$ is closed, it suffices to prove that $R(D_1^*)$ is closed. Since

$$D_1^* = \begin{pmatrix} I & 0 \\ 0 & B^* \end{pmatrix},$$

it follows that

$$\begin{aligned}\beta(D_1^*) &= \beta(B^*)\\ &= \alpha(B) < +\infty.\end{aligned}$$

Consequently, the range $R(D_1)$ is closed.

In a similar manner, it can be shown that $D_2 \in \Phi_+(H \oplus K)$, from which it follows that $\mathcal{M}_C$ is an upper semi-Fredholm relation.

Moreover, since C is a single-valued operator, we deduce that

$$i(\mathcal{M}_C) \leq 0.$$

In fact, we observe that

$$i(\mathcal{M}_C) = i\left(\begin{pmatrix} I & 0 \\ 0 & B \end{pmatrix}\begin{pmatrix} A & C \\ 0 & I \end{pmatrix}\right).$$

Hence, applying the index theorem for the product of linear relations as stated in Theorem 3.6.4 (iii), we obtain

$$\begin{aligned}i(\mathcal{M}_C) &= i\begin{pmatrix} I & 0 \\ 0 & B \end{pmatrix} + i\begin{pmatrix} A & C \\ 0 & I \end{pmatrix} - \dim\left(\begin{pmatrix} A & C \\ 0 & I \end{pmatrix}\begin{pmatrix} 0 \\ 0 \end{pmatrix} \bigcap \begin{pmatrix} I & 0 \\ 0 & B \end{pmatrix}^{-1}\begin{pmatrix} 0 \\ 0 \end{pmatrix}\right)\\ &= i\begin{pmatrix} I & 0 \\ 0 & B \end{pmatrix} + i\begin{pmatrix} I & C \\ 0 & I \end{pmatrix} + i\begin{pmatrix} A & 0 \\ 0 & I \end{pmatrix}\\ &= i(B) + i(A)\\ &\leq 0.\end{aligned}$$

This achieves the proof. Q.E.D.

As a consequence of Propositions 3.6.6, 3.6.7, 3.6.8 and 3.6.9, we obtain this fundamental result.

Theorem 3.6.13 *Let $A \in CB\mathcal{R}(H)$ and $B \in CB\mathcal{R}(K)$. Then,*

$$\mathcal{M}_C \in \Phi_{\pm}(H \oplus K) \quad \textit{for some } C \in \mathcal{L}(K, H)$$

if, and only if, $A \in \Phi_+(H)$, and one of the following conditions holds:

- *If $R(B)$ is closed:*
  - *Either $\alpha(B) < +\infty$ and $\alpha(A) + \alpha(B) \leq \beta(A) + \beta(B)$,*
  - *or $\alpha(B) = \beta(A) = +\infty$.*
- *If $R(B)$ is not closed. Then, $\beta(A) = +\infty$.*

◇

Corollary 3.6.5 *Let $A \in CB\mathscr{R}(H)$ and $B \in CB\mathscr{R}(K)$. Then,*

$$\mathscr{M}_C \in \Phi_-^+(H \oplus K) \ \textit{ for some } C \in \mathscr{L}(K, H)$$

if and only if $B \in \Phi_-(K)$, and one of the following conditions holds:

- *If $R(A)$ is closed:*
  - *Either $\beta(A) < +\infty$ and $\beta(A) + \beta(B) \leq \alpha(A) + \alpha(B)$,*
  - *or $\beta(A) = \alpha(B) = +\infty$.*
- *If $R(A)$ is not closed. Then, $\alpha(B) = +\infty$.*

◇

Proof This result is derived by reversing the roles of $\mathscr{M}_C$ and its adjoint $\mathscr{M}_C^*$. Q.E.D.

By jointly applying Theorem 3.6.13 and Corollary 3.6.5, we conclude the following result.

Corollary 3.6.6 *Let $A \in CB\mathscr{R}(H)$ and $B \in CB\mathscr{R}(K)$. Then,*

$$\mathscr{M}_C \textit{ is a Weyl relation for some } C \in \mathscr{L}(K, H)$$

if and only if $A \in \Phi_+(H)$, $B \in \Phi_-(K)$, and one of the following conditions holds:

$$\begin{cases} \beta(A) < +\infty, \ \alpha(B) < +\infty, \ \textit{and } \alpha(A) + \alpha(B) = \beta(A) + \beta(B); \\ \textit{or } \alpha(B) = \beta(A) = +\infty. \end{cases}$$

◇

3.6.5 Fredholmness of 3 × 3 Matrices Linear Relations

Theorem 3.6.14 *(i) If A, E and K are Fredholm linear relation, then the matrix $\mathscr{T}$, given in (2.15.16), is a Fredholm linear relation and if A, E are everywhere defined, then*

$$i(\mathscr{T}) = i(A) + i(E) + i(K).$$

(ii) Let

$$\mathscr{M}_C = \begin{pmatrix} A_1 & B & C \\ 0 & A_2 & D \\ 0 & 0 & A_3 \end{pmatrix}.$$

If $A_1 \in \Phi(X)$, $A_2 \in \Phi(Y)$ and $A_3 \in \Phi(Z)$, then for all $B \in \Phi(Y, X)$, $C \in \Phi(Z, Y)$ and $D \in \Phi(Z, Y)$, we have

$$i(\mathscr{M}_C) = i(A_1) + i(A_2) + i(A_3).$$ ◇

Proof (i) Suppose that A, E and K are Fredholm linear relations. Using Lemma 2.12.2 (i), it follows that $Q_A(A)$, $Q_E(E)$ and $Q_K(K)$ are a Fredholm operators. Then,

$$\begin{pmatrix} Q_A(A) & 0 & 0 \\ 0 & Q_E(E) & 0 \\ 0 & 0 & Q_K(K) \end{pmatrix}$$

is a Fredholm operator. Now, using Lemma 2.12.2 (i), we deduce that

$$\begin{pmatrix} A & 0 & 0 \\ 0 & E & 0 \\ 0 & 0 & K \end{pmatrix}$$

is a Fredholm linear relation. On the other hand,

$$\begin{aligned} \begin{pmatrix} A & 0 & 0 \\ 0 & E & 0 \\ 0 & 0 & K \end{pmatrix} &= \begin{pmatrix} A & 0 & 0 \\ 0 & I & 0 \\ 0 & 0 & I \end{pmatrix} \begin{pmatrix} I & 0 & 0 \\ 0 & E & 0 \\ 0 & 0 & I \end{pmatrix} \begin{pmatrix} I & 0 & 0 \\ 0 & I & 0 \\ 0 & 0 & K \end{pmatrix} \\ &= \begin{pmatrix} I & 0 & 0 \\ 0 & E & 0 \\ 0 & 0 & I \end{pmatrix} \mathscr{M} \end{aligned}$$

where

$$\mathscr{M} = \begin{pmatrix} I & 0 & 0 \\ 0 & E & 0 \\ 0 & 0 & I \end{pmatrix} \begin{pmatrix} I & 0 & 0 \\ 0 & I & 0 \\ 0 & 0 & K \end{pmatrix}.$$

Then,

$$\begin{aligned} i(\mathscr{T}) = & \; i(A) + i(\mathscr{M}) - \dim \left(\mathscr{M} \begin{pmatrix} 0 \\ 0 \\ 0 \end{pmatrix} \bigcap \begin{pmatrix} A^{-1} & 0 & 0 \\ 0 & I & 0 \\ 0 & 0 & I \end{pmatrix} \begin{pmatrix} 0 \\ 0 \\ 0 \end{pmatrix} \right) \\ = & \; i(A) + i(\mathscr{M}) - \dim \left(\begin{pmatrix} 0 \\ E(0) \\ K(0) \end{pmatrix} \bigcap \begin{pmatrix} A^{-1}(0) \\ 0 \\ 0 \end{pmatrix} \right) \\ = & \; i(A) + i(\mathscr{M}). \end{aligned}$$

A similar proof shows that $i(\mathscr{M}) = i(E) + i(K)$. Consequently,

$$i(\mathscr{T}) = i(A) + i(E) + i(K).$$

(ii) The matrix linear relation $\mathcal{M}_C$ admits the following decomposition

$$\mathcal{M}_C = \begin{pmatrix} I & 0 & 0 \\ 0 & I & 0 \\ 0 & 0 & A_3 \end{pmatrix} \begin{pmatrix} I & 0 & C \\ 0 & I & D \\ 0 & 0 & I \end{pmatrix} \begin{pmatrix} I & 0 & 0 \\ 0 & A_2 & 0 \\ 0 & 0 & I \end{pmatrix} \begin{pmatrix} I & B & 0 \\ 0 & I & 0 \\ 0 & 0 & I \end{pmatrix} \begin{pmatrix} A_1 & 0 & 0 \\ 0 & I & 0 \\ 0 & 0 & I \end{pmatrix}.$$

Since $A_1 \in \Phi_+(X_1)$, $A_2 \in \Phi_+(X_2)$ and $A_3 \in \Phi_+(X_3)$, then

$$\begin{pmatrix} A_1 & 0 & 0 \\ 0 & I & 0 \\ 0 & 0 & I \end{pmatrix},$$

$$\begin{pmatrix} I & 0 & 0 \\ 0 & A_2 & 0 \\ 0 & 0 & I \end{pmatrix}$$

and

$$\begin{pmatrix} I & 0 & 0 \\ 0 & I & 0 \\ 0 & 0 & A_3 \end{pmatrix}$$

are both an upper semi-Fredholm relations. Then, by using Lemma 2.7.2, we get

$$i(\mathcal{M}_C) = \; i(A_3) + i(\mathcal{A}) - i\left(\mathcal{A}\begin{pmatrix} 0 \\ 0 \\ 0 \end{pmatrix} \bigcap \begin{pmatrix} I & 0 & 0 \\ 0 & I & 0 \\ 0 & 0 & A_3 \end{pmatrix}^{-1} \begin{pmatrix} 0 \\ 0 \\ 0 \end{pmatrix}\right),$$

where

$$\mathcal{A} = \begin{pmatrix} I & 0 & C \\ 0 & I & D \\ 0 & 0 & I \end{pmatrix} \begin{pmatrix} I & 0 & 0 \\ 0 & A_2 & 0 \\ 0 & 0 & I \end{pmatrix} \begin{pmatrix} I & B & 0 \\ 0 & I & 0 \\ 0 & 0 & I \end{pmatrix} \begin{pmatrix} A_1 & 0 & 0 \\ 0 & I & 0 \\ 0 & 0 & I \end{pmatrix}.$$

Then,

$$\begin{aligned} i(\mathcal{M}_C) &= \; i(A_3) + i(\mathcal{A}) - \dim\left(\begin{pmatrix} A_1(0) \\ A_2(0) \\ 0 \end{pmatrix} \bigcap \begin{pmatrix} I & 0 & 0 \\ 0 & I & 0 \\ 0 & 0 & A_3^{-1} \end{pmatrix} \begin{pmatrix} 0 \\ 0 \\ 0 \end{pmatrix}\right) \\ &= \; i(A_3) + i(\mathcal{A}) - \dim\left(\begin{pmatrix} A_1(0) \\ A_2(0) \\ 0 \end{pmatrix} \bigcap \begin{pmatrix} 0 \\ 0 \\ A_3^{-1}(0) \end{pmatrix}\right) \\ &= \; i(A_3) + i(\mathcal{A}). \end{aligned}$$

A similar proof gives

$$i(\mathscr{A}) = i(A_2) + i(A_1).$$

Hence,

$$i(\mathscr{M}_C) = i(A_3) + i(A_2) + i(A_1).$$

This achieves the proof. Q.E.D.

3.6.6 Closed Fredholm and Semi-Fredholm Perturbation Classes

Let X and Y be Banach spaces. Then,
(i) The set of *upper semi-Fredholm perturbations* of linear relations is defined as

$$\mathscr{PR}\big(\Phi_+(X,Y)\big) = \big\{S \in \mathscr{LR}(X,Y) : S \text{ is continuous and } U + S \in \Phi_+(X,Y),$$
$$\text{for every } U \in \Phi_+(X,Y),\ \mathscr{D}(U) \subseteq \mathscr{D}(S),\ \text{and } U(0) \supseteq S(0)\big\}.$$

(ii) The set of *lower semi-Fredholm perturbations* of linear relations is given by

$$\mathscr{PR}\big(\Phi_-(X,Y)\big) = \big\{S \in \mathscr{LR}(X,Y) : S \text{ is continuous and } U + S \in \Phi_-(X,Y),$$
$$\text{for every } U \in \Phi_-(X,Y),\ \mathscr{D}(U) \subseteq \mathscr{D}(S) \text{and } U(0) \supseteq S(0)\big\}.$$

(iii) The set of *Fredholm perturbations* of linear relations is defined as

$$\mathscr{PR}\big(\Phi(X,Y)\big) = \big\{S \in \mathscr{LR}(X,Y) : S \text{ is continuous and } U + S \in \Phi(X,Y),$$
$$\text{for every } U \in \Phi(X,Y),\ \mathscr{D}(U) \subseteq \mathscr{D}(S),\ \text{and } U(0) \supseteq S(0)\big\}.$$

When $X = Y$, we write

$$\mathscr{PR}(\Phi_+(X)) := \mathscr{PR}(\Phi_+(X,X)), \quad \mathscr{PR}(\Phi_-(X)) := \mathscr{PR}(\Phi_-(X,X)),$$

and

$$\mathscr{PR}(\Phi(X)) := \mathscr{PR}(\Phi(X,X)).$$

Proposition 3.6.10 *Let X and Y be Banach spaces, and let $U_1, U_2 \in L\mathscr{R}(X,Y)$.*
(i) If $U_1 \in \mathscr{PR}\big(\Phi_+(X,Y)\big)$ and $U_2 \in \mathscr{PR}\big(\Phi_+(X,Y)\big)$, then

$$U_1 \pm U_2 \in \mathscr{PR}\big(\Phi_+(X,Y)\big).$$

(ii) If $U_1 \in \mathscr{PR}\big(\Phi_-(X,Y)\big)$ and $U_2 \in \mathscr{PR}\big(\Phi_-(X,Y)\big)$, then

$$U_1 \pm U_2 \in \mathscr{PR}\big(\Phi_-(X,Y)\big).$$

(iii) If $U_1 \in \mathscr{PR}\big(\Phi(X,Y)\big)$ and $U_2 \in \mathscr{PR}\big(\Phi(X,Y)\big)$, then

$$U_1 \pm U_2 \in \mathscr{PR}\big(\Phi(X,Y)\big).$$

◇

Proof Let $A \in \Phi_+(X,Y)$ be such that

$$\mathscr{D}(A) \subset \mathscr{D}(U_1 \pm U_2) = \mathscr{D}(U_1) \cap \mathscr{D}(U_2),$$

and

$$A(0) \supset (U_1 \pm U_2)(0) = U_1(0) + U_2(0).$$

From the first inclusion, we deduce that $\mathscr{D}(A) \subset \mathscr{D}(U_1)$, and from the second, it follows that $A(0) \supset U_1(0)$.

Assuming $U_1 \in \mathscr{PR}(\Phi_+(X,Y))$, then

$$A + U_1 \in \Phi_+(X,Y),$$

with

$$\mathscr{D}(A + U_1) = \mathscr{D}(A) \subset \mathscr{D}(U_2),$$

and

$$(U_1 + A)(0) \supset U_2(0).$$

Using these relations and the fact that $U_2 \in \mathscr{PR}(\Phi_+(X,Y))$, we conclude that

$$A + U_1 \pm U_2 \in \Phi_+(X,Y).$$

Therefore,

$$U_1 \pm U_2 \in \mathscr{PR}(\Phi_+(X,Y)).$$

The demonstration for the other cases proceeds in the same manner as part(i). Q.E.D.

Theorem 3.6.15 *Let $U \in C\mathscr{R}(X,Y)$ and $V \in L\mathscr{R}(X,Y)$. Then,*
(i) If $U \in \Phi_+(X,Y)$ and $V \in \mathscr{PR}(\Phi_+(X,Y))$, then $U + V \in \Phi_+(X,Y)$ and

$$i(U+V) = i(U).$$

(ii) If $U \in \Phi_-(X,Y)$ and $V \in \mathscr{PR}(\Phi_-(X,Y))$, then $U + V \in \Phi_-(X,Y)$ and

$$i(U+V) = i(U).$$

(iii) If $U \in \Phi(X,Y)$ *and* $V \in \mathcal{PR}(\Phi(X,Y))$*, then* $U + V \in \Phi(X,Y)$ *and*

$$i(U+V) = i(U).$$

◇

Proof (i) Suppose that $U \in \Phi_+(X,Y)$ and $V \in \mathcal{PR}(\Phi_+(X,Y))$. Since U is closed, then by invoking Proposition 2.11.1, we obtain that $U(0)$ is closed. Therefore, $(U+V)(0) = U(0)$ is closed. Furthermore, $Q_{U+V} = Q_U$. Thus,

$$\begin{aligned} Q_{U+V}(U+V) &= Q_U(U+V) \\ &= Q_U U + Q_U V. \end{aligned}$$

Since $V(0) \subset U(0)$, it follows that

$$\begin{aligned} \overline{V(0)} &\subset \overline{U(0)} \\ &= U(0). \end{aligned}$$

Additionally, applying Lemma 2.1.1 (ii), we get

$$Y/\overline{V(0)} \Big/ U(0)/\overline{V(0)} \equiv Y/U(0),$$

and $Q_U = Q_R Q_V$, where $R = U(0)/\overline{V(0)}$. This allows us to deduce that

$$Q_{U+V}(U+V) = Q_U U + Q_R Q_V V. \tag{3.6.10}$$

By using Proposition 2.11.1, we know that $Q_U U$ is closed and single-valued. Clearly, $Q_R Q_V V$ is single-valued and continuous, so applying Proposition 2.11.1 again, we infer that $Q_U U + Q_R Q_V V$ is closed. In light of (3.6.10), it follows that $Q_{U+V}(U+V)$ is closed and single-valued. Hence, by applying Proposition 2.11.1, we conclude that $U+V$ is closed.

Let us now distinguish two situations for V.

First Case. If $\|V\| < \gamma(U)$, then according to Lemma 3.6.9 (i), we have $U \in \Phi_+(X,Y)$ if and only if $Q_U U$ is single-valued and upper semi-Fredholm. Since $\mathcal{D}(Q_U U) = \mathcal{D}(U) \subset \mathcal{D}(V) = \mathcal{D}(Q_R Q_V V)$ and

$$\begin{aligned} \|Q_R Q_V V\| &\leq \|Q_V V\| \\ &= \|V\| \\ &< \gamma(U) \\ &= \gamma(Q_U U), \end{aligned}$$

then by Corollary 3.6.3, $Q_U U + Q_R Q_V V$ is upper semi-Fredholm and

$$i(Q_U U + Q_R Q_V V) = i(Q_U U).$$

Using (3.6.10), we deduce that $Q_{U+V}(U+V) \in \Phi_+(X, Y/(U+V)(0))$ with

$$i(Q_U U) = i\left(Q_{U+V}(U+V)\right).$$

By applying Lemma 3.6.9 (i), we conclude that $U+V \in \Phi_+(X,Y)$ and

$$i(U) = i(U+V).$$

<u>Second Case</u>. If $\|V\| \geq \gamma(U)$, then since $V \in \mathscr{PR}\,(\Phi_+(X,Y))$, we have

$$\begin{aligned} \|\lambda V\| &:= \|Q_{\lambda V}\lambda V\| \\ &= |\lambda|\|Q_V V\| \\ &= |\lambda|\|V\|. \end{aligned}$$

Since $\lambda V(0) = V(0)$ and $\mathscr{D}(\lambda V) = \mathscr{D}(V)$ for all $\lambda \in \mathbb{C}$, it follows that $\lambda V \in \mathscr{PR}\,(\Phi_+(X,Y))$. Therefore,

$$U + \lambda V \in \Phi_+(X,Y) \tag{3.6.11}$$

for all $\lambda \in \mathbb{C}$. Hence, $U+V \in \Phi_+(X,Y)$. Now, we show that

$$i(U+V) = i(U).$$

To do so, consider the function

$$\begin{aligned} \varphi : \mathbb{I} := [0,1] &\longrightarrow \mathbb{Z} \cup \{-\infty\} \\ \lambda &\longrightarrow \varphi(\lambda) = i(U+\lambda V) \end{aligned}$$

and demonstrate its continuity. Let $\lambda_0 \in \mathbb{I}$ be arbitrary. Then for $\lambda \in \mathbb{I}$ such that

$$|\lambda - \lambda_0| < \frac{\gamma(U+\lambda_0 V)}{\|V\|},$$

and using (3.6.11) along with Corollary 3.6.3 (iv), we define $A := U + \lambda_0 V - (\lambda_0 V - \lambda V) \in \Phi_+(X,Y)$ and

$$i(A) = i(U+\lambda_0 V). \tag{3.6.12}$$

Since $V(0) \subset U(0)$, we have

$$\begin{aligned} Q_U &= Q_{U+\lambda V} \\ &= Q_{U+\lambda_0 V} \\ &= Q_{U+\lambda V+\lambda_0 V-\lambda_0 V} \end{aligned}$$

and $Q_U = Q_R Q_V$ (as in (3.6.10)). Hence,

$$\begin{aligned} Q_V(U + \lambda V + \lambda_0 V - \lambda_0 V) &= Q_V(U + \lambda V) + \lambda_0 Q_V V - \lambda_0 Q_V V \\ &= Q_V(U + \lambda V), \end{aligned}$$

and

$$\begin{aligned} Q_U A &= Q_A A \\ &= Q_R Q_V A \\ &= Q_R Q_V (U + \lambda V) \\ &= Q_U(U + \lambda V). \end{aligned}$$

Therefore, we obtain

$$Q_A A = Q_{U+\lambda V}(U + \lambda V). \tag{3.6.13}$$

From Lemma 3.6.9 (i) and Eqs. (3.6.12) and (3.6.13), we conclude

$$\begin{aligned} i(Q_A A) &= i(A) \\ &= i(Q_{U+\lambda V}(U + \lambda V)) \\ &= i(U + \lambda V) \\ &= i(U + \lambda_0 V). \end{aligned}$$

Thus,

$$i(U + \lambda_0 V) = i(U + \lambda V).$$

Let $\varepsilon > 0$; then there exists $\delta := \frac{\gamma(U+\lambda_0 V)}{\|V\|}$ such that for all $\lambda \in \mathbb{I}$ with $|\lambda - \lambda_0| < \delta$, we have

$$0 = |i(U + \lambda V) - i(U + \lambda_0 V)| < \varepsilon.$$

So, φ is continuous. Hence, $\varphi(\mathbb{I})$ is a connected subset of $\mathbb{Z} \cup \{-\infty\}$, and thus consists of a single constant value. It follows that

$$\begin{aligned} i(U) &= \varphi(0) \\ &= \varphi(1) \\ &= i(U + V). \end{aligned}$$

This completes the proof of part (i).
(ii) Let $U \in \Phi_-(X, Y)$ and $V \in \mathscr{PR}(\Phi_-(X, Y))$. Then for all $\lambda \in \mathbb{C}$, $U + \lambda V \in \Phi_-(X, Y)$, which is equivalent to $(U + \lambda V)^* \in \Phi_+(Y^*, X^*)$. Now observe that

$$\begin{aligned} (U + \lambda V)^* &= U^* + (\lambda V)^* \\ &= U^* + \lambda V^* \end{aligned}$$

(clearly, $(\lambda V)^* = \lambda V^*$ for $\lambda \neq 0$), so $U^* + \lambda V^* \in \Phi_+(Y^*, X^*)$. Furthermore, since V^* is continuous, then λV^* is also continuous, and $\lambda V^*(0) = V^*(0) \subset U^*(0)$ with $\mathscr{D}(U^*) \subset \mathscr{D}(V^*) = \mathscr{D}(\lambda V^*)$. Thus, we may apply (i) and obtain that $i(U^*) = i(U^* + V^*)$ with U^* and $U^* + V^* \in \Phi_+(Y^*, X^*)$. Consequently,

$$\begin{aligned} -i(U) &= i(U^*) \\ &= i(U^* + V^*) \\ &= i((U+V)^*) \\ &= -i(U+V). \end{aligned}$$

Therefore, $i(U) = i(U+V)$, as required.
(iii) The proof of (iii) follows from the same argument as above. Q.E.D.

Lemma 3.6.13 *Let $U \in C\mathscr{R}(X)$ and $V \in L\mathscr{R}(X)$ be continuous operators such that $V(0) \subset U(0)$ and $\mathscr{D}(U) \subset \mathscr{D}(V)$. Then:*

$$U \in \Phi_+(X) \quad \textit{if and only if} \quad U + V - V \in \Phi_+(X).$$

In this case, their indices coincide:

$$i(U) = i(U + V - V).$$

$$U \in \Phi_-(X) \quad \textit{if and only if} \quad U + V - V \in \Phi_-(X).$$

Similarly, the indices satisfy:

$$i(U) = i(U + V - V).$$

◇

Proof We initially observe that by Proposition 2.11.2, $U + V$ and $U + V - V$ are closed. Moreover, we possess
(a) $(U + V - V)(0) = U(0)$ and $Q_{U+V-V} = Q_{U(0)/V(0)} Q_V$. Indeed,

$$\begin{aligned} (U + V - V)(0) &= U(0) + V(0) \\ &= U(0), \end{aligned}$$

thus

$$Q_{U+V-V} = Q_U$$

and by Lemma 2.1.1, we deduce that

$$Q_U = Q_{U(0)/V(0)} Q_V.$$

(b) $Q_{U+V-V}(U+V-V) = Q_U(U)$. Indeed,

$$\begin{aligned} Q_V(U+V-V) &= Q_V(U) + Q_V(V) - Q_V(V) \\ &= Q_V(U), \end{aligned}$$

so, by (a), we conclude (b). Now, merging (b) with Lemma 3.6.9, we derive the statement. Q.E.D.

3.6.7 Perturbation by Demicompact Linear Relation

Proposition 3.6.11 [99, Theorem 2.3] *Let $U : \mathscr{D}(U) \subset X \to X$ and $V : \mathscr{D}(V) \subset X \to X$ be densely defined closed linear operators such that $\mathscr{D}(U) \subset \mathscr{D}(V)$ and the operator $V - U$ is closed. If U is V-demicompact, then the operator $V - U$ is an upper semi-Fredholm operator.* $\diamond$

Theorem 3.6.16 *Let X be a Banach space and $U \in L\mathscr{R}(X)$. Then:*
(i) If $\dim(U(0)) < \infty$ and $I - U \in \mathscr{F}_+(X)$, then U is a demicompact linear relation.
(ii) If $\dim(U(0)) < \infty$ and U is a demicompact linear relation, then $I - U \in \mathscr{F}_+(X)$.
(iii) If $U \in C\mathscr{R}(X)$, $\dim(U(0)) < \infty$, and U is a demicompact linear relation, then $I - U \in \Phi_+(X)$. $\diamond$

Proof (i) Assume $I - U \in \mathscr{F}_+(X)$. By applying Lemma 3.6.1, we deduce

$$Q_{I-U}(I-U) = Q_U(I-U) \in \mathscr{F}_+(X, X/\overline{U(0)}).$$

Hence, invoking Theorem 3.6.3, there exist a bounded linear operator A and a bounded finite-rank projection operator P such that

$$AQ_U(I-U) = I_{\mathscr{D}(U)} - P.$$

Let $(x_n)_n \subset \mathscr{D}(U)$ be a bounded sequence satisfying

$$Q_U(I-U)x_n \to y \in X/\overline{U(0)}.$$

Then,

$$AQ_U(I-U)x_n \to Ay,$$

and consequently,

$$x_n - Px_n \to Ay.$$

Since P is compact, by Theorem 3.5.3 it follows that P is demicompact. Therefore, the sequence $(x_n)_n$ admits a convergent subsequence.
(ii) Suppose $I - U \notin \mathscr{F}_+(X)$. Then, by Theorem 3.6.2, there exists a sequence $(x_n)_n \subset \mathscr{D}(U)$ with unit norm such that $(x_n)_n$ has no Cauchy subsequence and

$$\lim_{n\to\infty} \|(I-U)x_n\| = 0.$$

Note that

$$\|(I-U)x_n\| = \|Q_{I-U}(I-U)x_n\| = \|Q_U(I-U)x_n\|.$$

Thus, the sequence $(Q_U(I-U)x_n)_n$ converges to zero. Since U is demicompact, it follows that $(Q_U x_n)_n$ contains a convergent subsequence. By Lemma 2.9.1, $(x_n)_n$ also possesses a convergent subsequence, contradicting the premise that $(x_n)_n$ has no Cauchy subsequence. Consequently, $I - U \in \mathscr{F}_+(X)$.
(iii) The assertion follows immediately from part (ii) and Proposition 2.11.2. .Q.E.D.

Theorem 3.6.17 *Let $U \in L\mathscr{R}(X)$ with $\dim(U(0)) < \infty$, and let W be a selection of U. Then, U is a demicompact linear relation if and only if W is a demicompact operator.* ◇

Proof Let U be a demicompact linear relation. By applying Theorem 3.6.16 (ii), we deduce that $I - U \in \Phi_+(X)$. Since $I - W$ is a selection of $I - U$ and $\dim(U(0)) < \infty$, it follows from Lemma 3.6.3 that $I - W \in \Phi_+(X)$. Then, by Theorem 3.6.3, there exist $A \in B\mathscr{R}(X)$ and a finite rank projection K such that

$$A(I-W) = I - K.$$

Let $(x_n)_n$ be a bounded sequence in $\mathscr{D}(W)$ such that $(I-W)x_n \to y$. Applying A, we have

$$A(I-W)x_n \to Ay,$$

which implies

$$(I-K)x_n \to Ay.$$

Since K is compact, the sequence $(Kx_n)_n$ admits a convergent subsequence, and consequently, $(x_n)_n$ also has a convergent subsequence.

Conversely, suppose W is a demicompact operator. Using Theorem 3.6.16, we obtain $I - W \in \Phi_+(X)$. Given that $I - W$ is a selection of $I - U$ and $\dim(U(0)) < \infty$, Lemma 3.6.3 implies $I - U \in \Phi_+(X)$. Applying Theorem 3.6.3 again, there exist $A \in \mathscr{L}(X)$ and a finite rank projection K such that

$$A(I-U) = I - K.$$

Let $(x_n)_n$ be a bounded sequence in $\mathscr{D}(U)$ with

$$Q_U(I-U)x_n \to y.$$

Since $Q_{A(I-U)}AQ_U^{-1}$ is bounded, it follows that

$$Q_{A(I-U)}AQ_U^{-1}Q_U(I-U)x_n \to Q_{A(I-U)}AQ_U^{-1}y.$$

Noting that $\overline{U(0)} \subset \mathscr{D}(A)$ and $A \in \mathscr{L}(X)$, we have

$$Q_{I-K}A(I-U)x_n \to Q_{I-K}AQ_U^{-1}y,$$

which is equivalent to

$$Q_K(I-K)x_n \to Q_KAQ_U^{-1}y.$$

Since K is compact, the sequence $(x_n)_n$ has a convergent subsequence. Q.E.D.

Theorem 3.6.18 *Let $U \in C\mathscr{R}(X)$ be with* $\dim U(0) < \infty$. *If $\frac{1}{\mu}U$ is demicompact for each $\mu \in [1, +\infty[$, then $\mu - U \in \Phi(X)$.* ◇

Proof Let S be a selection of U. Since $\frac{1}{\mu}U$ is demicompact for each $\mu \in [1, +\infty[$, then by using Theorem 3.6.16, we get $\mu - U \in \Phi_+(X)$. By using Lemma 3.6.3, we have $\mu - S \in \Phi(X)$. We shall prove that the map

$$\begin{aligned} \varphi : [1, +\infty[&\longrightarrow \mathbb{Z} \\ \mu &\longrightarrow i(\mu - S) \end{aligned}$$

is continuous in μ. For this, let $\mu, \mu_0 \in [1, +\infty[$ be arbitrary but fixed such that

$$|\mu - \mu_0| < \gamma(\mu - S).$$

Then,

$$\begin{aligned} i(\mu - S) &= i(\mu - S - \mu + \mu_0) \\ &= i(\mu_0 - S). \end{aligned}$$

Let $\varepsilon > 0$, then there exists $\delta := \gamma(\mu - S)$ such that, if $\mu, \mu_0 \in [1, +\infty[$ with $|\mu - \mu_0| < \delta$, then

$$|i(\mu - S) - i(\mu_0 - S)| = |0| = 0 < \varepsilon.$$

So, $\varphi(\mu)$ is continuous. Now, we know that every continuous mapping of a connected in $\mathbb{Z}$ is constant. If $\mu \to +\infty$, then

$$\begin{aligned} i\left(\left(I - \frac{1}{\mu}S\right)\mu\right) &= i\left(I - \frac{1}{\mu}S\right) \\ &= i(I) \\ &= 0. \end{aligned}$$

Showing that $i(\mu - S) = 0$ for each $\mu \in [1, +\infty[$. We conclude that

$$\alpha(\mu - S) = \beta(\mu - S) < \infty.$$

So, $\mu - S$ is a Fredholm linear operator. Now, by using Lemma 3.6.3, we conclude that $\mu - U \in \Phi(X)$. Q.E.D.

Theorem 3.6.19 *Let $U : \mathscr{D}(U) \subset X \longrightarrow X$ be a closed linear relation and $S : \mathscr{D}(S) \subset X \longrightarrow X$ be a continuous and a closed linear relation with $\mathscr{D}(U) \subset \mathscr{D}(S)$ and $S(0) \subset U(0)$. If $Q_S - Q_U$ is compact, then $S - U \in \Phi_+(X)$ if, and only if, U is S-demicompact.* ◇

Proof By using Proposition 2.11.2, we have $S - U$ is closed. If $S - U \in \Phi_+(X)$, then by using both $S(0) \subset U(0)$ and Lemma 3.6.9 (i), we get

$$Q_{S-U}(S - U) = Q_U(S - U) \in \Phi_+(X, X/U(0)). \tag{3.6.14}$$

In view of Theorem 3.6.3, there exist a bounded linear operator A and a bounded finite rank projection operator P such that

$$AQ_T(S - U) = I_{\mathscr{D}(U)} - P. \tag{3.6.15}$$

Let $(x_n)_n$ be a bounded sequence of $\mathscr{D}(U)$ such that $(Q_U(S - U)x_n)_n$ converges to $Q_U x$. Then, $(AQ_U(S - U)x_n)_n$ converges to $AQ_U x$. So, by (3.6.15), we have $(x_n - Px_n)_n$ converges to $AQ_U x$. Since P is compact and single-valued, then $(x_n)_n$ has a convergent subsequence. Finally, $(Q_U Sx_n)_n$ has a convergent subsequence. Conversely, let U be S-demicompact and $Q_S - Q_U$ be a compact linear relation. Using Proposition 3.6.11, we find that $Q_U U$ is $Q_U S$-demicompact. The latter implies and using Proposition 3.6.11, we obtain that $Q_U S - Q_U U$ is an upper semi-Fredholm single-valued. On the other side, since $S(0) \subset U(0)$, then

$$Q_U S - Q_U U = Q_U(S - U).$$

By using (3.6.14), we have $Q_{S-U}(S - U)$ is an upper semi-Fredholm single-valued linear operator. Using Lemma 3.6.9 (i), we obtain $S - U$ is an upper semi-Fredholm linear relation. Q.E.D.

Proposition 3.6.12 *Let $\mu \in \mathbb{C}^*$ and let $U : \mathscr{D}(U) \subset X \longrightarrow X$ and $S : \mathscr{D}(S) \subset X \longrightarrow X$ be two linear relations with $S(0) \subset U(0)$ and $\mathscr{D}(U) \subset \mathscr{D}(S)$. If $\mu S - U \in \mathscr{F}_+(X)$, then U is μS-demicompact.* ◇

Proof Suppose that $\mu S - U \in \mathscr{F}_+(X)$, then

$$Q_{\mu S-U}(\mu S - U) = Q_U(\mu S - U) \in \mathscr{F}_+(X, X/\overline{U(0)}).$$

Hence, there exists a bounded linear operator A and a bounded finite rank projection operator P such that

$$AQ_U(\mu S - U) = I_{\mathscr{D}(T)} - P. \tag{3.6.16}$$

Let $(x_n)_n$ be a bounded sequence of $\mathcal{D}(U)$ such that $(Q_U(\mu S - U)x_n)_n$ converges to $Q_U x$. Then, $(AQ_U(\mu S - U)x_n)_n$ converges to $AQ_U x$. So, by using (3.6.16), we have $(x_n - Px_n)_n$ converges to $AQ_U x$. Since P is a compact linear operator, then $(x_n)_n$ has a convergent subsequence. Finally, $(Q_U \mu S x_n)_n$ has a convergent subsequence. Q.E.D.

Proposition 3.6.13 *Let $\mu \in \mathbb{C}^*$ and let $V : \mathcal{D}(V) \subset X \longrightarrow X$ and $S : \mathcal{D}(S) \subset X \longrightarrow X$ be two linear relations with $S(0) \subset V(0)$, $\mathcal{D}(V) \subset \mathcal{D}(S)$ and $\overline{V(0)}$ is compact. If V is a μS-demicompact linear relation, then $\mu S - V \in \mathcal{F}_+(X)$.* ◇

Proof We suppose that $\mu S - V \notin \mathcal{F}_+(X)$. Then, there exists a sequence $(x_n)_n \in \mathcal{D}(V)$ of norm one such that $(x_n)_n$ has no Cauchy subsequence and

$$\lim_{n\to\infty} \|(\mu S - V)x_n\| = 0.$$

Since

$$\begin{aligned} \|(\mu S - V)x_n\| &= \|Q_{\mu S - V}(\mu S - V)x_n\|, \\ &= \|Q_V(\mu S - V)x_n\| \text{ (since } S(0) \subset V(0)), \end{aligned}$$

then $(Q_V(\mu S - V)x_n)_n$ converge to 0. Since V is μS-demicompact, then $(Q_V \mu S x_n)_n$ has a convergent subsequence. Using the fact that $\overline{V(0)}$ is compact, then $(x_n)_n$ has a convergent subsequence. This contradicts the fact $(x_n)_n$ has no Cauchy subsequence. Finally, $\mu S - V \in \mathcal{F}_+(X)$. Q.E.D.

Theorem 3.6.20 *Let $\mu \in \mathbb{C}^*$, and consider a closed linear relation $U : \mathcal{D}(U) \subset X \to X$ and a continuous, closed linear relation $W : \mathcal{D}(W) \subset X \to X$ such that $\mathcal{D}(U) \subset \mathcal{D}(W)$ and $W(0) \subset U(0)$. Suppose that the operator $|\mu|W - U$ is closed. If U is $|\mu|W$-demicompact and the difference $Q_{|\mu|W} - Q_U$ is compact, then $|\mu|W - U$ is an upper semi-Fredholm relation.* ◇

Proof Assume U is $|\mu|W$-demicompact and that $Q_{|\mu|W} - Q_U$ is a compact linear relation. By applying Lemma 3.5.5, we deduce that $Q_U U$ is $Q_U|\mu|W$-demicompact. This consequently implies that $Q_U|\mu|W - Q_U U$ is a single-valued upper semi-Fredholm operator. Furthermore,

$$Q_{|\mu|W-U}(|\mu|W - U) = Q_T(|\mu|W - U).$$

We observe that $Q_{|\mu|W-U}(|\mu|W - U)$ constitutes a single-valued upper semi-Fredholm linear operator. Therefore, $|\mu|W - U$ is an upper semi-Fredholm relation. Q.E.D.

Proposition 3.6.14 *Let $\alpha \in (0, 1)$, $\mu \in \mathbb{C}^*$, and consider continuous, closed linear relations*

$$U : \mathcal{D}(U) \subset X \to X \quad \textit{and} \quad W : \mathcal{D}(W) \subset X \to X$$

such that $\mathscr{D}(U) \subset \mathscr{D}(W)$, $W(0) \subset U(0)$, and $W \neq 0$. Suppose that W is relatively bounded with respect to U with U-bound strictly less than 1. If the operator $|\mu|W - \alpha U$ is closed, the difference $Q_{|\mu|W} - Q_U$ is compact, and αU is $|\mu|W$-demicompact, then the operator $|\mu|W - U$ is a Fredholm linear relation, and its index satisfies $i(|\mu|W - U) = i(|\mu|W)$. ◇

Proof Let αU be a $|\mu|W$-demicompact operator and suppose that $Q_{|\mu|W} - Q_U$ is a compact linear relation. By invoking Lemma 3.5.5, we deduce that $Q_U U$ is $Q_U|\mu|W$-demicompact. Consequently, the difference $Q_U|\mu|W - Q_U U$ is a Fredholm linear operator and satisfies

$$\begin{aligned} i(Q_U|\mu|W - Q_U U) &= i(Q_U|\mu|W) \\ &= i(|\mu|W). \end{aligned}$$

Moreover, observe that

$$Q_{|\mu|W-U}(|\mu|W - U) = Q_U(|\mu|W - U).$$

Therefore, $Q_{|\mu|W-U}(|\mu|W - U)$ is a Fredholm linear operator. It follows that $|\mu|W - U$ is a Fredholm linear relation with index

$$i(|\mu|W - U) = i(|\mu|W).$$

Q.E.D.

3.7 Essential Spectra of Linear Relation

3.7.1 *Definitions and Preliminary Results*

Definition 3.7.1 Let $V \in C\mathscr{R}(X)$. The essential resolvent sets of V, where we define by $\rho_{ei}(V)$, $i = 1, 2, 3, 4, 5, 6, ap, \delta$ as follows

$$\begin{aligned} \rho_{e1}(V) &= \{\lambda \in \mathbb{C} \text{ such that } \lambda - V \in \Phi_+(X)\}, \\ \rho_{e2}(V) &= \{\lambda \in \mathbb{C} \text{ such that } \lambda - V \in \Phi_-(X)\}, \\ \rho_{e3}(V) &= \left\{\lambda \in \mathbb{C} \text{ such that } \lambda - V \in \Phi_+(X) \bigcup \Phi_-(X)\right\}, \\ \rho_{e4}(V) &= \left\{\lambda \in \mathbb{C} \text{ such that } \lambda - V \in \Phi_+(X) \bigcap \Phi_-(X)\right\}, \\ \rho_{e5}(V) &= \left\{\lambda \in \mathbb{C} \text{ such that } \lambda - V \in \Phi_+(X) \bigcap \Phi_-(X) \text{ and } i(\lambda - V) = 0\right\}, \\ \rho_{e6}(V) &= \bigcup \rho_{e3}^{(n)}(V), \text{ where } \rho_{e3}^{(n)}(V) \text{ is a component of } \rho_{e3}(V) \text{and } \rho_{e3}^{(n)}(V) \bigcap \rho_{e3}^{(n)}(V) \neq \emptyset. \\ \rho_{eap}(V) &= \{\lambda \in \mathbb{C} \text{ such that } \lambda - V \in \Phi_+(X) \text{ and } i(\lambda - V) \leq 0\}, \\ \rho_{e\delta}(V) &= \{\lambda \in \mathbb{C} \text{ such that } \lambda - V \in \Phi_-(X) \text{ and } i(\lambda - V) \geq 0\}. \end{aligned}$$

There are several definitions of the essential spectra, most of which constitute an enlargement of the continuous spectrum. Let us define the following sets: the essential spectra $\sigma_{ei}(V)$, $i = 1, 2, 3, 4, 5, 6, ap, \delta$ are the complement in $\mathbb{C}$ of respective of the essential resolvent sets $\rho_{ei}(V)$. ◇

In general those spectra satisfy the following inclusions

$$\sigma_{e1}(V)\bigcap\sigma_{e2}(V)=\sigma_{e3}(V)\subset\sigma_{e4}(V)\subset\sigma_{e5}(V)\subset\sigma_{e6}(V)\subset\sigma(V),$$

and

$$\sigma_{e1}(V)\subset\sigma_{eap}(V),\ \sigma_{e2}(V)\subset\sigma_{e\delta}(V).$$

Remark 3.7.1 Let X be a Banach space and let $V\in C\mathscr{R}(X)$. Then,

(i) $\sigma_{e5}(V)=\sigma_{e4}(V)\bigcup\{\lambda\in\mathbb{C}$ such that $i(V-\lambda)\neq 0\}$.
(ii) $\sigma_{eap}(V)=\sigma_{e1}(V)\bigcup\{\lambda\in\mathbb{C}$ such that $i(V-\lambda)>0\}$.
(iii) $\sigma_{e\delta}(V)=\sigma_{e2}(V)\bigcup\{\lambda\in\mathbb{C}$ such that $i(V-\lambda)<0\}$. ◇

Remark 3.7.2 Let X be a Banach space and $V\in L\mathscr{R}(X)$. Then, (i) $\lambda\notin\sigma_{e5}(V)$ if, and only if, $\lambda-V\in\Phi(X)$ and $i(\lambda-V)=0$.
(ii) $\lambda\notin\sigma_{eap}(V)$ if, and only if, $\lambda-V\in\Phi_+(X)$ and $i(\lambda-V)\leq 0$.
(iii) $\lambda\notin\sigma_{e\delta}(V)$ if, and only if, $\lambda-V\in\Phi_-(X)$ and $i(\lambda-V)\geq 0$. ◇

Theorem 3.7.1 *Let X be a Banach space, $V\in C\mathscr{R}(X)$ and set*

$$\mathscr{K}_V(X)=\{K\in K\mathscr{R}(X) \text{ such that } \mathscr{D}(V)\subset\mathscr{D}(K), \text{ and } K(0)\subset V(0)\}.$$

Then,

$$(i)\ \sigma_{e5}(V)=\bigcap_{K\in\mathscr{K}_V(X)}\sigma(V+K).$$

$$(ii)\ \sigma_{eap}(V)=\bigcap_{K\in\mathscr{K}_V(X)}\sigma_{ap}(V+K).$$

$$(iii)\ \sigma_{e\delta}(V)=\bigcap_{K\in\mathscr{K}_V(X)}\sigma_{\delta}(V+K).$$ ◇

Proof (i) Let $\lambda\notin\sigma_{e5}(V)$. Then, by using Remark 3.7.2, we have $\lambda-V\in\Phi(X)$ and $i(\lambda-V)=0$. Hence, $n=\alpha(\lambda-V)=\beta(\lambda-V)<\infty$. So, there exists an everywhere defined single-valued B with $\dim R(B)\leq\alpha(V)$ such that $V-B$ is bijective. Furthermore, let

$$Bx:=\sum_{i=1}^{n}x_i'(x)y_i,\quad x\in X,$$

where $\{x_1,\ldots,x_n\}$ is a basis of $N(\lambda-V)$. Choose linear functionals $x_1',\ldots,x_n'$ such that $x_i'(x_j)=\delta_{ij}$ and choose $y_1,\ldots,y_n\in X$ such that $[y_1],\ldots,[y_n]\in X/R(\lambda-V)$ are linearly independent ($n=\alpha(\lambda-V)=\beta(\lambda-V)$). Hence, it is clear that B is a bounded finite rank operator. Since $B(0)=\{0\}\subset(\lambda-V)(0)=V(0)$ and $B\in\mathscr{K}(X)$, then $B\in\mathscr{K}_V(X)$ and also

$$(\lambda-V)-B=\lambda-(V+B)$$

is bijective. Therefore, $\lambda \in \rho(V+K)$. This shows that $\lambda \notin \bigcap_{K\in\mathcal{K}_V(X)}\sigma(V+K)$. Conversely, we suppose $\lambda \notin \bigcap_{K\in\mathcal{K}_V(X)}\sigma(V+K)$, then there exists $K \in \mathcal{K}_V(X)$ such that $\lambda \in \rho(V+K)$. Hence, $\lambda-(V+K) \in \Phi(X)$ and

$$i(\lambda-(V+K))=0.$$

Using Remark 2.7.1, we get $\lambda - V \in \Phi(X)$ and

$$i(\lambda-V)=i(\lambda-(V+K))=0.$$

We conclude that $\lambda \notin \sigma_{e5}(V)$. The statements for the $\sigma_{eap}(V)$ and $\sigma_{e\delta}(V)$ can be proved similarly. Q.E.D.

Proposition 3.7.1 *(i) If $\rho_{ei}^{(n)}(V)$ is a component of $\rho_{ei}(V)$, $i=1,2,3,4$, then $\alpha(\lambda-V)$ and $\beta(\lambda-V)$ have constant values, n_1 and n_2, respectively, $n_1,\ n_2 \in \mathbb{N}\bigcup\{\infty\}$, except perhaps at isolated points, where $\alpha(\lambda-V)>n_1$ and $\beta(\lambda-V)>n_2$.*
(ii) $\sigma_{ei}(V),\ i=1,2,3,4,5$ are closed.
(iii) The index is constant in $\rho_{ei}(\cdot),\ i=1,2,3,4,5$. ◇

Proof (i) We first prove the result for the quantities $\alpha(\lambda-V)$. Since any component of an open set in $\mathbb{C}$ is open, we have that $\rho_{ei}^{(n)}(V)$ are open sets. We first consider the case $\rho_{e1}^{(n)}(V)$. If $\alpha(\lambda-V)=\infty$ for all $\lambda \in \rho_{e1}^{(n)}(V)$, then we are done. Now, suppose $\alpha(\lambda-T)<\infty$ for some $\lambda \in \rho_{e1}^{(n)}(V)$. Define $\alpha(\lambda)=\alpha(\lambda-V)$ and choose λ_0 such that $\alpha(\lambda_0)=n_1$ is the smallest nonnegative integer attained by $\alpha(\lambda)$ on $\rho_{e1}^{(n)}(V)$. Suppose $\alpha(\lambda')\neq n_1$ for some λ'. Since $\rho_{e3}^{(n)}(V)$ is connected, then there exists an arc Λ in $\rho_{e3}^{(n)}(V)$ with endpoints λ_0 and λ'. Since $\lambda-V\in\Phi_+(X)\bigcup\Phi_-(X)$, it follows from Theorem 3.6.6 that for each $\mu\in\Lambda$, there exists an open ball B_μ contained in $\rho_{e3}^{(n)}(V)$ such that $\alpha(\lambda)$ is constant on $B_\mu\backslash\{\mu\}$. Since Λ is compact, there exists a finite set of points $\lambda_1,\lambda_2,\ldots,\lambda_n=\lambda'$ such that $B_{\lambda_0}, B_{\lambda_1},\ldots,B_{\lambda_n}$ cover Λ and for $0\leq i\leq n-1$,

$$B_{\lambda_i}\bigcap B_{\lambda_{i+1}}\neq\emptyset. \tag{3.7.1}$$

It follows from Theorem 2.10.1 that $\alpha(\lambda)\leq\alpha(\lambda_0)$ for λ sufficiently close to λ_0. Thus, since $\alpha(\lambda_0)$ is the minimum value attained by $\alpha(\lambda)$ on $\rho_{e1}^{(n)}(V)$, it follows that $\alpha(\lambda)=\alpha(\lambda_0)$ for λ sufficiently close to λ_0. Since $\alpha(\lambda)$ is constant for all $\lambda\neq\lambda_0$ in B_{λ_0}, this constant must be $\alpha(\lambda_0)$. Similarly $\alpha(\lambda)$ is constant on $B_{\lambda_i}\backslash\{\lambda_i\}$ for $1\leq i\leq n$. Thus, by (3.7.1) that $\alpha(\lambda)=\alpha(\lambda_0)$ for all $B_{\lambda'}\backslash\{\lambda'\}$ and $\alpha(\lambda')>n_1$. To see that the result holds for $\beta(\lambda-V)$, we pass to the adjoint of T and apply the above, and the equality

$$\alpha(\lambda-V^*)=\beta(\lambda-V).$$

The proofs for $\rho_{e1}^{(n)}(T)$, $\rho_{e1}^{(n)}(V)$ and $\rho_{e4}^{(n)}(V)$ are similar.
(ii) For $i=5$. Let $\lambda_0\in\rho_{e5}(V)$. Then, $R(V-\lambda_0)$ is closed. Hence, $V-\lambda_0$ is open. Then, according to Proposition 2.15.3 (i), we have $\gamma(V-\lambda_0)>0$. Thus,

$$\gamma(V-\lambda_0)>0.$$

Consider $\mu\in\mathbb{C}$ such that

$$|\mu-\lambda_0|<\gamma(V-\lambda_0).$$

Then, according to Corollary 3.6.3, $\lambda_0-\mu-\lambda_0+V\in\Phi(X)$ and

$$i(\lambda_0-\mu-\lambda_0+V)=0.$$

By using Proposition 2.7.2 allows us to conclude that

$$\lambda_0-\mu-\lambda_0+V=\mu-V.$$

Hence, $\mu-T\in\Phi(X)$ and

$$i(\mu-V)=0.$$

Therefore, $\rho_{e5}(V)$ is open. For the other cases, the same arguments have been used. (iii) Let λ_1 and λ_2 be any two points in $\rho_{ei}(V)$, $i=1,2,3,4,5$ which are connected by a smooth curve Γ whose points are all in $\rho_{ei}(T)$, $i=1,2,3,4,5$. Since $\rho_{ei}(V)$, $i=1,2,3,4,5$ are an open sets, then for each $\lambda\in\Gamma$, by using Corollary 3.6.3, there exists an $\varepsilon>0$ such that, for all $\mu\in\mathbb{C}$, $0<|\mu-\lambda|<\varepsilon$, $\mu\in\rho_{ei}(V)$, $i=1,2,3,4,5$ and

$$i(\mu-V)=i(\lambda-V).$$

By using the Heine–Borel theorem, there exists a finite number of such sets which cover Γ. Since each of these sets overlaps with, at least, another set and since $i(\mu-V)$ is constant on each one, we see that $i(\lambda_1-V)=i(\lambda_2-V)$. Q.E.D.

Proposition 3.7.2 *$\lambda\in\rho_{e6}(T)$ if, and only if, $\lambda\in\rho_{e5}(T)$ and a deleted neighborhood of λ lies in $\rho(T)$.* ◇

Proof Suppose $\lambda\in\rho_{e6}(T)$. Then, λ lies in a component $\rho_{e3}^{(n)}(T)$ of $\rho_{e1}(T)$ which intersects $\rho(T)$. Let C be such a component. Clearly $C\bigcap\rho(T)$ is open. Since $\mu\in C\bigcap\rho(T)$ implies

$$\begin{aligned}\alpha(\mu-T)=&\ \beta(\mu-T)\\ =&\ i(\mu-T)\\ =&\ 0.\end{aligned}$$

It follows from Theorem 3.6.5 (iii) and (iv) that

$$i(\lambda-T)=0$$

for $\lambda\in C$ when λ is sufficiently close to μ, and, hence for all $\lambda\in C$. Applying Proposition 2.11.2, we see that

$$\alpha(\lambda - T) = \beta(\lambda - T) = 0$$

for all except some isolated points, say λ_j, where $\alpha(\lambda_j - T) > 0$ and $\beta(\lambda_j - T) > 0$. Thus, if $\lambda \in \rho_{e5}(T)$, then either $\lambda \in \rho(T)$ or λ is one of these isolated points in $\rho_{e5}(T)$. The converse may be proved in a similar. This achieves the proof. Q.E.D.

Proposition 3.7.3 *Let $T \in C\mathscr{R}(X)$ be such that $\rho(T) \neq \emptyset$. Then,*

(i) If $\rho_{e5}(T)$ is connected, then $\sigma_{e1}(T) = \sigma_{eap}(T)$, $\sigma_{e2}(T) = \sigma_{e\delta}(T)$ and $\sigma_{e4}(T) = \sigma_{e5}(T)$.
(ii) If $\rho_{e5}(T)$ is connected, then $\rho_{e5}(T) = \rho_{e6}(T)$. ◇

Proof (i) It is easy to check that

$$\sigma_{e1}(T) \subset \sigma_{eap}(T).$$

We prove that $\sigma_{eap}(T) \subset \sigma_{e1}(T)$. Let $\lambda \notin \sigma_{e1}(T)$, then $\lambda \in \Phi_{+T}$. On the other hand,

$$\Phi_{+T} = \Phi_T \bigcup (\Phi_{+T} \setminus \Phi_T).$$

Hence, we will discuss these two cases:

First case: If $\lambda \in \Phi_T$, then
$$i(\lambda - T) = 0.$$

Indeed, let $\lambda_0 \in \rho(T)$, then $\lambda_0 \in \Phi_T$ and $i(T - \lambda_0) = 0$. Using, Proposition 2.11.2 (iii), we have $i(\lambda - T)$ is constant on any component of Φ_T. Therefore, $\rho(T) \subset \Phi_T$, so that $i(\lambda - T) = 0$ for all $\lambda \in \Phi_T$. Then, it follows from, Corollary 3.7.1 (ii) that $\lambda \notin \sigma_{eap}(T)$.

Second case: If $\lambda \in (\Phi_{+T} \setminus \Phi_T)$, then $\alpha(\lambda - T) < \infty$ and $\beta(\lambda - T) = +\infty$. So, $i(\lambda - T) = -\infty < 0$. Hence, it follows from Corollary 3.7.1 (ii) that $\lambda \notin \sigma_{eap}(T)$. Thus, we obtain from the above that $\sigma_{eap}(T) \subset \sigma_{e1}(T)$. Hence, we obtain the second inclusion from the above two cases. By following the same reasoning, we get the two equalities.
(ii) Since $\rho(T) \subset \rho_{e5}(T)$, it follows from the hypothesis and Proposition 2.11.2 (i) that
$$\alpha(\lambda - T) = \beta(\lambda - T) = 0$$

for all $\lambda \in \rho_{e4}(T)$ except perhaps at isolated points, i.e., a deleted neighborhood of λ lies in $\rho(T)$. The result follows from Proposition 3.7.2. This achieves the proof. Q.E.D.

As a direct consequence of [110, Remarks 3.1 and 3.3], we infer the following result:

Lemma 3.7.1 *Let X be a Banach space and let $T \in C\mathscr{R}(X)$. Then,*

$$\sigma_{ei}(T + K) = \sigma_{ei}(T), \text{ for all } K \in K\mathscr{R}(X) \text{ and } i = 1, \cdots, 4. \qquad \diamond$$

3.7.2 Pencil Essential Spectra of Multivalued Linear Operators

Let X be a normed vector space, $U \in C\mathscr{R}(X)$, and V be a continuous linear relation such that $V \neq 0$, $V(0) \subset U(0)$, and $\mathscr{D}(V) \supset \mathscr{D}(U)$. The pencil essential resolvent sets of U, where we define by $\rho_{ei,S}(U)$, $i = 1, 2, 3, 4, 5, 6$ are as follows

$$\begin{aligned}
\rho_{e1,V}(U) &= \{\lambda \in \mathbb{C} \text{ such that } \lambda V - U \in \Phi_+(X)\}, \\
\rho_{e2,V}(U) &= \{\lambda \in \mathbb{C} \text{ such that } \lambda V - U \in \Phi_-(X)\}, \\
\rho_{e3,V}(U) &= \left\{\lambda \in \mathbb{C} \text{ such that } \lambda V - U \in \Phi_+(X) \bigcup \Phi_-(X)\right\}, \\
\rho_{e4,V}(U) &= \left\{\lambda \in \mathbb{C} \text{ such that } \lambda V - U \in \Phi_+(X) \bigcap \Phi_-(X)\right\}, \\
\rho_{e5,V}(U) &= \left\{\lambda \in \mathbb{C} \text{ such that } \lambda V - U \in \Phi_+(X) \bigcap \Phi_-(X) \text{ and } i(\lambda V - U) = 0\right\}, \text{ and} \\
\rho_{e6,V}(U) &= \left\{\lambda \in \mathbb{C} \text{ such that } \lambda \in \rho_{e5,V}(U) \text{ such that all scalars near } \lambda \text{ are in } \rho_V(U)\right\}.
\end{aligned}$$

The pencil essential spectra $\sigma_{ei,V}(U)$, $i = 1, 2, 3, 4, 5, 6$ are the complement in $\mathbb{C}$ of respective $\rho_{ei,V}(U)$.

Definition 3.7.2 Let X be a normed vector space, $U \in C\mathscr{R}(X)$. The pencil essential spectra is a continuous linear relation such that $V \neq 0$, $V(0) \subset U(0)$, and $\mathscr{D}(V) \supset \mathscr{D}(U)$. We define

$$\sigma_{eap,V}(U) = \bigcap_{K \in \mathscr{K}_T(X)} \sigma_{ap,V}(U + K)$$

and

$$\sigma_{e\delta,V}(U) = \bigcap_{K \in \mathscr{K}_T(X)} \sigma_{\delta,V}(U + K).$$ ◇

Remark 3.7.3 Note that if $V = I$, then we find the usual definition of spectra and the essential spectra of a linear relation. ◇

Theorem 3.7.2 *Let $U \in C\mathscr{R}(X)$ and $V \in B\mathscr{R}(X)$ be such that $V(0) \subset U(0)$ and $\|V\| \neq 0$. Then,*

(i) $\sigma_{ei,V}(U)$, $i = 1, 2, 3, 4, 5$ are closed.
(ii) The index is constant in any component of $\rho_{ei,V}(\cdot)$, $i = 1, 2, 3, 4, 5$. ◇

Proof (i) For $i = 5$. Let $\lambda_0 \in \rho_{e5,V}(U)$. Then, $R(U - \lambda_0 V)$ is closed. Hence, $U - \lambda_0 V$ is open. Then, according to Proposition 2.15.3 (i), we have

$$\gamma(U - \lambda_0 V) > 0.$$

Thus,

$$\frac{\gamma(U - \lambda_0 V)}{\|V\|} > 0.$$

Consider $\mu \in \mathbb{C}$ such that

$$|\mu - \lambda_0| < \frac{\gamma(U - \lambda_0 V)}{\|V\|}.$$

Then, according to Corollary 3.6.3, $\lambda_0 S - \mu S - \lambda_0 S + U \in \Phi(X)$ and

$$i(\lambda_0 V - \mu V - \lambda_0 V + U) = 0.$$

Since $\lambda_0 V(0) \subset V(0) = (\mu V - U)(0)$ and $\mathcal{D}(\mu V - U) = \mathcal{D}(U) \subset \mathcal{D}(V)$, then the use Proposition 2.7.2 allows us to conclude that

$$\lambda_0 V - \mu V - \lambda_0 V + U = \mu V - U.$$

Hence, $\mu V - U \in \Phi(X)$ and

$$i(\mu V - U) = 0.$$

Therefore, $\rho_{e5}(U)$ is open. For the other cases, the same arguments have been used.

(ii) Let λ_1 and λ_2 be any two points in $\rho_{ei,V}(U)$, $i = 1, 2, 3, 4, 5$ which are connected by a smooth curve Γ whose points are all in $\rho_{ei,V}(U)$, $i = 1, 2, 3, 4, 5$. Since $\rho_{ei,V}(U)$, $i = 1, 2, 3, 4, 5$ are an open set, then for each $\lambda \in \Gamma$, by using Corollary 3.6.3, there exists an $\varepsilon > 0$ such that, for all $\mu \in \mathbb{C}$, $0 < |\mu - \lambda| < \varepsilon$, $\mu \in \rho_{ei,V}(U)$, $i = 1, 2, 3, 4, 5$ and

$$i(\mu V - U) = i(\lambda V - U).$$

By using the Heine–Borel theorem, there exists a finite number of such sets which cover Γ. Since each of these sets overlaps with, at least, another set and since $i(\mu V - U)$ is constant on each one, we see that

$$i(\lambda_1 V - U) = i(\lambda_2 V - U).$$

Q.E.D.

Remark 3.7.4 Let X be a normed space and $V \in L\mathcal{R}(X)$. Then, $\|V\| = 0$ if, and only if, $S(B_{\mathcal{D}(V)}) \subset \overline{V(0)}$. In fact, if $\|V\| = 0$, then by using Lemma 2.10.2, we have $\|Vx\| = 0$, for all $x \in B_{\mathcal{D}(V)}$. Hence, for all $x \in B_{\mathcal{D}(V)}$ and $y \in Sx$, we have $\text{dist}(y, U(0)) = 0$. So, for all $x \in B_{\mathcal{D}(V)}$ and $y \in Sx$, we have $\text{dist}(y, \overline{V(0)}) = 0$. This implies that for all $x \in B_{\mathcal{D}(V)}$, $Vx \subset \overline{V(0)}$. Thus, $V(B_{\mathcal{D}(V)}) \subset \overline{V(0)}$. Conversely, by using both the same above reasoning and Lemma 2.10.2, we find the result. ◇

Theorem 3.7.3 *Let $U \in C\mathcal{R}(X)$ and $V \in B\mathcal{R}(X)$ be such that $V(0) \subset U(0)$, $V \neq 0$, and $V \neq U$. Then,*

$$\sigma_{e5,V}(U) = \bigcap_{K \in \mathcal{K}_U(X)} \sigma_S(U + K).$$ ◇

Proof Let $\lambda \notin \sigma_{e5,V}(U)$. Then, $\lambda V - U \in \Phi(X)$ and

$$i(\lambda V - U) = 0.$$

Hence,

$$n = \alpha(\lambda V - U) = \beta(\lambda V - U) < \infty.$$

So, there exists an everywhere defined single-valued K with dim $R(K) \leq \alpha(\lambda S - U)$ such that $\lambda V - (U + K)$ is bijective, where K is defined by

$$Kx := \sum_{i=1}^{n} x_i(x) y_i, \quad x \in X,$$

$\{x_1, \ldots, x_n\}$ is a basis of $N(\lambda V - U)$, and $x_1, \cdots, x_n$ are linear functionals such that $x_i(x_j) = \delta_{ij}$. Choose $y_1, \ldots, y_n \in X$ such that $[y_1], \ldots, [y_n] \in X/R(\lambda S - T)$ are linearly independent ($n = \alpha(\lambda V - U) = \beta(\lambda V - U) < \infty$). Hence, it is clear that K is a bounded finite rank operator. Since

$$\begin{aligned} K(0) &= \{0\} \\ &\subset (\lambda V - U)(0) \\ &= \lambda V(0) - U(0) \\ &= U(0), \end{aligned}$$

then $K \in \mathscr{K}_U(X)$. It is clear that

$$(\lambda V - U) - K = \lambda V - (U + K)$$

is bijective. Therefore, $\lambda \in \rho_V(U + K)$. This shows that

$$\lambda \notin \bigcap_{K \in \mathscr{K}_U(X)} \sigma_V(U + K).$$

So,

$$\bigcap_{K \in \mathscr{K}_T(X)} \sigma_V(U + K) \subset \sigma_{e5,V}(U). \tag{3.7.2}$$

To prove the inverse inclusion of (3.7.2). Suppose $\lambda \notin \bigcap_{K \in \mathscr{K}_U(X)} \sigma_V(U + K)$, then there exists $K \in \mathscr{K}_U(X)$ such that $\lambda \in \rho_V(U + K)$. Hence, $\lambda V - (U + K) \in \Phi(X)$ and

$$i(\lambda V - (U + K)) = 0.$$

Since $K \in \mathscr{K}_U(X)$, it follows from Proposition 2.7.2 that $\lambda V - (U + K) + K \in \Phi(X)$ and

$$i(\lambda S - (U + K) + K) = i(\lambda S - (U + K)) = 0.$$

So, by using Proposition 2.7.2, we infer that $\lambda V - U \in \Phi(X)$ and

$$i(\lambda V - U) = 0.$$

We conclude $\lambda \notin \sigma_{e5,S}(U)$. Hence,

$$\sigma_{e5,V}(U) \subset \bigcap_{K \in \mathscr{K}_U(X)} \sigma_V(U + K).$$

This completes the proof. Q.E.D.

Theorem 3.7.4 *Let $U \in C\mathscr{R}(X)$ and $V \in B\mathscr{R}(X)$ be such that $S(0) \subset U(0)$, $S \neq 0$, and $S \neq U$. Then,*

(*i*) $\sigma_{e5,S}(U) = \sigma_{e4,S}(U) \bigcup \{\lambda \in \mathbb{C}$ *such that* $i(\lambda V - U) \neq 0\}$.
(*ii*) $\sigma_{eap,V}(U) = \sigma_{e1,S}(U) \bigcup \{\lambda \in \mathbb{C}$ *such that* $i(\lambda S - T) > 0\}$.
(*iii*) $\sigma_{e\delta,V}(U) = \sigma_{e2,V}(U) \bigcup \{\lambda \in \mathbb{C}$ *such that* $i(\lambda V - U) < 0\}$. ◇

Proof (*i*) Let $\lambda \notin \sigma_{e4,V}(U) \bigcup \{\lambda \in \mathbb{C}$ such that $i(\lambda V - U) \neq 0\}$. Then, $\lambda V - U \in \Phi(X)$ and

$$i(\lambda V - U) = 0.$$

Hence, $\lambda \notin \sigma_{e5,V}(U)$. So,

$$\sigma_{e5,V}(U) \subset \sigma_{e4,V}(U) \bigcup \{\lambda \in \mathbb{C} \text{ such that } i(\lambda V - U) \neq 0\}. \quad (3.7.3)$$

To prove the inverse inclusion of (3.7.3). Suppose $\lambda \notin \sigma_{e5,V}(U)$, then by using Theorem 3.7.3, there exists $K \in \mathscr{K}_U(X)$ such that $\lambda S - (T + K)$ is bijective. Moreover, by virtue of Proposition 2.7.2, we infer that

$$\lambda S - T = \lambda S - (T + K) + K.$$

So, by using Theorem 3.6.15 (*iii*), $\lambda S - T \in \Phi(X)$ and

$$i(\lambda S - T) = 0.$$

Thus, $\lambda \notin \sigma_{e4,S}(T)$ and $\lambda \notin \sigma_{e4,S}(T) \bigcup \{\lambda \in \mathbb{C}$ such that $i(\lambda S - T) \neq 0\}$. Hence,

$$\sigma_{e4,S}(T) \bigcup \{\lambda \in \mathbb{C} \text{ such that } i(\lambda S - T) \neq 0\} \subset \sigma_{e5,S}(T).$$

(*ii*) For all $\lambda \in \mathbb{C}$, we have $\lambda S(0) = S(0) \subset T(0)$ and $\overline{\mathscr{D}(T)} \subset \mathscr{D}(\lambda S) = X$. So, by using Proposition 2.11.2, we infer that $\lambda S - T$ is closed.
(*iii*) Let $\lambda \notin \sigma_{eap,S}(T)$. Then, there exists $K \in \mathscr{K}_T(X)$ such that

$$\lambda S - T - K = (\lambda S - T) - K$$

is bounded below. Since $\lambda S - T$ is closed, $K(0) \subset T(0) = (\lambda S - T)(0)$ and $\mathscr{D}(\lambda S - T) = \mathscr{D}(T) \subset \mathscr{D}(K)$, then by using Proposition 2.11.2, we have $\lambda S - (T + K)$ is closed. Since $\lambda S - (T + K)$ is one-to-one with closed range and open, then $\lambda S - (T + K) \in \Phi_{+}(X)$ and

$$\begin{aligned} i((\lambda S - T) - K) &= \alpha(\lambda S - T - K) - \beta(\lambda S - T - K) \\ &= 0 - \beta(\lambda S - T - K) \\ &\leq 0. \end{aligned}$$

Therefore, $\lambda S - (T + K) \in \Phi_{+}(X)$ and

$$i(\lambda S - (T + K)) \leq 0.$$

By Lemma 3.6.13 (i), we have $\lambda S - (T + K) + K = (\lambda S - T) + K - K \in \Phi_{+}(X)$ and

$$i(\lambda S - (T + K) + K) = i(\lambda S - (T + K)) \leq 0.$$

Since $K(0) \subset T(0) = (\lambda S - T)(0)$ and $\mathscr{D}(\lambda S - T) = \mathscr{D}(T) \subset \mathscr{D}(K)$, and by using Proposition 2.7.2, we have $\lambda S - T \in \Phi_{+}(X)$ and

$$i(\lambda S - T) \leq 0.$$

Therefore,

$$\lambda \notin \sigma_{e1,S}(T) \bigcup \{\lambda \in \mathbb{C} \text{ such that } i(\lambda S - T) > 0\}.$$

Conversely, let $\lambda \notin \sigma_{e1,S}(T) \bigcup \{\lambda \in \mathbb{C} \text{ such that } i(\lambda S - T) > 0\}$, then $\lambda S - T \in \Phi_{+}(X)$ and

$$i(\lambda S - T) \leq 0.$$

We define the operator K by

$$K : X \longrightarrow Kx := \sum_{i=1}^{n} x_i(x) y_i, \quad x \in X,$$

where $\{x_1, \ldots, x_n\}$ is a basis of $N(\lambda S - T)$ and $x_1, \ldots, x_n$ are a linear functionals such that $x_i(x_j) = \delta_{ij}$. Choose $y_1, \ldots, y_n \in X$ such that $[y_1], \ldots, [y_n] \in X/R(\lambda S - T)$ are linearly independent (such elements exist since $n \leq \beta(\lambda S - T)$). It is clear that K is continuous, and so that K is a bounded finite rank operator. So, $K \in \mathscr{K}_T(X)$ and also $(\lambda S - T) - K = \lambda S - (T + K)$ is one-to-one. Since $\lambda S - T \in \Phi_{+}(X)$, we have $\lambda S - (T + K) \in \Phi_{+}(X)$ with the same index that $\lambda S - T$. In this situation, we have $K \in \mathscr{K}_T(X)$, $\lambda S - (T + K)$ is one-to-one and closed with closed range. Hence, $\lambda S - (T + K)$ is open. So, $\lambda \in \rho_{ap,S}(T + K)$. Therefore, $\lambda \notin \sigma_{eap,S}(T)$.
The proofs of (iii) is similar to the previous one. Q.E.D.

Corollary 3.7.1 *Let $T \in C\mathcal{R}(X)$ and $S \in B\mathcal{R}(X)$ be such that $S(0) \subset T(0)$ and $S \neq 0$. Then,*
(i) $\lambda \notin \sigma_{eap,S}(T)$ if, and only if, $\lambda S - T \in \Phi_+(X)$ and $i(\lambda S - T) \leq 0$.
(ii) $\lambda \notin \sigma_{e\delta,S}(T)$ if, and only if, $\lambda S - T \in \Phi_-(X)$ and $i(\lambda S - T) \geq 0$. ◇

Proposition 3.7.4 *Let $V \in C\mathcal{R}(X)$ and $W \in B\mathcal{R}(X)$ be such that $W(0) \subset V(0)$, $W \neq 0$, and the resolvent set $\rho_W(V)$ is nonempty. If the set $\Phi_{V,W}$ is connected, then:*
(i) The essential spectra satisfy

$$\sigma_{e1,W}(V) = \sigma_{eap,W}(V) \quad \textit{and} \quad \sigma_{e2,W}(V) = \sigma_{e\delta,W}(V).$$

(ii) Moreover, if the set $\rho_{e4,W}(V)$ is connected and $\rho_W(V) \neq \emptyset$, then the equality

$$\sigma_{e4,W}(V) = \sigma_{e5,W}(V)$$

holds.

◇

Proof (i) It is straightforward to verify that

$$\sigma_{e1,W}(V) \subset \sigma_{eap,W}(V).$$

For the reverse inclusion, consider $\mu \in \rho_{e1,W}(V)$. Then,

$$\mu \in \Phi_{+V,W} = \Phi_{V,W} \cup (\Phi_{+V,W} \setminus \Phi_{V,W}).$$

Thus, we analyze these two scenarios:
First scenario: If $\mu \in \Phi_{V,W}$, then

$$i(\mu W - V) = 0.$$

Indeed, choose $\mu_0 \in \rho_W(V)$, then $\mu_0 \in \Phi_{V,W}$ and $i(\mu_0 W - V) = 0$. By Theorem 3.7.2 (ii), the index function $i(\mu W - V)$ remains invariant on each connected component of $\Phi_{V,W}$. Hence, $\rho_W(V) \subset \Phi_{V,W}$, and

$$i(\mu W - V) = 0$$

for every $\mu \in \Phi_{V,W}$. This establishes that $\mu \in \rho_{eap,W}(V)$.
Second scenario: If $\mu \in (\Phi_{+V,W} \setminus \Phi_{V,W})$, then the ascent $\alpha(\mu W - V)$ is finite but the descent $\beta(\mu W - V) = +\infty$. Thus,

$$i(\mu W - V) = -\infty < 0.$$

Therefore, the second inclusion follows from the preceding two situations. A parallel argument applies to the second equality, completing the demonstration of (i).

(ii) Obviously,

$$\sigma_{e4,W}(V) \subset \sigma_{e5,W}(V).$$

The converse inclusion can be established by proving that

$$\rho_{e4,W}(V) \cap \sigma_{e5,W}(V) = \emptyset.$$

Suppose, for contradiction, that

$$\rho_{e4,W}(V) \cap \sigma_{e5,W}(V) \neq \emptyset.$$

Then, there exists $\lambda_0 \in \mathbb{C}$ such that

$$\lambda_0 \in \rho_{e4,W}(V) \cap \sigma_{e5,W}(V).$$

Take $\lambda_1 \in \rho_W(V)$, which implies

$$\lambda_1 W - V \in \Phi_+(X) \cap \Phi_-(X),$$

and

$$i(\lambda_1 W - V) = 0.$$

Since $\rho_{e4,W}(V)$ is connected, Theorem 3.7.2 (ii) guarantees that the index $i(\lambda W - V)$ is constant on each connected component of $\rho_{e4,W}(V)$. Hence,

$$i(\lambda_1 W - V) = i(\lambda_0 W - V) = 0.$$

This contradicts the assumption that $\lambda_0 \in \sigma_{e5,W}(V)$, disproving the initial supposition.

Q.E.D.

3.8 Local Spectrum of a Linear Relation

In the following propositions we gather some elementary properties of $\sigma_T(x)$.

Proposition 3.8.1 *Let $Q \in L\mathcal{R}(X)$. Then the following properties hold:*
(i) For every $x \in Q(0)$, the spectrum of x with respect to Q is empty, i.e.,

$$\sigma_Q(x) = \emptyset.$$

(ii) For any vectors $x, y \in X$ and scalars $\alpha, \beta \in \mathbb{C}$, the spectrum satisfies the inclusion

$$\sigma_Q(\alpha x + \beta y) \subseteq \sigma_Q(x) \cup \sigma_Q(y).$$

(iii) For all $x \in X$, the spectrum relative to $-Q$ is the set of negatives of the spectrum relative to Q:

$$\sigma_{-Q}(x) = \{-\lambda : \lambda \in \sigma_Q(x)\}.$$

(iv) Given any subset $F \subseteq \mathbb{C}$, we have the equivalence

$$\sigma_{\lambda I+Q}(x) \subseteq F \iff \sigma_Q(x) \subseteq F \setminus \{\lambda\}.$$

In particular,

$$\sigma_{\lambda I-Q}(x) \subseteq \{0\} \iff \sigma_Q(x) \subseteq \{\lambda\}.$$

◇

Proof (i) It suffices to demonstrate that for every $x \in Q(0)$, the resolvent set $\rho_Q(x)$ equals the entire complex plane $\mathbb{C}$. Take any $\lambda \in \mathbb{C}$. Consider the zero analytic function $f \equiv 0$ defined on $\mathbb{C}$, which satisfies

$$(\mu I - Q)f(\mu) = x + Q(0)$$

for all $x \in Q(0)$ and all $\mu \in \mathbb{C}$. Hence, λ belongs to $\rho_Q(x)$.
(ii) It suffices to show that

$$\rho_Q(x) \cap \rho_Q(y) \subseteq \rho_Q(\alpha x + \beta y)$$

for every $x, y \in X$ and scalars $\alpha, \beta \in \mathbb{C}$. If $\alpha = \beta = 0$, the claim is trivial. Otherwise, let $\delta \in \rho_Q(x) \cap \rho_Q(y)$. Then there exist open neighborhoods U_δ and V_δ along with analytic mappings $f_{\delta,x} : U_\delta \to X$ and $g_{\delta,y} : V_\delta \to X$ such that

$$(\mu I - Q)f_{\delta,x}(\mu) = x + Q(0), \quad \forall \mu \in U_\delta$$

and

$$(\mu I - Q)g_{\delta,y}(\mu) = y + Q(0), \quad \forall \mu \in V_\delta.$$

Set $W_\delta = U_\delta \cap V_\delta$ and define the analytic function $h : W_\delta \to X$ by

$$h(\mu) = \alpha f_{\delta,x}(\mu) + \beta g_{\delta,y}(\mu).$$

Then, for all $\mu \in W_\delta$,

$$(\mu I - Q)h(\mu) = (\alpha x + \beta y) + Q(0).$$

Consequently, $\delta \in \rho_Q(\alpha x + \beta y)$.
(iii) Let $\lambda \in \rho_{-Q}(x)$. Then, there exists an open neighborhood U_λ and an analytic function $f : U_\lambda \to X$ such that

$$(\mu I - (-Q))f(\mu) = x + Q(0)$$

for all $\mu \in U_\lambda$. By applying the change of variable $\nu = -\mu$, we get

$$(\nu I - Q)g(\nu) = x + Q(0), \quad \forall \nu \in U_{-\lambda},$$

where $g(\nu) = -f(-\nu)$ is an analytic function defined on an open neighborhood $U_{-\lambda}$ of $-\lambda$. Hence, $-\lambda \in \rho_Q(x)$.
(iv) Similarly, it can be shown that

$$\mathbb{C} \setminus F \subseteq \rho_{\lambda I+Q}(x) \quad \text{if and only if} \quad \mathbb{C} \setminus (F \setminus \{\lambda\}) \subseteq \rho_Q(x).$$

For the "if" part, let $\lambda_0 \in \mathbb{C} \setminus (F \setminus \{\lambda\})$. Then $\lambda_0 + \lambda \in \mathbb{C} \setminus F \subseteq \rho_{\lambda I+Q}(x)$. Thus, there exists an open neighborhood $U_{\lambda_0+\lambda}$ and an analytic function

$$f_{\lambda_0+\lambda,x} : U_{\lambda_0+\lambda} \to X$$

such that

$$(\mu I - (\lambda I + Q))f_{\lambda_0+\lambda,x}(\mu) = x + Q(0), \quad \forall \mu \in U_{\lambda_0+\lambda}.$$

By the substitution $\nu = \mu - \lambda$, it follows that

$$(\nu I - Q)g(\nu) = x + Q(0), \quad \forall \nu \in U_{\lambda_0},$$

where $g(\nu) = f(\nu + \lambda)$ is analytic on an open neighborhood U_{λ_0} of λ_0. Therefore, $\lambda_0 \in \rho_Q(x)$.
Conversely, for the "only if" part, let $\lambda_0 \in \mathbb{C} \setminus F$. Then

$$\lambda_0 - \lambda \in \mathbb{C} \setminus (F \setminus \{\lambda\}) \subseteq \rho_Q(x).$$

Hence, there exists an open neighborhood $U_{\lambda_0-\lambda}$ and an analytic function

$$f_{\lambda_0-\lambda,x} : U_{\lambda_0-\lambda} \to X$$

such that

$$((\mu + \lambda)I - (\lambda I + Q))f_{\lambda_0-\lambda,x}(\mu) = x + Q(0), \quad \forall \mu \in U_{\lambda_0-\lambda}.$$

By letting $\nu = \mu + \lambda$, we get

$$(\nu I - (\lambda I + Q))g(\nu) = x + Q(0), \quad \forall \nu \in U_{\lambda_0},$$

where $g(\nu) = f(\nu - \lambda)$ is analytic on an open neighborhood U_{λ_0} of λ_0. Therefore, $\lambda_0 \in \rho_{\lambda I+Q}(x)$.

Finally, by choosing $F = \{0\}$ and replacing Q by $-Q$, the particular case follows readily. Q.E.D.

Lemma 3.8.1 *Let X be a Banach space and U be an open subset of $\mathbb{C}$. If $T \in C\mathscr{R}(X)$, $x \in \mathscr{D}(T)$ and $f \in H(U, \mathscr{D}(T))$ satisfying*

$$(\lambda - T)f(\lambda) = x + T(0), \textit{ for all } \lambda \in U, \tag{3.8.1}$$

then $\rho_T(x) = \rho_T(f(\lambda))$, for all $\lambda \in U$. ◇

Proof Let us assume that $\lambda \in \rho_T(x)$. We will consider two cases.

First case. Let $\omega \in \rho_T(x)\backslash U$. Then, we may choose an open neighborhood W of ω such that $\lambda \notin W$ and an analytic function $h : W \longrightarrow \mathscr{D}(T)$ for which

$$(\mu - T)h(\mu) = x + T(0), \text{ for all } \mu \in W. \tag{3.8.2}$$

Let $k : W \longrightarrow \mathscr{D}(T)$ be defined by

$$k(\mu) = \frac{h(\mu) - f(\lambda)}{\lambda - \mu}, \text{ for all } \mu \in W \text{ and } \lambda \in U.$$

Then, $k \in H(W, \mathscr{D}(T))$. Hence, for all $\mu \in W$, we can write

$$\begin{aligned}(\mu - T)k(\mu) &= (\mu - T)\left(\frac{h(\mu) - f(\lambda)}{\lambda - \mu}\right), \\ &= \frac{1}{\lambda - \mu}\left((\mu - T)h(\mu) - (\mu - \lambda)f(\lambda) - (\lambda - T)f(\lambda)\right) \\ &= f(\lambda) + \frac{1}{\lambda - \mu}\left((\mu - T)h(\mu) - (\lambda - T)f(\lambda)\right).\end{aligned}$$

By referring to (3.8.1) and (3.8.2), we obtain

$$\begin{aligned}(\mu - T)k(\mu) &= f(\lambda) + \frac{1}{\lambda - \mu}\left(x + T(0) - x - T(0)\right) \\ &= f(\lambda) + T(0).\end{aligned}$$

Thus, we conclude that $\omega \in \rho_T(f(\lambda))$.

Second case. Let $\lambda \in U$. Consider the function $g : U \longrightarrow \mathscr{D}(T)$ which defined by

$$g(\mu) = \frac{f(\mu) - f(\lambda)}{\lambda - \mu} \text{ for all } \mu \in U\backslash\{\lambda\}$$

and

$$g(\lambda) = f'(\lambda).$$

Then, $g \in H(U, \mathscr{D}(T))$. Hence, for all $\mu \in U\backslash\{\lambda\}$, we can write

$$\begin{aligned}(\mu - T)g(\mu) = &\ (\mu - T)\left(\frac{f(\mu) - f(\lambda)}{\lambda - \mu}\right),\\ = &\ \frac{1}{\lambda - \mu}\left((\lambda - \mu)f(\lambda) + (\mu - T)f(\mu) - (\lambda - T)f(\lambda)\right)\\ = &\ f(\lambda) + \frac{1}{\lambda - \mu}\left((\mu - T)f(\mu) - (\lambda - T)f(\lambda)\right).\end{aligned}$$

It follows from (3.8.1) that

$$\begin{aligned}(\mu - T)g(\mu) = &\ f(\lambda) + \frac{1}{\lambda - \mu}\left(x + T(0) - x - T(0)\right),\\ = &\ f(\lambda) + \frac{1}{\lambda - \mu}T(0).\end{aligned}$$

By linearity of T, we have

$$(\mu - T)g(\mu) = f(\lambda) + T(0).$$

This implies from Proposition 2.8.1 (i) that

$$\begin{aligned}Q_T(\mu - T)g(\mu) = &\ Q_T(f(\lambda) + T(0)),\\ = &\ Q_T f(\lambda) + Q_T T(0),\\ = &\ Q_T f(\lambda).\end{aligned}$$

The fact that $T \in C\mathscr{R}(X)$ implies from Propositions 2.11.2 and 2.11.1 (iv) that $Q_T(\mu - T) = Q_{\mu-T}(\mu - T)$ is a closed operator. Since $g(\mu) \in \mathscr{D}(\mu - T) = \mathscr{D}(T)$, for all $\mu \in U\backslash\{\lambda\}$, $g(\mu) \to g(\lambda)$ as $\mu \to \lambda$ and $Q_T(\mu - T)g(\mu) = Q_T f(\lambda) \to Q_T f(\lambda)$ as $\mu \to \lambda$, then $g(\lambda) \in \mathscr{D}(T)$ and

$$Q_T(\mu - T)g(\lambda) = Q_T f(\lambda).$$

At this level, we have

$$\begin{aligned}Q_T(\lambda - T)g(\lambda) = &\ Q_T(\lambda - \mu + \mu - T)g(\lambda)\\ = &\ Q_T(\lambda - \mu)g(\lambda) + Q_T(\mu - T)g(\lambda)\\ = &\ Q_T(\lambda - \mu)g(\lambda) + Q_T f(\lambda).\end{aligned}$$

Using the fact $\lim\limits_{\mu \longrightarrow \lambda} Q_T(\lambda - \mu)g(\lambda) = 0$, we infer that

$$Q_T(\lambda - T)g(\lambda) = Q_T f(\lambda).$$

This leads to

$$Q_T\left((\lambda - T)g(\lambda) - f(\lambda)\right) = 0. \tag{3.8.3}$$

Applying Q_T^{-1} to both sides of (3.8.3), we have

$$Q_T^{-1}Q_T\left((\lambda - T)g(\lambda) - f(\lambda)\right) = Q_T^{-1}(0).$$

The fact that $Q_T^{-1}(0) = T(0)$ implies from Proposition 2.8.1 (ii) that

$$((\lambda - T)g(\lambda) - f(\lambda))\bigcap \mathscr{D}(Q_T) + T(0) = T(0).$$

Since Q_T is bounded operator, then

$$(\lambda - T)g(\lambda) + T(0) = f(\lambda) + T(0).$$

Hence, we deduce that

$$(\lambda - T)g(\lambda) = f(\lambda) + T(0).$$

Thus,

$$(\mu - T)g(\mu) = f(\lambda) + T(0) \text{ for all } \mu \in U.$$

This shows that

$$U \subset \rho_T\left(f(\lambda)\right). \tag{3.8.4}$$

Conversely, let $\omega \in \rho_T\left(f(\lambda)\right)\backslash U$, we may choose an open neighborhood W of ω such that $\lambda \notin W$ and an analytic function $g : W \longrightarrow \mathscr{D}(T)$ for which

$$(\mu - T)g(\mu) = f(\lambda) + T(0), \text{ for all } \mu \in W. \tag{3.8.5}$$

Let $h : W \longrightarrow \mathscr{D}(T)$ be defined by

$$h(\mu) = g(\mu) + \frac{f(\lambda)}{\lambda - \mu}, \text{ for all } \mu \in W.$$

Then, $h \in H(W, \mathscr{D}(T))$. Hence, it follows from (3.8.1) and (3.8.5) that for all $\mu \in W$, we have

$$\begin{aligned}(\mu - T)h(\mu) &= (\mu - T)g(\mu) + \frac{1}{\lambda - \mu}\left((\mu - T)f(\lambda)\right), \\ &= f(\lambda) + T(0) + \frac{1}{\lambda - \mu}\left(x + T(0) + (\mu - \lambda)f(\lambda)\right) \\ &= \frac{1}{\lambda - \mu}x + T(0).\end{aligned}$$

This leads to

$$\rho_T\left(f(\lambda)\right)\backslash U \subset \rho_T\left(\frac{1}{\lambda - \mu}x\right).$$

Then, according to (iv) of Proposition 3.4.1, we obtain

$$\rho_T(f(\lambda))\backslash U \subset \rho_T\left(\frac{1}{\lambda-\mu}x\right) = \rho_T(x).$$

Finally, the use of (3.8.4) allows us to conclude that $\rho_T(f(\lambda)) \subset \rho_T(x)$. Q.E.D.

Proposition 3.8.2 *Let $T \in B\mathscr{R}(X)$, $x \in X$ and $\mathscr{U}$ be an open subset of $\mathbb{C}$. If there exists an analytic function $f_x : \mathscr{U} \longrightarrow X$ such that*

$$(\mu I - T)f(\mu) = x + T(0)$$

for all $\mu \in \mathscr{U}$, then $\mathscr{U} \subseteq \rho_T(f(\lambda))$ for all $\lambda \in \mathscr{U}$. If, moreover, T has a continuous selection A such that $T(0) \subset N(A)$, then

$$\sigma_T(x) = \sigma_T(f(\lambda)),\ \textit{for all } \lambda \in \mathscr{U}. \qquad \diamond$$

Proof Let $\lambda \in \mathscr{U}$. Define

$$h(\mu) = \begin{cases} \dfrac{f(\lambda)-f(\mu)}{\mu-\lambda} & \text{if } \mu \neq \lambda \\ -f'(\lambda) & \text{if } \mu = \lambda. \end{cases}$$

Then, $h(\cdot)$ is an analytic function on $\mathscr{U}$ and, we have for all $\mu \in \mathscr{U}\backslash\{\lambda\}$

$$\begin{aligned}(\mu I - T)h(\mu) &= \frac{1}{\mu-\lambda}[((\mu-\lambda)I + (\lambda I - T))(f(\lambda)) - x + T(0)] \\ &= f(\lambda) + T(0).\end{aligned}$$

Therefore,

$$(\mu I - T)h(\mu) = f(\lambda) + T(0)$$

holds for all $\mu \in \mathscr{U}\backslash\{\lambda\}$. On the other hand, we have

$$(\mu I - T)f(\mu) = x + T(0).$$

Then,

$$\mu Q_T(f(\mu)) - Q_T T f(\mu) = Q_T x.$$

Now, as Q_T is a bounded operator and by derivation on μ, we get

$$Q_T(f(\mu)) + \mu Q_T(f'(\mu)) - Q_T T f(\mu) = 0.$$

Thus, for $\mu = \lambda$, we have

$$-(\lambda I - T)f'(\lambda) = f(\lambda) + T(0).$$

Hence, the equality

$$(\mu I - T)h(\mu) = f(\lambda) + T(0)$$

holds for all $\mu \in \mathscr{U}$. Therefore,

$$\mathscr{U} \subseteq \rho_T(f(\lambda)).$$

To show the equality $\sigma_T(x) = \sigma_T(f(\lambda))$, we begin by proving the inclusion $\rho_T(x) \subseteq \rho_T(f(\lambda))$. Let $w \in \rho_T(x)$. If $w \in \mathscr{U}$, then $w \in \rho_T(f(\lambda))$ for all $\lambda \in \mathscr{U}$, by the first part of the proof. If $w \in \rho_T(x) \backslash \mathscr{U}$, then there exist an open neighborhood W which not contains λ and an analytic function $g : W \longrightarrow X$ such that

$$(\mu I - T)g(\mu) = x + T(0)$$

for all $\mu \in W$. Define

$$K(\mu) = \frac{g(\lambda) - g(\mu)}{\mu - \lambda}, \quad \text{for all} \quad \mu \in W.$$

$K(\cdot)$ is an analytic function on W and, we have

$$(\mu I - T)K(\mu) = f(\lambda) + T(0)$$

for all $\mu \in W$. Thus, $w \in \rho_T(f(\lambda))$. So,

$$\rho_T(x) \subseteq \rho_T(f(\lambda)).$$

It remains to prove the reverse inclusion. Let $\nu \in \rho_T(f(\lambda))$. Then, there exist an open neighborhood $\mathscr{V}$ of ν and an analytic function $h : \mathscr{V} \longrightarrow X$ such that

$$(\mu I - T)h(\mu) = f(\lambda) + T(0)$$

is satisfied for all $\mu \in \mathscr{V}$. By hypothesis, A is a continuous selection of T, then the function h_A defined on $\mathscr{V}$ by

$$h_A(\mu) = (\lambda I - A)h(\mu)$$

is analytic on $\mathscr{V}$. Besides, since $T(0) \subset N(A)$, then for all $\mu \in \mathscr{V}$

$$\begin{aligned}(\mu I - T)h_A(\mu) &= (\lambda I - A)(\mu I - T)h(\mu) \\ &= (\lambda I - A + T - T)f(\lambda) + T(0) - AT(0) \\ &= x + T(0).\end{aligned}$$

Thus, $\nu \in \rho_T(x)$ and hence $\rho_T(f(\lambda)) \subseteq \rho_T(x)$ which ends the proof. Q.E.D.

Proposition 3.8.3 *Let $T, S \in B\mathscr{R}(X)$ be such that $ST = TS$, $S(0) \subset N(T)$ and S have a linear continuous selection S_1. Then, for all $y \in Sx$, we have*

$$\sigma_T(y) \subset \sigma_T(x).$$ ◇

Proof We shall prove that $\rho_T(x) \subset \rho_T(y)$ for all $y \in Sx$. Let $y \in Sx$ and $\lambda \in \rho_T(x)$. Then, there exist an open neighborhood U_λ and an analytic function $f : U_\lambda \longrightarrow X$ such that

$$(\nu I - T)f(\nu) = x + T(0),$$

for all $\nu \in U_\lambda$. Then,

$$S(\nu I - T)f(\nu) = S(x + T(0)).$$

Using the fact that $TS = ST$, $S(0) \subset N(T)$, we get

$$(\nu I - T)Sf(\nu) = Sx + ST(0).$$

Thus,

$$(\nu I - T)(S_1 f(\nu) + S(0)) = Sx + ST(0).$$

So,

$$(\nu I - T)S_1 f(\nu) + ST(0) = y + S(0) + ST(0).$$

Hence,

$$(\nu I - T)S_1 f(\nu) = y + T(0).$$

Then, $\lambda \in \rho_T(y)$ which ends the proof. Q.E.D.

Definition 3.8.1 Let $T \in L\mathscr{R}(X)$. We say that T verifies the stabilization criteria if $T(0) = T^2(0)$. ◇

We denote by $\mathscr{S}\mathscr{T}\mathscr{R}_1(X)$ the set of all linear relations satisfying the stabilization criteria.

Lemma 3.8.2 *Let $R, S \in B\mathscr{R}(X)$. Then, $S(0) \subset N(SR)$ and $R(0) \subset N(RS)$ if, and only if, SR and RS are both in $\mathscr{S}\mathscr{T}\mathscr{R}_1(X)$.* ◇

Proof For the only if part, we have by hypothesis,

$$SRS(0) = SR(0)$$

and

$$RSR(0) = RS(0).$$

Then,

$$SRSR(0) = SR(0).$$

Thus, $SR \in \mathscr{S}\mathscr{T}\mathscr{R}_1(X)$. In a similar way, we can prove that $RS \in \mathscr{S}\mathscr{T}\mathscr{R}_1(X)$. For the if part, we have

$$SRSR(0) = SR(0).$$

So,

$$SRS(0) \subset SR(0).$$

Then, for all $y \in S(0)$, we have

$$SR(y) \subset SR(0).$$

So,

$$SR(y) = SR(0).$$

Thus,

$$SRS(0) = SR(0)$$

and, we conclude that

$$S(0) \subset N(SR).$$

By the same way we can prove that $R(0) \subset N(RS)$. Q.E.D.

In the next theorem, we study the relation between the local spectrums of the linear relations SR and RS.

Theorem 3.8.1 *Let S, $R \in B\mathscr{R}(X)$ be such that SR and RS are both in $\mathscr{S}\mathscr{T}\mathscr{R}_1(X)$. If S and R have continuous linear selection S_1 and R_1, respectively, then for all $x \in X$ and $y \in Sx$, we have*

$$\sigma_{SR}(y) \subseteq \sigma_{RS}(x) \subseteq \sigma_{SR}(y) \bigcup \{0\}. \tag{3.8.6}$$

◇

Proof Let be beginning by showing the first inclusion. It is equivalent to prove that $\rho_{RS}(x) \subseteq \rho_{SR}(y)$. Let $\lambda \in \rho_{RS}(x)$. Then, there exist an open neighborhood U_λ and an analytic function $f : U_\lambda \longrightarrow X$ such that

$$(\nu I - RS)f(\nu) = x + RS(0), \ \forall \ \nu \in U_\lambda.$$

Then,

$$S(\nu I - RS)f(\nu) = Sx + SRS(0).$$

Using Lemma 3.8.2 and (3.8.6), we get

$$(\nu I - SR)Sf(\nu) = Sx + SRS(0).$$

Thus,

$$(\nu I - SR)(S_1 f(\nu) + S(0)) = Sx + SRS(0).$$

Since $(\nu I - SR)S(0) \subseteq (\nu I - SR)(0)$, then

$$(\nu I - SR)S_1 f(\nu) = y + SR(0)$$

for all $\nu \in U_\lambda$. So, we conclude that $\lambda \in \rho_{SR}(y)$ and therefore,

$$\rho_{RS}(x) \subseteq \rho_{SR}(y).$$

In order to show the second inclusion, it suffices to prove that $\rho_{SR}(y) \bigcap \mathbb{C}^* \subset \rho_{RS}(x)$. Let $\lambda \in \rho_{SR}(y) \bigcap \mathbb{C}^*$. Then, there exist an open neighborhood $V_\lambda \subset \mathbb{C}^*$ and an analytic function $f : V_\lambda \longrightarrow X$ such that

$$(\mu I - SR) f(\mu) = y + SR(0), \ \forall \mu \in V_\lambda.$$

Let h be the function defined by

$$h(\mu) = \frac{1}{\mu}(x + R_1 f(\mu)).$$

Hence,

$$\begin{aligned}
(\mu I - RS)h(\mu) &= x - \frac{1}{\mu}RS(x) + \frac{1}{\mu}(\mu R_1 - RSR_1)f(\mu) \\
&= x - \frac{1}{\mu}RS(x) + \frac{1}{\mu}R[(\mu I - SR_1)f(\mu)] \\
&\subset x - \frac{1}{\mu}R(y) + \frac{1}{\mu}R(y) + RS(0) \\
&\subset x + RS(0).
\end{aligned}$$

Thus,

$$(\mu I - RS)h(\mu) = x + RS(0),$$

for all $\mu \in V_\lambda$. Therefore, we have $\lambda \in \rho_{RS}(x)$ and we obtain the desired inclusion. Q.E.D.

3.8.1 *Local Spectral Subspace of Linear Relations*

Let X be a Banach space and let $T \in C\mathscr{R}(X)$. For every subset F of $\mathbb{C}$ the local analytical or spectral subspace of T associated with F is the set

$$X_T(F) := \{x \in X \text{ such that } \sigma_T(x) \subseteq F\}.$$

Obviously, if $F_1 \subseteq F_2$, then

$$X_T(F_1) \subseteq X_T(F_2).$$

Let us notice that, according to Proposition 3.4.1 (iii), if $T \in C\mathscr{R}(X)$, then

$$T(0) \subset X_T(\emptyset). \tag{3.8.7}$$

We begin with the following proposition, which gives some properties of the local spectral subspace.

Proposition 3.8.4 *Let $G \in B\mathscr{R}(X)$ and let F be a subset of $\mathbb{C}$. Then,*

(i) The set $X_G(F)$ forms a linear subspace of X. (ii) The subspace $X_G(F)$ is hyper-invariant under G; that is, for every bounded operator S commuting with G (i.e., $GS = SG$), it holds that

$$S(X_G(F)) \subseteq X_G(F).$$

(iii) Furthermore, if $G \in \mathscr{S}\mathscr{T}\mathscr{R}_1(X)$ and admits a continuous selection, then

$$G(X_G(F)) \subseteq X_G(F). \qquad \diamondsuit$$

Proof (i) According to the initial part of Proposition 3.8.1, we have $G(0) \subset X_G(F)$. Hence, $X_G(F)$ is nonempty. Take any $x, y \in X_G(F)$. By definition, $\sigma_G(x) \subseteq F$ and $\sigma_G(y) \subseteq F$. Using the latter part of Proposition 3.8.1, it follows that for all scalars $\alpha, \beta \in \mathbb{C}$,

$$\sigma_G(\alpha x + \beta y) \subseteq \sigma_G(x) \cup \sigma_G(y) \subseteq F.$$

Therefore, $\alpha x + \beta y \in X_G(F)$, which shows that $X_G(F)$ is a linear subspace of X.

(ii) Let $S \in \mathscr{L}(X)$ be a bounded operator commuting with G, i.e., $SG = GS$. For any $x \in X_G(F)$ and any $\lambda \notin \sigma_G(x)$, there exists an open neighborhood U_λ of λ and an analytic map $f : U_\lambda \to X$ such that

$$(\mu I - G)f(\mu) = x + G(0), \quad \forall \mu \in U_\lambda.$$

Applying S to both sides, and using commutativity, we get

$$(\mu I - G)(Sf(\mu)) = Sx + G(0), \quad \forall \mu \in U_\lambda.$$

Since S is bounded, Sf remains analytic on U_λ. This implies $\lambda \notin \sigma_G(Sx)$, and consequently,

$$\sigma_G(Sx) \subseteq \sigma_G(x) \subseteq F.$$

Hence, $Sx \in X_G(F)$, demonstrating that $S(X_G(F)) \subseteq X_G(F)$.

(iii) Given that G has a continuous linear selection and that $G(0) \subseteq N(G)$, the assertion follows directly from Proposition 3.8.3. Q.E.D.

Let us study some basic properties of the local analytical subspace:

Proposition 3.8.5 *Let X be a Banach space, $T \in C\mathscr{R}(X)$ and $F \subset \mathbb{C}$. Then,*

(i) $X_T(F_1) \subset X_T(F_2)$, for all $F_1 \subset F_2 \subset \mathbb{C}$.
(ii) $X_T(F \bigcap \sigma(T)) = X_T(F)$.
(iii) $X_{\lambda+T}(F) = X_T(F \backslash \{\lambda\})$, for all $\lambda \in \mathbb{C}$.
(iv) If $x \in \mathscr{D}(T)$ satisfies $(\lambda - T)x \subset X_T(F)$, for some $\lambda \in F$, then $x \in X_T(F)$. ◇

Proof (i) Let $F_1 \subset F_2 \subset \mathbb{C}$ and $x \in X_T(F_1)$. Then,

$$\begin{aligned}\sigma_T(x) &\subseteq F_1 \\ &\subset F_2.\end{aligned}$$

As a result, $X_T(F_1) \subset X_T(F_2)$, as desired.
(ii) Since $F \bigcap \sigma(T) \subset \sigma(T)$, then by (i), we infer that

$$X_T(F \bigcap \sigma(T)) \subset X_T(F).$$

Conversely, let $x \in X_T(F)$. Then, $\sigma_T(x) \subseteq F$. This implies from Proposition 3.4.1 (ii) that

$$\begin{aligned}\sigma_T(x) = &\ \sigma_T(x) \bigcap \sigma(T) \\ \subseteq &\ F \bigcap \sigma(T).\end{aligned}$$

Hence, we conclude that $x \in X_T(F \bigcap \sigma(T))$.
(iii) Let us assume that $x \in X_{\lambda+T}(F)$. Then,

$$\sigma_{\lambda+T}(x) \subseteq F.$$

By referring to Proposition 3.4.1 (vi), we obtain

$$\sigma_T(x) \subseteq F \backslash \{\lambda\}.$$

This implies that $x \in X_T(F \backslash \{\lambda\})$. It is shown that

$$X_{\lambda+T}(F) \subset X_T(F \backslash \{\lambda\}).$$

The converse inclusion holds by substituting λ, F and T respectively, for $-\lambda$, $F - \{\lambda\}$ and $T + \lambda$ in $X_{\lambda+T}(F) \subset X_T(F \setminus \{\lambda\})$.

(iv) Let us assume that $x \in \mathscr{D}(T)$ satisfies $(\lambda - T)x \subset X_T(F)$, for some $\lambda \in F$. Then, $\mathbb{C}\backslash F \subset \rho_T(y)$ for all $y \in (\lambda - T)x$. Let $\omega \in \mathbb{C}\backslash F$. Then, we may choose an open neighborhood U of ω in $\mathbb{C}$ such that $\lambda \notin U$ and an analytic function $f : U \longrightarrow \mathscr{D}(T)$ which satisfies

$$(\mu - T)f(\mu) = y + T(0), \text{ for all } \mu \in U.$$

Hence, by referring to (i) of Theorem 2.8.1, we infer that

$$(\mu - T)f(\mu) = (\lambda - T)x, \text{ for all } \mu \in U. \tag{3.8.8}$$

By using the fact that $\lambda \neq \mu$, we can defined $g : U \longrightarrow \mathscr{D}(T)$ by

$$g(\mu) = \frac{f(\mu) - x}{\lambda - \mu}, \text{ for all } \mu \in U.$$

Therefore, we have $g \in H(U, \mathscr{D}(T))$ and

$$\begin{aligned}
(\mu - T)g(\mu) &= (\mu - T)\left(\frac{f(\mu) - x}{\lambda - \mu}\right), \\
&= \frac{1}{\lambda - \mu}\left((\mu - T)(f(\mu) - x)\right), \\
&= \frac{1}{\lambda - \mu}\left((\mu - T)f(\mu) - (\mu - \lambda + \lambda - T)x)\right), \\
&= \frac{1}{\lambda - \mu}\left((\mu - T)f(\mu) - (\mu - \lambda)x - (\lambda - T)x\right), \\
&= x + \frac{1}{\lambda - \mu}\left((\mu - T)f(\mu) - (\lambda - T)x)\right).
\end{aligned}$$

This implies from (3.8.8) that

$$\begin{aligned}
(\mu - T)g(\mu) &= x + \frac{1}{\lambda - \mu}\left((\lambda - T)x - (\lambda - T)x)\right), \\
&= x + T(0).
\end{aligned}$$

Thus, we conclude that $w \in \rho_T(x)$. This is equivalent to saying that $\mathbb{C}\backslash F \subset \rho_T(x)$. As a result, $x \in X_T(F)$, as desired. Q.E.D.

Theorem 3.8.2 *Let X be a Banach space and let $T \in C\mathscr{R}(X)$ such that*

$$T(0) \bigcap \mathscr{D}(T) = \{0\}. \tag{3.8.9}$$

If $N(\lambda_0 - T) \bigcap X_T(\emptyset) \subset T(0)$, then T has the SVEP at λ_0. ◇

Proof We can suppose that $\lambda_0 = 0$. Since $0 \in X_T(\emptyset)$, then $\rho_T(0) = \mathbb{C}$. This implies that $f : \mathbb{D}_\varepsilon \longrightarrow \mathscr{D}(T)$ such that

$$(\lambda - T)f(\lambda) = T(0), \text{ for all } \lambda \in \mathbb{D}_\varepsilon. \tag{3.8.10}$$

We assume that

$$f(\lambda) = \sum_{n=0}^{\infty} \lambda^n x_n \text{ for all } (x_n)_n \subset \mathscr{D}(T). \tag{3.8.11}$$

It follows from (3.8.10) and (3.8.11) that

$$\begin{aligned} T(0) &= Tf(0) \\ &= Tx_0. \end{aligned} \tag{3.8.12}$$

By referring to Remark 2.8.1 (i) that $x_0 \in N(T)$. Moreover, in view of Lemma 3.8.1, Proposition 3.4.1 (iii) and (3.8.10) implies that

$$\begin{aligned} \sigma_T(x_0) &= \sigma_T(f(0)) \\ &= \sigma_T(0) \\ &= \emptyset. \end{aligned}$$

This leads to $x_0 \in X_T(\emptyset)$. Based on the assumption that $N(T) \bigcap X_T(\emptyset) \subset T(0)$, we infer that

$$x_0 \in T(0). \tag{3.8.13}$$

For $0 < |\lambda| < \varepsilon$, we have

$$\begin{aligned} (\lambda - T)f(\lambda) &= (\lambda - T)\sum_{n=0}^{\infty} \lambda^n x_n, \\ &= \sum_{n=0}^{\infty} \lambda^{n+1} x_n + \lambda x_0 - T\left(x_0 + \sum_{n=1}^{\infty} \lambda^n x_n\right). \end{aligned}$$

Hence, by using Proposition 2.8.1 (i) and (3.8.10), we get that

$$T(0) = \sum_{n=1}^{\infty} \lambda^{n+1} x_n + \lambda x_0 - Tx_0 - T\left(\sum_{n=1}^{\infty} \lambda^n x_n\right).$$

This implies from (3.8.12) that

$$T(0) = \sum_{n=1}^{\infty} \lambda^{n+1} x_n + \lambda x_0 + T(0) - T\left(\sum_{n=1}^{\infty} \lambda^n x_n\right).$$

By the linearity of T and $\lambda \neq 0$, we infer that

$$T(0) = \lambda \left(\sum_{n=1}^{\infty} \lambda^n x_n + x_0 + T(0) - T\left(\sum_{n=1}^{\infty} \lambda^{n-1} x_n \right) \right).$$

Then, multiplying by λ^{-1}, we infer that

$$\begin{aligned} T(0) &= x_0 + T(0) + (\lambda - T)\left(\sum_{n=1}^{\infty} \lambda^{n-1} x_n \right) \\ &= x_0 + T(0) + (\lambda - T)\left(\sum_{n=0}^{\infty} \lambda^{n} x_{n+1} \right). \end{aligned} \quad (3.8.14)$$

It follows from both (3.8.13) and (3.8.14) that

$$\begin{aligned} (\lambda - T)\left(\sum_{n=0}^{\infty} \lambda^{n} x_{n+1} \right) + (\lambda - T)(0) &= T(0) \\ (\lambda - T)g(\lambda) &= T(0), \end{aligned} \quad (3.8.15)$$

where

$$g(\lambda) = \sum_{n=0}^{\infty} \lambda^n x_{n+1}.$$

By continuity, (3.8.15) holds for all $|\lambda| < \varepsilon$. Clearly, $g(\lambda) \in H(\mathbb{D}_\varepsilon, \mathscr{D}(T))$. It follows from (3.8.15) that

$$\begin{aligned} Tx_1 &= Tg(0) \\ &= T(0). \end{aligned}$$

By referring to Remark 2.8.1 (i), we infer that $x_1 \in N(T)$. Again by using Lemma 3.8.1, Proposition 3.4.1 (iii) and (3.8.15), we have $\sigma_T(x_1) = \emptyset$ which yields $x_1 \in X_T(\emptyset)$. Therefore,

$$x_1 \in N(T) \bigcap X_T(\emptyset) \subset T(0).$$

By iterating this procedure, we infer that $x_n \in T(0)$ for all $n \geq 0$. It follows from Lemma 2.6.1 that

$$\sum_{n=0}^{N} \lambda^n x_n \in T(0).$$

The fact that $T \in C\mathscr{R}(X)$ implies from Proposition 2.11.1 (iv) that $T(0)$ is closed. Since

$$\sum_{n=0}^{N} \lambda^n x_n \in T(0),$$

then $f(\lambda) \in T(0)$. Finally, the use of the assumption $T(0) \bigcap \mathscr{D}(T) = \{0\}$ allows us to conclude that $f \equiv 0$. This is equivalent to saying that T has the SVEP at 0. Q.E.D.

The following example shows that there are closed linear relations satisfying the condition (3.8.9).

Example 3.8.1 Let $X = \mathbb{R}^2$ and $T \in L\mathscr{R}(X)$ be defined by

$$G(T) = \left\{ \left(\begin{pmatrix} x \\ 0 \end{pmatrix}, \begin{pmatrix} x \\ y \end{pmatrix} \right) \text{ such that } x,\ y \in \mathbb{R} \right\}.$$

It is immediately to see that $G(T)$ is closed which implies that $T \in C\mathscr{R}(X)$. Moreover, we have $\mathscr{D}(T) = \mathbb{R} \times \{0\}$ and $T(0) = \{0\} \times \mathbb{R}$. This enables us to conclude that

$$T(0) \bigcap \mathscr{D}(T) = \left\{ \begin{pmatrix} 0 \\ 0 \end{pmatrix} \right\}.$$

Lemma 3.8.3 *Let X be a Banach space, $T \in BC\mathscr{R}(X)$ and $F \subset \mathbb{C}$ be closed. Then, $X_T(F) \subset (\lambda - T)X_T(F)$, for all $\lambda \notin F$.* ◇

Proof Let $\lambda \in \mathbb{C}\backslash F$ and $x \in X_T(F)$. Then, $\mathbb{C}\backslash F \subset \rho_T(x)$. Hence, the fact $\mathbb{C}\backslash F$ is open implies that there exists an analytic function $f : \mathbb{C}\backslash F \longrightarrow X$ which satisfies

$$(\lambda - T)f(\lambda) = x + T(0), \text{ for all } \lambda \in \mathbb{C}\backslash F.$$

It follows from Theorem 2.8.1 (i) that

$$x \in (\lambda - T)f(\lambda).$$

Combining the fact that $\mathbb{C}\backslash F \subset \rho_T(x)$, and Lemma 3.8.1, we infer that

$$\mathbb{C}\backslash F \subset \rho_T(f(\lambda)).$$

This leads to $f(\lambda) \in X_T(F)$ which yields

$$(\lambda - T)f(\lambda) \subset (\lambda - T)X_T(F).$$

Hence, we conclude $x \in (\lambda - T)X_T(F)$. Q.E.D.

Theorem 3.8.3 *Let $T \in C\mathscr{R}(H)$ and $Q \in \mathscr{L}(H)$ be such that $Q^2 = I$. If we take $F \subset \mathbb{C}$ be closed, then*

$$H_T(F) = Q\left(H_{T^Q}(F^c)\right).$$ ◇

Proof We propose to show that

$$H_T(F) \subseteq Q\left(H_{T^Q}(F^c)\right). \tag{3.8.16}$$

Let $x \in H_T(F)$, then there exists $f : \mathbb{C}\backslash F \longrightarrow \mathcal{D}(T)$ such that

$$(\lambda - T)f(\lambda) = x + T(0), \quad \text{for all } \lambda \in \mathbb{C}\backslash F.$$

The fact that $Q^2 = I$ implies that

$$(\lambda - T)Q(Qf)(\lambda) = x + T(0).$$

Since $T(0) \subset H = \mathcal{D}(Q)$, then

$$\begin{aligned} Q(\lambda - T)Q(Qf)(\lambda) &= Q(x + T(0)) \\ &= Qx + QT(0) \\ &= Qx + QTQ(0) \ \ (\text{as } Q(0) = \{0\}). \end{aligned}$$

This implies that

$$(\lambda - T)^Q(Qf)(\lambda) = Qx + T^Q(0), \quad \text{for all } \lambda \notin F. \tag{3.8.17}$$

Let us assume that $g(\overline{\lambda}) = Qf(\lambda)$, for all $\overline{\lambda} \notin F^c$. Then, by using Lemma 2.10.8 (ii), we infer that $g \in \mathcal{H}(F^c, H)$. Hence, by (3.8.17), we obtain

$$(\overline{\lambda} - T)^Q g(\overline{\lambda}) = Qx + T^Q(0), \quad \text{for all } \overline{\lambda} \notin F^c.$$

On the other hand, we have

$$\begin{aligned} (\overline{\lambda} - T)^Q &= Q(\overline{\lambda} - T)Q \\ &= Q\overline{\lambda}Q - QTQ \\ &= \overline{\lambda}\underbrace{QQ}_{=I} - QTQ \\ &= \overline{\lambda} - T^Q. \end{aligned}$$

Hence,

$$(\overline{\lambda} - T^Q)g(\overline{\lambda}) = Qx + T^Q(0), \quad \text{for all } \overline{\lambda} \notin F^c.$$

This is equivalent to saying that $Qx \in (H_{T^Q}(F^c))$. Based on the hypothesis $Q^2 = I$, we get $x \in Q\,(H_{T^Q}(F^c))$.

Conversely, by substituting T for T^Q and F for F^c in (3.8.16), we get

$$H_{T^Q}(F^c) \subseteq Q(H_{(T^Q)^Q}(F)).$$

The use of Remark 2.11.10 (i) makes us to conclude that

$$H_{TQ}(F^c) \subseteq Q\,(H_T(F))\,.$$

Therefore,

$$\begin{aligned} Q\left(H_{TQ}(F^c)\right) \subseteq &\ Q^2\,(H_T(F)) \\ = &\ H_T(F). \end{aligned}$$

This completes the proof. Q.E.D.

3.8.2 *The Single-Valued Extension Property*

Let X be a Banach space and let $T \in C\mathcal{R}(X)$. T is said to have the single-valued extension property at $\lambda \in \mathbb{C}$, abbreviated SVEP at λ, if for every neighborhood U of λ the only analytic function $f : U \longrightarrow \mathcal{D}(T)$ which satisfies the equation

$$(\mu - T)f(\mu) = T(0)$$

is the constant function $f \equiv 0$. T is said to have the SVEP, if T has the SVEP at every $\lambda \in \mathbb{C}$.

Proposition 3.8.6 *Let X be a Banach space and let $T \in C\mathcal{R}(X)$.*

(i) If $x \in X$ and T has the SVEP at $\lambda \in \rho_T(x)$, then there exists a neighborhood U of λ and an unique analytic function $f : U \longrightarrow \mathcal{D}(T)$ satisfying the equation

$$(\mu - T)f(\mu) = x + T(0), \textit{ for all } \mu \in U.$$

(ii) The SVEP is transmitted under translations, i.e., T has SVEP if, and only if, $\lambda - T$ has SVEP, for all $\lambda \in \mathbb{C}$. ◇

Proof (i) Let U be an open neighborhood of λ in $\mathbb{C}$. We assume there exist two analytic functions $f : U \longrightarrow \mathcal{D}(T)$ and $g : U \longrightarrow \mathcal{D}(T)$ which satisfies

$$(\mu - T)f(\mu) = x + T(0) \text{ and } (\mu - T)g(\mu) = x + T(0), \text{ for all } \mu \in U.$$

Let $h : U \longrightarrow \mathcal{D}(T)$ be defined by

$$h(\mu) = f(\mu) - g(\mu).$$

Hence, $h(\cdot)$ is an analytic function and for all $\mu \in U$, we have

$$\begin{aligned}(\mu - T)h(\mu) &= (\mu - T)(f(\mu) - g(\mu)),\\ &= (\mu - T)f(\mu) - (\mu - T)g(\mu),\\ &= x + T(0) - x + T(0),\\ &= T(0). \end{aligned} \tag{3.8.18}$$

The fact that T has the SVEP implies from (3.8.18) that $h \equiv 0$. As a result, $f \equiv g$, as desired.
(ii) Let λ, $\lambda_0 \in \mathbb{C}$. Assume that T has the SVEP at λ_0. Show that $\lambda + T$ has the SVEP at $\lambda + \lambda_0$. We consider an open neighborhood U of $\lambda_0 + \lambda$ and an analytic function $f : U \longrightarrow \mathscr{D}(T)$ satisfying

$$(\mu - (\lambda + T))f(\mu) = T(0), \text{ for all } \mu \in U. \tag{3.8.19}$$

Consider $g : U\backslash\{\lambda\} \longrightarrow \mathscr{D}(T)$ defined by

$$g(\xi) = f(\xi + \lambda).$$

Then, g is an analytic function and $U\backslash\{\lambda\}$ is an open neighborhood of λ_0. Hence, it follows from (3.8.19), for all $\xi \in U\backslash\{\lambda\}$, that

$$(\xi - T)g(\xi) = T(0).$$

Using the fact that T has the SVEP at λ_0, we infer that $g \equiv 0$. This implies that $f \equiv 0$. This enables us to conclude that $\lambda + T$ has the SVEP at $\lambda + \lambda_0$, for all $\lambda \in \mathbb{C}$. This achieves the proof. Q.E.D.

Theorem 3.8.4 *Let $\lambda \in \mathbb{C}$ and $T \in C\mathscr{R}(H)$. Then, T has SVEP at λ if, and only if, T^Q has SVEP at $\overline{\lambda}$.* ◇

Proof Let us assume that T has SVEP at λ. Then, there exists the analytic function $f : \mathbb{D}(\overline{\lambda}, r) \longrightarrow \mathscr{D}\left(T^Q\right)$, for which $(\overline{\mu} - T^Q)f(\overline{\mu}) = T^Q(0)$ for all $\overline{\mu} \in \mathbb{D}(\overline{\lambda}, r)$. Hence,

$$(\overline{\mu} - QTQ)\, f\,(\overline{\mu}) = QT(0).$$

This implies from the fact that $Q^2 = I$ that

$$QT(0) = (\overline{\mu}QQ - QTQ)\, f\,(\overline{\mu})\,.$$

Based on the hypothesis $Q \in \mathscr{L}(H)$, we have

$$QT(0) = Q\,(\overline{\mu} - T)\,(Qf)\,(\overline{\mu})\,.$$

Hence,

$$(\overline{\mu} - T)\,(Qf)\,(\overline{\mu}) = T(0).$$

Define

$$h(\mu) = (Qf)\,(\overline{\mu})$$

for $\mu \in \mathbb{D}\,(\lambda, r)$. The use of Lemma 2.10.8 (ii) makes us conclude that h is an analytic function on $\mathbb{D}\,(\lambda, r)$. This implies that

$$(\mu - T)\,h\,(\mu) = T(0) \quad \text{for } \mu \in \mathbb{D}\,(\lambda, r)\,.$$

The fact that T has SVEP at λ implies that $h(\mu) = 0$ for all $\mu \in \mathbb{D}\,(\lambda, r)$, which yields that $Qf\,(\overline{\mu}) = 0$. Therefore,

$$f\,(\overline{\mu}) = Q^2 f\,(\overline{\mu}) = 0.$$

So, T^Q has SVEP at $\overline{\lambda}$.

Conversely, by substituting T for T^Q, we obtain from Remark 2.11.10 (i) that $T = (T^Q)^Q$ has SVEP at λ. Q.E.D.

Remark 3.8.1 Let $\lambda \in \mathbb{C}$ and $T \in C\mathscr{R}(H)$.

(i) T have SVEP if, and only if, T^Q have SVEP.
(ii) If $T^* = T^Q$, then T has SVEP at λ if, and only if, T^* has SVEP at $\overline{\lambda}$. ◇

Proposition 3.8.7 ([113]) (i) *Let* $T \in C\mathscr{R}(H)$. *Then,* T *have SVEP at* λ *if, and only if,* $\lambda - T$ *have SVEP at* 0, *for every* $\lambda \in \mathbb{C}$.
(ii) *Let* $T \in C\mathscr{R}(H)$ *be onto. If* T *have SVEP at* 0, *then* T *is one-to-one.* ◇

Theorem 3.8.5 ([113]) *Let* X *be a Banach space and let* $T \in C\mathscr{R}(X)$ *be with closed range. If* T *is not one-to-one, then* T *does not have the SVEP at* 0. ◇

Theorem 3.8.6 *Let* $T \in C\mathscr{R}(H)$ *be such that* $\mathscr{D}(T) \subset T(0)^{\perp}$. *If* T *is one-to-one, then* T *have SVEP at* 0. ◇

Proof Based on the assumption $N(T) = \{0\}$, we infer that

$$N(T) \bigcap X_T(\emptyset) \subset \{0\}.$$

This implies that

$$N(T) \bigcap X_T(\emptyset) \subset T(0).$$

Now, since T is closed, then by using Proposition 2.11.1, we have $T(0)$ is closed. This leads us to write H as follows

$$H = T(0) \bigoplus T(0)^{\perp}.$$

Then, the fact that $\mathscr{D}(T) \subset T(0)^{\perp}$ implies that

$$T(0)\bigcap \mathscr{D}(T) \subset T(0)\bigcap T(0)^{\perp} = \{0\}.$$

Consequently,

$$T(0)\bigcap \mathscr{D}(T) = \{0\}.$$

Hence, by referring to [113, Theorem 3.1], we deduce that T has SVEP at 0. Q.E.D.

As a direct consequence of Proposition 3.8.7 (ii) and Theorem 3.8.6, we infer the following result.

Corollary 3.8.1 *Let $T \in C\mathscr{R}(H)$ be onto and $\mathscr{D}(T) \subset T(0)^{\perp}$. Then, T have SVEP at 0 if, and only if T is one-to-one.* ◇

Theorem 3.8.7 *Let X be a Banach space and let $T \in C\mathscr{R}(X)$ be with closed range. If T is not one-to-one, then T does not has the SVEP at 0.* ◇

Proof Since T is not one-to-one, then there exists $x_0 \in \mathscr{D}(T)$, $\|x_0\| = 1$ and $0 \in Tx_0$. By Theorem 2.8.1 (i), we have $Tx_0 = T(0)$. According $T \in C\mathscr{R}(X)$ and $R(T)$ is closed, we infer from (iii) of Theorem 2.11.1 that T is open. This implies that T^{-1} is continuous. Therefore, there exists $M > 0$ such that

$$\|T^{-1}y\| \leq M\|y\|, \text{ for all } y \in R(T). \tag{3.8.20}$$

For any $y \in R(T)$, there is a $x \in \mathscr{D}(T)$ such that $x \in T^{-1}y$. It follows from Lemma 2.10.2 that

$$\begin{aligned}\|T^{-1}y\| &= \operatorname{dist}(x, T^{-1}(0)) \\ &= \inf_{z\in T^{-1}(0)} \|x - z\|,\end{aligned}$$

which yields

$$\|x\| \leq \|T^{-1}y\| + \varepsilon,$$

by the arbitrariness of ε. In view of (3.8.20) implies that

$$\|x\| \leq M\|y\| + \varepsilon.$$

Choose $(x_n)_n \subset \mathscr{D}(T)$ such that $x_{n-1} \in Tx_n$. Then, we obtain

$$\|x_n\| \leq M\|x_{n-1}\| + \varepsilon.$$

Hence, the fact that ε is arbitrary implies that

$$\|x_n\| \leq M\|x_{n-1}\|.$$

Let us assume that

$$f(\lambda) = \sum_{n=0}^{\infty} \lambda^n x_n.$$

Since

$$\begin{aligned}\|x_n\| &\le M\|x_{n-1}\| \\ &\vdots \\ &\le M^n \|x_0\| \\ &\le M^n,\end{aligned}$$

then the sum converges for $|\lambda| < M^{-1}$. Consequently, f is analytic in $\mathbb{D}_{M^{-1}}$. The fact that $T \in C\mathscr{R}(X)$ implies from Proposition 2.11.2 that $\lambda - T$ is closed. Moreover, we have

$$f_N(\lambda) = \sum_{n=0}^{N} \lambda^n x_n \in \mathscr{D}(T).$$

Hence, for $|\lambda| < M^{-1}$, we have

$$\begin{aligned}(\lambda - T) f_N(\lambda) &= (\lambda - T) \sum_{n=0}^{N} \lambda^n x_n \\ &= \sum_{n=0}^{N} \lambda^{n+1} x_n - T\left(\sum_{n=0}^{N} \lambda^n x_n\right) \\ &= \sum_{n=0}^{N} \lambda^{n+1} x_n - T\left(x_0 + \sum_{n=1}^{N} \lambda^n x_n\right).\end{aligned}$$

By virtue of Proposition 2.8.1 (i), we obtain

$$(\lambda - T) f_N(\lambda) = \sum_{n=0}^{N} \lambda^{n+1} x_n - T x_0 - T\left(\sum_{n=1}^{N} \lambda^n x_n\right).$$

This implies from $T x_0 = T(0)$ that

$$\begin{aligned}(\lambda - T) f_N(\lambda) &= \sum_{n=0}^{N} \lambda^{n+1} x_n - T(0) - T\left(\sum_{n=1}^{N} \lambda^n x_n\right), \\ &= \sum_{n=0}^{N} \lambda^{n+1} x_n - T\left(\sum_{n=1}^{N} \lambda^n x_n\right).\end{aligned} \tag{3.8.21}$$

We discuss two cases.
First case. If $\lambda = 0$, then by using (3.8.21), we infer that $Tf_N(0) = T(0)$.
Second case. If $0 < |\lambda| < M^{-1}$, then it follows from (3.8.21) that

$$(\lambda - T)f_N(\lambda) = \sum_{n=0}^{N} \lambda^{n+1} x_n - \sum_{n=1}^{N} \lambda^n T(x_n).$$

At this level, the fact that $x_{n-1} \in Tx_n$ implies from (i) of Theorem 2.8.1 (i) that

$$\begin{aligned}(\lambda - T)f_N(\lambda) &= \sum_{n=0}^{N} \lambda^{n+1} x_n - \sum_{n=1}^{N} \lambda^n (x_{n-1} + T(0)),\\ &= \sum_{n=0}^{N} \lambda^{n+1} x_n - \sum_{n=1}^{N} \lambda^n x_{n-1} - \sum_{n=1}^{N} \lambda^n T(0),\\ &= \sum_{n=0}^{N} \lambda^{n+1} x_n - \sum_{n=0}^{N-1} \lambda^{n+1} x_n - T\left(\left(\sum_{n=1}^{N} \lambda^n\right) 0\right),\\ &= \lambda^{N+1} x_N + T(0).\end{aligned}$$

Consequently, we obtain

$$(\lambda - T)f_N(\lambda) = \lambda^{N+1} x_N + T(0), \text{ for all } |\lambda| < M^{-1}.$$

Now, combining the fact that $\lambda - T$ is closed, $(f_N)_N \subset \mathscr{D}(T)$, $f(\lambda) = \lim_{N\to\infty} f_N(\lambda)$ and Remark 2.11.6, we deduce that $f(\lambda) \in \mathscr{D}(T)$. This implies, for all $\lambda \in \mathbb{D}_{M^{-1}}$, that

$$\begin{aligned}(\lambda - T)f(\lambda) &= (\lambda - T)\lim_{N\to\infty} f_N(\lambda)\\ &= \lim_{N\to\infty} ((\lambda - T)f_N(\lambda))\\ &= \lim_{N\to\infty} (\lambda^{N+1} x_N) + T(0).\end{aligned} \tag{3.8.22}$$

Since $\lim_{N\to\infty} \|x_N \lambda^{N+1}\| \leq \lim_{N\to\infty} M^N |\lambda|^{N+1} = 0$, for all $\lambda \in \mathbb{D}_{M^{-1}}$, then by referring to (3.8.22), we obtain

$$(\lambda - T)f(\lambda) = T(0).$$

$$(\lambda - T)f(\lambda) = T(0) \text{ for all } \lambda \in \mathbb{D}_{M^{-1}}.$$

This shows that T does not has the SVEP at 0. Q.E.D.

Theorem 3.8.8 *Let X be a Banach space and let $T \in BC\mathscr{R}(X)$ be such that $T(0)$ is the kernel of continuous projection P defined on $R(T)$. Assume that $T(0) \in N(PT)$. If T has the SVEP at λ_0, then $N(\lambda_0 - T) \bigcap X_T(\emptyset) \subset T(0)$.* ◇

Proof We can suppose that $\lambda_0 = 0$. Assume that, for some $x \in N(T)$, we have $\sigma_T(x) = \emptyset$. Then, $0 \in \rho_T(x)$ which implies that there is an open disc $\mathbb{D}_\varepsilon$ and an analytic function $f : \mathbb{D}_\varepsilon \longrightarrow X$ such that

$$(\lambda - T)f(\lambda) = x + T(0), \text{ for all } \lambda \in \mathbb{D}_\varepsilon. \tag{3.8.23}$$

Since $A = PT$ is continuous selection of T, then from $\mathscr{D}(T) = X$ that A is a bounded selection of T. Hence, it follows from (3.8.23) that

$$(\lambda - A)f(\lambda) + T(0) = x + T(0), \text{ for all } \lambda \in \mathbb{D}_\varepsilon. \tag{3.8.24}$$

For all $\lambda \in \mathbb{D}_\varepsilon$, we have

$$\begin{aligned}(\lambda - T)(Af(\lambda)) &= (\lambda - A)Af(\lambda) + T(0), \\ &= A(\lambda - A)f(\lambda) + T(0).\end{aligned}$$

Using the fact $T(0) \subset N(A)$, we infer that

$$(\lambda - T)(Af(\lambda)) = A(\lambda - A)f(\lambda) + AT(0) + T(0).$$

By virtue of Proposition 2.8.2 for all $\lambda \in \mathbb{D}_\varepsilon$, we obtain

$$\begin{aligned}(\lambda - T)(Af(\lambda)) &= A((\lambda - A)f(\lambda) + T(0)) + T(0). \\ &= A((\lambda - T)f(\lambda)) + T(0).\end{aligned}$$

Consequently, in view of (3.8.23), we get

$$\begin{aligned}(\lambda - T)(Af(\lambda)) &= A(x + T(0)) + T(0), \\ &= Ax + T(0), \\ &= Tx.\end{aligned} \tag{3.8.25}$$

Since $x \in N(T)$, then by Remark 2.8.1 (i), we have

$$Tx = T(0).$$

This implies from (3.8.25) that

$$(\lambda - T)(Af(\lambda)) = T(0). \tag{3.8.26}$$

Based on the assumption T has the SVEP at 0, we can conclude from (3.8.26) that $Af \equiv 0$. Thus, by virtue of (3.8.24), we infer that

$$\begin{aligned} T(0) &= -A\ f(0) + T(0) \\ &= x + T(0). \end{aligned}$$

Finally, the use of Theorem 2.8.1 (i) allows us to conclude that $x \in T(0)$. This achieves the proof. Q.E.D.

Theorem 3.8.9 *Let X be a Banach space and let $T \in C\mathscr{R}(X)$ be such that $\mathscr{D}(T) \bigcap T(0) = \{0\}$. Then, T has the SVEP if, and only if, $X_T(\emptyset) = T(0)$.* ◇

Proof Let us assume that T has the SVEP. Let $x \in X_T(\emptyset)$. Then, $\rho_T(x) = \mathbb{C}$ which yields, there exists an analytic function $f : \mathbb{C} \longrightarrow \mathscr{D}(T)$ for which

$$(\lambda - T)f(\lambda) = x + T(0), \text{ for all } \lambda \in \mathbb{C}. \tag{3.8.27}$$

Since $f(\lambda) = (\lambda - T)^{-1}x$, for all $\lambda \in \rho(T)$ and $\lim_{|\lambda| \to \infty} \|(\lambda - T)^{-1}\| = 0$, then f is a bounded analytic function on $\mathbb{C}$. It follows from Liouville's theorem that f is constant. Using the fact that

$$\lim_{|\lambda| \to \infty} (\lambda - T)^{-1}x = 0,$$

we get f is null. By referring to (3.8.27), we obtain

$$\begin{aligned} (\lambda - T)f(\lambda) &= (\lambda - T)(0) \\ &= x + T(0), \text{ for all } \lambda \in \mathbb{C}. \end{aligned}$$

Hence, we deduce that $x \in T(0)$. Finally, the use of Eq. (3.8.7) allows us to conclude that

$$X_T(\emptyset) = T(0),$$

whenever T has the SVEP. Conversely, let us assume that $X_T(\emptyset) = T(0)$ and let $f : \mathbb{C} \longrightarrow \mathscr{D}(T)$ be an analytic function such that

$$(\lambda - T)f(\lambda) = T(0), \text{ for all } \lambda \in \mathbb{C}.$$

This implies that $\lambda \in \rho_T(0)$ for all $\lambda \in \mathbb{C}$. Consequently, $\sigma_T(0) = \emptyset$. Hence, it follows from Lemma 3.8.1 that

$$\sigma_T(f(\lambda)) = \emptyset.$$

This leads to $f(\lambda) \in X_T(\emptyset) = T(0)$. The fact that $T(0) \bigcap \mathscr{D}(T) = \{0\}$ and $f(\lambda) \in \mathscr{D}(T)$ implies that $f \equiv 0$. Hence, T has the SVEP. Q.E.D.

3.8.3 Relationship Between the Spectrum and the Defect Spectrum of a Closed Linear Relation on a Hilbert Space Having SVEP

In the following result, we discuss a relationship between the spectrum and the defect spectrum of a closed linear relation on a Hilbert space having SVEP.

Corollary 3.8.2 *Let $T \in C\mathscr{R}(H)$. Then,*

(i) If T have SVEP and $\mathscr{D}(T) \subset T(0)^{\perp}$, then $\sigma(T) = \sigma_{\delta}(T)$.
(ii) If T^* have SVEP and $\mathscr{D}(T^*) \subset T^*(0)^{\perp}$, then $\sigma(T) = \sigma_{ap}(T)$. ◇

Remark 3.8.2 Let $T \in C\mathscr{R}(H)$. If $T(0) = \mathscr{D}(T)^{\perp}$ and T, T^* have SVEP, then $\sigma_{ap}(T) = \sigma_{\delta}(T)$. Indeed, based on the hypothesis $T(0) = \mathscr{D}(T)^{\perp}$, we infer that $\mathscr{D}(T) \subset \overline{\mathscr{D}(T)} = T(0)^{\perp}$. Since T have SVEP, then by applying Corollary 3.8.2 (i), we obtain

$$\sigma(T) = \sigma_{\delta}(T).$$

Since T is closed, then by using Proposition 2.11.1, we infer that $T(0)$ is closed, which yields that

$$\overline{T}(0) = T(0) = \mathscr{D}(T)^{\perp}. \tag{3.8.28}$$

Again from (3.1.4), we have $\mathscr{D}(T^*)^{\perp} = \overline{T}(0)$ and $\mathscr{D}(T)^{\perp} = T^*(0)$. This implies from (3.8.28) that $\mathscr{D}(T^*)^{\perp} = T^*(0)$. Consequently, $\mathscr{D}(T^*) \subset \overline{\mathscr{D}(T^*)} = T^*(0)^{\perp}$. Hence, the fact that T^* have SVEP implies from Corollary 3.8.2 (ii) that

$$\sigma(T) = \sigma_{ap}(T).$$

Thus, we conclude that $\sigma_{ap}(T) = \sigma_{\delta}(T)$. ◇

Remark 3.8.3 (i) Let $T \in B\mathscr{R}(X)$. Then, T has the SVEP if, and only if, $S_T = \emptyset$ (the empty set).
(ii) It is known that $\sigma(T) = S_T \bigcup \sigma_{\delta}(T)$. In particular, if T has the SVEP, we obtain $\sigma(T) = \sigma_{\delta}(T)$. ◇

Let us define the following set

$$\Xi(T) = \{\lambda \in \mathbb{C} \text{ such that } T \text{ does not have SVEP at } \lambda\}.$$

Then, $\lambda \notin \Xi(T)$ if, and only if, T have SVEP at λ. The following results consists of establishing some results on spectral theory which will be needed in the sequel.

Theorem 3.8.10 *Let $T \in C\mathscr{R}(H)$. Then,*

(i) If $\mathscr{D}(T) \subset T(0)^{\perp}$, then $\sigma(T) = \sigma_{\delta}(T) \bigcup \Xi(T)$.
(ii) If $\mathscr{D}(T^*) \subset T^*(0)^{\perp}$, then $\sigma(T) = \sigma_{ap}(T) \bigcup \left\{\lambda \in \mathbb{C} \text{ such that } \overline{\lambda} \in \Xi(T^*)\right\}$. ◇

Proof (i) The fact that $\mathscr{D}(T) \subset T(0)^{\perp}$, $(\lambda - T)(0) = T(0)$ and $\mathscr{D}(\lambda - T) = \mathscr{D}(T)$ implies that

$$\mathscr{D}(\lambda - T) \subset (\lambda - T)(0)^{\perp}. \tag{3.8.29}$$

Our purpose is to show that $\sigma(T) \subset \sigma_{\delta}(T) \bigcup \Xi(T)$. Let us assume that $\lambda \notin [\sigma_{\delta}(T) \bigcup \Xi(T)]$. Then, T have SVEP at λ. It follows from (i) of Proposition 3.8.7 that $\lambda - T$ have SVEP at 0. Since $\lambda - T$ is onto, then by using (3.8.29) and Corollary 3.8.1, we infer that $\lambda - T$ is one-to-one. This is equivalent to saying from (3.1.4) that

$$R((\lambda - T)^{*})^{\perp} = N(\lambda - T) = \{0\}. \tag{3.8.30}$$

The use of Theorem 2.15.1 makes us conclude from the hypothesis $R(\lambda - T) = H$ is closed that $R((\lambda - T)^{*})$ is so. This implies from (3.1.3) and (3.8.30) that $R(\overline{\lambda} - T^{*}) = H$. In addition, from (3.1.3) and (3.1.4), we have

$$\begin{aligned} \{0\} &= R(\lambda - T)^{\perp} \\ &= N((\lambda - T)^{*}) \\ &= N(\overline{\lambda} - T^{*}). \end{aligned}$$

This leads us to $\overline{\lambda} - T^{*}$ is one-to-one. Consequently, by (3.1.1), we get $\overline{\lambda} \notin \sigma(T^{*})$. Finally, the use of (3.1.5) allows us to conclude that $\lambda \notin \sigma(T)$.

Conversely, suppose that $\lambda \notin \sigma(T)$. Then, by using (3.1.5), we obtain $\overline{\lambda} \notin \sigma(T^{*})$. This implies from (3.1.3) and (3.1.1) that

$$R((\lambda - T)^{*}) = H$$

and

$$N((\lambda - T)^{*}) = \{0\}.$$

Hence, it follows from (3.1.4) that

$$N(\overline{\lambda - T}) = \{0\}$$

and

$$R(\lambda - T)^{\perp} = \{0\}. \tag{3.8.31}$$

By the closedness of $\lambda - T$, we infer that $N(\lambda - T) = \{0\}$. Now, since $R((\lambda - T)^{*})$ is closed, then by Theorem 2.15.1, we obtain $R(\lambda - T)$ is so. This implies from (3.8.31) that $\lambda - T$ is onto. In view of (3.8.29), Corollary 3.8.1 and Proposition 3.8.7 (i) implies that T have SVEP at λ. This enables us to conclude that $\lambda \notin [\sigma_{\delta}(T) \bigcup \Xi(T)]$.

(ii) Since T^{*} is closed and $\mathscr{D}(T^{*}) \subset T^{*}(0)^{\perp}$, then we infer from (i) that

$$\sigma(T^{*}) = \sigma_{\delta}(T^{*}) \bigcup \Xi(T^{*}).$$

By referring to Lemma 3.2.3 and (3.1.5), we deduce that

$$\{\overline{\lambda} \text{ such that } \lambda \in \sigma(T)\} = \{\overline{\lambda} \text{ such that } \lambda \in \sigma_{ap}(T)\} \bigcup \Xi(T^*).$$

This shows that $\sigma(T) = \sigma_{ap}(T) \bigcup \left\{\lambda \in \mathbb{C} \text{ such that } \overline{\lambda} \in \Xi(T^*)\right\}$. Q.E.D.

Now, for $T \in C\mathscr{R}(H)$, we define the new sets by means $\Xi(T)$ as follows

$$\begin{aligned} \Xi_+(T) &= \{\lambda \in \mathbb{C} \text{ such that } \lambda - T \notin \Phi_+(H) \text{ or } \lambda \in \Xi(T)\} \\ \Xi_-(T) &= \{\lambda \in \mathbb{C} \text{ such that } \lambda - T \notin \Phi_-(H) \text{ or } \lambda \in \Xi(T)\} \\ \Xi_\pm(T) &= \{\lambda \in \mathbb{C} \text{ such that } \lambda - T \notin \Phi_\pm(H) \text{ or } \lambda \in \Xi(T)\} \\ \Xi_0(T) &= \{\lambda \in \mathbb{C} \text{ such that } \lambda - T \notin \Phi(H) \text{ or } \lambda \in \Xi(T)\}. \end{aligned}$$

Remark 3.8.4 Let us supply the Hilbert space $H \times K$ equipped with the norm

$$\|(h,k)\|^2 = \|h\|^2 + \|k\|^2,\ h \in H,\ k \in K.$$

Then, the norms $\|(\cdot,\cdot)\|$ and $\|(\cdot,\cdot)\|_1$ are equivalent, where $\|(h,k)\|_1 = \|h\| + \|k\|,\ h \in H,\ k \in K$. ◇

Proposition 3.8.8 *Let H be an infinite-dimensional Hilbert space and let $T \in L\mathscr{R}(H \times H)$ be defined by*

$$T\begin{pmatrix} x \\ y \end{pmatrix} = \begin{pmatrix} x \\ y \end{pmatrix} + T\begin{pmatrix} 0 \\ 0 \end{pmatrix}, \tag{3.8.32}$$

where

$$T\begin{pmatrix} 0 \\ 0 \end{pmatrix} = \left\{\begin{pmatrix} x \\ y \end{pmatrix} \in H \times H \text{ such that } xy = 0\right\}.$$

(i) The identity operator I on $H \times H$ is a selection of T.
(ii) T is a closed linear relation on $H \times H$.
(iii) $\lambda - T \in \Phi_\pm(H \times H)$ if, and only if, $\lambda - I \in \Phi_\pm(H \times H)$.
(iv) $1 \in \Xi_\pm(T)$. ◇

Proof (i) It follows from (3.8.32) and (2.14.1) that I is a selection of T.
(ii) Since $\mathscr{D}(T) = H \times H$ and I is a continuous selection of T, we infer from (2.14.1) that T is bounded. It remains to prove from Theorem 2.11.1 (iv) that $T\begin{pmatrix} 0 \\ 0 \end{pmatrix}$ is closed. Let us assume that $\begin{pmatrix} x_n \\ y_n \end{pmatrix} \in T\begin{pmatrix} 0 \\ 0 \end{pmatrix}$ such that

$$\begin{pmatrix} x_n \\ y_n \end{pmatrix} \to \begin{pmatrix} x \\ y \end{pmatrix}, \quad \text{as } n \to \infty.$$

This implies that $x_n y_n = 0$ and $x_n y_n \to xy$, as $n \to \infty$. According to the uniqueness of the limit of $(x_n y_n)_n$, we infer that $xy = 0$. This yields that $T\begin{pmatrix} 0 \\ 0 \end{pmatrix}$ is closed.

(iii) Let $\lambda - T \in \Phi_{\pm}(H \times H)$. Then, by using Definition 2.14.1, we have $\lambda - T + T - I \in \Phi_{\pm}(H \times H)$. At this level, we show that $T - T$ is compact. By referring to Proposition 2.8.2, we have

$$\begin{aligned} Q_{T-T}(T-T) &= \ Q_T(T-T) \\ &= \ Q_T T - Q_T T \\ &= \ 0 \text{ (by Proposition 2.9.1)}, \end{aligned}$$

where 0 is a null operator. This implies that

$$\overline{Q_{T-T}(T-T)(\mathscr{B})} = \{0\},$$

where $\mathscr{B} = \{(x, y) \in H \times H : \|(x, y)\| \leq 1\}$. Thus, it is compact. Hence, we can conclude that $\lambda - I \in \Phi_{\pm}(H \times H)$. Conversely, the same reasoning as before leads to the result.

(iv), we have

$$\{\lambda \in \mathbb{C} \text{ such that } \lambda - T \notin \Phi_{\pm}(H \times H)\} = \{1\}.$$

Indeed, since $\lambda - I = (\lambda - 1)I$ invertible if, and only if, $\lambda \neq 1$, then $\{1\} = \sigma(I)$. Hence,

$$\{\lambda \in \mathbb{C} \text{ such that } \lambda - I \notin \Phi_{\pm}(H \times H)\} \subset \{1\}.$$

Now, suppose that

$$1 \notin \{\lambda \in \mathbb{C} \text{ such that } \lambda - I \notin \Phi_{\pm}(H \times H)\}.$$

Then, $I - I = 0 \in \Phi_{\pm}(H \times H)$, which is a contradiction since $\alpha(0) = \infty$ and $\beta(0) = \infty$. Therefore, by using (iii), we infer that

$$\{\lambda \in \mathbb{C} \text{ such that } \lambda - T \notin \Phi_{\pm}(H \times H)\} = \{1\}.$$

It is clear that

$$\{\lambda \in \mathbb{C} \text{ such that } \lambda - T \notin \Phi_{\pm}(H \times H)\} \subset \Xi_{\pm}(T).$$

As a result, $1 \in \Xi_{\pm}(T)$. Q.E.D.

Remark 3.8.5 Let $T \in C\mathscr{R}(H)$. Then,

(i) $\Xi_+(T) = \sigma_1(T) \bigcup \Xi(T)$, $\Xi_-(T) = \sigma_2(T) \bigcup \Xi(T)$, $\Xi_\pm(T) = \sigma_3(T) \bigcup \Xi(T)$ and $\Xi_0(T) = \sigma_4(T) \bigcup \Xi(T)$.

(ii) The fact that $\Xi_\pm(T) = \Xi_-(T) \bigcap \Xi_+(T) \subset \Xi_0(T)$ implies from Proposition 3.8.8 that $\Xi_-(T)$, $\Xi_+(T)$ and $\Xi_0(T)$ are nonempty sets. ◇

3.8.4 The Analytical Core of Closed Multivalued Linear Operator

Definition 3.8.2 Let X be a Banach space. The analytical core of $T \in C\mathscr{R}(X)$ is the set $\mathscr{K}(T)$ for all $x \in \mathscr{D}(T)$ for which there is a sequence $(x_n)_n$ and $\delta > 0$ such that

(i) $x = x_0$.
(ii) $Tx_{n+1} = x_n + T(0)$, for all $n \geq 0$.
(iii) $\|x_n\| \leq \delta^n \|x\|$, for all $n \geq 0$.

We now establish the relationship between the correlation analytic core and the local spectral subspace of a relation T.

Theorem 3.8.11 *Let X be a Banach space and $T \in BC\mathscr{R}(X)$. Then,*

$$\mathscr{K}(T) = X_T(\mathbb{C} \setminus \{0\}) = \{x \in X \textit{ such that } 0 \in \rho_T(x)\}.$$ ◇

Proof Let prove the first inclusion $\mathscr{K}(T) \subseteq X_T(\mathbb{C} \setminus \{0\})$. Let $x \in \mathscr{K}(T)$. Then, there exist a sequence $(x_n)_n \subset X$ and a positive scalar a such that $x = x_0$, $x_n \in Tx_{n+1}$ and

$$\text{dist}(x_n, T(0) \bigcap N(T)) \leq a^n \text{dist}(x, T(0) \bigcap N(T)),$$

$\forall n \in \mathbb{N}$. Let $b > a$. From the last inequality, we deduce that for all $n \geq 1$, there exists $\alpha_n \in T(0) \bigcap N(T)$ such that

$$\|x_n - \alpha_n\| \leq b^n \|x\|.$$

Let $(y_n)_n$ be the sequence defined by $y_0 = x$, and for all $n \geq 1$, $y_n = x_n - \alpha_n$. Then,

$$\|y_n\| \leq b^n \|x\|.$$

Let f be the analytic function $f : B(0, \frac{1}{b}) \longrightarrow X$ defined by

$$f(\lambda) = -\sum_{n \geq 1} \lambda^{n-1} y_n.$$

We can easily verify that $y_n \in Ty_{n+1}$ for all $n \in \mathbb{N}$. Moreover, by Lemma 2.10.7, for all $\lambda \in B(0, \frac{1}{b})$, we have

$$\begin{aligned}(\lambda I - T)f(\lambda) &= \sum_{n\geq 1} \lambda^{n-1} y_{n-1} + T(0) - \lambda \sum_{n\geq 1} \lambda^{n-1} y_n \\ &= y_0 + T(0) = x + T(0).\end{aligned}$$

Thus, $0 \in \rho_T(x)$ and then $x \in X_T(\mathbb{C} \setminus \{0\})$, which provide the desired inclusion. Moving to the reverse inclusion. Let $x \in X_T(\mathbb{C} \setminus \{0\})$. Then, $0 \in \rho_T(x)$ which implies that there exist an open disc $\mathbb{D}(0, \varepsilon)$ and an analytic function $f : \mathbb{D}(0, \varepsilon) \longrightarrow X$ such that the equation

$$(\mu I - T)f(\mu) = x + T(0)$$

holds for all $\mu \in \mathbb{D}(0, \varepsilon)$. Since f is an analytic function, then there exists a sequence $(u_n)_{n\geq 1} \subset X$ such that

$$f(\lambda) = -\sum_{n\geq 1} \lambda^{n-1} u_n, \quad \text{for all } \lambda \in \mathbb{D}(0, \varepsilon).$$

Clearly, $f(0) = -u_1$. Then,

$$T(u_1) = x + T(0)$$

and so $x \in Tu_1$. Take $u_0 = x$ and let prove that $u_n \in Tu_{n+1}$ for all $n \in \mathbb{N}$. The proof is given by induction. For $n = 0$, we have $u_0 \in Tu_1$. Suppose that this property is true until the order n, and let prove it for the order $n + 1$. For all $\lambda \in \mathbb{D}(0, \varepsilon)$, we have

$$\begin{aligned}(\lambda I - T)f(\lambda) &= -\sum_{k\geq 1} \lambda^k u_k + T\left(\sum_{k\geq 1} \lambda^{k-1} u_k\right) \\ &= -\sum_{k=1}^{n+1} \lambda^k u_k - \sum_{k\geq n+2} \lambda^k u_k + T\left(\sum_{k=1}^{n+2} \lambda^{k-1} u_k\right) + T\left(\sum_{k\geq n+3} \lambda^{k-1} u_k\right).\end{aligned}$$

So,

$$Tu_1 + \lambda^{n+1}(Tu_{n+2} - u_{n+1}) - \sum_{k\geq n+2} \lambda^k u_k + T\left(\sum_{k\geq n+3} \lambda^{k-1} u_k\right) = Tu_1.$$

Since

$$\lambda^{n+1} T\left(\sum_{k\geq n+3} \lambda^{k-(n+2)} u_k\right) \subset T\left(\lambda^{n+1} \sum_{k\geq n+3} \lambda^{k-(n+2)} u_k\right),$$

then

$$(Tu_{n+2} - u_{n+1}) - \sum_{k\geq n+2} \lambda^{k-(n+1)} u_k + T\left(\sum_{k\geq n+3} \lambda^{k-(n+2)} u_k\right) \subset T(0).$$

If we take $\lambda = 0$, then we find that $Tu_{n+2} - u_{n+1} \subset T(0)$ and hence, $u_{n+1} \in Tu_{n+2}$. Consequently, $u_n \in Tu_{n+1}$, for all $n \in \mathbb{N}$. It remains to prove that there exists a positive scalar which verify the second condition of $\mathscr{K}(T)$. If $x \in T(0) \bigcap N(T)$, then there is nothing to prove. If not, since the series

$$-\sum_{n\geq 1} \lambda^{n-1} u_n$$

converges, then $|\lambda|^{n-1}\|u_n\| \to 0$ as $n \to \infty$, for all $|\lambda| < \varepsilon$. In particular,

$$\frac{1}{\mu^{n-1}}\|u_n\| \to 0$$

as $n \to \infty$, for $\mu > \frac{1}{\varepsilon}$. Take $\mu_0 > \frac{1}{\varepsilon}$. So, there exists $c > 0$ such that

$$\|u_n\| \leq c\mu_0^{n-1}.$$

Hence, we obtain that

$$\|u_n\| \leq \left(\mu_0 + \frac{c}{\operatorname{dist}(x, T(0)\bigcap N(T))}\right)^n \operatorname{dist}(x, T(0)\bigcap N(T)).$$

Thus, we have

$$\operatorname{dist}(u_n, T(0)\bigcap N(T)) \leq \|u_n\| \leq b^n \operatorname{dist}(x, T(0)\bigcap N(T))$$

with $b = \mu_0 + \frac{c}{\operatorname{dist}(x, T(0)\bigcap N(T))}$. So, $x \in \mathscr{K}(T)$ and therefore,

$$X_T(\mathbb{C}\setminus\{0\}) \subseteq \mathscr{K}(T)$$

which ends the proof. Q.E.D.

Theorem 3.8.12 *Let $T \in C\mathscr{R}(H)$ be such that $\|T\| < \infty$. Then,*

$$H_{T^*}(\{0\}) \subseteq \mathscr{K}(T)^{\perp}.$$ ◇

Proof Let $x \in H_{T^*}(\{0\})$ and fix an arbitrary $y \in \mathscr{K}(T)$. Our purpose is to show that $\langle x, y\rangle = 0$. First we consider the an analytic function $g : \mathbb{D}_r \longrightarrow \mathscr{D}(T)$ satisfying

$$(\lambda - T)g(\lambda) = y + T(0), \text{ for all } \lambda \in \mathbb{D}_r.$$

Since $x \in H_{T^*}(\{0\})$, then there exists an analytic function $f : \mathbb{C}\backslash\{0\} \longrightarrow \mathscr{D}(T^*)$ such that $(\overline{\mu} - T^*)f(\overline{\mu}) = x + T^*(0)$, for all $\mu \in \mathbb{C}\backslash\{0\}$. Since the functions f and g are defined in $\mathbb{D}_r\backslash\{0\}$ and for $\lambda \in \mathbb{D}_r\backslash\{0\}$, we have

$$(\lambda - T)g(\lambda) = y + (\lambda - T)(0) \tag{3.8.33}$$

and

$$(\overline{\lambda} - T^*)f(\overline{\lambda}) = x + (\overline{\lambda} - T^*)(0).$$

This implies that

$$(\lambda - T)^* f(\overline{\lambda}) = x + (\lambda - T)^*(0), \text{ for all } \lambda \in \mathbb{D}_r\backslash\{0\}. \tag{3.8.34}$$

In view of both (3.8.33) and (3.8.34) implies that $\left(f(\overline{\lambda}), x\right) \in G\left((\lambda - T)^*\right)$ and $(g(\lambda), y) \in G\,(\lambda - T)$. The implies that

$$\langle x, g(\lambda)\rangle = \left\langle f(\overline{\lambda}), y\right\rangle.$$

Let us define that

$$h(\overline{\lambda}) = \begin{cases} \left\langle f(\overline{\lambda}), y\right\rangle & \text{if } \lambda \notin \mathbb{D}_r \\ \langle x, g(\lambda)\rangle & \text{if } \lambda \in \mathbb{D}_r. \end{cases}$$

It is clear that $h(\cdot)$ is an analytic function on $\mathbb{C}$. We assume that $h(\overline{\lambda}) = \left\langle f(\overline{\lambda}), y\right\rangle$. Since $f(\overline{\mu}) = (\overline{\mu} - T^*)^{-1}x$ for all $\overline{\mu} \in \rho(T^*)$, then $f(\overline{\mu}) \to 0$ as $|\overline{\mu}| \to +\infty$. This implies that $h(\overline{\mu}) \to 0$ as $|\overline{\mu}| \to +\infty$. By the vector-valued version of Liouville's theorem, we obtain $h = 0$. Now, based on $\|T\| < \infty$, implies that $\|T^*\| < \infty$. Hence, by using Lemma 3.1.2, we obtain

$$\overline{\mu}(\overline{\mu} - T^*)^{-1}x \to x \text{ as } |\overline{\mu}| \to +\infty.$$

Thus,

$$\begin{aligned} \langle x, y\rangle &= \left\langle \lim_{|\overline{\mu}| \to +\infty} \overline{\mu}(\overline{\mu} - T^*)^{-1}x, y\right\rangle \\ &= \left\langle \lim_{|\overline{\mu}| \to +\infty} \overline{\mu} f(\overline{\mu}), y\right\rangle \text{ (since } \overline{\mu} \in \rho(T^*)) \\ &= \lim_{|\overline{\mu}| \to +\infty} \overline{\mu}\,\langle f(\overline{\mu}), y\rangle \\ &= \lim_{|\overline{\mu}| \to +\infty} \overline{\mu} h(\overline{\mu}) \\ &= 0. \end{aligned}$$

This shows that $x \in \mathscr{K}(T)^{\perp}$. The case $h(\overline{\lambda}) = \langle x, g(\lambda)\rangle$ is handled similarly. Q.E.D.

As a direct consequence of Lemma 2.15.9, Theorems 3.8.3 and 3.8.12, we infer the following result.

Corollary 3.8.3 *Let $T \in C\mathscr{R}(H)$ be such that $\|T\| < \infty$. If $T^* = T^{\mathcal{Q}}$, then*

$$H_{T^{\mathcal{Q}}}(\{0\}) = \mathcal{Q}\left(H_T(\mathbb{C}\backslash\{0\})\right) \subseteq \mathscr{K}(T)^{\perp}. \qquad \diamond$$

3.8.5 Glocal Spectral Subspace

Let X be a Banach space and let $F \subset \mathbb{C}$ be a closed subset. If $T \in C\mathscr{R}(X)$, then we define the glocal spectral subspace $\mathscr{X}_T(F)$ as the set of all $x \in X$ such that there exists an analytic function $f : \mathbb{C} \setminus F \longrightarrow X$ checking

$$(\lambda I - T)f(\lambda) = x + T(0) \quad for\ all\ \lambda \in \mathbb{C} \setminus F.$$

Remark 3.8.6 Let X be a Banach space, $F \subset \mathbb{C}$ be a closed subset and let $T \in C\mathscr{R}(X)$. Then,

(i) $\mathscr{X}_T(F) \subset X_T(F)$.
(ii) Moreover, if T has the SVEP, then $\mathscr{X}_T(F) = X_T(F)$. $\diamond$

Now, let us study some basic properties of glocal spectral subspace of closed linear relations.

Proposition 3.8.9 *Let X be a Banach space, $T \in C\mathscr{R}(X)$ and let $F \subset \mathbb{C}$ be a closed subset.*

(i) $\mathscr{X}_T(F_1) \subset \mathscr{X}_T(F_2)$, for every closed subset $F_1 \subset F_2 \subset \mathbb{C}$.
(ii) For all $x \in T(0)$, we have $x \in \mathscr{X}_T(\emptyset)$.
(iii) $\mathscr{X}_T(F)$ is a linear subspace of X. $\diamond$

Proof (i) Let us assume that $x \in \mathscr{X}_T(F_1)$. Then, there exists an analytic function $f : \mathbb{C}\backslash F_1 \longrightarrow \mathscr{D}(T)$ such that

$$(\lambda - T)f(\lambda) = x + T(0), \text{ for all } \lambda \in \mathbb{C}\backslash F_1.$$

Using the fact that $F_1 \subset F_2$, we infer that the function g which a restriction of f on $\mathbb{C}\backslash F_2$, is analytic and

$$(\lambda - T)g(\lambda) = x + T(0), \text{ for all } \lambda \in \mathbb{C}\backslash F_2.$$

Hence, we conclude that $x \in \mathscr{X}_T(F_2)$.
(ii) Let $x \in T(0)$ and f be a null analytic function defined on $\mathbb{C}$. Then, by using Theorem 2.8.1 (i), we infer that

$$(\lambda - T)f(\lambda) = x + T(0), \text{ for all } \lambda \in \mathbb{C}.$$

This shows that $x \in \mathscr{X}_T(\emptyset)$.
(iii) It follows from (i) and (ii) that

$$T(0) \subset \mathscr{X}_T(F). \tag{3.8.35}$$

This implies that $\mathscr{X}_T(F) \neq \emptyset$. Let us assume $x \in \mathscr{X}_T(F)$ and $\alpha \in \mathbb{C}\backslash\{0\}$. Then, there exists an analytic function $f : \mathbb{C}\backslash F \longrightarrow \mathscr{D}(T)$ such that

$$(\lambda - T)f(\lambda) = x + T(0), \text{ for all } \lambda \in \mathbb{C}\backslash F.$$

Since $\alpha f \in H(\mathbb{C}\backslash F, \mathscr{D}(T))$, then for all $\lambda \in \mathbb{C}\backslash F$, we obtain

$$\begin{aligned} (\lambda - T)(\alpha f)(\lambda) &= \alpha((\lambda - T)f(\lambda)), \\ &= \alpha(x + T(0)), \\ &= \alpha x + T(0). \end{aligned}$$

This implies that $\alpha x \in \mathscr{X}_T(F)$, for all $\alpha \in \mathbb{C}\backslash\{0\}$. The use of (3.8.35) makes us conclude that $\alpha x \in \mathscr{X}_T(F)$, for all $\alpha \in \mathbb{C}$. Let $x, y \in \mathscr{X}_T(F)$. Then, there exists $f : \mathbb{C}\backslash F \longrightarrow X$ and $g : \mathbb{C}\backslash F \longrightarrow X$ such that

$$(\lambda - T)f(\lambda) = x + T(0)$$

and

$$(\lambda - T)g(\lambda) = y + T(0), \text{ for all } \lambda \in \mathbb{C}\backslash F.$$

Let $h : \mathbb{C}\backslash F \longrightarrow X$ be given by

$$h(\lambda) = (f + g)(\lambda), \text{ for all } \lambda \in \mathbb{C}\backslash F.$$

Then, $h(\cdot)$ is analytic function and satisfies

$$\begin{aligned} (\lambda - T)h(\lambda) &= (\lambda - T)f(\lambda) + (\lambda - T)g(\lambda), \\ &= x + T(0) + y + T(0) \\ &= x + y + T(0). \end{aligned}$$

This enables us to conclude that $x + y \in \mathscr{X}_T(F)$. Q.E.D.

Some basic properties of the glocal spectral subspace are gathered in the following proposition.

Proposition 3.8.10 *Let $T \in B\mathscr{R}(X)$ and $F \subset \mathbb{C}$ be a closed subset. Then,*

(i) $\mathscr{X}_T(F)$ is a subspace of X and $\mathscr{X}_T(F) \subset X_T(F)$,
(ii) $\mathscr{X}_T(F)$ is a linear hyper-invariant subspace for T, i.e., for every bounded operator S such that $TS = ST$, we have $S(\mathscr{X}_T(F)) \subseteq \mathscr{X}_T(F)$,
(iii) If, moreover, $T \in \mathscr{S}\mathscr{T}\mathscr{R}_1(X)$ and have a continuous selection, then

$$T(\mathscr{X}_T(F)) \subseteq \mathscr{X}_T(F). \qquad \diamondsuit$$

Proof Proceeding as in the proof of Proposition 3.8.4, we obtain the desired results. Q.E.D.

Theorem 3.8.13 *Let $T \in BC\mathscr{R}(X)$. Then,*

$$\mathscr{X}_T(\mathbb{D}(0, \varepsilon)) \supset H_\varepsilon(T) + T(0),$$

where

$$H_\varepsilon(T) := \left\{x \in X : \exists (x_n) \subset X \ \text{such that} \ x = x_0, \ x_{n+1} \in Tx_n \ \text{and} \ \limsup_{n\to\infty} \|x_n\|^{\frac{1}{n}} \leq \varepsilon\right\}.$$

If, moreover, $\sigma(T)$ is bounded, then the equality holds. $\diamondsuit$

Proof Let $x \in H_\varepsilon(T)$. Then, there exists $(x_n)_n \subset X$ such that $x = x_0, \ x_{n+1} \in Tx_n$ and

$$\limsup_{n\to\infty} \|x_n\|^{1/n} \leq \varepsilon.$$

Thus, the series defined by

$$f(\lambda) := \sum_{n\geq 1} \lambda^{-n} x_{n-1}$$

converges uniformly on $\mathbb{C} \setminus \mathbb{D}(0, \varepsilon)$. So, f is analytic on $\mathbb{C} \setminus \mathbb{D}(0, \varepsilon)$. Besides, using Lemma 2.11.2, we get for all λ such that $|\lambda| > \varepsilon$

$$\begin{aligned}(\lambda I - T)f(\lambda) &= \sum_{n\geq 0} \lambda^{-n} x_n - \sum_{n\geq 1} \lambda^{-n} x_n - T(0)\\ &= x + T(0).\end{aligned}$$

Then, $x \in \mathscr{X}_T(\mathbb{D}(0, \varepsilon))$. On the other hand, since $T(0) \subset \mathscr{X}_T(\mathbb{D}(0, \varepsilon))$ and $\mathscr{X}_T(\mathbb{D}(0, \varepsilon))$ is a subspace of X, then the inclusion

$$H_\varepsilon(T) + T(0) \subseteq \mathscr{X}_T(\mathbb{D}(0, \varepsilon))$$

is satisfied. Conversely, let $x \in \mathscr{X}_T(\mathbb{D}(0, \varepsilon))$. Then, there exists an analytic function $f : \mathbb{C} \setminus \mathbb{D}(0, \varepsilon) \longrightarrow X$ such that

$$(\lambda I - T)f(\lambda) = x + T(0)$$

holds for every $\lambda \in \mathbb{C} \setminus \mathbb{D}(0, \varepsilon)$. By assumption, we have $\sigma(T)$ is bounded. Hence, the set $V := (\mathbb{C} \setminus \mathbb{D}(0, \varepsilon)) \bigcap \rho(T)$ is not empty, open and the function f is analytic on V. Besides, for all $\lambda \in V$, we have

$$f(\lambda) = (\lambda I - T)^{-1} x. \tag{3.8.36}$$

Indeed, we have

$$(\lambda I - T) f(\lambda) = x + T(0).$$

Hence,

$$f(\lambda) + N(\lambda I - T) = (\lambda I - T)^{-1} x + N(\lambda I - T).$$

So,

$$f(\lambda) = (\lambda I - T)^{-1} x$$

and thus, we get the result. Now, let $\lambda \in V$. Using (3.8.36), we have

$$\lim_{|\lambda| \to \infty} f(\lambda) = 0.$$

Let consider the analytic function g defined by

$$g(\mu) := \begin{cases} f(\frac{1}{\mu}) \text{ if } 0 \neq \mu \in \mathbb{D}(0, 1/\varepsilon), \\ 0 \quad\quad \text{if } \mu = 0. \end{cases}$$

Since g is analytic on $\mathbb{D}(0, 1/\varepsilon)$ and $g(0) = 0$, then there exists a sequence $(x_n)_n \subset X$ such that $x_0 = 0$ and

$$g(\mu) = \sum_{n \geq 0} \mu^n x_n$$

holds even for all $\mathbb{D}(0, 1/\varepsilon)$. This shows that the radius of convergence of the power series representing $g(\mu)$ is greater then $1/\varepsilon$. Hence,

$$\limsup_{n \to \infty} \|x_n\|^{1/n} \leq \varepsilon.$$

Besides, we have

$$\begin{aligned} f(\lambda) &= g\left(\frac{1}{\lambda}\right) \\ &= \sum_{n \geq 0} \lambda^{-n} x_n. \end{aligned}$$

From Lemma 2.11.3, for all $n \in \mathbb{N}$, there exists $\alpha_{n+1} \in T x_n$ such that for all λ with $|\lambda| > \varepsilon$, we have

$$\begin{aligned}(\lambda I - T)f(\lambda) &= \sum_{n\geq 0}\lambda^{-n+1}x_n - \sum_{n\geq 0}\lambda^{-n}\alpha_{n+1} + T(0)\\ &= \sum_{n\geq 0}\lambda^{-n}(x_{n+1} - \alpha_{n+1}) + T(0).\end{aligned}$$

Thus,

$$Q_T(x) = \sum_{n\geq 0}\lambda^{-n}Q_T(x_{n+1} - \alpha_{n+1}).$$

Hence,

$$Q_T(x_1 - \alpha_1) = Q_T(x)$$

and

$$Q_T(x_{n+1} - \alpha_{n+1}) = 0, \quad n \geq 1.$$

So, $x = x_1 + \alpha$, with $\alpha \in T(0)$ and $x_{n+1} \in Tx_n, \quad n \geq 1$. Now, let prove that $x_1 \in H_\varepsilon(T)$. For this, consider the sequence $(y_n)_n$ defined by $y_n := x_{n+1}$. We have $x_1 = y_0$ and $y_{n+1} = x_{n+2} \in Tx_{n+1} = Ty_n$. Besides, we have

$$\begin{aligned}\limsup_{n\to\infty}\|y_n\|^{1/n} &\leq \limsup_{n\to\infty}\|x_{n+1}\|^{1/n}\\ &\leq \varepsilon.\end{aligned}$$

So, $x_1 \in H_\varepsilon(T)$ and therefore, $\mathscr{X}_T(\mathbb{D}(0,\varepsilon)) \subseteq H_\varepsilon(T) + T(0)$. Q.E.D.

As a consequence of Theorem 3.8.13, we prove that the quasi-nilpotent part of a linear relation may be characterized in terms of the glocal spectral subspace as follows.

Corollary 3.8.4 *For every $T \in BC\mathscr{R}(X)$, we have*

$$\mathscr{X}_T(\{0\}) \supset H_0(T) + T(0).$$

If, moreover, $\sigma(T)$ is bounded, then the equality holds. ◇

In the following result, we shall find the defect spectrum by means the glocal spectral subspace.

Proposition 3.8.11 *Let X be a Banach space, $T \in BC\mathscr{R}(X)$ be such that T has the SVEP and let $F \subset \mathbb{C}$ be a closed subset. If $\mathscr{X}_T(F) = X$, then $\sigma_\delta(T) \subseteq F$.* ◇

Proof Let us assume that $\lambda \notin F$. The fact that $\mathscr{X}_T(F) = X$ and T has the SVEP implies from Remark 3.8.6 (ii) that $X_T(F) = X$. Hence, it follows from Lemma 3.8.3 that $X \subset (\lambda - T)X$. Therefore, we infer that $\lambda - T$ is onto. This enables us to conclude that $\sigma_\delta(T) \subseteq F$. Q.E.D.

Theorem 3.8.14 *Let X be a Banach space. If $T \in C\mathscr{R}(X)$ has a continuous selection A, then*

$$\mathscr{X}_T(\overline{\mathbb{D}_\varepsilon}) = \left\{x \in X \text{ such that } \limsup_{n\to\infty}\|A^n x\|^{1/n} \leq \varepsilon\right\}. \quad ◇$$

Proof Let $x \in X$ such that $\delta := \limsup_{n \to \infty} \|A^n x\|^{1/n}$. The series

$$f(\lambda) = \sum_{n=1}^{\infty} \lambda^{-n} A^{n-1}x, \text{ for all } \lambda \in \mathbb{C}\backslash\overline{\mathbb{D}_\varepsilon}$$

converges locally uniformly, so it defines an analytic vector-valued function on the open set $\mathbb{C}\backslash\overline{\mathbb{D}_\varepsilon}$. Let

$$f_N(\lambda) = \sum_{n=1}^{N} \lambda^{-n} A^{n-1}x, \text{ for all } \lambda \in \mathbb{C}\backslash\overline{\mathbb{D}_\varepsilon}.$$

Let us prove that $f(\lambda) \in \mathscr{D}(T)$. If $M < N$, then

$$\begin{aligned} \|Tf_N(\lambda) - Tf_M(\lambda)\| &= \left\| T\left(\sum_{n=1}^{N} \lambda^{-n} A^{n-1}x - \sum_{n=1}^{M} \lambda^{-n} A^{n-1}x\right)\right\|, \\ &= \left\| T\left(\sum_{n=M+1}^{N} \lambda^{-n} A^{n-1}x\right)\right\|. \end{aligned}$$

The fact that A is a selection of T implies that

$$\begin{aligned} \|Tf_N(\lambda) - Tf_M(\lambda)\| &= \left\| A\left(\sum_{n=M+1}^{N} \lambda^{-n} A^{n-1}x\right) + T(0)\right\|, \\ &\leq \|A\| \left\|\sum_{n=M+1}^{N} \lambda^{-n} A^{n-1}x\right\| + \|T(0)\|. \end{aligned}$$

Hence, it follows from (ii) of Lemma 2.10.2 that

$$\|Tf_N(\lambda) - Tf_M(\lambda)\| \leq \|A\| \left\|\sum_{n=M+1}^{N} \lambda^{-n} A^{n-1}x\right\|.$$

This implies that

$$\|Tf_M(\lambda) - Tf_N(\lambda)\| \to 0 \text{ when } M,\ N \to \infty. \tag{3.8.37}$$

The fact that

$$\|f_N(\lambda) - f_M(\lambda)\| = \left\|\sum_{n=M+1}^{N} \lambda^{-n} A^{n-1}x\right\| \to 0 \text{ when } M,\ N \to \infty$$

implies from (3.8.37) that

$$\|f_N(\lambda) - f_M(\lambda)\|_T \to 0 \text{ when } M,\ N \to \infty.$$

This leads to $(f_N)_N$ is a Cauchy sequence. Therefore, we deduce from Remark 2.11.6 that $\lim_{N\to\infty} f_N = f(\lambda) \in \mathscr{D}(T)$. For all $\lambda \in \mathbb{C}\backslash\overline{\mathbb{D}_\varepsilon}$, we have

$$\begin{aligned}(\lambda - T)f_N(\lambda) &= (\lambda - T)\left(\sum_{n=1}^{N}\lambda^{-n}A^{n-1}x\right),\\ &= \sum_{n=1}^{N}\lambda^{-(n-1)}\,A^{n-1}x - T\left(\sum_{n=1}^{N}\lambda^{-n}A^{n-1}x\right),\\ &= \sum_{n=1}^{N}\lambda^{-(n-1)}\,A^{n-1}x - \sum_{n=1}^{N}\lambda^{-n}A(A^{n-1}x) + T(0),\\ &= \sum_{n=0}^{N-1}\lambda^{-n}\,A^{n}x - \sum_{n=1}^{N}\lambda^{-n}A^{n}x + T(0)\\ &= x + T(0) - \lambda^{-N}A^{N}x.\end{aligned}$$

Since $f(\lambda)$ converges for all $\lambda \in \mathbb{C}\backslash\overline{\mathbb{D}_\varepsilon}$, then $\lambda^{-N}A^N x \to 0$ when $n \to \infty$. This implies that

$$(\lambda - T)f(\lambda) = x + T(0),\ \text{for all } \lambda \in \mathbb{C}\backslash\overline{\mathbb{D}_\varepsilon}.$$

This enables us to conclude that $x \in \mathscr{X}_T(\overline{\mathbb{D}_\varepsilon})$. Conversely, let $x \in \mathscr{X}_T(\overline{\mathbb{D}_\varepsilon})$. Then, there exists an analytic function $f : \mathbb{C}\backslash\overline{\mathbb{D}_\varepsilon} \longrightarrow \mathscr{D}(T)$ such that

$$(\lambda - T)f(\lambda) = x + T(0) \text{ for all } \lambda \in \mathbb{C}\backslash\overline{\mathbb{D}_\varepsilon}.$$

This leads to

$$(\lambda - A)f(\lambda) + T(0) = x + T(0) \text{ for all } \lambda \in \mathbb{C}\backslash\overline{\mathbb{D}_\varepsilon}.$$

For $|\lambda| > \max\{\varepsilon, \|A\|\}$, we infer that

$$\begin{aligned}f(\lambda) &= (\lambda - A)^{-1}x\\ &= \sum_{n=1}^{\infty}\lambda^{-n}A^{n-1}x,\end{aligned}$$

and therefore, $\lim_{|\lambda|\to\infty} f(\lambda) = 0$. We consider the analytic function $g : \mathbb{D}_{\varepsilon^{-1}} \longrightarrow \mathscr{D}(T)$ which that defined by

$$g(\mu) = \begin{cases} f\left(\frac{1}{\lambda}\right) & \text{if } \mu \in \mathbb{D}_{\frac{1}{\varepsilon}}\backslash\{0\} \\ 0 & \text{if } \mu = 0. \end{cases}$$

Consequently,

$$g(\mu) = \sum_{n=1}^{\infty} \mu^n A^{n-1} x, \text{ for all } |\mu| < \frac{1}{\max\{\varepsilon, \|A\|\}}.$$

The final part of the proof is almost identical to that [8, Theorem 27.2.1]. Q.E.D.

3.9 Some Properties of the Onto Spectrum of a Linear Relation

3.9.1 Characterization of the Onto Spectrum

The onto spectrum may be characterized by means of the local spectrum as follows.

Theorem 3.9.1 *Let $T \in BC\mathscr{R}(X)$. Then,*

(i) $\sigma_\delta(T) = \bigcup_{x \in X} \sigma_T(x)$,

(ii) The set $\{x \in X$ such that $\sigma_T(x) = \sigma_\delta(T)\}$ is of the second category in X. ◇

Proof (*i*) First observe that the desired result is equivalent to $\mathbb{C} \setminus \sigma_\delta(T) = \bigcap_{x \in X} \rho_T(x)$. For the opposite inclusion, let $\lambda \in \bigcap_{x \in X} \rho_T(x)$. Then, $\lambda \in \rho_T(x)$ for all $x \in X$. Let $x \in X$. There exists an open neighborhood U_λ of λ and an analytic function $f_x : U_\lambda \longrightarrow X$ such that

$$(\mu I - T) f_x(\mu) = x + T(0) \ \forall \ \mu \in U_\lambda.$$

This implies that

$$x \in (\lambda I - T) f_x(\lambda) \subset R(\lambda I - T) \text{ for all } x \in X.$$

Then, $\lambda I - T$ is onto. Thus, $\lambda \in \mathbb{C} \setminus \sigma_\delta(T)$. For the direct inclusion, let $\lambda \in \mathbb{C} \setminus \sigma_\delta(T)$. Then, $\lambda I - T$ is onto. So, by using Theorem 3.8.11, we have $\mathscr{K}(\lambda I - T) = X$ and we obtain that $0 \in \rho_{\lambda I - T}(x)$. So, there exist an open neighborhood U_0 of 0 and an analytic function $f : U_0 \longrightarrow X$ such that

$$((\mu - \lambda)I + T) f(\mu) = x + T(0) \ \forall \mu \in U_0.$$

Therefore,

$$(\delta I - T) g(\delta) = x + T(0) \ \forall \delta \in U_\lambda,$$

where U_λ is the open neighborhood of λ given by $U_\lambda = \lambda - U_0$ and g is the analytic function defined on U_λ by

$$g(\delta) = -f(\lambda - \delta).$$

Therefore, $\lambda \in \rho_T(x)$ for all $x \in X$ which ends the proof of (i).
(ii) Let E be a dense countable subset of $\sigma_\delta(T)$. Then, for each $\lambda \in E$, we have $(\lambda I - T)X \neq X$. We claim that $(\lambda I - T)X$ is of the first category in X. Indeed, let us suppose that $(\lambda I - T)X$ is of the second category in X. We show that $(\lambda I - T)X = X$, which is absurd. To do this, it suffices to prove that $\lambda I - T$ is an open mapping. Let U be the open ball in X with center 0 and radius $r > 0$. We prove that $(\lambda I - T)(U)$ contains a neighborhood of 0 in X. Define,

$$U_n := \{x \in X \text{ such that} \|x\| < 2^{-n} r\} \quad (n = 0, 1, 2, \ldots).$$

We note that $U_1 \supset U_2 - U_2$. So,

$$(\lambda I - T)U_1 \supset (\lambda I - T)U_2 - (\lambda I - T)U_2.$$

Hence,

$$\overline{(\lambda I - T)U_1} \supset \overline{(\lambda I - T)U_2 - (\lambda I - T)U_2} \supset \overline{(\lambda I - T)U_2} - \overline{(\lambda I - T)U_2}. \tag{3.9.1}$$

On the other hand, we have

$$(\lambda I - T)X = \bigcup_{k=1}^{\infty} k(\lambda I - T)(U_2).$$

As, the union of countably many first category sets is first category and as $(\lambda I - T)X$ is of the second category, then at least one $k(\lambda I - T)(U_2)$ is of the second category of X and so, $(\lambda I - T)(U_2)$ is of the second category. Hence, $int(\overline{(\lambda I - T)(U_2)}) \neq \emptyset$. Thus, by (3.9.1), there is some neighborhood W of 0 in X such that

$$W \subset \overline{(\lambda I - T)(U_1)}.$$

We shall now show that $\overline{(\lambda I - T)(U_1)} \subset (\lambda I - T)(U)$. Fix $y_1 \in \overline{(\lambda I - T)(U_1)}$. As, what just proved for U_1 holds by the same way for U_2, so $\overline{(\lambda I - T)(U_2)}$ contains a neighborhood of 0. Thus,

$$\left(y_1 - \overline{(\lambda I - T)(U_2)}\right) \bigcap (\lambda I - T)(U_1) \neq \emptyset.$$

Hence, there exists $\alpha_1 \in (\lambda I - T)(x_1)$ with $x_1 \in U_1$ such that

$$y_2 = y_1 - \alpha_1 \in \overline{(\lambda I - T)(U_2)}.$$

Proceeding by induction we can construct the sequences $(\alpha_n)_{n\geq 1}$, $(x_n)_{n\geq 1}$ and $(y_n)_{n\geq 1}$ such that for all $n \geq 1$, $x_n \in U_n$, $\alpha_n \in (\lambda I - T)(x_n)$ and

$$y_{n+1} = y_n - \alpha_n \in \overline{(\lambda I - T)(U_{n+1})}.$$

Since, $\|x_n\| < \frac{r}{2^n}$, then

$$\sum_{n\geq 1} x_n$$

converges. Let

$$x = \sum_{n=1}^{\infty} x_n.$$

Then, $\|x\| < r$ and so, $x \in U$. By the construction of the sequences $(y_n)_{n\geq 1}$ and $(\alpha_n)_{n\geq 1}$, we have

$$\begin{aligned} \sum_{n=1}^{m} Q_T \alpha_n &= \sum_{n=1}^{m} Q_T(y_n - y_{n+1}) \\ &= Q_T y_1 - Q_T y_{m+1}. \end{aligned}$$

So,

$$\sum_{n=1}^{m} Q_T(\lambda I - T)x_n = Q_T y_1 - Q_T y_{m+1}.$$

Thus,

$$Q_T(\lambda I - T)\left(\sum_{n=1}^{m} x_n\right) = Q_T y_1 - Q_T y_{m+1}.$$

Now, we claim that $Q_T y_m \to 0$ as $m \to \infty$. Indeed, let $\alpha \in (\lambda I - T)(U_1)$. Then, there exists $\beta \in U_1$ such that $\alpha \in (\lambda I - T)(\beta)$. Hence,

$$\begin{aligned} \text{dist}(\alpha, T(0)) &= \|(\lambda I - T)(\beta)\| \\ &\leq \|(\lambda I - T)\|\|\beta\|. \end{aligned}$$

Thus, we obtain

$$(\lambda I - T)(U_1) \subset \left\{\alpha \in X \text{ such that } \text{dist}(\alpha, T(0)) \leq \|(\lambda I - T)\|\frac{r}{2^n}\right\}.$$

But, the map $\alpha \longrightarrow \text{dist}(\alpha, T(0))$ is continuous on X. Then, the set

$$\left\{\alpha \in X \text{ such that } \text{dist}(\alpha, T(0)) \leq \|(\lambda I - T)\|\frac{r}{2^n}\right\}$$

is closed and, we get

$$y_n \in \overline{(\lambda I - T)(U_n)} \subset \left\{\alpha \in X \text{ such that } \operatorname{dist}(\alpha, T(0)) \leq \|(\lambda I - T)\|\frac{r}{2^n}\right\}.$$

So,

$$\operatorname{dist}(y_n, T(0)) \leq \|(\lambda I - T)\|\frac{r}{2^n}.$$

Therefore,

$$\|Q_T y_n\| \to 0 \text{ as } n \to \infty.$$

Then, we conclude that

$$Q_T(\lambda I - T)(x) = Q_T(y_1)$$

and hence $y_1 \in (\lambda I - T)(x)$ which confirm our claim. Now, since, for all $\lambda \in E$, we have $(\lambda I - T)X$ is of the first category, then

$$F = \bigcup_{\lambda \in E} (\lambda I - T)X$$

is also of the first category. Thus, $X \backslash F$ is of the second category. We claim that for all $x \in X \backslash F$, $\sigma_\delta(T) \subset \sigma_T(x)$. Indeed, let $x \in X \backslash F$, then $E \subset \sigma_T(x)$. So,

$$\begin{aligned}\sigma_\delta(T) &= \overline{E}\\ &\subset \overline{\sigma_T(x)}\\ &= \sigma_T(x).\end{aligned}$$

Hence,

$$\sigma_\delta(T) = \sigma_T(x).$$

Therefore, $\{x \in X \text{ such that } \sigma_T(x) = \sigma_\delta(T)\}$ is of the second category in X.Q.E.D.

We can now state the main result of this section which is a sufficient condition to ensure that the onto spectrum $\sigma_\delta(T)$ is not cluster at a point λ.

Theorem 3.9.2 *Let* $T \in BC\mathscr{R}(X)$. *If* $X = H_0(\lambda I - T) + \mathscr{K}(\lambda I - T)$, *then* $\sigma_\delta(T)$ *does not cluster at* λ. ◇

Proof Without loss of generality, we can suppose that $\lambda = 0$. Suppose that $0 \in \sigma_\delta(T)$ and $X = H_0(T) + \mathscr{K}(T)$. Then, every $x \in X$ may be written as $x = x_1 + x_2$, where $x_1 \in H_0(T)$ and $x_2 \in \mathscr{K}(T)$. From Corollary 3.8.4, we have $\sigma_T(x_1) \subseteq \{0\}$. Therefore,

$$\begin{aligned}\sigma_T(x) &\subseteq \sigma_T(x_1) \bigcup \sigma_T(x_2)\\ &\subseteq \{0\} \bigcup \sigma_T(x_2).\end{aligned}$$

Now, by Theorem 3.8.11 and since $\sigma_T(x_2)$ is closed, we conclude that 0 is isolated in $\sigma_T(x)$ for any $x \in X$. Using Theorem 3.9.1, we conclude that there exists $x_0 \in X$ such that

$$\sigma_T(x_0) = \sigma_\delta(T).$$

Hence, 0 is isolated in $\sigma_\delta(T)$. Q.E.D.

We close this section by stating the next corollary which is related to Theorems 3.2.1 and 3.9.2.

Corollary 3.9.1 *Let $T \in BC\mathscr{R}(X)$. If $X = H_0(\lambda I - T) + \mathscr{K}(\lambda I - T)$, then $\sigma_T(x)$ does not cluster at λ for every $x \in X$.* ◇

3.9.2 Some Results of an Upper Triangular Linear Relation Matrix $\mathscr{M}_C$ Involving SVEP

The basic objective of this subsection resides in displaying and assessing some results of an upper triangular linear relation matrix $\mathscr{M}_C$ involving SVEP. In addition, we exhibit some necessary conditions such that $\mathscr{M}_C$ has the SVEP.

Proposition 3.9.1 *Let $A \in L\mathscr{R}(X)$, $B \in L\mathscr{R}(Y)$ and $C \in L\mathscr{R}(Y, X)$. If $A(0) \subset C(0)$, then for all $x \in V$, we have*

$$S_B \bigcup \sigma_A(x) = S_B \bigcup \sigma_{\mathscr{M}_C}\left(x \bigoplus 0\right),$$

where $\mathscr{M}_C$ is given in (2.9.1). ◇

Proof Let $\lambda \notin S_B \bigcup \sigma_{\mathscr{M}_C}\left(x \bigoplus 0\right)$. Then, there exist an open neighborhood V of λ in $\mathbb{C}$ and a nonzero analytic operator function $f : V \longrightarrow X \bigoplus Y$ satisfying

$$(\mathscr{M}_C - \mu)f(\mu) = \left(x \bigoplus 0\right) + \mathscr{M}_C(0), \quad \text{for all } \mu \in V.$$

Let $f_1 : V \longrightarrow X$ and $f_2 : V \longrightarrow Y$ be analytic operators functions such that $f = f_1 \bigoplus f_2$. Hence, we get

$$(A - \mu)f_1(\mu) + Cf_2(\mu) = x + A(0) \text{ and } (B - \mu)f_2(\mu) = B(0), \quad \text{for all } \mu \in V.$$

Since $\lambda \notin S_B$, we can deduce that $f_2 = 0$ on V. Moreover, based on the assumption $A(0) \subset C(0)$, we get

$$(A - \mu)f_1(\mu) = x + A(0), \quad \text{for all } \mu \in V.$$

Consequently, we obtain $\lambda \notin \sigma_A(x)$. The reverse inclusion is clear. Q.E.D.

The following result can be easily derived from Proposition 3.9.1.

Corollary 3.9.2 *Let $A \in L\mathscr{R}(X)$, $B \in L\mathscr{R}(Y)$ and $C \in L\mathscr{R}(Y, X)$. If B has the SVEP and $A(0) \subset C(0)$, then for all $x \in V$, we have*

$$\sigma_A(x) = \sigma_{\mathscr{M}_C}\left(x \bigoplus 0\right),$$

where $\mathscr{M}_C$ is given in (2.9.1). ◇

Proof By referring to Remark 3.8.5, we infer that B has the SVEP if, and only if, $S_B = \emptyset$. As a result, $\sigma_A(x) = \sigma_{\mathscr{M}_C}\left(x \bigoplus 0\right)$, as desired. Q.E.D.

Proposition 3.9.2 *Let $A \in L\mathscr{R}(X)$, $B \in L\mathscr{R}(Y)$ and $C \in L\mathscr{R}(Y, X)$ be satisfying $A(0) \subset C(0)$. If A has a selection E which satisfies $A(0) \subset R(E)$ and B has the SVEP, then*

$$\sigma_\delta(\mathscr{M}_C) = \sigma_\delta(A) \bigcup \sigma_\delta(B),$$

where $\mathscr{M}_C$ is given in (2.9.1). ◇

Proof Relying upon Corollary 3.9.2, we obtain

$$\sigma_A(x) = \sigma_{\mathscr{M}_C}\left(x \bigoplus 0\right).$$

Thus,

$$\begin{aligned}\sigma_\delta(A) &= \bigcup_{y \in X} \sigma_A(y) \\ &= \bigcup_{y \in X} \sigma_{\mathscr{M}_C}\left(y \bigoplus 0\right) \\ &\subset \sigma_\delta(\mathscr{M}_C).\end{aligned}$$

On the other side, by using Remark 3.4.2 (ii), we get

$$\begin{aligned}\sigma_\delta(B) &= \bigcup_{z \in Y} \sigma_B(z) \\ &= \bigcup_{z \in Y} \sigma_{\mathscr{M}_C}\left(0 \bigoplus z\right) \\ &\subset \sigma_\delta(\mathscr{M}_C).\end{aligned}$$

Hence, we have

$$\sigma_\delta(A) \bigcup \sigma_\delta(B) \subset \sigma_\delta(\mathscr{M}_C).$$

Conversely, it follows from Corollary 3.4.1 that

$$\sigma_\delta(\mathscr{M}_C) \subset \sigma_\delta(A)\bigcup\sigma_\delta(B).$$

As a result, $\sigma_\delta(\mathscr{M}_C) = \sigma_\delta(A)\bigcup\sigma_\delta(B)$, as desired. Q.E.D.

Theorem 3.9.3 *Let $A \in L\mathscr{R}(X)$, $B \in L\mathscr{R}(Y)$ and $C \in L\mathscr{R}(Y, X)$ be satisfy $A(0) \subset C(0)$. If A has a selection E satisfying $A(0) \subset R(E)$ and B has the SVEP, then*

$$\sigma(\mathscr{M}_C) = \sigma(A)\bigcup\sigma(B),$$

where $\mathscr{M}_C$ is given in (2.9.1). ◇

Proof It follows from Proposition 3.9.2 that

$$\sigma_\delta(\mathscr{M}_C) = \sigma_\delta(A)\bigcup\sigma_\delta(B).$$

Based on the assumption B which has the SVEP, we can deduce that

$$\sigma_\delta(B) = \sigma(B).$$

Hence, we have

$$\sigma_\delta(A)\bigcup\sigma(B) \subset \sigma(\mathscr{M}_C).$$

This is equivalent to

$$S_A\bigcup\sigma_\delta(A)\bigcup\sigma(B) \subset S_A\bigcup\sigma(\mathscr{M}_C).$$

Therefore, by using Corollary 3.9.2, we get

$$\begin{aligned} S_A\bigcup\sigma_\delta(A)\bigcup\sigma(B) &= \sigma(A)\bigcup\sigma(B) \\ &\subset S_A\bigcup\sigma(\mathscr{M}_C). \end{aligned}$$

Since $S_A \subset S_{\mathscr{M}_C} \subset \sigma(\mathscr{M}_C)$, it follows that

$$\sigma(A)\bigcup\sigma(B) \subset \sigma(\mathscr{M}_C).$$

Conversely, referring to Remark 3.4.2 (iii), we infer that $\sigma_\delta(\mathscr{M}_C) \subset \sigma_\delta(A)\bigcup\sigma_\delta(B)$, where $\mathscr{M}_C$ is given in (2.9.1). Q.E.D.

Proposition 3.9.3 *Let $A \in L\mathscr{R}(X)$, $B \in L\mathscr{R}(Y)$ and $C \in L\mathscr{R}(Y, X)$ be satisfy $A(0) \subset C(0)$. If A and B have the SVEP, then $\mathscr{M}_C$ also has the SVEP, where $\mathscr{M}_C$ is given in (2.9.1).* ◇

Proof Let $f : V \longrightarrow X\bigoplus Y$ be an analytic operator function on an open neighborhood V of λ in $\mathbb{C}$ satisfying

$$(\mathcal{M}_C - \mu)f(\mu) = \mathcal{M}_C(0), \quad \text{for all } \mu \in V.$$

Let $f = f_1 \bigoplus f_2$ be analytic operators functions, with $f_1 : V_\lambda \longrightarrow X$ and $f_2 : V_\lambda \longrightarrow Y$. Therefore,

$$(A - \mu)f_1(\mu) + Cf_2(\mu) = A(0)$$

and

$$(B - \mu)f_2(\mu) = B(0), \quad \text{for all } \mu \in V_\lambda.$$

Since B has the SVEP, then $f_2(\mu) = 0$ on V. Hence, we obtain

$$(A - \mu)f_1(\mu) = A(0), \quad \text{for all } \mu \in V.$$

The fact that A has the SVEP, allows us to conclude that $f_1(\mu) = 0$ on V. Thus, $\mathcal{M}_C$ has the SVEP. Q.E.D.

Corollary 3.9.3 *Let $A \in L\mathcal{R}(X)$, $B \in L\mathcal{R}(Y)$ and $C \in L\mathcal{R}(Y, X)$ be satisfying $A(0) \subset C(0)$. If A has a selection E satisfying $A(0) \subset R(E)$, A and B have the SVEP, then*

$$\sigma(\mathcal{M}_C) = \sigma(A) \bigcup \sigma(B),$$

where $\mathcal{M}_C$ is given in (2.9.1). ◇

Proof On the one hand, using Proposition 3.9.2, we deduce that

$$\sigma_\delta(\mathcal{M}_C) = \sigma_\delta(A) \bigcup \sigma_\delta(B). \tag{3.9.2}$$

On the other hand, based on the assumption that A and B have the SVEP, we infer that $S_A = S_B = \emptyset$, which implies

$$\sigma_\delta(A) = \sigma(A) \quad \text{and} \quad \sigma_\delta(B) = \sigma(B). \tag{3.9.3}$$

Therefore, again grounded on the assumption that A and B have the SVEP, and according to Proposition 3.9.3, we conclude that $\mathcal{M}_C$ also has the SVEP. This means that

$$\sigma_\delta(\mathcal{M}_C) = \sigma(\mathcal{M}_C). \tag{3.9.4}$$

Thus, combining (3.9.2), (3.9.3) and (3.9.4), we get

$$\sigma(\mathcal{M}_C) = \sigma(A) \bigcup \sigma(B).$$

This achieves the proof. Q.E.D.

Proposition 3.9.4 *Let $A \in L\mathscr{R}(X)$, $B \in L\mathscr{R}(Y)$ and $C \in L\mathscr{R}(Y, X)$. If A has a selection E satisfying $A(0) \subset R(E)$ and $A(0) \subset C(0)$, then*

$$S_B \bigcup \sigma_\delta(\mathscr{M}_C) = \sigma_\delta(A) \bigcup \sigma(B),$$

where $\mathscr{M}_C$ is given in (2.9.1). ◇

Proof Applying Corollary 3.4.1, we have

$$\sigma_\delta(\mathscr{M}_C) \subset \sigma_\delta(A) \bigcup \sigma_\delta(B).$$

This implies that

$$S_B \bigcup \sigma_\delta(\mathscr{M}_C) \subset \sigma_\delta(A) \bigcup \sigma(B).$$

For the inverse inclusion, it follows from Proposition 3.9.1 that

$$S_B \bigcup \sigma_A(x) = S_B \bigcup \sigma_{\mathscr{M}_C}\left(x \bigoplus 0\right), \quad \text{for all } x \in V.$$

Therefore,

$$S_B \bigcup \sigma_\delta(A) \subset S_B \bigcup \sigma_\delta(\mathscr{M}_C).$$

Since $\sigma_\delta(B) \subset \sigma_\delta(\mathscr{M}_C)$, then we get

$$\begin{aligned} \sigma_\delta(B) \bigcup S_B \bigcup \sigma_\delta(A) = & \ \sigma_\delta(A) \bigcup \sigma(B) \\ \subset & \ S_B \bigcup \sigma_\delta(\mathscr{M}_C). \end{aligned}$$

This achieves the proof. Q.E.D.

We are now in a position to provide another characterization of our main result using the adjoint of linear relations matrices $\mathscr{M}_C$.

Theorem 3.9.4 *Let $A \in BC\mathscr{R}(X)$, $B \in BC\mathscr{R}(Y)$ and let C be a bounded linear operator. If A has a selection E satisfying $A(0) \subset R(E)$, then*

$$\left(S_{A^*} \bigcap S_B\right) \bigcup \sigma(\mathscr{M}_C) = \sigma(A) \bigcup \sigma(B),$$

where $\mathscr{M}_C$ is given in (2.9.1). ◇

Proof We divide this proof into two steps.

Step1. We prove that

$$S_B \bigcup \sigma(\mathscr{M}_C) = \sigma(A) \bigcup \sigma(B). \tag{3.9.5}$$

Hence, Remark 3.4.2 (iii) and Proposition 3.9.4 yield that

$$S_A \bigcup \sigma_\delta(A) \bigcup \sigma(B) \subset S_B \bigcup \sigma_\delta(\mathcal{M}_C) \bigcup S_{\mathcal{M}_C}.$$

As a consequence,

$$\sigma(A) \bigcup \sigma(B) \subset S_B \bigcup \sigma(\mathcal{M}_C).$$

Conversely, since $\sigma(\mathcal{M}_C) \subset \sigma(A) \bigcup \sigma(B)$, it follows that

$$S_B \bigcup \sigma(\mathcal{M}_C) \subset \sigma(A) \bigcup \sigma(B).$$

Step2. In view of Theorem 2.15.3, the adjoint of $\mathcal{M}_C$ is

$$\mathcal{M}_C{}^* = \begin{pmatrix} A^* & 0 \\ C^* & B^* \end{pmatrix}.$$

Proceeding in the same way, we obtain

$$S_{A^*} \bigcup \sigma(\mathcal{M}_C^*) = \sigma(A^*) \bigcup \sigma(B^*).$$

Using Proposition 2.11.2, we get

$$S_{A^*} \bigcup \sigma(\mathcal{M}_C) = \sigma(A) \bigcup \sigma(B). \tag{3.9.6}$$

Therefore, combining (3.9.5) and (3.9.6), we get the desired result. Q.E.D.

In what follows, we shall present a new condition under which we have the desired equality.

Corollary 3.9.4 *Let $A \in BC\mathcal{R}(X)$, $B \in BC\mathcal{R}(Y)$ and C be a bounded linear operator. If A has a selection E satisfying $A(0) \subset R(E)$ and $S_{A^*} \bigcap S_B = \emptyset$, then*

$$\sigma(\mathcal{M}_C) = \sigma(A) \bigcup \sigma(B).$$ ◇

In the following result, we shall give some necessary conditions such that $\mathcal{M}_C$ has the SVEP.

Theorem 3.9.5 *Let $A \in L\mathcal{R}(X)$, $B \in L\mathcal{R}(Y)$ and $C \in L\mathcal{R}(Y, X)$ be such that the onto spectrum of A has an emply interior and $A(0) \subset C(0)$. If A has a selection E satisfying $A(0) \subset R(E)$, then $S_{\mathcal{M}_C} = \emptyset$ if, and only if, $S_A = S_B = \emptyset$, where $\mathcal{M}_C$ is given in (2.9.1). In particular, $\mathcal{M}_C$ has the SVEP if, and only if, A and B have the SVEP.* ◇

Proof If A and B have the SVEP, then by using Proposition 3.9.3, we obtain $\mathcal{M}_C$ has the SVEP. Conversely, since

$$\sigma(B) \subset \sigma_\delta(A) \bigcup \sigma(\mathcal{M}_C)$$

and

$$\begin{aligned} S_A &\subset S_{\mathcal{M}_C} \\ &\subset S_A \bigcup S_B, \end{aligned}$$

we deduce that $S_A = S_B = \emptyset$. This achieves the proof. Q.E.D.

The following proposition provides a characterization for closedness of local analytical subspace.

Proposition 3.9.5 *Let $A \in L\mathcal{R}(X)$, $B \in L\mathcal{R}(Y)$, $C \in L\mathcal{R}(Y, X)$ be satisfying $A(0) \subset C(0)$ and let F be a closed set satisfying $S_B \subset F$. If the local analytical subspace $(X \bigoplus Y)_{\mathcal{M}_C}(F)$ is closed, then $X_A(F)$ is closed, where $\mathcal{M}_C$ is given in (2.9.1).* ◇

Proof Let $(x_n)_n$ be a sequence of elements of $X_A(F)$ which converges to $x \in X$. In order to demonstrate that $X_A(F)$ is closed, we need to show that $x \in X_A(F)$. Applying Proposition 3.9.1, we get

$$S_B \bigcup \sigma_A(x_n) = S_B \bigcup \sigma_{\mathcal{M}_C}\left(x_n \bigoplus 0\right), \quad \text{for all } x \in X.$$

This implies that

$$\sigma_{\mathcal{M}_C}\left(x_n \bigoplus 0\right) \subset F \bigcup S_B.$$

In addition, it follows from the assumption $S_B \subset F$ that

$$\sigma_{\mathcal{M}_C}\left(x_n \bigoplus 0\right) \subset F.$$

Since $(X \bigoplus Y)_{\mathcal{M}_C}(F)$ is closed, we get $\sigma_{\mathcal{M}_C}\left(x \bigoplus 0\right) \subset F$. Moreover, we conclude that

$$\sigma_A(x) \subset S_B \bigcup \sigma_A(x) = S_B \bigcup \sigma_{\mathcal{M}_C}\left(x \bigoplus 0\right).$$

Hence, we infer that

$$\sigma_A(x) \subset S_B \bigcup F \subset F.$$

This means that $x \in X_A(F)$. Q.E.D.

3.9.3 Local Spectral Properties of $\mathcal{L}_0$

Now, our main purpose is to characterize the local spectral and the spectral properties of the block matrices $\mathcal{L}_0$ of linear relations, where the entries are unbounded linear relations.

Theorem 3.9.6 *Let $A \in C\mathcal{R}(H)$, $D \in C\mathcal{R}(K)$ and let B, C be continuous linear relations from K into H and from H into K, respectively. Suppose that $\mathcal{D}(A) \subset N(C)$, $\mathcal{D}(D) \subset N(B)$, $B(0) \subset A(0)$ and $C(0) \subset D(0)$. Then,*

$$\rho_A(x) \bigcap \rho_D(y) = \rho_{\mathcal{L}_0}\begin{pmatrix} x \\ y \end{pmatrix}, \quad \textit{for all } \begin{pmatrix} x \\ y \end{pmatrix} \in H \times K,$$

where $\mathcal{L}_0$ is given in (2.13.1). ◇

Proof Since $\mathcal{D}(A) \subset N(C) \subset \mathcal{D}(C)$ and $\mathcal{D}(D) \subset N(B) \subset \mathcal{D}(B)$, then by referring to Remark 2.13.2, we infer that $\mathcal{L}_0$ is closed and $\mathcal{D}(\mathcal{L}_0) = \mathcal{D}(A) \times \mathcal{D}(D)$. Suppose that $\lambda \in \rho_A(x) \bigcap \rho_D(y)$. Then, there exist two open neighborhoods U_1, U_2 of λ in $\mathbb{C}$ and two analytic functions $f : U_1 \longrightarrow \mathcal{D}(A)$, $g : U_2 \longrightarrow \mathcal{D}(D)$ such that for all $\mu \in U_1 \bigcap U_2$, we have

$$(\mu - A)f(\mu) = x + A(0) \tag{3.9.7}$$

and

$$(\mu - D)g(\mu) = y + D(0). \tag{3.9.8}$$

Let $\begin{pmatrix} f \\ g \end{pmatrix} : U_1 \bigcap U_2 \longrightarrow \mathcal{D}(A) \times \mathcal{D}(D) = \mathcal{D}(\mathcal{L}_0)$. Hence, $\begin{pmatrix} f \\ g \end{pmatrix}$ is an analytic function and for all $\mu \in U_1 \bigcap U_2$, we have

$$\begin{aligned} (\mu - \mathcal{L}_0)\begin{pmatrix} f(\mu) \\ g(\mu) \end{pmatrix} &= \begin{pmatrix} \mu - A & B \\ C & \mu - D \end{pmatrix}\begin{pmatrix} f(\mu) \\ g(\mu) \end{pmatrix} \\ &= \begin{pmatrix} (\mu - A)f(\mu) + Bg(\mu) \\ Cf(\mu) + (\mu - D)g(\mu) \end{pmatrix} \\ &= \begin{pmatrix} x + A(0) + Bg(\mu) \\ Cf(\mu) + y + D(0) \end{pmatrix} \quad \text{(by (3.9.7) and (3.9.8))}. \end{aligned}$$

The fact that $f(\mu) \in \mathcal{D}(A) \subset N(C)$ and $g(\mu) \in \mathcal{D}(D) \subset N(B)$ implies from Theorem 2.8.1 (i) that

$$Bg(\mu) = B(0) \text{ and } Cf(\mu) = C(0), \quad \text{for all } \mu \in U_1 \bigcap U_2.$$

Hence, we deduce that

$$\begin{aligned} (\mu - \mathcal{L}_0)\begin{pmatrix} f(\mu) \\ g(\mu) \end{pmatrix} &= \begin{pmatrix} x + A(0) + B(0) \\ y + C(0) + D(0) \end{pmatrix} \\ &= \begin{pmatrix} x \\ y \end{pmatrix} + \begin{pmatrix} A(0) + B(0) \\ C(0) + D(0) \end{pmatrix} \\ &= \begin{pmatrix} x \\ y \end{pmatrix} + \mathcal{L}_0\begin{pmatrix} 0 \\ 0 \end{pmatrix}. \end{aligned}$$

This shows that

$$\rho_A(x)\bigcap \rho_B(y) \subset \rho_{\mathscr{L}_0}\begin{pmatrix} x \\ y \end{pmatrix}.$$

Conversely, let us assume that $\lambda \in \rho_{\mathscr{L}_0}\begin{pmatrix} x \\ y \end{pmatrix}$. Then, there exist an open neighborhood U and an analytic function $\begin{pmatrix} f \\ g \end{pmatrix} : U \longrightarrow \mathscr{D}(\mathscr{L}_0) = \mathscr{D}(A) \times \mathscr{D}(D)$ such that

$$(\mu - \mathscr{L}_0)\begin{pmatrix} f(\mu) \\ g(\mu) \end{pmatrix} = \begin{pmatrix} x \\ y \end{pmatrix} + \mathscr{L}_0\begin{pmatrix} 0 \\ 0 \end{pmatrix}, \quad \text{for all} \quad \mu \in U.$$

This implies that

$$(\mu - A)f(\mu) + Bg(\mu) = x + A(0) + B(0), \quad \text{for all} \quad \mu \in U, \tag{3.9.9}$$

and

$$(\mu - D)g(\mu) + Cf(\mu) = y + D(0) + C(0), \quad \text{for all} \quad \mu \in U. \tag{3.9.10}$$

Based on the assumption $B(0) \subset A(0)$, we infer that $A(0) = A(0) + B(0)$. This implies from (3.9.9) that

$$(\mu - A)f(\mu) + Bg(\mu) = x + A(0).$$

Then, from the fact that $f(\mu) \in \mathscr{D}(A) \subset N(B)$ for all $\mu \in U$, we deduce that

$$\begin{aligned} x + A(0) = &\ (\mu - A)f(\mu) + B(0) \\ = &\ (\mu - A)[f(\mu) + 0] + B(0) \\ = &\ (\mu - A)f(\mu) + (\mu - A)(0) + (\mu - B)(0) \ \text{(by the linearity of } \mu - A) \\ = &\ (\mu - A)f(\mu) + (\mu - A)(0) \ \text{(as } (\mu - A)(0) = (\mu - A)(0) + (\mu - B)(0)) \\ = &\ (\mu - A)f(\mu). \end{aligned}$$

Therefore, $\lambda \in \rho_A(x)$. By using (3.9.10) and the same reasoning as before leads to $\lambda \in \rho_D(x)$. As a result, $\rho_{\mathscr{L}_0}\begin{pmatrix} x \\ y \end{pmatrix} \subset \rho_A(x)\bigcap \rho_D(y)$, as desired. Q.E.D.

We illustrate Theorem 3.9.6 with the following example:

Example 3.9.1 Let

$$F = \left\{\begin{pmatrix} x \\ y \end{pmatrix} \in \mathbb{R}^2 \text{ such that } xy = 0\right\}$$

be a linear subspace of $\mathbb{R}^2$. Assume that

$$\begin{aligned} T : \mathbb{R}^2 &\longrightarrow \mathbb{R}^2 \\ \begin{pmatrix} x \\ y \end{pmatrix} &\longrightarrow \begin{pmatrix} x \\ y \end{pmatrix} + F. \end{aligned}$$

Let us consider the following matrix

$$\mathfrak{D} = \begin{pmatrix} T & 0 \\ 0 & 2T \end{pmatrix}.$$

Then, the local resolvent set of $\mathfrak{D}$ at $\begin{pmatrix} x \\ y \end{pmatrix} \in \mathbb{R}^2$ is not empty. Indeed, since $\dim \mathscr{D}(T) < \infty$, then by Proposition 2.10.4 (ii), we obtain T is continuous. Moreover, we have

$$\dim \left(T \begin{pmatrix} 0 \\ 0 \end{pmatrix} \right) < \infty.$$

This implies from Theorem 2.11.1 (iv) that T is closed. Consequently, by Lemma 2.11.6, we infer that $2T$ is closed. Hence, by referring to [114, Corollary 3.4 (i)], we deduce that $\mathfrak{D}$ is closed. The spectrum of T is associated with the equation

$$(\lambda - T) \begin{pmatrix} z \\ t \end{pmatrix} = (\lambda - T) \begin{pmatrix} 0 \\ 0 \end{pmatrix},$$

for all $\begin{pmatrix} z \\ t \end{pmatrix} \neq \begin{pmatrix} 0 \\ 0 \end{pmatrix}$. This implies that

$$\begin{pmatrix} (\lambda - 1)z \\ (\lambda - 1)t \end{pmatrix} - F = F.$$

The fact that F is a linear subspace implies that

$$(\lambda - 1) \begin{pmatrix} z \\ t \end{pmatrix} \in F,$$

which yields that

$$(\lambda - 1)^2 zt = 0.$$

Hence, $\lambda = 1$. This is equivalent to saying that $\rho(T) = \mathbb{C} \backslash \{1\}$. Therefore, we can conclude from the same reasoning that $\rho(2T) = \mathbb{C} \backslash \{2\}$. By virtue of [113, Proposition 3.1 (ii)], we obtain

$$\rho(T) \bigcap \rho(2T) \subset \rho_T(x) \bigcap \rho_{2T}(y).$$

This implies from Theorem 3.9.6 that

$$\rho(T)\bigcap\rho(2T) \subset \rho_{\mathfrak{D}}\begin{pmatrix} x \\ y \end{pmatrix}, \quad \text{for all } \begin{pmatrix} x \\ y \end{pmatrix} \in \mathbb{R}^2.$$

Hence,

$$\mathbb{C}\backslash\{1, 2\} \subset \rho_{\mathfrak{D}}\begin{pmatrix} x \\ y \end{pmatrix}, \quad \text{for all } \begin{pmatrix} x \\ y \end{pmatrix} \in \mathbb{R}^2.$$

Proposition 3.9.6 *Let $A \in C\mathcal{R}(H)$, $D \in C\mathcal{R}(K)$ and let B, C be continuous linear relations from K into H and from H into K, respectively. Suppose that $\mathcal{D}(A) \subset N(C)$, $\mathcal{D}(D) \subset N(B)$, $B(0) \subset A(0)$ and $C(0) \subset D(0)$. For every subset F of $\mathbb{C}$ the local analytical subspace of $\mathcal{L}_0$, given in (2.13.1), associated with F is the set*

$$X_{\mathcal{L}_0}(F) = X_A(F) \times X_D(F). \qquad \diamond$$

Proof The local analytical subspace of $\mathcal{L}_0$ associated with F is given by

$$\begin{aligned} X_{\mathcal{L}_0}(F) &= \left\{\begin{pmatrix} x \\ y \end{pmatrix} \in H \times K \text{ such that } \sigma_{\mathcal{L}_0}\begin{pmatrix} x \\ y \end{pmatrix} \subset F\right\} \\ &= \left\{\begin{pmatrix} x \\ y \end{pmatrix} \in H \times K \text{ such that } \sigma_A(x)\bigcup\sigma_D(y) \subset F\right\} \text{ (by Theorem 3.9.6)} \\ &= \{x \in H \text{ such that } \sigma_A(x) \subset F\} \times \{y \in K \text{ such that } \sigma_D(y) \subset F\}. \end{aligned}$$

This is equivalent to saying that $X_{\mathcal{L}_0}(F) = X_A(F) \times X_D(F)$. Q.E.D.

In the assumptions of Theorem 3.9.6 and Lemma 2.8.5, the condition that the domain of a linear relation included in the kernel of the other is satisfied for some linear relations. We illustrate this in the following example.

Example 3.9.2 Let H be an infinite-dimensional Hilbert space and let $A \in L\mathcal{R}(H \times H)$ be defined by

$$G(A) = \left\{\left[\begin{pmatrix} x \\ 0 \end{pmatrix}, \begin{pmatrix} x \\ y \end{pmatrix}\right] \text{ such that } x,\ y \in H\right\}.$$

It is obvious that $G(A)$ is closed. Then, A is closed. Moreover, it is easy to see that $\mathcal{D}(A) = H \times \{0\}$. Let

$$F = \left\{\begin{pmatrix} x \\ y \end{pmatrix} \in H \times H \text{ such that } xy = 0\right\}$$

be a linear subspace of $H \times H$. Assume that

$$\begin{aligned} T : H \times H &\longrightarrow H \times H \\ \begin{pmatrix} x \\ y \end{pmatrix} &\longrightarrow \begin{pmatrix} x \\ y \end{pmatrix} + F, \end{aligned}$$

where

$$\begin{pmatrix} x \\ y \end{pmatrix} + F = \left\{ \begin{pmatrix} x \\ y \end{pmatrix} + \begin{pmatrix} z \\ t \end{pmatrix} \text{ such that } \begin{pmatrix} z \\ t \end{pmatrix} \in F \right\}.$$

It is clear that

$$F = T \begin{pmatrix} 0 \\ 0 \end{pmatrix} \neq \left\{ \begin{pmatrix} 0 \\ 0 \end{pmatrix} \right\}.$$

Hence, by referring to Proposition 3.8.8, we have $T \in C\mathscr{R}(H \times H)$. Now, let us assume that $C = T^{-1}$. It follows that C is closed. It remains to prove that C is continuous. By Proposition 2.15.3, it is sufficient to show that $\mathscr{D}(C) = R(T)$ is closed. It follows from Proposition 3.8.8 (i) and Definition 2.14.1 that

$$\begin{aligned} R(T) =& \ R(I) + T \begin{pmatrix} 0 \\ 0 \end{pmatrix} \\ =& \ H \times H + F \\ =& \ H \times H \ \text{ (as } F \subset H \times H). \end{aligned}$$

Moreover, we have

$$\begin{aligned} N(C) =& \ C^{-1} \begin{pmatrix} 0 \\ 0 \end{pmatrix} \\ =& \ T \begin{pmatrix} 0 \\ 0 \end{pmatrix} \\ =& \ F. \end{aligned}$$

Our purpose is to verify that

$$\mathscr{D}(A) \subset N(C). \tag{3.9.11}$$

Let $\begin{pmatrix} x \\ y \end{pmatrix} \in \mathscr{D}(A)$, then $x \in H$ and $y = 0$. This implies that $xy = 0$. This is equivalent to saying that $\begin{pmatrix} x \\ y \end{pmatrix} \in F$. Thus, $\begin{pmatrix} x \\ y \end{pmatrix} \in N(C)$. Hence, (3.9.11) holds.

Remark 3.9.1 For $A \in C\mathscr{R}(H)$, $D \in C\mathscr{R}(K)$ and B, C are continuous linear relations from K into H and from H into K, respectively. The conditions $\mathscr{D}(A) \subset N(C)$ and $\mathscr{D}(D) \subset N(B)$ play a crucial role in the sequel. ◇

Corollary 3.9.5 *Let $A \in L\mathscr{R}(H)$, $B \in L\mathscr{R}(K, H)$, $C \in L\mathscr{R}(H, K)$ and $D \in L\mathscr{R}(K)$. Suppose that $\mathscr{D}(A) \subset N(C)$, $\mathscr{D}(D) \subset N(B)$, $B(0) \subset A(0)$ and $C(0) \subset D(0)$. Then,*

(i) $P\sigma(\mathscr{L}_0) = P\sigma(A) \bigcup P\sigma(D)$, where $\mathscr{L}_0$ is given in (2.13.1),
(ii) $\sigma_\delta(\mathscr{L}_0) = \sigma_\delta(A) \bigcup \sigma_\delta(D)$,
(iii) $\sigma_r(\mathscr{L}_0) = \sigma_r(A) \bigcup \sigma_r(D)$. ◇

Proposition 3.9.7 *Let $A \in C\mathscr{R}(H)$, $D \in C\mathscr{R}(K)$ and let B, C be continuous linear relations from K into H and from H into K, respectively. Then, $\mathscr{L}_0$, given in (2.13.1), have a SVEP if A and D have a SVEP and either (i) or (ii) holds*

(i) $B(0) \subset A(0)$, $C(0) \subset D(0)$ and $\mathscr{D}(A) \subset N(C)$.
(ii) $B(0) \subset A(0)$, $C(0) \subset D(0)$ and $\mathscr{D}(D) \subset N(B)$. ◇

Proof It follows from Remark 2.13.2 that $\mathscr{L}_0$ is closed. Let $\lambda \in \mathbb{C}$ and U be an open neighborhood of λ in $\mathbb{C}$. Let $\begin{pmatrix} f \\ g \end{pmatrix} : U \longrightarrow \mathscr{D}(\mathscr{L}_0)$ be an analytic function such that

$$(\mu - \mathscr{L}_0)\begin{pmatrix} f(\mu) \\ g(\mu) \end{pmatrix} = \mathscr{L}_0\begin{pmatrix} 0 \\ 0 \end{pmatrix}, \text{ for all } \mu \in U.$$

Then,

$$(\mu - A)f(\mu) + Bg(\mu) = A(0) + B(0)$$

and

$$(\mu - D)g(\mu) + Cf(\mu) = D(0) + C(0).$$

Let us assume that (i) holds. Then, we obtain from (2.8.15) and (2.8.16) that

$$(\mu - A)f(\mu) + Bg(\mu) = A(0) \text{ and } (\mu - D)g(\mu) + Cf(\mu) = D(0). \quad (3.9.12)$$

The fact that $\mathscr{D}(A) \subset N(C)$ implies from Theorem 2.8.1 (i) that $Cf(\mu) = C(0)$. Hence, by using (2.8.16) and the linearity of $\mu - D$ for all $\mu \in U$, we get $(\mu - D)g(\mu) = D(0)$, for all $\mu \in U$. Based on the assumption D have SVEP, we infer that g is null function. By referring to (3.9.12), we deduce that

$$(\mu - A)f(\mu) + B(0) = A(0).$$

Thus, from the same reasoning as above, we conclude that f is null function. This shows that $\mathscr{L}_0$ have SVEP at λ. The same reasoning if we assume that (ii) holds. Q.E.D.

Theorem 3.9.7 *Let $A \in C\mathscr{R}(H)$, $D \in C\mathscr{R}(K)$ and let B, C be continuous linear relations from K into H and from H into K, respectively. Suppose that $\mathscr{D}(A) \subset N(C)$, $\mathscr{D}(D) \subset N(B)$, $B(0) \subset A(0)$ and $C(0) \subset D(0)$. Then, $\mathscr{L}_0$ have SVEP if, and only if, A and D have SVEP, where $\mathscr{L}_0$ is given in (2.13.1).* ◇

Proof It follows from Remark 2.13.2 that $\mathscr{L}_0$ is closed. By Proposition 3.9.7, it is sufficient to prove that A and D have SVEP if $\mathscr{L}_0$ have SVEP. For all $\lambda \in \mathbb{C}$, let us assume that U be an open neighborhood of λ in $\mathbb{C}$ and $f : U \longrightarrow \mathscr{D}(A)$, $g : U \longrightarrow \mathscr{D}(D)$ be analytic functions such that

$$(\mu - A)f(\mu) = A(0) \text{ and } (\mu - D)g(\mu) = D(0), \text{ for all } \mu \in U.$$

Then,

$$(\mu - A)f(\mu) + B(0) = A(0) + B(0)$$

and

$$(\mu - D)g(\mu) + C(0) = D(0) + C(0).$$

This implies from the fact that $\mathscr{D}(A) \subset N(C)$ and $\mathscr{D}(D) \subset N(B)$ that

$$(\mu - A)f(\mu) + Bg(\mu) = A(0) + B(0)$$

and

$$(\mu - D)g(\mu) + Cf(\mu) = D(0) + C(0).$$

Moreover, $\begin{pmatrix} f \\ g \end{pmatrix} : U \longrightarrow \mathscr{D}(\mathscr{L}_0)$ is an analytic function. Hence, we infer that

$$(\mu - \mathscr{L}_0)\begin{pmatrix} f(\mu) \\ g(\mu) \end{pmatrix} = \mathscr{L}_0\begin{pmatrix} 0 \\ 0 \end{pmatrix}, \text{ for all } \mu \in U.$$

The fact that $\mathscr{L}_0$ have SVEP implies that $\begin{pmatrix} f \\ g \end{pmatrix}$ is null. This is equivalent to saying that f and g are null. Thus, A and D have SVEP at λ. Q.E.D.

Proposition 3.9.8 *Let $A \in C\mathscr{R}(H)$, $D \in C\mathscr{R}(K)$ and let B, C be continuous linear relations from K into H and from H into K, respectively. Suppose that $\mathscr{D}(A) \subset N(C)$, $\mathscr{D}(D) \subset N(B)$, $B(0) \subset A(0)$ and $C(0) \subset D(0)$. If $\begin{pmatrix} x \\ y \end{pmatrix} \in H \times K$, A and D have SVEP at $\lambda \in \rho_A(x) \bigcap \rho_D(y)$, then there exists a neighborhood U of λ and an unique analytic function $h = \begin{pmatrix} f \\ g \end{pmatrix} : U \longrightarrow \mathscr{D}(\mathscr{L}_0)$ satisfying the equation*

$$(\mu - \mathscr{L}_0)h(\mu) = x + \mathscr{L}_0(0), \text{ for all } \mu \in U,$$

where $\mathscr{L}_0$ is given in (2.13.1). ◇

Proof It follows from Remark 2.13.2 that $\mathscr{L}_0$ is closed. By referring to Theorem 3.9.6, we have

$$\rho_A(x) \bigcap \rho_D(y) = \rho_{\mathscr{L}_0}\begin{pmatrix} x \\ y \end{pmatrix}.$$

This implies from Theorem 3.9.7 that $\mathscr{L}_0$ have SVEP at $\lambda \in \rho_{\mathscr{L}_0}\begin{pmatrix} x \\ y \end{pmatrix}$. Hence, by using [113, Proposition 3.2 (i)], we infer that there exists a neighborhood U of λ and an unique analytic function $h = \begin{pmatrix} f \\ g \end{pmatrix} : U \longrightarrow \mathscr{D}(\mathscr{L}_0)$ satisfying the equation

$$(\mu - \mathscr{L}_0)h(\mu) = x + \mathscr{L}_0(0), \text{ for all } \mu \in U.$$

Q.E.D.

Theorem 3.9.8 *Let $A \in C\mathscr{R}(H)$, $D \in C\mathscr{R}(K)$ and let B, C be continuous linear relations from K into H and from H into K, respectively. Suppose that $\mathscr{D}(A) \subset \left[\mathscr{D}(C) \bigcap A(0)^{\perp}\right]$, $\mathscr{D}(D) \subset \left[\mathscr{D}(B) \bigcap D(0)^{\perp}\right]$, $B(0) \subset A(0)$ and $C(0) \subset D(0)$. Then,*

$$\sigma(\mathscr{L}_0) = \sigma_{\delta}(\mathscr{L}_0) \bigcup \Xi(\mathscr{L}_0),$$

where $\mathscr{L}_0$ is given in (2.13.1). ◇

Proof It follows from Remark 2.13.2 that $\mathscr{L}_0$ is closed and $\mathscr{D}(\mathscr{L}_0) = \mathscr{D}(A) \times \mathscr{D}(D)$. The fact that $\mathscr{D}(A) \subset A(0)^{\perp}$ and $\mathscr{D}(D) \subset D(0)^{\perp}$ implies that

$$\begin{aligned}
\mathscr{D}(\mathscr{L}_0) &\subset A(0)^{\perp} \times D(0)^{\perp} \\
&\subset [A(0) \times D(0)]^{\perp} \quad \text{(by Remark 2.1.2 } (ii)) \\
&\subset [(A(0) + B(0)) \times (C(0) + D(0))]^{\perp} \quad \text{(by (2.8.15) and (2.8.16))} \\
&\subset \mathscr{L}_0 \begin{pmatrix} 0 \\ 0 \end{pmatrix}^{\perp}.
\end{aligned}$$

Hence, the use of Theorem 3.8.9 allows us to conclude that $\sigma(\mathscr{L}_0) = \sigma_{\delta}(\mathscr{L}_0) \bigcup \Xi(\mathscr{L}_0)$. This achieves the proof. Q.E.D.

The linear relations considered in the following example shows that the conditions of Theorem 3.9.8 hold.

Example 3.9.3 Consider the linear relations A and C given in Example 3.9.2. Recall that A is closed, $\mathscr{D}(A) = H \times \{0\}$, C is continuous and $N(C) = F$. Moreover, we have $A(0) = \{0\} \times H$. This implies that

$$\begin{aligned}
A(0)^{\perp} &= (\{0\} \times H)^{\perp} \\
&= H \times \{0\} \quad \text{(by Lemma 2.1.2 } (ii)) \\
&= \mathscr{D}(A).
\end{aligned}$$

It follows from (3.9.11) that

$$A(0)^{\perp} \subset N(C).$$

Consequently,

$$\mathscr{D}(A) = A(0)^{\perp} \bigcap N(C).$$

This enables us to conclude that $\mathscr{D}(A) = A(0)^{\perp} \bigcap \mathscr{D}(C)$.

As a direct consequence of Theorems 3.9.7 and 3.9.8, we infer the following result.

Corollary 3.9.6 *Let $A \in C\mathscr{R}(H)$, $D \in C\mathscr{R}(K)$ and let B, C be continuous linear relations from K into H and from H into K, respectively. Suppose that $\mathscr{D}(A) \subset$*

$[N(C) \bigcap A(0)^{\perp}]$, $\mathscr{D}(D) \subset [N(B) \bigcap D(0)^{\perp}]$, $B(0) \subset A(0)$ and $C(0) \subset D(0)$. If A and D have SVEP, then

$$\sigma(\mathscr{L}_0) = \sigma_{\delta}(\mathscr{L}_0),$$

where $\mathscr{L}_0$ is given in (2.13.1). ◇

Theorem 3.9.9 *Let $A \in C\mathscr{R}(H)$, $D \in C\mathscr{R}(K)$ and let B, C be continuous linear relations from K into H and from H into K, respectively. Suppose that $\mathscr{D}(A) \subset N(C)$, $\mathscr{D}(D) \subset N(B)$, $[B(0) \bigcap \mathscr{D}(A)^{\perp}] \subset A(0)$ and $[C(0) \bigcap \mathscr{D}(D)^{\perp}] \subset D(0)$. Then,*

(i) $\sigma(\mathscr{L}_0) \bigcup \{\lambda \in \mathbb{C}$ such that $\overline{\lambda} \notin \Xi(A^*) \bigcap \Xi(D^*)\} \subset \sigma(A) \bigcup \sigma(D)$, where $\mathscr{L}_0$ is given in (2.13.1).
(ii) $\sigma(A) \bigcup \sigma(D) \subset \sigma_{ap}(\mathscr{L}_0) \bigcup \{\lambda \in \mathbb{C}$ such that $\overline{\lambda} \notin \Xi(A^*) \bigcap \Xi(D^*)\}$. ◇

Proof It follows from Remark 2.13.2 that $\mathscr{L}_0$ is closed. By referring to (3.1.4), we have $A^*(0) = \mathscr{D}(A)^{\perp}$ and $A(0) = \mathscr{D}(A^*)^{\perp}$. Hence, based on the assumption $\mathscr{D}(A)^{\perp} \subset A(0)$, we infer that $A^*(0) \subset \mathscr{D}(A^*)^{\perp}$. This yields for all $\lambda \in \mathbb{C}$ that

$$\mathscr{D}(\overline{\lambda} - A^*) = \mathscr{D}(A^*) \subset \overline{\mathscr{D}(A^*)} \subset A^*(0)^{\perp} = (\overline{\lambda} - A^*)(0)^{\perp}. \tag{3.9.13}$$

A same reasoning as before leads to

$$\mathscr{D}(\overline{\lambda} - D^*) = \mathscr{D}(D^*) \subset \overline{\mathscr{D}(D^*)} \subset D^*(0)^{\perp} = (\overline{\lambda} - D^*)(0)^{\perp}. \tag{3.9.14}$$

(i) Let us assume that $\lambda \notin \sigma(A) \bigcup \sigma(D)$. Then, by using (3.1.1), we have

$$N(\lambda - A) = \{0\},\ R(\lambda - A) = H,\ N(\lambda - D) = \{0\} \text{ and } R(\lambda - D) = K. \tag{3.9.15}$$

It follows from Lemma 2.8.5 that

$$N(\lambda - \mathscr{L}_0) = \left\{ \begin{pmatrix} 0 \\ 0 \end{pmatrix} \right\}$$

and

$$R(\lambda - \mathscr{L}_0) = H \times K.$$

This implies from (3.1.1) that $\lambda \notin \sigma(\mathscr{L}_0)$. On the other hand, by the closdeness of $\lambda - A$, (3.1.3), (3.1.4) and (3.9.15), we infer that

$$N(\overline{\lambda} - A^*) = \{0\} \text{ and } R(\overline{\lambda} - A^*)^{\perp} = \{0\}. \tag{3.9.16}$$

Moreover, we have $R(\lambda - A)$ is closed. Then, by using Theorem 2.15.1 and (3.1.3), we obtain that $R(\overline{\lambda} - A^*)$ is closed. This implies from (3.9.16) that $\overline{\lambda} - A^*$ is onto. Hence, by using (3.9.13) and (3.9.16), we infer from Corollary 3.8.1 that $\overline{\lambda} - A^*$

have SVEP at 0. The use of Proposition 3.8.7 (i) allows us to conclude that A^* have SVEP at $\overline{\lambda}$. A similar reasoning as above gives D^* have SVEP at $\overline{\lambda}$. Therefore,

$$\lambda \notin \sigma(\mathscr{L}_0) \bigcup \left\{\lambda \in \mathbb{C} \text{ such that } \overline{\lambda} \notin \Xi(A^*) \bigcap \Xi(D^*)\right\}.$$

(ii) Suppose that $\lambda \notin \sigma_{ap}(\mathscr{L}_0) \bigcup \left\{\lambda \in \mathbb{C} \text{ such that } \overline{\lambda} \notin \Xi(A^*) \bigcap \Xi(D^*)\right\}$. Then, by using Proposition 3.3.1, we infer that $\lambda - \mathscr{L}_0$ is one-to-one and open. It follows from Lemma 2.8.5 (i) that $N(\lambda - A) = \{0\}$ and $N(\lambda - D) = \{0\}$. This implies from (3.1.3) and (3.1.4) that

$$R(\overline{\lambda} - A^*)^{\perp} = \{0\}$$

and

$$R(\overline{\lambda} - D^*)^{\perp} = \{0\}.$$

Consequently,

$$\overline{R\left(\overline{\lambda} - A^*\right)} = H \text{ and } \overline{R\left(\overline{\lambda} - D^*\right)} = K. \tag{3.9.17}$$

Now, by the closedness and the openness of $\lambda - \mathscr{L}_0$, we infer from Theorem 2.11.1 (ii) that $R(\lambda - \mathscr{L}_0)$ is closed. This leads us from Lemma 2.8.5 (ii) to

$$\begin{aligned} R(\lambda - A) \times R(\lambda - D) &= \overline{R(\lambda - A) \times R(\lambda - D)} \\ &= \overline{R(\lambda - A)} \times \overline{R(\lambda - D)}, \end{aligned}$$

which yields $R(\lambda - A)$ and $R(\lambda - D)$ are closed. This implies from Theorem 2.15.1 and (3.9.17) that $\overline{\lambda} - A^*$ and $\overline{\lambda} - D^*$ are onto. The fact that A^* and D^* have SVEP at $\overline{\lambda}$ implies from Proposition 3.8.7 (i) that $\overline{\lambda} - A^*$ and $\overline{\lambda} - D^*$ have SVEP at 0. This leads from (3.9.13), (3.9.14) and Corollary 3.8.1 that $\overline{\lambda} - A^*$ and $\overline{\lambda} - D^*$ are one-to-one. By virtue of (3.1.3) and (3.1.4), we obtain

$$R(\lambda - A)^{\perp} = \{0\}$$

and

$$R(\lambda - D)^{\perp} = \{0\}.$$

This implies from the fact that $R(\lambda - A)$ and $R(\lambda - D)$ are closed that $R(\lambda - A) = H$ and $R(\lambda - D) = K$. Hence, by (3.1.1), we conclude that $\lambda \notin \sigma(A) \bigcup \sigma(D)$. Q.E.D.

Remark 3.9.2 Note that Example 3.9.3 satisfy the assumptions of the Theorem 3.9.9. ◇

Example 3.9.4 Consider the same linear relations A and T given in Proposition 3.8.8 and Example 3.9.2. Recall that A is closed, T has a continuous selection I, which yields from Proposition 2.14.3 that T is continuous. In addition, $\mathscr{D}(A) = H \times \{0\}$. This implies that

$$\begin{aligned}\mathcal{D}(A)^{\perp} &= (H \times \{0\})^{\perp} \\ &= H^{\perp} \times \{0\}^{\perp} \quad \text{(by Lemma 2.1.2}(ii)) \\ &= \{0\} \times H \\ &\subset T(0) \\ &= \left\{ \begin{pmatrix} x \\ y \end{pmatrix} \in H \times H \text{ such that } xy = 0 \right\}.\end{aligned}$$

The fact that $A(0) = \{0\} \times H$ implies that $\mathcal{D}(A)^{\perp} \bigcap T(0) = A(0)$.

As an immediate consequence of Remark 3.2.3 and Theorem 3.9.9, we have.

Corollary 3.9.7 *Let $A \in C\mathcal{R}(H)$, $D \in C\mathcal{R}(K)$ and let B, C be continuous linear relations from K into H and from H into K, respectively. Suppose that $\mathcal{D}(A) \subset N(C)$, $\mathcal{D}(D) \subset N(B)$, $[B(0) \bigcap \mathcal{D}(A)^{\perp}] \subset A(0)$ and $[C(0) \bigcap \mathcal{D}(D)^{\perp}] \subset D(0)$. If A^* or D^* have SVEP, then*

$$\sigma(\mathcal{L}_0) = \sigma(A) \bigcup \sigma(D),$$

where $\mathcal{L}_0$ is given in (2.13.1). ◇

Theorem 3.9.10 *Let $A \in C\mathcal{R}(H)$, $D \in C\mathcal{R}(K)$, $B \in L\mathcal{R}(K, H)$ and $C \in L\mathcal{R}(H, K)$ be four continuous linear relations such that $\mathcal{D}(A^*) \subset \mathcal{D}(A) \subset \mathcal{D}(C)$, $\mathcal{D}(D^*) \subset \mathcal{D}(D) \subset \mathcal{D}(B)$, $B(0) \subset A(0)$ and $C(0) \subset D(0)$. Then,*

$$\sigma(\mathcal{L}_0^*) = \sigma_\delta(\mathcal{L}_0^*) \bigcup \Xi(\mathcal{L}_0^*),$$

where $\mathcal{L}_0$ is given in (2.13.1). ◇

Proof we have

$$\begin{aligned}\mathcal{L}_0^* \begin{pmatrix} 0 \\ 0 \end{pmatrix} &= \begin{pmatrix} A^*(0) + C^*(0) \\ B^*(0) + D^*(0) \end{pmatrix} \\ &= \begin{pmatrix} \mathcal{D}(A)^{\perp} + \mathcal{D}(C)^{\perp} \\ \mathcal{D}(B)^{\perp} + \mathcal{D}(D)^{\perp} \end{pmatrix} \quad \text{(by (3.1.4)).}\end{aligned}$$

Since $\mathcal{D}(A) \subset \mathcal{D}(C)$ and $\mathcal{D}(D) \subset \mathcal{D}(B)$, then

$$\mathcal{D}(C)^{\perp} \subset \mathcal{D}(A)^{\perp}$$

and

$$\mathcal{D}(B)^{\perp} \subset \mathcal{D}(D)^{\perp}.$$

This implies that

$$\mathcal{D}(A)^{\perp} = \mathcal{D}(A)^{\perp} + \mathcal{D}(C)^{\perp}$$

and

$$\mathscr{D}(D)^{\perp} = \mathscr{D}(B)^{\perp} + \mathscr{D}(D)^{\perp}.$$

Hence, we infer that

$$\begin{aligned} \mathscr{L}_0^* \begin{pmatrix} 0 \\ 0 \end{pmatrix} &= \mathscr{D}(A)^{\perp} \times \mathscr{D}(D)^{\perp} \\ &= (\mathscr{D}(A) \times \mathscr{D}(D))^{\perp} \text{ (by Lemma 2.1.2 } (ii)). \end{aligned} \tag{3.9.18}$$

Based on the assumptions $\mathscr{D}(A^*) \subset \mathscr{D}(A)$ and $\mathscr{D}(D^*) \subset \mathscr{D}(D)$, we obtain from (2.15.3) that

$$\begin{aligned} \mathscr{D}(\mathscr{L}_0^*) &\subset \mathscr{D}(A) \times \mathscr{D}(D) \\ &\subset \overline{\mathscr{D}(A) \times \mathscr{D}(D)} \\ &\subset \mathscr{L}_0^* \begin{pmatrix} 0 \\ 0 \end{pmatrix}^{\perp} \text{ (by (3.9.18)).} \end{aligned}$$

The above implies from Theorem 3.8.9 that

$$\sigma(\mathscr{L}_0) = \sigma_{ap}(\mathscr{L}_0) \bigcup \{\lambda \in \mathbb{C} \text{ such that } \overline{\lambda} \in \Xi(\mathscr{L}_0^*)\}.$$

Finally, the use of Lemma 3.2.3 and (3.1.5) gives the wanted result and achieves the proof. Q.E.D.

As a direct consequence of Theorems 3.9.7 and 3.9.10, we infer the following result.

Corollary 3.9.8 *Let $A \in C\mathscr{R}(H)$, $D \in C\mathscr{R}(K)$, $B \in L\mathscr{R}(K, H)$ and $C \in L\mathscr{R}(H, K)$ be four continuous linear relation such that $\mathscr{D}(A^*) \subset \mathscr{D}(A) \subset N(C) \bigcap R(B)^{\perp}$, $\mathscr{D}(D^*) \subset \mathscr{D}(D) \subset N(B) \bigcap R(C)^{\perp}$, $B(0) \subset A(0)$ and $C(0) \subset D(0)$. If A^* and D^* have SVEP, then*

$$\sigma(\mathscr{L}_0^*) = \sigma_{\delta}(\mathscr{L}_0^*).$$

◇

Theorem 3.9.11 *Let $A \in L\mathscr{R}(X)$, $D \in L\mathscr{R}(Y)$ and $B \in L\mathscr{R}(Y, X)$ be satisfying $A(0) \subset B(0)$. Then, there exists $x \in X \backslash \{0\}$ such that for all $\lambda \in \rho_A(x)$, there exists $C \in L\mathscr{R}(X, Y)$ for which $\lambda \in P\sigma(\mathscr{L}_0)$, where $\mathscr{L}_0$ is given in (2.13.1).* ◇

Proof Let $B \in L\mathscr{R}(Y, X)$. Then, there exists $z \in Y$ such that

$$Bz = x + B(0).$$

Let $\lambda \in \rho_A(x)$. Therefore, there exists $x_\lambda \in X$ satisfying

$$(A - \lambda)x_\lambda = x + A(0).$$

For this x_λ, there exists $\varphi \in X^*$ such that $\varphi(x_\lambda) = \|x_\lambda\|$. Hence,

$$(D - \lambda)z = -Cx_\lambda + D(0).$$

If we put $C = \frac{1}{\|x_\lambda\|}\varphi(\cdot)(D - \lambda)z$, we get

$$(\mathcal{L}_0 - \lambda)\left((-x_\lambda)\bigoplus z\right) = \mathcal{L}_0(0).$$

Thus, we conclude that $\lambda \in P\sigma(\mathcal{L}_0)$.

Corollary 3.9.9 *Let $A \in L\mathcal{R}(X)$, $D \in L\mathcal{R}(Y)$ and $B \in L\mathcal{R}(Y, X)$ be satisfying $A(0) \subset B(0)$. Then, for all $\lambda \in \mathbb{B}\backslash\sigma_\delta(A)$, there exists $C \in L\mathcal{R}(X, Y)$ for which $\lambda \in P\sigma(\mathcal{L}_0)$.* ◇

Proof Since $\sigma_\delta(T) = \bigcup_{x\in X}\sigma_T(x)$, then $\lambda \in \mathbb{C}\backslash\sigma_\delta(A)$ implying that $\lambda \in \rho_A(x)$ for all $x \in X$. It follows from Theorem 3.9.11 that $\lambda \in P\sigma(\mathcal{L}_0)$. Q.E.D.

Corollary 3.9.10 *Let $A \in L\mathcal{R}(X)$, $D \in L\mathcal{R}(Y)$ and $B \in L\mathcal{R}(Y, X)$ be satisfying $A(0) \subset B(0)$. Then, for all $\lambda \in \rho(A)$, there exists $C \in L\mathcal{R}(X, Y)$ for which $\lambda \in P\sigma(\mathcal{L}_0)$.* ◇

Proof Since $\sigma_\delta(T) = \bigcup_{x\in X}\sigma_T(x)$, then $\lambda \in \mathbb{C}\backslash\sigma_\delta(A)$ implying that $\lambda \in \rho_A(x)$ for all $x \in X$. It follows from Theorem 3.9.11 that $\lambda \in P\sigma(\mathcal{L}_0)$. Q.E.D.

Corollary 3.9.11 *Let $A \in L\mathcal{R}(X)$, $D \in L\mathcal{R}(Y)$ and $B \in L\mathcal{R}(Y, X)$ be satisfying $A(0) \subset B(0)$. Then, for all $\lambda \in \rho(A)$, there exists $C \in L\mathcal{R}(X, Y)$ for which $\lambda \in P\sigma(\mathcal{L}_0)$.* ◇

3.9.4 Local Spectral of 3 × 3 Linear Relations Matrices

Consider the upper triangular 3 × 3-block matrix of linear relations

$$\mathcal{T} = \begin{pmatrix} T_{11} & T_{12} & T_{13} \\ 0 & T_{22} & T_{23} \\ 0 & 0 & T_{33} \end{pmatrix} \tag{3.9.19}$$

on $X_1 \bigoplus X_2 \bigoplus X_3$, where $T_{ii} \in L\mathcal{R}(X_i)$, and $T_{ij} \in L\mathcal{R}(X_j, X_i)$ for all $i \in \{1, 2, 3\}$, $j \in \{2, 3\}$ and X_1, X_2, X_3 are complex Banach spaces. In the beginning of this subsection, we discuss some properties of the local spectral theory of $\mathcal{T}$, given in (3.9.19), involving the onto spectrum and the analytic residuum of its diagonal entries. In the further proposition, we illustrate a new result of the local spectrum of $\mathcal{T}$ in terms of the onto spectrum.

Proposition 3.9.9 *Let $T_{ii} \in B\mathscr{R}(X_i)$, and $T_{ij} \in B\mathscr{R}(X_j, X_i)$ for all $i \in \{1,2,3\}$ and $j \in \{2,3\}$ such that $T_{ii}(0) \subset T_{ij}(0)$ for $j \neq i \in \{1,2\}$. If T_{11} (resp. T_{22}) has a selection E_1 (resp. E_2) which satisfies $T_{11}(0) \subset R(E_1)$ (resp. $T_{22}(0) \subset R(E_2)$), then for all $x = \begin{pmatrix} y \\ z \\ w \end{pmatrix} \in X_1 \times X_2 \times X_3$, we have*

$$\sigma_{\mathscr{T}}(x) \subset \sigma_\delta(T_{11}) \bigcup \sigma_\delta(T_{22}) \bigcup \sigma_{T_{33}}(w),$$

where the matrix $\mathscr{T}$ is given in (3.9.19). ◇

Proof Let $\lambda_0 \notin \sigma_\delta(T_{11}) \bigcup \sigma_\delta(T_{22}) \bigcup \sigma_{T_{33}}(w)$. Then, there exist three open neighborhoods V_1, V_2 and V_3 of λ_0 and an analytic operator function $f_3 : V_3 \longrightarrow X_3$ such that

$$(T_{33} - \lambda) f_3(\lambda) = w + T_{33}(0),$$

for all $\lambda \in V_3$, $V_2 \bigcap \sigma_\delta(T_{22}) = \emptyset$ and $V_1 \bigcap \sigma_\delta(T_{11}) = \emptyset$. Let

$$V = \bigcap_{i \in \{1,2,3\}} V_i.$$

Hence, we get

$$(T_{33} - \lambda) f_3(\lambda) = w + T_{33}(0), \quad \text{for all } \lambda \in V.$$

In addition, by referring to Theorem 2.14.1, we infer that there exist two analytic functions $f_2 : V \longrightarrow X_2$ and $f_1 : V \longrightarrow X_1$ such that

$$(T_{22} - \lambda) f_2(\lambda) = z - T_{23} f_1(\lambda) + T_{22}(0),$$

and

$$(T_{11} - \lambda) f_3(\lambda) = y - T_{12} f_2(\lambda) - T_{13} f_1(\lambda) + T_{11}(0),$$

for all $\lambda \in V$. Furthermore, the nonzero analytic function

$$\begin{pmatrix} f_3 \\ f_2 \\ f_1 \end{pmatrix} : V \longrightarrow X_3 \times X_2 \times X_1$$

defined by

$$\begin{pmatrix} f_3 \\ f_2 \\ f_1 \end{pmatrix}(\lambda) = \begin{pmatrix} f_3(\lambda) \\ f_2(\lambda) \\ f_1(\lambda) \end{pmatrix}$$

such that

$$(\mathscr{T}-\lambda)\begin{pmatrix} f_3(\lambda) \\ f_2(\lambda) \\ f_1(\lambda) \end{pmatrix} = \begin{pmatrix} y+T_{11} \\ z+T_{22} \\ w+T_{33} \end{pmatrix}$$
$$= x+\mathscr{T}(0), \quad \text{for all } \lambda \in V.$$

Therefore, $\lambda_0 \notin \sigma_{\mathscr{T}}(x)$. Q.E.D.

As a direct consequence of Proposition 3.9.9, we get the following result.

Corollary 3.9.12 *Let $T_{ii} \in B\mathscr{R}(X_i)$ and $T_{ij} \in B\mathscr{R}(X_j, X_i)$ for all $i \in \{1,2,3\}$ and $j \in \{2,3\}$ be such that $T_{ii}(0) \subset T_{ij}(0)$ for $j \neq i \in \{1,2\}$. If T_{11} (resp. T_{22}) has a selection E_1 (resp. E_2) which satisfies $T_{11}(0) \subset R(E_1)$ (resp. $T_{22}(0) \subset R(E_2)$)). Then,*

$$\sigma_\delta(\mathscr{T}) \subset \bigcup_{i\in\{1,2,3\}} \sigma_\delta(T_{ii}),$$

where the matrix $\mathscr{T}$ is given in (3.9.19). ◇

Proposition 3.9.10 *Let $T_{ii} \in B\mathscr{R}(X_i)$ and $T_{ij} \in B\mathscr{R}(X_j, X_i)$, for all $i \in \{1,2,3\}$, $j = 2,3$ and $i \neq j \in \{1,2\}$ be such that $\mathscr{D}(T_{33}) \subset N(T_{23})$ and $T_{23}(0) \subset T_{22}(0)$. Then,*

$$\sigma_{T_{33}}(w)\bigcup\sigma_{T_{22}}(z) \subset \sigma_{\mathscr{T}}\begin{pmatrix} y \\ z \\ w \end{pmatrix},$$

for all $y \in \mathscr{D}(T_{11})$, $z \in \mathscr{D}(T_{22})$, and $w \in \mathscr{D}(T_{33})$, where the matrix $\mathscr{T}$ is given in (3.9.19). ◇

Proof Let $\lambda_0 \notin \sigma_{\mathscr{T}}\begin{pmatrix} y \\ z \\ w \end{pmatrix}$. Then, there exist an open neighborhood V_{λ_0} and a nonzero analytic function $f : V_{\lambda_0} \longrightarrow \mathscr{D}(T_{11}) \times \mathscr{D}(T_{22}) \times \mathscr{D}(T_{33})$ such that

$$(\mathscr{T}-\lambda)f(\lambda) = \begin{pmatrix} y \\ z \\ w \end{pmatrix} + \mathscr{T}(0), \quad \text{for all } \lambda \in V_{\lambda_0}.$$

Hence, for $f = \begin{pmatrix} f_1 \\ f_2 \\ f_3 \end{pmatrix}$ with $f_1 : V_{\lambda_0} \longrightarrow \mathscr{D}(T_{11})$, $f_2 : V_{\lambda_0} \longrightarrow \mathscr{D}(T_{22})$ and $f_3 : V_{\lambda_0} \longrightarrow \mathscr{D}(T_{33})$ which are analytic functions, we have

$$\begin{cases} (T_{11}-\lambda)f_1(\lambda) + T_{12}f_2(\lambda) + T_{13}f_3(\lambda) = y + T_{11}(0) \\ (T_{22}-\lambda)f_2(\lambda) + T_{23}f_3(\lambda) = z + T_{22}(0), \\ (T_{33}-\lambda)f_3(\lambda) = w + T_{33}(0), \quad \text{for all } \lambda \in V_{\lambda_0}. \end{cases} \tag{3.9.20}$$

So, $\lambda \notin \sigma_{T_{33}}(w)$. This allows us to deduce that

$$\sigma_{T_{33}}(w) \subset \sigma_{\mathscr{T}} \begin{pmatrix} y \\ z \\ w \end{pmatrix}.$$

Moreover, by using the fact that $f_3(\lambda) \in \mathscr{D}(T_{33}) \subset N(T_{23})$, we get

$$T_{23} f_3(\lambda) = T_{23}(0). \tag{3.9.21}$$

By combining (3.9.20) and (3.9.21), we obtain

$$\begin{aligned} z + T_{22}(0) &= (T_{22} - \lambda) f_2(\lambda) + T_{23}(0) \\ &= (T_{22} - \lambda)(f_2(\lambda) + 0) + T_{23}(0) \\ &= (T_{22} - \lambda) f_2(\lambda) + (T_{22} - \lambda)(0) + (T_{23} - \lambda)(0), \quad \text{for all } \lambda \in V_{\lambda_0}. \end{aligned}$$

Since, $T_{23}(0) \subset T_{22}(0)$, then

$$(T_{22} - \lambda) f_2(\lambda) = z + T_{22}(0), \quad \text{for all } \lambda \in V_{\lambda_0}.$$

As a consequence, $\sigma_{T_{22}}(z) \subset \sigma_{\mathscr{T}} \begin{pmatrix} y \\ z \\ w \end{pmatrix}$. Q.E.D.

Remark 3.9.3 It is clear that

$$\begin{aligned} S_{T_{11}} &\subset S_{\mathscr{T}} \\ &\subset \bigcup_{i \in \{1,2,3\}} S_{T_{ii}}, \end{aligned}$$

where the matrix $\mathscr{T}$ is given in (3.9.19). Hence, by using of Corollary 3.9.12, we obtain that

$$\sigma(\mathscr{T}) \subset \bigcup_{i \in \{1,2,3\}} \sigma(T_{ii}).$$

◇

Proposition 3.9.11 *Let $T_{ii} \in B\mathscr{R}(X_i)$ and $T_{ij} \in B\mathscr{R}(X_j, X_i)$, for all $i \in \{1, 2, 3\}$ and $j \in \{2, 3\}$ be such that $T_{ii}(0) \subset T_{ij}(0)$ for $j \neq i \in \{1, 2\}$. Then, for all $x \in V_{\lambda_0}$,*

$$S_{T_{22}} \bigcup S_{T_{33}} \bigcup \sigma_{T_{11}}(x) = S_{T_{22}} \bigcup S_{T_{33}} \bigcup \sigma_{\mathscr{T}} \begin{pmatrix} x \\ 0 \\ 0 \end{pmatrix},$$

where the matrix $\mathscr{T}$ is given in (3.9.19). ◇

Proof Let $\lambda_0 \notin S_{T_{ii}} \bigcup \sigma_{\mathscr{T}}\begin{pmatrix} x \\ 0 \\ 0 \end{pmatrix}$, for all $i \in \{2, 3\}$. Then, there exist an open neighborhood V_{λ_0} of λ_0 in $\mathbb{C}$ and a nonzero analytic function $f_i : V_{\lambda_0} \longrightarrow X_1 \times X_2 \times X_3$ such that

$$(\mathscr{T} - \lambda) f_i(\lambda) = \begin{pmatrix} x \\ 0 \\ 0 \end{pmatrix} + \mathscr{T}(0), \quad \text{for all } \lambda \in V_{\lambda_0}.$$

Let $f_1 : V_{\lambda_0} \longrightarrow X_1$, $f_2 : V_{\lambda_0} \longrightarrow X_2$ and $f_3 : V_{\lambda_0} \longrightarrow X_3$ be analytic functions such that $f_i = \begin{pmatrix} f_1 \\ f_2 \\ f_3 \end{pmatrix}$. Thus,

$$\begin{cases} (T_{11} - \lambda) f_1(\lambda) + T_{12} f_2(\lambda) + T_{13} f_3(\lambda) = x + T_{11}(0), \\ (T_{22} - \lambda) f_2(\lambda) + T_{23} f_3(\lambda) = T_{22}(0), \\ (T_{33} - \lambda) f_3(\lambda) = T_{33}(0), \qquad \text{for all } \lambda \in V_{\lambda_0}. \end{cases} \tag{3.9.22}$$

Since $\lambda_0 \notin S_{T_{33}}$, it follows from (3.9.22) that $f_3 = 0$ on V_{λ_0}. In addition, using both of hypotheses $\lambda_0 \notin S_{T_{22}}$ and $T_{22}(0) \subset T_{23}(0)$ allows us to conclude that $f_2 = 0$ on V_{λ_0}. Again, by referring to (3.9.22), we infer that

$$(T_{11} - \lambda) f_1(\lambda) = x + T_{11}(0) \ \text{ for all } \lambda \in V_{\lambda_0}.$$

Thus, $\lambda_0 \notin \sigma_{T_{11}}(x)$. It is easy to see the converse inclusion. Q.E.D.

Corollary 3.9.13 *Let $T_{ii} \in B\mathscr{R}(X_i)$ and $T_{ij} \in B\mathscr{R}(X_j, X_i)$, for all $i \in \{1, 2, 3\}$ and $j \in \{2, 3\}$ be such that $T_{ii}(0) \subset T_{ij}(0)$ for $j \neq i \in \{1, 2\}$. If T_{22} and T_{33} have the SVEP. Then, for all $x \in V_{\lambda_0}$,*

$$\sigma_{T_{11}}(x) = \sigma_{\mathscr{T}}\begin{pmatrix} x \\ 0 \\ 0 \end{pmatrix},$$

where the matrix $\mathscr{T}$ is given in (3.9.19). ◇

3.9.5 Spectral Properties of 3 × 3 Linear Relations Matrices Through the SVEP

We begin this subsection by giving a new characterization of the onto spectrum of $\mathscr{T}$, given in (3.9.19), by means of SVEP. We first need the following result on the onto spectrum of $\mathscr{T}$.

Lemma 3.9.1 *Let $T_{ii} \in B\mathscr{R}(X_i)$ and $T_{ij} \in B\mathscr{R}(X_j, X_i)$, for all $i \in \{1, 2, 3\}$ and $j \in \{2, 3\}$ be such that $T_{ii}(0) \subset T_{ij}(0)$ for $j \neq i \in \{1, 2\}$ and $\mathscr{D}(T_{33}) \subset N(T_{23})$. If T_{11} (resp. T_{22}) has a selection E_1 (resp. E_2) which satisfies $T_{11}(0) \subset R(E_1)$ (resp. $T_{22}(0) \subset R(E_2)$) and T_{33} has the SVEP, then*

$$\sigma_\delta(\mathscr{T}) = \bigcup_{i\in\{1,2,3\}} \sigma_\delta(T_{ii}),$$

where the matrix $\mathscr{T}$ is given in (3.9.19). ◇

Proof By applying Corollary 3.9.13, we can deduce that

$$\sigma_{T_{11}}(x) = \sigma_{\mathscr{T}}\begin{pmatrix} x \\ 0 \\ 0 \end{pmatrix}. \tag{3.9.23}$$

Since $\sigma_\delta(T) = \bigcup\limits_{x\in X} \sigma_T(x)$, then

$$\begin{aligned} \sigma_\delta(T_{11}) &= \bigcup_{y\in\mathscr{D}(T_{11})} \sigma_{T_{11}}(y) \\ &= \bigcup_{y\in\mathscr{D}(T_{11})} \sigma_{\mathscr{T}}\begin{pmatrix} y \\ 0 \\ 0 \end{pmatrix}, \quad \text{(from (3.9.23))} \\ &\subset \sigma_\delta(\mathscr{T}). \end{aligned}$$

In addition, by referring to Proposition 3.9.10, we infer that

$$\begin{aligned} \sigma_\delta(T_{33}) &\subset \bigcup_{w\in\mathscr{D}(T_{33})} \sigma_{T_{33}}\begin{pmatrix} 0 \\ 0 \\ w \end{pmatrix} \\ &\subset \bigcup_{w\in\mathscr{D}(T_{33})} \sigma_{\mathscr{T}}\begin{pmatrix} 0 \\ 0 \\ w \end{pmatrix} \\ &\subset \sigma_\delta(\mathscr{T}). \end{aligned}$$

Again, by using of Proposition 3.9.10, we obtain

$$\sigma_\delta(T_{22}) \subset \sigma_\delta(\mathscr{T}).$$

So, we get

$$\bigcup_{i\in\{1,2,3\}} \sigma_\delta(T_{ii}) \subset \sigma_\delta(\mathscr{T}).$$

Conversely, by applying Corollary 3.9.12, we can deduce that

$$\sigma_\delta(\mathscr{T}) \subset \sigma_\delta(T_{11}) \bigcup \sigma_\delta(T_{22}) \bigcup \sigma_\delta(T_{33}).$$

This end the proof. Q.E.D.

One of our main results is to find an equality between the spectrum of $\mathscr{T}$, given in (3.9.19), and the spectrum of its diagonal entries invelvig SVEP which is the following.

Theorem 3.9.12 *Let $T_{ii} \in B\mathscr{R}(X_i)$ and $T_{ij} \in B\mathscr{R}(X_j, X_i)$, for all $i \in \{1, 2, 3\}$ and $j \in \{2, 3\}$ be such that $T_{ii}(0) \subset T_{ij}(0)$ for $j \neq i \in \{1, 2\}$ and $\mathscr{D}(T_{33}) \subset N(T_{23})$. If T_{11} (resp. T_{22}) has a selection E_1 (resp. E_2) which satisfies $T_{11}(0) \subset R(E_1)$ (resp. $T_{22}(0) \subset R(E_2)$) and T_{33} has the SVEP, then*

$$\sigma(\mathscr{T}) = \bigcup_{i\in\{1,2,3\}} \sigma(T_{ii}),$$

where the matrix $\mathscr{T}$ is given in (3.9.19). ◇

Proof In view of Lemma 3.9.1, we have

$$\sigma_\delta(\mathscr{T}) = \bigcup_{i\in\{1,2,3\}} \sigma_\delta(T_{ii}).$$

Using the fact that T_{33} has the SVEP, we infer that

$$\bigcup_{i\in\{1,2\}} \sigma_\delta(T_{ii}) \bigcup \sigma(T_{33}) \subset \sigma(\mathscr{T}).$$

Accordingly,

$$\bigcup_{i\in\{1,2\}} S_{T_{ii}} \bigcup_{i\in\{1,2\}} \sigma_\delta(T_{ii}) \bigcup \sigma(T_{33}) \subset \bigcup_{i\in\{1,2\}} S_{T_{ii}} \bigcup \sigma(\mathscr{T}).$$

Also, applying Corollary 3.9.13, we have

$$\begin{aligned}\bigcup_{i\in\{1,2\}} S_{T_{ii}} \bigcup_{i\in\{1,2\}} \sigma_\delta(T_{ii}) \bigcup \sigma(T_{33}) &= \bigcup_{i\in\{1,2,3\}} \sigma(T_{ii}) \\ &\subset \bigcup_{i\in\{1,2\}} S_{T_{ii}} \bigcup \sigma(\mathscr{T}).\end{aligned}$$

Furthermore, since $S_{T_{11}} \bigcup S_{T_{22}} \subset S_{\mathscr{T}}$, then

$$\bigcup_{i\in\{1,2,3\}} \sigma(T_{ii}) \subset \sigma(\mathscr{T}).$$

Conversely, it follows from Remark 3.9.3, we get

$$\sigma_\delta(\mathcal{T}) \subset \bigcup_{i\in\{1,2,3\}} \sigma_\delta(T_{ii}). \qquad \text{Q.E.D.}$$

In what follows, under some new conditions, we prove that the upper triangular matrix of linear relations also has the SVEP.

Proposition 3.9.12 *Let $T_{ii} \in B\mathcal{R}(X_i)$ and $T_{ij} \in B\mathcal{R}(X_j, X_i)$, for all $i \in \{1, 2, 3\}$ and $j = 2, 3$ be such that $T_{ii}(0) \subset T_{ij}(0)$ for $j \neq i \in \{1, 2\}$. If T_{ii} have the SVEP, for all $i \in \{1, 2, 3\}$, then $\mathcal{T}$ also has the SVEP, where the matrix $\mathcal{T}$ is given in (3.9.19).* ◇

Proof Let $f : V_{\lambda_0} \longrightarrow X_1 \times X_2 \times X_3$ be an analytic function on an open neighborhood V_{λ_0} of λ_0 in $\mathbb{C}$ satisfying

$$(\mathcal{T} - \lambda) f(\lambda) = \mathcal{T}(0), \quad \text{for all } \lambda \in V.$$

Let $f = \begin{pmatrix} f_1 \\ f_2 \\ f_3 \end{pmatrix}$ be with $f_1 : V_{\lambda_0} \longrightarrow X_1$, $f_2 : V_{\lambda_0} \longrightarrow X_2$ and $f_3 : V_{\lambda_0} \longrightarrow X_3$ are analytic functions. Hence, we get

$$\begin{cases} (T_{11} - \lambda) f_1(\lambda) + T_{12} f_2(\lambda) + T_{13} f_3(\lambda) = T_{11}(0) \\ (T_{22} - \lambda) f_2(\lambda) + T_{23} f_3(\lambda) = T_{22}(0), \\ (T_{33} - \lambda) f_3(\lambda) = T_{33}(0), \qquad \text{for all } \lambda \in V_{\lambda_0}. \end{cases}$$

Based on the fact that T_{33} has the SVEP, we can conclude that $f_3(\lambda) = 0$, for all $\lambda \in V$. Therefore, we have

$$(T_{22} - \lambda) f_2(\lambda) = T_{22}(0), \quad \text{for all } \mu \in V.$$

The fact that T_{22} has the SVEP and $T_{22}(0) \subset T_{23}(0)$, allows us to deduce that $f_2(\lambda) = 0$ on V. So, we obtain

$$(T_{11} - \lambda) f_1(\lambda) = T_{11}(0).$$

Again, since T_{11} has the SVEP, $T_{11}(0) \subset T_{12}(0)$ and $T_{11}(0) \subset T_{13}(0)$, then $f_1(\lambda) = 0$ on V. Consequently, $\mathcal{T}$ has the SVEP. Q.E.D.

An immediate consequence of the Proposition 3.9.12 is the following.

Corollary 3.9.14 *Let $T_{ii} \in B\mathcal{R}(X_i)$ and $T_{ij} \in B\mathcal{R}(X_j, X_i)$, for all $i \in \{1, 2, 3\}$ and $j = 2, 3$ be such that $T_{ii}(0) \subset T_{ij}(0)$ for $j \neq i \in \{1, 2\}$. If T_{ii} have the SVEP, for all $i \in \{1, 2, 3\}$, then*

$$\sigma(\mathcal{T}) = \bigcup_{i\in\{1,2,3\}} \sigma(T_{ii}), \tag{3.9.24}$$

where the matrix $\mathcal{T}$ is given in (3.9.19). ◇

Proof By referring to Theorem 3.9.1, we infer that

$$\sigma_\delta(\mathscr{T}) = \bigcup_{i\in\{1,2,3\}} \sigma_\delta(T_{ii}). \tag{3.9.25}$$

Also, since T_{ii} have the SVEP, for all $i \in \{1, 2, 3\}$, we can obtain that

$$\begin{aligned} S_{T_{11}} &= S_{T_{22}} \\ &= S_{T_{33}} \\ &= \emptyset. \end{aligned}$$

This is equivalent to

$$\sigma_\delta(T_{11}) = \sigma(T_{11}), \quad \sigma_\delta(T_{22}) = \sigma(T_{22}) \quad \text{and} \quad \sigma_\delta(T_{33}) = \sigma(T_{33}). \tag{3.9.26}$$

In view of Proposition 3.9.12, we deduce that $\mathscr{T}$ also has the SVEP. This leads to

$$\sigma_\delta(\mathscr{T}) = \sigma(\mathscr{T}). \tag{3.9.27}$$

As a consequence, by combining (3.9.25), (3.9.26) and (3.9.27), we conclude that (3.9.24) holds. Q.E.D.

Proposition 3.9.13 *Let $T_{ii} \in B\mathscr{R}(X_i)$ and $T_{ij} \in B\mathscr{R}(X_j, X_i)$ for all $i \in \{1, 2, 3\}$ and $j \in \{2, 3\}$ be such that $T_{ii}(0) \subset T_{ij}(0)$ for $j \neq i \in \{1, 2\}$. If T_{11} (resp. T_{22}) has a selection E_1 (resp. E_2) which satisfies $T_{11}(0) \subset R(E_1)$ (resp. $T_{22}(0) \subset R(E_2)$). Then,*

$$S_{T_{22}} \bigcup S_{T_{33}} \bigcup \sigma_\delta(\mathscr{T}) = \sigma_\delta(T_{11}) \bigcup \sigma(T_{22}) \bigcup \sigma(T_{33}),$$

where the matrix $\mathscr{T}$ is given in (3.9.19). ◇

Proof By using Corollary 3.9.12, we obtain

$$\sigma_\delta(\mathscr{T}) \subset \bigcup_{\in\{1,2,3\}} \sigma_\delta(T_{ii}).$$

Hence,

$$S_{T_{22}} \bigcup S_{T_{33}} \bigcup \sigma_\delta(\mathscr{T}) \subset S_{T_{22}} \bigcup S_{T_{33}} \bigcup \bigcup_{\in\{1,2,3\}} \sigma_\delta(T_{ii}).$$

Consequently,

$$S_{T_{22}} \bigcup S_{T_{33}} \bigcup \sigma_\delta(\mathscr{T}) \subset \sigma_\delta(T_{11}) \bigcup \bigcup_{\in\{2,3\}} \sigma(T_{ii}).$$

Conversely, by referring to Proposition 3.9.11, we infer that

$$S_{T_{22}} \bigcup S_{T_{33}} \bigcup \sigma_{T_{11}}(x) = S_{T_{22}} \bigcup S_{T_{33}} \bigcup \sigma_{\mathscr{T}} \begin{pmatrix} x \\ 0 \\ 0 \end{pmatrix}.$$

Therefore,

$$S_{T_{22}} \bigcup S_{T_{33}} \bigcup \sigma_{\delta}(T_{11}) \subset S_{T_{22}} \bigcup S_{T_{33}} \bigcup \sigma_{\delta}(\mathscr{T}).$$

Using the fact that $\sigma_{\delta}(T_{22}) \subset \sigma_{\delta}(\mathscr{T})$ and $\sigma_{\delta}(T_{33}) \subset \sigma_{\delta}(\mathscr{T})$, we can deduce that

$$\begin{aligned} \sigma_{\delta}(T_{22}) \bigcup \sigma_{\delta}(T_{33}) \bigcup S_{T_{22}} \bigcup S_{T_{33}} \bigcup \sigma_{\delta}(T_{11}) = \; & \sigma_{\delta}(T_{11}) \bigcup \sigma(T_{22}) \bigcup \sigma(T_{33}) \\ \subset \; & S_{T_{22}} \bigcup S_{T_{33}} \bigcup \sigma_{\delta}(\mathscr{T}). \end{aligned}$$

This complete the proof. Q.E.D.

In the following theorem, we will study a new characterization of $\sigma(\mathscr{T})$ by means of its adjoint $\mathscr{T}^*$, where the matrix $\mathscr{T}$ is given in (3.9.19).

Theorem 3.9.13 *Let $T_{ii} \in B\mathscr{R}(X_i)$ and $T_{ij} \in B\mathscr{R}(X_j, X_i)$ for all $i \in \{1, 2, 3\}$ and $j \in \{2, 3\}$ be such that $T_{ii}(0) \subset T_{ij}(0)$ for $j \neq i \in \{1, 2\}$. If T_{11} (resp. T_{22}) has a selection E_1 (resp. E_2) which satisfies $T_{11}(0) \subset R(E_1)$ (resp. $T_{22}(0) \subset R(E_2)$). Then,*

$$\left(S_{T_{11}^*} \bigcup S_{T_{22}} \bigcup S_{T_{33}}\right) \bigcup \sigma(\mathscr{T}) = \bigcup_{i \in \{1,2,3\}} \sigma(T_{ii}),$$

where the matrix $\mathscr{T}$ is given in (3.9.19). ◇

Proof In the one hand, we show that

$$\left(S_{T_{22}} \bigcup S_{T_{33}}\right) \bigcup \sigma(\mathscr{T}) = \bigcup_{i \in \{1,2,3\}} \sigma(T_{ii}).$$

Then, by referring to Remark 3.9.3 and Proposition 3.9.13, we infer that

$$\left[S_{T_{11}} \bigcup \sigma_{\delta}(T_{11})\right] \bigcup_{i \in \{2,3\}} \sigma(T_{ii}) \subset S_{T_{22}} \bigcup S_{T_{33}} \bigcup \sigma_{\delta}(\mathscr{T}) \bigcup S_{\mathscr{T}}.$$

It follows that

$$\bigcup_{i \in \{1,2,3\}} \sigma(T_{ii}) \subset S_{T_{22}} \bigcup S_{T_{33}} \bigcup \sigma(\mathscr{T}).$$

In the other hand, in view of Theorem 2.15.6, we get the adjoint of $\mathscr{T}$ is

$$\mathscr{T}^* = \begin{pmatrix} T_{11}^* & 0 & 0 \\ T_{12}^* & T_{22}^* & 0 \\ T_{13}^* & T_{23}^* & T_{33}^* \end{pmatrix}.$$

Using the same reasoning as before, we have

$$S_{T_{11}^*} \bigcup \sigma(\mathscr{T}^*) = \bigcup_{i \in \{1,2,3\}} \sigma(T_{ii}^*).$$

In view of Proposition 3.1.1 (iii), we obtain

$$S_{T_{11}^*} \bigcup \sigma(\mathscr{T}) = \bigcup_{i \in \{1,2,3\}} \sigma(T_{ii})$$

which proves the statement. Q.E.D.

Corollary 3.9.15 *Let $T_{ii} \in B\mathscr{R}(X_i)$ and $T_{ij} \in B\mathscr{R}(X_j, X_i)$ for all $i \in \{1, 2, 3\}$ and $j \in \{2, 3\}$ be such that $T_{ii}(0) \subset T_{ij}(0)$ for $j \neq i \in \{1, 2\}$. If T_{11} (resp. T_{22}) has a selection E_1 (resp. E_2) which satisfies $T_{11}(0) \subset R(E_1)$ (resp. $T_{22}(0) \subset R(E_2)$) and $S_{T_{11}^*} \bigcup S_{T_{22}} \bigcup S_{T_{33}} = \emptyset$. Then,*

$$\sigma(\mathscr{T}) = \bigcup_{i \in \{1,2,3\}} \sigma(T_{ii}),$$

where the matrix $\mathscr{T}$ is given in (3.9.19). ◇

Moving forward, we show an equivalence between the SVEP of $\mathscr{T}$ and the SVEP of its diagonal entries.

Theorem 3.9.14 *Let $T_{ii} \in B\mathscr{R}(X_i)$ and $T_{ij} \in B\mathscr{R}(X_j, X_i)$ for all $i \in \{1, 2, 3\}$ and $j \in \{2, 3\}$ be such that $T_{ii}(0) \subset T_{ij}(0)$ for $j \neq i \in \{1, 2\}$. If T_{11} (resp. T_{22}) has a selection E_1 (resp. E_2) which satisfies $T_{11}(0) \subset R(E_1)$ (resp. $T_{22}(0) \subset R(E_2)$). Then, $S_{\mathscr{T}} = \emptyset$ if, and only if, $S_{T_{11}} = S_{T_{22}} = S_{T_{33}} = \emptyset$, where the matrix $\mathscr{T}$ is given in (3.9.19). In particular, $\mathscr{T}$ has the SVEP if, and only if, T_{ii} have the SVEP, for all $i \in \{1, 2, 3\}$.* ◇

Proof Suppose that $S_{\mathscr{T}} = \emptyset$, then $\mathscr{T}$ has the SVEP. Hence, $\sigma(\mathscr{T}) = \sigma_\delta(\mathscr{T})$. Based on the fact that

$$\sigma_\delta(T_{22}) \subset \sigma_\delta(\mathscr{T})$$

and

$$S_{T_{22}} \subset S_{\mathscr{T}},$$

we can conclude that $S_{T_{11}} = \emptyset$. The same reasoning, we can deduce that

$$S_{T_{11}} = S_{T_{33}} = \emptyset.$$

Conversely, if T_{ii} for all $i \in \{1, 2, 3\}$ have the SVEP, then using Proposition 3.9.12, we obtain $\mathscr{T}$ which has the SVEP. Q.E.D.

We finish this section by giving a new characterization of the local spectrum of 3×3-block matrix of linear relations.

Theorem 3.9.15 *Let $T_{ii} \in B\mathcal{R}(X_i)$ and $T_{ij} \in B\mathcal{R}(X_j, X_i)$, for all $i \in \{1,2,3\}$ and $j \in \{2,3\}$ be such that $T_{12}(0) \bigcap T_{13}(0) \subset T_{11}(0)$, $T_{21}(0) \bigcap T_{23}(0) \subset T_{22}(0)$ and $T_{31}(0) \bigcap T_{32}(0) \subset T_{33}(0)$. If $\mathcal{D}(T_{11}) \subset N(T_{23}) \bigcap N(T_{31})$, $\mathcal{D}(T_{22}) \subset N(T_{12}) \bigcap N(T_{32})$, and $\mathcal{D}(T_{33}) \subset N(T_{13}) \bigcap N(T_{23})$, then for all $\begin{pmatrix} x \\ y \\ z \end{pmatrix} \in X_1 \times X_2 \times X_3$, we have*

$$\sigma_{\mathcal{T}} \begin{pmatrix} x \\ y \\ z \end{pmatrix} = \sigma_{T_{11}}(x) \bigcup \sigma_{T_{22}}(y) \bigcup \sigma_{T_{33}}(z),$$

where the matrix $\mathcal{T}$ is given in (3.9.19). ◇

Proof First, it should be noted that the result is equivalent to show that

$$\rho_{T} \begin{pmatrix} x \\ y \\ z \end{pmatrix} = \rho_{T_{11}}(x) \bigcap \rho_{T_{22}}(y) \bigcap \rho_{T_{33}}(z),$$

for all $\begin{pmatrix} x \\ y \\ z \end{pmatrix} \in X_1 \times X_2 \times X_3$. Let $\lambda \in \rho_{T_{11}}(x) \bigcap \rho_{T_{22}}(y) \bigcap \rho_{T_{33}}(z)$, then there exist three open neighborhoods V_1, V_2 and V_3 of $\lambda \in \mathbb{C}$ and three analytic functions $f_1 : V_1 \longrightarrow X_1$, $f_2 : V_2 \longrightarrow X_2$ and $f_3 : V_3 \longrightarrow X_3$ such that for all $\lambda_0 \in V_1 \bigcap V_2 \bigcap V_3$, we have

$$\begin{cases} (T_{11} - \lambda_0) f_1(\lambda_0) = x + T_{11}(0), \\ (T_{22} - \lambda_0) f_2(\lambda_0) = x + T_{22}(0), \\ (T_{33} - \lambda_0) f_3(\lambda_0) = x + T_{33}(0). \end{cases} \tag{3.9.28}$$

Now, let $\begin{pmatrix} f_1 \\ f_2 \\ f_3 \end{pmatrix} : V_1 \bigcap V_2 \bigcap V_3 \longrightarrow X_1 \times X_2 \times X_3$. Then, $\begin{pmatrix} f_1 \\ f_2 \\ f_3 \end{pmatrix}$ is an analytic function and for all $\lambda_0 \in V_1 \bigcap V_2 \bigcap V_3$, we obtain

$$\begin{aligned} (\mathcal{T} - \lambda_0) \begin{pmatrix} f_1(\lambda_0) \\ f_2(\lambda_0) \\ f_3(\lambda_0) \end{pmatrix} &= \begin{pmatrix} T_{11} - \lambda_0 & T_{12} & T_{13} \\ T_{21} & T_{22} - \lambda_0 & T_{23} \\ T_{31} & T_{32} & T_{33} - \lambda_0 \end{pmatrix} \begin{pmatrix} f_1(\lambda_0) \\ f_2(\lambda_0) \\ f_3(\lambda_0) \end{pmatrix} \\ &= \begin{pmatrix} (T_{11} - \lambda_0) f_1(\lambda_0) + T_{12} f_2(\lambda_0) + T_{13} f_3(\lambda_0) \\ T_{21} f_1(\lambda_0) + (T_{22} - \lambda_0) f_2(\lambda_0) + T_{23} f_3(\lambda_0) \\ T_{31} f_1(\lambda_0) + T_{32} f_2(\lambda_0) + (T_{33} - \lambda_0) f_3(\lambda_0) \end{pmatrix} \end{aligned}$$

$$= \begin{pmatrix} x + T_{11}(0) + T_{12}f_2(\lambda_0) + T_{13}f_3(\lambda_0) \\ T_{21}f_1(\lambda_0) + y + T_{22}(0) + T_{23}f_3(\lambda_0) \\ T_{31}f_1(\lambda_0) + T_{32}f_2(\lambda_0) + y + T_{33}(0) \end{pmatrix} \quad \text{(from (3.9.28))}.$$

Based on the hypotheses $\mathscr{D}(T_{11}) \subset N(T_{23}) \bigcap N(T_{31})$, $\mathscr{D}(T_{22}) \subset N(T_{12}) \bigcap N(T_{32})$, and $\mathscr{D}(T_{33}) \subset N(T_{13}) \bigcap N(T_{23})$ and by referring to Theorem 2.8.1, we infer that

$$\begin{aligned}
(\mathscr{T} - \lambda_0) \begin{pmatrix} f_1(\lambda_0) \\ f_2(\lambda_0) \\ f_3(\lambda_0) \end{pmatrix} &= \begin{pmatrix} x + T_{11}(0) + T_{12}(0) + T_{13}(0) \\ T_{21}(0) + y + T_{22}(0) + T_{23}(0) \\ T_{31}(0) + T_{32}(0) + y + T_{33}(0) \end{pmatrix} \\
&= \begin{pmatrix} T_{11}(0) + T_{12}(0) + T_{13}(0) \\ T_{21}(0) + T_{22}(0) + T_{23}(0) \\ T_{31}(0) + T_{32}(0) + T_{33}(0) \end{pmatrix} + \begin{pmatrix} x \\ y \\ z \end{pmatrix} \\
&= \mathscr{T} \begin{pmatrix} 0 \\ 0 \\ 0 \end{pmatrix} + \begin{pmatrix} x \\ y \\ z \end{pmatrix}.
\end{aligned}$$

Hence,

$$\rho_{T_{11}}(x) \bigcap \rho_{T_{22}}(y) \bigcap \rho_{T_{33}}(z) \subset \rho_T \begin{pmatrix} x \\ y \\ z \end{pmatrix}.$$

For the reverse inclusion, let us suppose that $\lambda \in \rho_T \begin{pmatrix} x \\ y \\ z \end{pmatrix}$, then there exist an open neighborhoods V of $\lambda \in \mathbb{C}$ and an analytic function

$$\begin{pmatrix} f_1 \\ f_2 \\ f_3 \end{pmatrix} : V \longrightarrow X_1 \times X_2 \times X_3$$

such that for all $\lambda_0 \in V$, we have

$$(\mathscr{T} - \lambda_0) \begin{pmatrix} f_1(\lambda_0) \\ f_2(\lambda_0) \\ f_3(\lambda_0) \end{pmatrix} = \begin{pmatrix} x \\ y \\ z \end{pmatrix} + \mathscr{T} \begin{pmatrix} 0 \\ 0 \\ 0 \end{pmatrix}.$$

This is equivalent to

$$\begin{cases} (T_{11} - \lambda_0)f_1(\lambda_0) + T_{12}f_2(\lambda_0) + T_{13}f_3(\lambda_0) = x + T_{11}(0) + T_{12}(0) + T_{13}(0), \\ T_{21}f_1(\lambda_0) + (T_{22} - \lambda_0)f_2(\lambda_0) + T_{23}f_3(\lambda_0) = T_{21}(0) + y + T_{22}(0) + T_{23}(0), \\ T_{31}f_1(\lambda_0) + T_{32}f_2(\lambda_0) + (T_{33} - \lambda_0)f_3(\lambda_0) = T_{31}(0) + T_{32}(0) + y + T_{33}(0). \end{cases}$$

Since, $T_{12}(0) \bigcap T_{13}(0) \subset T_{11}(0)$, then

$$(T_{11} - \lambda_0) f_1(\lambda_0) + T_{12} f_2(\lambda_0) + T_{13} f_3(\lambda_0) = x + T_{11}(0).$$

Also, it follows from $f_2(\lambda_0) \in \mathcal{D}(T_{22}) \subset N(T_{12})$ and $f_3(\lambda_0) \in \mathcal{D}(T_{33}) \subset N(T_{13})$ that

$$\begin{aligned} x + T_{11}(0) &= (T_{11} - \lambda_0) f_1(\lambda_0) + T_{12} f_2(\lambda_0) + T_{13} f_3(\lambda_0) \\ &= (T_{11} - \lambda_0) f_1(\lambda_0) + T_{12}(0) + T_{13}(0) \\ &= (T_{11} - \lambda_0) f_1(\lambda_0) + (T_{11} - \lambda_0)(0) + (T_{12} - \lambda_0)(0) + (T_{13} - \lambda_0)(0) \\ &= (T_{11} - \lambda_0) f_1(\lambda_0). \end{aligned}$$

Thus, $\lambda \in \rho_{T_{11}}(x)$. The same reasoning allows us to conclude that $\lambda \in \rho_{T_{22}}(y)$ and $\lambda \in \rho_{T_{33}}(z)$. Q.E.D.

3.10 The Jeribi Essential Spectrum

3.10.1 Definition

Definition 3.10.1 Let X be a Banach space and let $A \in C\mathcal{R}(X)$. The Jeribi essential spectrum is defined by

$$\sigma_j(A) := \bigcap_{K \in \mathcal{W}_*(X)} \sigma(A + K),$$

where $\mathcal{W}_*(X)$ stands for each one of the sets $WK\mathcal{R}(X)$ and $\mathcal{S}\mathcal{S}\mathcal{R}(X)$, where $WK\mathcal{R}(X)$ (resp. $\mathcal{S}\mathcal{S}\mathcal{R}(X)$) is the families of all weakly compact linear relations on X (resp. the set of strictly singular linear relations on X). ◇

For more important contributions concerning the Jeribi essential spectrum and its applications to transport operators were due to A. Jeribi and his collaborators (see the bibliography at the end of the chapter [103, 115, 116]).

3.10.2 Open Question

Can we prove the same results, obtained in this chapter, for the study of the Jeribi essential spectrum.

References

1. Erdélyi, I., Wang, S.W.: A local spectral theory for closed operators. In: London Mathematical Society Lecture Note Series, vol. 105. Cambridge University Press, Cambridge (1985)
2. Laursen, K.B., Neumann, M.M.: An introduction to local spectral theory. In: London Mathematical Society Monographs. New Series, vol. 20. The Clarendon Press, Oxford University Press, New York (2000)
3. Dunford, N.: Spectral theory. II. Resolutions of the identity. Pac. J. Math. **2**, 559–614 (1952)
4. Finch, J.K.: The single-valued extension property on a Banach space. Pac. J. Math. **58**(1), 61–69 (1975)
5. Aiena, P., Triolo, S.: Fredholm spectra and Weyl type theorems for Drazin invertible operators. Mediterr. J. Math. **13**(6), 4385–4400 (2016)
6. Aiena, P., Triolo, S.: Local spectral theory for Drazin invertible operators. J. Math. Anal. Appl. **435**(1), 414–424 (2016)
7. Aiena, P., Triolo, S.: Weyl-type theorems on Banach spaces under compact perturbations. Mediterr. J. Math. **15**(3), Paper No. 126, 18 (2018)
8. Dales, H.G., Aiena, P., Eschmeier, J., Laursen, K., Willis, G.A.: Introduction to Banach algebras, operators, and harmonic analysis. In: London Mathematical Society Student Texts, vol. 57. Cambridge University Press, Cambridge (2003)
9. Aiena, P., Trapani, C., Triolo, S.: SVEP and local spectral radius formula for unbounded operators. Filomat **28**(2), 263–273 (2014)
10. Cross, R.: Multivalued linear operators. In: Monographs and Textbooks in Pure and Applied Mathematics, vol. 213, pp. x+335. Marcel Dekker, Inc., New York (1998). ISBN: 0-8247-0219-0
11. Lajnef, M., Mnif, M.: Isolated spectral points of a linear relation. Monatsh. Math. **191**, 595–614 (2020)
12. Reed, M., Simon, B.: Methods of Modern Mathematical Physics: Analysis of Operators, vol. 4. Academic, New York (1978)
13. Lajnef, M., Mnif, M.: On generalized Drazin invertible linear relations. Rocky Mountain J. Math. **50**(4), 1387–1408 (2020)
14. Ammar, A., Jeribi, A.: Spectral Theory of Multivalued Linear Operators, p. 314 (English). CRC Press, Boca Raton, FL (2021). ISBN 9781771889667
15. Aiena, P.: Fredholm and Local Spectral Theory, with Applications to Multipliers, pp. xiv+444. Kluwer Academic Publishers, Dordrecht (2004). ISBN: 1-4020-1830-4
16. Chamkha, Y., Mnif, M.: Browder spectra of upper triangular matrix linear relations. Publ. Math. Debrecen **82**(3–4), 569–590 (2013)
17. Ammar, A., Fakhfakh, S., Jeribi, A.: Demicompact operators and ellipsoid spectra on quaternionic Hilbert spaces. Rev. R. Acad. Cienc. Exactas Fís. Nat. Ser. A Mat. RACSAM **114**(3), Paper No. 149, 10 (2020)
18. Jeribi, A., Krichen, B., Zitouni, A.: Properties of demicompact operators, essential spectra and some perturbation results for block operator matrices with applications. Linear Multilinear Algebra **68**(12), 2506–2522 (2020)
19. Ammar, A., Daoud, H., Jeribi, A.: Pseudospectra of multivalued linear operators. (English) Zbl 07361671; Jeribi, A. (ed.).: Operator theory. In: Proceedings of the International Conference on Operator Theory, pp. 85–100. De Gruyter. De Gruyter Proc. Math., Hammamet, Tunisia, Berlin (2021)
20. Benkhaled, H., Elleuch, A., Jeribi, A.: Demicompactness properties for uniformly continuous cosine families. Mediterr. J. Math. **19**(4), 157 (2022)
21. Jeribi, A., Krichen, B., Salhi, M.: Characterization of Essential Spectra Involving Weakly Demicompact Operators on Banach Spaces. Bull. Iranian Math. Soc. **48**(4), 1513–1538 (2022)
22. Charfi, S., Jeribi, A., Walha, I.: Riesz basis property of families of nonharmonic exponentials and application to a problem of a radiation of a vibrating structure in a light fluid. Numer. Funct. Anal. Optim. **32**(4), 370–382 (2011)

23. Charfi, S., Jeribi, A., Walha, I.: Essential spectra, matrix operator and applications. Acta Appl. Math. **111**(3), 319–337 (2010)
24. Chelly, S., Jeribi, A., Krichen, B.: Stability of relative essential spectra of the transport operator in L^1-space by means of measures of noncompactness and relative weak demicompactness. Monatsh. Math. **198**(4), 653–674 (2022)
25. Ammar, A., Fakhfakh, S., Jeribi, A.: Fredholm theory for demicompact linear relations. Appl. Gen. Topol. **23**(2), 425–436 (2022)
26. Benkhaled, H., Elleuch, A., Jeribi, A.: Uniformly continuous cosine families properties around weak demicompactness concept. Hacet. J. Math. Stat. **51**(6), 1674–1679 (2022)
27. Ammar, A., Jeribi, A., Saadaoui, B.: Demicompactness, selection of linear relation and application to multivalued matrix. Filomat **36**(7), 2215–2235 (2022)
28. Jeribi, A., Krichen, B., Salhi, M.: Weak demicompactness involving measures of weak noncompactness and invariance of the essential spectrum. Filomat **36**(9), 3051–3073 (2022)
29. Benkhaled, H., Hajji, M., Jeribi, A.: L-weakly and M-weakly demicompact operators on Banach lattices. Filomat **36**(13), 4319–4329 (2022)
30. Chelly, S., Jeribi, A., Krichen, B.: Demicompactness perturbation in Banach algebras and some stability results. Georgian Math. J. **30**(1), 53–63 (2023)
31. Jeribi, A.: Linear Operators and Their Essential Pseudospectra. CRC Press, Boca Raton (2018)
32. Jeribi, A., (ed.).: Operator theory. In: Proceedings of the International Conference on Operator Theory. Hammamet, Tunisia (2018). (English) Zbl 1458.47004 De Gruyter Proceedings in Mathematics Series Profile, pp. x, 216. De Gruyter, Berlin (2021). ISBN 978-3-11-059686-1/hbk; 978-3-11-059819-3/ebook
33. Jeribi, A.: Perturbation theory for linear operators. In: Denseness and Bases with Applications, p. 509. Springer, Singapore (2021). ISBN: 978-981-16-2527-5; 978-981-16-2528-2
34. Jeribi, A., Krichen, B., Dhahri, M.Z.: Essential spectra of some matrix operators involving-relatively bounded entries and an application. Linear Multilinear Algebra **64**(8), 1654–1668 (2016)
35. Jeribi, A., Krichen, B., Mefteh, B.: Existence of solutions of a two-dimensional boundary value problem for a system of nonlinear equations arising in growing cell populations. J. Biol. Dyn. **7**(1), 218–232 (2013)
36. Jeribi, A., Krichen, B., Salhi, M.: Characterization of relatively demicompact operators by means of measures of noncompactness. J. Korean Math. Soc. **55**(4), 877–895 (2018)
37. Abdelmoumen, B., Dehici, A., Jeribi, A., Mnif, M.: Some new properties in Fredholm theory, Schechter essential spectrum, and application to transport theory. J. Inequal. Appl. Art. ID 852676, 14 (2008)
38. Abdelmoumen, B., Jedidi, O., Jeribi, A.: Time asymptotic description of an abstract Cauchy problem solution and application to transport equation. Appl. Math. **59**(1), 53–67 (2014)
39. Abdelmoumen, B., Jeribi, A., Mnif, M.: Invariance of the Schechter essential spectrum under polynomially compact operators perturbation. Extracta Math. **26**(1), 61–73 (2011)
40. Abdelmoumen, B., Jeribi, A., Mnif, M.: Time asymptotic description of the solution to an abstract Cauchy problem and application to transport equation. Math. Z. **268**(3–4), 837–869 (2011)
41. Abdelmoumen, B., Jeribi, A., Mnif, M.: Measure of weak noncompactness, some new properties in Fredholm theory, characterization of the Schechter essential spectrum and application to transport operators. Ric. Mat. **61**(2), 321–340 (2012)
42. Abdelmoumen, B., Jeribi, A., Mnif, M.: On graph measures in Banach spaces and description of essential spectra of multidimensional transport equation. Acta Math. Sci. Ser. B (Engl. Ed.) **32**(5), 2050–2064 (2012)
43. Abdmouleh, F., Ammar, A., Jeribi, A.: Stability of the pencils essential spectra on a Banach space. Math. Slovaca **63**(2), 299–320 (2013)
44. Abdmouleh, F., Ammar, A., Jeribi, A.: A characterization of the pseudo-Browder essential spectra of linear operators and application to a transport equation. J. Comput. Theor. Transp. **44**(3), 141–153 (2015)

45. Abdmouleh, F., Charfi, S., Jeribi, A.: On a characterization of the essential spectra of the sum and the product of two operators. J. Math. Anal. Appl. **386**(1), 83–90 (2012)
46. Abdmouleh, F., Àlvarez, T., Ammar, A., Jeribi, A.: Spectral mapping theorem for Rakočević and Schmoeger essential spectra of a multivalued linear operator. Mediterr. J. Math. **12**(3), 1019–1031 (2015)
47. Abdmouleh, F., Jeribi, A.: Gustafson, Weidman, Kato, Wolf, Schechter, Browder, Rakočević and Schmoeger essential spectra of the sum of two bounded operators and application to a transport operator. Math. Nachr. **284**(2–3), 166–176 (2011)
48. Abdmouleh, F., Jeribi, A.: Symmetric family of Fredholm operators of indices zero, stability of essential spectra and application to transport operators. J. Math. Anal. Appl. **364**(2), 414–423 (2010)
49. Charfi, S., Jeribi, A., Moalla, R.: Essential spectra of operator matrices and applications. Math. Methods Appl. Sci. **37**(4), 597–608 (2014)
50. Jeribi, A.: Quelques remarques sur le spectre de Weyl et applications. (French) [Some remarks on the Weyl spectrum and applications] C. R. Acad. Sci. Paris Sér. I Math. **327**(5), 485–490 (1998)
51. Ammar, A., Boutaf, F.Z., Jeribi, A.: Extended eigenvalues of a closed linear operator. Filomat **36**(13), 4277–4291 (2022)
52. Ammar, A., Boutaf, F.Z., Jeribi, A.: New results on perturbations of p-adic linear operators. Georgian Math. J. **30**(2), 173–184 (2023)
53. Ammar, A., Boutaf, F.Z., Jeribi, A.: Extended eigenvalues of 2×2 block operator matrices. Filomat **37**(5), 1377–1389 (2023)
54. Ammar, A., Bouzayeni, F., Jeribi, A.: Perturbation of unbounded linear operators by-relative boundedness. Ric. Mat. **67**(2), 433–445 (2018)
55. Ammar, A., Daoud, H., Jeribi, A.: Pseudospectra and essential pseudospectra of multivalued linear relations. Mediterr. J. Math. **12**(4), 1377–1395 (2015)
56. Ammar, A., Daoud, H., Jeribi, A.: Demicompact and k-D-set-contractive multivalued linear operators. Mediterr. J. Math. **15**(2), Art. 41, 18 (2018)
57. Ammar, A., Daoud, H., Jeribi, A.: The index of a linear relation under strictly singular perturbations. Adv. Oper. Theory **7**(1), Paper No. 11 (2022)
58. Ammar, A., Dhahri, M.Z., Jeribi, A.: Some properties of the M-essential spectra of closed linear operator on a Banach space. Funct. Anal. Approx. Comput. **7**(1), 15–28 (2015)
59. Ammar, A., Dhahri, M.Z., Jeribi, A.: A characterization of S-essential spectrum by means of measure of non-strict-singularity and application. Azerb. J. Math. **5**(1), 73–88 (2015)
60. Ammar, A., Dhahri, M.Z., Jeribi, A.: Stability of essential approximate point spectrum and essential defect spectrum of linear operator. Filomat **29**(9), 1983–1994 (2015)
61. Ammar, A., Fakhfakh, S., Jeribi, A.: A New stability of the S essential spectrum of multivalued linear operators. Int. J. Anal. Appl. **14**, 1–8 (2017)
62. Aimar, M.T., Intissar, A., Jeribi, A.: On an unconditional basis of generalized eigenvectors of the nonself-adjoint Gribov operator in Bargmann space. J. Math. Anal. Appl. **231**(2), 588–602 (1999)
63. Ammar, A.: A characterization of some subsets of essential spectra of a multivalued linear operator. Complex Anal. Oper. Theory **11**(1), 175–196 (2017)
64. Ammar, A.: Some results on semi-Fredholm perturbations of multivalued linear operators. In: Linear and Multilinear Algebra, pp. 1563–5139 (2017)
65. Ammar, A., Benharrat, M., Jeribi, A., Messirdi, B.: On the Kato, semi-regular and essentially semi-regular spectra. Funct. Anal. Approx. Comput. **6**(2), 9–22 (2014)
66. Ammar, A., Bouchama, C., Jeribi, A.: Some properties of extended eigenvalues for operators pair. Filomat **37**(6), 1927–1940 (2023)
67. Ammar, A., Bouchekoua, A., Jeribi, A.: Pseudospectra in a non-Archimedean Banach space and essential pseudospectra in $\mathbb{E}_{\omega}$. Filomat **33**(12), 3961–3975 (2019)
68. Ammar, A., Bouchekoua, A., Jeribi, A.: Some approximation results in a non-Archimedean Banach space. Funct. Anal. Approx. Comput. **12**(1), 33–50 (2020)

69. Ammar, A., Fakhfakh, S., Jeribi, A.: Shechter spectra and relatively demicompact linear relations. Commun. Korean Math. Soc. **35**(2), 499–516 (2020)
70. Ammar, A., Heraiz, T., Jeribi, A.: Essential approximate point and essential defect spectrum of a sequence of linear operators in Banach spaces. Methods Funct. Anal. Topol. **25**(4), 373–380 (2019)
71. Ammar, A., Jedidi, O., Jeribi, A.: Stability of some essential spectra of closed linear relations on Banach spaces. Afr. Mat. **29**(7–8), 987–993 (2018)
72. Ammar, A., Jeribi, A.: A characterization of the essential pseudospectra and application to a transport equation. Extracta Math. **28**(1), 95–112 (2013)
73. Ammar, A., Bouchekoua, A., Lazrag, N.: Coupled numerical range for a linear relation and a linear operator on Hilbert spaces (preprint)
74. Ammar, A., Boukettaya, B., Jeribi, A.: Stability of the S-left and S-right essential spectra of a linear operator. Acta Math. Sci. Ser. B (Engl. Ed.) **34**(6), 1922–1934 (2014)
75. Ammar, A., Boukettaya, B., Jeribi, A.: A note on the essential pseudospectra and application. Linear Multilinear Algebra **64**(8), 1474–1483 (2016)
76. Abdelhedi, B., Boubaker, W., Moalla, N.: Joint essential numerical spectrum and Jeribi essential numerical spectrum of linear operators in Banach spaces. Commun. Korean Math. Soc. **38**(4), 1029–1044 (2023)
77. Dharmarha, P., Kumari, S.: Properties of Fredholm, Weyl and Jeribi essential S-spectra in a right quaternionic Hilbert space. Rev. Math. Phys. **35**(9), Paper No. 2350021, 20 (2023)
78. Moalla, N., Selmi, W.: On a characterization of Jeribi, Rakocevic, Schechter, Schmoeger and Wolf essential spectra of a 3×3 block operator matrices with non diagonal domain and application. Oper. Matrices **16** (2022)
79. Ammar, A., Boukettaya, B., Jeribi, A.: Essentially semi-regular linear relations. Facta Univ. Ser. Math. Inform. **32**(4), 421–434 (2017)
80. Ammar, A., Boukettaya, B., Jeribi, A.: Relationship between the left-right Jeribi essential spectra and the left-right Weyl essential spectra. Lib. Math. (N.S.) **39**(2), 13–26 (2019)
81. Ammar, A., Jeribi, A.: A characterization of the essential pseudospectra on a Banach space. J. Arab. Math. **2**, 139–145 (2013)
82. Jeribi, A.: Problems in finite element methods, Aubin Nitsche?s Duality Process, Nodal Methods and Friedrichs Systems. book series: Infosys Science Foundation Series, Springer, Singapore (2024). ISBN: 978-981-97-5709-1
83. Ammar, A., Jeribi, A.: Analyse. Cours et exercices corrigés MPSI-PCSI, édition Cépaduès, Toulouse, France (2024)
84. Jeribi, A.: Théorie des points fixes pour la topologie faibleâŁ¯: exercices et problèmes corrigés, édition Cépaduès. Toulouse, France (2023)
85. Jeribi, A.: Time asymptotic behaviour for unbounded linear operator arising in growing cell populations. Nonlinear Anal. Real World Appl. **4**(5), 667–688 (2003)
86. Benhamadou, M., Jeribi, A.: Analyse numérique matricielle - Méthodes et algorithmes, exercices et problèmes corrigés, Edition Ellipses, p. 680 (2020). ISBN 978-2-340-03803-5
87. Jeribi, A., Krichen, B.: Nonlinear functional analysis in Banach spaces and Banach algebrasâŁ¯: fixed point theory under weak topology for nonlinear operators and block operator matrices with applications. In: Monographs and Research Notes in Mathematics. Taylor-Francis (2015)
88. Jeribi, A., Hammami, M.A., Masmoudi, A., (ed.).: Applied mathematics in Tunisia. In: Proceedings of the International Conference on Ad- vances in Applied Mathematics (ICAAM) held in Hammamet, December 16-19, 2013. Edited by Aref Jeribi, Mohamed Ali Hammami and Afif Masmoudi. Springer Proceedings in Mathematics and Statistics, vol. 131, pp. xix+397. Springer, Cham, 2015. ISBN : 978-3-319-18041-0 ; 978-3-319-18040-3
89. Jeribi, A., Kaddachi, N.: Topological Fixed-point Theory in Suitable Banach Algebras with Applications. Springer, Singapore (2025)
90. Elleuch, A., Jeribi, A.: New description of the structured essential pseudospectra. Indag. Math. (N.S.) **27**(1), 368–382 (2016)

91. Elleuch, A., Jeribi, A.: On a characterization of the structured Wolf, Schechter and Browder essential pseudospectra. Indag. Math. (N.S.) **27**(1), 212–224 (2016)
92. Ellouz, H., Feki, I., Jeribi, A.: On a Riesz basis of exponentials related to the eigenvalues of an analytic operator and application to a non-selfadjoint problem deduced from a perturbation method for sound radiation. J. Math. Phys. **54**(11), 112101, 15 (2013)
93. Ellouz, H., Feki, I., Jeribi, A.: On a Riesz basis of exponentials related to a family of analytic operators and application. J. Pseudo-Differ. Oper. Appl. **10**(4), 999–1014 (2019)
94. Ellouz, H., Feki, I., Jeribi, A.: On the asymptotic behavior of the eigenvalues of an analytic operator in the sense of Kato and applications. Serdica Math. J. **45**(1), 55–88 (2019)
95. Ellouz, H., Feki, I., Jeribi, A.: Non-orthogonal fusion frames of an analytic operator and application to a one-dimensional wave control system. Mediterr. J. Math. **16**(2), Paper No. 52, 23 (2019)
96. Ellouz, H., Feki, I., Jeribi, A.: On the basis property of root vectors related to a non-self-adjoint analytic operator and applications. J. Math. Anal. **45**, 59–86 (2022)
97. Dehici, A., Latrach, K., Jeribi, A.: On a transport operator arising in growing cell populations. Spectral analysis. In: Advances in Mathematics Research, vol. 1, pp. 159–185. Adv. Math. Res., 1, Nova Sci. Publ., Hauppauge, NY (2002)
98. Chaker, W., Ghribi, A., Jeribi, A., Krichen, B.: Some fixed point theorems for orbitally-(p,q)-quasi-contraction mappings. In: Applied Mathematics in Tunisia, vol. 131, pp. 153–160. Springer Proc. Math. Stat., Springer, Cham (2015)
99. Chaker, W., Jeribi, A., Krichen, B.: Demicompact linear operators, essential spectrum and some perturbation results. Math. Nachr. **288**(13), 1476–1486 (2015)
100. Chaker, W., Ghribi, A., Jeribi, A., Krichen, B.: Fixed point theorem for quasi-contractive mapping in partial cone metric space. Afr. Mat. **26**(5–6), 1153–1159 (2015)
101. Chaker, W., Jeribi, A.: Common fixed point of single-valued and multivalued mappings. Afr. Mat. **25**(3), 617–622 (2014)
102. Jeribi, A.: Une nouvelle caractérisation du spectre essentiel et application. C. R. Acad. Sci. Paris, Série I **331**, 525–530 (2000)
103. Damak, M., Jeribi, A., Moalla, N.: Essential spectra of some matrix operators and application to two-group transport operators with general boundary conditions. J. Math. Anal. Appl. **323**(2), 1071–1090 (2006)
104. Damak, M., Jeribi, A.: On the essential spectra of matrix operators and applications. Electron. J. Differ. Equ. (11), 16 (2007)
105. Jeribi, A.: A characterization of the essential spectrum and applications. J. Boll. dell. Union. Mat. Ital. J. Boll. dell. Union. Mat. Ital. (8) 5-B, 805–825 (2002)
106. Jeribi, A.: A characterization of the Schechter essential spectrum on Banach spaces and applications. J. Math. Anal. Appl. **271**, 343–358 (2002)
107. Jeribi, A.: On the Schechter essential spectrum on Banach spaces and application. Facta Univ. Ser. Math. Inform. **17**, 35–55 (2002)
108. Jeribi, A.: Some remarks on the Schechter essential spectrum and applications to transport equations. J. Math. Anal. Appl. **275**(1), 222–237 (2002)
109. Zhao, L., Xu, J., Huang, J.: Gustafson, Jeribi, Weidmann and Wolf essential spectra and generator properties for one-sided coupled operator matrices, Linear and Multilinear Algebra (2025)
110. Álvarez, T., Ammar, A., Jeribi, A.: On the essential spectra of some matrix of linear relations. Math. Methods Appl. Sci. **37**(5), 620–644 (2014)
111. Álvarez, T.: On regular linear relations. Acta. Math. Sin. **28**, 183–194 (2012)
112. Cao, X.H., Meng, B.: Essential approximate point spectra and Weyl's theorem for operator matrices. J. Math. Anal. Appl. **304**, 759–771 (2005)
113. Ammar, A., Bouchekoua, A., Jeribi, A.: The ε-pseudospectra and the essential ε-pseudospectra of linear relations. J. Pseudo-Differ. Oper. Appl. **11**(2), 879–915 (2020)
114. Ammar, A., Fakhfakh, S., Jeribi, A.: Stability of the essential spectrum of the diagonally and off-diagonally dominant Block Matrix Linear Relations. J. Pseudo-Differ. Oper. Appl. **7**(4), 493–509 (2016)

115. Jeribi, A.: Spectral Theory and Applications of Linear Operators and Block Operator Matrices. Springer, New York (2015)
116. Ammar, A., Djaidja, N., Jeribi, A.: The essential spectrum of a sequence of linear operators in Banach spaces. Int. J. Anal. Appl. 1–7 (2017)

Chapter 4
Pseudospectra and Numerical Range of Linear Relations

The concept of numerical range of an operator on Banach and Hilbert spaces treated by several researchers. There are various manner to determine the last concept of closed linear operators in a Hilbert space. Among them we are interested of the numerical range of a closed linear operator which is the range of the restriction to the unit sphere of the quadratic form associated with it. This is a convex subset of the complex plane whose closure contains its spectrum. In this chapter, we present the new idea of the numerical range of two linear relations on Hilbert spaces. After that, we examines the aspects of the coupled numerical range for a linear relation and a linear operator on Hilbert spaces. Besides, we study its properties. After that, we establish some results describing the relationship between the various concepts of spectral theory and the numerical range.

4.1 Pseudospectra of Multivalued Linear Operator

In the beginning of this section, let us recall some results of pseudospectra of linear relations which are developed by Ammar, Daoud, and Jeribi in [1, 2] see also [3–15, 15–17]. Let X be a normed spaces, $U \in L\mathscr{R}(X)$ and $\varepsilon > 0$. The pseudospectra of T is defined by

$$\sigma_\varepsilon(U) = \sigma(U) \bigcup \left\{ \lambda \in \mathbb{C} \text{ such that } \|(\lambda - \widetilde{U})^{-1}\|_{L\mathscr{R}(\widetilde{X})} > \frac{1}{\varepsilon} \right\},$$

where $\widetilde{U}$ is the completion of U, given in (2.11.1). ◇

Lemma 4.1.1 ([1, Proposition 3.1]) *Giving $U \in L\mathscr{R}(X)$ and $\varepsilon > 0$. Then,*

$$\bigcap_{\varepsilon > 0} \sigma_\varepsilon(U) = \sigma(U).$$ ◇

A. Ammar and A. Jeribi, *Spectral Theory of Block Multivalued Operator Matrices*, Infosys Science Foundation Series, https://doi.org/10.1007/978-981-95-0397-1_4

Proposition 4.1.1 ([1, Remark 3.1 and Theorem 3.2]) Giving $U \in C\mathscr{R}(X)$ and $\varepsilon > 0$. Then, we have

(i) $\sigma_\varepsilon(U) = \sigma(U) \bigcup \left\{\lambda \in \mathbb{C} \text{ such that } \|(\lambda - U)^{-1}\|_{L\mathscr{R}(X)} > \frac{1}{\varepsilon}\right\}$.

(ii) $\sigma_\varepsilon(U^*) = \sigma_\varepsilon(U)$. ◇

Proposition 4.1.2 ([1, Remark 3.2]) Giving $U \in C\mathscr{R}(X)$ and $\varepsilon > 0$. Then,

$$\sigma_\varepsilon(U) = \bigcup_{\substack{\|V\|_{L\mathscr{R}(X)}<\varepsilon \\ V(0)\subset U(0) \\ \mathscr{D}(V)\supset\mathscr{D}(U)}} \sigma(V+U).$$ ◇

Proposition 4.1.3 *If T be one-to-one with dense range. Then,*

$$\sigma_\varepsilon(U) \subset \{\lambda \in \mathbb{C} \text{ such that } |\lambda| > \gamma(U) - \varepsilon\},$$

where $\gamma(\cdot)$ is the minimum modulus of U. ◇

Proof We are going to discuss three cases:

First case. If $\gamma(U) < \varepsilon$, then we obtain the result.

Second case. If $\gamma(U) = \varepsilon$. Then, U is open. Therefore, $0 \in \rho(T)$. Moreover,

$$\|U^{-1}\| = \frac{1}{\varepsilon}.$$

Hence, $0 \notin \sigma_\varepsilon(U)$. As a consequence,

$$\sigma_\varepsilon(U) \subset \{\lambda \in \mathbb{C} \text{ such that } |\lambda| > 0\}.$$

Third case: If $\gamma(U) > \varepsilon$, then T is open and one-to-one with dense range. Now, for $\lambda \in \mathbb{C}$ such that $|\lambda| \leq \gamma(T) - \varepsilon$, then

$$|\lambda| < \gamma(U).$$

Applying Lemma 2.10.11, we get the relation $\lambda - U$ is open, one-to-one with dense range. So, $\lambda \in \rho(U)$.
For $x \in \mathscr{D}(U)$, we obtain

$$\begin{aligned} \|Ux\| &= \|Ux - \lambda x + \lambda x\| \\ &\leq \|Ux - \lambda x\| + \|\lambda x\|. \end{aligned}$$

Thus, for $x \in \mathscr{D}(U)$, we get

$$\begin{aligned} \|(U-\lambda)x\| &\geq \|Ux\| - |\lambda|\|x\| \\ &\geq (\gamma(U) - |\lambda|)\|x\|. \end{aligned}$$

On the other hand,

$$\gamma(\lambda - U) \geq \gamma(U) - |\lambda|$$

and

$$\gamma(U) - |\lambda| \geq \gamma(U) - \gamma(U) + \varepsilon.$$

So,

$$\|(\lambda - U)^{-1}\| \leq \frac{1}{\varepsilon}.$$

As a result, $\lambda \in \rho_\varepsilon(U)$. Then,

$$\sigma_\varepsilon(U) \subset \{\lambda \in \mathbb{C} \text{ such that } |\lambda| > \gamma(U) - \varepsilon\}.$$

This completes the proof. Q.E.D.

Lemma 4.1.2 *Let $V \in C\mathscr{R}(X)$ and $\varepsilon > 0$. If $\eta \notin \sigma(V)$, then $\eta \in \sigma_\varepsilon(V)$ if, and only if, there exists $x \in \mathscr{D}(V)$ such that $\|(\eta - V)x\| < \varepsilon\|x\|$.* ◇

Proof If we take $\eta \in \sigma_\varepsilon(V)$ and $\eta \notin \sigma(V)$, then

$$\|(\eta - V)^{-1}\| > \frac{1}{\varepsilon}.$$

Hence, there exists a non-zero vector $y \in X$ such that

$$\|(\eta - V)^{-1}y\| > \frac{1}{\varepsilon}\,\|y\|. \tag{4.1.1}$$

Focus $x := (\eta - V)^{-1}y$, then $x \in \mathscr{D}(V)$ and

$$\begin{aligned}(\eta - V)x &= (\eta - V)(\eta - V)^{-1}y\\ &= y + (\eta - V)(0).\end{aligned}$$

Referring to Lemma 2.8.1, we infer that $y \in (\eta - V)x$. In addition,

$$(\eta - V)(0) = V(0).$$

It follows from Lemma 2.8.1, $0 \in V(0)$ from Lemma 2.10.2 that

$$\begin{aligned}\|(\eta - V)x\| &= \operatorname{dist}(y, (\eta - V)(0))\\ &= \operatorname{dist}(y, V(0))\\ &\leq \operatorname{dist}(y, 0)\\ &\leq \|y\|.\end{aligned} \tag{4.1.2}$$

So, from both (4.1.1) and (4.1.2), we get

$$\begin{aligned}\|x\| &> \frac{1}{\varepsilon}\|y\| \\ &\geq \frac{1}{\varepsilon}\|(\eta - V)x\|.\end{aligned}$$

Therefore,

$$\|(\eta - V)x\| < \varepsilon\|x\|.$$

Conversely, suppose that there exists $x \in \mathscr{D}(V)$ such that

$$\|(\eta - V)x\| < \varepsilon\|x\|.$$

Since $\eta \in \rho(V)$, then $\eta - V$ is one-to-one, open and, we obtain that

$$\begin{aligned}\gamma(\eta - V)\|x\| &\leq \|(\eta - V)x\| \\ &< \varepsilon\|x\|.\end{aligned}$$

Then,

$$0 < \gamma(\eta - V) < \varepsilon.$$

By Lemma 2.10.2, we get

$$\gamma(\eta - V) = \|(\eta - V)^{-1}\|^{-1}.$$

Thus,

$$\|(\eta - T)^{-1}\| > \frac{1}{\varepsilon}.$$

This completes the proof. Q.E.D.

Theorem 4.1.1 *Let $U \in C\mathscr{R}(X)$ and $\varepsilon > 0$. Then, the succeeding conditions are equivalent.*

$(i)\eta \in \sigma_\varepsilon(V)$.
(ii) There exists a continuous linear relation U satisfying $\mathscr{D}(U) \supset \mathscr{D}(V)$, $U(0) \subset V(0)$, $\|U\| < \varepsilon$ such that $\eta \in \sigma(V + U)$.
(iii) Whether $\eta \in \sigma(V)$ or $\|(\eta - V)^{-1}\| > \dfrac{1}{\varepsilon}$. ◇

Proof $(i) \Longrightarrow (ii)$ Suppose that $\eta \in \sigma_\varepsilon(V)$. We will discuss these two cases:

First case. If $\eta \in \sigma(V)$, then we may focus $U = 0$.

Second case : If $\eta \notin \sigma(V)$, then by using Lemma 4.1.2, there exists $x_0 \in \mathscr{D}(V)$, $\|x_0\| = 1$ such that

$$\|(\eta - V)x_0\| < \varepsilon.$$

It follows from Theorem 2.1.3, that there exists $x' \in X^*$ such that $\|x'\| = 1$ and

$$x'(x_0) = \|x_0\|.$$

We define the relation $U : X \longrightarrow X$ by

$$U(x) := x'(x)(\lambda - V)x_0.$$

It becomes clear that $\mathscr{D}(U) = X$ and $U(0) = 0$ which leads to U is single-valued everywhere defined. Furthermore, for $x \in X$, we get

$$\begin{aligned} \|Ux\| &\leq \|x'(x)(\lambda - V)x_0)\| \\ &\leq \|x'\|\|x\|\|(\lambda - V)x_0\| \\ &\leq \varepsilon\|x\|. \end{aligned}$$

This equivalent to $U \in B\mathscr{R}(X)$ and $\|U\| \leq \varepsilon$. Moreover, we get

$$\begin{aligned} (\lambda - (V + U))x_0 &= (\eta - V)x_0 - Ux_0 \\ &= (\eta - V)x_0 - x'(x_0)(\eta - V)x_0 \\ &= (\eta - (V + U))(0), \end{aligned}$$

then $0 \neq x_0 \in N(\eta - (V + U))$. So, $\eta - (V + U)$ is not one-to-one. Hence, $\eta \in \sigma(V + U)$.
$(ii) \Longrightarrow (iii)$ We derive a contradiction from the assumption that $\eta \notin \sigma(V)$ and $\|(\eta - V)^{-1}\| \leq \frac{1}{\varepsilon}$. By Lemma 2.10.2, we get $\eta \in \rho(V)$ and $\gamma(\eta - V) \geq \varepsilon$. Therefore, the use of Remark 3.2.1 (i), implies $\eta - V$ is one-to-one, open with dense range. In addition, $U(0) \subset V(0) = (\eta - V)(0)$, $\mathscr{D}(U) \supset \mathscr{D}(V) = \mathscr{D}(\eta - V)$ and $\|U\| < \varepsilon \leq \gamma(\eta - V)$. Therefore, using Lemma 2.10.11, we infer that $\eta - V - U$ is one-to-one, open with dense range. Thus, $\eta \in \rho(V + U)$. This is a contradiction.
$(iii) \Longrightarrow (i)$ By the uniform reasoning like back, we achieve the result. Q.E.D.

Theorem 4.1.2 *Let $U \in C\mathscr{R}(X)$, and $\varepsilon > 0$. If $\lambda \notin \sigma_\varepsilon(U)$, then*

$$\|(\lambda - U)^{-1}\| \geq \frac{1}{dist(\lambda, \sigma_\varepsilon(U)) + \varepsilon}. \qquad \diamond$$

Proof Let $\lambda \notin \sigma_\varepsilon(U)$. Assume that

$$\|(\lambda - U)^{-1}\| < \frac{1}{\text{dist}(\lambda, \sigma_\varepsilon(U)) + \varepsilon}.$$

Then,

$$\gamma(\lambda - U) > \text{dist}(\lambda, \sigma_\varepsilon(U)) + \varepsilon.$$

As

$$\text{dist}(\lambda, \sigma_\varepsilon(U)) = \inf\{|z - \lambda| \text{ such that } z \in \sigma_\varepsilon(U)\},$$

for η, $0 < \eta \leq \varepsilon$, there exists $z_\eta \in \sigma_\varepsilon(U)$ such that

$$|\lambda - z_\eta| < \text{dist}(\lambda, \sigma_\varepsilon(U)) + \eta. \tag{4.1.3}$$

Thus,

$$\begin{aligned}|\lambda - z_\eta| &< \text{dist}(\lambda, \sigma_\varepsilon(U)) + \varepsilon \\ &< \gamma(\lambda - U).\end{aligned}$$

But $(z_\eta - \lambda)I(0) = 0 \subset (\lambda - U)(0)$, $\mathscr{D}((z_\eta - \lambda)I) = \mathscr{D}(I) = X \supset \mathscr{D}((\lambda - U)) = \mathscr{D}(U)$ and $\lambda - T$ is one-to-one open with dense range (as $\lambda \in \rho(U)$), so by referring to Lemma 2.10.11, we infer that $\lambda - U + z_\eta - \lambda = z_\eta - U$ is one-to-one open with dense range. Hence, $z_\eta \in \rho(U)$. But $z_\eta \in \sigma_\varepsilon(U)$, so

$$\|(z_\eta - U)^{-1}\| > \frac{1}{\varepsilon},$$

i.e., $\gamma(z_\eta - U) < \varepsilon$. In addition, applying Lemma 2.10.10,

$$\begin{aligned}\gamma(z_\eta - U) &= \gamma(\lambda - U + z_\eta - \lambda) \\ &\geq \gamma(\lambda - U) - |z_\eta - \lambda|.\end{aligned}$$

Then,

$$\gamma(\lambda - U) \leq \gamma(z_\eta - U) + |z_\eta - \lambda|.$$

This implies from (4.1.3) that for all $0 < \eta \leq \varepsilon$

$$\gamma(\lambda - U) < \varepsilon + \text{dist}(\lambda, \sigma_\varepsilon(U)) + \eta.$$

Consequently,

$$\gamma(\lambda - U) \leq \varepsilon + \text{dist}(\lambda, \sigma_\varepsilon(U)).$$

This is a contradiction. Therefore,

$$\|(\lambda - U)^{-1}\| \geq \frac{1}{\text{dist}(\lambda, \sigma_\varepsilon(U)) + \varepsilon}.$$ Q.E.D.

Corollary 4.1.1 *Let $V \in C\mathscr{R}(X)$, and $\varepsilon > 0$. Then,*

$$\{\lambda \in \mathbb{C} \text{ such that } dist(\lambda, \sigma(V)) < \varepsilon\} \subset \sigma_\varepsilon(V).$$ ◇

Proof Let $\lambda \notin \sigma_\varepsilon(T)$. So, by Theorem 4.1.2, we get

$$\begin{aligned}\frac{1}{\text{dist}(\lambda, \sigma(T))} &\leq \|(\lambda - T)^{-1}\| \\ &\leq \frac{1}{\varepsilon}.\end{aligned}$$

Hence,

$$\mathrm{dist}(\lambda, \sigma(T)) \geq \varepsilon.$$

This completes the proof. Q.E.D.

Proposition 4.1.4 *Let $U \in L\mathscr{R}(X)$ and $\varepsilon > 0$. Then, for any η, $\vartheta \in \mathbb{C}$ with $\vartheta \neq 0$, we have*

$$\sigma_\varepsilon(\eta + \vartheta U) = \eta + \vartheta \sigma_{\varepsilon/|\vartheta|}(U). \qquad \diamond$$

Proof For $\eta, \mu \in \mathbb{C}$ and $\vartheta \in \mathbb{C}\backslash\{0\}$, we have

$$\mu\eta - \vartheta \widetilde{U} = \vartheta \left(\frac{\mu\eta}{\vartheta} - \widetilde{U} \right), \tag{4.1.4}$$

First case. If $\mu \in \sigma(\eta + \vartheta U)$, then in view of (4.1.4), we obtain that

$$\begin{aligned} \frac{\mu\eta}{\vartheta} &\in \sigma(\widetilde{U}) \\ &= \sigma(U) \\ &\subset \sigma_{\varepsilon/|\vartheta|}(U). \end{aligned}$$

This implies that $\mu \in \eta + \vartheta\ \sigma_{\varepsilon/|\vartheta|}(U)$.

Second case. If $\mu \in \sigma_\varepsilon(\eta + \vartheta U)$ and $\mu \notin \sigma(\eta + \vartheta U)$, then by applying (4.1.4), we get

$$\begin{aligned} \left\| \left(\frac{\mu\eta}{\vartheta} - \widetilde{V} \right)^{-1} \right\| &= \left\| \left(\vartheta^{-1}(\mu\eta - \vartheta \widetilde{V}) \right)^{-1} \right\| \\ &= |\vartheta| \left\| \left(\mu\eta - \vartheta \widetilde{V} \right)^{-1} \right\| \\ &> \frac{|\vartheta|}{\varepsilon}. \end{aligned}$$

Then, we conclude that $\frac{\mu\eta}{\vartheta} \in \sigma_{\varepsilon/|\vartheta|}(V)$. Therefore, $\mu \in \eta + \vartheta\sigma_{\varepsilon/|\vartheta|}(V)$. As a consequence

$$\sigma_\varepsilon(\eta + \vartheta U) \subset \eta + \vartheta\ \sigma_{\varepsilon/|\vartheta|}(V).$$

Conversely, by the uniform reasoning like back, we achieve the result. Q.E.D.

Proposition 4.1.5 *Let $U \in C\mathscr{R}(X)$ and $\varepsilon > 0$. Then, for $\delta > 0$, we have*

$$\mathbb{D}(0, \delta) + \sigma_\varepsilon(V) \subset \sigma_{\varepsilon+\delta}(V). \qquad \diamond$$

Proof Let $\eta \in \mathbb{D}(0, \delta) + \sigma_\varepsilon(V)$. Then, there exists $\eta_1 \in \mathbb{D}(0, \delta)$ and $\eta_2 \in \sigma_\varepsilon(V)$ such that

$$\eta = \eta_1 + \eta_2.$$

Suppose that $\eta \notin \sigma_{\varepsilon+\delta}(V)$, then $\eta_1 + \eta_2 \in \rho(V)$ and

$$\|(\eta_1 + \eta_2 - V)^{-1}\| \leq \frac{1}{\varepsilon + \delta}.$$

This leads to $\eta_1 + \eta_2 - V$ is one-to-one, open with dense range and

$$\gamma(\eta_1 + \eta_2 - V) \geq \varepsilon + \delta.$$

The fact that $\eta_1 \in \mathbb{D}(0, \delta)$ implies that

$$|\eta_1| \leq \delta < \varepsilon + \delta \leq \gamma(\eta_1 + \eta_2 - V).$$

By referring to Lemma 2.10.11, we infer that $\eta_1 + \eta_2 - V - \eta_1 = \eta_2 - V$ is one-to-one and open with dense range. Thus, $\eta_2 \in \rho(V)$. By means of Lemma 2.10.10, we obtain

$$\begin{aligned} \gamma(\eta_2 - V) = & \ \gamma(\eta_2 - \eta_1 + \eta_1 - V) \\ \geq & \ \gamma(\eta_1 + \eta_2 - V) - |\eta_1| \\ \geq & \ \varepsilon. \end{aligned}$$

Therefore,

$$\|(\eta_2 - V)^{-1}\| \leq \frac{1}{\varepsilon}.$$

So, it is to be concluded that $\eta_2 \notin \sigma_\varepsilon(V)$, which is a contradiction. Q.E.D.

Proposition 4.1.6 *Let $V \in C\mathscr{R}(X)$, $\varepsilon > 0$ and $U \in L\mathscr{R}(X)$ be such that $U(0) \subset V(0)$, $\mathscr{D}(U) \supset \mathscr{D}(V)$ and $\|U\| < \varepsilon$, we have*

$$\sigma_{\varepsilon-\|U\|}(V) \subset \sigma_\varepsilon(U + V) \subset \sigma_{\varepsilon+\|U\|}(V). \qquad \diamond$$

Proof Let $\lambda \notin \sigma_{\varepsilon+\|U\|}(V)$. So, $\lambda \in \rho(T)$ and

$$\|(\lambda - V)^{-1}\| \leq \frac{1}{\varepsilon + \|U\|}.$$

Then, $\lambda - V$ is one-to-one, open with dense range and

$$\gamma(\lambda - V) \geq \varepsilon + \|U\|.$$

Since $\|U\| < \varepsilon + \|U\| \leq \gamma(\lambda - V)$, then by Lemma 2.10.11, we infer that $\lambda - V - U$ is one-to-one and open with dense range. Hence, $\lambda \in \rho(V + U)$. In addition, by using Lemma 2.10.10, we get

$$\begin{aligned} \gamma(\lambda - V - U) &\geq \gamma(\lambda - V) - \|U\| \\ &\geq \varepsilon + \|U\| - \|U\| \\ &\geq \varepsilon. \end{aligned}$$

Therefore,

$$\|(\lambda - V - U)^{-1}\| \leq \frac{1}{\varepsilon}.$$

Thus, $\lambda \notin \sigma_{\varepsilon}(U + V)$. So,

$$\sigma_{\varepsilon}(U + V) \subset \sigma_{\varepsilon+\|U\|}(V).$$

Let $\lambda \notin \sigma_{\varepsilon}(U + V)$. Then, $\lambda \in \rho(V + U)$ and

$$\|(\lambda - U - V)^{-1}\| \leq \frac{1}{\varepsilon}.$$

Hence, $\lambda - V - U$ is one-to-one, open with dense range and

$$\gamma(\lambda - V - U) \geq \varepsilon.$$

Since $\|U\| < \gamma(\lambda - V - U)$, then it follows from Lemma 2.10.11 that

$$\lambda - V - U + U = \lambda - V$$

is one-to-one, open with dense range. Thus, $\lambda \in \rho(V)$. On the other hand, applying Lemma 2.10.10, we get

$$\begin{aligned}\gamma(\lambda - V) &= \gamma(\lambda - V - U + U)\\ &\geq \gamma(\lambda - V - U) - \|U\|\\ &\geq \varepsilon - \|U\|.\end{aligned}$$

Therefore,

$$\|(\lambda - V)^{-1}\| \leq \frac{1}{\varepsilon - \|U\|}.$$

Consequently, $\lambda \notin \sigma_{\varepsilon-\|U\|}(V)$. Hence,

$$\sigma_{\varepsilon+\|U\|}(U) \subset \sigma_{\varepsilon}(U + V).$$

This achieves the proof. Q.E.D.

Proposition 4.1.7 *Let $U \in C\mathcal{R}(X)$. Suppose that $R \in \mathcal{L}(X)$ such that $0 \in \rho(R)$. Then,*

$$\sigma(RUR^{-1}) = \sigma(U)$$

and, for $\varepsilon > 0$, we have

$$\sigma_{\varepsilon/\|R\|\|R^{-1}\|}(U) \subset \sigma_{\varepsilon}(RUR^{-1}) \subset \sigma_{\|R\|\|R^{-1}\|\varepsilon}(U). \quad \diamond$$

Proof Let us first prove that RUR^{-1} is closed. Using the fact that $0 \in \rho(R)$ and R is closed, we obtain the range of R is closed. Since $\alpha(R) = 0 < \infty$, $\gamma(R) > 0$ and as R is one-to-one and open and by means of Proposition 2.11.3, we infer that RU is closed. In addition, since R^{-1} is single-valued and bounded, so RTR^{-1} is closed. Then, RTR^{-1} is closed. Since the fact R and R^{-1} are single-valued and bounded, we obtain that

$$\begin{aligned} \lambda - RUR^{-1} &= \lambda - RUR^{-1} \\ &= R(\lambda R^{-1} - UR^{-1}) \\ &= R(\lambda - U)R^{-1}. \end{aligned} \tag{4.1.5}$$

Now, let $\lambda \in \rho(U)$. So, $(\lambda - U)^{-1}$ is a bounded single-valued. This leads to $\lambda - U$ is a bounded below and surjective. Since RUR^{-1} is closed, then by Proposition 2.11.3 and (4.1.5), we get $\lambda - RUR^{-1}$ is also closed, bounded below, surjective. Therefore, $\lambda \in \rho(RUR^{-1})$. Conversely, if $\lambda \in \rho(RUR^{-1})$, then $(\lambda - RUR^{-1})^{-1}$ is a bounded single-valued. This implies that $\lambda - RUR^{-1}$ is a bounded below and surjective. Since R and R^{-1} are bounded single-valued, then by (4.1.5), we obtain

$$\lambda - U = R^{-1}(\lambda - RUR^{-1})R. \tag{4.1.6}$$

Because of the fact that RUR^{-1} is closed implies by using of Propositions 2.11.2 and 2.11.3, and (4.1.6) that $\lambda - T$ is also closed, bounded below, onto. Then, $\lambda \in \rho(T)$, which implies the first result. From (4.1.5) and (4.1.6), we infer that

$$(\lambda - U)^{-1} = R^{-1}(\lambda - RTR^{-1})^{-1}V$$

and

$$(\lambda - RUR^{-1})^{-1} = R(\lambda - U)^{-1}R^{-1}.$$

Therefore,

$$\begin{aligned} \|(\lambda - RUR^{-1})^{-1}\| &= \|R(\lambda - U)^{-1}R^{-1}\| \\ &\leq \|R(\lambda - U)^{-1}\| \, \|R^{-1}\| \\ &\leq \|R\| \, \|(\lambda - U)^{-1}\| \|R^{-1}\| \\ &\leq \|R\|\|R^{-1}\|\|(\lambda - U)^{-1}\|. \end{aligned}$$

In the same way,

$$\|(\lambda - U)^{-1}\| \leq \|R\|\|R^{-1}\|\|(\lambda - RUR^{-1})^{-1}\|.$$

For $\lambda \in \sigma_{\varepsilon/\|R\|\|R^{-1}\|}(U)$, we get

$$\lambda \in \sigma(U)$$

or

$$\|(\lambda - U)^{-1}\| > \frac{\|R\|\|R^{-1}\|}{\varepsilon}.$$

Hence,

$$\lambda \in \sigma(RUR^{-1})$$

or

$$\begin{aligned}\|(\lambda - RUR^{-1})^{-1}\| &\geq \frac{1}{\|R\|\|R^{-1}\|}\|(\lambda - U)^{-1}\| \\ &> \frac{1}{\varepsilon}.\end{aligned}$$

Therefore,

$$\lambda \in \sigma_\varepsilon(RUR^{-1}).$$

Thus,

$$\sigma_{\varepsilon/\|R\|\|R^{-1}\|}(U) \subset \sigma_\varepsilon(RUR^{-1}).$$

On the other hand, for $\lambda \in \sigma_\varepsilon(RUR^{-1})$, we obtain

$$\lambda \in \sigma(RUR^{-1})$$

or

$$\|(\lambda - RUR^{-1})^{-1}\| > \frac{1}{\varepsilon}.$$

So,

$$\lambda \in \sigma(U)$$

or

$$\begin{aligned}\|(\lambda - U)^{-1}\| &\geq \frac{1}{\|R\|\|R^{-1}\|}\|(\lambda - RUR^{-1})^{-1}\| \\ &> \frac{1}{\|R\|\|R^{-1}\|\varepsilon}.\end{aligned}$$

As a consequence,

$$\lambda \in \sigma_{\|R\|\|R^{-1}\|\varepsilon}(U).$$

As a result, $\sigma_\varepsilon(RUR^{-1}) \subset \sigma_{\|R\|\|R^{-1}\|\varepsilon}(U)$. Q.E.D.

4.1.1 The ε-Pseudospectra of Linear Relations

The goal of this subsection is to introduce the ε-pseudospectra of linear relations and to establish its properties. Let X be a normed space, $T \in L\mathscr{R}(X)$ and $\varepsilon > 0$. The ε-pseudospectrum of T is defined by

$$\Sigma_\varepsilon(T) = \sigma(T) \bigcup \left\{ \lambda \in \mathbb{C} \text{ such that } \|(\lambda - \widetilde{T})^{-1}\| \geq \frac{1}{\varepsilon} \right\},$$

where $\widetilde{T}$ is the completion of T, given in (2.11.1).

Remark 4.1.1 Let X be a normed space, $T \in L\mathscr{R}(X)$ and $\varepsilon > 0$. Then,

(i) $\Sigma_\varepsilon(T)\backslash\sigma_\varepsilon(T) = \left\{ \lambda \in \mathbb{C} \text{ such that } \|(\lambda - \widetilde{T})^{-1}\| = \frac{1}{\varepsilon} \right\}$.
(ii) $\Sigma_\varepsilon(T)$ is a closed subset of the complex plane.
(iii) If $0 < \varepsilon_1 < \varepsilon_2$, then $\sigma(T) \subset \Sigma_{\varepsilon_1}(T) \subset \Sigma_{\varepsilon_2}(T)$.
(iv) Let X be a nonzero normed space, $\varepsilon > 0$ and let T be a continuous and densely defined linear relation in $L\mathscr{R}(X)$. Then,

$$\sigma_\varepsilon(T) \neq \emptyset$$

and

$$\Sigma_\varepsilon(T) \neq \emptyset.$$

Indeed, since $\sigma(T) \subset \sigma_\varepsilon(T) \subset \Sigma_\varepsilon(T)$, then by referring to Theorem 3.2.1, we infer that $\sigma_\varepsilon(T) \neq \emptyset$ and $\Sigma_\varepsilon(T) \neq \emptyset$.
(v) If X is a Banach space and $T \in C\mathscr{R}(X)$, then

$$\Sigma_\varepsilon(T) = \sigma(T) \bigcup \left\{ \lambda \in \mathbb{C} \text{ such that } \|(\lambda - T)^{-1}\| \geq \frac{1}{\varepsilon} \right\}.$$

Indeed, since X is a Banach space and $T \in C\mathscr{R}(X)$, hence by Remark 2.11.4 (ii), $(\lambda - \widetilde{T})^{-1} = (\lambda - T)^{-1}$. This implies that

$$\Sigma_\varepsilon(T) = \sigma(T) \bigcup \left\{ \lambda \in \mathbb{C} \text{ such that } \|(\lambda - T)^{-1}\| \geq \frac{1}{\varepsilon} \right\}.$$

(vi) Let X be a normed vector space, $T \in L\mathscr{R}(X)$ and $\varepsilon > 0$. Then,

(i_1) $\Sigma_\varepsilon(T)\backslash\sigma_\varepsilon(T) = \left\{ \lambda \in \mathbb{C} \text{ such that } \|(\lambda - \widetilde{T})^{-1}\| = \frac{1}{\varepsilon} \right\}$.
(i_2) $\Sigma_\varepsilon(T)$ is a closed subset of the complex plane.
(i_3) If $0 < \varepsilon_1 < \varepsilon_2$, then $\sigma(T) \subset \Sigma_{\varepsilon_1}(T) \subset \Sigma_{\varepsilon_2}(T)$.

(vii) Let X be a nonzero normed space, $\varepsilon > 0$ and let T be a continuous and densely defined linear relation in $L\mathscr{R}(X)$. Hence, $\sigma(T) \neq \emptyset$. So,

$$\sigma_\varepsilon(T) \neq \emptyset$$

and

$$\Sigma_\varepsilon(T) \neq \emptyset.$$

$(viii)$ If X is a Banach space and $T \in C\mathscr{R}(X)$, then

$$\Sigma_\varepsilon(T) = \sigma(T) \bigcup \left\{\lambda \in \mathbb{C} \text{ such that } \|(\lambda - T)^{-1}\| \geq \frac{1}{\varepsilon}\right\}. \qquad \diamond$$

As a direct consequence of Theorem 3.1.1 and Corollary [18, Corollary 3.1], we have the following result.

Corollary 4.1.2 *Let $T \in L\mathscr{R}(H)$, $T^2 \in C\mathscr{R}(H)$, and $\mathscr{D}(T^2)$ be closed. Then,*

$$\Sigma_\varepsilon(T) = \overline{\sigma_\varepsilon(T)}. \qquad \diamond$$

Theorem 4.1.3 *Let $T \in C\mathscr{R}(X)$ and $\varepsilon > 0$. Then,*

$(i) \sigma_\varepsilon(T) = \sigma_\varepsilon(T^*)$.
$(ii) \Sigma_\varepsilon(T) = \Sigma_\varepsilon(T^*)$. $\diamond$

Proof (i) Let $\lambda \in \rho_\varepsilon(T^*)$. Then, $\lambda \in \rho(T^*)$ and

$$\|(\lambda - T^*)^{-1}\| \leq \frac{1}{\varepsilon}.$$

The use of Remark 3.2.1 implies that $\lambda \in \rho(T)$. Hence, $(\lambda - T)^{-1}$ is continuous. It follows from Proposition 2.15.1 (vi) that

$$\begin{aligned} \|(\lambda - T)^{-1}\| &= \|(\lambda - T^*)^{-1}\| \\ &\leq \frac{1}{\varepsilon}. \end{aligned}$$

Then, $\lambda \in \rho_\varepsilon(T)$. Conversely, if $\lambda \in \rho_\varepsilon(T)$, then

$$\|(\lambda - T)^{-1}\| \leq \frac{1}{\varepsilon}.$$

By using Proposition 2.15.1 (vi), we get

$$\begin{aligned} \|(\lambda - T)^{-1}\| &= \|(\lambda - T^*)^{-1}\| \\ &\leq \frac{1}{\varepsilon}. \end{aligned}$$

Moreover, $\lambda \in \rho(T^*)$ by using of Remark 3.2.1 (iii). Hence, $\lambda \in \rho_\varepsilon(T^*)$.

(ii) Let us suppose that $\lambda \in \Sigma_\varepsilon(T)$. So, by Remark 4.1.1 $(vi)(i_1)$, we obtain that

$$\lambda \in \sigma_\varepsilon(T) \bigcup \left\{\lambda \in \mathbb{C} \text{ such that } \|(\lambda - T)^{-1}\| = \frac{1}{\varepsilon}\right\}.$$

So, there are two cases:

First case. Let $\lambda \in \sigma_\varepsilon(T)$. Hence, by using (i) and Remark 4.1.1 (vi), we conclude that $\lambda \in \Sigma_\varepsilon(T^*)$.

Second case. Let us suppose that $\lambda \notin \sigma_\varepsilon(T)$ and

$$\|(\lambda - T)^{-1}\| = \frac{1}{\varepsilon}.$$

So, $\lambda \in \rho(T)$,

$$\|(\lambda - T)^{-1}\| \leq \frac{1}{\varepsilon}$$

and

$$\|(\lambda - T)^{-1}\| = \frac{1}{\varepsilon}.$$

This leads to $(\lambda - T)^{-1}$ is a bounded linear operator and

$$\|(\lambda - T)^{-1}\| = \frac{1}{\varepsilon}.$$

Then, by using of Proposition 2.15.1 (vi), we have

$$\begin{aligned}\|(\lambda - T^*)^{-1}\| &= \|(\lambda - T)^{-1}\| \\ &= \frac{1}{\varepsilon}.\end{aligned}$$

Thus, $\lambda \in \Sigma_\varepsilon(T^*)$. This proves that

$$\Sigma_\varepsilon(T) \subset \Sigma_\varepsilon(T^*). \tag{4.1.7}$$

By the same reasoning as (4.1.7), we prove the reverse inclusion. Q.E.D.

Proposition 4.1.8 *Let X be a Banach space, $T \in C\mathcal{R}(X)$ and $\varepsilon > 0$. Then,*

$$\Sigma_\varepsilon(T) = \sigma(T) \bigcup \left\{\lambda \in \mathbb{C} \text{ such that } \exists\ (x_n) \subset \mathcal{D}(T),\ \|x_n\| = 1 \text{ and } \lim_{n\to\infty} \|(\lambda - T)x_n\| \leq \varepsilon\right\}.$$

◇

Proof Let $\lambda \in \left\{\lambda \in \mathbb{C} \text{ such that } \exists\ (x_n) \subset \mathcal{D}(T),\ \|x_n\| = 1 \text{ and } \lim_{n\to\infty} \|(\lambda - T)x_n\| \leq \varepsilon\right\}$ and $\lambda \notin \sigma(T)$. Since $\lambda \in \rho(T)$, then $(\lambda - T)^{-1}$ is a bounded operator which implies from Lemma 2.10.9 that

$$\gamma(\lambda - T) = \|(\lambda - T)^{-1}\|^{-1} < \infty.$$

Applying Lemma 2.10.2 (i), we obtain that

$$\begin{aligned}\|(\lambda - T)x_n\| &= \operatorname{dist}((\lambda - T)x_n, 0) \\ &\geq \gamma(\lambda - T) \operatorname{dist}\left(x_n, (\lambda - T)^{-1}(0)\right).\end{aligned}$$

Using the fact that $(\lambda - T)^{-1}$ is an operator, we get $(\lambda - T)^{-1}(0) = 0$. Hence,

$$\begin{aligned}\|(\lambda - T)x_n\| &\geq \gamma(\lambda - T)\,\mathrm{dist}(x_n, 0)\\ &\geq \gamma(\lambda - T)\|x_n\|\\ &\geq \gamma(\lambda - T)\ \text{as}\|x_n\| = 1.\end{aligned}$$

Together, from Lemma 2.10.9, we infer that

$$\|(\lambda - T)x_n\| \geq \|(\lambda - T)^{-1}\|^{-1}.$$

Therefore, $\|(\lambda - T)^{-1}\| \geq \frac{1}{\varepsilon}$. As a consequence, $\lambda \in \Sigma_\varepsilon(T)$. Conversely, we suppose that $\lambda \in \rho(T)$ and

$$\|(\lambda - T)^{-1}\| \geq \frac{1}{\varepsilon}.$$

The use of Lemma 2.10.2 (ii) implies that for every $n \in \mathbb{N}\backslash\{0\}$, there exists $(y_n)_n$ such that $\|y_n\| \leq 1$ and

$$\begin{aligned}\|(\lambda - T)^{-1}\| - \frac{1}{n} &\leq \|(\lambda - T)^{-1}y_n\|\\ &\leq \|(\lambda - T)^{-1}\|.\end{aligned}$$

Consequently,

$$\begin{aligned}\lim_{n\to+\infty} \|(\lambda - T)^{-1}y_n\| &= \lim_{n\to+\infty} \|(\lambda - T)^{-1}\|\\ &= \|(\lambda - T)^{-1}\|\\ &\geq \frac{1}{\varepsilon}.\end{aligned} \tag{4.1.8}$$

Since $(\lambda - T)^{-1}$ is an operator, hence we can take

$$x_n = \|(\lambda - T)^{-1}y_n\|^{-1}(\lambda - T)^{-1}y_n.$$

Therefore, $(x_n)_n \subset \mathcal{D}(T)$, $\|x_n\| = 1$ and

$$y_n \in \|(\lambda - T)^{-1}y_n\|\,(\lambda - T)x_n.$$

Thus,

$$\begin{aligned}\|(\lambda - T)^{-1}y_n\|\,\|(\lambda - T)x_n\| &= \mathrm{dist}\left(y_n, \|(\lambda - T)^{-1}y_n\|\,(\lambda - T)(0)\right)\\ &= \mathrm{dist}\left(y_n, (\lambda - T)(0)\right).\end{aligned}$$

Consequently,

$$\begin{aligned} \|(\lambda - T)x_n\| &\leq \|(\lambda - T)^{-1}y_n\|^{-1}\|y_n\| \\ &\leq \|(\lambda - T)^{-1}y_n\|^{-1}. \end{aligned}$$

Hence, by applying (4.1.8), we obtain

$$\lim_{n\to+\infty} \|(\lambda - T)x_n\| \leq \varepsilon.$$

This leads to

$$\Sigma_\varepsilon(T) = \sigma(T) \bigcup \left\{\lambda \in \mathbb{C} \text{ such that } \exists\ (x_n) \subset \mathscr{D}(T),\ \|x_n\| = 1 \text{ and } \lim_{n\to\infty} \|(\lambda - T)x_n\| \leq \varepsilon\right\}.$$

This achieves the proof. Q.E.D.

Proposition 4.1.9 *Let X be a normed vector space and $T \in L\mathscr{R}(X)$. Then,*

$$\bigcap_{\varepsilon>0} \sigma_\varepsilon(T) = \sigma(T) = \bigcap_{\varepsilon>0} \Sigma_\varepsilon(T).$$ ◇

Proof It is clear that $\sigma(T) \subset \sigma_\varepsilon(T) \subset \Sigma_\varepsilon(T)$ for all $\varepsilon > 0$, then

$$\begin{aligned} \sigma(T) &\subset \bigcap_{\varepsilon>0} \sigma_\varepsilon(T) \\ &\subset \bigcap_{\varepsilon>0} \Sigma_\varepsilon(T). \end{aligned} \tag{4.1.9}$$

Conversely, if $\lambda \notin \sigma(T)$, then $\lambda \in \rho(T)$. So, $(\lambda - \widetilde{T})^{-1}$ is a bounded linear operator, where $\widetilde{T}$ is given in (2.11.1). Hence, there exists $\varepsilon > 0$ such that

$$\|(\lambda - \widetilde{T})^{-1}\| < \frac{1}{\varepsilon}.$$

Thus, $\lambda \notin \Sigma_\varepsilon(T)$. Therefore, $\lambda \notin \bigcap_{\varepsilon>0} \Sigma_\varepsilon(T)$. Consequently,

$$\bigcap_{\varepsilon>0} \Sigma_\varepsilon(T) \subset \sigma(T). \tag{4.1.10}$$

As a result, from (4.1.9) and (4.1.10), we can conclude that

$$\begin{aligned} \bigcap_{\varepsilon>0} \sigma_\varepsilon(T) &= \sigma(T) \\ &= \bigcap_{\varepsilon>0} \Sigma_\varepsilon(T). \end{aligned}$$

This completes the proof. Q.E.D.

Now, let us study some properties of ε-pseudospectra of closed linear relations.

Proposition 4.1.10 *Let X be a normed space and let $T \in L\mathscr{R}(X)$. Then,*

$$\sigma(T) = \bigcap_{\varepsilon>0} \Sigma_\varepsilon(T). \qquad \diamond$$

Proof It follows from Remark 4.1.1 (iii) that

$$\sigma(T) \subset \bigcap_{\varepsilon>0} \Sigma_\varepsilon(T).$$

Conversely, let us assume that $\lambda \notin \sigma(T)$. Then, $\lambda \in \rho(T)$ which yields $(\lambda - \widetilde{T})^{-1}$ is a bounded linear operator. This leads to there exists $M > 0$ such that

$$\|(\lambda - \widetilde{T})^{-1}\| \leq M.$$

Taking $\varepsilon < \dfrac{1}{M}$, we obtain that

$$\|(\lambda - \widetilde{T})^{-1}\| < \frac{1}{\varepsilon}.$$

Hence, $\lambda \notin \bigcap_{\varepsilon>0} \Sigma_\varepsilon(T)$. Q.E.D.

In this section, we study the characterization of ε-pseudospectra and establish the pseudospectra of a sequence of closed linear relations in a Banach space.

Theorem 4.1.4 *Let X be a Banach space, $\varepsilon > 0$ and let $T \in C\mathscr{R}(X)$. Then,*

$$\bigcup_{\substack{\|S\|\leq\varepsilon \\ S(0)\subset T(0) \\ \mathscr{D}(S)\supset\mathscr{D}(T)}} \sigma(T+S) \subset \Sigma_\varepsilon(T). \qquad \diamond$$

Proof We derive a contradiction from the assumption that $\lambda \notin \Sigma_\varepsilon(T)$. Then, $\lambda \in \rho(T)$ and

$$\|(\lambda - T)^{-1}\| < \frac{1}{\varepsilon}.$$

This leads to $\lambda - T$ is one-to-one, open with dense range and

$$\gamma(\lambda - T) > \varepsilon. \tag{4.1.11}$$

Let $S \in L\mathscr{R}(X)$ be satisfy $\|S\| \leq \varepsilon$, $S(0) \subset T(0)$ and $\mathscr{D}(S) \supset \mathscr{D}(T)$. Then, by using (4.1.11), we infer that

$$\|S\| < \gamma(\lambda - T).$$

Thus, by applying Lemma 2.10.11, we deduce that $\lambda - T - S$ is one-to-one and open with dense range. This is equivalent to saying that $\lambda \in \rho(T+S)$. Q.E.D.

Remark 4.1.2 (i) We should notice that if X is a Banach space and $T \in C\mathscr{R}(X)$, then

$$\bigcup_{\substack{\|S\|\leq\varepsilon \\ S(0)\subset T(0) \\ \mathscr{D}(S)\supset\mathscr{D}(T)}} \sigma(T+S) \subset \Sigma_\varepsilon(T).$$

In fact, it suffices to consider the following example: let $X = l_p(\mathbb{N})$ be the space of sequences $x : \mathbb{N} \longrightarrow \mathbb{C}$ summable with a power $p \in [1, \infty)$ with the standard norm and $\varepsilon > 0$. Let us consider the linear relation T defined by

$$G(T) = \{(x, y) \in X \times X \text{ such that } x(n) = \varepsilon\, y(n-1),\ n \geq 2\}.$$

It is clear that $T = \varepsilon L^{-1}$, where L is the left shift single-valued operator defined by

$$L(x(n)) = x(n+1),\quad n \geq 1 \text{ and } x \in X.$$

We know that L is bounded operator with $\|L\| = 1$. Then, $\mathscr{D}\left(\varepsilon^{-1}L\right) = X$ and

$$\begin{aligned}\left\|\varepsilon^{-1}L\right\| &= \varepsilon^{-1}\,\|L\| \\ &= \varepsilon^{-1} \\ &< \infty.\end{aligned}$$

This leads to $\varepsilon^{-1}L$ is bounded operator. Therefore, $\varepsilon^{-1}L$ is closed which yields T is closed. In addition,

$$\|T^{-1}\| = \frac{1}{\varepsilon}.$$

Then, we deduce that

$$0 \in \Sigma_\varepsilon(T).$$

Now, we prove that

$$0 \notin \bigcup_{\substack{\|S\|\leq\varepsilon \\ S(0)\subset T(0) \\ \mathscr{D}(S)\supset\mathscr{D}(T)}} \sigma(T+S). \tag{4.1.12}$$

Since $N(T) = (\varepsilon^{-1}L)(0) = \varepsilon^{-1}L(0) = \{0\}$, $\gamma(T) = \|T^{-1}\|^{-1} = \varepsilon > 0$ and $R(T) = X$, then T is one-to-one and open with dense range. Let $S \in L\mathscr{R}(X)$ be satisfy that $\|S\| \leq \varepsilon$, $S(0) \subset T(0)$ and $\mathscr{D}(S) \supset \mathscr{D}(T)$. By using Lemma 2.10.11, we infer that $T + S$ is one-to-one and open with dense range which is leads to $0 \in \rho(T+S)$. Consequently, (4.1.12) holds.
(ii) Note that

$$\overline{\sigma_\varepsilon(T)} \subset \Sigma_\varepsilon(T). \tag{4.1.13}$$

In fact, we get

$$\sigma_\varepsilon(T) \subset \Sigma_\varepsilon(T).$$

Since $\Sigma_\varepsilon(T)$ is closed, then we have

$$\overline{\sigma_\varepsilon(T)} \subset \Sigma_\varepsilon(T).$$

(iii) The opposite inclusion of the above theorem (Theorem 4.1.4) is false. ◇

Now, the aim is to show the equality in (4.1.13). It is well known that if the resolvent norm of the closed linear operator T acting in a Banach space cannot be constant on an open set, hence

$$\overline{\sigma_\varepsilon(T)} = \Sigma_\varepsilon(T).$$

Note that, even if X is Banach space, $T \in C\mathscr{R}(X)$ and $\lambda \in \rho(T)$, we have $(\lambda - T)^{-1}$ is bounded operator in a Banach space X.

Theorem 4.1.5 *Let X be a Banach space, $\varepsilon > 0$, and $T \in C\mathscr{R}(X)$ be such that the resolvent norm of T cannot be constant on an open set. Then,*

$$\Sigma_\varepsilon(T) \subset \overline{\sigma_\varepsilon(T)}.$$ ◇

Proof It is clear that

$$\Sigma_\varepsilon(T)\backslash \left\{\lambda \in \mathbb{C} \text{ such that} \|(\lambda - T)^{-1}\| = \frac{1}{\varepsilon}\right\} \subset \overline{\sigma_\varepsilon(T)}.$$

Hence, it is sufficient to show that

$$\left\{\lambda \in \mathbb{C} \text{ such that } \|(\lambda - T)^{-1}\| = \frac{1}{\varepsilon}\right\} \subset \overline{\sigma_\varepsilon(T)}.$$

Let us suppose that $\|(\lambda - T)^{-1}\| = \frac{1}{\varepsilon}$. Hence, there exists μ belongs to the neighborhood of λ and $\mu \in \rho(T)$ such that

$$\begin{aligned} \|(\mu - T)^{-1}\| &> \|(\lambda - T)^{-1}\| \\ &= \frac{1}{\varepsilon}. \end{aligned}$$

This implies that $\mu \in \sigma_\varepsilon(T)$. So, we can deduce that $\lambda \in \overline{\sigma_\varepsilon(T)}$. Q.E.D.

As an immediate consequence of Remark 4.1.2 and Theorem 4.1.5, we obtain the following result

Corollary 4.1.3 *Let X be a Banach space, $\varepsilon > 0$, and $T \in C\mathscr{R}(X)$ be such that the resolvent norm of T cannot be constant on an open set. Then,*

$$\Sigma_\varepsilon(T) = \overline{\sigma_\varepsilon(T)}. \qquad \diamondsuit$$

The following result is devoted to give some equivalent definitions of ε-pseudospectra.

Theorem 4.1.6 *Let X be a Banach space, $\varepsilon > 0$, and $T \in C\mathscr{R}(X)$ be such that the resolvent norm of T cannot be constant on an open set. Then, the following propositions are all equivalent*

$$(i)\lambda \in \overline{\bigcup_{\substack{\|S\|\leq\varepsilon \\ S(0)\subset T(0) \\ \mathscr{D}(S)\supset\mathscr{D}(T)}} \sigma(T+S)}.$$

$$(ii)\lambda \in \sigma(T)\bigcup\left\{\lambda \in \mathbb{C} \text{ such that} \|(\lambda - T)^{-1}\| \geq \frac{1}{\varepsilon}\right\}.$$

$$(iii)\lambda \in \overline{\sigma(T)\bigcup\{\lambda \in \mathbb{C} \text{ such that } \exists\, x \in \mathscr{D}(T),\ \|x\| = 1 \text{ and } \|(\lambda - T)x\| < \varepsilon\}}. \ \diamondsuit$$

Proof $(i) \Longrightarrow (ii)$ Since $\Sigma_\varepsilon(T)$ is closed, then the use of Theorem 4.1.4 gives the wanted inclusion and achieves the proof of $(i) \Longrightarrow (ii)$.

$(ii) \Longrightarrow (iii)$ Let us suppose that $\lambda \in \sigma(T)\bigcup\left\{\lambda \in \mathbb{C} \text{ such that} \|(\lambda - T)^{-1}\| \geq \frac{1}{\varepsilon}\right\}$. Then, we will discuss the following two cases:

First case. If $\lambda \in \sigma(T)$, then the result is trivial.

Second case. If $\lambda \in \Sigma_\varepsilon(T)\backslash\sigma(T)$, then $\lambda \in \rho(T)$ and

$$\|(\lambda - T)^{-1}\| \geq \frac{1}{\varepsilon}.$$

We obtain that there exists μ belongs to the neighborhood of λ such that

$$\|(\mu - T)^{-1}\| > \|(\lambda - T)^{-1}\|$$

which yields

$$\|(\mu - T)^{-1}\| > \frac{1}{\varepsilon}.$$

This leads to there exists $y \in X$ such that $\|y\| = 1$ and

$$\|(\mu - T)^{-1}y\| > \frac{1}{\varepsilon}.$$

Since $\mu \in \rho(T)$, so $(\mu - T)^{-1}$ is an operator. Put

$$x = \|(\mu - T)^{-1}y\|^{-1}(\mu - T)^{-1}y,$$

then $x \in \mathscr{D}(T)$, $\|x\| = 1$ and $y \in \|(\mu - T)^{-1}y\|\ (\mu - T)x$. We obtain that

$$\|(\mu - T)^{-1}y\| \|(\mu - T)x\| = \text{dist}\,(y, (\mu - T)(0))\,.$$

As a consequence,

$$\begin{aligned}\|(\mu - T)x\| &= \|(\mu - T)^{-1}y\|^{-1}\,\text{dist}\,(y, (\mu - T)(0)) \\ &\leq \|(\mu - T)^{-1}y\|^{-1}\|y\| \\ &< \varepsilon.\end{aligned}$$

This enables us to deduce that

$$\lambda \in \overline{\{\lambda \in \mathbb{C} \text{ such that there is } x \in \mathscr{D}(T),\ \|x\| = 1 \text{ and } \|(\lambda - T)x\| < \varepsilon\}}.$$

$(iii) \Longrightarrow (i)$ Let $\lambda \in \sigma(T) \bigcup \{\lambda \in \mathbb{C}$ such that $\exists\, x \in \mathscr{D}(T)$, $\|x\| = 1$ and $\|(\lambda - T)x\| < \varepsilon\}$. We notice the existence of two cases.

First case. Let $\lambda \in \sigma(T)$. Then, we may put $S = 0$.

Second case. Let $\lambda \in \{\lambda \in \mathbb{C}$ such that $\exists\, x \in \mathscr{D}(T)$, $\|x\| = 1$ and $\|(\lambda - T)x\| < \varepsilon\}$ and $\lambda \in \rho(T)$. Then, there exists $x' \in X^*$ such that $\|x'\| = 1$ and $x'(x) = 1$. Now, we consider the following linear relation

$$S(y) = x'(y)\,(\lambda - T)x.$$

Then, $S(0) = 0$ and $\mathscr{D}(S) = X$. Let $y \in X$, we have

$$\begin{aligned}\|S(y)\| &\leq \|x'\|\|y\|\|(\lambda - T)x\| \\ &\leq \varepsilon\|y\|,\end{aligned}$$

which leads to $\|S\| \leq \varepsilon$. Moreover,

$$\begin{aligned}(\lambda - T - S)x &= (\lambda - T)(0) \\ &= (\lambda - T)(0) - S(0) \\ &= (\lambda - T - S)(0),\end{aligned}$$

and by applying Lemma 4.1.1, we infer that $0 \in (\lambda - T - S)x$. This leads to $0 \neq x \in N(\lambda - T - S)$. So, $\lambda - T - S$ is not one-to-one. Hence, $\lambda \in \sigma(T + S)$. This achieves the proof. Q.E.D.

As a direct consequence of Theorem 4.1.6, we have the following result.

Corollary 4.1.4 *Let X be a Banach space, $\varepsilon > 0$, and $T \in C\mathscr{R}(X)$ be such that the resolvent norm of T cannot be constant on an open set. Then,*

$$\overline{\bigcup_{\substack{\|S\| \leq \varepsilon \\ S(0) \subset T(0) \\ \mathscr{D}(S) \supset \mathscr{D}(T)}} \sigma(T + S)} = \Sigma_\varepsilon(T). \qquad \diamondsuit$$

The following result proves the relation between the pseudospectra of a closed linear relation on a Banach space and the ε-pseudospectra of all perturbed linear relations.

Proposition 4.1.11 *Let X be a Banach space, $T \in C\mathcal{R}(X)$ and δ, $\varepsilon > 0$. For any continuous $S \in L\mathcal{R}(X)$ such that $S(0) \subset T(0)$ and $\mathcal{D}(S) \supset \overline{\mathcal{D}(T)}$. Then,*

$$\sigma_{\varepsilon}(T) \subset \Sigma_{\varepsilon+\|S\|}(T+S) \subset \sigma_{\varepsilon+\delta+2\|S\|}(T). \qquad \diamond$$

Proof By using of Proposition 2.11.2 and Remark 4.1.1 (vi), we obtain that

$$\sigma_{\varepsilon}(T) \subset \Sigma_{\varepsilon+\|S\|}(T+S).$$

Now, we shall prove

$$\Sigma_{\varepsilon+\|S\|}(T+S) \subset \sigma_{\varepsilon+\delta+2\|S\|}(T).$$

Let $\lambda \notin \sigma_{\varepsilon+\delta+2\|S\|}(T)$. Hence, $\lambda \in \rho(T)$ and

$$\|(\lambda - T)^{-1}\| \leq \frac{1}{\varepsilon+\delta+2\|S\|},$$

which leads to $\lambda - T$ is open, one-to-one with dense range and

$$\begin{aligned} \gamma(\lambda - T) \geq &\ \varepsilon+\delta+2\|S\| \\ > &\ \|S\|. \end{aligned}$$

Since $S(0) \subset T(0) = \overline{T(0)}$ and $\mathcal{D}(S) \supset \overline{\mathcal{D}(T)} \supset \mathcal{D}(T)$, then from (ii) and (iii) of Lemma 2.10.11 that $\lambda - T - S$ is open, one-to-one with dense range and

$$\begin{aligned} \gamma(\lambda - T - S) \geq &\ \gamma(\lambda - T) - \|S\| \\ \geq &\ \varepsilon+\delta+2\|S\| - \|S\| \\ > &\ \varepsilon + \|S\|. \end{aligned}$$

This implies that $\lambda \in \rho(S+T)$ and

$$\|(\lambda - T - S)^{-1}\| < \frac{1}{\varepsilon+\|S\|}.$$

Moreover, $T+S$ is closed. Indeed, using the fact that $S(0) \subset T(0)$ and $\mathcal{D}(S) \supset \overline{\mathcal{D}(T)}$, so by using (i) of Lemma 2.10.11, we deduce that $T+S$ is closed. This enables us to conclude that

$$\Sigma_{\varepsilon+\|S\|}(T+S) \subset \sigma_{\varepsilon+\delta+2\|S\|}(T). \qquad \text{Q.E.D.}$$

Corollary 4.1.5 *Let X be a Banach space, $T \in C\mathcal{R}(X)$ and $\varepsilon > 0$. For any continuous $S \in L\mathcal{R}(X)$ such that $S(0) \subset T(0)$ and $\mathcal{D}(S) \supset \overline{\mathcal{D}(T)}$. Then,*

$$\Sigma_\varepsilon(T) \subset \Sigma_{\varepsilon+\|S\|}(T+S) \subset \Sigma_{\varepsilon+2\|S\|}(T). \qquad \diamond$$

Proof Since $S(0) \subset T(0)$ and $\mathscr{D}(S) \supset \overline{\mathscr{D}(T)}$, then by using Lemma 2.10.11, we get $T + S$ is closed. Let us suppose that $\lambda \notin \Sigma_{\varepsilon+2\|S\|}(T)$. So, $\lambda \in \rho(T)$ and

$$\|(\lambda - T)^{-1}\| < \frac{1}{\varepsilon + 2\|S\|}.$$

This implies that $\lambda - T$ is open, one-to-one with dense range and

$$\gamma(\lambda - T) > \varepsilon + 2\|S\|. \tag{4.1.14}$$

Since $\|S\| < \gamma(\lambda - T)$, then we have from (4.1.14) that $\lambda - T - S$ is open, one-to-one with dense range. In addition, from Lemma 2.10.11 (iii), we can deduce that

$$\begin{aligned}\gamma(\lambda - T - S) >\; & \gamma(\lambda - T) - \|S\| \\ >\; & \varepsilon + 2\|S\| - \|S\| \\ >\; & \varepsilon + \|S\|.\end{aligned}$$

Then, $\lambda \in \rho(T + S)$ and

$$\|(\lambda - T - S)^{-1}\| < \frac{1}{\varepsilon + \|S\|},$$

which is equivalent to

$$\lambda \notin \Sigma_{\varepsilon+\|S\|}(T + S).$$

Now, we show $\Sigma_\varepsilon(T) \subset \Sigma_{\varepsilon+\|S\|}(T + S)$. Let us suppose that $\lambda \notin \Sigma_{\varepsilon+\|S\|}(T + S)$, then $\lambda \in \rho(T + S)$ and

$$\|(\lambda - T - S)^{-1}\| < \frac{1}{\varepsilon + \|S\|}.$$

This leads to $\lambda - T - S$ is open, one-to-one with dense range and

$$\begin{aligned}\gamma(\lambda - T - S) >\; & \varepsilon + \|S\| \\ >\; & \|S\|.\end{aligned}$$

Combining the fact that $S(0) \subset T(0) = \overline{T(0)}$, $\mathscr{D}(S) \supset \mathscr{D}(T)$ and (ii) of Lemma 2.10.11, we infer that $\lambda - T - S + S$ is open and one-to-one with dense range. By using (iv) of Lemma 2.10.11, we can write

$$\lambda - T = \lambda - T - S + S.$$

Finally, in view Lemma 2.10.11 (iii), we have

$$\begin{aligned}\gamma(\lambda - T) \geq & \ \gamma(\lambda - T - S) - \|S\| \\ > & \ \varepsilon.\end{aligned}$$

Thus, $\lambda \notin \Sigma_\varepsilon(T)$. This proves that $\Sigma_\varepsilon(T) \subset \Sigma_{\varepsilon+\|S\|}(T + S)$. Q.E.D.

Now, the goal is achieved the equality in (4.1.13). It is well known that if the resolvent norm of the closed linear operator T acting in a Banach space cannot be constant on an open set, then

$$\overline{\sigma_\varepsilon(T)} = \Sigma_\varepsilon(T) \text{ (see [19]).}$$

Note that, even if X is Banach space, $T \in C\mathscr{R}(X)$ and $\lambda \in \rho(T)$, we have $(\lambda - T)^{-1}$ is bounded operator in a Banach space X. Consequently, we can assume the following hypothesis:

$$(H1) : \begin{cases} \text{The resolvent norm of a linear relation acting} \\ \text{in a Banach space cannot be constant on an open set.} \end{cases}$$

Theorem 4.1.7 *Let X be a Banach space, $T \in C\mathscr{R}(X)$ be satisfying the hypothesis $(H1)$ and $\varepsilon > 0$. Then,*

$$\Sigma_\varepsilon(T) \subset \overline{\sigma_\varepsilon(T)}.$$

◇

Proof It is clear that

$$\Sigma_\varepsilon(T) \backslash \left\{ \lambda \in \mathbb{C} \text{ such that } \|(\lambda - T)^{-1}\| = \frac{1}{\varepsilon} \right\} \subset \overline{\sigma_\varepsilon(T)}.$$

Hence, it is sufficient to prove that

$$\left\{ \lambda \in \mathbb{C} \text{ such that } \|(\lambda - T)^{-1}\| = \frac{1}{\varepsilon} \right\} \subset \overline{\sigma_\varepsilon(T)}.$$

Suppose that $\|(\lambda - T)^{-1}\| = \dfrac{1}{\varepsilon}$. Then, by applying $(H1)$, we infer that there exists μ belongs to the neighborhood of λ and $\mu \in \rho(T)$ such that

$$\begin{aligned}\|(\mu - T)^{-1}\| > & \ \|(\lambda - T)^{-1}\| \\ = & \ \frac{1}{\varepsilon}.\end{aligned}$$

This implies that $\mu \in \sigma_\varepsilon(T)$. Thus, we deduce that $\lambda \in \overline{\sigma_\varepsilon(T)}$. Q.E.D.

As an immediate consequence of Remark 4.1.2 and Theorem 4.1.7, we get:

Corollary 4.1.6 *Let X be a Banach space, $T \in C\mathscr{R}(X)$ be satisfying the hypothesis $(H1)$ and $\varepsilon > 0$. Then,*

$$\Sigma_\varepsilon(T) = \overline{\sigma_\varepsilon(T)}. \qquad \diamondsuit$$

The following theorems is devoted to give a characterization of the ε-pseudospectra.

Theorem 4.1.8 *Let X be a Banach space, $T \in C\mathscr{R}(X)$ be satisfying the hypothesis $(H1)$ and $\varepsilon > 0$. Then,*

$$\Sigma_\varepsilon(T) = \overline{\bigcup_{\substack{\|S\|\leq\varepsilon \\ S(0)\subset T(0) \\ \mathscr{D}(S)\supset\mathscr{D}(T)}} \sigma(T+S)}. \qquad \diamondsuit$$

Proof By using of Proposition 4.1.2 and Theorem 4.1.4, we obtain that

$$\begin{aligned}\sigma_\varepsilon(T) &\subset \bigcup_{\substack{\|S\|\leq\varepsilon \\ S(0)\subset T(0) \\ \mathscr{D}(S)\supset\mathscr{D}(T)}} \sigma(T+S) \\ &\subset \Sigma_\varepsilon(T).\end{aligned}$$

Combining the fact that $\Sigma_\varepsilon(T)$ is closed and Corollary 4.1.6, we infer that

$$\begin{aligned}\Sigma_\varepsilon(T) &\subset \overline{\bigcup_{\substack{\|S\|\leq\varepsilon \\ S(0)\subset T(0) \\ \mathscr{D}(S)\supset\mathscr{D}(T)}} \sigma(T+S)} \\ &\subset \Sigma_\varepsilon(T).\end{aligned}$$

This achieves the proof. Q.E.D.

Theorem 4.1.9 *Let X be a Banach space and $\varepsilon > 0$. Let $T \in C\mathscr{R}(X)$ be satisfying the hypothesis $(H1)$. Then,*

$$\Sigma_\varepsilon(T) = \sigma(T) \bigcup \overline{\{\lambda \in \mathbb{C} \text{ such that } \exists\, x \in \mathscr{D}(T),\ \|x\| = 1 \text{ and } \|(\lambda - T)x\| < \varepsilon\}}. \diamondsuit$$

Proof Let us assume that $\lambda \in \Sigma_\varepsilon(T)$. Then, we will discuss the following two cases:
First case. If $\lambda \in \sigma(T)$, then

$$\lambda \in \sigma(T) \bigcup \overline{\{\lambda \in \mathbb{C} \text{ such that } \exists\, x \in \mathscr{D}(T),\ \|x\| = 1 \text{ and } \|(\lambda - T)x\| < \varepsilon\}}.$$

Second case. If $\lambda \in \Sigma_\varepsilon(T)\backslash\sigma(T)$, then $\lambda \in \rho(T)$ and

$$\|(\lambda - T)^{-1}\| \geq \frac{1}{\varepsilon}.$$

By using the hypothesis $(H1)$, we infer that there exists μ belongs to the neighborhood of λ such that

$$\|(\mu - T)^{-1}\| > \|(\lambda - T)^{-1}\|$$

which implies that

$$\|(\mu - T)^{-1}\| > \frac{1}{\varepsilon}.$$

This leads to there exists $y \in X$ such that $\|y\| = 1$ and

$$\|(\mu - T)^{-1}y\| > \frac{1}{\varepsilon}.$$

Since $\mu \in \rho(T)$, so $(\mu - T)^{-1}$ is an operator. Put

$$x = \|(\mu - T)^{-1}y\|^{-1}(\mu - T)^{-1}y,$$

hence $x \in \mathscr{D}(T)$, $\|x\| = 1$ and $y \in \|(\mu - T)^{-1}y\|\ (\mu - T)x$. Now, by applying (i) of Lemma 2.10.2 to the linear relation $\|(\mu - T)^{-1}y\|\ (\mu - T)$, we obtain that

$$\big\| \|(\mu - T)^{-1}y\|\ (\mu - T)x \big\| = \operatorname{dist}\left(y, \|(\mu - T)^{-1}y\|\ (\mu - T)(0)\right).$$

Since $\|(\mu - T)^{-1}y\| > 0$, then by the linearity of $\mu - T$, we get

$$\begin{aligned} \|(\mu - T)x\| &= \|(\mu - T)^{-1}y\|^{-1} \operatorname{dist}(y, (\mu - T)(0)) \\ &\leq \|(\mu - T)^{-1}y\|^{-1} \|y\| \\ &< \varepsilon. \end{aligned}$$

Thus,

$$\lambda \in \overline{\{\lambda \in \mathbb{C} \text{ such that } \exists\, x \in \mathscr{D}(T),\ \|x\| = 1 \text{ and } \|(\lambda - T)x\| < \varepsilon\}}.$$

Conversely, let $\lambda \in \sigma(T) \bigcup \{\lambda \in \mathbb{C}$ such that $\exists\, x \in \mathscr{D}(T)$, $\|x\| = 1$ and $\|(\lambda - T)x\| < \varepsilon\}$. We notice the existence of two cases.

First case. Let $\lambda \in \sigma(T)$. Then, we may put $S = 0$.

Second case. Let $\lambda \in \{\lambda \in \mathbb{C}$ such that $\exists\, x \in \mathscr{D}(T)$, $\|x\| = 1$ and $\|(\lambda - T)x\| < \varepsilon\}$ and $\lambda \in \rho(T)$. It follows from Theorem 2.1.3 that there exists $x^* \in X^*$ such that $\|x^*\| = 1$ and $x^*(x) = 1$. Now, we consider the following linear relation

$$S(y) = x^*(y)(\lambda - T)x.$$

Let $z \in S(0)$. Then, $z \in 0\ (\lambda - T)x$ which yields $(x, z) \in G(0\ (\lambda - T))$. It follows from Remark 2.7.2 (ii) that $z = 0$. Consequently, S is an operator, $S(0) = 0 \in T(0)$

and $\mathscr{D}(S) = X$. Let $y \in X$, we have

$$\begin{aligned}\|S(y)\| &\leq \|x^*\|\|y\|\|(\lambda - T)x\| \\ &\leq \varepsilon\|y\|,\end{aligned}$$

which implies that $\|S\| \leq \varepsilon$. Moreover, we have

$$\begin{aligned}(\lambda - T - S)x &= (\lambda - T)(0) \\ &= (\lambda - T)(0) - S(0) \\ &= (\lambda - T - S)(0).\end{aligned}$$

by using Lemma 4.1.1 (i), we have $0 \in (\lambda - T - S)x$. This implies that $x \in N(\lambda - T - S)$ but $x \neq 0$. Consequently, $\lambda - T - S$ is not one-to-one. Then, we deduce that $\lambda \in \sigma(T + S)$. Q.E.D.

The following result proves the relation between the pseudospectra of a closed linear relation on a Banach space and the ε-pseudospectra of all perturbed linear relations:

Proposition 4.1.12 *Let X be a Banach space, $T \in C\mathscr{R}(X)$ and δ, $\varepsilon > 0$. For any continuous $S \in L\mathscr{R}(X)$ such that $S(0) \subset T(0)$ and $\mathscr{D}(S) \supset \overline{\mathscr{D}(T)}$. Then,*

(i) $\sigma_\varepsilon(T) \subset \Sigma_{\varepsilon+\|S\|}(T + S) \subset \sigma_{\varepsilon+\delta+2\|S\|}(T)$.
(ii) $\Sigma_\varepsilon(T) \subset \Sigma_{\varepsilon+\|S\|}(T + S) \subset \Sigma_{\varepsilon+2\|S\|}(T)$. ◇

Proof Since $\mathscr{D}(S) \supset \overline{\mathscr{D}(T)}$, then by using Remark 2.7.1 (i), we obtain

$$\mathscr{D}(\lambda - T - S) = \mathscr{D}(\lambda - T) \bigcap \mathscr{D}(S) \subset \mathscr{D}(S) \tag{4.1.15}$$

and

$$\begin{aligned}\mathscr{D}(\lambda - T) &= \mathscr{D}(\lambda) \bigcap \mathscr{D}(T) \\ &= \mathscr{D}(T) \subset \mathscr{D}(S).\end{aligned} \tag{4.1.16}$$

Based on the assumption $S(0) \subset T(0)$, we infer from Lemma 2.6.1 (i) that

$$\begin{aligned}S(0) &\subset S(0) + 0 && (4.1.17)\\ &\subset S(0) + T(0) \\ &= (\lambda - T - S)(0) && (4.1.18)\end{aligned}$$

and

$$\begin{aligned}S(0) &\subset T(0) \\ &= (\lambda - T)(0).\end{aligned} \tag{4.1.19}$$

Since $S(0) \subset T(0)$ and $\mathscr{D}(S) \supset \overline{\mathscr{D}(T)}$, then it follows from Proposition 2.11.2 that $T + S \in C\mathscr{R}(X)$.

(i) Let $\lambda \notin \Sigma_{\varepsilon+\|S\|}(T + S)$. Hence, by Remark 4.1.1 (v), we get $\lambda \in \rho(T + S)$ and

$$\|(\lambda - T - S)^{-1}\| < \frac{1}{\varepsilon + \|S\|}.$$

It follows from Lemma 2.10.9 that $\lambda - T - S$ is open, one-to-one with dense range and

$$\gamma(\lambda - T - S) > \varepsilon + \|S\|.$$

Since $T + S \in C\mathscr{R}(X)$, then by Remark 2.11.9 and Proposition 2.11.2, we obtain that $\lambda - T - S \in C\mathscr{R}(X)$. Applying Proposition 2.11.1 (ii), we get that $(\lambda - T - S)(0)$ is closed which implies from (4.1.18) that

$$\begin{aligned} S(0) \subset\ & (\lambda - T - S)(0) \\ =\ & \overline{(\lambda - T - S)(0)}. \end{aligned} \tag{4.1.20}$$

Combining the fact that $\|S\| < \gamma(\lambda - T - S)$, (4.1.15), (4.1.20), Lemma 2.10.11 and Proposition 2.7.2, we infer that $\lambda - T = \lambda - T - S + S$ is open, one-to-one with dense range. Finally, the use of Lemmas 2.10.9 and 2.10.10 allows us to deduce that

$$\begin{aligned} \|(\lambda - T)^{-1}\|^{-1} =\ & \gamma(\lambda - T) \\ & = \gamma(\lambda - T - S + S) \\ \geq\ & \gamma(\lambda - T - S) - \|S\| \\ >\ & \varepsilon + \|S\| - \|S\| \\ >\ & \varepsilon. \end{aligned}$$

Therefore, $\lambda \in \rho(T)$ and

$$\|(\lambda - T)^{-1}\| < \frac{1}{\varepsilon}.$$

Thus, $\lambda \notin \sigma_\varepsilon(T)$. This shows that

$$\sigma_\varepsilon(T) \subset \Sigma_{\varepsilon+\|S\|}(T + S).$$

Now, we will show

$$\Sigma_{\varepsilon+\|S\|}(T + S) \subset \sigma_{\varepsilon+\delta+2\|S\|}(T).$$

Let us assume that $\lambda \notin \sigma_{\varepsilon+\delta+2\|S\|}(T)$. Then, by using Proposition 4.1.1 (i), we get $\lambda \in \rho(T)$ and

$$\|(\lambda - T)^{-1}\| \leq \frac{1}{\varepsilon + \delta + 2\|S\|}$$

which implies from Lemma 2.10.9 that $\lambda - T$ is open, one-to-one with dense range and

$$\gamma(\lambda - T) \geq \varepsilon + \delta + 2\|S\|.$$

Consequently, $\gamma(\lambda - T) > \|S\|$. In view of Proposition 2.11.2 and Proposition 2.11.1 (ii) implies that $(\lambda - T)(0)$ is closed. Hence, it follows from (4.1.19) that

$$\begin{aligned} S(0) \subset\ & T(0) \\ =\ & (\lambda - T)(0) \\ =\ & \overline{(\lambda - T)(0)} \end{aligned} \tag{4.1.21}$$

Since $\gamma(\lambda - T) > \|S\|$, then by (4.1.16), (4.1.21) and Lemma 2.10.11, we obtain that $\lambda - T - S$ is open, one-to-one with dense range. Therefore, by applying Lemma 2.10.10, we obtain

$$\begin{aligned} \gamma(\lambda - T - S) \geq\ & \gamma(\lambda - T) - \|S\| \\ \geq\ & \varepsilon + \delta + 2\|S\| - \|S\| \\ >\ & \varepsilon + \|S\|. \end{aligned}$$

Finally, by Lemma 2.10.9, we conclude that

$$\|(\lambda - T)^{-1}\|^{-1} > \varepsilon + \|S\|.$$

This leads to $\lambda \in \rho(S + T)$ and

$$\|(\lambda - T - S)^{-1}\| < \frac{1}{\varepsilon + \|S\|}.$$

This implies that

$$\Sigma_{\varepsilon + \|S\|}(T + S) \subset \sigma_{\varepsilon + \delta + 2\|S\|}(T).$$

(ii) We will show

$$\Sigma_{\varepsilon}(T) \subset \Sigma_{\varepsilon + \|S\|}(T + S).$$

Let us suppose that $\lambda \notin \Sigma_{\varepsilon + \|S\|}(T + S)$. Then, by using Remark 4.1.1 (v), we have $\lambda \in \rho(T + S)$ and

$$\|(\lambda - T - S)^{-1}\| < \frac{1}{\varepsilon + \|S\|}.$$

This implies from Lemma 2.10.9 that $\lambda - T - S$ is open, one-to-one with dense range and

$$\gamma(\lambda - T - S) > \varepsilon + \|S\|.$$

Consequently,

$$\gamma(\lambda - T - S) > \|S\|.$$

Hence, it follows from (4.1.15), (4.1.20), Lemma 2.10.11 and Proposition 2.7.2 that $\lambda - T = \lambda - T - S + S$ is open and one-to-one with dense range. Finally, applying Lemma 2.10.10, we obtain

$$\begin{aligned}\gamma(\lambda - T) &\geq \gamma(\lambda - T - S) - \|S\| \\ &> \varepsilon.\end{aligned}$$

This leads from Lemma 2.10.9 that

$$\|(\lambda - T)^{-1}\| < \frac{1}{\varepsilon}.$$

Therefore, by virtue of Remark 4.1.1, we infer that $\lambda \notin \Sigma_\varepsilon(T)$. Now, we will show

$$\Sigma_{\varepsilon+\|S\|}(T + S) \subset \Sigma_{\varepsilon+2\|S\|}(T).$$

Let $\lambda \notin \Sigma_{\varepsilon+2\|S\|}(T)$. Then, by Remark 4.1.1 (v) and Lemma 2.10.9, we infer that $\lambda - T$ is open, one-to-one with dense range and

$$\gamma(\lambda - T) > \varepsilon + 2\|S\|.$$

Since $\gamma(\lambda - T) > \|S\|$, then by (4.1.16), (4.1.21) and Lemma 2.10.11, we have $\lambda - T - S$ is open, one-to-one with dense range. Then, by referring to Lemmas 2.10.10 and 2.10.9, we obtain

$$\begin{aligned}\|(\lambda - T - S)^{-1}\|^{-1} &= \gamma(\lambda - T - S) \\ &\geq \gamma(\lambda - T) - \|S\| \\ &> \varepsilon + 2\|S\| - \|S\| \\ &> \varepsilon + \|S\|.\end{aligned}$$

Thus, we conclude that $\lambda \notin \Sigma_{\varepsilon+\|S\|}(T + S)$. This proves that

$$\Sigma_{\varepsilon+\|S\|}(T + S) \subset \Sigma_{\varepsilon+2\|S\|}(T).$$ Q.E.D.

4.1.2 *The Pseudospectra and the ε-Pseudospectra of a Sequence of Closed Linear Relations*

The goal of this subsection is to study the relation between the pseudospectra (resp. ε-pseudospectra) of the sequence T_n and its limit.

Theorem 4.1.10 *Let X be a Banach space, $(T_n)_n$ be a sequence of closed linear relations from X into X, $T \in C\mathscr{R}(X)$ be such that $\|T_n - T\| \to 0$ when $n \to \infty$, $\mathscr{D}(T_n) = \mathscr{D}(T)$ and there exists $n_1 \in \mathbb{N}$ such that $T(0) = T_n(0)$, for all $n \geq n_1$. Then,*

(i) For every $\varepsilon_3 > 0$, there exists $\varepsilon_1, \varepsilon_2 > 0$ such that $\varepsilon_1 < \varepsilon_2 < \varepsilon_3$, and there exists a positive integers $n_2 \geq n_1$ such that

$$\sigma_{\varepsilon_1}(T) \subset \sigma_{\varepsilon_2}(T_n) \subset \sigma_{\varepsilon_3}(T), \text{ for all } n \geq n_2.$$

(ii) To every real number $\varepsilon > 0$, there exists a positive integers $n_2 \geq n_1$ such that

$$\Sigma_\varepsilon(T) = \Sigma_\varepsilon(T_n), \text{ for all } n \geq n_2. \qquad \diamond$$

Proof Using the fact $\mathscr{D}(T_n) = \mathscr{D}(T)$ for all $n \geq n_1$, we infer from (i) of Remark 2.7.1 that

$$\begin{aligned} \mathscr{D}(\lambda - T_n) &= \mathscr{D}(T_n) \\ &= \mathscr{D}(T_n) \bigcap \mathscr{D}(T) \\ &= \mathscr{D}(T_n - T). \end{aligned} \tag{4.1.22}$$

It follows from Proposition 2.11.2 that $\lambda - T_n$ is closed. This implies from Proposition 2.11.1 (ii) that $(\lambda - T_n)(0)$ is closed. Since $T(0) = T_n(0)$ for all $n \geq n_1$, then by the linearity of (T_n), we obtain

$$\begin{aligned} (T - T_n)(0) &= T_n(0) \\ &= (\lambda - T_n)(0) \\ &= \overline{(\lambda - T_n)(0)}. \end{aligned} \tag{4.1.23}$$

(i) We will prove that

$$\sigma_{\varepsilon_2}(T_n) \subset \sigma_{\varepsilon_3}(T), \text{ for all } n \geq n_2.$$

Assume that $\lambda \notin \sigma_{\varepsilon_3}(T)$. Hence, by Remark 4.1.1 (i), we get $\lambda \in \rho(T)$ and

$$\|(\lambda - T)^{-1}\| \leq \frac{1}{\varepsilon_3}.$$

The use of Lemma 2.10.9 allows us to conclude that $\lambda - T$ is open, one-to-one with dense range and $\gamma(\lambda - T) \geq \varepsilon_3$. This implies that there exists $0 < \varepsilon_2 < \varepsilon_3$ such that $\gamma(\lambda - T) > \varepsilon_2$. Since $\|T_n - T\| \to 0$ as $n \to \infty$, then for any $\epsilon > 0$, there exists $n_0 \in \mathbb{N}$ such that

$$\|T_n - T\| < \epsilon, \text{ for all } n \geq n_0.$$

In particular, for $\epsilon = \gamma(\lambda - T) - \varepsilon_2$, we have

$$\|T_n - T\| < \gamma(\lambda - T) - \varepsilon_2 \text{ for all } n \geq n_0.$$

Consequently,

$$\|T_n - T\| < \gamma(\lambda - T) \text{ for all } n \geq n_0. \tag{4.1.24}$$

By referring to (4.1.22), (4.1.23), (4.1.24), Lemma 2.10.11 and Proposition 2.7.2, we deduce that

$$\lambda - T_n = \lambda - T - T_n + T$$

is open and one-to-one with dense range for all $n \geq n_1$. Let $n_2 = \max\{n_0, n_1\}$. Hence, by applying Lemma 2.10.9 and Lemma 2.10.10, we infer, for all $n \geq n_2$, that

$$\begin{aligned} \|(\lambda - T_n)^{-1}\|^{-1} &= \gamma(\lambda - T_n) \\ &\geq \gamma(\lambda - T) - \|T_n - T\| \\ &\geq \gamma(\lambda - T) - \gamma(\lambda - T) + \varepsilon_2 \\ &\geq \varepsilon_2. \end{aligned}$$

By using of Remark 4.1.1 (i), we deduce that

$$\sigma_{\varepsilon_2}(T_n) \subset \sigma_{\varepsilon_3}(T), \text{ for all } n \geq n_2.$$

We will prove that

$$\sigma_{\varepsilon_1}(T) \subset \sigma_{\varepsilon_2}(T_n), \text{ for all } n \geq n_2.$$

Let us suppose that $\lambda \notin \sigma_{\varepsilon_2}(T_n)$, for all $n \geq 0$. Then, by Remark 4.1.1 (i), we have $\lambda \in \rho(T_n)$ and

$$\|(\lambda - T_n)^{-1}\| \leq \frac{1}{\varepsilon_2}.$$

This leads from Lemma 2.10.9 that $\lambda - T_n$ is open, one-to-one with dense range and

$$\gamma(\lambda - T_n) \geq \varepsilon_2.$$

This implies that there exists $0 < \varepsilon_1 < \varepsilon_2$ such that

$$\gamma(\lambda - T_n) > \varepsilon_1.$$

Since $\|T_n - T\| \to 0$ as $n \to \infty$, then for any $\epsilon > 0$, there exists $n_0 \in \mathbb{N}$ such that

$$\|T_n - T\| < \epsilon, \text{ for all } n \geq n_0.$$

In particular, for $\epsilon = \gamma(\lambda - T_n) - \varepsilon_1$, we have

$$\|T_n - T\| < \gamma(\lambda - T_n) - \varepsilon_1, \text{ for all } n \geq n_0. \tag{4.1.25}$$

This leads to

$$\|T_n - T\| < \gamma(\lambda - T_n), \text{ for all } n \geq n_0. \tag{4.1.26}$$

By virtue of (4.1.22), (4.1.23), (4.1.26), Lemma 2.10.11 and Proposition 2.7.2, we deduce that

$$\lambda - T = \lambda - T_n + T_n - T$$

is open and one-to-one with dense range for all $n \geq n_1$. It remains to prove that $\gamma(\lambda - T) \geq \varepsilon_1$. Let $n_2 = \max\{n_0, n_1\}$. By using of (4.1.25) and Lemma 2.10.10, we deduce for all $n \geq n_2$, that

$$\begin{aligned}\gamma(\lambda - T) \geq\ & \gamma(\lambda - T_n) - \|T_n - T\| \\ \geq\ & \gamma(\lambda - T_n) - \gamma(\lambda - T_n) + \varepsilon_1 \\ \geq\ & \varepsilon_1.\end{aligned}$$

Hence, by using Lemma 2.10.9, we deduce that $\lambda \in \rho(T)$ and

$$\|(\lambda - T)^{-1}\|^{-1} \geq \varepsilon_1$$

which implies that $\lambda \notin \sigma_{\varepsilon_1}(T)$. This proves that

$$\sigma_{\varepsilon_1}(T) \subset \sigma_{\varepsilon_2}(T_n), \text{ for all } n \geq n_2.$$

(ii) Let $\lambda \notin \Sigma_\varepsilon(T_n)$, for all $n \geq 0$. Hence, by using Remark 4.1.1 (v) and Lemma 2.10.9, we have $\lambda \in \rho(T_n)$ and $\gamma(\lambda - T_n) > \varepsilon$. This leads to $\lambda - T_n$ is open, one-to-one with dense range and

$$\gamma(\lambda - T_n) > \varepsilon.$$

Since $\|T_n - T\| \to 0$ as $n \to 0$, so for any $\epsilon > 0$, there exists $n_0 \in \mathbb{N}$ such that $\|T_n - T\| < \epsilon$, for all $n \geq n_0$. In particular, for $\epsilon = \gamma(\lambda - T_n) - \varepsilon$, we have

$$\begin{aligned}\|T_n - T\| <\ & \gamma(\lambda - T_n) - \varepsilon \\ <\ & \gamma(\lambda - T_n), \text{ for all } n \geq n_0.\end{aligned}$$

Hence, it follows from (4.1.22), (4.1.23), (4.1.26), Lemma 2.10.11 and Proposition 2.7.2 that

$$\lambda - T = \lambda - T_n + T_n - T$$

is open and one-to-one with dense range. Let $n_2 = \max\{n_0, n_1\}$. Hence, by Lemma 2.10.10, for all $n \geq n_2$, we deduce that

$$\begin{aligned}\gamma(\lambda - T) \geq\ & \gamma(\lambda - T_n) - \|T_n - T\| \\ \geq\ & \gamma(\lambda - T_n) - \gamma(\lambda - T_n) + \varepsilon\end{aligned}$$

$$> \varepsilon.$$

Finally, the use of Proposition 2.11.2, Remark 4.1.1 (v) and Lemma 2.10.9 allows us to conclude that $\lambda \notin \Sigma_\varepsilon(T)$. This shows that

$$\Sigma_\varepsilon(T) \subset \Sigma_\varepsilon(T_n),$$

for all $n \geq n_2$. The reverse inclusion is analogous. Q.E.D.

The following result can be directly derived from Propositions 4.1.6, 4.1.12 and Theorem 4.1.11.

Corollary 4.1.7 *Let X be a Banach space, $\varepsilon, \delta_1, \delta_2 > 0$, (T_n) be a sequence of closed linear relations from X into X and $T \in C\mathcal{R}(X)$ be such that $\|T_n - T\| \to 0$ when $n \to \infty$. We suppose that there exists $n_1 \in \mathbb{N}$ such that $\mathcal{D}(T_n) = \mathcal{D}(T)$ and $T(0) = T_n(0)$, for all $n \geq n_1$. Then, for any continuous $S \in L\mathcal{R}(X)$ such that $S(0) \subset T(0)$ and $\mathcal{D}(S) \supset \overline{\mathcal{D}(T)}$, there exists a positive integers $n_2 \geq n_1$ such that*

$$\sigma_\varepsilon(T) \subset \sigma_{\varepsilon+\delta_1+\|S\|}(T_n + S) \subset \sigma_{\varepsilon+\delta_1+\delta_2+2\|S\|}(T), \text{ for all } n \geq n_2,$$

and

$$\Sigma_\varepsilon(T) \subset \Sigma_{\varepsilon+\|S\|}(T_n + S) \subset \Sigma_{\varepsilon+2\|S\|}(T), \text{ for all } n \geq n_2. \quad \diamond$$

The purpose of this subsection is to establish the relation between the pseudospectra (resp. ε-pseudospectra) of the sequence T_n and its limit.

Theorem 4.1.11 *Let X be a Banach space, $(T_n)_n$ be a sequence of closed linear relations from X into X, $T \in C\mathcal{R}(X)$ be such that $\|T_n - T\| \to 0$ when $n \to \infty$, $\mathcal{D}(T_n) = \mathcal{D}(T)$ and there exists $n_1 \in \mathbb{N}$ such that $T(0) = T_n(0)$, for all $n \geq n_1$. Then,*

(i) For every triplet $(\varepsilon_1, \varepsilon_2, \varepsilon_3)$ of real numbers with $0 < \varepsilon_1 < \varepsilon_2 < \varepsilon_3$, there exists a positive integers $n_2 \geq n_1$ such that

$$\sigma_{\varepsilon_1}(T) \subset \sigma_{\varepsilon_2}(T_n) \subset \sigma_{\varepsilon_3}(T), \text{ for all } n \geq n_2.$$

(ii) For every real number $\varepsilon > 0$, there exists a positive integers $n_2 \geq n_1$ such that

$$\Sigma_\varepsilon(T) = \Sigma_\varepsilon(T_n), \text{ for all } n \geq n_2. \quad \diamond$$

Proof (i) First, we show that $\sigma_{\varepsilon_1}(T) \subset \sigma_{\varepsilon_2}(T_n)$. Let us suppose that $\lambda \notin \sigma_{\varepsilon_2}(T_n)$, for all $n \geq 0$. Hence, $\lambda \in \rho(T_n)$ and

$$\|(\lambda - T_n)^{-1}\| \leq \frac{1}{\varepsilon_2}.$$

This implies that $\lambda - T_n$ is open, one-to-one with dense range and

$$\gamma(\lambda - T_n) \geq \varepsilon_2.$$

Using the fact that $0 < \varepsilon_1 < \varepsilon_2$, we have $\gamma(\lambda - T_n) > \varepsilon_1$. Since $\|T_n - T\| \to 0$ when $n \to \infty$ and let $n_0 \in \mathbb{N}$ be such that

$$\begin{aligned}\|T_n - T\| &< \gamma(\lambda - T_n) - \varepsilon_1 \\ &< \gamma(\lambda - T_n), \text{ for all } n \geq n_0.\end{aligned}$$

Since

$$\begin{aligned}\mathscr{D}(\lambda - T_n) &= \mathscr{D}(T_n) \\ &= \mathscr{D}(T_n) \bigcap \mathscr{D}(T) \\ &= \mathscr{D}(T_n - T)\end{aligned}$$

and

$$\begin{aligned}(T - T_n)(0) &= T_n(0) \\ &= (\lambda - T_n)(0) \\ &= \overline{(\lambda - T_n)(0)},\end{aligned}$$

for all $n \geq n_1$, then by Lemma 2.10.11 (ii), we conclude that $\lambda - T_n + T_n - T$ is open and one-to-one with dense range. Applying Proposition 2.7.2, we can write

$$\lambda - T = \lambda - T - T_n + T_n$$

for all $n \geq n_1$. Now, we shall prove that $\gamma(\lambda - T) \geq \varepsilon_1$. Let $n_2 = \max\{n_0, n_1\}$. By Lemma 2.10.10, we conclude, for all $n \geq n_2$, that

$$\begin{aligned}\gamma(\lambda - T) &\geq \gamma(\lambda - T_n) - \|T_n - T\| \\ &\geq \gamma(\lambda - T_n) - \gamma(\lambda - T_n) + \varepsilon_1 \\ &\geq \varepsilon_1.\end{aligned}$$

Therefore, $\lambda \in \rho(T)$ and

$$\gamma(\lambda - T) \geq \varepsilon_1$$

which implies that $\lambda \notin \sigma_{\varepsilon_1}(T)$. This proves that

$$\sigma_{\varepsilon_1}(T) \subset \sigma_{\varepsilon_2}(T_n), \text{ for all } n \geq n_2.$$

A similar reasoning allows us to reach $\sigma_{\varepsilon_2}(T_n) \subset \sigma_{\varepsilon_3}(T)$, for all $n \geq n_2$.
(ii) Let $\lambda \notin \Sigma_\varepsilon(T_n)$, for all $n \geq 0$. Then, $\lambda \in \rho(T_n)$ and

$$\|(\lambda - T_n)^{-1}\| < \frac{1}{\varepsilon}.$$

This leads to $\lambda - T_n$ is open, one-to-one with dense range and

$$\gamma(\lambda - T_n) > \varepsilon.$$

Since $\|T_n - T\| \to 0$ as $n \to 0$, then there is $n_0 \in \mathbb{N}$ such that

$$\begin{aligned}\|T_n - T\| &< \gamma(\lambda - T_n) - \varepsilon \\ &< \gamma(\lambda - T_n), \text{ for all } n \geq n_0.\end{aligned}$$

Since $\mathcal{D}(\lambda - T_n) = \mathcal{D}(T_n - T)$ and $(T - T_n)(0) = \overline{(\lambda - T_n)(0)}$, for all $n \geq n_1$, then by referring to Proposition 2.7.2, we infer that

$$\lambda - T = \lambda - T_n + T_n - T$$

is open and one-to-one with dense range. Let $n_2 = \max\{n_0, n_1\}$. Hence, by Lemma 2.10.10, for all $n \geq n_2$, we deduce that

$$\begin{aligned}\gamma(\lambda - T) &\geq \gamma(\lambda - T_n) - \|T_n - T\| \\ &\geq \gamma(\lambda - T_n) - \gamma(\lambda - T_n) + \varepsilon \\ &> \varepsilon.\end{aligned}$$

Hence, we have $\lambda \notin \Sigma_\varepsilon(T)$. This proves that $\Sigma_\varepsilon(T) \subset \Sigma_\varepsilon(T_n)$, for all $n \geq n_2$. The opposite inclusion is analogous. Q.E.D.

The following result can be directly derived from Proposition 4.1.6, Corollary 4.1.5 and Theorem 4.1.11.

Corollary 4.1.8 *Let X be a Banach space, ε, δ_1, $\delta_2 > 0$, $(T_n)_n$ be a sequence of closed linear relations from X into X and $T \in C\mathcal{R}(X)$ be such that $\|T_n - T\| \to 0$ when $n \to \infty$. We suppose that there exists $n_1 \in \mathbb{N}$ such that $\mathcal{D}(T_n) = \mathcal{D}(T)$ and $T(0) = T_n(0)$, for all $n \geq n_1$. Then, for any continuous $S \in L\mathcal{R}(X)$ such that $S(0) \subset T(0)$ and $\mathcal{D}(S) \supset \overline{\mathcal{D}(T)}$, there exists a positive integers $n_2 \geq n_1$ such that*

$$\sigma_\varepsilon(T) \subset \sigma_{\varepsilon+\delta_1+\|S\|}(T_n + S) \subset \sigma_{\varepsilon+\delta_1+\delta_2+2\|S\|}(T), \text{ for all } n \geq n_2,$$

and

$$\Sigma_\varepsilon(T) \subset \Sigma_{\varepsilon+\|S\|}(T_n + S) \subset \Sigma_{\varepsilon+2\|S\|}(T), \text{ for all } n \geq n_2. \qquad \diamond$$

4.2 Essential Pseudospectra of Linear Relations

4.2.1 Some Properties of Pseudospectra of a Linear Relation

In this section, we establish some properties of essential pseudospectra of a linear relation and study some results of perturbation on the context of linear relations. Let us assume that $(\mathbb{K}, |\cdot|)$ be a complete filed and $(X, \|\cdot\|)$ be a Banach space, where $\mathbb{K} = \mathbb{C}$, $|\cdot|$ is absolute value, X is a vector space over a complex filed and $\|\cdot\|$ is a norm on X. For simplicity, we use $\|\cdot\|$ for both the norm on X and the relation norm on $L\mathscr{R}(X, Y)$ (i.e., $\|\cdot\| := \|\cdot\|_{L\mathscr{R}(X,Y)}$).

Definition 4.2.1 Let $\varepsilon > 0$ and $T \in C\mathscr{R}(X)$. We define the essential pseudospectra of T by

$$\begin{aligned}
\sigma_{e1,\varepsilon}(T) &= \mathbb{C}\backslash\left\{\lambda \in \mathbb{C} \text{ such that } \lambda - T + S \in \Phi_+(X), \text{ for all } S \in \mathscr{J}_T(X)\right\},\\
\sigma_{e2,\varepsilon}(T) &= \mathbb{C}\backslash\left\{\lambda \in \mathbb{C} \text{ such that } \lambda - T + S \in \Phi_-(X), \text{ for all } S \in \mathscr{J}_T(X)\right\},\\
\sigma_{e3,\varepsilon}(T) &= \mathbb{C}\backslash\left\{\lambda \in \mathbb{C} \text{ such that } \lambda - T + S \in \Phi_\pm(X), \text{ for all } S1\mathscr{J}_T(X)\right\},\\
\sigma_{e4,\varepsilon}(T) &= \mathbb{C}\backslash\left\{\lambda \in \mathbb{C} \text{ such that } \lambda - T + S \in \Phi(X), \text{ for all } S \in \mathscr{J}_T(X)\right\},\\
\sigma_{e5,\varepsilon}(T) &= \bigcap_{K \in \mathscr{K}_T(X)} \sigma_\varepsilon(T + K),
\end{aligned}$$

where $\mathscr{J}_T(X) = \{S \in L\mathscr{R}(X)$ is continuous such that $\|S\| < \varepsilon$, $S(0) \subset T(0)$ and $\mathscr{D}(S) \supset \mathscr{D}(T)\}$. ◇

Remark 4.2.1 Let $\varepsilon > 0$ and $T \in C\mathscr{R}(X)$, then one may order pseudospectra as

$$\sigma_{e3,\varepsilon}(T) = \sigma_{e1,\varepsilon}(T)\bigcap\sigma_{e2,\varepsilon}(T) \subset \sigma_{e4,\varepsilon}(T) \subset \sigma_{e5,\varepsilon}(T).$$ ◇

Theorem 4.2.1 *Let $T \in C\mathscr{R}(X)$ and $\varepsilon > 0$. Then, $\sigma_{ei,\varepsilon}(T)$ is a closed set with $i \in \{1, 2, 3, 4\}$.* ◇

Proof Since $T \in C\mathscr{R}(X)$ and $S \in \mathscr{J}_T(X)$, then we have $\lambda - T - S \in C\mathscr{R}(X)$ for all $\lambda \in \mathbb{C}$.

<u>For $i = 1$</u>: Let $\lambda \notin \sigma_{e1,\varepsilon}(T)$, then for all $S \in \mathscr{J}_T(X)$, we get $\lambda - T - S \in \Phi_+(X)$. Since $\lambda - T - S$ is closed and $R(\lambda - T - S)$ is closed, then applying Proposition 2.15.3 (i), we have $\gamma(\lambda - T - S) > 0$. Let $r > 0$ be such that $r < \gamma(\lambda - T - S)$, and let $\mu \in \mathbb{D}(\lambda, r)$, then $|\lambda - \mu| < r < \gamma(\lambda - T - S)$. Therefore,

$$\mu - T - S = \lambda - T - S - (\lambda - \mu) \in \Phi_+(X).$$

Thus, $\mu \notin \sigma_{e1,\varepsilon}(T)$ which leads to that $\mu \in \mathbb{C}\backslash\sigma_{e1,\varepsilon}(T)$. This allows us to conclude that

$$\mathbb{D}(\lambda, r) \subset \mathbb{C}\backslash\sigma_{e1,\varepsilon}(T).$$

So, $\sigma_{e1,\varepsilon}(T)$ is a closed. The same reasoning as above, we get $\sigma_{ei,\varepsilon}(T)$ is a closed set with $i \in \{2, 3, 4\}$. Q.E.D.

Proposition 4.2.1 *Let $T \in C\mathscr{R}(X)$. Then,*

(i) If $0 < \varepsilon_1 < \varepsilon_2$, then $\sigma_{ei}(T) \subset \sigma_{ei,\varepsilon_1}(T) \subset \sigma_{ei,\varepsilon_2}(T)$, with $i \in \{1, 2, 3, 4\}$.
(ii) Let $\varepsilon > 0$, then $\sigma_{ei,\varepsilon}(T) \subset \sigma_\varepsilon(T)$, with $i \in \{1, 2, 3, 4\}$.
(iii) $\bigcap\limits_{\varepsilon>0} \sigma_{ei,\varepsilon}(T) = \sigma_{ei}(T)$, with $i \in \{1, 2, 3, 4\}$. ◇

Proof Since $T \in C\mathscr{R}(X)$ and $S \in \mathscr{I}_T(X)$, then $\lambda - T - S \in C\mathscr{R}(X)$ for all $\lambda \in \mathbb{C}$.

(i) For $i = 1$: Let $\lambda \notin \sigma_{e1,\varepsilon_2}(T)$, then for all $S \in \mathscr{I}_T(X)$, we have $\lambda - T - S \in \Phi_+(X)$. Since $\varepsilon_1 < \varepsilon_2$, then for all continuous $S \in L\mathscr{R}(X)$ such that $S(0) \subset T(0)$, $\mathscr{D}(S) \supset \overline{\mathscr{D}(T)}$ and $\|S\| < \varepsilon_1$, we obtain that $\lambda - T - S \in \Phi_+(X)$. Therefore, we conclude that $\lambda \notin \sigma_{e1,\varepsilon_1}(T)$. Thus,

$$\sigma_{e1,\varepsilon_1}(T) \subset \sigma_{e1,\varepsilon_2}(T).$$

Now, we claim that $\sigma_{e1}(T) \subset \sigma_{e1,\varepsilon_1}(T)$. Assume that $\lambda \notin \sigma_{e1,\varepsilon_2}(T)$ which is equivalent to $\lambda - T - S \in \Phi_+(X)$ for all $S \in \mathscr{I}_T(X)$. In particular, for $S = 0$, then $\lambda - T - S \in \Phi_+(X)$. So, we obtain that $\lambda \notin \sigma_{e1}(T)$.
For $i = 2$: Let $\lambda \notin \sigma_{e2,\varepsilon_2}(T)$, then for all $S \in \mathscr{I}_T(X)$, we have $\lambda - T - S \in \Phi_-(X)$. From Theorem 3.6.5 (ii), we have $(\lambda - T - S)^* \in \Phi_+(X^*)$. Since S is continuous, then by Proposition 2.15.1, we have

$$(\lambda - T - S)^* = \lambda - T^* - S^*.$$

By referring to Proposition 2.15.1, we infer that $\|S\| = \|S^*\|$. Thus, $\lambda - T^* - S^* \in \Phi_+(X^*)$. Since $\mathscr{D}(T) \subset \mathscr{D}(S)$ and $S(0) \subset T(0)$, we have

$$\begin{aligned} S^*(0) = & \ \mathscr{D}(S)^\perp \\ \subset & \ \mathscr{D}(T)^\perp \\ = & \ T^*(0), \end{aligned}$$

and

$$\begin{aligned} \mathscr{D}(T^*) = & \ T(0)^\perp \\ \subset & \ S(0)^\perp \\ = & \ \mathscr{D}(S^*). \end{aligned}$$

Then, $\lambda - T^* - S^* \in \Phi_+(X^*)$, for all continuous linear relation S^* such that $S^*(0) \subset T^*(0)$, $\mathscr{D}(S^*) \supset \overline{\mathscr{D}(T^*)}$ and $\|S^*\| \leq \varepsilon_2$. Therefore, $\lambda \notin \sigma_{e1,\varepsilon_2}(T^*)$. By referring to the case that $i = 1$, we deduce that $\lambda \notin \sigma_{e1,\varepsilon_1}(T^*)$. This leads to $\lambda - T^* - S^* \in \Phi_+(X^*)$, for all continuous linear relation S^* such that $\|S^*\| \leq \varepsilon_1$. Therefore, $\lambda - T - S \in \Phi_-(X)$, for all $S \in \mathscr{I}_T(X)$. Hence, $\lambda \notin \sigma_{e2,\varepsilon_1}(T)$. Now, let us suppose that

$\lambda \notin \sigma_{e2,\varepsilon_1}(T)$, then $\lambda - T - S \in \Phi_-(X)$, for all $S \in \mathcal{J}_T(X)$. Since $0 \in \mathcal{J}_T(X)$, then $\lambda \notin \sigma_{e2}(T)$.

For $i = 3$: Since $\sigma_{e3}(T) = \sigma_{e1}(T) \bigcap \sigma_{e2}(T)$, $\sigma_{e3,\varepsilon_1}(T) = \sigma_{e1,\varepsilon_1}(T) \bigcap \sigma_{e2,\varepsilon_1}(T)$ and $\sigma_{e3,\varepsilon_2}(T) = \sigma_{e1,\varepsilon_2}(T) \bigcap \sigma_{e2,\varepsilon_2}(T)$, then

$$\begin{aligned}\sigma_{e3}(T) &\subset \sigma_{e3,\varepsilon_1}(T)\\ &\subset \sigma_{e3,\varepsilon_2}(T).\end{aligned}$$

For $i = 4$: This assertion is immediately deduced from (i) and (ii).

(ii) Let $\lambda \notin \sigma_\varepsilon(T)$. Then, $\lambda \in \rho(T)$ and

$$\|(\lambda - T)^{-1}\| \leq \frac{1}{\varepsilon}.$$

This leads to $\lambda - T$ is one-to-one, open with dense range and $\gamma(\lambda - T) \geq \varepsilon$. Now, if we take $S \in \mathcal{J}_T(X)$, then

$$\|S\| \leq \varepsilon \leq \gamma(\lambda - T).$$

Thus, $\lambda - T - S$ is one-to-one, open with dense range. This is equivalent to $\alpha(\lambda - T - S) = 0$, $\beta(\lambda - T - S) = 0$ and $R(\lambda - T - S)$ is closed. Therefore, $\lambda - T - S \in \Phi(X)$. Therefore, $\lambda \in \sigma_{ei,\varepsilon}(T)$, for all $i \in \{1, 2, 3, 4\}$.

(iii) For $i = 1$: From (i) and (ii), we have $\sigma_{e1}(T) \subset \sigma_\varepsilon(T)$ for all $\varepsilon > 0$. Hence,

$$\sigma_{e1}(T) \subset \bigcap_{\varepsilon>0} \sigma_\varepsilon(T).$$

Conversely, let $\lambda \notin \sigma_{e1}(T)$, then $\lambda - T \in \Phi_+(X)$. Therefore, $R(\lambda - T)$ is closed. Hence,

$$\gamma(\lambda - T) > 0.$$

Let ε be such that $0 < \varepsilon \leq \gamma(\lambda - T)$ and let $S \in \mathcal{J}_T(X)$. Since $\gamma(\lambda - T) \geq \varepsilon > \|S\|$, then

$$\alpha(\lambda - T - S) \leq \alpha(\lambda - T) < \infty.$$

Hence, $\lambda - T - S$ is open, and $R(\lambda - T - S)$ is closed. So, $\lambda \notin \sigma_{e1,\varepsilon}(T)$. As a result, we deduce that $\lambda \notin \bigcap_{\varepsilon>0} \sigma_{e1,\varepsilon}(T)$.

By using the same reasoning as above, we get the next inclusion. Q.E.D.

Theorem 4.2.2 *Let $T \in C\mathscr{R}(X)$ and $\varepsilon > 0$. Then, the following properties are equivalent*

(i) $\lambda \notin \sigma_{e5,\varepsilon}(T)$.
(ii) For all continuous linear relations $S \in L\mathscr{R}(X)$ *such that* $\mathscr{D}(S) \supset \mathscr{D}(T)$, $S(0) \subset T(0)$ *and* $\|S\| < \varepsilon$, *we have* $T + S - \lambda \in \Phi(X)$ *and* $i(T + S - \lambda) = 0$.
(iii) For all $D \in \mathscr{L}(X)$ *such that* $\|D\| < \varepsilon$, *we have* $T + D - \lambda \in \Phi(X)$ *and* $i(T + D - \lambda) = 0$. ◇

Proof *(i)* $\Longrightarrow$ *(ii)* Let $\lambda \notin \sigma_{e5,\varepsilon}(T)$. Then, there exists $K \in \mathscr{K}_T(X)$ such that $\lambda \notin \sigma_\varepsilon(T + K)$. By Theorem 4.1.1, for all continuous linear relations $S \in L\mathscr{R}(X)$ such that $\mathscr{D}(S) \supset \mathscr{D}(T + K) = \mathscr{D}(T) \bigcap \mathscr{D}(K) = \mathscr{D}(T)$ as $\mathscr{D}(K) \supset \mathscr{D}(T)$, $S(0) \subset (T + K)(0) = T(0)$ as $K(0) \subset T(0)$ and $\|S\| < \varepsilon$, we get $\lambda \in \rho(T + S + K)$. Hence, $T + S + K - \lambda$ is open, one-to-one with dense range. Since K is compact, then K is continuous. This leads to $S + K - \lambda$ is continuous. In addition, $(S + K - \lambda)(0) \subset T(0)$ and $\mathscr{D}(S + K - \lambda) = \mathscr{D}(S) \bigcap \mathscr{D}(K)$. Since T is closed, then by referring to Proposition 2.11.2, we infer that $T + S + K - \lambda$ is closed. Therefore, $R(T + S + K - \lambda)$ is closed. So, $R(T + S + K - \lambda) = X$. Hence, $T + S + K - \lambda \in \Phi(X)$ and $i(T + S + K - \lambda) = 0$, for all continuous linear relations $S \in L\mathscr{R}(X)$ such that $\mathscr{D}(S) \supset \mathscr{D}(T)$, $S(0) \subset T(0)$ and $\|S\| < \varepsilon$. It follows from Proposition 2.7.2 for all continuous linear relations $S \in L\mathscr{R}(X)$ such that $\mathscr{D}(S) \supset \mathscr{D}(T)$, $S(0) \subset T(0)$ and $\|S\| < \varepsilon$ that $T + S - \lambda \in \Phi(X)$ and $i(T + S - \lambda) = 0$.
(ii) $\Longrightarrow$ *(iii)* is trivial.
(iii) $\Longrightarrow$ *(i)* We assume that for all $D \in \mathscr{L}(X)$ such that $\|D\| < \varepsilon$, we get $T + D - \lambda \in \Phi(X)$ and $i(T + D - \lambda) = 0$. Applying Proposition 2.15.1, $N((T + D - \lambda)^*) = R(T + D - \lambda)^\perp$. Let $n = \alpha(T + D - \lambda) = \beta(T + D - \lambda)$, $\{x_1, \ldots, x_n\}$ be basis for $N(T + D - \lambda)$ and $\{y'_1, \ldots, y'_n\}$ be basis for the $N((T + D - \lambda)^*)$. Then, there are functionals $x'_1, \ldots, x'_n \in X^*$ (the adjoint space of X) and elements $y_1, \ldots, y_n$ such that

$$x'_j(x_k) = \delta_{jk} \quad 1 \leq j, k \leq n$$

and

$$y'_j(y_k) = \delta_{jk}, \quad 1 \leq j, k \leq n,$$

where $\delta_{jk} = 0$ if $j \neq k$ and $\delta_{jk} = 1$ if $j = k$. The single-valued operator K is defined by

$$Kx = \sum_{k=1}^{n} x'_k(x) y_k, \ x \in X.$$

K is bounded, since $\mathscr{D}(K) = X$ and

$$\|Kx\| \leq \|x\| \left(\sum_{k=1}^{n} \|x'_k\| \|y_k\| \right).$$

Furthermore, the range of K is contained in a finite subspace of X. Hence, K is a finite rank relation in X. Therefore, K is compact. We show that

$$N(T+D-\lambda)\bigcap N(K)=\{0\} \quad \text{and} \quad R(T+D-\lambda)\bigcap R(K)=\{0\}, \qquad (4.2.1)$$

for all $D \in \mathscr{L}(X)$ such that $\|D\| < \varepsilon$. Let $x \in N(T+D-\lambda)$, then

$$x=\sum_{k=1}^{n}\alpha_k x_k.$$

Therefore, $x'_j(x)=\alpha_j,\ 1 \le j \le n$. On the other hand, if $x \in N(K)$, then $x'_j(x)=0,\ 1 \le j \le n$. This proves the first relation in (4.2.1). The second inclusion is similar. In fact, if $y \in R(K)$, then

$$y=\sum_{k=1}^{n}\alpha_k y_k.$$

Hence,

$$y'_j(y)=\alpha_j,\ 1 \le j \le n.$$

But, if $y \in R(T+D-\lambda)$, then

$$y'_j(y)=0,\ 1 \le j \le n.$$

This gives the second relation in (4.2.1). On the other hand $K \in \mathscr{K}_T(X)=\mathscr{K}_{T+D-\lambda}(X)$ since $K(0)\subset T(0)=(T+D-\lambda)(0)$ and $\mathscr{D}(K)\supset\mathscr{D}(T)=\mathscr{D}(T+D-\lambda)$. We conclude from Corollary 3.6.3 (vii) that $T+D+K-\lambda \in \Phi(X)$ and

$$i(T+D+K-\lambda)=0.$$

If $x \in N(T+D+K-\lambda)$, then $0 \in Tx+Dx+Kx-\lambda x$. So, $-Kx \in (T+D-\lambda)x$. Thus, $Kx \in R(K)\bigcap R(T+D-\lambda)=\{0\}$. Hence, $Kx=0$ and $0 \in (T+D-\lambda)x$. This implies that $x \in N(T+D-\lambda)\bigcap N(K)$. Therefore, $x=0$. Consequently,

$$\alpha(T+D+K-\lambda)=0.$$

In the some way, one proves that $R(T+D+K-\lambda)=X$. The fact that $T+D+K-\lambda$ is closed and from Theorem 2.11.1 (ii), we deduce that $T+D+K-\lambda$ is open. Moreover, it is one-to-one with dense range and $\lambda \in \rho(T+D+K)$ for all continuous single-valued operators $D \in L\mathscr{R}(X)$ such that $\mathscr{D}(D)\supset\mathscr{D}(T)$ and $\|D\|<\varepsilon$. This leads to $\lambda \in \rho(T+D+K)$ for all $D \in \mathscr{L}(X)$ such that $\|D\|<\varepsilon$. This implies that

$$\lambda \notin \bigcup_{\substack{\|D\|<\varepsilon \\ \mathscr{D}(D)\supset\mathscr{D}(T)}} \sigma_\varepsilon(T+K+D).$$

It follow from Theorem 4.1.1 that $\lambda \notin \sigma_\varepsilon(T+K)$. This implies that

$$\lambda \notin \bigcap_{K \in \mathcal{K}_T(X)} \sigma_\varepsilon(T+K).$$

Hence, $\lambda \notin \sigma_{e5,\varepsilon}(T)$. Q.E.D.

Remark 4.2.2 It follow immediately, from Theorem 4.2.2 that for $T \in C\mathcal{R}(X)$ and $\varepsilon > 0$, we have

$$\begin{aligned} \sigma_{e5,\varepsilon}(T) &= \bigcup_{\substack{\|D\|<\varepsilon \\ \mathcal{D}(D)\supset\mathcal{D}(T)}} \sigma_{e5}(T+D) \\ &= \bigcup_{\substack{\|S\|<\varepsilon \\ S(0)\subset T(0) \\ \mathcal{D}(S)\supset\mathcal{D}(T)}} \sigma_{e5}(T+S). \end{aligned}$$

◇

Theorem 4.2.3 *Let $T \in C\mathcal{R}(X)$ and $\varepsilon > 0$. Then,*

$$\sigma_{e5,\varepsilon}(T) = \bigcap_{P \in \mathcal{PR}(\Phi(X))} \sigma_\varepsilon(T+P).$$ ◇

Proof Let $\mathcal{O} := \bigcap_{P \in \mathcal{PR}(\Phi(X))} \sigma_\varepsilon(T+P)$. Since $\mathcal{K}_T(X) \subset \mathcal{PR}(\Phi(X))$, we have $\mathcal{O} \subset \sigma_{e5,\varepsilon}(T)$. Conversely, let $\lambda \notin \mathcal{O}$, so there exists $P \in \mathcal{PR}(\Phi(X))$ such that $\lambda \notin \sigma_\varepsilon(T+P)$. Since P is continuous, hence by Proposition 2.11.2, we obtain that $T+P$ is closed. Then, by Theorem 4.1.1, we get $\lambda \in \rho(T+S+P)$ for all continuous linear relations $S \in L\mathcal{R}(X)$ such that $S(0) \subset (T+P)(0) = T(0)$, $\mathcal{D}(S) \supset \mathcal{D}(T+P) = \mathcal{D}(T) \bigcap \mathcal{D}(P) = \mathcal{D}(T)$ and $\|S\| < \varepsilon$. Using the fact that $(S+P-\lambda)(0) \subset T(0)$, $\mathcal{D}(S+P-\lambda) = \mathcal{D}(S) \bigcap \mathcal{D}(P) \supset \mathcal{D}(T)$ and T is closed, we infer Proposition 2.11.2 that $T+S+P-\lambda$ is closed, and $T+S+P-\lambda$ is open as $\lambda \in \rho(T+P+S)$. So, from Theorem 2.11.1 (ii), $R(T+S+P-\lambda)$ is closed. Therefore, $T+S+P-\lambda$ is one-to-one and onto. So, $T+S+P-\lambda \in \Phi(X)$ and $i(T+S+P-\lambda) = 0$. Since $-P \in \mathcal{PR}(\Phi(X))$, then by using Theorem 3.6.15, we obtain $T+S+P-P-\lambda \in \Phi(X)$ and

$$i(T+S-P+P-\lambda) = 0.$$

The fact that $-P(0) = P(0) \subset (T+S+P-\lambda)(0)$ and $\mathcal{D}(-P) = \mathcal{D}(P) \supset \mathcal{D}(T+S+P-\lambda) = \mathcal{D}(T) \bigcap \mathcal{D}(S) \bigcap \mathcal{D}(P)$ implies from Proposition 2.7.2 that

$$T+S-P+P-\lambda = T+S-\lambda.$$

Hence, for all continuous linear relations $S \in L\mathscr{R}(X)$ such that $S(0) \subset T(0)$, $\mathscr{D}(S) \supset \mathscr{D}(T)$ and $\|S\| < \varepsilon$, we have $T + S - \lambda \in \Phi(X)$ and $i(T + S - \lambda) = 0$. Finally, in view of Theorem 4.2.2, we can conclude that $\lambda \notin \sigma_{e5,\varepsilon}(T)$. Q.E.D.

Corollary 4.2.1 *Let $T \in C\mathscr{R}(X)$, $\varepsilon > 0$ and $\mathscr{J}(X)$ be a subset of $L\mathscr{R}(X)$. If $\mathscr{K}_T(X) \subset \mathscr{J}(X) \subset \mathscr{P}\mathscr{R}(\Phi(X))$, then*

$$\sigma_{e5,\varepsilon}(T) = \bigcap_{J \in \mathscr{J}(X)} \sigma_\varepsilon(T + J). \qquad \diamond$$

Proof It is clear to see that

$$\begin{aligned}\sigma_{e5,\varepsilon}(T) &= \bigcap_{P \in \mathscr{P}\mathscr{R}(\Phi(X))} \sigma_\varepsilon(T + P) \\ &\subset \bigcap_{J \in \mathscr{J}(X)} \sigma_\varepsilon(T + J) \\ &\subset \bigcap_{K \in \mathscr{K}_T(X)} \sigma_\varepsilon(T + K).\end{aligned}$$

This achieves the proof. Q.E.D.

Remark 4.2.3 By the definition of essential pseudospectra, and both Theorem 4.2.3 and Corollary 4.2.1 that for $T \in C\mathscr{R}(X)$ and $\varepsilon > 0$, we have

(i) $\sigma_{e5,\varepsilon}(T + K) = \sigma_{e5,\varepsilon}(T)$ for all $K \in \mathscr{K}_T(X)$.
(ii) $\sigma_{e5,\varepsilon}(T + P) = \sigma_{e5,\varepsilon}(T)$ for all $P \in \mathscr{P}\mathscr{R}(\Phi(X))$.
(iii) If $\mathscr{K}_T(X) \subset \mathscr{J}(X) \subset \mathscr{P}\mathscr{R}(\Phi(X))$ and if for all $J_1, J_2 \in \mathscr{J}(X)$, $J_1 \pm J_2 \in \mathscr{J}(X)$, then $\sigma_{e5,\varepsilon}(T + J) = \sigma_{e5,\varepsilon}(T)$, for all $J \in \mathscr{J}(X)$. $\diamond$

Proposition 4.2.2 *Let $T \in C\mathscr{R}(X)$. Then,*

(i) If $0 < \varepsilon_1 < \varepsilon_2$, then $\sigma_{e5}(T) \subset \sigma_{e5,\varepsilon_1}(T) \subset \sigma_{e5,\varepsilon_2}(T)$.
(ii) For $\varepsilon > 0$, $\sigma_{e5,\varepsilon}(T) \subset \sigma_\varepsilon(T)$.
(iii) $\bigcap_{\varepsilon>0} \sigma_{e5,\varepsilon(T)} = \sigma_{e5}(T)$. $\diamond$

Proof (i) If $\lambda \notin \sigma_{e5,\varepsilon_2}(T)$, then applying Theorem 4.2.2 and for all continuous linear relations $S \in L\mathscr{R}(X)$ such that $\mathscr{D}(S) \supset \mathscr{D}(T)$, $S(0) \subset T(0)$ and $\|S\| < \varepsilon_2$, we get $T + S - \lambda \in \Phi(X)$ and $i(T + S - \lambda) = 0$. Hence, for all continuous linear relations $S \in L\mathscr{R}(X)$ such that $\mathscr{D}(S) \supset \mathscr{D}(T)$, $S(0) \subset T(0)$ and $\|S\| < \varepsilon_1$, we have $T + S - \lambda \in \Phi(X)$ and $i(T + S - \lambda) = 0$. So, $\lambda \notin \sigma_{e5,\varepsilon_1}(T)$. On the other hand, if $\lambda \notin \sigma_{e5,\varepsilon_1}(T)$, then for all continuous linear relations $S \in L\mathscr{R}(X)$ such that $\mathscr{D}(S) \supset \mathscr{D}(T)$, $S(0) \subset T(0)$ and $\|S\| < \varepsilon_1$, we obtain that $T + S - \lambda \in \Phi(X)$ and $i(T + S - \lambda) = 0$. In particular for $S = 0$, we get $\lambda \notin \sigma_{e5}(T)$.

(ii) $\sigma_{e5,\varepsilon}(T) = \bigcap_{K \in \mathscr{K}_T(X)} \sigma_\varepsilon(T + K) \subset \sigma_\varepsilon(T + K)$ for all $K \in \mathscr{K}_T(X)$. In particular, $K = 0$.

(iii) From (i), $\sigma_{e5}(T) \subset \sigma_{e5,\varepsilon}(T)$ for all $\varepsilon > 0$. Therefore,

$$\sigma_{e5}(T) \subset \bigcap_{\varepsilon>0} \sigma_{e5,\varepsilon}(T).$$

Conversely, if $\lambda \notin \sigma_{e5}(T)$, then $T - \lambda \in \Phi(X)$ and $i(T - \lambda) = 0$. Hence, $R(T - \lambda)$ is closed and by applying Theorem 2.11.1 (ii), we have $T - \lambda$ is open. So,

$$\gamma(T - \lambda) > 0.$$

Let ε be such that $0 < \varepsilon \leq \gamma(T - \lambda)$ and let D be a single-valued linear relation in $L\mathscr{R}(X)$ such that

$$\mathscr{D}(D) \supset \mathscr{D}(T)$$

and

$$\|D\| < \varepsilon \leq \gamma(T - \lambda).$$

In view of Proposition 2.15.3 (iii),

$$\gamma(T - \lambda) = \gamma((T - \lambda)^*),$$

we infer that

$$\|D\| < \varepsilon \leq \gamma((T - \lambda)^*).$$

Thus, from Corollary 3.6.3 (iii), $T + D - \lambda \in \Phi_-(X)$. Now, by Theorem 2.10.1,

$$\alpha(T + D - \lambda) \leq \alpha(T - \lambda) < \infty.$$

So, $T + D - \lambda \in \Phi_+(X)$. Hence, $T + D - \lambda \in \Phi(X)$. Furthermore,

$$i(T + D - \lambda) = i(T - \lambda) = 0.$$

Therefore, by referring to Theorem 4.2.2, we get $\lambda \notin \sigma_{e5,\varepsilon}(T)$. So, $\lambda \notin \bigcap_{\varepsilon>0} \sigma_{e5,\varepsilon}(T)$. It follows that

$$\bigcap_{\varepsilon>0} \sigma_{e5,\varepsilon}(T) \subset \sigma_{e5}(T).$$

This achieves the proof. Q.E.D.

Theorem 4.2.4 *Let $T \in C\mathscr{R}(X)$ and let $\varepsilon > 0$. Then, $\sigma_{e5,\varepsilon}(T)$ is a closed set.* ◇

Proof Let $\lambda \in \rho_{e5,\varepsilon}(T)$, then $\lambda \notin \sigma_{e5,\varepsilon}(T)$. Let D be a single-valued continuous linear relation such that

$$\mathscr{D}(D) \supset \mathscr{D}(T)$$

and

$$\|D\| < \varepsilon.$$

By using Theorem 4.2.2, we have $T + D - \lambda \in \Phi(X)$ and

$$i(T + D - \lambda) = 0.$$

Therefore, $R(T + D - \lambda)$ is closed and $T + D - \lambda$ is open. So,

$$\gamma(T + D - \lambda) > 0.$$

Thus, by Proposition 2.15.1,

$$\gamma(T + D - \lambda) = \gamma((T + D - \lambda)^*).$$

Let $r > 0$ be such that $r < \gamma(T + D - \lambda)$, and let $\mu \in \mathbb{D}(\lambda, r)$. Then,

$$\begin{aligned} |\lambda - \mu| &< r \\ &< \gamma(T + D - \lambda) \\ &= \gamma((T + D - \lambda)^*). \end{aligned}$$

Then,

$$T + D - \lambda + \lambda - \mu = T + D - \mu \in \Phi_-(X).$$

Also,

$$\alpha(T + D - \mu) \leq \gamma(T + D - \lambda) < \infty$$

as $T + D - \lambda \in \Phi(X)$. Therefore, $T + D - \mu \in \Phi_+(X)$. Thus, $T + D - \mu \in \Phi(X)$. On the other hand,

$$\begin{aligned} i(T + D - \mu) &= i(T + D - \lambda + \lambda - \mu) \\ &= i(T + D - \lambda) \\ &= 0. \end{aligned}$$

Then, for all continuous single-valued operators $D \in L\mathscr{R}(X)$ such that $\mathscr{D}(D) \supset \mathscr{D}(T)$ and $\|D\| < \varepsilon$, we have $T + D - \mu \in \Phi(X)$ and

$$i(T + D - \mu) = 0.$$

Applying Theorem 4.2.2, we have $\mu \in \rho_{e5,\varepsilon}(T)$. Hence, there exists $r > 0$ such that

$$\mathbb{D}(\lambda, r) \subset \rho_{e5,\varepsilon}(T).$$

So, $\rho_{e5,\varepsilon}(T)$ is an open set. Q.E.D.

Theorem 4.2.5 *Let $T \in C\mathscr{R}(X)$ and $\varepsilon > 0$. Then, $\sigma_{e5,\varepsilon}(T) = \sigma_{e5,\varepsilon}(T^*)$.* ◇

Proof Let $K \in \mathscr{K}_T(X)$, then K is compact. Hence, it is continuous by Lemma 3.5.1 and, by using Proposition 3.5.2, we have K^* is compact. Moreover, $\mathscr{D}(K) \supset \mathscr{D}(T)$. So, by Proposition 2.15.1,

$$\begin{aligned} K^*(0) = & \ \mathscr{D}(K)^{\perp} \\ \subset & \ T^*(0) \\ = & \ \mathscr{D}(T)^{\perp}, \end{aligned}$$

and $K(0) \subset T(0)$ since K is continuous. Then,

$$\begin{aligned} \mathscr{D}(K^*) = & \ K(0)^{\perp} \\ \supset & \ T(0)^{\perp} \\ = & \ \overline{\mathscr{D}(T^*)} \\ \supset & \ \mathscr{D}(T^*). \end{aligned}$$

Therefore,

$$\mathscr{K}_T(X) \subset \{K \in L\mathscr{R}(X) \text{ such that } K \text{ is continuous and } K^* \in \mathscr{K}_{T^*}(X^*)\}.$$

Now, let $K \in \{K \in L\mathscr{R}(X)$ such that K is continuous and $K^* \in \mathscr{K}_{T^*}(X^*)\}$, then by using Proposition 3.5.2, we have K is compact. Using the fact that $K^*(0) = \mathscr{D}(K)^{\perp} \subset T^*(0) = \mathscr{D}(T)^{\perp}$, then $\mathscr{D}(K) \supset \mathscr{D}(T)$, also

$$\begin{aligned} K(0) \subset & \ \overline{K(0)} \\ = & \ \mathscr{D}(K^*)^{\top} \\ \subset & \ \mathscr{D}(T^*)^{\top} \\ = & \ T(0) \end{aligned}$$

as T is closed. Thus, $K \in \mathscr{K}_T(X)$. Therefore,

$$\mathscr{K}_T(X) = \{K \in L\mathscr{R}(X) \text{ such that } K \text{ is continuous and } K^* \in \mathscr{K}_{T^*}(X^*)\}. \quad (4.2.2)$$

On the other hand, for $K \in \mathscr{K}_T(X)$, $T + K$ is closed. Applying Theorem 4.1.3, we conclude that

$$\sigma_\varepsilon(T + K) = \sigma_\varepsilon((T + K)^*).$$

But $\mathscr{D}(K) \supset \mathscr{D}(T)$ and K is continuous, then by Proposition 2.15.1, we have

$$(T + K)^* = T^* + K^*.$$

Then,

$$\sigma_\varepsilon(T+K)=\sigma_\varepsilon(T^*+K^*) \text{ for all } K\in\mathscr{K}_T(X).$$

Hence, using Eq. (4.2.2), we get

$$\begin{aligned}\sigma_{e5,\varepsilon}(T) &= \bigcap_{K\in\mathscr{K}_T(X)}\sigma_\varepsilon(T+K)\\ &= \bigcap_{\substack{K^*\in\mathscr{K}_{T^*}(X^*)\\ K\in L\mathscr{R}(X)\\ K \text{ continuous}}}\sigma_\varepsilon(T^*+K^*)\\ &\supset \bigcap_{\substack{K^*\in\mathscr{K}_{T^*}(X^*)\\ K\in L\mathscr{R}(X)}}\sigma_\varepsilon(T^*+K^*)\\ &= \sigma_{e5,\varepsilon}(T^*).\end{aligned}$$

Let

$$\mathscr{O}:=\bigcap_{\substack{K^*\in\mathscr{K}_{T^*}(X^*)\\ K\in L\mathscr{R}(X)\\ K \text{ continuous}}}\sigma_\varepsilon(T^*+K^*),$$

so

$$\mathscr{O}=\sigma_{e5,\varepsilon}(T)\supset\sigma_{e5,\varepsilon}(T^*). \tag{4.2.3}$$

On the other hand, let $\lambda\in\mathscr{O}$, then for all $K\in L\mathscr{R}(X)$, K is continuous such that $K^*\in\mathscr{K}_{T^*}(X^*)$, we have $\lambda\in\sigma_\varepsilon(T^*+K^*)$. Let $K\in L\mathscr{R}(X)$ be such that $K^*\in\mathscr{K}_{T^*}(X)$, then $\widetilde{K}$ is compact, where $\widetilde{K}$ is given in (2.11.1). Since $G(\widetilde{K})$ is the completion of $G(K)$, therefore

$$\begin{aligned}G(\widetilde{K}) &= \overline{G(K)}\\ &= G(\overline{K}).\end{aligned}$$

Hence, $\widetilde{K}=\overline{K}$ is compact. Hence, $\overline{K}$ is continuous. In addition,

$$\begin{aligned}G(\overline{K}^*) &= G(-\overline{K}^{-1})^\perp\\ &= \overline{G(-K^{-1})}^\perp\\ &= G(-K^{-1})^\perp\\ &= G(K^*).\end{aligned}$$

Hence, $\overline{K}^*=K^*$. So, $\overline{K}$ is in $L\mathscr{R}(X)$, $\overline{K}$ is continuous and $\overline{K}^*=K^*\in\mathscr{K}_{T^*}(X^*)$, since $\lambda\in\mathscr{O}$, so $\lambda\in\sigma_\varepsilon(T^*+\overline{K}^*)=\sigma_\varepsilon(T^*+K^*)$. We deduce that if $\lambda\in\mathscr{O}$, for all $K\in L\mathscr{R}(X)$ such that $K^*\in\mathscr{K}_{T^*}(X^*)$ and $\lambda\in\sigma_\varepsilon(T^*+K^*)$, then

$$\lambda \in \bigcap_{\substack{K^* \in \mathcal{K}_{T^*}(X^*) \\ K \in L\mathcal{R}(X)}} \sigma_\varepsilon(T^* + K^*) = \sigma_{e5,\varepsilon}(T^*).$$

Hence, $\mathcal{O} \subset \sigma_{e5,\varepsilon}(T^*)$. From (4.2.3), we have

$$\begin{aligned}\mathcal{O} &= \sigma_{e5,\varepsilon}(T) \\ &= \sigma_{e5,\varepsilon}(T^*).\end{aligned}$$

This achieves the proof. Q.E.D.

4.2.2 Stability of Essential Pseudospectra of Linear Relations

Proposition 4.2.3 *Let $T \in L\mathcal{R}(X)$ and $\varepsilon > 0$, then for any α, $\beta \in \mathbb{C}$ with $\beta \neq 0$, we have*

$$\sigma_{e5,\varepsilon}(\alpha + \beta T) = \alpha + \sigma_{w,\varepsilon/|\beta|}(T)\beta. \qquad \diamond$$

Proof Let α, $\beta \in \mathbb{C}$ be with $\beta \neq 0$. Then,

$$\begin{aligned}\sigma_{e5,\varepsilon}(\alpha + \beta T) &= \bigcap_{K \in \mathcal{K}_T(X)} \sigma_\varepsilon(\alpha + \beta T + K) \\ &= \bigcap_{K \in \mathcal{K}_T(X)} \sigma_\varepsilon(\beta(\beta^{-1}\alpha + T + \beta^{-1}K)).\end{aligned}$$

It is simply to verify that $\{\beta^{-1}K : K \in \mathcal{K}_T(X)\} = \mathcal{K}_T(X)$. Indeed, if $K \in \mathcal{K}_T(X)$, then $(\beta^{-1}K)(0) = K(0) \subset T(0)$ and $\mathcal{D}(\beta^{-1}K) = \mathcal{D}(K) \supset \mathcal{D}(T)$. In addition, since K is compact, then

$$\begin{aligned}\overline{Q_{\beta^{-1}K}\beta^{-1}KB_X} &= \overline{\beta^{-1}Q_K KB_X} \\ &= \beta^{-1}\overline{Q_K KB_X}\end{aligned}$$

is compact. Therefore, $\beta^{-1}K$ is compact. Hence, $\beta^{-1}K \in \mathcal{K}_T(X)$. Thus,

$$\{\beta^{-1}K \text{ such that } K \in \mathcal{K}_T(X)\} \subset \mathcal{K}_T(X).$$

Conversely, by the same way, if $K \in \mathcal{K}_T(X)$, $\beta K \in \mathcal{K}_T(X)$, then

$$\mathcal{K}_T(X) \subset \{\beta^{-1}K \text{ such that } K \in \mathcal{K}_T(X)\}.$$

Hence,

$$\sigma_{e5,\varepsilon}(\alpha I + \beta T) = \bigcap_{K \in \mathscr{K}_T(X)} \sigma_\varepsilon(\beta(\beta^{-1}\alpha I + T + K))$$
$$= \bigcap_{K \in \mathscr{K}_T(X)} \sigma_\varepsilon(\alpha I + \beta(T + K)).$$

Applying Proposition 4.1.4, we get

$$\sigma_{e5,\varepsilon}(\alpha I + \beta T) = \bigcap_{K \in \mathscr{K}_T(X)} (\alpha + \sigma_{\varepsilon/|\beta|}(T + K)\beta)$$
$$= \alpha + \left(\bigcap_{K \in \mathscr{K}_T(X)} \sigma_{\varepsilon/|\beta|}(T + K) \right) \beta.$$
$$= \alpha + \sigma_{w,\frac{\varepsilon}{|\beta|}}(T)\beta.$$

This achieves the proof. Q.E.D.

Proposition 4.2.4 *Let X be a Banach space, $T \in C\mathscr{R}(X)$, and $\varepsilon > 0$. Then, for $\delta > 0$, we have*

$$\sigma_{e5,\varepsilon}(T) \subset \mathbb{D}(0, \delta) + \sigma_{e5,\varepsilon}(T) \subset \sigma_{e5,\varepsilon+\delta}(T). \qquad \diamond$$

Proof Let $K \in \mathscr{K}_T(X)$, then K is compact. Therefore, it is continuous (by Lemma 3.5.1). In view of Proposition 2.11.2, we have $T + K$ is closed. By Proposition 4.1.5,

$$\mathbb{D}(0, \delta) + \sigma_\varepsilon(T + K) \subset \sigma_{\varepsilon+\delta}(T + K).$$

Hence,

$$\mathbb{D}(0, \delta) + \bigcap_{K \in \mathscr{K}_T(X)} \sigma_\varepsilon(T + K) \subset \bigcap_{K \in \mathscr{K}_T(X)} \sigma_{\varepsilon+\delta}(T + K).$$

Therefore,

$$\mathbb{D}(0, \delta) + \sigma_{e5,\varepsilon}(T) \subset \sigma_{e5,\varepsilon+\delta}(T).$$ Q.E.D.

Proposition 4.2.5 *Let X be a Banach space and $T \in C\mathscr{R}(X)$, so for any $\varepsilon > 0$ and $S \in L\mathscr{R}(X)$ such that $S(0) \subset T(0)$, $\mathscr{D}(S) \supset \mathscr{D}(T)$ and $\|S\| < \varepsilon$, we get*

$$\sigma_{e5,\varepsilon-\|S\|}(T) \subset \sigma_{e5,\varepsilon}(T + S) \subset \sigma_{e5,\varepsilon+\|S\|}(T). \qquad \diamond$$

Proof Let $K \in \mathscr{K}_T(X)$, then K is compact. Hence, it is continuous (by Lemma 3.5.1). Using Proposition 2.11.2, $T + K$ is closed. Furthermore, let $S \in L\mathscr{R}(X)$ be such that $S(0) \subset T(0) = (T + K)(0)$, $\mathscr{D}(S) \supset \mathscr{D}(T) = \mathscr{D}(T + K)$ and $\|S\| < \varepsilon$. Then, by Proposition 4.1.6, we get

$$\sigma_{\varepsilon-\|S\|}(T + K) \subset \sigma_\varepsilon(T + S + K)$$
$$\subset \sigma_{\varepsilon+\|S\|}(T + K).$$

Hence,

$$\bigcap_{K\in\mathcal{K}_T(X)}\sigma_{\varepsilon-\|S\|}(T+K)\subset\bigcap_{K\in\mathcal{K}_T(X)}\sigma_{\varepsilon}(T+S+K)$$
$$\subset\bigcap_{K\in\mathcal{K}_T(X)}\sigma_{\varepsilon+\|S\|}(T+K).$$

Since $T(0)=(T+S)(0)$ and $\mathcal{D}(T)=\mathcal{D}(T+S)$, then

$$\mathcal{K}_T(X)=\mathcal{K}_{(T+S)}(X).$$

Therefore,

$$\bigcap_{K\in\mathcal{K}_T(X)}\sigma_{\varepsilon-\|S\|}(T+K)\subset\bigcap_{K\in\mathcal{K}_{T+S}(X)}\sigma_{\varepsilon}(T+S+K)$$
$$\subset\bigcap_{K\in\mathcal{K}_T(X)}\sigma_{\varepsilon+\|S\|}(T+K).$$

Thus,

$$\begin{aligned}\sigma_{e5,\varepsilon-\|S\|}(T)&\subset\sigma_{e5,\varepsilon}(T+S)\\&\subset\sigma_{e5,\varepsilon+\|S\|}(T).\end{aligned}$$

This completes the proof. Q.E.D.

Theorem 4.2.6 *Let X be a Banach space, $\varepsilon>0$, $T\in C\mathcal{R}(X)$ and $V\in\mathcal{L}(X)$ be such that $0\in\rho(V)$. Let $k=\|V\|\|V^{-1}\|$ and $S=VTV^{-1}$. Then,*

$$\sigma_{e5,\varepsilon/k}(T)\subset\sigma_{e5,\varepsilon}(S)\subset\sigma_{e5,k\varepsilon}(T).$$

In particular

$$\sigma_{e5}(S)=\sigma_{e5}(T).$$ ◇

Proof Let $K\in\mathcal{K}_T(X)$, then K is compact. Hence, it is continuous (by Lemma 3.5.1). Applying Proposition 2.11.2, $T+K$ is closed. Since V and V^{-1} are bounded single-valued, together with Proposition 2.7.3, we have

$$\begin{aligned}S+VKV^{-1}&=VTV^{-1}+VKV^{-1}\\&=V(TV^{-1}+KV^{-1})\\&=V(T+K)V^{-1}.\end{aligned}$$

From Proposition 4.1.7, we have for any $\varepsilon>0$,

$$\begin{aligned}\sigma_{\varepsilon/k}(T+K)&\subset\sigma_{\varepsilon}(S+VKV^{-1})\\&\subset\sigma_{k\varepsilon}(T+K).\end{aligned}$$

Thus,

$$\bigcap_{K\in\mathscr{K}_T(X)} \sigma_{\varepsilon/k}(T+K) \subset \bigcap_{K\in\mathscr{K}_T(X)} \sigma_{\varepsilon}(S+VKV^{-1})$$
$$\subset \bigcap_{K\in\mathscr{K}_T(X)} \sigma_{k\varepsilon}(T+K).$$

But $\{VKV^{-1}$ such that $K \in \mathscr{K}_T(X)\} = \mathscr{K}_S(X)$. Indeed, if $K \in \mathscr{K}_T(X)$, then $K(0) \subset T(0)$. Hence,

$$\begin{aligned} \lambda VKV^{-1}(0) = &\ VK(0) \\ \subset &\ VT(0) \\ = &\ VTV^{-1}(0) \\ = &\ S(0). \end{aligned}$$

In addition, since V is bounded, $\mathscr{D}(VK) = \{x \in X$ such that $Kx \bigcap \mathscr{D}(V) \neq \emptyset\} = \mathscr{D}(K)$ and $\mathscr{D}(VT) = \{x \in X$ such that $Tx \bigcap \mathscr{D}(V) \neq \emptyset\} = \mathscr{D}(T)$, so

$$\begin{aligned} \mathscr{D}(VKV^{-1}) = &\ \{x \in X \text{ such that } V^{-1}x \bigcap \mathscr{D}(VK) \neq \emptyset\} \\ \supset &\ \{x \in X \text{ such that } V^{-1}x \bigcap \mathscr{D}(VT) \neq \emptyset\} \\ = &\ \mathscr{D}(VTV^{-1}) \\ = &\ \mathscr{D}(S). \end{aligned}$$

On the other hand, K is compact, V is continuous, and $V(0) = \{0\} \subset \mathscr{D}(K)$, so VK is precompact. Therefore,

$$\overline{\Gamma}_0(VK) = 0.$$

Further, since V^{-1} is single-valued ($0 \in \rho(V)$), so

$$\overline{\Gamma}_0(VKV^{-1}) \leq \overline{\Gamma}_0(VK)\overline{\Gamma}_0(V^{-1}) = 0.$$

Hence, VKV^{-1} is precompact and, we have X is complete. So, VKV^{-1} is compact. Then,

$$\left\{VKV^{-1} \text{ such that } K \in \mathscr{K}_T(X)\right\} \subset \mathscr{K}_S(X).$$

In the similar way, if $K \in \mathscr{K}_S(X)$, hence $K(0) \subset S(0)$. Thus,

$$\begin{aligned} V^{-1}KV(0) = &\ V^{-1}K(0) \\ \subset &\ V^{-1}S(0) \\ = &\ V^{-1}SV(0) \\ = &\ T(0). \end{aligned}$$

Moreover, since V^{-1} is bounded as $0 \in \rho(V)$, $\mathscr{D}(V^{-1}K) = \{x \in X \text{ such that } Kx \bigcap \mathscr{D}(V^{-1}) \neq \emptyset\} = \mathscr{D}(K)$ and $\mathscr{D}(VS) = \{x \in X \text{ such that } Sx \bigcap \mathscr{D}(V^{-1}) \neq \emptyset\} = \mathscr{D}(S)$, so

$$\begin{aligned} \mathscr{D}(V^{-1}KV) = & \{x \in X \text{ such that } Vx \bigcap \mathscr{D}(V^{-1}K) \neq \emptyset\} \\ \supset & \{x \in X \text{ such that } Vx \bigcap \mathscr{D}(V^{-1}S) \neq \emptyset\} \\ = & \mathscr{D}(V^{-1}SV) \\ = & \mathscr{D}(T). \end{aligned}$$

In addition, K is compact, V^{-1} is continuous and $V^{-1}(0) = \{0\} \subset \mathscr{D}(K)$ (as $0 \in \rho(V)$), so the use of Theorem 3.5.2 implies that $V^{-1}K$ is precompact. Then,

$$\overline{\Gamma}_0(V^{-1}K) = 0.$$

Moreover, since V is single-valued, then

$$\overline{\Gamma}_0(V^{-1}KV) \leq \overline{\Gamma}_0(V^{-1}K)\overline{\Gamma}_0(V) = 0.$$

Then, $V^{-1}KV$ is precompact and, we have X is complete. Therefore, $V^{-1}KV$ is compact. Hence, $V^{-1}KV \in \mathscr{K}_T(X)$. Thus,

$$\mathscr{K}_S(X) \subset \{VKV^{-1} \text{ such that } K \in \mathscr{K}_T(X)\}.$$

Then,

$$\begin{aligned} \bigcap_{K \in \mathscr{K}_T(X)} \sigma_{\varepsilon/k}(T+K) \subset & \bigcap_{K \in \mathscr{K}_S(X)} \sigma_{\varepsilon}(S+K) \\ \subset & \bigcap_{K \in \mathscr{K}_T(X)} \sigma_{k\varepsilon}(T+K). \end{aligned}$$

So, for any $\varepsilon > 0$

$$\begin{aligned} \sigma_{e5,\varepsilon/k}(T) \subset & \ \sigma_{e5,\varepsilon}(S) \\ \subset & \ \sigma_{e5,k\varepsilon}(T). \end{aligned}$$

Moreover,

$$\begin{aligned} \bigcap_{\varepsilon>0} \sigma_{e5,\varepsilon/k}(T) \subset & \bigcap_{\varepsilon>0} \sigma_{e5,\varepsilon}(S) \\ \subset & \bigcap_{\varepsilon>0} \sigma_{e5,k\varepsilon}(T). \end{aligned}$$

From the proof of Proposition 4.1.7, S is closed. Then,

$$\begin{aligned}\sigma_{e5}(T) &\subset \sigma_{e5}(S)\\ &\subset \sigma_w(T).\end{aligned}$$

This achieves the proof. Q.E.D.

Theorem 4.2.7 *Let X be a Banach space, $T \in C\mathscr{R}(X)$, and $\varepsilon > 0$. Then,*

(i) If $\lambda \notin \sigma_{e5}(T)$, then

$$\|(\lambda - T)^{-1}\| \geq \frac{1}{dist(\lambda, \sigma_{e5}(T))}.$$

(ii) If $\lambda \notin \sigma_{e5,\varepsilon}(T)$, then

$$\|(\lambda - T)^{-1}\| > \frac{1}{dist(\lambda, \sigma_{e5,\varepsilon}(T)) + \varepsilon}.$$ ◇

Proof Let $\lambda \in \rho_{e5}(T)$. Then,

$$\text{dist}(\lambda, \sigma_{e5}(T)) = \inf\{|z - \lambda| \text{ such that } z \in \sigma_{e5}(T)\}.$$

Hence, for all $\eta > 0$, there exists $z_\eta \in \sigma_{e5}(T)$ such that

$$|\lambda - z_\eta| < \text{dist}(\lambda, \sigma_{e5}(T)) + \eta.$$

Assume that $|\lambda - z_\eta| < \gamma(\lambda - T)$. Using the fact that $\lambda - T \in \Phi(X)$ and $i(\lambda - T) = 0$, $(z_\eta - \lambda)I(0) = 0 \subset (\lambda - T)(0)$, $\mathscr{D}((z_\eta - \lambda)I) = \mathscr{D}(I) = X \supset \overline{\mathscr{D}((\lambda - T))} = \overline{\mathscr{D}(T)}$ and $|(z_\eta - \lambda)I| = |\lambda - z_\eta| < \gamma(\lambda - T)$, then we can conclude that $\lambda - T + z_\eta - \lambda = z_\eta - T \in \Phi(X)$ and

$$\begin{aligned} i(z_\eta - T) &= i(\lambda - T + z_\eta - \lambda)\\ &= i(\lambda - T)\\ &= 0.\end{aligned}$$

Hence, $z_\eta \in \rho_{e5}(T)$. This is a contradiction. Thus,

$$|\lambda - z_\eta| \geq \gamma(\lambda - T)$$

for all $\eta > 0$. Consequently,

$$\begin{aligned}\gamma(\lambda - T) &\leq |\lambda - z_\eta|\\ &< \text{dist}(\lambda, \sigma_{e5}(T)) + \eta \text{ for all } \eta > 0.\end{aligned}$$

Therefore,

$$\gamma(\lambda - T) \leq \text{dist}(\lambda, \sigma_{e5}(T)).$$

As a result,

$$\|(\lambda - T)^{-1}\| \geq \frac{1}{\text{dist}(\lambda, \sigma_{e5,\varepsilon}(T))}.$$

(ii) Let $\lambda \notin \sigma_{e5,\varepsilon}(T)$. Suppose that

$$\|(\lambda - T)^{-1}\| \leq \frac{1}{\text{dist}(\lambda, \sigma_{e5,\varepsilon}(T)) + \varepsilon},$$

i.e.,

$$\gamma(\lambda - T) \geq \text{dist}(\lambda, \sigma_{e5,\varepsilon}(T)) + \varepsilon.$$

As $\text{dist}(\lambda, \sigma_{e5,\varepsilon}(T)) = \inf\{|z - \lambda| \text{ such that } z \in \sigma_{e5,\varepsilon}(T)\}$, then for η, $0 < \eta \leq \varepsilon$, there exists $z_\eta \in \sigma_{e5,\varepsilon}(T)$ such that

$$|\lambda - z_\eta| < \text{dist}(\lambda, \sigma_{e5,\varepsilon}(T)) + \eta.$$

Hence,

$$\begin{aligned} |\lambda - z_\eta| &< \text{dist}(\lambda, \sigma_{e5,\varepsilon}(T)) + \varepsilon \\ &< \gamma(\lambda - T). \end{aligned}$$

Let S be a linear relation such that $S(0) \subset T(0)$, $\mathcal{D}(S) \supset \mathcal{D}(T)$ and $\|S\| < \varepsilon$. If $\lambda \notin \sigma_{e5,\varepsilon}(T)$, then by Theorem 4.2.2, we get $\lambda - T - S \in \Phi(X)$ and

$$i(\lambda - T - S) = 0.$$

In addition, T is closed, $(\lambda - S)(0) = S(0) \subset T(0)$, $\mathcal{D}((\lambda - S) = \mathcal{D}(S) \supset \mathcal{D}(T)$, and $\lambda - S$ is continuous, then by referring to Proposition 2.11.2, we infer that $\lambda - T - S$ is closed. Now, assume that $|\lambda - z_\eta| < \gamma(\lambda - T - S)$. Since $(z_\eta - \lambda)I(0) = 0 \subset (\lambda - T - S)(0)$, $\mathcal{D}((z_\eta - \lambda)I) = \mathcal{D}(I) = X \supset \overline{\mathcal{D}((\lambda - T - S))} = \overline{\mathcal{D}(T)}$ and $|(z_\eta - \lambda)I| = |\lambda - z_\eta| < \gamma(\lambda - T - S)$, then $\lambda - T - S + z_\eta - \lambda = z_\eta - T - S \in \Phi(X)$ and

$$\begin{aligned} i(z_\eta - T - S) &= i(\lambda - T - S + z_\eta - \lambda) \\ &= i(\lambda - T - S) \\ &= 0. \end{aligned}$$

So, $z_\eta \in \rho_{e5,\varepsilon}(T)$. This is a contradiction. Thus,

$$\gamma(\lambda - T - S) \leq |\lambda - z_\eta|$$

for all linear relations S such that $S(0) \subset T(0)$, $\mathscr{D}(S) \supset \mathscr{D}(T)$ and $\|S\| < \varepsilon$. In particular, we put $S = 0$. Hence,

$$\begin{aligned}\gamma(\lambda - T) &\leq |\lambda - z_\eta| \\ &< \operatorname{dist}(\lambda, \sigma_{e5,\varepsilon}(T)) + \eta.\end{aligned}$$

The fact that $\eta \leq \varepsilon$ implies that

$$\gamma(\lambda - T) < \operatorname{dist}(\lambda, \sigma_{e5,\varepsilon}(T)) + \varepsilon.$$

This is a contradiction. Then,

$$\|(\lambda - T)^{-1}\| > \frac{1}{\operatorname{dist}(\lambda, \sigma_{e5,\varepsilon}(T)) + \varepsilon}.$$

Q.E.D.

4.3 The ε-Essential Pseudospectra of Linear Relations

The goal of this section is to introduce and establish the ε-essential pseudospectra of linear relations in a Banach space.

Definition 4.3.1 Let X be a Banach space, $\varepsilon > 0$ and let $T \in C\mathscr{R}(X)$. The ε-essential pseudospectra sets of T are defined as follows

$$\begin{aligned}
\Sigma_{e1,\varepsilon}(T) &= \Upsilon_1 \bigcup \left\{\lambda \in \mathbb{C} \text{ such that } \|(\lambda - T)^{-1}\| \geq \frac{1}{\varepsilon}\right\}, \\
\Sigma_{e2,\varepsilon}(T) &= \Upsilon_2 \bigcup \left\{\lambda \in \mathbb{C} \text{ such that } \|(\lambda - T)^{-1}\| \geq \frac{1}{\varepsilon}\right\}, \\
\Sigma_{e3,\varepsilon}(T) &= \left(\Upsilon_1 \bigcap \Upsilon_2\right) \bigcup \left\{\lambda \in \mathbb{C} \text{ such that } \|(\lambda - T)^{-1}\| \geq \frac{1}{\varepsilon}\right\}, \\
\Sigma_{e4,\varepsilon}(T) &= \Upsilon_1 \bigcup \Upsilon_2 \bigcup \left\{\lambda \in \mathbb{C} \text{ such that } \|(\lambda - T)^{-1}\| \geq \frac{1}{\varepsilon}\right\}, \\
\Sigma_{e5,\varepsilon}(T) &= \Upsilon_1 \bigcup \Upsilon_2 \bigcup \Upsilon_3 \bigcup \left\{\lambda \in \mathbb{C} \text{ such that } \|(\lambda - T)^{-1}\| \geq \frac{1}{\varepsilon}\right\},
\end{aligned}$$

where $\Upsilon_1 = \{\lambda \in \mathbb{C} \text{ such that } \alpha(\lambda - T) = \infty\}$, $\Upsilon_2 = \{\lambda \in \mathbb{C} \text{ such that } \beta(\lambda - T) = \infty\}$ and $\Upsilon_3 = \{\lambda \in \mathbb{C} \text{ such that } \alpha(\lambda - T) \neq \beta(\lambda - T)\}$. ◇

Remark 4.3.1 Let X be a Banach space and let $T \in C\mathscr{R}(X)$. Then,

(i) If $0 < \varepsilon_1 < \varepsilon_2$, then $\Sigma_{ei,\varepsilon_1}(T) \subset \Sigma_{ei,\varepsilon_2}(T)$ for $i = 1, \ldots, 5$.
(ii) If $\varepsilon > 0$, then $\Sigma_{e3,\varepsilon}(T) = \Sigma_{e1,\varepsilon}(T) \bigcap \Sigma_{e2,\varepsilon}(T)$.
(iii) If $\varepsilon > 0$, then $\Sigma_{e4,\varepsilon}(T) = \Sigma_{e1,\varepsilon}(T) \bigcup \Sigma_{e2,\varepsilon}(T)$.

(iv) If $\varepsilon > 0$, then $\Sigma_{e5,\varepsilon}(T) = \Sigma_{e4,\varepsilon}(T) \bigcup \{\lambda \in \mathbb{C}$ such that $\alpha(\lambda - T) \neq \beta(\lambda - T)\}$. ◇

Lemma 4.3.1 *Let X be a Banach space and let $T \in C\mathscr{R}(X)$. If $\varepsilon > 0$, then*

$$\Sigma_{ei,\varepsilon}(T) \subset \Sigma_\varepsilon(T) \text{ for } i = 1, \ldots, 4. \quad \diamond$$

Proof Let us suppose that $\lambda \in \Sigma_\varepsilon(T)$. So, by Remark 4.1.1 (v), we have $\lambda \in \rho(T)$ and

$$\|(\lambda - T)^{-1}\| < \frac{1}{\varepsilon}.$$

This implies that $(\lambda - T)^{-1}$ is a bounded operator which implies $\lambda - T$ is one-to-one and $\mathscr{D}((\lambda - T)^{-1}) = X$. Hence, $N(\lambda - T) = \{0\}$ and

$$R(\lambda - T) = \mathscr{D}((\lambda - T)^{-1}) = X.$$

Consequently, $\alpha(\lambda - T)$ and $\beta(\lambda - T)$ are finite. Thus, $\lambda \notin (\Sigma_{e1,\varepsilon}(T) \bigcap \Sigma_{e2,\varepsilon}(T))$. This is equivalent to saying that

$$\Sigma_{e1,\varepsilon}(T) \bigcup \Sigma_{e2,\varepsilon}(T) \subset \Sigma_\varepsilon(T).$$

The use of Remark 4.3.1 (ii) and (iii) makes us conclude that $\Sigma_{ei,\varepsilon}(T) \subset \Sigma_\varepsilon(T)$ for $i = 1, \ldots, 4$. Q.E.D.

Example 4.3.1 Let H be the operator defined on $L^2(\mathbb{R})$ by

$$H = \frac{d^2}{dx^2} + ix.$$

Then, for $i = 1, \ldots, 5$, we have

$$\left\{\lambda \in \mathbb{C} \text{ such that } \max\left\{c_1, \left(\log\frac{1}{\varepsilon}\right)^{\frac{2}{3}} c_2\right\} \leq Re\lambda\right\} \subset \sigma_{ei,\varepsilon}(H),$$

where $c_1, c_2 > 0$. Indeed, the operator H is closed and non self-adjoint when considered on its maximal domain

$$\mathscr{D}(H) = \left\{\psi \in L^2(\mathbb{R}) \text{ such that } -\psi'' + i\psi \in L^2(\mathbb{R})\right\}.$$

Moreover, we have $\rho(H) = \mathbb{C}$ and

$$\left\{\lambda \in \mathbb{C} \text{ such that } \max\left\{c_1, \left(\log\frac{1}{\varepsilon}\right)^{\frac{2}{3}} c_2\right\} \leq Re\lambda\right\} \subset \left\{\lambda \in \mathbb{C} \text{ such that} \|(\lambda - H)^{-1}\| \geq \frac{1}{\varepsilon}\right\},$$

where $c_1, c_2 > 0$. Hence, for $i = 1, \ldots, 5$, we have

$$\left\{\lambda \in \mathbb{C} \text{ such that } \max\left\{c_1, \left(\log\frac{1}{\varepsilon}\right)^{\frac{2}{3}} c_2\right\} \leq Re\lambda\right\} \subset \Sigma_{ei,\varepsilon}(H).$$

Proposition 4.3.1 *Let X be a Banach space, $T \in C\mathcal{R}(X)$ and $\varepsilon > 0$. Then,*
(i) If $\lambda \notin \Sigma_{e1,\varepsilon}(T)$, then $\lambda - T - S \in \Phi_+(X)$, for all $S \in \mathcal{J}_T(X)$, where $\mathcal{J}_T(X) := \{S \in L\mathcal{R}(X)$ such that $S(0) \subset T(0)$, $\mathcal{D}(S) \supset \overline{\mathcal{D}(T)}$ and $\|S\| \leq \varepsilon\}$.
(ii) If $\lambda \notin \Sigma_{e2,\varepsilon}(T)$, then $\lambda - T - S \in \Phi_-(X)$, for all $S \in \mathcal{J}_T(X)$,
(iii) If $\lambda \notin \Sigma_{e3,\varepsilon}(T)$, then $\lambda - T - S \in \Phi_\pm(X)$, for all $S \in \mathcal{J}_T(X)$,
(iv) If $\lambda \notin \Sigma_{e4,\varepsilon}(T)$, then $\lambda - T - S \in \Phi(X)$, for all $S \in \mathcal{J}_T(X)$. ◇

Proof (i) Let $\lambda \notin \Sigma_{e1,\varepsilon}(T)$. Then, $\alpha(\lambda - T) < \infty$ and

$$\|(\lambda - T)^{-1}\| < \frac{1}{\varepsilon}.$$

In view of Lemma 2.10.9, we obtain

$$\gamma(\lambda - T) > \varepsilon$$

which leads to $\lambda - T$ is open. Combining the fact that $T \in C\mathcal{R}(X)$ and Proposition 2.11.2, we have $\lambda - T$ is closed. Then, applying Theorem 2.11.1 (ii), we obtain that $R(\lambda - T)$ is closed. Consequently, $\lambda - T \in \Phi_+(X)$. Let $S \in \mathcal{J}_T(X)$, then

$$\|S\| < \gamma(\lambda - T).$$

Finally, by Lemma 3.6.8, we can deduce that

$$\lambda - T - S \in \Phi_+(X).$$

(ii) Let $\lambda \notin \Sigma_{e2,\varepsilon}(T)$. Then, $\beta(\lambda - T) < \infty$ and

$$\gamma(\lambda - T) > \varepsilon.$$

By virtue of Proposition 2.11.2, we obtain $\lambda - T$ is closed. Now, since $\lambda - T$ is open and Theorem 2.11.1 (ii), then $R(\lambda - T)$ is closed. Thus, $\lambda - T \in \Phi_-(X)$. Let $S \in \mathcal{J}_T(X)$. Since $\lambda - T$ is open and $\|S\| < \gamma(\lambda - T)$, then by Proposition 2.15.3 (iv), we have

$$\|S\| < \gamma((\lambda - T^*)).$$

Finally, by using of Lemma 3.6.8, we can conclude that

$$\lambda - T - S \in \Phi_-(X).$$

(iii) Let us suppose that $\lambda \notin \Sigma_{e3,\varepsilon}(T)$. Therefore, it follows from (ii) of Remark 4.3.1 that $\lambda \notin (\Sigma_{e2,\varepsilon}(T) \bigcap \Sigma_{e1,\varepsilon}(T))$. In view of (i) and (ii) implies that

$$\lambda - T - S \in \Phi_-(X) \bigcup \Phi_+(X) = \Phi_\pm(X), \text{ for all } S \in \mathscr{J}_T(X).$$

(iv) Let us assume that $\lambda \notin \Sigma_{e4,\varepsilon}(T)$. Then, by using (iii) of Remark 4.3.1, we infer that $\lambda \notin (\Sigma_{e2,\varepsilon}(T) \bigcup \Sigma_{e1,\varepsilon}(T))$. The use of (i) and (ii) makes us conclude that

$$\lambda - T - S \in \Phi_-(X) \bigcap \Phi_+(X) = \Phi(X), \text{ for all } S \in \mathscr{J}_T(X). \qquad \text{Q.E.D.}$$

The following result gives a characterization of the essential ε-pseudospectra of a closed linear relations by means of Fredholm relations and minimum modulus.

Proposition 4.3.2 *Let X be a Banach space, $T \in C\mathscr{R}(X)$ and $\varepsilon > 0$. Then,*

(i) $\lambda \notin \Sigma_{e1,\varepsilon}(T)$ if, and only if, $\lambda - T \in \Phi_+(X)$ and $\gamma(\lambda - T) > \varepsilon$.
(ii) $\lambda \notin \Sigma_{e2,\varepsilon}(T)$ if, and only if, $\lambda - T \in \Phi_-(X)$ and $\gamma(\lambda - T) > \varepsilon$. ◇

Proof (i) Let $\lambda \notin \Sigma_{e1,\varepsilon}(T)$. Then, by using Lemma 2.10.9, we get $\alpha(\lambda - T) < \infty$ and $\gamma(\lambda - T) > \varepsilon$. Since $\gamma(\lambda - T) > 0$, then by using of Proposition 2.10.4 (ii), we infer that $\lambda - T$ is open. It follows from Proposition 2.11.2 and Theorem 2.11.1 (ii) that $R(\lambda - T)$ is closed. Hence, we conclude that $\lambda - T \in \Phi_+(X)$ and $\gamma(\lambda - T) > \varepsilon$. Conversely, let $\lambda - T \in \Phi_+(X)$ and $\gamma(\lambda - T) > \varepsilon$. Then, $\alpha(\lambda - T) < \infty$ and $\gamma(\lambda - T) > \varepsilon$. This implies in view of Lemma 2.10.9 that

$$\|(\lambda - T)^{-1}\| < \frac{1}{\varepsilon}.$$

Thus, $\lambda \notin \Sigma_{e1,\varepsilon}(T)$.
(ii) Let $\lambda \notin \Sigma_{e2,\varepsilon}(T)$. so, it follows from Lemma 2.10.9 that $\beta(\lambda - T) < \infty$ and $\gamma(\lambda - T) > \varepsilon$. Since $\gamma(\lambda - T) > 0$ then, by Proposition 2.10.4 (ii), we infer that $\lambda - T$ is open. By virtue of Proposition 2.11.2 and Theorem 2.11.1 (ii), we infer that $R(\lambda - T)$ is closed. Therefore, we deduce that $\lambda - T \in \Phi_-(X)$ and $\gamma(\lambda - T) > \varepsilon$. Conversely, let $\lambda - T \in \Phi_-(X)$ and $\gamma(\lambda - T) > \varepsilon$. Then, $\beta(\lambda - T) < \infty$ and $\gamma(\lambda - T) > \varepsilon$. From Lemma 2.10.9, we obtain that

$$\|(\lambda - T)^{-1}\| < \frac{1}{\varepsilon}.$$

Thus, $\lambda \notin \Sigma_{e2,\varepsilon}(T)$. Q.E.D.

4.3.1 The Essential ε-Pseudospectra of Linear and a Sequence Linear Relations

The goal of this subsection is to introduce and study the essential ε-pseudospectra of linear relations in a Banach space. Let $T \in C\mathscr{R}(X)$ and $\varepsilon > 0$. Consider $\mathscr{J}_{T,\varepsilon}(X)$ the set

$$\mathscr{J}_{T,\varepsilon}(X) := \{S \in L\mathscr{R}(X) \text{ such that } S(0) \subset T(0),\ \mathscr{D}(S) \supset \overline{\mathscr{D}(T)} \text{ and } \|S\| \le \varepsilon\}.$$

Remark 4.3.2 Let X be a Banach space and let $T \in C\mathscr{R}(X)$. Then,

(i) If $0 < \varepsilon_1 < \varepsilon_2$, then $\Sigma_{ei,\varepsilon_1}(T) \subset \Sigma_{ei,\varepsilon_2}(T)$ for $i = 1, \ldots, 5$.
(ii) If $\varepsilon > 0$, then $\Sigma_{ei,\varepsilon}(T) \subset \Sigma_{\varepsilon}(T)$ for $i = 1, \ldots, 5$.
(iii) If $\varepsilon > 0$, then $\Sigma_{e3,\varepsilon}(T) = \Sigma_{e1,\varepsilon}(T) \bigcap \Sigma_{e2,\varepsilon}(T)$.
(iv) If $\varepsilon > 0$, then $\Sigma_{e4,\varepsilon}(T) = \Sigma_{e1,\varepsilon}(T) \bigcup \Sigma_{e2,\varepsilon}(T)$.
(v) If $\varepsilon > 0$, then $\Sigma_{e5,\varepsilon}(T) = \Sigma_{e4,\varepsilon}(T) \bigcup \{\lambda \in \mathbb{C}$ such that $\alpha(\lambda - T) \neq \beta(\lambda - T)\}$. ◇

Proposition 4.3.3 *Let X be a Banach space, $T \in C\mathscr{R}(X)$ and $\varepsilon > 0$. Then,*

(i) If $\lambda \notin \Sigma_{e1,\varepsilon}(T)$, then $\lambda - T - S \in \Phi_+(X)$, for all $S \in \mathscr{J}_{T,\varepsilon}(X)$,
(ii) If $\lambda \notin \Sigma_{e2,\varepsilon}(T)$, then $\lambda - T - S \in \Phi_-(X)$, for all $S \in \mathscr{J}_{T,\varepsilon}(X)$,
(iii) If $\lambda \notin \Sigma_{e3,\varepsilon}(T)$, then $\lambda - T - S \in \Phi_{\pm}(X)$, for all $S \in \mathscr{J}_{T,\varepsilon}(X)$,
(iv) If $\lambda \notin \Sigma_{e4,\varepsilon}(T)$, then $\lambda - T - S \in \Phi(X)$, for all $S \in \mathscr{J}_{T,\varepsilon}(X)$,
(v) If $\lambda \notin \Sigma_{e5,\varepsilon}(T)$, then $\lambda - T - S \in \Phi(X)$ and $i(\lambda - T - S) = 0$, for all $S \in \mathscr{J}_{T,\varepsilon}(X)$. ◇

Proof Since $T \in C\mathscr{R}(X)$, then by Proposition 2.11.2, we get that for all $S \in \mathscr{J}_{T,\varepsilon}(X)$ and $\lambda \in \mathbb{C}$, we have $\lambda - T - S \in C\mathscr{R}(X)$.

(i) Let $\lambda \notin \Sigma_{e1,\varepsilon}(T)$. So, $\alpha(\lambda - T) < \infty$ and

$$\|(\lambda - T)^{-1}\| < \frac{1}{\varepsilon}.$$

In view of Lemma 2.10.9, we obtain $\gamma(\lambda - T) > \varepsilon$ which leads to $\lambda - T$ is open. Combining the fact that $T \in C\mathscr{R}(X)$ and Proposition 2.11.2 (i), we obtain that $\lambda - T$ is closed. Therefore, applying Theorem 2.11.1 (ii), we have $R(\lambda - T)$ is closed. Thus, $\lambda - T \in \Phi_+(X)$. It is clear for any $S \in \mathscr{J}_{T,\varepsilon}(X)$ that $\|S\| < \gamma(\lambda - T)$. Finally, in view of Corollary 3.6.3 (iv), we conclude that

$$\lambda - T - S \in \Phi_+(X).$$

(ii) Let $\lambda \notin \Sigma_{e2,\varepsilon}(T)$. So, $\beta(\lambda - T) < \infty$ and $\gamma(\lambda - T) > \varepsilon$. By virtue of Proposition 2.10.11 (i), then $\lambda - T$ is closed. Now, since $\lambda - T$ is open and by Theorem 2.11.1 (ii), we infer that $R(\lambda - T)$ is closed. Therefore, $\lambda - T \in \Phi_-(X)$. Since $\|S\| < \gamma(\lambda - T)$ for any $S \in \mathscr{J}_{T,\varepsilon}(X)$, then it follows from Corollary 3.6.3 (ii) that

$$\lambda - T - S \in \Phi_-(X).$$

(iii) Let us suppose that $\lambda \notin \Sigma_{e3,\varepsilon}(T)$. Then, by (iii) of Remark 4.3.2, we get $\lambda \notin \Sigma_{e2,\varepsilon}(T) \bigcap \Sigma_{e1,\varepsilon}(T)$. it follows from (i) and (ii) that

$$\lambda - T - S \in \Phi_-(X) \bigcup \Phi_+(X) = \Phi_{\pm}(X), \text{ for all } S \in \mathscr{J}_{T,\varepsilon}(X).$$

(iv) Let us suppose that $\lambda \notin \Sigma_{e4,\varepsilon}(T)$. Then, applying (iv) of Remark 4.3.2, we obtain that $\lambda \notin \Sigma_{e2,\varepsilon}(T) \bigcup \Sigma_{e1,\varepsilon}(T)$. The use of (i) and (ii) makes us deduce that

$$\lambda - T - S \in \Phi_{-}(X) \bigcap \Phi_{+}(X) = \Phi(X), \text{ for all } S \in \mathscr{I}_{T,\varepsilon}(X).$$

(v) Let us suppose that $\lambda \notin \Sigma_{e5,\varepsilon}(T)$. Then, by (v) of Remark 4.3.2, we conclude that $\lambda \notin \Sigma_{e4,\varepsilon}(T)$ and $i(\lambda - T) = 0$. Let $S \in \mathscr{I}_{T,\varepsilon}(X)$. It follows from (iv) that $\lambda - T - S \in \Phi(X)$. Since $\|S\| \leq \varepsilon$, then

$$i(\lambda - T - S) = i(\lambda - T) = 0.$$

This achieves the proof of (v). Q.E.D.

The following result gives a characterization of the essential ε-pseudospectra of a closed linear relations by means of Fredholm relations and minimum modulus.

Proposition 4.3.4 *Let X be a Banach space, $T \in C\mathscr{R}(X)$ and $\varepsilon > 0$. Then,*

(i) $\lambda \notin \Sigma_{e1,\varepsilon}(T)$ if, and only if, $\lambda - T \in \Phi_{+}(X)$ and $\gamma(\lambda - T) > \varepsilon$.
(ii) $\lambda \notin \Sigma_{e2,\varepsilon}(T)$ if, and only if, $\lambda - T \in \Phi_{-}(X)$ and $\gamma(\lambda - T) > \varepsilon$. ◇

Proof (i) Let $\lambda \notin \Sigma_{e1,\varepsilon}(T)$. Then, $\alpha(\lambda - T) < \infty$ and $\gamma(\lambda - T) > \varepsilon$. The fact that $\gamma(\lambda - T) > 0$ implies that $\lambda - T$ is open. By using the item (i) of Lemma 2.10.11 and (ii) of Theorem 2.11.1, we infer that $R(\lambda - T)$ is closed. Therefore, $\lambda - T \in \Phi_{+}(X)$ and $\gamma(\lambda - T) > \varepsilon$. Conversely, let $\lambda - T \in \Phi_{+}(X)$ and $\gamma(\lambda - T) > \varepsilon$. Thus, $\alpha(\lambda - T) < \infty$ and $\gamma(\lambda - T) > \varepsilon$. As a consequence $\lambda \notin \Sigma_{e1,\varepsilon}(T)$.
The proof of the assertion (ii) can be checked in the same way as (i). Q.E.D.

From Proposition 4.3.4 and Remark 4.3.2, we can conclude the following corollary.

Corollary 4.3.1 *Let X be a Banach space, $T \in C\mathscr{R}(X)$ and $\varepsilon > 0$. Then,*

(i) $\lambda \notin \Sigma_{e3,\varepsilon}(T)$ if, and only if, $\lambda - T \in \Phi_{\pm}(X)$ and $\gamma(\lambda - T) > \varepsilon$.
(ii) $\lambda \notin \Sigma_{e4,\varepsilon}(T)$ if, and only if, $\lambda - T \in \Phi(X)$ and $\gamma(\lambda - T) > \varepsilon$.
(iii) $\lambda \notin \Sigma_{e5,\varepsilon}(T)$ if, and only if, $\lambda - T \in \Phi(X)$, $i(\lambda - T) = 0$ and $\gamma(\lambda - T) > \varepsilon$. ◇

Lemma 4.3.2 *Let X be a Banach space, $T \in C\mathscr{R}(X)$ and let $\varepsilon > 0$. Then,*

(i) $\Sigma_{e1,\varepsilon}(T) = \Sigma_{e2,\varepsilon}(T^*)$
(ii) $\Sigma_{e2,\varepsilon}(T) = \Sigma_{e1,\varepsilon}(T^*)$.
(iii) $\Sigma_{ei,\varepsilon}(T) = \Sigma_{ei,\varepsilon}(T^*)$, $i = 3, 4, 5$. ◇

Proof (i) Let us suppose that $\lambda \notin \Sigma_{e1,\varepsilon}(T)$. So, applying (i) of Proposition 4.3.4, we get $\lambda - T \in \Phi_{+}(X)$ and $\gamma(\lambda - T) > \varepsilon$. In view of Theorem 3.6.5 (ii) and Proposition 2.15.1 (iii), we have $\lambda - T^* \in \Phi_{-}(X^*)$. Combining the fact that $\lambda - T$ is open and Propositions 2.15.3 (i), (iii), and 2.15.1 (iii), we obtain that

$$\gamma(\lambda - T^*) > \varepsilon.$$

Therefore, $\lambda \notin \Sigma_{e2,\varepsilon}(T^*)$. Thus,

$$\Sigma_{e1,\varepsilon}(T) \subset \Sigma_{e2,\varepsilon}(T^*).$$

A similar reasoning as above gives

$$\Sigma_{e2,\varepsilon}(T^*) \subset \Sigma_{e1,\varepsilon}(T).$$

(ii) Let us assume that $\lambda \notin \Sigma_{e1,\varepsilon}(T^*)$. Hence, from (i) of Proposition 4.3.4, we infer that $\lambda - T^* \in \Phi_+(X^*)$ and $\gamma(\lambda - T^*) > \varepsilon$. Applying Theorem 3.6.5 (i), we have $\lambda - T \in \Phi_-(X)$. By using of (iii) of Theorem 2.11.1, we get that $\lambda - T$ is open. So, according to Propositions 2.15.3 (i), (iii), we obtain that $\gamma(\lambda - T) > \varepsilon$. Therefore, $\lambda \notin \Sigma_{e2,\varepsilon}(T)$. Then,

$$\Sigma_{e2,\varepsilon}(T) \subset \Sigma_{e1,\varepsilon}(T^*).$$

Conversely, a same reasoning as before leads to the result.

(iii) For $i = 3$, by (i), (ii) and Remark 4.3.2 (iii), we conclude that

$$\begin{aligned}\Sigma_{e3,\varepsilon}(T) &= \Sigma_{e1,\varepsilon}(T) \bigcap \Sigma_{e2,\varepsilon}(T)\\ &= \Sigma_{e2,\varepsilon}(T^*) \bigcap \Sigma_{e1,\varepsilon}(T^*)\\ &= \Sigma_{e3,\varepsilon}(T^*).\end{aligned}$$

In the same way, one checks easily

$$\Sigma_{e4,\varepsilon}(T) = \Sigma_{e4,\varepsilon}(T^*). \tag{4.3.1}$$

For $i = 5$, let us suppose that $\lambda \notin \Sigma_{e5,\varepsilon}(T)$. This leads to by using of (v) of Remark 4.3.2 that $\lambda \notin \Sigma_{e4,\varepsilon}(T)$ and $i(\lambda - T) = 0$. In view of (4.3.1), we have $\lambda \notin \Sigma_{e4,\varepsilon}(T^*)$. It remains to prove the following

$$i(\lambda - T^*) = -i(\lambda - T) = 0.$$

Since $\lambda \notin \Sigma_{e4,\varepsilon}(T)$, then by applying of Corollary 4.3.1, we obtain that $\lambda - T \in \Phi(X)$ and $\gamma(\lambda - T) > \varepsilon$. This is equivalent to $R(\lambda - T)$ is closed and $\lambda - T$ is open. Therefore, by using of Theorem 3.6.5 (ii) and Proposition 2.15.1 (iii), we can deduce that

$$\alpha(\lambda - T^*) = \beta(\lambda - T)$$

and

$$\alpha(\lambda - T) = \beta(\lambda - T^*).$$

This implies that

$$i(\lambda - T^*) = -i(\lambda - T) = 0.$$

Again, by (v) of Remark 4.3.2, we conclude that $\lambda \notin \Sigma_{e5,\varepsilon}(T^*)$. Thus,

$$\Sigma_{e5,\varepsilon}(T^*) \subset \Sigma_{e5,\varepsilon}(T).$$

Conversely, by the same reasoning as before, we obtain the result. Q.E.D.

4.3.2 The Essential ε-Pseudospectra of a Sequence of Linear Relations

The goal of this subsection is to study the essential ε-pseudospectra of a sequence of linear relations in a Banach space.

Theorem 4.3.1 *Let X be a Banach space, $(T_n)_n$ be a sequence of closed linear relations from X into X and $T \in C\mathscr{R}(X)$. We suppose that $\|T_n - T\| \to 0$ when $n \to \infty$. If there exists $n_1 \in \mathbb{N}$ such that $T(0) = T_n(0)$ and $\mathscr{D}(T) = \mathscr{D}(T_n)$, for all $n \geq n_1$. Then, for every $\varepsilon > 0$, there exist ε_1, $\varepsilon_2 > 0$ and a positive integers n_2 such that*

$$\Sigma_{e1,\varepsilon_1}(T) \subset \Sigma_{e1,\varepsilon_2}(T_n) \subset \Sigma_{e1,\varepsilon}(T), \text{ for all } n \geq n_2. \qquad \diamondsuit$$

Proof Let us suppose that $\lambda \notin \sigma_{e1,\varepsilon}(T)$. Then, $\alpha(\lambda - T) < \infty$ and $\|(\lambda - T)^{-1}\| < \frac{1}{\varepsilon}$. By Lemma 2.10.9, we have $\gamma(\lambda - T) > \varepsilon$. Using the fact that $\|T_n - T\| \to 0$ when $n \to \infty$, then there exists $n_0 \in \mathbb{N}$ such that

$$\|T_n - T\| < \|P\|^{-1}(\gamma(\lambda - T) - \varepsilon_3), \text{ for all } n \geq n_0,$$

where P is any continuous projection defined on $\mathscr{D}(\lambda - T)$ with kernel $N(\lambda - T)$. On the other hand, for all $n \geq n_0$, we can obtain that

$$\begin{aligned} \|T_n - T\| &< \gamma(\lambda - T) - \varepsilon_3 \\ &< \gamma(\lambda - T). \end{aligned}$$

By using Proposition 2.7.2, we can write

$$\lambda - T_n = \lambda - T + T - T_n, \quad \text{for all } n \geq n_1. \tag{4.3.2}$$

In view of Theorem 2.10.1 (i) and (4.3.2) implies that $\alpha(\lambda - T_n) \leq \alpha(\lambda - T)$, for all $n \geq n_1$. Consequently, $\alpha(\lambda - T_n) < \infty$, for all $n \geq n_1$. Let $n_2 = \max\{n_0, n_1\}$. Now, by using of Proposition 3.6.4 (i) and (4.3.2), we deduce, for all $n \geq n_2$, that

$$\gamma(\lambda - T_n) \geq \|P\|^{-1}\gamma(\lambda - T) - \|T_n - T\|$$

$$> \ \varepsilon\|P\|^{-1}.$$

Therefore, we conclude that $\lambda \notin \Sigma_{e1,\varepsilon_2}(T)$, where $\varepsilon_2 = \varepsilon\|P\|^{-1}$. Let us prove that

$$\Sigma_{e1,\varepsilon_1}(T) \subset \Sigma_{e1,\varepsilon_2}(T_n), \text{ for all } n \geq n_2.$$

Suppose that $\lambda \notin \Sigma_{e1,\varepsilon_2}(T_n)$. Hence, $\alpha(\lambda - T_n) < \infty$ and $\|(\lambda - T_n)^{-1}\| < \frac{1}{\varepsilon_2}$. By referring to Lemma 2.10.9, we infer that $\gamma(\lambda - T_n) > \varepsilon_2$. Since $\|T_n - T\| \to 0$ when $n \to \infty$, so there exists $n_0 \in \mathbb{N}$ such that

$$\|T_n - T\| < \|P_1\|^{-1}(\gamma(\lambda - T_n) - \varepsilon_2), \text{ for all } n \geq n_0,$$

where P_1 is any continuous projection defined on $\mathscr{D}(\lambda - T_n)$ with kernel $N(\lambda - T_n)$. Further, for all $n \geq n_0$, we get that

$$\begin{aligned}\|T_n - T\| &< \ \gamma(\lambda - T_n) - \varepsilon_2 \\ &< \ \gamma(\lambda - T_n).\end{aligned}$$

By using Proposition 2.7.2, we can write

$$\lambda - T = \lambda - T - T_n + T_n, \text{ for all } n \geq n_1. \tag{4.3.3}$$

In view of Theorem 2.10.1 (i) and (4.3.3) implies that $\alpha(\lambda - T) \leq \alpha(\lambda - T_n)$, for all $n \geq n_1$. Consequently, $\alpha(\lambda - T) < \infty$, for all $n \geq n_1$. Let $n_2 = \max\{n_0, n_1\}$. Now, by using Lemma 3.6.4 (i) and (4.3.3), we conclude, for all $n \geq n_2$, that

$$\begin{aligned}\gamma(\lambda - T) \geq \ & \|P_1\|^{-1}\gamma(\lambda - T_n) - \|T_n - T\| \\ > \ & \varepsilon_2\|P_1\|^{-1}.\end{aligned}$$

Therefore, $\lambda \notin \Sigma_{e1,\varepsilon_1}(T)$, where $\varepsilon_1 = \varepsilon_2\|P_1\|^{-1} = \varepsilon\|P\|^{-1}\|P_1\|^{-1}$. Q.E.D.

Theorem 4.3.2 *Let X be a Banach space, $(T_n)_n$ be a sequence of closed linear relations from X into X and $T \in C\mathscr{R}(X)$. We suppose that $\|T_n - T\| \to 0$ when $n \to \infty$. Then,*

(i) If there exists $n_1 \in \mathbb{N}$ such that $\overline{\mathscr{D}(T)} \subset \mathscr{D}(T_n) \subset \mathscr{D}(T)$ and $T(0) = T_n(0)$, for all $n \geq n_1$, then for every $\varepsilon_2 > 0$, there exist a positive real number ε_1, and a positive integer n_2 such that

$$\Sigma_{e2,\varepsilon_1}(T_n) \subset \Sigma_{e2,\varepsilon_2}(T), \textit{ for all } n \geq n_2.$$

(ii) If there exists $n_1 \in \mathbb{N}$ such that $\overline{\mathscr{D}(T_n)} \subset \mathscr{D}(T) \subset \mathscr{D}(T_n)$ and $T(0) = T_n(0)$, for all $n \geq n_1$, then for every $\varepsilon_2 > 0$, there exist a positive real number ε_1, and a positive integer n_2 such that

$$\Sigma_{e2,\varepsilon_1}(T) \subset \Sigma_{e2,\varepsilon_2}(T_n), \textit{ for all } n \geq n_2.$$

◇

Proof (i) Let $\lambda \notin \Sigma_{e2,\varepsilon_2}(T)$. Then, applying (ii) of Proposition 4.3.4, we obtain that $\lambda - T \in \Phi_-(X)$ and $\gamma(\lambda - T) > \varepsilon_2$. Since $\|T_n - T\| \to 0$ when $n \to \infty$, so there exists $n_0 \in \mathbb{N}$ such that

$$\|T_n - T\| < \|P\|^{-1}\left(\gamma(\lambda - T) - \varepsilon_2\right), \text{ for all } n \geq n_0,$$

where P is any continuous projection defined on $\mathscr{D}((\lambda - T)^*)$ with kernel $N((\lambda - T)^*)$. In view of Theorem 3.6.5 (i) and since $\lambda - T \in \Phi_-(X)$, we get that $(\lambda - T)^* \in \Phi_+(X^*)$. This leads to $(\lambda - T)^*$ is open and $\alpha((\lambda - T)^*) < \infty$. On the other hand, for all $n \geq n_0$, we infer that

$$\begin{aligned} \|T_n - T\| &< \gamma(\lambda - T) - \varepsilon_2 \\ &< \gamma(\lambda - T). \end{aligned} \tag{4.3.4}$$

Since $\lambda - T$ is open, then by virtue of Proposition 2.15.3 (iii) and (4.3.4), we obtain

$$\|T_n - T\| < \gamma((\lambda - T)^*).$$

Let $n_2 = \max\{n_0, n_1\}$. Combining both the fact that $(T_n - T)(0) \subset (\lambda - T)(0)$ and $\mathscr{D}(T) \subset \mathscr{D}(T_n - T)$, for all $n \geq n_2$, and Corollary 3.6.3 (ii), we infer that $\lambda - T + T - T_n \in \Phi_-(X)$. Using the fact that for all $n \geq n_1$, $\mathscr{D}(\lambda - T_n) = \mathscr{D}(T_n) \subset \mathscr{D}(T)$, $T(0) = T_n(0) = (\lambda - T_n)(0)$, and in view of Lemma 2.10.11 (iv), we can write

$$\lambda - T_n = \lambda - T_n - T + T, \text{ for all } n \geq n_2.$$

Consequently, $\lambda - T_n \in \Phi_-(X)$. Now, by virtue of Proposition 3.6.4, for all $n \geq n_2$, we obtain

$$\begin{aligned} \gamma(\lambda - T_n) &\geq \|P\|^{-1}\gamma(\lambda - T) - \|T_n - T\| \\ &\geq \|P\|^{-1}\gamma(\lambda - T) - \|P\|^{-1}\left(\gamma(\lambda - T) - \varepsilon_2\right) \\ &> \varepsilon_1, \end{aligned}$$

where $\varepsilon_1 = \|P\|^{-1}\varepsilon_2$. Finally, by Proposition 4.3.4, we obtain that $\lambda \notin \Sigma_{e2,\varepsilon_1}(T_n)$, for all $n \geq n_2$.
The same reasoning as (i), we may achieve the proof of (ii). Q.E.D.

As an immediate consequence of Theorems 4.3.1 and 4.3.2.

Corollary 4.3.2 *Let X be a Banach space, $\varepsilon > 0$, $(T_n)_n$ be a sequence of closed and bounded linear relations from X into X and $T \in BC\mathscr{R}(X)$. We suppose that $\|T_n - T\| \to 0$ when $n \to \infty$. If there exists $n_1 \in \mathbb{N}$ such that $T(0) = T_n(0)$, for all $n \geq n_1$, then for every triplet $(\varepsilon_1, \varepsilon_2, \varepsilon_3)$ such that $0 < \varepsilon_1 < \varepsilon_2 < \varepsilon_3$, there exists a positive integer n_2 such that*

$$\Sigma_{ei,\varepsilon_1}(T) \subset \Sigma_{ei,\varepsilon_2}(T_n) \subset \Sigma_{ei,\varepsilon_3}(T), \textit{ for all } n \geq n_2, \textit{ for } i = 1, \ldots, 4. \quad \diamond$$

4.4 Pencil Pseudospectra and Pencil Essential Pseudospectra of Linear Relations

Definition 4.4.1 We define the S-pseudospectra of T by

$$\sigma_{\varepsilon,S}(T) = \sigma_S(T) \bigcup \left\{ \lambda \in \mathbb{C} \text{ such that } \|(\lambda S - T)^{-1}\| > \frac{1}{\varepsilon} \right\}.$$

We denote the S-pseudoresolvent set of T

$$\rho_{\varepsilon,S}(T) = \mathbb{C}\backslash\sigma_{\varepsilon,S}(T) = \rho_S(T) \bigcap \left\{ \lambda \in \mathbb{C} \text{ such that } \|(\lambda S - T)^{-1}\| \leq \frac{1}{\varepsilon} \right\}. \quad \diamond$$

4.4.1 Pencil Pseudospectra of Linear Relations

The aim of this section is to introduce and study the pencil pseudospectra of multivalued linear operator and establish some properties.

Lemma 4.4.1 *Let X be a Banach space, $\varepsilon > 0$, $T \in C\mathscr{R}(X)$ and let $S \in L\mathscr{R}(X)$ be such that $S(0) \subset T(0)$ and $\mathscr{D}(S) \supset \mathscr{D}(T)$. If $\lambda \notin \sigma_S(T)$, then $\lambda \in \sigma_{\varepsilon,S}(T)$ if, and only if, there exists $x \in X$ such that $\|(\lambda S - T)x\| < \varepsilon\|x\|$.* $\diamond$

Proof Let $\lambda \in \sigma_{\varepsilon,S}(T)\backslash\sigma_S(T)$, then

$$\|(\lambda S - T)^{-1}\| > \frac{1}{\varepsilon}.$$

The fact that $(\lambda S - T)^{-1}$ is bounded operator, then there exists a non-zero vector $y \in X$ such that

$$\|(\lambda S - T)^{-1}y\| > \frac{1}{\varepsilon}\, \|y\|. \tag{4.4.1}$$

Putting $x := (\lambda S - T)^{-1}y$, then $y \in (\lambda S - T)x$. On the other hand,

$$\begin{aligned}(\lambda S - T)(0) &= \lambda S(0) - T(0)\\ &= T(0) \ \ (\text{as } S(0) \subset T(0)).\end{aligned}$$

Applying 2.10.2, we have

$$\begin{aligned}\|(\lambda S - T)x\| &= \operatorname{dist}(y, (\lambda S - T)(0))\\ &= \operatorname{dist}(y, T(0))\\ &\leq \inf_{z \in T(0)} \|y - z\|\\ &= \|y\|.\end{aligned} \tag{4.4.2}$$

From (4.4.1) and (4.4.2), we have

$$\begin{aligned}\|x\| &> \frac{1}{\varepsilon}\|y\| \\ &\geq \frac{1}{\varepsilon}\|(\lambda S - T)x\|.\end{aligned}$$

Hence,

$$\|(\lambda S - T)x\| < \varepsilon\|x\|.$$

Conversely, suppose that there exists $x \in X$ such that

$$\|(\lambda S - T)x\| < \varepsilon\|x\|.$$

Since $\lambda \in \rho_S(T)$, then $\lambda S - T$ is one-to-one. This leads to

$$\begin{aligned}\gamma(\lambda S - T)\|x\| &\leq \|(\lambda S - T)x\| \\ &< \varepsilon\|x\|.\end{aligned}$$

So, $0 < \gamma(\lambda S - T) < \varepsilon$. By Lemma 2.10.2, we obtain that

$$\gamma(\lambda S - T) = \|(\lambda S - T)^{-1}\|^{-1}.$$

Then, $\lambda \in \sigma_{\varepsilon,S}(T)$. Q.E.D.

Theorem 4.4.1 *Let X be a Banach space, $\varepsilon > 0$ and let $T \in C\mathscr{R}(X)$. Then, $\lambda \in \sigma_{\varepsilon,S}(T)$ if, and only if, there exists a continuous linear relation B satisfying $\mathscr{D}(B) \supset \mathscr{D}(T)$, $B(0) \subset T(0)$, and $\|B\| < \varepsilon$ such that*

$$\lambda \in \sigma_S(T + B).$$ ◇

Proof Firstly, suppose that $\lambda \in \sigma_{\varepsilon,S}(T)$. We will discuss two cases.

First case. If $\lambda \in \sigma_S(T)$, then we may put $B = 0$.

Second case. If $\lambda \notin \sigma_S(T)$, then by Lemma 4.4.1 there exists $x_0 \in X$, $\|x_0\| = 1$ such that

$$\|(\lambda S - T)x_0\| < \varepsilon.$$

In view of Theorem 2.1.3, there exists $x' \in X^*$ such that $\|x'\| = 1$ and $x'(x_0) = \|x_0\|$. We can define the relation $B : X \longrightarrow X$ by

$$B(x) := x'(x)(\lambda S - T)x_0.$$

It is clear that B is everywhere defined and single-valued. Therefore,

$$\begin{aligned}\|Bx\| &= \|x'(x)(\lambda S - T)x_0)\| \\ &\leq \|x'\| \|x\| \|(\lambda S - T)x_0\|.\end{aligned}$$

For $x \neq 0$, we have

$$\frac{\|Bx\|}{\|x\|} \leq \|(\lambda S - T)x_0\|.$$

So, $\|B\| < \varepsilon$. In addition,

$$\begin{aligned}(\lambda S - (T + B))x_0 &= (\lambda S - T)x_0 - Bx_0 \\ &= (\lambda S - T)x_0 - x'(x_0)(\lambda S - T)x_0 \\ &= (\lambda S - (T + B))(0).\end{aligned}$$

Hence,

$$0 \neq x_0 \in N(\lambda S - (T + B)).$$

Then, $\lambda S - (T + B)$ is not one-to-one. Then, $\lambda \in \sigma_S(T + B)$. For the reverse inclusion, we derive a contradiction from the assumption that $\lambda \notin \sigma_{\varepsilon,S}(T)$, so $\lambda \in \rho_S(T)$ and $\gamma(\lambda S - T) \geq \varepsilon$. Thus, $\lambda S - T$ is one-to-one, open with dense range. Moreover,

$$\begin{aligned}B(0) &\subset T(0) \\ &= (\lambda S - T)(0),\end{aligned}$$

$$\begin{aligned}\mathscr{D}(B) &\supset \mathscr{D}(T) \\ &= \mathscr{D}(\lambda S - T)\end{aligned}$$

and

$$\|B\| < \varepsilon \leq \gamma(\lambda S - T).$$

Applying Lemma 2.10.11, $\lambda S - T - B$ is one-to-one, open with dense range. Hence, $\lambda \in \rho_S(T + B)$ and this is a contradiction. Q.E.D.

Remark 4.4.1 It follows, immediately, from Theorem 4.4.1, that for $T \in C\mathscr{R}(X)$ and $\varepsilon > 0$,

$$\sigma_{\varepsilon,S}(T) = \bigcup_{\substack{\|B\| < \varepsilon \\ B(0) \subset T(0) \\ \mathscr{D}(B) \supset \mathscr{D}(T)}} \sigma_S(T + B).$$

◇

Proposition 4.4.1 *Let X be a Banach space, $\varepsilon > 0$, $T \in C\mathscr{R}(X)$ and let $S \in L\mathscr{R}(X)$ be such that $S(0) \subset T(0)$ and $\mathscr{D}(S) \supset \mathscr{D}(T)$. Then,*

$$\bigcap_{\varepsilon>0} \sigma_{\varepsilon,S}(T) = \sigma_S(T). \qquad \diamond$$

Proof It is clear that $\sigma_S(T) \subset \sigma_{\varepsilon,S}(T)$ for all $\varepsilon > 0$. Hence,

$$\sigma_S(T) \subset \bigcap_{\varepsilon>0} \sigma_{\varepsilon,S}(T).$$

Conversely, if $\lambda \notin \sigma_S(T)$, then $\lambda \in \rho_S(T)$. Therefore, $(\lambda S - T)^{-1}$ is a bounded linear operator. Then, there exists $\varepsilon > 0$ such that

$$\|(\lambda S - T)^{-1}\| \leq \frac{1}{\varepsilon}.$$

As a consequence, $\lambda \notin \sigma_{\varepsilon,S}(T)$ and

$$\lambda \notin \bigcap_{\varepsilon>0} \sigma_{\varepsilon,S}(T). \qquad \text{Q.E.D.}$$

Theorem 4.4.2 *Let $T \in C\mathscr{R}(X)$ and S be a continuous linear relation such that $S(0) \subset T(0)$ and $\mathscr{D}(S) \supset \overline{\mathscr{D}(T)}$. Then,*

$$\sigma_S(T) = \sigma_{S^*}(T^*). \qquad \diamond$$

Proof Let $\lambda \in \rho_S(T)$, then $\lambda S - T$ is one-to-one, open with dense range. Using Proposition 2.15.1 (iv), we have

$$\begin{aligned} N((\lambda S - T)^*) &= (R(\lambda S - T))^{\perp} \\ &= (\overline{R(\lambda S - T)})^{\perp} \\ &= X^{\perp} \\ &= \{0\}. \end{aligned}$$

By referring to Proposition 2.15.3 (i), (iii), it follows that

$$\gamma((\lambda S - T)^*) = \gamma(\lambda S - T) > 0$$

and Proposition 2.15.1 (iv),

$$\begin{aligned} R((\lambda S - T)^*) &= N(\lambda S - T)^{\perp} \\ &= \{0\}^{\perp} \\ &= X^*. \end{aligned}$$

Then, $(\lambda S - T)^*$ is one-to-one, open with dense range. It remains to show that $(\lambda S - T)^* = \lambda S^* - T^*$. Applying Proposition 2.15.1 (vi), λS is continuous, and

$\mathscr{D}(\lambda S) = \mathscr{D}(S) \supset \mathscr{D}(T)$. This implies that

$$\begin{aligned}(\lambda S - T)^* &= (\lambda S)^* - T^* \\ &= \lambda S^* - T^*. \end{aligned} \tag{4.4.3}$$

Hence, from (4.4.3), we get $\lambda \in \rho_{S^*}(T^*)$. Then,

$$\sigma_{S^*}(T^*) \subset \sigma_S(T).$$

Conversely, let $\lambda \in \rho_{S^*}(T^*)$, then $\lambda S^* - T^* = (\lambda S - T)^*$ is one-to-one, open with dense range. The use of Proposition 2.15.1 (i) implies that $(\lambda S - T)^*$ is closed. So, by applying (4.4.3), we obtain that $\lambda S^* - T^*$ is a closed linear relation. Since $(\lambda S - T)^*$ is open and from Theorem 2.11.1 (ii), $R((\lambda S - T)^*)$ is closed. This leads to $R((\lambda S - T)^*) = X^*$. In addition, by referring to Proposition 2.15.1, we infer that

$$\begin{aligned}\overline{R(\lambda S - T)} &= (R(\lambda S - T)^{\perp})^{\top} \\ &= N((\lambda S - T)^*)^{\top} \\ &= \{0\}^{\top} \\ &= X.\end{aligned}$$

On the other hand, by using of Proposition 2.15.1, we get

$$\begin{aligned}N(\overline{\lambda S - T}) &= R((\lambda S - T)^*)^{\top} \\ &= X^{*\top} \\ &= \{0\}.\end{aligned}$$

Since $\lambda S - T$ is closed (see Proposition 2.11.2), then

$$N(\overline{\lambda S - T}) = N(\lambda S - T) = \{0\}.$$

Hence,

$$R((\lambda S - T)^*) = \overline{R((\lambda S - T)^*)} = X^*.$$

Moreover, based on the assumption that $\lambda S - T$ is closed, we can deduce that

$$(\lambda S - T)^{**} = \lambda S - T.$$

Hence, $\lambda S - T$ is one-to-one, open with dense range. Consequently, $\lambda \in \rho_S(T)$. Thus, $\sigma_S(T) \subset \sigma_{S^*}(T^*)$. Q.E.D.

Theorem 4.4.3 *Let $\varepsilon > 0$, $T \in C\mathscr{R}(X)$ and let $S \in L\mathscr{R}(X)$ be such that $S(0) \subset T(0)$ and $\mathscr{D}(S) \supset \mathscr{D}(T)$. Then,*

$$\sigma_{\varepsilon,S}(T) = \sigma_{\varepsilon,S^*}(T^*). \qquad \diamond$$

Proof Firstly, it is clear from the proof of Theorem 4.4.2 that $(\lambda S - T)^* = \lambda S^* - T^*$. Now, let $\lambda \in \rho_{\varepsilon,S^*}(T^*)$. Hence, $\lambda \in \rho_{S^*}(T^*)$ and

$$\|((\lambda S - T)^*)^{-1}\| \leq \frac{1}{\varepsilon}.$$

Applying Theorem 4.4.2, we infer that $\lambda \in \rho_S(T)$. Then, $(\lambda S - T)^{-1}$ is a continuous single-valued. Using Proposition 2.15.1, we get

$$\begin{aligned}\|(\lambda S - T)^{-1}\| &= \|((\lambda S - T)^{-1})^*\| \\ &\leq \frac{1}{\varepsilon}.\end{aligned}$$

Then, $\lambda \in \rho_{\varepsilon,S}(T)$. Conversely, if $\lambda \in \rho_{\varepsilon,S}(T)$, then

$$\|(\lambda S - T)^{-1}\| \leq \frac{1}{\varepsilon}.$$

By Proposition 2.15.1, we have

$$\begin{aligned}\|(\lambda S - T)^{-1}\| &= \|(\lambda S^* - T^*)^{-1}\| \\ &\leq \frac{1}{\varepsilon}.\end{aligned}$$

Moreover, $\lambda \in \rho_{S^*}(T^*)$ by using Theorem 4.4.2. Hence, $\lambda \in \rho_{\varepsilon,S^*}(T^*)$. Q.E.D.

Proposition 4.4.2 *Let $T \in C\mathscr{R}(H, K)$ and $A \in \mathscr{L}(H, K)$ be such that $0 \in \rho(A)$. Set $k = \dfrac{\|A\|}{\|A^{-1}\|}$. Then,*

(i) $\sigma_A(T) = \sigma(TA^{-1}) = \sigma(A^{-1}T)$.
(ii) $\sigma_{\varepsilon\|A\|^{-1}}(A^{-1}T) \subset \sigma_{\varepsilon,A}(T) \subset \sigma_{\varepsilon\|A^{-1}\|^{-1}}(TA^{-1}) \subset \sigma_{\varepsilon k}(T, A)$. ◇

Proof *(i)* For $\lambda \in \mathbb{C}$, we infer from the fact that A is one-to-one and Lemma 2.6.3 that

$$\lambda A - T = \lambda A - TA^{-1}A.$$

Since A is an operator, we conclude from Proposition 2.8.2 *(iii)* that

$$\lambda A - T = \left(\lambda - TA^{-1}\right) A. \tag{4.4.4}$$

Since $0 \in \rho(A)$, then by (4.4.4), we conclude that

$$\sigma_A(T) = \sigma(TA^{-1}).$$

For $\lambda \in \mathbb{C}$, we obtain from Remark 2.8.1 *(ii)* that

$$\lambda A - T = \lambda A - AA^{-1}T.$$

Since $\mathscr{D}(A) = H$, then by Proposition 2.8.2 (iii), we infer that

$$\lambda A - T = A\left(\lambda - A^{-1}T\right). \tag{4.4.5}$$

Hence, the fact that $0 \in \rho(A)$ leads to (4.4.5) that

$$\sigma_A(T) = \sigma(A^{-1}T).$$

(ii) First, we will prove

$$\sigma_{\varepsilon\|A\|^{-1}}(A^{-1}T) \subset \sigma_{\varepsilon,A}(T). \tag{4.4.6}$$

Let $\lambda \in \sigma_{\varepsilon\|A\|^{-1}}(A^{-1}T)$. We discuss two cases.

First case. If $\lambda \in \sigma(A^{-1}T)$, then by (i), we have $\lambda \in \sigma_A(T)$. This implies that $\lambda \in \sigma_{\varepsilon,A}(T)$.

Second case. If $\lambda \in \sigma_{\varepsilon\|A\|^{-1}}(A^{-1}T)\backslash\sigma(A^{-1}T)$, then $\lambda \in \rho(A^{-1}T) = \rho_A(T)$ and

$$\begin{aligned}
\frac{1}{\varepsilon\|A\|^{-1}} &< \left\|\left(\lambda - A^{-1}T\right)^{-1}\right\| \\
&< \left\|\left(\lambda A^{-1}A - A^{-1}T\right)^{-1}\right\| \quad \text{(from Lemma 2.6.3)} \\
&< \left\|\left[A^{-1}\left(\lambda A - T\right)\right]^{-1}\right\| \quad \text{(by Proposition 2.8.2 } (iii)\text{)} \\
&< \left\|(\lambda A - T)^{-1}A\right\| \\
&< \left\|(\lambda A - T)^{-1}\right\| \|A\| \quad \text{(by Corollary 2.10.1).}
\end{aligned}$$

Consequently,

$$\left\|(\lambda A - T)^{-1}\right\| > \frac{1}{\varepsilon}.$$

Hence, (4.4.6) holds. Second, we will show

$$\sigma_{\varepsilon,A}(T) \subset \sigma_{\varepsilon\|A^{-1}\|^{-1}}(TA^{-1}). \tag{4.4.7}$$

Let $\lambda \in \sigma_{\varepsilon,A}(T)$. We notice the existence of two cases.

First case. If $\lambda \in \sigma_A(T)$, then by (i), we have $\lambda \in \sigma(TA^{-1})$. This leads to $\lambda \in \sigma_{\varepsilon\|A^{-1}\|^{-1}}(TA^{-1})$.

Second case. If $\lambda \in \sigma_{\varepsilon,A}(T)\backslash\sigma_A(T)$, then $\lambda \in \rho_A(T) = \rho(TA^{-1})$ and

$$\begin{aligned}
\frac{1}{\varepsilon} &< \left\|(\lambda A - T)^{-1}\right\| \\
&< \left\|\left(\lambda A - TA^{-1}A\right)^{-1}\right\| \quad \text{(from Lemma 2.8.1)}
\end{aligned}$$

$$
\begin{aligned}
&< \left\| \left[\left(\lambda - TA^{-1} \right) A \right]^{-1} \right\| \text{ (by Proposition 2.8.2 } (iii)) \\
&< \left\| A^{-1} \left(\lambda - TA^{-1} \right)^{-1} \right\| \\
&< \left\| A^{-1} \right\| \left\| \left(\lambda - TA^{-1} \right)^{-1} \right\| \text{ (by Corollary 2.10.1).}
\end{aligned}
$$

This implies that

$$\left\| \left(\lambda - TA^{-1} \right)^{-1} \right\| > \frac{1}{\varepsilon \left\| A^{-1} \right\|^{-1}}.$$

Hence, (4.4.7) holds. Finally, we will prove that

$$\sigma_{\varepsilon \|A^{-1}\|^{-1}}(TA^{-1}) \subset \sigma_{\varepsilon k}(T, A). \tag{4.4.8}$$

Let $\lambda \in \sigma_{\varepsilon \|A^{-1}\|^{-1}}(TA^{-1})$. Then, there exists two cases.

First case. If $\lambda \in \sigma(TA^{-1})$, then by (i), we have $\lambda \in \sigma_A(T)$. Hence, $\lambda \in \sigma_{\varepsilon k}(T, A)$.

Second case. If $\lambda \in \sigma_{\varepsilon \|A^{-1}\|^{-1}}(TA^{-1}) \backslash \sigma(TA^{-1})$, then $\lambda \in \rho(TA^{-1}) = \rho_A(T)$ and

$$
\begin{aligned}
\frac{1}{\varepsilon \|A^{-1}\|^{-1}} &< \left\| \left(\lambda - TA^{-1} \right)^{-1} \right\| \\
&< \left\| \left(\lambda AA^{-1} - TA^{-1} \right)^{-1} \right\| \text{ (from Remark 2.8.1 } (ii)) \\
&< \left\| \left[(\lambda A - T) A^{-1} \right]^{-1} \right\| \text{ (by Proposition 2.8.2 } (iii)) \\
&< \left\| A (\lambda A - T)^{-1} \right\| \\
&< \|A\| \left\| (\lambda A - T)^{-1} \right\| \text{ (by Corollary 2.10.1).}
\end{aligned}
$$

This leads to deduce that

$$\left\| (\lambda A - T)^{-1} \right\| > \frac{1}{\varepsilon k}.$$

It is show that (4.4.8) holds. Q.E.D.

Proposition 4.4.3 *Let $T \in C\mathscr{R}(X)$ and $\varepsilon > 0$. If T is one-to-one with dense range and $\|S\| \neq 0$, then*

$$\sigma_{\varepsilon,S}(T) \subset \left\{ \lambda \in \mathbb{C} \text{ such that } |\lambda| > \frac{\gamma(T) - \varepsilon}{\|S\|} \right\}. \qquad \diamond$$

Proof If $\gamma(T) < \varepsilon$, then there is nothing to prove. If $\gamma(T) = \varepsilon > 0$, then $T = T - 0\,S$ is open. Therefore, $0 \in \rho_S(T)$. In addition,

$$
\begin{aligned}
\|T^{-1}\| &= \|(0\,S - T)^{-1}\| \\
&= \frac{1}{\varepsilon}.
\end{aligned}
$$

So, $0 \notin \sigma_{\varepsilon,S}(T)$. Hence,

$$\sigma_{\varepsilon,S}(T) \subset \{\lambda \in \mathbb{C} \text{ such that } |\lambda| > 0\}.$$

Now, assume that $\gamma(T) > \varepsilon$. Then, T is open, one-to-one and onto. On the other hand, for $\lambda \in \mathbb{C}$ such that $|\lambda| \leq \frac{\gamma(T)-\varepsilon}{\|S\|}$, we have

$$\begin{aligned} \|\lambda S\| &= |\lambda|\|S\| \\ &\leq \gamma(T) - \varepsilon. \end{aligned}$$

Therefore, $\|\lambda S\| < \gamma(T)$. Using Lemma 2.10.11, the relation $\lambda S - T$ is open, one-to-one with dense range, i.e., $\lambda \in \rho_S(T)$. For $x \in \mathscr{D}(T)$, since $S(0) \subset T(0)$, we have

$$\begin{aligned} \|(\lambda S - T)x\| &= \|(T - \lambda S)x\| \\ &\geq \|Tx\| - \|\lambda Sx\| \\ &\geq (\gamma(T) - |\lambda|\|S\|)\|x\|. \end{aligned}$$

Consequently,

$$\begin{aligned} \gamma(\lambda S - T) &\geq \gamma(T) - |\lambda|\|S\| \\ &\geq \gamma(T) - \gamma(T) + \varepsilon \\ &= \varepsilon. \end{aligned}$$

So,

$$\|(\lambda S - T)^{-1}\| \leq \frac{1}{\varepsilon}.$$

Therefore,

$$\lambda \in \rho_{\varepsilon,S}(T).$$

Thus,

$$\sigma_{\varepsilon,S}(T) \subset \left\{\lambda \in \mathbb{C} \text{ such that } |\lambda| > \frac{\gamma(T) - \varepsilon}{\|S\|}\right\}.$$

Q.E.D.

4.4.2 Pencil Essential Pseudospectra of Linear Relations

In this section, we introduce the pencil essential pseudospectra of a closed linear relation, establish some properties and study some results of perturbation on the context of linear relations.

Definition 4.4.2 Let T be a linear relation in $C\mathscr{R}(X)$ and $\varepsilon > 0$. The pencil essential pseudospectra of T is the set

$$\sigma_{e5,\varepsilon,S}(T) = \bigcap_{K \in \mathscr{K}_T(X)} \sigma_{\varepsilon,S}(T+K).$$

We define the pencil essential pseudoresolvent set by

$$\rho_{e5,\varepsilon,S}(T) = \mathbb{C} \backslash \sigma_{e5,\varepsilon,S}(T). \qquad \diamond$$

Theorem 4.4.4 *The following properties are equivalent*

(i) $\lambda \notin \sigma_{e5,\varepsilon,S}(T)$.
(ii) For all continuous linear relations $B \in L\mathscr{R}(X)$ *such that* $\mathscr{D}(B) \supset \mathscr{D}(T)$, $B(0) \subset T(0)$ *and* $\|B\| < \varepsilon$, *we have* $T + B - \lambda S \in \Phi(X)$ *and* $i(T + B - \lambda S) = 0$.
(iii) For all continuous single-valued operators $D \in L\mathscr{R}(X)$ *such that* $\mathscr{D}(D) \supset \mathscr{D}(T)$ *and* $\|D\| < \varepsilon$, *we have* $T + D - \lambda S \in \Phi(X)$ *and* $i(T + D - \lambda S) = 0$. $\diamond$

Proof $(i) \Longrightarrow (ii)$ Let $\lambda \notin \sigma_{e5,\varepsilon,S}(T)$. Then, there exists $K \in \mathscr{K}_T(X)$ such that $\lambda \notin \sigma_{\varepsilon,S}(T+K)$. Applying Theorem 4.4.1, for all continuous linear relations $B \in L\mathscr{R}(X)$ such that $\mathscr{D}(B) \supset \mathscr{D}(T+K) = \mathscr{D}(T) \bigcap \mathscr{D}(K) = \mathscr{D}(T)$ as $\mathscr{D}(K) \supset \mathscr{D}(T)$, $B(0) \subset (T+K)(0) = T(0)$ as $K(0) \subset T(0)$ and $\|B\| < \varepsilon$, we get that $\lambda \in \rho_S(T + B + K)$. Hence, $T + B + K - \lambda S$ is open, one-to-one with dense range. But T is closed, K is compact, then K is continuous. Then, $B + K - \lambda S$ is continuous. In addition, $(B + K - \lambda S)(0) \subset T(0)$, and by using Proposition 2.11.2, we have $T + B + K - \lambda S$ is closed. Then, $R(T + B + K - \lambda S)$ is closed. Therefore, $R(T + B + K - \lambda S) = X$. So, $T + B + K - \lambda S \in \Phi(X)$ and $i(T + B + K - \lambda S) = 0$, for all continuous linear relations $B \in L\mathscr{R}(X)$ such that $\mathscr{D}(B) \supset \mathscr{D}(T)$, $B(0) \subset T(0)$ and $\|B\| < \varepsilon$. It comes from Proposition 2.7.2 and Theorem 2.11.4 for all continuous linear relations $B \in L\mathscr{R}(X)$ such that $\mathscr{D}(B) \supset \mathscr{D}(T)$, $B(0) \subset T(0)$ and $\|B\| < \varepsilon$, we get $T + B - \lambda S \in \Phi(X)$ and $i(T + B - \lambda S) = 0$.

$(ii) \Longrightarrow (iii)$ is trivial.

$(iii) \Longrightarrow (i)$ Let D be a continuous single-valued operators in $L\mathscr{R}(X)$ such that $\mathscr{D}(D) \supset \mathscr{D}(T)$ and $\|D\| < \varepsilon$, then $T + D - \lambda S \in \Phi(X)$ and $i(T + D - \lambda S) = 0$. By referring Proposition 2.15.1, we infer that $N((T + D - \lambda S)^*) = R(T + D - \lambda S)^{\perp}$. Let $n = \alpha(T + D - \lambda S) = \beta(T + D - \lambda S)$, $\{x_1, \ldots, x_n\}$ be basis for $N(T + D - \lambda S)$ and $\{y_1', \ldots, y_n'\}$ be basis for $N((T + D - \lambda S)^*)$. Then, there are functionals $x_1', \ldots, x_n' \in X^*$ and elements $y_1, \ldots, y_n$ such that

$$x_j'(x_k) = \delta_{jk} \quad \text{and} \quad y_j'(y_k) = \delta_{jk}, \quad 1 \leq j, k \leq n,$$

where $\delta_{jk} = 0$ if $j \neq k$ and $\delta_{jk} = 1$ if $j = k$. The single-valued operator K is defined by

$$Kx = \sum_{k=1}^{n} x_k'(x) y_k, \; x \in X.$$

K is bounded, since $\mathscr{D}(K) = X$ and

$$\|Kx\| \leq \|x\| \left(\sum_{k=1}^{n} \|x'_k\| \|y_k\| \right).$$

Furthermore, the range of K is contained in a finite subspace of X. So, K is a finite rank relation in X. Then, K is compact. Let $x \in N(T + D - \lambda S)$, then

$$x = \sum_{k=1}^{n} \alpha_k x_k.$$

Thus, $x'_j(x) = \alpha_j,\ 1 \leq j \leq n$. On the other hand, if $x \in N(K)$, so $x'_j(x) = 0,\ 1 \leq j \leq n$. This proves that $N(T + D - \lambda S) \bigcap N(K) = \{0\}$. Now, if $y \in R(K)$, then

$$y = \sum_{k=1}^{n} \alpha_k y_k.$$

Hence,

$$y'_j(y) = \alpha_j,\ 1 \leq j \leq n.$$

Now, if $y \in R(T + D - \lambda S)$, then

$$y'_j(y) = 0,\ 1 \leq j \leq n.$$

So,

$$N(T + D - \lambda S) \bigcap N(K) = \{0\}$$

and

$$R(T + D - \lambda S) \bigcap R(K) = \{0\}.$$

On the other hand, $K \in \mathscr{K}_T(X) = \mathscr{K}_{T+D-\lambda S}(X)$ since $K(0) \subset T(0) = (T + D - \lambda S)(0)$ and $\mathscr{D}(K) \supset \mathscr{D}(T) = \mathscr{D}(T + D - \lambda S)$. Thus, we conclude that $T + D + K - \lambda S \in \Phi(X)$ and

$$i(T + D + K - \lambda S) = 0.$$

If $x \in N(T + D + K - \lambda S)$, then $0 \in Tx + Dx + Kx - \lambda Sx$. Therefore, $-Kx \in (T + D - \lambda S)x$. Then, $Kx \in R(K) \bigcap R(T + D - \lambda S) = \{0\}$. Hence, $Kx = 0$ and $0 \in (T + D - \lambda S)x$. This implies that $x \in N(T + D - \lambda S) \bigcap N(K)$. So, $x = 0$. Consequently,

$$\alpha(T + D + K - \lambda S) = 0.$$

In the some way, one proves that $R(T + D + K - \lambda S) = X$. The fact that $T + D + K - \lambda S$ is closed by Proposition 2.11.2, then $\lambda \in \rho_S(T + D + K)$ for all continuous single-valued operators $D \in L\mathscr{R}(X)$ such that

$$\mathscr{D}(D) \supset \mathscr{D}(T)$$

and

$$\|D\| < \varepsilon.$$

But from the proof of Theorem 4.4.1($(i) \Longrightarrow (ii)$), if $\lambda \in \sigma_{\varepsilon,S}(T+K)$, then there exists a continuous single-valued operator $D \in L\mathscr{R}(X)$ satisfying $\mathscr{D}(D) \supset \mathscr{D}(T+K) = \mathscr{D}(T)$ and $\|D\| < \varepsilon$ such that $\lambda \in \sigma_S(T+K+D)$. Hence, $\lambda \notin \sigma_{\varepsilon,S}(T+K)$. Since $K \in \mathscr{K}_T(X)$, then

$$\lambda \notin \bigcap_{K \in \mathscr{K}_T(X)} \sigma_{\varepsilon,S}(T+K).$$

Thus, $\lambda \notin \sigma_{e5,\varepsilon,S}(T)$. Q.E.D.

Remark 4.4.2 It follow immediately, from Theorem 4.4.4 that for $T \in C\mathscr{R}(X)$ and $\varepsilon > 0$,

$$\begin{aligned}\sigma_{e5,\varepsilon,S}(T) &= \bigcup_{\substack{\|D\|<\varepsilon \\ \mathscr{D}(D)\supset\mathscr{D}(T)}} \sigma_{e5,S}(T+D) \\ &= \bigcup_{\substack{\|B\|<\varepsilon \\ B(0)\subset T(0) \\ \mathscr{D}(B)\supset\mathscr{D}(T)}} \sigma_{e5,S}(T+B).\end{aligned}$$

◇

Proposition 4.4.4 *Let $T \in C\mathscr{R}(X)$. Then,*

(i) If $0 < \varepsilon_1 < \varepsilon_2$, then $\sigma_{e5,S}(T) \subset \sigma_{e5,\varepsilon_1,S}(T) \subset \sigma_{e5,\varepsilon_2,S}(T)$.
(ii) For $\varepsilon > 0$, $\sigma_{e5,\varepsilon,S}(T) \subset \sigma_{\varepsilon,S}(T)$.
(iii) $\bigcap_{\varepsilon>0} \sigma_{e5,\varepsilon,S}(T) = \sigma_{e5,S}(T)$. ◇

Proof (i) If $\lambda \notin \sigma_{e5,\varepsilon_2,S}(T)$, then by Theorem 4.4.4 for all continuous linear relations $B \in L\mathscr{R}(X)$ such that $\mathscr{D}(B) \supset \mathscr{D}(T)$, $B(0) \subset T(0)$ and $\|B\| < \varepsilon_2$, we get that $T + B - \lambda S \in \Phi(X)$ and

$$i(T + B - \lambda S) = 0.$$

Then, for all continuous linear relations $B \in L\mathscr{R}(X)$ such that $\mathscr{D}(B) \supset \mathscr{D}(T)$, $B(0) \subset T(0)$ and $\|B\| < \varepsilon_1$, we obtain that $T + B - \lambda S \in \Phi(X)$ and

$$i(T + B - \lambda S) = 0.$$

So, $\lambda \notin \sigma_{\varepsilon_1,S}(T)$. On other hand, if $\lambda \notin \sigma_{e5,\varepsilon_1,S}(T)$, then for all continuous linear relations $B \in L\mathscr{R}(X)$ such that $\mathscr{D}(B) \supset \mathscr{D}(T)$, $B(0) \subset T(0)$ and $\|B\| < \varepsilon_1$, we get

$T + B - \lambda S \in \Phi(X)$ and

$$i(T + B - \lambda S) = 0.$$

In particular $B = 0$, so $\lambda \notin \sigma_{e5,S}(T)$.

(ii) $\sigma_{e5,\varepsilon,S}(T) = \bigcap_{K \in \mathscr{K}_T(X)} \sigma_{\varepsilon,S}(T + K) \subset \sigma_{\varepsilon,S}(T + K)$ for all $K \in \mathscr{K}_T(X)$. In particular, $K = 0$.

(iii) From (i), $\sigma_{e5,S}(T) \subset \sigma_{e5,\varepsilon,S}(T)$ for all $\varepsilon > 0$, we have

$$\sigma_{e5,S}(T) \subset \bigcap_{\varepsilon>0} \sigma_{e5,\varepsilon,S}(T).$$

For the reverse inclusion, if $\lambda \notin \sigma_{e5,S}(T)$, then $T - \lambda S \in \Phi(X)$ and

$$i(T - \lambda S) = 0.$$

Hence, $R(T - \lambda S)$ is closed. So, $T - \lambda S$ is open. Then,

$$\gamma(T - \lambda S) > 0.$$

Let ε be such that $0 < \varepsilon \leq \gamma(T - \lambda S)$ and let D be a single-valued linear relation in $L\mathscr{R}(X)$ such that $\mathscr{D}(D) \supset \mathscr{D}(T)$ and $\|D\| < \varepsilon \leq \gamma(T - \lambda S)$. Using $\gamma(T - \lambda S) = \gamma((T - \lambda S)^*)$, we have

$$\|D\| < \varepsilon \leq \gamma((T - \lambda S)^*).$$

Hence, $T + D - \lambda S \in \Phi_-(X)$. Therefore,

$$\alpha(T + D - \lambda S) \leq \alpha(T - \lambda S) < \infty.$$

Consequently, $T + D - \lambda S \in \Phi_+(X)$ and $T + D - \lambda S \in \Phi(X)$. In addition, we get

$$i(T + D - \lambda S) = i(T - \lambda S) = 0.$$

Then, by using Theorem 4.4.4, we infer that

$$\lambda \notin \sigma_{e5,\varepsilon,S}(T).$$

Then, $\lambda \notin \bigcap_{\varepsilon>0} \sigma_{e5,\varepsilon,S}(T)$. This implies that

$$\bigcap_{\varepsilon>0} \sigma_{e5,\varepsilon,S}(T) \subset \sigma_{e5,S}(T).$$

Q.E.D.

Theorem 4.4.5 *Let X be a Banach space, $\varepsilon > 0$, $T \in C\mathscr{R}(X)$ and let $S \in L\mathscr{R}(X)$ be such that $S(0) \subset T(0)$ and $\mathscr{D}(S) \supset \mathscr{D}(T)$. Then,*

$$\sigma_{e5,\varepsilon,S}(T) = \sigma_{e5,\varepsilon,S^*}(T^*).$$

◇

Proof Let $K \in \mathscr{K}_T(X)$, then K is compact. Therefore, it is continuous by Lemma 3.5.1 and from Proposition 3.5.2, K^* is compact. Moreover, by Lemma 2.15.2, we have $K^*(0) \subset T^*(0)$ and $\mathscr{D}(K^*) \supset \mathscr{D}(T^*)$. Hence,

$$\mathscr{K}_T(X) \subset \{K \in L\mathscr{R}(X) \text{ such that } K \text{ is continuous and } K^* \in \mathscr{K}_{T^*}(X^*)\}.$$

Now, let $K \in \{K \in L\mathscr{R}(X) \text{ such that } K \text{ is continuous and } K^* \in \mathscr{K}_{T^*}(X^*)\}$, then K is compact and

$$\begin{aligned} K^*(0) &= \mathscr{D}(K)^{\perp} \\ &\subset T^*(0) \\ &= \mathscr{D}(T)^{\perp}. \end{aligned}$$

So,

$$\mathscr{D}(K) \supset \mathscr{D}(T),$$

also

$$\begin{aligned} K(0) &\subset \overline{K}(0) \\ &= \mathscr{D}(K^*)^{\top} \\ &\subset \mathscr{D}(T^*)^{\top} \\ &= T(0) \end{aligned}$$

as T is closed. Then, $K \in \mathscr{K}_T(X)$. Hence,

$$\mathscr{K}_T(X) = \{K \in L\mathscr{R}(X) \text{ such that } K \text{ is continuous and } K^* \in \mathscr{K}_{T^*}(X^*)\}. \quad (4.4.9)$$

On the other hand for $K \in \mathscr{K}_T(X)$, $T + K$ is closed (by Proposition 2.11.2). By using Theorem 4.4.3, we have

$$\sigma_{\varepsilon,S}(T + K) = \sigma_{\varepsilon,S^*}((T + K)^*).$$

But $\mathscr{D}(K) \supset \mathscr{D}(T)$ and K is continuous, then

$$(T + K)^* = T^* + K^*.$$

Hence,

$$\sigma_{\varepsilon,S}(T + K) = \sigma_{\varepsilon,S^*}(T^* + K^*) \text{ for all } K \in \mathscr{K}_T(X).$$

Therefore, by using (4.4.9), we have

$$\begin{aligned}
\sigma_{e5,\varepsilon,S}(T) &= \bigcap_{K\in\mathcal{K}_T(X)} \sigma_{\varepsilon,S}(T+K) \\
&= \bigcap_{\substack{K^*\in\mathcal{K}_{T^*}(X^*)\\ K\in L\mathcal{R}(X)\\ K\, continuous}} \sigma_{\varepsilon,S^*}(T^*+K^*) \\
&\supset \bigcap_{\substack{K^*\in\mathcal{K}_{T^*}(X^*)\\ K\in L\mathcal{R}(X)}} \sigma_{\varepsilon,S^*}(T^*+K^*) \\
&= \sigma_{e5,\varepsilon,S^*}(T^*).
\end{aligned}$$

Let

$$\mathcal{O} := \bigcap_{\substack{K^*\in\mathcal{K}_{T^*}(X^*)\\ K\in L\mathcal{R}(X)\\ K\ continuous}} \sigma_{\varepsilon,S^*}(T^*+K^*),$$

then

$$\mathcal{O} = \sigma_{e5,\varepsilon,S}(T) \supset \sigma_{e5,\varepsilon,S^*}(T^*). \tag{4.4.10}$$

On the other hand, let $\lambda \in \mathcal{O}$, then for all $K \in L\mathcal{R}(X)$, K is continuous such that $K^* \in \mathcal{K}_{T^*}(X^*)$, $\lambda \in \sigma_{\varepsilon,S^*}(T^*+K^*)$. Let $K \in L\mathcal{R}(X)$ be such that $K^* \in \mathcal{K}_{T^*}(X^*)$. Hence, $\widetilde{K}$ is compact, where $\widetilde{T}$ is given in (2.11.1). Since $G(\widetilde{K})$ is the completion of $G(K)$, then

$$\begin{aligned}
G(\widetilde{K}) &= \overline{G(K)} \\
&= G(\overline{K}).
\end{aligned}$$

Hence, $\widetilde{K} = \overline{K}$ is compact and hence, continuous. Furthermore,

$$\begin{aligned}
\lambda G(\overline{K}^*) &= G(-\overline{K}^{\,-1})^{\perp} \\
&= \overline{G(-K^{-1})}^{\perp} \\
&= G(-K^{-1})^{\perp} \\
&= G(K^*).
\end{aligned}$$

Hence,

$$\overline{K}^* = K^*.$$

Thus, $\overline{K}$ in $L\mathcal{R}(X)$, $\overline{K}$ is continuous and $\overline{K}^{\,*} = K^* \in \mathcal{K}_{T^*}(X^*)$. Since $\lambda \in \mathcal{O}$, then $\lambda \in \sigma_{\varepsilon,S^*}(T^* + \overline{K}^*) = \sigma_{\varepsilon,S^*}(T^*+K^*)$. We conclude that if $\lambda \in \mathcal{O}$, for all $K \in L\mathcal{R}(X)$ such that $K^* \in \mathcal{K}_{T^*}(X^*)$, then $\lambda \in \sigma_{\varepsilon,S^*}(T^*+K^*)$. So,

$$\lambda \in \bigcap_{\substack{K^* \in \mathscr{K}_{T^*}(X^*) \\ K \in L\mathscr{R}(X)}} \sigma_{\varepsilon,S^*}(T^* + K^*) = \sigma_{e5,\varepsilon,S^*}(T^*).$$

Hence,

$$\mathscr{O} \subset \sigma_{e5,\varepsilon,S^*}(T^*).$$

By using (4.4.10), we have

$$\begin{aligned} \mathscr{O} &= \sigma_{e5,\varepsilon,S}(T) \\ &= \sigma_{e5,\varepsilon,S^*}(T^*). \end{aligned}$$

This achieves the proof. Q.E.D.

Theorem 4.4.6 *Let X be a Banach space, $\varepsilon > 0$, $T \in C\mathscr{R}(X)$ and let $S \in L\mathscr{R}(X)$ be such that $S(0) \subset T(0)$ and $\mathscr{D}(S) \supset \mathscr{D}(T)$. Then,*

$$\sigma_{e5,\varepsilon,S}(T) = \bigcap_{P \in \mathscr{P}\mathscr{R}(\Phi(X))} \sigma_{\varepsilon,S}(T + P). \qquad \diamond$$

Proof Let $\mathscr{O} := \bigcap_{P \in \mathscr{P}\mathscr{R}(\Phi(X))} \sigma_{\varepsilon,S}(T + P)$. The fact that $\mathscr{K}_T(X) \subset \mathscr{P}\mathscr{R}(\Phi(X))$, we obtain that

$$\mathscr{O} \subset \sigma_{e5,\varepsilon,S}(T).$$

Conversely, let $\lambda \notin \mathscr{O}$, then there exists $P \in \mathscr{P}\mathscr{R}(\Phi(X))$ such that $\lambda \notin \sigma_{\varepsilon,S}(T + P)$. But P is continuous, then by using of Proposition 2.11.2, $T + P$ is closed. Therefore, by Theorem 4.4.1, we can see that $\lambda \in \rho_S(T + B + P)$ for all continuous linear relations $B \in L\mathscr{R}(X)$ such that $B(0) \subset (T + P)(0) = T(0)$, $\mathscr{D}(B) \supset \mathscr{D}(T + P) = \mathscr{D}(T) \bigcap \mathscr{D}(P) = \mathscr{D}(T)$ and $\|B\| < \varepsilon$. But $(B + P - \lambda S)(0) \subset T(0)$, $\mathscr{D}(B + P - \lambda S) = \mathscr{D}(B) \bigcap \mathscr{D}(P) \supset \mathscr{D}(T)$ and T is closed, applying Proposition 2.11.2, $T + B + P - \lambda S$ is closed. Therefore, $T + B + P - \lambda S$ is one-to-one and onto. Hence,

$$T + B + P - \lambda S \in \Phi(X) \ \text{ and } \ i(T + B + P - \lambda S) = 0.$$

Since $-P \in \mathscr{P}\mathscr{R}(\Phi(X))$, $-P(0) = P(0) \subset (T + B + P - \lambda S)(0)$ and $\mathscr{D}(-P) = \mathscr{D}(P) \supset \mathscr{D}(T + B + P - \lambda S) = \mathscr{D}(T) \bigcap \mathscr{D}(B) \bigcap \mathscr{D}(P) \bigcap \mathscr{D}(S)$, then $-P$ is in $\mathscr{P}\mathscr{R}(\Phi(X))$. Applying Theorem 3.6.15, we deduce that for all continuous linear relations $B \in L\mathscr{R}(X)$ such that $B(0) \subset T(0)$, $\mathscr{D}(B) \supset \mathscr{D}(T)$ and $\|B\| < \varepsilon$,

$$T + B - \lambda S \in \Phi(X) \ \text{ and } \ i(T + B - \lambda S) = 0.$$

Finally, Theorem 4.4.4 proves that $\lambda \notin \sigma_{e5,\varepsilon,S}(T)$. Q.E.D.

Corollary 4.4.1 *Let $\mathscr{J}(X)$ be a subset of $L\mathscr{R}(X)$. If $\mathscr{K}_T(X) \subset \mathscr{J}(X) \subset \mathscr{P}\mathscr{R}(\Phi(X))$, then*

$$\sigma_{e5,\varepsilon,S}(T) = \bigcap_{J \in \mathscr{J}(X)} \sigma_{\varepsilon,S}(T + J). \qquad \diamond$$

Proof It is easy to see

$$\begin{aligned}\lambda\sigma_{e5,\varepsilon,S}(T) &= \bigcap_{P\in\mathscr{P}\mathscr{R}(\Phi(X))} \sigma_{\varepsilon,S}(T+P)\\ &\subset \bigcap_{J\in\mathscr{J}(X)} \sigma_{\varepsilon,S}(T+J)\\ &\subset \bigcap_{K\in\mathscr{K}_T(X)} \sigma_{\varepsilon,S}(T+K).\end{aligned}$$

This achieves the proof. Q.E.D.

Remark 4.4.3 It follows, by the definition of pencil essential pseudospectra, Theorem 4.4.6 and Corollary 4.4.1 that

(i) $\sigma_{e5,\varepsilon,S}(T+K) = \sigma_{e5,\varepsilon,S}(T)$ for all $K \in \mathscr{K}_T(X)$.
(ii) $\sigma_{e5,\varepsilon,S}(T+P) = \sigma_{e5,\varepsilon,S}(T)$ for all $P \in \mathscr{P}\mathscr{R}(\Phi(X))$.
(iii) If for all $J_1, J_2 \in \mathscr{J}(X)$, $J_1 \pm J_2 \in \mathscr{J}(X)$, then $\sigma_{e5,\varepsilon,S}(T+J) = \sigma_{e5,\varepsilon,S}(T)$ for all $J \in \mathscr{J}(X)$ such that $\mathscr{K}_T(X) \subset \mathscr{J}(X) \subset \mathscr{P}\mathscr{R}(\Phi(X))$. ◇

4.5 Numerical Range of a Linear Relation

Let X be a complex Hilbert space and let $A \in L\mathscr{R}(X)$. Then, the numerical range of A is the set

$$W(A) := \{\langle x, y\rangle \text{ such that } x \in \mathscr{D}(A), \quad y \in Ax, \quad \text{and } \|x\| = 1\}.$$

The numerical range of a linear relation is always convex in $\mathbb{C}$. If A is a bounded linear operator in X, then the numerical range of A is the set

$$W(A) := \{\langle Ax, x\rangle \text{ such that } x \in \mathbb{S}_X\},$$

where

$$\mathbb{S}_X := \{x \in X \text{ such that } \|x\| = 1\}$$

is the unit sphere in X. The following properties of $W(A)$ are immediate

$$W(\alpha I + \beta A) = \alpha + \beta W(A) \text{ for } \alpha,\ \beta \in \mathbb{C},$$

$$W(A^*) = \{\overline{\lambda},\ \lambda \in W(A)\}, \text{ and}$$

$$W(U^*AU) = W(A), \text{ for any unitary } U.$$

By the well-known Toeplitz–Hausdorff theorem, the numerical range is a convex subset of $\mathbb{C}$ and it satisfies the, so-called, spectral inclusion property

$$\sigma_p(A) \subset W(A)$$

and

$$\sigma(A) \subset \overline{W(A)}.$$

Note that $W(A)$ is closed if $\dim X < \infty$. Moreover, the resolvent of A can be estimated in terms of the distance to the numerical range,

$$\|(\lambda - A)^{-1}\| \leq \frac{1}{\operatorname{dist}(\lambda, W(A))}, \quad \lambda \notin \overline{W(A)}, \tag{4.5.1}$$

where $\operatorname{dist}(\lambda, W(A))$ denotes the distance between λ to the set $W(A)$. If a point $\lambda \in \overline{W(A)}$ is a corner of the numerical range (i.e., $W(A)$ lies in a sector with vertex λ and angle less than π), then $\lambda \in \sigma(A)$. If, in addition, $\lambda \in W(A)$, then $\lambda \in \sigma_p(A)$. The estimate (4.5.1) leads to if $\lambda \in \sigma_p(A)$ is a boundary point of $W(A)$, hence there are no associated vectors at λ.

For many applications it is useful to consider unbounded operators. Suppose that $A \in \mathscr{L}(\mathscr{D}(A), X)$ is a closed linear operator with domain $\mathscr{D}(A)$ densely and continuously embedded in X. Its numerical range is then defined by

$$W(A) := \left\{ \langle Ax, x \rangle \text{ such that } x \in \mathscr{D}(A) \bigcap \mathbb{S}_X \right\}.$$

4.5.1 The Numerical Range of a Linear Relation T Associated with A

Let $T \in L\mathscr{R}(H, K)$ and let $A : \mathscr{D}(A) \subset H \longrightarrow K$ be a linear operator on Hilbert spaces H and K. To study the properties of the numerical range, coupled numerical range for a linear relation T and a linear operator A, we are need to define this notion, which is done in the following definition.

Definition 4.5.1 Let H and K be Hilbert spaces. Let $T \in L\mathscr{R}(H, K)$ and $A : \mathscr{D}(A) \subset H \longrightarrow K$ be a linear operator. The numerical range of T associated with A is defined as the set

$$W(T, A) = \left\{ \langle y, Ax \rangle \text{ such that } x \in \mathscr{D}(A) \bigcap \mathscr{D}(T), \ y \in Tx \text{ and } \|Ax\| = 1 \right\}. \diamondsuit$$

Proposition 4.5.1 *Let H and K be Hilbert spaces.*

(i) If $T \in L\mathscr{R}(H)$ and $A = I$, then $W(T, A) = W(T)$.
(ii) If $T \in L\mathscr{R}(H, K)$ and $A : \mathscr{D}(A) \subset H \longrightarrow K$ is a linear operator, then

$$W(T, A) = W(TA^{-1}).$$

(iii) If $T \in C\mathscr{R}(H, K)$ and $A \in \mathscr{C}(H, K)$ are everywhere defined, then

$$W(T, A) \subset \overline{\mathbb{D}}_{\|TA^{-1}\|},$$

where $\mathscr{C}(H, K)$ is a class of all closed linear operators from H into K. ◇

Proof (i) If $A = I$, then it is easy to verify that $W(T, A) = W(T)$.
(ii) Suppose that $\lambda \in W(T, A)$, then there exist $x \in \mathscr{D}(A) \bigcap \mathscr{D}(T)$ and $y \in Tx$ satisfying

$$\lambda = \langle y, Ax \rangle, \ \|Ax\| = 1. \tag{4.5.2}$$

The fact that $(Ax, x) \in G(A^{-1})$ and $(x, y) \in G(T)$ implies from the Subsection 2.8.1 that $(Ax, y) \in G(TA^{-1})$. This leads to (4.5.2) that $\lambda \in W(TA^{-1})$. Hence,

$$W(T, A) \subset W(TA^{-1}).$$

Conversely, let us assume that $\lambda \in W(TA^{-1})$, then there exists $(x, y) \in G(TA^{-1})$ satisfying

$$\lambda = \langle y, x \rangle, \ \|x\| = 1. \tag{4.5.3}$$

It follows from Subsection 2.8.1 that there exists $z \in \mathscr{D}(A) \bigcap \mathscr{D}(T)$ such that $(x, z) \in G(A^{-1})$ and $(z, y) \in G(T)$. This implies that $x = Az$ and $(z, y) \in G(T)$. Hence, there exist $z \in \mathscr{D}(A) \bigcap \mathscr{D}(T)$ and $(z, y) \in G(T)$ satisfies from (4.5.3) that

$$\lambda = \langle y, Az \rangle, \ \|Az\| = 1.$$

This implies that $\lambda \in W(T, A)$.
(iii) Let $\lambda \in W(T, A)$. From (ii), we get $\lambda \in W(TA^{-1})$. Hence, there exists $(x, y) \in G(TA^{-1})$ satisfying $\lambda = \langle y, x \rangle$, where $\|x\| = 1$. This yields from Cauchy–Schwarz inequality that

$$\begin{aligned} |\lambda| &= |\langle y, x \rangle| \\ &\leq \|y\|\|x\| \\ &\leq \|y\| \quad (\text{as } \|x\| = 1). \end{aligned} \tag{4.5.4}$$

Since $y \in TA^{-1}x$, so by using Lemma 2.10.2, we infer that

$$\|TA^{-1}x\| = \inf_{y \in TA^{-1}x} \|y\|.$$

Using the definition of the lower bound, for any given $\eta > 0$, we obtain that

$$\|y\| < \|TA^{-1}x\| + \eta. \tag{4.5.5}$$

Since $A^{-1}(0) = N(A) = \mathscr{D}(T)$, then by Corollary 2.10.1, we obtain

$$\|TA^{-1}\| \leq \|T\|\|A^{-1}\|.$$

Based on the hypothesis $\mathscr{D}(T)$ is closed and $T \in C\mathscr{R}(H)$, we infer that

$$\|T\| < \infty.$$

The fact that $\gamma(A) > 0$ implies from Lemma 2.10.9 that $\|A^{-1}\| < \infty$. Hence,

$$\|TA^{-1}\| < \infty.$$

For any given $\eta > 0$, we deduce from (4.5.4) and (4.5.5) that

$$\begin{aligned} |\lambda| &\leq \|TA^{-1}x\| + \eta \\ &\leq \|TA^{-1}\| + \eta. \end{aligned}$$

By arbitrariness of η, we conclude that

$$|\lambda| \leq \|TA^{-1}\|.$$

This equivalent to saying that

$$W(T, A) \subset \overline{\mathbb{D}}_{\|TA^{-1}\|}.$$

Q.E.D.

Remark 4.5.1 *(i)* Let $T \in L\mathscr{R}(H, K)$ and $A : \mathscr{D}(A) \subset H \longrightarrow K$ be a linear operator such that $N(T) \not\subseteq N(A)$. Then,

$$0 \in W(T, A).$$

Indeed, since $\{0\} \subsetneq A(0) = N(A^{-1}) \subset N(TA^{-1})$, then TA^{-1} is not one-to-one. This implies that there exists $x \in \mathscr{D}(TA^{-1})$ such that $\|x\| = 1$ and $0 \in TA^{-1}x$. Thus, $(x, 0) \in G(TA^{-1})$ with $\|x\| = 1$. Hence, $\langle 0, x\rangle \in W(TA^{-1})$. Consequently, $0 \in W(TA^{-1})$. Finally, using Proposition 4.5.1 *(ii)*, we have $0 \in W(T, A)$.
(ii) Let $T \in L\mathscr{R}(H, K)$ be continuous and $A \in L\mathscr{R}(H, K)$ such that $\gamma(A) > 0$ and $N(A) \subset \mathscr{D}(T)$. Then,

$$\overline{W(T, A)} \subset \overline{\mathbb{D}}_{\|TA^{-1}\|}.$$

Moreover, if $\dim(W(T, A)) < \infty$, then $W(T, A)$ is compact. ◇

Proposition 4.5.2 *Let T, $S \in L\mathscr{R}(H, K)$, $A : \mathscr{D}(A) \subset H \longrightarrow K$ be a linear operator and $\alpha_1, \alpha_2, \alpha_3 \in \mathbb{C}$ such that $|\alpha_3| = 1$. Then,*

(i) $W(\alpha_1 T, A) = \alpha_1 W(T, A)$.
(ii) $W(\alpha_1 S + \alpha_2 T, A) \subset \alpha_1 W(S, A) + \alpha_2 W(T, A)$.
(iii) $W(T, \alpha_3 A) = \overline{\alpha}_3 W(T, A)$. ◇

Proof (i) If $\alpha_1 = 0$, then

$$W(0T, A) = \{\langle 0, Ax\rangle : x \in \mathscr{D}(A) \text{ and } \|Ax\| = 1\} = \{0\}.$$

This implies that

$$W(\alpha_1 T, A) = \alpha_1 W(T, A).$$

Suppose that $\alpha_1 \in \mathbb{C}\backslash\{0\}$ and $\lambda \in W(\alpha_1 T, A)$. Hence, there exists $x \in \mathscr{D}(A) \bigcap \mathscr{D}(\alpha_1 T)$ and $y \in \alpha_1 Tx$ satisfying

$$\lambda = \langle y, Ax\rangle \ \text{ with } \|Ax\| = 1. \tag{4.5.6}$$

The fact that $(x, y) \in G(\lambda_1 T)$ implies from Subsection 2.8.1 that $\alpha_1^{-1} y \in Tx$. This yields that $\left\langle \alpha_1^{-1} y, Ax\right\rangle \in W(T, A)$. It follows from (4.5.6) that

$$\begin{aligned} \lambda = &\ \alpha_1 \left\langle \alpha_1^{-1} y, Ax\right\rangle \\ \subset &\ \alpha_1 W(T, A). \end{aligned}$$

Hence,

$$W(\alpha_1 T, A) \subset \alpha_1 W(T, A).$$

Conversely, assume that $\alpha_1 \in \mathbb{C}\backslash\{0\}$ and $\lambda \in \alpha_1 W(T, A)$. Then, $\alpha_1^{-1}\lambda \in W(T, A)$. This implies that there exists $x \in \mathscr{D}(A) \bigcap \mathscr{D}(T)$ and $y \in Tx$ satisfying

$$\alpha_1^{-1}\lambda = \langle y, Ax\rangle$$

with $\|Ax\| = 1$. Therefore,

$$\begin{aligned} \lambda = &\ \alpha_1 \alpha_1^{-1}\lambda \\ = &\ \alpha_1 \langle y, Ax\rangle \\ = &\ \langle \alpha_1 y, Ax\rangle. \end{aligned} \tag{4.5.7}$$

The use of the hypothesis $(x, y) \in G(T)$ leads to deduce from Subsection 2.8.1 that $(x, \alpha_1 y) \in G(\lambda_1 T)$. Hence, by the fact that $x \in \mathscr{D}(A) \bigcap \mathscr{D}(T)$, we conclude from (4.5.7) that $\lambda \in W(\alpha_1 T, A)$.

(ii) If $\alpha_1 = \alpha_2 = 0$, then it is clear that

$$W(\alpha_1 S + \alpha_2 T, A) \subset \alpha_1 W(S, A) + \alpha_2 W(T, A).$$

Now, assume that $\alpha_1 \neq 0$ and $\alpha_2 \neq 0$ and $\lambda \in W(\alpha_1 S + \alpha_2 T, A)$. Then, there exist $x \in \mathscr{D}(\alpha_1 S + \alpha_2 T) \bigcap \mathscr{D}(A) = \mathscr{D}(S) \bigcap \mathscr{D}(T) \bigcap \mathscr{D}(A)$ and $y \in (\alpha_1 S + \alpha_2 T)x$ satisfying

$$\lambda = \langle y, x\rangle, \tag{4.5.8}$$

with $\|Ax\| = 1$. The fact that $(x, y) \in G(\alpha_1 S + \alpha_2 T)$ implies from Remark 2.7.1 (ii) that there exists $z \in K$ such that $(x, z) \in G(\alpha_1 S)$ and $(x, y - z) \in G(\alpha_2 T)$. This leads from Subsection 2.8.1 that $\left(x, \alpha_1^{-1} z\right) \in G(S)$ and $\left(x, \alpha_2^{-1}(y - z)\right) \in G(T)$. Since $x \in \mathscr{D}(S) \bigcap \mathscr{D}(T) \bigcap \mathscr{D}(A)$ and $\|Ax\| = 1$, then we can conclude that

$$\left\langle \alpha_1^{-1} z, x \right\rangle \in W(S, A)$$

and

$$\left\langle \alpha_2^{-1}(y - z), x \right\rangle \in W(T, A).$$

Thus, the use of (4.5.8) makes us deduce that

$$\begin{aligned} \lambda &= \langle y - z + z, x \rangle \\ &= \langle y - z, x \rangle + \langle z, x \rangle \\ &= \alpha_1 \left\langle \alpha_1^{-1} z, x \right\rangle + \alpha_2 \left\langle \alpha_2^{-1}(y - z), x \right\rangle \\ &\subset \alpha_1 W(S, A) + \alpha_2 W(T, A). \end{aligned}$$

(iii) Let $\lambda \in W(T, \alpha_3 A)$. Then, there exist $(x, y) \in G(T)$, $x \in \mathscr{D}(\alpha_3 A) = \mathscr{D}(A)$ such that $\|\alpha_3 Ax\| = 1$ and $\lambda = \langle y, \alpha_3 Ax \rangle$. Based on the assumption $|\alpha_3| = 1$, we infer that

$$\begin{aligned} 1 &= \|\alpha_3 Ax\| \\ &= |\alpha_3| \|Ax\| \\ &= \|Ax\|. \end{aligned}$$

This implies that $\langle y, Ax \rangle \in W(T, A)$. Moreover, we have

$$\lambda = \overline{\alpha}_3 \langle y, Ax \rangle.$$

Hence, $\lambda \in \overline{\alpha}_3 W(T, A)$. Conversely, let $\lambda \in \overline{\alpha}_3 W(T, A)$. Then, $(\overline{\alpha_3})^{-1} \lambda \in W(T, A)$. This implies that there exist $(x, y) \in G(T)$ and $x \in \mathscr{D}(A)$ satisfying

$$(\overline{\alpha_3})^{-1} \lambda = \langle y, Ax \rangle,$$

with $\|Ax\| = 1$. Consequently,

$$\begin{aligned} \lambda &= \overline{\alpha}_3 \langle y, Ax \rangle \\ &= \langle y, \alpha_3 (Ax) \rangle \\ &= \langle y, (\alpha_3 A) x \rangle. \end{aligned}$$

Since $x \in \mathscr{D}(A) = \mathscr{D}(\alpha_3 A)$ and $\|\alpha_3 Ax\| = |\alpha_3| \|Ax\| = 1$, then $\lambda \in W(T, \alpha_3 A)$. This achieves the proof. Q.E.D.

4.5.2 The Pseudo-Spectral Theory Associated with the Numerical Range

The goal of this section is to study same results of the classical pseudospectral theory associated with the numerical range.

Theorem 4.5.1 *Let $T \in C\mathscr{R}(H, K)$, $A \in \mathscr{L}(H, K)$ and $\varepsilon > 0$. Then,*

$$\sigma_{\varepsilon\|A^{-1}\|}(T, A) \subset \sigma_\varepsilon(TA^{-1}) \subset \overline{W(T, A)} + \mathbb{D}_\varepsilon,$$

where $\mathbb{D}_\varepsilon$ denotes the open disc centered at 0 and with radius ε. ◇

Proof By using Proposition 4.4.2 (ii), we have $\sigma_{\varepsilon\|A^{-1}\|}(T, A) \subset \sigma_\varepsilon(TA^{-1})$. It remains to prove

$$\sigma_\varepsilon(TA^{-1}) \subset \overline{W(T, A)} + \mathbb{D}_\varepsilon.$$

Let $\lambda \in \sigma_\varepsilon(TA^{-1})$. Then, from [20, Lemma 3.1], we have $\lambda \in \sigma(TA^{-1}) \bigcup U_\varepsilon$, where $U_\varepsilon = \big\{\lambda \in \mathbb{C}$ such that there exists $x_0 \in \mathscr{D}(TA^{-1})$ with $\|(\lambda - TA^{-1})x_0\| < \varepsilon\|x_0\|\big\}$. Then, we will discuss the following two cases:

<u>First case.</u> If $\lambda \in \sigma(TA^{-1})$, then by Theorem 4.5.6, we obtain $\lambda \in \overline{W(TA^{-1})}$. By using of Proposition 4.5.1 (ii), we obtain that $\lambda \in \overline{W(T, A)}$. This implies that

$$\begin{aligned}\sigma(TA^{-1}) =&\ \sigma(TA^{-1}) + \{0\}\\ \subset&\ \overline{W(T, A)} + \mathbb{D}_\varepsilon.\end{aligned}$$

<u>Second case.</u> If $\lambda \in U_\varepsilon$, then there exists $x_0 \in \mathscr{D}(TA^{-1})$ such that

$$\|(\lambda - TA^{-1})x_0\| < \varepsilon\|x_0\|. \tag{4.5.9}$$

We suppose that $x_0 = 0$. Then, by (4.5.9), we obtain $0 < 0$. This contradiction implies that $x_0 \neq 0$. Putting $x = \dfrac{x_0}{\|x_0\|}$. Then, $x \in \mathscr{D}(TA^{-1})$ such that $\|x\| = 1$ and

$$\|(\lambda - TA^{-1})x\| < \varepsilon. \tag{4.5.10}$$

For $y \in TA^{-1}x$, we obtain

$$\begin{aligned}\lambda =&\ \lambda\langle x, x\rangle - \langle y, x\rangle + \langle y, x\rangle\\ =&\ \langle \lambda x - y, x\rangle + \langle y, x\rangle\\ \in&\ \langle \lambda x - y, x\rangle + W(TA^{-1}).\end{aligned} \tag{4.5.11}$$

On the other hand, from Cauchy–Schwarz inequality, we get

$$\begin{aligned}\lambda\,|\langle \lambda x - y, x\rangle| \le\ & \|\lambda x - y\|\|x\| \\ =\ & \|\lambda x - y\|. \end{aligned} \tag{4.5.12}$$

The fact that $(x, y) \in G(TA^{-1})$ and $(x, \lambda x) \in G(\lambda I)$ implies from Remark 2.7.1 (ii) that $(x, \lambda x - y) \in G(\lambda - TA^{-1})$. Hence, for all $\varepsilon_1 > 0$, we get that

$$\|\lambda x - y\| \le \|(\lambda - TA^{-1})x\| + \varepsilon_1.$$

By referring to (4.5.12), we infer that

$$\begin{aligned}|\langle \lambda x - y, x\rangle| \le\ & \|(\lambda - TA^{-1})x\| + \varepsilon_1 \\ <\ & \varepsilon + \varepsilon_1 \quad \text{(from (4.5.10))}. \end{aligned} \tag{4.5.13}$$

By arbitrariness of ε_1, we deduce that

$$|\langle \lambda x - y, x\rangle| < \varepsilon.$$

Therefore, from (4.5.11) and (4.5.13), we conclude that $\lambda \in W(TA^{-1}) + \mathbb{D}_\varepsilon$. This leads to

$$\begin{aligned}U_\varepsilon \subset\ & W(TA^{-1}) + \mathbb{D}_\varepsilon \\ \subset\ & \overline{W(TA^{-1})} + \mathbb{D}_\varepsilon. \end{aligned}$$

By referring to Proposition 4.5.1 (ii), we infer that

$$U_\varepsilon \subset \overline{W(T, A)} + \mathbb{D}_\varepsilon.$$

This shows that

$$\sigma_\varepsilon(TA^{-1}) \subset \overline{W(T, A)} + \mathbb{D}_\varepsilon. \qquad \text{Q.E.D.}$$

Theorem 4.5.2 *Let $T \in C\mathcal{R}(H, K)$ be with $\mathcal{D}(T^2)$ is closed, $A \in \mathcal{L}(H, K)$ and $\varepsilon > 0$. Then,*

$$\Sigma_\varepsilon(TA^{-1}) \subset \overline{W(T, A)} + \overline{\mathbb{D}}_\varepsilon.$$

Moreover, if $0 \in \rho(A)$, then

$$\Sigma_{\varepsilon\|A^{-1}\|}(T, A) \subset \overline{W(T, A)} + \overline{\mathbb{D}}_\varepsilon. \qquad \diamond$$

Proof It follows from Theorem 4.5.1 that

$$\begin{aligned}\sigma_\varepsilon(TA^{-1}) \subset\ & \overline{W(T, A)} + \mathbb{D}_\varepsilon \\ \subset\ & \overline{W(T, A)} + \overline{\mathbb{D}}_\varepsilon. \end{aligned}$$

Since $\overline{W(T, A)}$ is a closed and $\overline{\mathbb{D}}_\varepsilon$ is compact, then

$$\overline{\overline{W(T, A)} + \overline{\mathbb{D}}_\varepsilon} = \overline{W(T, A)} + \overline{\mathbb{D}}_\varepsilon.$$

Hence,

$$\overline{\sigma_\varepsilon(TA^{-1})} \subset \overline{W(T,A)} + \overline{\mathbb{D}}_\varepsilon.$$

This leads from Theorem 3.1.1 and [18, Corollary 3.1] that

$$\Sigma_\varepsilon(TA^{-1}) \subset \overline{W(T,A)} + \overline{\mathbb{D}}_\varepsilon.$$

This achieves the proof. Q.E.D.

Theorem 4.5.3 *Let $T \in C\mathcal{R}(H,K)$ and $A \in \mathcal{L}(H,K)$ be such that $N(T) \subset N(A)$ $R(A) = K$ and AT^{-1} have SVEP at 0. For a posit if integer $n \geq 1$, assume that $0 < n\,\varepsilon < \gamma(AT^{-1})$. Then,*

$$\sigma_\varepsilon(AT^{-1}) \subset \left(\mathcal{B}_{\delta(\varepsilon)}\left(W\left(T,A\right)\right)\right)^{-1},$$

where $\mathcal{B}_{\delta(\varepsilon)}\left(W\left(T,A\right)\right) = \{\lambda \in \mathbb{C}$ such that $dist\left(\lambda, W\left(T,A\right)\right) < \delta(\varepsilon)\}$ and $\delta(\varepsilon) = \dfrac{\varepsilon\,\left(\gamma\left(AT^{-1}\right)\right)^{-1}}{n^{-1}\gamma\left(AT^{-1}\right) - \varepsilon}$. ◇

Proof By referring to Proposition 4.5.1 (ii), we have

$$\mathcal{B}_{\delta(\varepsilon)}\left(W\left(TA^{-1}\right)\right) = \mathcal{B}_{\delta(\varepsilon)}\left(W\left(T,A\right)\right).$$

Let us denote $O = \left(\mathcal{B}_{\delta(\varepsilon)}\left(W(T,A)\right)\right)^{-1}$. Let $\lambda \in \mathbb{C}\backslash O$. We shall show that

$$\left\|\left(\lambda - AT^{-1}\right)y\right\| \geq \varepsilon\|y\| \text{ for all } y \in \mathcal{D}\left(AT^{-1}\right).$$

We discuss two cases.

First case. If $|\lambda| > n^{-1}\gamma(AT^{-1}) - \varepsilon > 0$, then λ^{-1} exists and $\lambda^{-1} \notin \mathcal{B}_{\delta(\varepsilon)}(W(TA^{-1}))$ which yields

$$\operatorname{dist}(\lambda^{-1}, W(TA^{-1})) \geq \delta(\varepsilon).$$

Let $(x,y) \in G(TA^{-1})$ and $\|x\| = 1$, then $\langle y, x\rangle \in W(TA^{-1})$. Hence,

$$\begin{aligned}
\delta(\varepsilon) \leq\ & \left|\lambda^{-1} - \langle y, x\rangle\right| \\
\leq\ & \left|\lambda^{-1}\langle x, x\rangle - \langle y, x\rangle\right| \\
\leq\ & \left|\langle \lambda^{-1}x - y, x\rangle\right| \\
\leq\ & \left\|\lambda^{-1}x - y\right\|\ \|x\| \text{ (by Cauchy-Schwarz inequality)} \\
\leq\ & \left\|\lambda^{-1}x - y\right\| \quad (\text{as } \|x\| = 1).
\end{aligned} \tag{4.5.14}$$

The fact that $(x,y) \in G(TA^{-1})$ and $(y, \lambda y) \in G(\lambda I)$ implies from Remark 2.7.1 (ii) that $(y, \lambda y - x) \in G(\lambda - AT^{-1})$. Now, we shall show that $\lambda - AT^{-1}$ is bounded below. Hence, for all $\varepsilon_0 > 0$,

$$\begin{aligned}\left\|(\lambda - AT^{-1})y\right\| &\geq \|\lambda y - x\| + |\lambda|\varepsilon_0 \\ &\geq |\lambda| \left\|y - \lambda^{-1}x\right\| + |\lambda|\varepsilon_0 \\ &\geq |\lambda| \left(\delta(\varepsilon) + \varepsilon_0\right) \quad \text{(from (4.5.14))}. \end{aligned} \tag{4.5.15}$$

Since $\gamma(AT^{-1}) > 0$, then by Proposition 2.15.3, we infer that AT^{-1} is open which yields that TA^{-1} is continuous. This implies from the fact that $(x, y) \in G(TA^{-1})$ that for every $\varepsilon_1 > 0$, we obtain

$$\begin{aligned}\|y\| &\leq \|y\| + \varepsilon_1 \\ &\leq \|TA^{-1}x\| \\ &\leq \|TA^{-1}\| \ \text{(since } TA^{-1} \text{ is continuous and } \|x\| = 1).\end{aligned}$$

It follows from Lemma 2.10.9 that

$$\left(\gamma(AT^{-1})\right)^{-1} \geq \|y\|.$$

Hence, by the arbitrariness of ε_0 and (4.5.15), we deduce that

$$\begin{aligned}\left\|(\lambda - AT^{-1})y\right\| &\geq |\lambda|\delta(\varepsilon) \\ &\geq \left(n^{-1}\gamma(AT^{-1}) - \varepsilon\right)\delta(\varepsilon) \\ &\geq \left(\gamma(AT^{-1})\right)^{-1}\gamma(AT^{-1})\left(n^{-1}\gamma(AT^{-1}) - \varepsilon\right)\delta(\varepsilon) \\ &\geq \left[\gamma(AT^{-1})\left(n^{-1}\gamma(AT^{-1}) - \varepsilon\right)\delta(\varepsilon)\right]\|y\|.\end{aligned}$$

<u>Second case.</u> Assume that $|\lambda| \leq n^{-1}\gamma(AT^{-1}) - \varepsilon$. Since $y \in \mathcal{D}(\lambda - AT^{-1})$, then $(y, y) \in G(I_{\mathcal{D}(\lambda - AT^{-1})})$. It follows from [21, Subsection I.1.3 (5)] that $y \in (\lambda - AT^{-1})^{-1}(\lambda - AT^{-1})y$. For $y \neq 0$, we have $z = \dfrac{y}{\|y\|} \in \mathcal{D}(\lambda - AT^{-1})$ and $\|z\| = 1$. Then, for all $\varepsilon_1 > 0$, we obtain

$$1 = \|z\| \leq \|(\lambda - AT^{-1})^{-1}(\lambda - AT^{-1})z\| + \varepsilon_1\|(\lambda - AT^{-1})^{-1}\|.$$

This implies from the fact that $(\lambda - AT^{-1})(0) \subset R(\lambda - AT^{-1})$ and Remark 2.10.1 that

$$1 \leq \|(\lambda - AT^{-1})^{-1}\|\|(\lambda - AT^{-1})z\| + \varepsilon_1\|(\lambda - AT^{-1})^{-1}\|.$$

Using of Lemma 2.10.9, we infer that

$$\begin{aligned}\gamma(\lambda - AT^{-1}) &= \|(\lambda - AT^{-1})^{-1}\|^{-1} \\ &\leq \|(\lambda - AT^{-1})z\| + \varepsilon_1.\end{aligned} \tag{4.5.16}$$

The fact that $N(T) \subset N(A)$ implies that AT^{-1} is an operator.

Now, we propose to prove that AT^{-1} is one-to-one. The fact that AT^{-1} is closed and $\gamma\left(AT^{-1}\right) > 0$ implies from Theorem 2.11.1 (ii) that $R\left(AT^{-1}\right)$ is closed. Moreover, AT^{-1} have SVEP at 0. By using [18, Theorem 3.4], we infer that AT^{-1} is one-to-one. On the other hand, we have $\gamma(AT^{-1}) > 0$. This leads from [?, Lemma 2.5] that

$$\gamma(AT^{-1} - \lambda) \geq \gamma(AT^{-1}) - |\lambda|.$$

This implies that

$$\begin{aligned}\gamma(AT^{-1} - \lambda) \geq & \ \gamma(AT^{-1}) - \left(\frac{1}{n}\gamma(AT^{-1}) - \varepsilon\right) \\ \geq & \left(1 - \frac{1}{n}\right)\gamma(AT^{-1}) + \varepsilon.\end{aligned}$$

By referring to (4.5.16), we deduce that

$$\left(1 - \frac{1}{n}\right)\gamma(AT^{-1}) + \varepsilon - \varepsilon_1 \leq \|(\lambda - AT^{-1})z\|.$$

Putting $\varepsilon_1 = \left(1 - \frac{1}{n}\right)\gamma(AT^{-1}) > 0$. Thus,

$$\|(\lambda - AT^{-1})z\| \geq \varepsilon.$$

Consequently,

$$\|(\lambda - AT^{-1})y\| \geq \varepsilon\|y\|, \quad \text{for all } y \in \mathscr{D}(AT^{-1})\backslash\{0\}. \tag{4.5.17}$$

It is clear that (4.5.17) holds for $y = 0$. This enables us to conclude that

$$\|(\lambda - AT^{-1})y\| \geq \varepsilon\|y\|, \quad \text{for all } y \in \mathscr{D}(AT^{-1}).$$

This shows that $\mathbb{C}\backslash O \subset \mathbb{C}\backslash\sigma_{ap}(AT^{-1})$ This implies that

$$\sigma_{ap}(AT^{-1})\bigcap\mathbb{C}\backslash O = \emptyset. \tag{4.5.18}$$

It remains to show that

$$\mathbb{C}\backslash O \subset \rho(AT^{-1}).$$

By referring to Propositions 4.5.1 (ii), we have $W(TA^{-1})$ is convex. Since $\mathbb{D}_{\delta(\varepsilon)}$ is convex, then

$$\mathscr{B}_{\delta(\varepsilon)}(W(TA^{-1})) = W(TA^{-1})\bigcap\mathbb{D}_{\delta(\varepsilon)}$$

is also. In addition, we have $\mathscr{B}_{\delta(\varepsilon)}(W(TA^{-1}))$ is bounded. Hence, $\mathbb{C}^*\backslash\mathscr{B}_{\delta(\varepsilon)}(W(TA^{-1}))$ is connected. Let us consider

$$\begin{aligned} \varphi : \mathbb{C}^* &\longrightarrow \mathbb{C}^* \\ z &\longrightarrow z^{-1}. \end{aligned}$$

It is clear that φ is continuous. This implies that

$$\begin{aligned} \lambda\varphi\left(\mathbb{C}^*\backslash\mathscr{B}_{\delta(\varepsilon)}\left(W\left(TA^{-1}\right)\right)\right) &= \left(\mathbb{C}^*\backslash\mathscr{B}_{\delta(\varepsilon)}\left(W\left(TA^{-1}\right)\right)\right)^{-1} \\ &= \mathbb{C}^*\backslash O \end{aligned}$$

is connected. On the other hand, the boundedness of $\mathscr{B}_{\delta(\varepsilon)}(W(TA^{-1})$ implies that a neighborhood around 0 belongs to $\mathbb{C}\backslash O = \mathbb{C}^*\backslash O \bigcup\{0\}$. This lead us to $0 \in \mathbb{C}\backslash O$ and $\mathbb{C}\backslash O$ is connected. This implies by using (4.5.18) and the fact that $\partial\sigma(AT^{-1}) \subset \sigma_{ap}(AT^{-1})$ that

$$\mathbb{C}\backslash O \subset \rho(AT^{-1})$$

or

$$\mathbb{C}\backslash O \subset \sigma(AT^{-1})\backslash\sigma_{ap}(AT^{-1}).$$

To verify that $\mathbb{C}\backslash O \subset \rho(AT^{-1})$, it is enough to show from the fact that $\mathbb{C}\backslash O$ is connected containing 0 that $0 \in \rho(AT^{-1})$. At this level, based on the hypotheses $\gamma\left(AT^{-1}\right) > 0$ and AT^{-1} is one-to-one, we shall show that $\overline{R(AT^{-1})} = K$. Moreover, we have $R\left(AT^{-1}\right)$ is closed. So, it remains to prove that AT^{-1} is onto. Suppose that $R(AT^{-1}) \subsetneq K$. Hence, there exists $k \in K$ such that $k \notin R\left(AT^{-1}\right)$. This implies for all $t \in \mathscr{D}\left(AT^{-1}\right)$ that $(t, k) \notin G\left(AT^{-1}\right)$, which yields that $(k, t) \notin G\left(TA^{-1}\right)$. By referring to Eq. (2.8.1), we infer for all $z \in \mathscr{D}(A)\bigcap\mathscr{D}(T) = \mathscr{D}(A)$ that $(k, z) \notin G\left(A^{-1}\right)$ or $\begin{pmatrix} z \\ t \end{pmatrix} \notin G\,(T)$. Assume that $(k, z) \notin G\left(A^{-1}\right)$. Hence, $(z, k) \notin G\,(A)$, which implies that $k \neq Az$. This leads to $k \notin R(A) = K$, which a contradiction. Now, assume that $\begin{pmatrix} z \\ t \end{pmatrix} \notin G\,(T)$. This implies that $t \notin R(T)$, which is impossible because it follows from Eq. (2.8.2) that $t \in \mathscr{D}\left(AT^{-1}\right) = T(\mathscr{D}(A)) \subset T(\mathscr{D}(T)) = R(T)$. This lead us to conclude that $R(AT^{-1}) = K$. Thus,

$$\mathbb{C}\backslash O \subset \rho(AT^{-1})\bigcap\left\{\lambda \in \mathbb{C} \text{ such that } \forall\, y \in \mathscr{D}(AT^{-1}),\ \text{we have } \|(\lambda - AT^{-1})y\| \geq \varepsilon\|y\|\right\}.$$

Finally, the use of [20, Lemma 3.1] allows us to conclude that

$$\sigma_\varepsilon(AT^{-1}) \subset \left(\mathscr{B}_{\delta(\varepsilon)}\left(W(T, A)\right)\right)^{-1}. \qquad \text{Q.E.D.}$$

Theorem 4.5.4 *Let* $T \in C\mathscr{R}(H, K)$, $S \in L\mathscr{R}(H, K)$ *be continuous and* $A \in \mathscr{L}(H, K)$ *such that* $0 \in \rho(A)$, $S(0) \subset T(0)$ *and* $\overline{\mathscr{D}(T)} \subset \mathscr{D}(S)$. *Then, for every* $\varepsilon, \delta > 0$, *we have*

$$\Sigma_\varepsilon\left(A^{-1}T + A^{-1}S\right) \subset \overline{W(T, A)} + \mathbb{D}_{\gamma(A)\|A\|\left(\varepsilon+\delta+\|A^{-1}S\|\right)}. \qquad \diamond$$

Proof Since $0 \in \rho(A)$, then $A^{-1} \in \mathscr{L}(K, H)$. The fact that $\|S\| < \infty$ and $S(0) \subset \mathscr{D}\left(A^{-1}\right) = K$ implies from Corollary 2.10.1 that

$$\left\|A^{-1}S\right\| \leq \left\|A^{-1}\right\| \|S\| < \infty.$$

This yields that $A^{-1}S$ is continuous. Based on the assumption $S(0) \subset T(0)$, we infer that

$$A^{-1}S(0) \subset A^{-1}T(0).$$

Since $\overline{\mathscr{D}(T)} \subset \mathscr{D}(S)$, then by Eq. (2.8.2), we get that

$$\begin{aligned}
\overline{\mathscr{D}\left(A^{-1}T\right)} &= \overline{T^{-1}\mathscr{D}\left(A^{-1}\right)} \\
&= \overline{T^{-1}K} \\
&= \overline{\mathscr{D}(T)} \subset \mathscr{D}(S) \\
&= \overline{S^{-1}K} \\
&= \overline{S^{-1}\mathscr{D}\left(A^{-1}\right)} \\
&= \overline{\mathscr{D}\left(A^{-1}S\right)}.
\end{aligned}$$

It follows from [18, Proposition 3.1] that

$$\Sigma_\varepsilon\left(A^{-1}T + A^{-1}S\right) \subset \sigma_{\varepsilon+\delta+\|A^{-1}S\|}\left(A^{-1}T\right).$$

This implies from Proposition 4.4.2 (ii) that

$$\Sigma_\varepsilon\left(A^{-1}T + A^{-1}S\right) \subset \sigma_{\gamma(A)\|A\|(\varepsilon+\delta+\|A^{-1}S\|)}\left(T^{-1}A\right).$$

Hence, the use of Theorem 4.5.1 allows us to deduce that

$$\Sigma_\varepsilon\left(A^{-1}T + A^{-1}S\right) \subset \overline{W(T, A)} + \mathbb{D}_{\gamma(A)\|A\|(\varepsilon+\delta+\|A^{-1}S\|)}.$$

This achieves the proof. Q.E.D.

Theorem 4.5.5 *Let $A \in \mathscr{L}(H, K)$ be such that $0 \in \rho(A)$. Let (T_n) be a sequence of closed linear relations from X into X, $T \in C\mathscr{R}(X)$ such that $\|T_n - T\| \to 0$ when $n \to \infty$, $\mathscr{D}(T_n) = \mathscr{D}(T)$ and there exists $n_1 \in \mathbb{N}$ such that $T(0) = T_n(0)$, for all $n \geq n_1$. Then, to every couple $(\varepsilon_1, \varepsilon_2)$ of real numbers with $0 < \varepsilon_1 < \varepsilon_2 < 0$, there exists a positive integers $n_2 \geq n_1$ such that*

$$\sigma_{\varepsilon_1\gamma(A)^{-1}}(T, A) \subset \sigma_{\varepsilon_2}\left(T_nA^{-1}\right) \subset \overline{W(T, A)} + \mathbb{D}_{\varepsilon_2}, \text{ for all } n \geq n_2. \qquad \diamond$$

Proof Now, we shall prove that

$$\sigma_{\varepsilon_1\gamma(A)^{-1}}(T, A) \subset \sigma_{\varepsilon_2}\left(T_nA^{-1}\right).$$

Since $0 \in \rho(A)$, then A^{-1} is an operator. This implies from Proposition 2.8.2 (iii) that

$$\left\|T_n A^{-1} - T A^{-1}\right\| = \left\|(T_n - T)\, A^{-1}\right\|.$$

It follows from Corollary 2.10.1 that

$$\left\|T_n A^{-1} - T A^{-1}\right\| \leq \|T_n - T\| \left\|A^{-1}\right\|.$$

Based on the assumption $\|T_n - T\| \to 0$ when $n \to \infty$ and A^{-1} is bounded, we infer that

$$\left\|T_n A^{-1} - T A^{-1}\right\| \to 0 \text{ when } n \to \infty.$$

Now, by referring to Eq. (2.8.2), we have

$$\begin{aligned} \mathscr{D}\left(T_n A^{-1}\right) &= A(\mathscr{D}(T_n)) \\ &= A(\mathscr{D}(T)) \\ &= \mathscr{D}\left(T A^{-1}\right). \end{aligned}$$

Moreover, for $n \geq n_1$, we have

$$\begin{aligned} T A^{-1}(0) &= T(0) \\ &= T_n(0) \\ &= T_n A^{-1}(0). \end{aligned}$$

Hence, by using [18, Theorem 3.4], we get that exists a positive integers $n_2 \geq n_1$ such that

$$\sigma_{\varepsilon_1}\left(T A^{-1}\right) \subset \sigma_{\varepsilon_2}\left(T_n A^{-1}\right), \text{ for all } n \geq n_2.$$

By using of Proposition 4.4.2 (ii), we conclude that

$$\sigma_{\varepsilon_1 \gamma(A)^{-1}}(T, A) \subset \sigma_{\varepsilon_2}\left(T_n A^{-1}\right).$$

Thus, by using Theorem 4.5.1 and [18, Theorem 3.4] that

$$\begin{aligned} \sigma_{\varepsilon_1 \gamma(A)^{-1}}(T, A) &\subset \sigma_{\varepsilon_2}\left(T_n A^{-1}\right) \\ &\subset \overline{W(T_n, A)} + \mathbb{D}_{\varepsilon_2}, \text{ for all } n \geq n_2. \end{aligned}$$

It remains to show that

$$W(T_n, A) \subset W(T, A).$$

Assume that $\lambda \in W(T_n, A)$, then there exist $x \in \mathscr{D}(A) \bigcap \mathscr{D}(T_n)$ and $y \in Tx$ satisfying $\lambda = \langle y, Ax\rangle$ and $\|Ax\| = 1$. The fact that $\mathscr{D}(T_n) = \mathscr{D}(T)$ implies that there exists $z \in K$ such that $(x, z) \in G(T)$. Hence, by Remark 2.7.1 (ii), we have

$(x, z - y) \in G(T - T_n)$. This implies that for every $\eta > 0$, we obtain

$$\begin{aligned}\|y - z\| &\leq \|(T_n - T)x\| + \eta \\ &\leq \|T_n - T\| + \eta.\end{aligned}$$

Since $\|T_n - T\| \to 0$ when $n \to \infty$, then by arbitrariness of η, we infer that $y = z$. Therefore, $(x, y) \in G(T)$. This enables us to conclude that $\lambda \in W(T, A)$. Q.E.D.

Theorem 4.5.6 *Let* $\varepsilon > 0$, $A \in \mathscr{L}(H, K)$ *be such that* $0 \in \rho(A)$. *Let* (T_n) *be a sequence of closed linear relations from* X *into* X, $T \in C\mathscr{R}(X)$ *such that* $\|T_n - T\| \to 0$ *when* $n \to \infty$, $\mathscr{D}(T_n) = \mathscr{D}(T)$ *and there exists* $n_1 \in \mathbb{N}$ *such that* $T(0) = T_n(0)$, *for all* $n \geq n_1$. *Then, there exists a positive integers* $n_2 \geq n_1$ *such that*

$$\sigma_{\varepsilon\|A^{-1}\|}(T, A) \subset \overline{W(T_n, A)} + \mathbb{D}_\varepsilon, \text{ for all } n \geq n_2. \qquad \diamond$$

Proof By using Theorem 4.5.1, we obtain

$$\sigma_{\varepsilon\|A^{-1}\|}(T, A) \subset \overline{W(T, A)} + \mathbb{D}_\varepsilon.$$

Now, we shall show that

$$W(T, A) \subset W(T_n, A).$$

Let $\lambda \in W(T, A)$, then there exist $x \in \mathscr{D}(A) \bigcap \mathscr{D}(T)$ and $y \in Tx$ satisfying

$$\lambda = \langle y, Ax \rangle$$

and

$$\|Ax\| = 1.$$

Based on the assumption $\mathscr{D}(T_n) = \mathscr{D}(T)$, we infer that there exists $z \in K$ such that $(x, z) \in G(T_n)$. Hence, by using Remark 2.7.1 (ii), we obtain $(x, z - y) \in G(T_n - T)$. This implies that for every $\eta > 0$, we obtain

$$\begin{aligned}\|y - z\| &\leq \|(T_n - T)x\| + \eta \\ &\leq \|T_n - T\| + \eta.\end{aligned}$$

Since $\|T_n - T\| \to 0$ when $n \to \infty$, then by arbitrariness of η, we get that $y = z$. Consequently, $(x, y) \in G(T_n)$. This shows that $\lambda \in W(T_n, A)$. Q.E.D.

4.5.3 The Numerical Range of Two Linear Relations

To study the properties of the numerical range for two linear relations, we are need to define this notion, which is done in the following definition.

Definition 4.5.2 Let H and K be Hilbert spaces. Let $A, T \in LR(H, K)$. The numerical range of the pair (T, A) is defined as the set

$$W(T, A) = \Big\{ \langle y, z\rangle : \text{there exists } x \in \mathscr{D}(A) \bigcap \mathscr{D}(T), \quad y \in Tx, \quad z \in Ax \text{ and } \|z\| = 1 \Big\}.$$ ◇

Proposition 4.5.3 *Let H and K be Hilbert spaces and let $A, T \in LR(H, K)$.*

(i) If $H = K$ and $A = I$, then $W(T, A) = W(T)$.
(ii) $W(T, A) = W(TA^{-1})$.
(iii) If T is continuous, $\gamma(A) > 0$ and $N(A) \subset \mathscr{D}(T)$, then $W(T, A) \subset \overline{\mathbb{D}}_{\|TA^{-1}\|}$. ◇

Proof The assertion (i) is trivially.
(ii) Suppose that $\lambda \in W(T, A)$, then there exists $x \in \mathscr{D}(A) \bigcap \mathscr{D}(T)$ and $y \in Tx$ and $z \in Ax$ satisfying

$$\lambda = \langle y, z\rangle, \ \|z\| = 1. \tag{4.5.19}$$

The fact that $(z, x) \in G(A^{-1})$ and $(x, y) \in G(T)$ implies that $(z, y) \in G(TA^{-1})$. This implies from (4.5.19) that $\lambda \in W(TA^{-1})$. Hence,

$$W(T, A) \subseteq W(TA^{-1}).$$

Conversely, let us assume that $\lambda \in W(TA^{-1})$, then there exists $(z, y) \in G(TA^{-1})$ satisfying

$$\lambda = \langle y, z\rangle, \ \|z\| = 1. \tag{4.5.20}$$

There exists $x \in \mathscr{D}(A) \bigcap \mathscr{D}(T)$ such that $(z, x) \in G(A^{-1})$ and $(x, y) \in G(T)$. This implies that $(x, z) \in G(A)$ and $(x, y) \in G(T)$. Hence, from (4.5.20) that $\lambda \in W(T, A)$.
(iii) Let $\lambda \in W(T, A)$. From (ii), we get $\lambda \in W(TA^{-1})$. Hence, there exists $(z, y) \in G(TA^{-1})$ satisfying $\lambda = \langle y, z\rangle$, where $\|z\| = 1$. This yields from Cauchy–Schwarz inequality that

$$\begin{aligned} |\lambda| &= |\langle y, z\rangle| \\ &\le \|y\|\|z\| \\ &\le \|y\| \quad (\text{as } \|z\| = 1). \end{aligned} \tag{4.5.21}$$

Since $y \in TA^{-1}z$, then that

$$\|TA^{-1}z\| = \inf_{y \in TA^{-1}z} \|y\|.$$

Using the definition of the lower bound, for any given $\eta > 0$, we obtain that

$$\|y\| < \|TA^{-1}z\| + \eta. \tag{4.5.22}$$

Since $A^{-1}(0) = N(A) \subset \mathscr{D}(T)$, then

$$\|TA^{-1}\| \leq \|T\|\|A^{-1}\|.$$

Based on the hypothesis T is continuous, we infer that $\|T\| < \infty$. The fact that $\gamma(A) > 0$ implies that $\|A^{-1}\| < \infty$. Hence,

$$\|TA^{-1}\| < \infty.$$

For any given $\eta > 0$, we deduce from (4.5.21) and (4.5.22) that

$$\begin{aligned} |\lambda| \leq & \ \|TA^{-1}z\| + \eta \\ \leq & \ \|TA^{-1}\| + \eta. \end{aligned}$$

By arbitrariness of η, we conclude that

$$|\lambda| \leq \|TA^{-1}\|.$$

This equivalent to saying that $W(T, A) \subset \overline{\mathbb{D}}_{\|TA^{-1}\|}$. Q.E.D.

Remark 1 Let $T \in LR(H, K)$ be continuous and $A \in LR(H, K)$ such that $\gamma(A) > 0$ and $N(A) \subset \mathscr{D}(T)$. Then,

$$\overline{W(T, A)} \subset \overline{\mathbb{D}}_{\|TA^{-1}\|}.$$

Moreover, if $\dim(W(T, A)) < \infty$, then $W(T, A)$ is compact. ◇

Remark 2 Let $A, T \in LR(H, K)$ such that $\{0\} \subsetneq A(0)$. Then,

$$0 \in W(T, A).$$

Indeed, since $\{0\} \subsetneq A(0) = N(A^{-1}) \subset N(TA^{-1})$, then TA^{-1} is not injective. This implies that there exists $x \in \mathscr{D}(TA^{-1})$ such that $\|x\| = 1$ and $0 \in TA^{-1}x$. Thus, $(x, 0) \in G(TA^{-1})$ with $\|x\| = 1$. Hence, $\langle 0, x\rangle \in W(TA^{-1})$. Consequently, $0 \in W(TA^{-1})$. Finally, the use of Proposition 4.5.3 (ii) allows us to conclude that $0 \in W(T, A)$. ◇

Proposition 4.5.4 *Let $A, T, S \in LR(H, K)$ and $\alpha_1, \alpha_2, \alpha_3 \in \mathbb{C}$ such that $|\alpha_3| = 1$. Then,*

(i) $W(\alpha_1 T, A) = \alpha_1 W(T, A)$.
(ii) $W(\alpha_1 S + \alpha_2 T, A) \subset \alpha_1 W(S, A) + \alpha_2 W(T, A)$.
(iii) $W(T, \alpha_3 A) = \overline{\alpha}_3 W(T, A)$. ◇

Proof (i) If $\alpha_1 = 0$, then

$$W(0T, A) = \{\langle 0, z\rangle : \text{ there exists } x \in \mathscr{D}(A), z \in Ax \text{ and } \|z\| = 1\} = \{0\}.$$

This implies that $W(\alpha_1 T, A) = \alpha_1 W(T, A)$.

Suppose that $\alpha_1 \in \mathbb{C}\backslash\{0\}$ and $\lambda \in W(\alpha_1 T, A)$. Then, there exists $x \in \mathscr{D}(A) \bigcap \mathscr{D}(\alpha_1 T)$, $y \in \alpha_1 T x$ and $z \in Ax$ satisfying

$$\lambda = \langle y, z \rangle \text{ with } \|z\| = 1. \tag{4.5.23}$$

The fact that $(x, y) \in G(\alpha_1 T)$ then $\alpha_1^{-1} y \in Tx$. This yields that $\left\langle \alpha_1^{-1} y, z \right\rangle \in W(T, A)$. It follows from (4.5.23) that

$$\lambda = \alpha_1 \left\langle \alpha_1^{-1} y, z \right\rangle \subset \alpha_1 W(T, A).$$

Hence,

$$W(\alpha_1 T, A) \subset \alpha_1 W(T, A).$$

Conversely, assume that $\alpha_1 \in \mathbb{C}\backslash\{0\}$ and $\lambda \in \alpha_1 W(T, A)$. Then, $\alpha_1^{-1}\lambda \in W(T, A)$. This implies that there exists $x \in \mathscr{D}(A) \bigcap \mathscr{D}(T)$, $y \in Tx$ and $z \in Ax$ satisfying

$$\alpha_1^{-1}\lambda = \langle y, z \rangle \text{ with } \|z\| = 1.$$

Therefore,

$$\begin{aligned} \lambda &= \alpha_1 \alpha_1^{-1} \lambda \\ &= \alpha_1 \langle y, z \rangle \\ &= \langle \alpha_1 y, z \rangle. \end{aligned} \tag{4.5.24}$$

The use of the hypothesis $(x, y) \in G(T)$, we have $(x, \alpha_1 y) \in G(\alpha_1 T)$. Hence, by the fact that $x \in \mathscr{D}(A) \bigcap \mathscr{D}(T)$, we conclude from (4.5.24) that $\lambda \in W(\alpha_1 T, A)$.

(ii) It is trivially when $\alpha_1 = \alpha_2 = 0$. Now, assume that $\alpha_1 \neq 0$, $\alpha_2 \neq 0$ and $\lambda \in W(\alpha_1 S + \alpha_2 T, A)$. Then, there exists $x \in \mathscr{D}(\alpha_1 S + \alpha_2 T) \bigcap \mathscr{D}(A) = \mathscr{D}(S) \bigcap \mathscr{D}(T) \bigcap \mathscr{D}(A)$, $y \in (\alpha_1 S + \alpha_2 T)x$ and $z \in Ax$ satisfying

$$\lambda = \langle y, z \rangle, \quad \text{with } \|z\| = 1. \tag{4.5.25}$$

The fact that $(x, y) \in G(\alpha_1 S + \alpha_2 T)$ implies that there exists $t \in K$ such that

$$(x, t) \in G(\alpha_1 S)$$

and

$$(x, y - t) \in G(\alpha_2 T).$$

This implies that $\left(x, \alpha_1^{-1} t\right) \in G(S)$ and $\left(x, \alpha_2^{-1}(y - t)\right) \in G(T)$. Since $x \in \mathscr{D}(S) \bigcap \mathscr{D}(T) \bigcap \mathscr{D}(A)$ and $\|z\| = 1$, then we can conclude that

$$\left\langle \alpha_1^{-1} t, z \right\rangle \in W(S, A)$$

and

$$\left\langle \alpha_2^{-1}(y-t), z \right\rangle \in W(T, A).$$

Thus, the use of (4.5.25) makes us deduce that

$$\begin{aligned} \lambda = & \ \langle t + y - t, z \rangle \\ = & \ \langle t, z \rangle + \langle y - t, z \rangle \\ = & \ \alpha_1 \left\langle \alpha_1^{-1} t, z \right\rangle + \alpha_2 \left\langle \alpha_2^{-1}(y-t), z \right\rangle \\ \subset & \ \alpha_1 W(S, A) + \alpha_2 W(T, A). \end{aligned}$$

(iii) Let $\lambda \in W(T, \alpha_3 A)$. Then, there exists $x \in \mathscr{D}(\alpha_3 A) \bigcap \mathscr{D}(T) = \mathscr{D}(A) \bigcap \mathscr{D}(T)$, $(x, y) \in G(T)$, $z \in (\alpha_3 A)x$ such that $\|z\| = 1$ and $\lambda = \langle y, z \rangle$. Based on the assumption $|\alpha_3| = 1$, we infer that $\alpha_3 \neq 0$. Now, the fact that $(x, z) \in G(\alpha_3 A)$ implies that $\left(x, \alpha_3^{-1} z\right) \in G(A)$. Moreover, we have

$$\begin{aligned} 1 = & \ \|z\| \\ = & \ \left|\alpha_3 \alpha_3^{-1} z\right| \\ = & \ |\alpha_3| \ \left\|\alpha_3^{-1} z\right\| \\ = & \ \left\|\alpha_3^{-1} z\right\| . \end{aligned}$$

This implies that $\langle y, \alpha_3^{-1} z \rangle \in W(T, A)$. Hence,

$$\begin{aligned} \lambda = & \ \langle y, \alpha_3 \alpha_3^{-1} z \rangle \\ = & \ \overline{\alpha}_3 \langle y, \alpha_3^{-1} z \rangle \in \overline{\alpha}_3 W(T, A). \end{aligned}$$

Conversely, let $\lambda \in \overline{\alpha_3} W(T, A)$ with $|\overline{\alpha_3}| = |\alpha_3| = 1$. Then, $(\overline{\alpha_3})^{-1} \lambda \in W(T, A)$. This implies that there exists $x \in \mathscr{D}(A) \bigcap \mathscr{D}(T)$, $(x, y) \in G(T)$ and $(x, z) \in G(A)$ satisfying

$$(\overline{\alpha_3})^{-1} \lambda = \langle y, z \rangle, \quad \text{with } \|z\| = 1.$$

Consequently,

$$\begin{aligned} \lambda = & \ \overline{\alpha_3} \langle y, z \rangle \\ = & \ \langle y, \alpha_3 z \rangle. \end{aligned}$$

Since $(x, z) \in G(A)$, we obtain $(x, \alpha_3 z) \in G(\alpha_3 A)$. Moreover, we have $\|\alpha_3 z\| = |\alpha_3| \|z\| = 1$. Hence, $\lambda \in W(T, \alpha_3 A)$. Q.E.D.

Proposition 4.5.5 *Let T, $A \in L\mathscr{R}(H)$ be such that $A(0) \subset N(T)$ and $\mathscr{D}(A) \subset R(T)$ and $G(T) \subset \mathscr{S}(G(A))$, where the operator $\mathscr{S}$ is defined by*

$$\begin{aligned}\mathscr{S} : H \times H &\longrightarrow H \times H\\ (x, y) &\longrightarrow (y, x).\end{aligned}$$

If T and A are self-adjoint, then $W(T, A)$ is a subset of the real axis. ◇

Proof Let us assume that $\lambda \in W(T, A)$. By referring to Proposition 4.5.3 (ii), we get $\lambda \in W(TA^{-1})$. Then, there exists $(z, y) \in G(TA^{-1})$ satisfying

$$\lambda = \langle y, z\rangle, \quad \text{with } \|z\| = 1.$$

The fact that $(z, y) \in G(TA^{-1})$ implies from Eq. (2.8.1) that there exists $x \in \mathscr{D}(A) \bigcap \mathscr{D}(T)$ such that $(z, x) \in G(A^{-1})$ and $(x, y) \in G(T)$. Based on the hypotheses $G(T) \subset \mathscr{S}(G(A)) = G(A^{-1})$, $R(A^{-1}) = \mathscr{D}(A) \subset R(T)$ and $N(A^{-1}) = A(0) \subset N(T) = T^{-1}(0)$, we infer from Proposition 2.7.3 that

$$G(T) = G(A^{-1}). \tag{4.5.26}$$

Since T and A are self-adjoint, then we deduce from (4.5.26) and Proposition 2.15.4 that

$$\begin{aligned}G(T^*) =& \ G(T)\\ =& \ G\left(A^{-1}\right)\\ =& \ G\left(\left(A^*\right)^{-1}\right)\\ =& \ G\left(\left(A^{-1}\right)^*\right).\end{aligned}$$

This implies that $(z, x) \in G(T^*)$ and $(x, y) \in G\left(\left(A^{-1}\right)^*\right)$. Again from Eq. (2.8.1), we get that $(z, y) \in G\left(\left(A^{-1}\right)^* T^*\right)$. The use of Theorem 2.15.2 makes us to conclude that

$$(z, y) \in G\left(\left(TA^{-1}\right)^*\right).$$

It follows from the definition of the adjoint relation on Hilbert space that

$$\langle z, y\rangle = \langle y, z\rangle, \text{ for all } (z, y) \in G\left(TA^{-1}\right).$$

Therefore,

$$\begin{aligned}\lambda =& \ \langle y, z\rangle\\ =& \ \langle z, y\rangle\\ =& \ \overline{\langle y, z\rangle}\\ =& \ \overline{\lambda}.\end{aligned}$$

This is equivalent to saying that $\lambda \in \mathbb{R}$. This achieves the proof. Q.E.D.

Corollary 4.5.1 *Let $T \in L\mathscr{R}(H)$ and $A : \mathscr{D}(A) \subset H \longrightarrow H$ be a linear operator such that $\mathscr{D}(A) \subset R(T)$ and $G(T) \subset \mathscr{S}(G(A))$, where the operator $\mathscr{S}$ is defined by*

$$\begin{aligned} \mathscr{S} : H \times H &\longrightarrow H \times H \\ (x, y) &\longrightarrow (y, x). \end{aligned}$$

If T and A are self-adjoint, then $W(T, A)$ is a subset of the real axis. ◇

Proposition 4.5.6 *(i) Let $T \in L\mathscr{R}(H)$ and $U \in \mathscr{L}(H)$ be such that $U^*U = UU^* = I$. Then,*

$$W(U^*TU) = W(T) = W(UTU^*).$$

(ii) Let $T \in C\mathscr{R}(H)$ be such that $\mathscr{D}(T) = T(0)^{\perp}$. Then,

$$W(T^*) = \{\overline{\lambda} \in \mathbb{C} \text{ such that } \lambda \in W(T)\}.$$ ◇

Proof (i) We shall show that

$$W(U^*TU) = W(T). \tag{4.5.27}$$

Let $\lambda \in W(U^*TU)$. Then, there exists $(x, y) \in G(U^*TU)$ such that $\|x\| = 1$ and $\lambda = \langle y, x\rangle$. By Theorem 2.8.1, we get that

$$U^*TUx = y + U^*TU(0).$$

Based on the assumption $UU^* = I$, we infer that

$$\begin{aligned} TUx &= UU^*TUx \\ &= U(y + U^*TU(0)) \\ &= Uy + UU^*TU(0) \text{ (by Proposition 2.8.2)} \\ &= Uy + TU(0) \\ &= Uy + T(0) \text{ (as } U(0) = 0). \end{aligned}$$

This implies from Theorem 2.8.1 that $(Ux, Uy) \in G(T)$. Moreover, we have

$$\begin{aligned} \lambda &= \langle y, x\rangle \\ &= \langle y, U^*Ux\rangle \\ &= \langle Uy, Ux\rangle \end{aligned}$$

and

$$\begin{aligned} \|Ux\| &= \sqrt{\langle Ux, Ux\rangle} \\ &= \sqrt{\langle x, U^*Ux\rangle} \\ &= \sqrt{\langle x, x\rangle} \end{aligned}$$

$$\begin{aligned} &= \|x\| \\ &= 1. \end{aligned}$$

Hence, $\langle Uy, Ux\rangle \in W(T)$. Thus, $\lambda \in W(T)$. Conversely, let $\lambda \in W(T)$. Then, there exists $(x, y) \in G(T)$ such that $\|x\| = 1$ and $\lambda = \langle y, x\rangle$. The fact that $UU^* = I$ implies that $(x, x) \in G(UU^*)$. Hence, $(x, y) \in G(TUU^*)$. By using Theorem 2.8.1, we infer from the fact that $U^*(0) = 0$ that

$$\begin{aligned} TUU^*x &= y + TUU^*(0) \\ &= y + TU(0). \end{aligned}$$

Hence, by Proposition 2.8.2, we conclude that

$$U^*TU(U^*x) = U^*y + U^*TU(0),$$

which yields from Theorem 2.8.1 that $(U^*x, U^*y) \in G(U^*TU)$. Moreover, we have

$$\langle U^*y, U^*x\rangle = \langle y, x\rangle$$

and

$$\left\|U^*x\right\| = \|x\| = 1.$$

Therefore, $\lambda \in W(U^*TU)$. Thus, (4.5.27) holds. Substituting T for UTU^* in (4.5.27) yields

$$W(T) = W(UTU^*).$$

(ii) We shall prove that

$$W(T) \subset \{\lambda \in \mathbb{C} \text{ such that } \overline{\lambda} \in W(T^*)\}. \tag{4.5.28}$$

Let $\lambda \in W(T)$. Then, there exists $(x, y) \in G(T)$ such that $\|x\| = 1$ and $\lambda = \langle y, x\rangle$. Based on the assumption $\mathscr{D}(T)$ is closed, we infer from Theorem 2.11.1 that $\|T\| < \infty$. Then, by using Proposition 2.15.1, we obtain

$$\mathscr{D}(T^*) = T(0)^{\perp}.$$

This implies from the fact that $\mathscr{D}(T) = T(0)^{\perp}$ that

$$\mathscr{D}(T^*) = \mathscr{D}(T).$$

Hence, there exists $z \in H$ such that $(x, z) \in G(T^*)$. This lead us to $\langle z, x\rangle \in W(T^*)$. Since $(x, y) \in G(T)$ and $(x, z) \in G(T^*)$, then

$$\langle x, y\rangle = \langle z, x\rangle.$$

Moreover, we have

$$\langle y, x\rangle = \overline{\langle x, y\rangle}.$$

Thus, (4.5.28) holds. Conversely, substituting T for T^* in (4.5.28) yields

$$W(T^*) \subset \{\overline{\lambda} \in \mathbb{C} \text{ such that } \lambda \in W(T^{**})\}.$$

Since T is closed, then by [22, p.16], we have $T^{**} = T$. Hence,

$$W(T^*) \subset \{\overline{\lambda} \in \mathbb{C} \text{ such that } \lambda \in W(T)\}.$$

Q.E.D.

Proposition 4.5.7 *If* $T, A \in L\mathscr{R}(H, K)$, *then* $W(T, A)$ *i s a convex set.* ◇

Proof Let us assume that λ_1 and λ_2 are two points in $W(T, A)$. We divide this proof into two cases.

First case. If $\lambda_1 = \lambda_2$, then we can conclude that

$$t\lambda_1 + (1-t)\lambda_1 = \lambda_1 \in W(T, A), \quad \text{for all } t \in [0, 1].$$

This is equivalent to saying that $W(T, A)$ is a convex set.

Second case. If $\lambda_1 \neq \lambda_2 \in W(T, A)$, then there exist $x_1, x_2 \in \mathscr{D}(A) \bigcap \mathscr{D}(T)$, $y_1 \in Tx_1$ and $y_2 \in Tx_2$, $z_1 \in Ax_1$ and $z_2 \in Ax_2$ such that $\|z_1\| = \|z_2\| = 1$ satisfying

$$\lambda_1 = \langle y_1, z_1\rangle$$

and

$$\lambda_2 = \langle y_2, z_2\rangle .$$

• Suppose that x_1 and x_2 are linearly dependent, i.e., $x_2 = \alpha x_1$, for some $\alpha \in \mathbb{C}$ with $|\alpha| = 1$. This implies that $(x_1, y_1) \in G(T)$ and $(\alpha x_1, y_2) \in G(T)$. The fact that $G(T)$ is a vector subspace of $H \times K$ implies that $(\alpha x_1, \alpha y_1) \in G(T)$, which yields that $(0, \alpha y_1 - y_2) \in G(T)$. Hence, for all $\beta \in \mathbb{C}$, we have $(0, \beta(\alpha y_1 - y_2)) \in G(T)$. Thus,

$$(x_1, \beta(\alpha y_1 - y_2) + y_1) \in G(T).$$

Based on the hypothesis $(z_1, x_1) \in G(A^{-1})$, we infer from Eq. (2.8.1) that

$$(z_1, \beta(\alpha y_1 - y_2) + y_1) \in G(TA^{-1}), \quad \text{with } \|z_1\| = 1.$$

This implies that $\langle \beta(\alpha y_1 - y_2) + y_1, z_1\rangle \in W(TA^{-1})$, which yields from Proposition 4.5.3 (ii) that

$$\langle \beta(\alpha y_1 - y_2) + y_1, z_1\rangle \in W(T, A), \quad \text{for all } \beta \in \mathbb{C}.$$

As a result,

$$\begin{aligned}\mathbb{C} = &\ \{\langle \beta(\alpha y_1 - y_2) + y_1, z_1\rangle \text{ such that } \beta \in \mathbb{C}\}\\ \subset &\ W(T, A)\\ \subset &\ \mathbb{C}.\end{aligned}$$

Therefore, $W(T, A) = \mathbb{C}$ is a convex set.

• Suppose that x_1 and x_2 are linearly independent. If λ_1 and λ_2 are distinct points in $W(T, A)$, then there exist $x_1, x_2 \in \mathscr{D}(A) \bigcap \mathscr{D}(T)$, $y_1 \in Tx_1$, $y_2 \in Tx_2$, $z_1 \in Ax_1$ and $z_2 \in Ax_2$ satisfying

$$\lambda_1 = \langle y_1, z_1\rangle$$

and

$$\lambda_2 = \langle y_2, z_2\rangle,$$

with $\|z_1\| = \|z_2\| = 1$. We must show that whenever λ is a point of the line segment joining λ_1 and λ_2, there exists an element $x_0 \in \mathscr{D}(A) \bigcap \mathscr{D}(T)$, $y_0 \in Tx_0$ and $z_0 \in Ax_0$, for which

$$\langle y_0, z_0\rangle = \lambda, \quad \text{with } \|z_0\| = 1.$$

We find x_0 as a linear combination of x_1 and x_2. For this purpose and based on the hypotheses $\mathscr{D}(A) \bigcap \mathscr{D}(T)$ and $G(T)$ are vector subspaces of H and $H \times K$, respectively, we can conclude, for all $\alpha_1, \alpha_2 \in \mathbb{C}$, that

$$(\alpha_1 x_1 + \alpha_2 x_2, \alpha_1 y_1 + \alpha_2 y_2) \in G(T)$$

and

$$(\alpha_1 x_1 + \alpha_2 x_2, \alpha_1 z_1 + \alpha_2 z_2) \in G(A).$$

Hence, we can consider the bilinary forms $B_i : \mathbb{C} \times \mathbb{C} \longrightarrow \mathbb{C}$, where $i = 1, 2$ defined by

$$\begin{aligned}B_1(\alpha_1, \alpha_2) = &\ \langle \alpha_1 y_1 + \alpha_2 y_2, \alpha_1 z_1 + \alpha_2 z_2\rangle\\ B_2(\alpha_1, \alpha_2) = &\ \langle \alpha_1 z_1 + \alpha_2 z_2, \alpha_1 z_1 + \alpha_2 z_2\rangle.\end{aligned}$$

It follows from the fact that $\|z_1\| = \|z_2\| = 1$ that

$$\begin{aligned}B_1(\alpha_1, \alpha_2) = &\ \alpha_1\overline{\alpha_1}\langle y_1, z_1\rangle + \alpha_1\overline{\alpha_2}\langle y_1, z_2\rangle + \alpha_2\overline{\alpha_1}\langle y_2, z_1\rangle + \alpha_2\overline{\alpha_2}\langle y_2, z_2\rangle\\ = &\ |\alpha_1|^2\lambda_1 + \alpha_1\overline{\alpha_2}\langle y_1, z_2\rangle + \alpha_2\overline{\alpha_1}\langle y_2, z_1\rangle + |\alpha_2|^2\lambda_2,\end{aligned}$$

and

$$\begin{aligned}B_2(\alpha_1, \alpha_2) = &\ \alpha_1\overline{\alpha_1}\langle z_1, z_1\rangle + \alpha_1\overline{\alpha_2}\langle z_1, z_2\rangle + \alpha_2\overline{\alpha_1}\langle z_2, z_1\rangle + \alpha_2\overline{\alpha_2}\langle z_2, z_2\rangle\\ = &\ |\alpha_1|^2\|z_1\|^2 + \alpha_1\overline{\alpha_2}\langle z_1, z_2\rangle + \alpha_2\overline{\alpha_1}\langle z_2, z_1\rangle + |\alpha_2|^2\|z_2\|^2\\ = &\ |\alpha_1|^2 + \alpha_1\overline{\alpha_2}\langle z_1, z_2\rangle + \alpha_2\overline{\alpha_1}\langle z_2, z_1\rangle + |\alpha_2|^2\\ = &\ |\alpha_1|^2 + |\alpha_2|^2 + 2\Re\left(\alpha_2\overline{\alpha_1}\langle z_2, z_1\rangle\right). \end{aligned} \tag{4.5.29}$$

We must show that B_1 assumes every values on the line segment joining λ_1 and λ_2 while $B_2 = 1$. First, we introduce a bilinary form $B(\alpha_1, \alpha_2)$ as follows:

$$B(\alpha_1, \alpha_2) = \frac{B_1(\alpha_1, \alpha_2) - \lambda_2 B_2(\alpha_1, \alpha_2)}{\lambda_1 - \lambda_2}.$$

A simple calculation shows that

$$B(\alpha_1, \alpha_2) = |\alpha_1|^2 + \overline{\alpha_1}\alpha_2 b_{12} + \alpha_1 \overline{\alpha_2} b_{21}, \tag{4.5.30}$$

where

$$b_{12} = \frac{\langle y_1, z_2\rangle - \lambda_2 \langle z_1, z_2\rangle}{\lambda_1 - \lambda_2}$$

and

$$b_{21} = \frac{\langle y_2, z_1\rangle - \lambda_2 \langle z_2, z_1\rangle}{\lambda_1 - \lambda_2}.$$

In order that B_1 should have the requisite behavior, it is necessary and sufficient that while $B_2 = 1$, the form B should take on every real value from 0 to 1 inclusive. We shall now exhibit values of α_1 and α_2 which bring about the desired result. Choose $\xi \in \mathbb{C}$ as follows:

$$\xi = \begin{cases} \pm 1 & \text{if } b_{21} = \overline{b_{12}} \\ \pm \dfrac{\overline{b_{12}} - b_{21}}{|\overline{b_{12}} - b_{21}|} & \text{if } b_{21} \neq \overline{b_{12}}. \end{cases}$$

So that $\Re(\xi\langle Ax_2, Ax_1\rangle) \geq 0$. Now, we will replace α_1 and α_2, respectively, by f and ξg. On the one hand, we obtain from (4.5.30) that

$$\begin{aligned} B(f, \xi g) &= f^2 + \overline{\xi} f g b_{21} + \xi f g b_{12} \\ &= f^2 + fg(\overline{\xi} b_{21} + \xi b_{12}). \end{aligned} \tag{4.5.31}$$

At this point, we need to calculate the value of $\theta = \overline{\xi} b_{21} + \xi b_{12}$. Then, we will consider two cases:

- If $b_{21} = \overline{b_{12}}$, then $\overline{\xi} = \xi = \pm 1$. It follows from (4.5.31) that

$$B(f, \xi g) = f^2 \pm fg(\overline{b_{12}} + b_{12}).$$

Hence, $\theta = \overline{b_{12}} + b_{12} \in \mathbb{R}$, which yields that $\theta = \overline{\theta}$.

- If $b_{21} \neq \overline{b_{12}}$, then

$$\xi = \pm \frac{\overline{b_{12}} - b_{21}}{|\overline{b_{12}} - b_{21}|}$$

and

$$\overline{\xi} = \pm \frac{b_{12} - \overline{b_{21}}}{|\overline{b_{12}} - b_{21}|}.$$

Hence,

$$\begin{aligned}\theta &= \frac{b_{21}(b_{12} - \overline{b_{21}}) + b_{12}(\overline{b_{12}} - b_{21})}{|\overline{b_{12}} - b_{21}|} \\ &= \frac{b_{21}b_{12} - b_{21}\overline{b_{21}} + b_{12}\overline{b_{12}} - b_{12}b_{21}}{|\overline{b_{12}} - b_{21}|} \\ &= \frac{|b_{12}|^2 - |b_{21}|^2}{|\overline{b_{12}} - b_{21}|} \in \mathbb{R}.\end{aligned}$$

Therefore, the real number θ is defined as follows:

$$\theta = \begin{cases} \pm(b_{21} + b_{12}) & \text{if } b_{21} = \overline{b_{12}} \\ \pm\dfrac{|b_{12}|^2 - |b_{21}|^2}{|\overline{b_{12}} - b_{21}|} & \text{if } b_{21} \neq \overline{b_{12}}. \end{cases}$$

On the other hand, we get from (4.5.29) that

$$\begin{aligned}B_2(f, \xi g) &= f^2 + |\xi|^2 g^2 + 2\Re\left(\xi f g \langle z_2, z_1\rangle\right) \\ &= f^2 + g^2 + 2fg\, \Re\left(\xi \langle z_2, z_1\rangle\right), \quad (\text{as } |\xi| = 1).\end{aligned}$$

Suppose that $\delta = \Re\left(\xi \langle z_2, z_1\rangle\right)$. We observe that $0 \leq \delta \leq 1$. Indeed, by Cauchy–Schwarz inequality, we infer that

$$\begin{aligned}|\delta| &= |\Re\left(\xi \langle z_2, z_1\rangle\right)| \\ &\leq |\xi \langle z_2, z_1\rangle| \\ &\leq |\xi|\, |\langle z_2, z_1\rangle| \\ &\leq |\xi|\, \|z_2\|^2 \|z_1\|^2 \\ &\leq 1 \;\; (\text{as } |\xi| = \|z_2\|^2 = \|z_1\|^2 = 1).\end{aligned}$$

Solving for $B_2(f, \xi g) = 1$, we get

$$g = -\delta f + \sqrt{1 - (1 - \delta^2) f^2},$$

which is a real valued for $f \in [-1, 1]$. Since g is a function of f, we can let $\widehat{B}(f) = B(f, \xi g)$, which yields from (4.5.31) that

$$\begin{aligned}\widehat{B}(f) &= f^2 + \theta f \left(-\delta f + \sqrt{1-(1-\delta^2)f^2}\right) \\ &= f^2 - \theta\delta f^2 + \theta f\sqrt{1-(1-\delta^2)f^2} \\ &= (1-\theta\delta) f^2 + \theta f\sqrt{1-(1-\delta^2)f^2}.\end{aligned}$$

For $f \in [0, 1]$, $\widehat{B}(f)$ is evidently a continuous real valued function with $\widehat{B}(0) = 0$ and $\widehat{B}(1) = 1$, so that $\widehat{B}(f)$ assumes all values between 0 and 1. This result shows that $W(T, A)$ is a convex set. Q.E.D.

4.5.4 The Spectral Theory Associated with the Numerical Range

The pencil point spectrum $\sigma_{p,A}(T)$ is defined by

$$\sigma_{p,A}(T) = \{\lambda \in \mathbb{C} : N(\lambda A - T) \neq \{0\}\}.$$

Theorem 4.5.7 *Let $T, A \in L\mathscr{R}(H, K)$. Then,*

(i) If $N(A)\bigcap \mathscr{D}(T) = \{0\}$, then $\sigma_{p,A}(T)\backslash\{0\} \subset W(T, A)$.
(ii)If A is one-to-one, then $\sigma_{ap,A}(T) \subset \overline{W(T, A)}$.
(iii) If A is onto and $\mathscr{D}(A) \subset \mathscr{D}(T)$, then $\sigma_{A,0}(T) \subset W(T, A)$. ◇

Proof (i) Let $\lambda \in \sigma_{p,A}(T)\backslash\{0\}$. Since $\mathscr{D}(\lambda A - T) = \mathscr{D}(A)\bigcap \mathscr{D}(T)$, then there exists $x \in \left(\mathscr{D}(A)\bigcap \mathscr{D}(T)\right)\backslash\{0\}$ such that $0 \in (\lambda A - T)x$. Consequently, $(x, 0) \in G(\lambda A - T)$. It follows from Remark 2.7.1 (ii) that there exists $y \in K$ such that $(x, \lambda^{-1}y) \in G(A)$ and $(x, y) \in G(T)$. Now, suppose that $y = 0$. Then, $(x, 0) \in G(A)$ which yields that $x \in N(A)$. Moreover, we have $x \in \mathscr{D}(T)$. Therefore, $x \in N(A)\bigcap \mathscr{D}(T) = \{0\}$. This contradiction implies that $y \neq 0$. We know that $G(T)$ and $G(A)$ are vector subspaces of $H \times K$, so we get

$$\left(\frac{x}{\|\lambda^{-1}y\|}, \lambda^{-1}\frac{y}{\|\lambda^{-1}y\|}\right) \in G(A)$$

and

$$\left(\frac{x}{\|\lambda^{-1}y\|}, \frac{y}{\|\lambda^{-1}y\|}\right) \in G(T).$$

Hence,

$$\left\|\lambda^{-1}\frac{y}{\|\lambda^{-1}y\|}\right\| = 1$$

and

$$\begin{aligned}\left\langle \frac{y}{\|\lambda^{-1}y\|}, \lambda^{-1}\frac{y}{\|\lambda^{-1}y\|}\right\rangle &= \overline{\lambda^{-1}}\frac{\|y\|^2}{\|\lambda^{-1}y\|^2}\\ &= \overline{\lambda^{-1}}\frac{\|y\|^2}{|\lambda^{-1}|^2\|y\|^2}\\ &= \lambda.\end{aligned}$$

This implies from the above that $\lambda \in W(T, A)$.

(ii) Let us suppose that $\lambda \in \sigma_{ap,A}(T)$. Then, there exists $(x_n) \subset \mathscr{D}(A)\bigcap\mathscr{D}(T)$ with $\|x_n\| = 1$ and

$$\lim_{n\to+\infty} \|(\lambda A - T)x_n\| = 0.$$

The fact that $(x_n) \subset \mathscr{D}(A)\bigcap\mathscr{D}(T)$ implies that there exist two sequence (y_n) and (z_n) in K such that $(x_n, y_n) \in G(T)$ and $(x_n, z_n) \in G(A)$. Now, assume that z_n is a zero sequence. Then, $(x_n, 0) \in G(A)$ which yields that $x_n \in N(A) = \{0\}$. This contradiction implies that z_n is a nonzero sequence. This implies from the fact that $G(T)$ and $G(A)$ are vector subspaces of $H \times K$ that

$$\left(\frac{x_n}{\|z_n\|}, \frac{y_n}{\|z_n\|}\right) \in G(T)$$

and

$$\left(\frac{x_n}{\|z_n\|}, \frac{z_n}{\|z_n\|}\right) \in G(A).$$

This lead us to $\left\|\frac{z_n}{\|z_n\|}\right\| = 1$ and

$$\left\langle\frac{y_n}{\|z_n\|}, \frac{z_n}{\|z_n\|}\right\rangle \in W(T, A).$$

On the one hand, we have

$$\begin{aligned}\left|\lambda - \left\langle\frac{y_n}{\|z_n\|}, \frac{z_n}{\|z_n\|}\right\rangle\right| &= \left|\lambda\left\langle\frac{z_n}{\|z_n\|}, \frac{z_n}{\|z_n\|}\right\rangle - \left\langle\frac{y_n}{\|z_n\|}, \frac{z_n}{\|z_n\|}\right\rangle\right|\\ &= \frac{1}{\|z_n\|^2}|\langle\lambda z_n - y_n, z_n\rangle|\\ &\leq \frac{1}{\|z_n\|^2}\|\lambda z_n - y_n\|\|z_n\| \text{ (by Cauchy-Schwarz inequality)}\\ &\leq \frac{1}{\|z_n\|}\|\lambda z_n - y_n\|. \end{aligned} \tag{4.5.32}$$

On the other hand, by using the fact that $(x_n, y_n) \in G(T)$ and $(x_n, \lambda z_n) \in G(\lambda A)$, we infer from Remark 2.7.1 (ii) that $(x_n, \lambda z_n - y_n) \in G(\lambda A - T)$. Hence, for all $\varepsilon > 0$, we have

$$\|\lambda z_n - y_n\| \leq \|(\lambda A - T)x_n\| + \varepsilon.$$

It follows from (4.5.32) that

$$\left|\left\langle \frac{y_n}{\|z_n\|}, \frac{z_n}{\|z_n\|} \right\rangle - \lambda\right| \leq \frac{1}{\|z_n\|} \left(\|(\lambda A - T)x_n\| + \varepsilon\right).$$

By arbitrariness of ε, we deduce that

$$\left|\left\langle \frac{y_n}{\|z_n\|}, \frac{z_n}{\|z_n\|} \right\rangle - \lambda\right| \to 0 \text{ as } n \to +\infty.$$

This yields that

$$\left\langle \frac{y_n}{\|z_n\|}, \frac{z_n}{\|z_n\|} \right\rangle \to \lambda \text{ as } n \to +\infty.$$

This equivalent to saying that $\lambda \in \overline{W(T, A)}$.

(iii) Let $\lambda \in \sigma_{A,0}(T)$, then

$$\overline{R(\lambda A - T)} \subsetneq K.$$

Since every closed subspace of a Hilbert space is topologically complemented by its orthogonal complement, then we can conclude that

$$K = \overline{R(\lambda A - T)} \bigoplus R(\lambda A - T)^{\perp}.$$

This leads to that there exists $z \in K$ such that $\|z\| = 1$ and $z \in R(\lambda A - T)^{\perp}$. The fact that A is onto implies that there exists $x \in \mathscr{D}(A)$ such that $(x, z) \in G(A)$. Based on the hypothesis $\mathscr{D}(A) \subset \mathscr{D}(T)$, we infer that there exists $y \in K$ such that $(x, y) \in G(T)$. Hence, by referring to Remark 2.7.1 (ii), we get $(x, \lambda z - y) \in G(\lambda A - T)$, which yields that $\lambda z - y \in R(\lambda A - T)$. This implies from the fact that $z \in R(\lambda A - T)^{\perp}$ that

$$\begin{aligned} 0 = &\ \langle \lambda z - y, z\rangle \\ = &\ \lambda\langle z, z\rangle - \langle y, z\rangle \\ = &\ \lambda - \langle y, z\rangle \quad (\text{as } \|z\| = 1). \end{aligned}$$

Therefore,

$$\lambda = \langle y, z\rangle.$$

This is equivalent to saying that $\lambda \in W(T, A)$. Q.E.D.

Corollary 4.5.2 *Let $T \in L\mathscr{R}(H, K)$ and $A : \mathscr{D}(A) \subset H \longrightarrow K$ be a linear operator.*

(i) If $N(A) \bigcap \mathscr{D}(T) = \{0\}$, then $\sigma_{p,A}(T) \subset W(T, A)$.
(ii) $\sigma_{ap,A}(T) \subset \overline{W(T, A)}$.

(iii) If A onto and $\mathscr{D}(A) \subset \mathscr{D}(T)$, then $\sigma_{A,0}(T) \subset W(T, A)$. ◇

Lemma 4.5.1 *Let $A, T \in L\mathscr{R}(H, K)$ be such that $0 < \|A\| < \infty$. Suppose that one of the following conditions holds*

(i) T is not one-to-one.
(ii) T is not open.
(iii) A is an operator.

Then, $\sigma_{ap,A}(T)$ is closed. ◇

Proof Our purpose is to prove that $\mathbb{C}\backslash\sigma_{ap,A}(T)$ is open set. In order to show it, let us suppose that $\lambda \in \mathbb{C}\backslash\sigma_{ap,A}(T)$. Hence, applying Proposition 3.3.1, we infer that there exists $c > 0$ such that

$$\|(\lambda A - T)x\| \geq c\|x\| \text{ for all } x \in \mathscr{D}(T)\bigcap\mathscr{D}(A). \tag{4.5.33}$$

(i) Suppose that $\lambda = 0$. Then, by Proposition 3.3.1, we obtain $0A - T = -T$ is one-to-one and open. This is in contradiction with the fact $\{0\} \subsetneq N(T) = N(-T)$. Hence, $\lambda \neq 0$. Then,

$$\begin{aligned} \mu A(0) \subset \; & A(0) \\ = \; & A(\lambda \times 0) \\ = \; & \lambda A(0), \text{ for } \mu \in \mathbb{C}. \end{aligned}$$

Consequently,

$$\begin{aligned} \mu A(0) = \; & \mu A(0) - 0 \\ \subset \; & \lambda A(0) - T(0). \end{aligned}$$

Moreover, we have

$$\begin{aligned} \mathscr{D}(\lambda A - T) = \; & \mathscr{D}(T)\bigcap\mathscr{D}(A) \\ \subset \; & \mathscr{D}(A) \\ = \; & \mathscr{D}(\mu A). \end{aligned}$$

Hence, by (4.5.33) and [23, Lemma 2.5], we deduce that

$$\begin{aligned} c\|x\| \leq \; & \|(\lambda - \mu)\, Ax + (\mu A - T)\, x\| \\ \leq \; & \|(\lambda - \mu)\, Ax\| + \|(\mu A - T)\, x\| \\ \leq \; & |\lambda - \mu| \; \|Ax\| + \|(\mu A - T)\, x\| \,. \end{aligned}$$

Based on the assumption $\|A\| < \infty$, we get from Proposition 2.10.4 (i) that

$$c\|x\| \leq |\lambda - \mu| \, \|A\| \, \|x\| + \|(\mu A - T)\, x\| \,.$$

This implies that

$$(c - |\lambda - \mu| \, \|A\|) \, \|x\| \leq \|(\mu A - T)\, x\| \,.$$

Assume that $|\lambda - \mu| < c \, \|A\|^{-1}$. This enables us to conclude from Proposition 3.3.1 that $\mu \in \mathbb{C}\backslash\sigma_{ap,A}(T)$. Hence,

$$\mathbb{D}\left(\lambda, c \, \|A\|^{-1}\right) \subset \mathbb{C}\backslash\sigma_{ap,A}(T).$$

This is equivalent to saying that $\mathbb{C}\backslash\sigma_{ap,A}(T)$ is open.

The proof of (ii) is completely similar to (i).

(iii) By using (4.5.33), we infer that

$$c\|x\| \leq \|(\lambda - \mu)\, Ax + (\mu A - T)\, x\| \,.$$

The final part of the proof is almost identical to that of (i). Q.E.D.

Proposition 4.5.8 *Let $A \in L\mathscr{R}(H, K)$ and $T \in C\mathscr{R}(H, K)$ be such that $0 < \|A\| < \infty$, $\mathscr{D}(T) \subset \mathscr{D}(A)$ and $R(A) = K$. Assume that the one of the conditions of Lemma 4.5.1 holds. Then,*

$$\partial\sigma_A(T) \subset \sigma_{ap,A}(T). \qquad \diamondsuit$$

Proof Assume that $\partial\sigma_A(T) \subset \mathbb{C}\backslash\sigma_{ap,A}(T)$. Then,

$$\partial\sigma_A(T) \bigcap \mathbb{C}\backslash\sigma_{ap,A}(T) \neq \emptyset,$$

which yields that there exists

$$\lambda \in \partial\sigma_A(T) \bigcap \mathbb{C}\backslash\sigma_{ap,A}(T).$$

By referring to Lemma 4.5.1, we have $\mathbb{C}\backslash\sigma_{ap,A}(T)$ is open. Then, we can consider a connected Ω of $\mathbb{C}\backslash\sigma_{ap,A}(T)$ containing λ. The fact that Ω is open and $\lambda \in \Omega$ implies that Ω is a neighborhood of λ. Consequently, $\lambda \in \partial\sigma_A(T) \bigcap \Omega$. Thus, Ω contains points of $\rho_A(T)$. Hence,

$$\rho_A(T) \bigcap \Omega \neq \emptyset,$$

which yields that there exists $\mu \in \rho_A(T) \bigcap \Omega$. Since $\|\mu A\| < \infty$, then by using Proposition 2.10.4 (i), we infer that μA is continuous. The fact that $T \in C\mathscr{R}(H, K)$,

$$\mu A(0) = A(0) \subset T(0) \text{ and } \mathscr{D}(T) \subset \mathscr{D}(\mu A) = \mathscr{D}(A) \qquad (4.5.34)$$

implies from [23, Proposition 3.1] that $\mu A - T \in C\mathcal{R}(H, K)$. This leads from Remark 3.3.1 that $(\mu A - T)^{-1} \in \mathcal{L}(K, H)$.

If $\lambda = \mu$, then $\lambda \in \rho_A(T)$, which is contradiction because $\lambda \in \partial\sigma_A(T)$.

We assume $\lambda \neq \mu$. The fact that $\lambda \in \mathbb{C}\backslash\sigma_{ap,A}(T)$ implies from Proposition 3.3.1 and Proposition 2.10.4 (i) that $(\lambda A - T)^{-1}$ is a continuous operator. At this point, we shall prove that

$$\mathcal{D}\left((\lambda A - T)^{-1}\right) = K.$$

We assume that $\mathcal{D}\left((\lambda A - T)^{-1}\right) \subsetneq K$. Then, there exists $y \in K$ such that $y \notin \mathcal{D}\left((\lambda A - T)^{-1}\right) = R\,(\lambda A - T)$. This implies for all $x \in \mathcal{D}(\lambda A - T) = \mathcal{D}(T) \bigcap \mathcal{D}(A)$ that $(x, y) \notin G\,((\lambda A - T))$. By using (4.5.34) and [23, Lemma 2.5], we obtain

$$(x, y) \notin G\,((\lambda - \mu)A + (\mu A - T)).$$

It follows from Remark 2.7.1 (ii) that $(x, z) \notin G\,(A)$ or $(x, y - (\lambda - \mu)z) \notin G\,(\mu A - T)$. But, $(x, z) \notin G\,(A)$ is impossible and $y - (\lambda - \mu)z \notin R\,(\mu A - T) = K$ is also impossible. This contradiction implies that $\mathcal{D}\left((\lambda A - T)^{-1}\right) = K$. Hence, $\lambda \in \rho_A(T)$, which is contradiction.

Remark 4.5.2 Let $T \in C\mathcal{R}(H)$. Then, $\sigma_{ap}(T)$ is closed and $\partial\sigma(T) \subset \sigma_{ap}(T)$. ◇

Corollary 4.5.3 *Let $T \in L\mathcal{R}(H, K)$ and let $A \in \mathcal{L}(H, K)$ be non null operator. Then,*

(i) $\sigma_{ap,A}(T)$ is closed.
(ii) If T is closed, then $\partial\sigma_A(T) \subset \sigma_{ap,A}(T)$. ◇

As a direct consequence of Theorem 4.5.7 and Proposition 4.5.8, we infer the following result.

Corollary 4.5.4 *Let $A \in L\mathcal{R}(H, K)$ and $T \in C\mathcal{R}(H, K)$ be such that $0 < \|A\| < \infty$, $\mathcal{D}(T) \subset \mathcal{D}(A)$, $N(A) = \{0\}$, $A(0) \subset T(0)$ and $R(A) = K$. Assume that the one of the conditions of Lemma 4.5.1 holds. Then,*

$$\partial\sigma_A(T) \subset \overline{W(T, A)}.$$ ◇

Theorem 4.5.8 *Let A, $T \in L\mathcal{R}(H, K)$ be such that $N(A) = \{0\}$ and $\gamma(A) > 0$. Then,*

(i) $\sigma_A(T) \subset \overline{W(T, A)}$.
(ii) Moreover, if $T \in C\mathcal{R}(H, K)$ and $\|A\| < \infty$, then

$$\sigma_{\varepsilon,A}(T) \subset \left\{\lambda \in \mathbb{C} \text{ such that } dist\left(\lambda, \overline{W(T, A)}\right) < \gamma(A)^{-1}\varepsilon\right\}.$$ ◇

Proof (i) Let $\lambda \in \sigma_A(T)$. To prove $\lambda \in \overline{W(T, A)}$, assume the contrary, i.e., let $\lambda \notin \overline{W(T, A)}$. Then, $r = \text{dist}(\lambda, \overline{W(T, A)}) = \text{dist}(\lambda, W(T, A)) > 0$. This implies that there exist $x \in \mathcal{D}(T) \bigcap \mathcal{D}(A)$, $y \in Tx$, $z \in Ax$ and $\|z\| = 1$ such that

$$\begin{aligned}
0 < r \leq & \ |\lambda - \langle y, z\rangle| \\
= & \ |\lambda \langle z, z\rangle - \langle y, z\rangle| \\
= & \ |\langle \lambda z - y, z\rangle| \\
\leq & \ \|\lambda z - y\| \|z\| \text{ (by Cauchy-Schwarz inequality)} \\
\leq & \ \|\lambda z - y\| \quad (\text{as } \|z\| = 1).
\end{aligned} \tag{4.5.35}$$

The use of Remark 2.7.1 (ii) makes us to infer that $(x, \lambda z - y) \in G(\lambda A - T)$. Hence, for all $\varepsilon_0 > 0$, we have

$$\|\lambda z - y\| \leq \|(\lambda A - T)x\| + \varepsilon_0.$$

Choose $0 < \varepsilon_0 < r$, we get from (4.5.35) that

$$0 < (r - \varepsilon_0)\|z\| \leq \|(\lambda A - T)x\|. \tag{4.5.36}$$

By using the hypotheses A is one-to-one and $\gamma(A) > 0$, we deduce from [21, Definition III.6.1] that there exists $c > 0$ such that $\|Ax\| \geq c\|x\|$. It follows from Theorem 2.8.1 (i) that

$$\begin{aligned}
c\|x\| \leq & \ \|z + A(0)\| \\
\leq & \ \|z\|.
\end{aligned}$$

This implies from (4.5.36) that

$$c(r - \varepsilon_0)\|x\| \leq \|(\lambda A - T)x\|,$$

where $c(r - \varepsilon) > 0$. Again from [21, Definition III.6.1], we conclude that $\lambda A - T$ is one-to-one and open. Now, we shall show that

$$\overline{R(\lambda A - T)} = K.$$

By referring to Proposition 2.15.1 (iv), it is sufficient to prove that $N((\lambda A - T)^*) = \{0\}$. Assume that $(u, \overline{\lambda} t) \in G(\overline{\lambda} A^*)$ and $(u, f) \in G(T^*)$, we infer from Remark 2.7.1 (ii) that $(u, \overline{\lambda} t - f) \in G(\overline{\lambda} A^* - T^*)$. This leads from Lemma 2.10.3 (i) that

$$(u, \overline{\lambda} t - f) \in G((\lambda A - T)^*).$$

Hence, for all $\varepsilon_1 > 0$, we have

$$\|\overline{\lambda} t - f\| \leq \|(\lambda A - T)^* u\| + \varepsilon_1. \tag{4.5.37}$$

The fact that $(x, \lambda z - y) \in G(\lambda A - T)$ implies that

$$\langle \overline{\lambda} t - f, x \rangle = \langle u, \lambda z - y \rangle.$$

By (4.5.37), we have

$$\begin{aligned} \varepsilon_1 + \|(\lambda A - T)^* u\| &\geq \|\overline{\lambda} t - f\| \|z\| \\ &\geq c \|\overline{\lambda} t - f\| \|x\| \\ &\geq c\, |\langle \overline{\lambda} t - f, x \rangle| \text{ (by Cauchy-Schwarz inequality)} \\ &= c\, |\langle u, \lambda z - y \rangle|. \end{aligned} \tag{4.5.38}$$

The generalized Riesz representation theorem asserts that one can denote that

$$\varphi_{\lambda z - y}(u) = \langle u, \lambda z - y \rangle,$$

where φ is a bounded linear functional defined on K and verifies the following property

$$\|\varphi_{\lambda z - y}\| = \|\lambda z - y\|. \tag{4.5.39}$$

For all $u \neq 0$, we conclude from (4.5.38) that

$$\begin{aligned} \varepsilon_1 + \|(\lambda A - T)^* u\| &\geq c\, |\varphi_{\lambda z - y}(u)| \\ &\geq c\, \frac{|\varphi_{\lambda z - y}(u)|}{\|u\|} \|u\|. \end{aligned}$$

Using the definition of the upper bound, for any given $\varepsilon_2 > 0$, we get from (4.5.38) that

$$\frac{|\varphi_{\lambda z - y}(u)|}{\|u\|} \geq \|\varphi_{\lambda z - y}\| - \varepsilon_2.$$

This implies that

$$\begin{aligned} \varepsilon_1 + \|(\lambda A - T)^* u\| &\geq c \|\varphi_{\lambda z - y}\| \|u\| - c\, \varepsilon_2 \|u\| \\ &\geq c \|\lambda z - y\| \|u\| - c\, \varepsilon_2 \|u\| \text{ (from (4.5.39))} \\ &\geq c\, (r - \varepsilon_2)\, \|u\| \text{ (from (4.5.35))}. \end{aligned} \tag{4.5.40}$$

In addition, if $u = 0$, we infer that (4.5.40) holds. Hence, by arbitrariness of ε_1, we conclude that $(\lambda A - T)^*$ is bounded below, which yields from [21, Definition III.6.1] that $(\lambda A - T)^*$ is one-to-one. Thus, $\overline{R(\lambda A - T)} = K$. Consequently, $\lambda \in \rho_A(T)$, which is a contradiction. Therefore, $\lambda \in \overline{W(T, A)}$.
(ii) We assume that

$$\operatorname{dist}\left(\lambda, \overline{W(T, A)}\right) \geq \gamma(A)^{-1} \varepsilon.$$

Then,

$$\operatorname{dist}(\lambda, W(T, A)) \geq \gamma(A)^{-1} \varepsilon$$

which yields that

$$\|\lambda - \langle y, z\rangle\| \geq \gamma(A)^{-1}\varepsilon$$

for all $y \in Tx$, $z \in Ax$ and $\|z\| = 1$. This lead us to

$$\begin{aligned}\gamma(A)^{-1}\varepsilon \leq\ & \|\langle \lambda z - y, z\rangle\| \\ \leq\ & \|\lambda z - y\|\,\|z\| \ \text{(by Cauchy-Schwarz inequality)} \\ \leq\ & \|\lambda z - y\| \ \text{(as } \|z\| = 1\text{)}.\end{aligned} \tag{4.5.41}$$

Since $y \in Tx$ and $z \in Ax$, then by using Remark 2.7.1 (ii), we infer that

$$(x, \lambda z - y) \in G(\lambda A - T).$$

Hence, for all $\varepsilon_0 > 0$, we have

$$\|\lambda z - y\| \leq \|(\lambda A - T)x\| + \varepsilon_0.$$

This implies from (4.5.41) that

$$\gamma(A)^{-1}\varepsilon - \varepsilon_0 \leq \|(\lambda A - T)x\|, \ \text{for all } 0 < \varepsilon_0 < \gamma(A)^{-1}\varepsilon. \tag{4.5.42}$$

Since

$$\operatorname{dist}\left(\lambda, \overline{W(T, A)}\right) \geq \gamma(A)^{-1}\varepsilon > 0,$$

then $\lambda \notin \overline{W(T, A)}$. It follows from (i) that $\lambda \notin \sigma_A(T)$ which yields that $\lambda A - T$ is one-to-one. Now, we suppose that $x = 0$. Then, by (4.5.42), we obtain

$$\gamma(A)^{-1}\varepsilon - \varepsilon_0 \leq \|(\lambda A - T)0\| = 0.$$

This contradiction implies that $x \neq 0$. Consequently,

$$\|x\| = \operatorname{dist}(x, N(\lambda A - T)) > 0.$$

By referring to Definition 2.10.2 and Lemma 2.10.9, we have

$$\left\|(\lambda A - T)^{-1}\right\|^{-1} = \inf\left\{\frac{\|(\lambda A - T)x\|}{\operatorname{dist}(x, N(\lambda A - T))} : x \in \mathscr{D}(\lambda A - T),\ x \notin \overline{N(\lambda A - T)}\right\}.$$

Using the definition of the lower bound, for any given $\varepsilon_1 > 0$, we have

$$\begin{aligned}\left\|(\lambda A - T)^{-1}\right\|^{-1} + \varepsilon_1 \geq\ & \frac{\|(\lambda A - T)x\|}{\|x\|} \\ \geq\ & \frac{\gamma(A)^{-1}\varepsilon - \varepsilon_0}{\|x\|}.\end{aligned} \tag{4.5.43}$$

Based on the hypothesis $\gamma(A) > 0$, we infer from Proposition 2.15.3 that A is open. This leads to A^{-1} is continuous. Moreover, A is one-to-one i.e., A^{-1} is an operator. Hence,

$$\begin{aligned}\|x\| &= \|A^{-1}z\| \\ &\leq \|A^{-1}\|\|z\| \\ &= \|A^{-1}\|.\end{aligned}$$

Therefore, by (4.5.43), we deduce that

$$\left\|(\lambda A - T)^{-1}\right\|^{-1} \geq \frac{\gamma(A)^{-1}\varepsilon - \varepsilon_0 - \varepsilon_1}{\|A^{-1}\|}.$$

Consequently, by arbitrariness of ε_0 and ε_1, we get that

$$\left\|(\lambda A - T)^{-1}\right\|^{-1} \geq \frac{\gamma(A)^{-1}\varepsilon}{\|A^{-1}\|}.$$

Finally, the use of Lemma 2.10.9 allows us to conclude that

$$\left\|(\lambda A - T)^{-1}\right\| \leq \frac{1}{\varepsilon}.$$

This achieves the proof. Q.E.D.

As a direct consequence of Theorem 4.5.7 and Proposition 4.5.8, we obtain the following result.

Corollary 4.5.5 *Let $T \in L\mathscr{R}(H, K)$ and let $A \in \mathscr{L}(H, K)$ be non null operator. Then,*

$$\partial\sigma_A(T) \subset \overline{W(T, A)}. \quad \diamond$$

Corollary 4.5.6 *Let $T \in L\mathscr{R}(H, K)$ and $A \in \mathscr{C}(H, K)$. If A is one-to-one and $\gamma(A) > 0$, then*

$$\sigma_A(T) \subset \overline{W(T, A)}. \quad \diamond$$

References

1. Ammar, A., Daoud, H., Jeribi, A.: Pseudospectra and essential pseudospectra of multivalued linear relations. Mediterr. J. Math. **12**(4), 1377–1395 (2015)
2. Ammar, A., Daoud, H., Jeribi, A.: The stability of pseudospectra and essential pseudospectra of linear relations. J. Pseudo-Differ. Oper. Appl. **7**(4), 473–491 (2016)
3. Ammar, A., Bouchekoua, A., Jeribi, A.: Some results of S-pseudospectra of bounded operators in a Hilbert space. In: (English) Zbl 07361674 Jeribi, A. (ed.) Operator Theory: Proceedings

of the International Conference on Operator Theory, Hammamet, Tunisia, April 30–May 3, 2018. De Gruyter, Berlin. De Gruyter Proceedings in Mathematics, pp. 121–129 (2021)

4. Ammar, A., Bouchekoua, A., Jeribi, A.: A characterization of S-pseudospectra of linear operators in a Hilbert space. Filomat **37**(5), 1331–1339 (2023)
5. Ammar, A., Bouchekoua, A., Jeribi, A.: New results of the pseudospectra of linear operators in a Banach space. Funct. Anal. Approx. Comput. **15**(1), 13–26 (2023)
6. Ammar, A., Daoud, H., Jeribi, A.: S-pseudospectra and S-essential pseudospectra. Mat. Vesnik **72**(2), 95–105 (2020)
7. Ammar, A., Jeribi, A., Lazrag, N.: Sequence of linear operators converging in the generalized sense and its essential pseudospectra. In: Jeribi, A. (ed.) (English) Zbl 07361676, Operator Theory: Proceedings of the International Conference on Operator Theory, Hammamet, Tunisia, April 30–May 3, 2018. De Gruyter, Berlin. De Gruyter Proceedings in Mathematics, Berlin, pp. 147–155 (2021)
8. Ammar, A., Jeribi, A., Mahfoudhi, K.: Some description of essential structured approximate and defect pseudospectrum. Korean J. Math. **28**(4), 673–697 (2020)
9. Ammar, A., Jeribi, A., Mahfoudhi, K.: The condition pseudospectrum of a operator pencil. Asian-Eur. J. Math. **14**(4), 2150057, 12 (2021)
10. Ammar, A., Jeribi, A., Mahfoudhi, K.: On the left (right) condition pseudospectrum of linear operators. Filomat **36**(10), 3207–3215 (2022)
11. Ammar, A., Jeribi, A., Mahfoudhi, K.: A new spectral approach in the matrix algebra: trace pseudospectrum. Bol. Soc. Parana. Mat. (3), 41 (2023)
12. Ammar, A., Jeribi, A., Zayed, M.: Some properties of the condition pseudospectrum of multivalued linear operators. Filomat **37**(5), 1625–1634 (2023)
13. Jeribi, A., Mahfoudhi, K.: Generalized Drazin-meromorphic pseudospectrum for a bounded linear operator on a Banach space. Rend. Circ. Mat. Palermo **72**(2), 1303–1314 (2023)
14. Chaker, W., Ghribi, A., Jeribi, A., Krichen, B.: Some fixed point theorems for orbitally-(p,q)-quasi-contraction mappings. In: Applied mathematics in Tunisia, Springer Proceedings in Mathematics and Statistics, vol. 131, pp. 153–160. Springer, Cham (2015)
15. Chaker, W., Jeribi, A., Krichen, B.: Demicompact linear operators, essential spectrum and some perturbation results. Math. Nachr. **288**(13), 1476–1486 (2015)
16. Chaker, W., Ghribi, A., Jeribi, A., Krichen, B.: Fixed point theorem for quasi-contractive mapping in partial cone metric space. Afr. Mat. **26**(5–6), 1153–1159 (2015)
17. Chaker, W., Jeribi, A.: Common fixed point of single-valued and multivalued mappings. Afr. Mat. **25**(3), 617–622 (2014)
18. Ammar, A., Bouchekoua, A., Jeribi, A.: The ε-pseudospectra and the essential ε-pseudospectra of linear relations. J. Pseudo-Differ. Oper. Appl. **11**(2), 879–915 (2020)
19. Chaitin-Chetelin, F., Harrabi, A.: About definitions of pseudospectra of closed operators in banach spaces, Technical Report TR/PA/98/08. CERFACS, Toulouse, France (1998)
20. Ammar, A., Dhahri, M.Z., Jeribi, A.: Some properties of upper triangular 3×3-block matrices of linear relations. Boll. Unione Mat. Ital. **8**(3), 189–204 (2015)
21. Cross, R.: Multivalued linear operators. In: Monographs and Textbooks in Pure and Applied Mathematics, vol. 213. Marcel Dekker, Inc., New York, x+335. ISBN: 0-8247-0219-0 (1998)
22. Arens, R.: Operational calculus of linear relations. Pacific J. Math. **11**, 9–23 (1961)
23. Álvarez, T., Ammar, A., Jeribi, A.: On the essential spectra of some matrix of linear relations. Math. Methods Appl. Sci. **37**(5), 620–644 (2014)

Chapter 5
Frobenius–Schur Factorization and Essential Spectra of Block Multivalued Operator Matrices

5.1 Frobenius–Schur Factorization for 2 × 2 Matrices Multivalued Linear Operator

5.1.1 Closability and Closure of $\mathscr{L}_0$

In this section, we are concerned with the Frobenius–Schur factorization of 2×2-block matrices multivalued linear operator $\mathscr{L}_0$, acting on the Banach space $X \times Y$, of the form:

$$\mathscr{L}_0 = \begin{pmatrix} A & B \\ C & D \end{pmatrix}, \tag{5.1.1}$$

where A, B, C and D are four closable linear relations with dense domains. We will first collect some hypotheses:

$(H2)$ The linear relation A is continuous, $\mathscr{D}(A)$ is closed, and $\dim(A(0)) < \infty$.
$(H3)$ The linear relation C is closable which satisfies $\mathscr{D}(A) \subset \mathscr{D}(C)$.
$(H4)$ The linear relation B is closable and $\overline{\mathscr{D}(B)} = Y$.
$(H5)$ The linear relation D is bounded and $D(0) \subset C(A - \mu I)^{-1}B(0) \subset C(0)$.

Remark 5.1.1 (i) From the assumption $(H2)$, we infer that $A - \mu I$ is closed.
(ii) Using Theorem 2.11.1 and the assumption $(H3)$, we deduce that the linear relation $F(\mu) = C(A - \mu I)^{-1}$ is bounded.
(iii) From the assumption $(H4)$, for $\mu \in \rho(A)$, the linear relation $(A - \mu I)^{-1}B$ is closable. It follows from Theorem 2.11.1 that the linear relation $E(\mu) = \overline{(A - \mu I)^{-1}B}$ is bounded.
(iv) From the assumption $(H5)$, for $\mu \in \rho(A)$, the linear relation $M(\mu) = D - C(A - \mu I)^{-1}B$ is closable. ◇

A. Ammar and A. Jeribi, *Spectral Theory of Block Multivalued Operator Matrices*, Infosys Science Foundation Series, https://doi.org/10.1007/978-981-95-0397-1_5

Theorem 5.1.1 *Let $\mu \in \rho(A)$. Suppose that the conditions $(H2)$–$(H5)$ are satisfied and $F(\mu)$ and $E(\mu)$ are single-valued, then*
(i) $\mathscr{L}_0$, given in (5.1.1), is closable and its closure $\mathscr{L}$ is given by

$$\mathscr{L} = \overline{\mathscr{L}_0} = \mu I + \begin{pmatrix} I & 0 \\ F(\mu) & I \end{pmatrix} \begin{pmatrix} A - \mu I & 0 \\ 0 & \overline{M(\mu)} - \mu I \end{pmatrix} \begin{pmatrix} I & E(\mu) \\ 0 & I \end{pmatrix}.$$

(ii) Let $\mu \in \rho(A)$, then for $\lambda \in \mathbb{C}$, we have

$$\lambda I - \mathscr{L} = UV(\lambda)W - (\lambda - \mu)\mathscr{I}(\mu)$$

where $U = \begin{pmatrix} I & 0 \\ F(\mu) & I \end{pmatrix}$, $W = \begin{pmatrix} I & E(\mu) \\ 0 & I \end{pmatrix}$, $V(\lambda) = \begin{pmatrix} \lambda I - A & 0 \\ 0 & \lambda I - \overline{M(\mu)} \end{pmatrix}$
and $\mathscr{I}(\mu) = \begin{pmatrix} 0 & E(\mu) \\ F(\mu) & F(\mu)E(\mu) \end{pmatrix}$. ◇

Proof (i) We start by determining the properties of the matrices

$$U = \begin{pmatrix} I & 0 \\ F(\mu) & I \end{pmatrix}$$

and

$$W = \begin{pmatrix} I & E(\mu) \\ 0 & I \end{pmatrix}.$$

(a) W is single-valued bounded, W is bijective and

$$W^{-1} = \begin{pmatrix} I & -E(\mu) \\ 0 & I \end{pmatrix}.$$

(b) U is single-valued bounded with $\mathscr{D}(H) = \mathscr{D}(E(\mu)) \times Y = X \times Y$, U is bijective and

$$U^{-1} = \begin{pmatrix} I & 0 \\ -F(\mu) & I \end{pmatrix}.$$

(c) Since $A - \mu I$ is continuous and $\mathscr{D}(A) = \mathscr{D}(A - \mu I)$ is closed. By Theorem 2.11.1 (iv), we obtain that $A - \mu I$ is closed. Since $M(\mu)$ is closable, then by Proposition 2.12.2, we have $M(\mu) - \mu I$ is closable. So,

$$V(\mu) = \begin{pmatrix} A - \mu I & 0 \\ 0 & M(\mu) - I \end{pmatrix}$$

is closable. The use of (a), (b), (c) and Theorem 2.12.3 allowed us to conclude that HKJ is closable and $\overline{UV(\mu)W} = U\overline{V(\mu)}W$. Then, applying Proposition 2.12.2, we have

$$\begin{aligned}\overline{\mathscr{L}_0 - \mu I} &= \mathscr{L} - \mu I \\ &= U\overline{V(\mu)}W.\end{aligned}$$

Hence,

$$\overline{\mathscr{L}_0} = \mu I + U\overline{V(\mu)}W.$$

It follows that the relation $\mathscr{L}_0$ is closable.

(ii) Since I is single-valued, then

$$\lambda I - \mathscr{L} = \mu I - \mathscr{L} + (\lambda - \mu)I.$$

So,

$$\begin{aligned}\lambda I - \mathscr{L} &= \begin{pmatrix} I & 0 \\ F(\mu) & I \end{pmatrix}\begin{pmatrix} \mu I - A & 0 \\ 0 & \mu I - \overline{M(\mu)} \end{pmatrix}\begin{pmatrix} I & E(\mu) \\ 0 & I \end{pmatrix} + (\lambda - \mu)I \\ &= \begin{pmatrix} I & 0 \\ F(\mu) & I \end{pmatrix}\begin{pmatrix} \mu I - A - \lambda I + \lambda I & 0 \\ 0 & \mu I - \overline{M(\mu)} - \lambda I + \lambda I \end{pmatrix}\begin{pmatrix} I & E(\mu) \\ 0 & I \end{pmatrix} + (\lambda - \mu)I \\ &= \begin{pmatrix} I & 0 \\ F(\mu) & I \end{pmatrix}\left[\begin{pmatrix} \lambda I - A & 0 \\ 0 & \lambda I - \overline{M(\mu)} \end{pmatrix} + \begin{pmatrix} (\mu - \lambda)I & 0 \\ 0 & (\mu - \lambda)I \end{pmatrix}\right]\begin{pmatrix} I & E(\mu) \\ 0 & I \end{pmatrix} + (\lambda - \mu)I.\end{aligned}$$

Since $\mathscr{D}(U) = X \times Y$ and W is single-valued, then we obtain

$$\begin{aligned}\lambda I - \mathscr{L} = &\begin{pmatrix} I & 0 \\ F(\mu) & I \end{pmatrix}\begin{pmatrix} \lambda I - A & 0 \\ 0 & \lambda I - \overline{M(\mu)} \end{pmatrix}\begin{pmatrix} I & E(\mu) \\ 0 & I \end{pmatrix} + \\ &\begin{pmatrix} I & 0 \\ F(\mu) & I \end{pmatrix}\begin{pmatrix} (\mu - \lambda)I & 0 \\ 0 & (\mu - \lambda)I \end{pmatrix}\begin{pmatrix} I & E(\mu) \\ 0 & I \end{pmatrix} + (\lambda - \mu)I.\end{aligned}$$

The use of Lemma 2.8.4 (iii) leads to

$$\begin{aligned}\lambda I - \mathscr{L} &= \begin{pmatrix} I & 0 \\ F(\mu) & I \end{pmatrix}\begin{pmatrix} \lambda I - A & 0 \\ 0 & \lambda I - \overline{M(\mu)} \end{pmatrix}\begin{pmatrix} I & E(\mu) \\ 0 & I \end{pmatrix} + \\ &\quad (\mu - \lambda)\begin{pmatrix} I & 0 \\ F(\mu) & I \end{pmatrix}\begin{pmatrix} I & E(\mu) \\ 0 & I \end{pmatrix} + (\lambda - \mu)I \\ &= \begin{pmatrix} I & 0 \\ F(\mu) & I \end{pmatrix}\begin{pmatrix} \lambda I - A & 0 \\ 0 & \lambda I - \overline{M(\mu)} \end{pmatrix}\begin{pmatrix} I & E(\mu) \\ 0 & I \end{pmatrix} + \\ &\quad (\mu - \lambda)\begin{pmatrix} I & E(\mu) \\ F(\mu) & F(\mu)E(\mu) + I \end{pmatrix} + (\lambda - \mu)I.\end{aligned}$$

By using Lemma 2.8.4 (i), we get

$$\lambda I - \mathscr{L} = \begin{pmatrix} I & 0 \\ F(\mu) & I \end{pmatrix}\begin{pmatrix} \lambda I - A & 0 \\ 0 & \lambda I - \overline{M(\mu)} \end{pmatrix}\begin{pmatrix} I & E(\mu) \\ 0 & I \end{pmatrix} - (\lambda - \mu)\begin{pmatrix} 0 & E(\mu) \\ F(\mu) & F(\mu)E(\mu) \end{pmatrix}.$$

Consequently,

$$\lambda I - \mathscr{L} = UV(\lambda)W - (\lambda - \mu)\mathscr{I}(\mu).$$

This achieves the proof. Q.E.D.

Concerning the multivalued linear operators $E(\mu)$, $F(\mu)$, $M(\mu)$, and A, we impose the following: conditions

$(H6)$ $Z(\mu)$ is a continuous selection of $F(\mu)$.
$(H7)$ $W(\mu)$ is a continuous selection of $E(\mu)$.
$(H8)$ For some $\mu \in \rho(A)$, $\mathscr{D}(V(\mu))$ contains the ranges of both

$$\begin{pmatrix} I & W(\mu) \\ 0 & I \end{pmatrix}$$

and

$$\begin{pmatrix} 0 & E(\mu) - E(\mu) \\ 0 & 0 \end{pmatrix}.$$

Theorem 5.1.2 *Let $\mu \in \rho(A)$. If the conditions $(H2)$–$(H8)$ are satisfied and $E(\mu)$ is single-valued, then*

(i)

$$\mu I - \mathscr{L}_0 = H_1 \begin{pmatrix} \mu I - A & 0 \\ 0 & \mu I - M(\mu) \end{pmatrix} W + \mathscr{H}(\mu),$$

where $\mathscr{L}_0$ is given in (5.1.1), H_1 is the lower-triangular multivalued operator matrix, given by

$$H_1 = \begin{pmatrix} I & 0 \\ Z(\mu) & I \end{pmatrix}$$

and

$$\mathscr{H}(\mu) = \begin{pmatrix} 0 & 0 \\ (F(\mu) - F(\mu))(\mu I - A) & (F(\mu) - F(\mu))B \end{pmatrix}.$$

(ii) If $\mathscr{H}(\mu)$ is continuous, then $\mathscr{L}_0$ is closable and its closure $\mathscr{L}$ is given by

$$\mathscr{L} = \overline{\mathscr{L}_0} = \mu I + H_1 V(\mu) J_1 + \mathscr{H}(\mu),$$

where

$$\mathscr{I}_1(\mu) = \begin{pmatrix} 0 & E(\mu) \\ Z(\mu) & Z(\mu)E(\mu) \end{pmatrix}.$$

(iii) Let $\mu \in \rho(A)$, then for $\lambda \in \mathbb{C}$, we have

$$\lambda I - \mathscr{L} = H_1 V(\lambda) J - (\lambda - \mu)\mathscr{I}_1(\mu) + \mathscr{H}(\mu).$$

◇

Proof (i) Since $F(\mu) = Z(\mu) + F(\mu) - F(\mu)$, then

$$\mu I - \mathscr{L}_0 = \begin{pmatrix} I & 0 \\ Z(\mu) + F(\mu) - F(\mu) & I \end{pmatrix} \begin{pmatrix} \mu I - A & 0 \\ 0 & \mu I - M(\mu) \end{pmatrix} \begin{pmatrix} I & E(\mu) \\ 0 & I \end{pmatrix}.$$

Using Lemma 2.8.4 (i), we get

$$\mu I - \mathscr{L}_0 = \left[\begin{pmatrix} I & 0 \\ Z(\mu) & I \end{pmatrix} + \begin{pmatrix} 0 & 0 \\ F(\mu) - F(\mu) & 0 \end{pmatrix} \right] \begin{pmatrix} \mu I - A & 0 \\ 0 & \mu I - M(\mu) \end{pmatrix} \begin{pmatrix} I & E(\mu) \\ 0 & I \end{pmatrix}.$$

Since $\begin{pmatrix} A(0) \\ C(0) \end{pmatrix} \subset N \begin{pmatrix} 0 & 0 \\ F(\mu) - F(\mu) & 0 \end{pmatrix} = X \times Y$, then

$$\begin{aligned} \mu I - \mathscr{L}_0 = & \begin{pmatrix} I & 0 \\ Z(\mu) & I \end{pmatrix} \begin{pmatrix} \mu I - A & 0 \\ 0 & \mu I - M(\mu) \end{pmatrix} \begin{pmatrix} I & E(\mu) \\ 0 & I \end{pmatrix} + \\ & \begin{pmatrix} 0 & 0 \\ F(\mu) - F(\mu) & 0 \end{pmatrix} \begin{pmatrix} \mu I - A & 0 \\ 0 & \mu I - M(\mu) \end{pmatrix} \begin{pmatrix} I & E(\mu) \\ 0 & I \end{pmatrix}. \end{aligned}$$

Let

$$T = \begin{pmatrix} I & 0 \\ Z(\mu) & I \end{pmatrix} \begin{pmatrix} \mu I - A & 0 \\ 0 & \mu I - M(\mu) \end{pmatrix} \begin{pmatrix} I & E(\mu) \\ 0 & I \end{pmatrix}$$

and let $\begin{pmatrix} x_1 \\ x_2 \end{pmatrix} \in X \times Y$. Then,

$$\begin{aligned} (\mu I - \mathscr{L}_0) \begin{pmatrix} x_1 \\ x_2 \end{pmatrix} = & T \begin{pmatrix} x_1 \\ x_2 \end{pmatrix} + \begin{pmatrix} 0 & 0 \\ F(\mu) - F(\mu) & 0 \end{pmatrix} \begin{pmatrix} \mu I - A & 0 \\ 0 & \mu I - M(\mu) \end{pmatrix} \begin{pmatrix} I & E(\mu) \\ 0 & I \end{pmatrix} \begin{pmatrix} x_1 \\ x_2 \end{pmatrix} \\ = & T \begin{pmatrix} x_1 \\ x_2 \end{pmatrix} + \begin{pmatrix} 0 & 0 \\ F(\mu) - F(\mu) & 0 \end{pmatrix} \begin{pmatrix} \mu I - A & 0 \\ 0 & \mu I - M(\mu) \end{pmatrix} \begin{pmatrix} x_1 + E(\mu) x_2 \\ x_2 \end{pmatrix} \\ = & T \begin{pmatrix} x_1 \\ x_2 \end{pmatrix} + \begin{pmatrix} 0 & 0 \\ F(\mu) - F(\mu) & 0 \end{pmatrix} \begin{pmatrix} (\mu I - A) x_1 + B x_2 \\ (\mu I - M(\mu)) x_2 \end{pmatrix} \\ = & T \begin{pmatrix} x_1 \\ x_2 \end{pmatrix} + \begin{pmatrix} 0 \\ (F(\mu) - F(\mu))(\mu I - A) x_1 + (F(\mu) - F(\mu)) B x_2 \end{pmatrix}. \end{aligned}$$

Therefore,

$$(\mu I - \mathscr{L}_0) \begin{pmatrix} x_1 \\ x_2 \end{pmatrix} = H_1 \begin{pmatrix} \mu I - A & 0 \\ 0 & \mu I - M(\mu) \end{pmatrix} W \begin{pmatrix} x_1 \\ x_2 \end{pmatrix} + \mathscr{H}(\mu) \begin{pmatrix} x_1 \\ x_2 \end{pmatrix}. \tag{5.1.2}$$

First, we have

$$\mathscr{D}(\mathscr{H}(\mu)) = \left\{ \begin{pmatrix} x_1 \\ x_2 \end{pmatrix} \in X \times Y \text{ such that } (\mu I - A) x_1 \in \mathscr{D}(F(\mu)) \text{ and } B x_2 \in \mathscr{D}(F(\mu)) \right\}.$$

Since $\mathscr{D}(F(\mu)) = X$, then $\mathscr{D}(\mathscr{H}(\mu)) = \mathscr{D}(A) \times \mathscr{D}(B)$. In the second place, we have $\mathscr{D}(Z(\mu)) = \mathscr{D}(F(\mu))$, whereas $\mathscr{D}(HKJ) = \mathscr{D}(H_1 KJ)$. Thus,

$$
\begin{aligned}
&\mathscr{D}\left(H_1\begin{pmatrix}\mu I-A & 0\\ 0 & \mu I-M(\mu)\end{pmatrix}W+\mathscr{H}(\mu)\right)\\
&=\mathscr{D}\left(H_1\begin{pmatrix}\mu I-A & 0\\ 0 & \mu I-M(\mu)\end{pmatrix}W\right)\bigcap\mathscr{D}(\mathscr{H}(\mu))\\
&=(\mathscr{D}(A)\times\mathscr{D}(B))\bigcap(\mathscr{D}(A)\times\mathscr{D}(B))\\
&=\mathscr{D}(A)\times\mathscr{D}(B).
\end{aligned}
$$

Therefore,

$$
\begin{aligned}
\lambda\mathscr{D}(\mu I-\mathscr{L}_0)&=\mathscr{D}(A)\times\mathscr{D}(B)\\
&=\mathscr{D}\left(H_1\begin{pmatrix}\mu I-A & 0\\ 0 & \mu I-M(\mu)\end{pmatrix}W+\mathscr{H}(\mu)\right).
\end{aligned}\tag{5.1.3}
$$

By Eqs. (5.1.2) and (5.1.3), we have

$$
\mu I-\mathscr{L}_0=H_1\begin{pmatrix}\mu I-A & 0\\ 0 & \mu I-M(\mu)\end{pmatrix}W+\mathscr{H}(\mu).
$$

(ii) We start by determining the properties of the matrices H_1 and W. It is easy to notice that W and H_1 are single-valued bounded bijective, where

$$
W^{-1}=\begin{pmatrix}I & -E(\mu)\\ 0 & I\end{pmatrix}
$$

and

$$
H_1^{-1}=\begin{pmatrix}I & 0\\ -Z(\mu) & I\end{pmatrix}.
$$

We have $\mu I-A$ is closed. On the other hand, we have $M(\mu)$ is closable. Then, by using Proposition 2.12.2, we have $\mu I-M(\mu)$ is closable. So,

$$
\begin{pmatrix}\mu I-A & 0\\ 0 & \mu I-M(\mu)\end{pmatrix}
$$

is closable and

$$
V(\mu)=\begin{pmatrix}\mu I-A & 0\\ 0 & \mu I-\overline{M(\mu)}\end{pmatrix}
$$

denotes its closure. Theorem 2.12.3 allowed us to conclude that

$$
H_1\begin{pmatrix}\mu I-A & 0\\ 0 & \mu I-M(\mu)\end{pmatrix}W
$$

is closable and

$$\overline{H_1\begin{pmatrix}\mu I - A & 0\\ 0 & \mu I - M(\mu)\end{pmatrix}W} = H_1 V(\mu) W.$$

Further,

$$\mathscr{D}\left(H_1\begin{pmatrix}\mu I - A & 0\\ 0 & \mu I - M(\mu)\end{pmatrix}W\right) \subset \mathscr{D}(\mathscr{H}(\mu))$$

and

$$\mathscr{H}(\mu)(0) \subset H_1\begin{pmatrix}\mu I - A & 0\\ 0 & \mu I - M(\mu)\end{pmatrix}W(0).$$

Then, we get

$$H_1\begin{pmatrix}\mu I - A & 0\\ 0 & \mu I - M(\mu)\end{pmatrix}W + \mathscr{H}(\mu)$$

is closable and

$$\begin{aligned}\overline{H_1\begin{pmatrix}\mu I - A & 0\\ 0 & \mu I - M(\mu)\end{pmatrix}W + \mathscr{H}(\mu)} &= \overline{H_1\begin{pmatrix}\mu I - A & 0\\ 0 & \mu I - M(\mu)\end{pmatrix}W} + \mathscr{H}(\mu)\\ &= H_1 V(\mu) W + \mathscr{H}(\mu).\end{aligned}$$

By applying Proposition 2.12.2, we have

$$\begin{aligned}\overline{\mathscr{L}_0 - \mu} &= \overline{\mathscr{L}_0} - \mu I\\ &= H_1 V(\mu) W + \mathscr{H}(\mu).\end{aligned}$$

So,

$$\overline{\mathscr{L}_0} = \mu I + H_1 V(\mu) W + \mathscr{H}(\mu).$$

It follows that the relation $\mathscr{L}_0$ is closable.
(iii) Since I is single-valued, then

$$\begin{aligned}\lambda I - \mathscr{L} &= \mu I - \mathscr{L} + (\lambda - \mu) I\\ &= H_1 V(u) W + \mathscr{H}(\mu) + (\lambda - \mu) I.\end{aligned}$$

Thus,

$$\begin{aligned}\lambda I - \mathscr{L} &= \begin{pmatrix} I & 0\\ Z(\mu) & I\end{pmatrix}\begin{pmatrix}\mu I - A & 0\\ 0 & \mu I - \overline{M(\mu)}\end{pmatrix}\begin{pmatrix} I & E(\mu)\\ 0 & I\end{pmatrix} + \mathscr{H}(\mu) + (\lambda - \mu) I,\\ &= H_1\begin{pmatrix}\mu I - A - \lambda I + \lambda I & 0\\ 0 & \mu I - \overline{M(\mu)} - \lambda I + \lambda I\end{pmatrix} J + \mathscr{H}(\mu) + (\lambda - \mu) I,\\ &= H_1\left[\begin{pmatrix}\lambda I - A & 0\\ 0 & \lambda I - \overline{M(\mu)}\end{pmatrix} + \begin{pmatrix}(\mu - \lambda) I & 0\\ 0 & (\mu - \lambda) I\end{pmatrix}\right] J + \mathscr{H}(\mu) + (\lambda - \mu) I.\end{aligned}$$

Since $\mathscr{D}(H_1) = X \times Y$ and W is single-valued, then

$$\lambda I - \mathscr{L} = H_1 \begin{pmatrix} \lambda I - A & 0 \\ 0 & \lambda I - \overline{M(\mu)} \end{pmatrix} J + H_1 \begin{pmatrix} (\mu-\lambda)I & 0 \\ 0 & (\mu-\lambda)I \end{pmatrix} J$$
$$+ \mathscr{H}(\mu) + (\lambda - \mu)I.$$

Show that

$$H_1 \begin{pmatrix} (\mu-\lambda)I & 0 \\ 0 & (\mu-\lambda)I \end{pmatrix} J + (\lambda-\mu)I = \mathscr{I}_1(\mu).$$

Indeed, let $\begin{pmatrix} x_1 \\ x_2 \end{pmatrix} \in X \times Y$. Then,

$$\begin{aligned} H_1 \begin{pmatrix} (\mu-\lambda)I & 0 \\ 0 & (\mu-\lambda)I \end{pmatrix} J \begin{pmatrix} x_1 \\ x_2 \end{pmatrix} &= (\mu-\lambda) H_1 \begin{pmatrix} x_1 + E(\mu)x_2 \\ x_2 \end{pmatrix} \\ &= (\mu-\lambda) \begin{pmatrix} x_1 + E(\mu)x_2 \\ Z(\mu)x_1 + Z(\mu)E(\mu) + x_2 \end{pmatrix} \\ &= (\mu-\lambda) \begin{pmatrix} I & E(\mu) \\ Z(\mu) & Z(\mu)E(\mu) + I \end{pmatrix} \begin{pmatrix} x_1 \\ x_1 \end{pmatrix}. \end{aligned}$$

Since

$$\mathscr{D}\left[\begin{pmatrix} (\mu-\lambda)I & (\mu-\lambda)E(\mu) \\ Z(\mu)(\mu-\lambda) & Z(\mu)(\mu-\lambda)E(\mu) + (\mu-\lambda)I \end{pmatrix}\right]$$

$$\begin{aligned} &= \mathscr{D}(Z(\mu)) \bigcap \mathscr{D}(E(\mu)) \\ &= \mathscr{D}\left\{\begin{pmatrix} x_1 \\ x_2 \end{pmatrix} \in X \times Y \text{ such that } x_1 + E(\mu)x_2 \in X \text{ and } x_2 \in Y\right\} \\ &= \mathscr{D}\left((\mu-\lambda)H_1 W\right), \end{aligned}$$

then

$$(\mu-\lambda)H_1 W + (\lambda-\mu)I = (\mu-\lambda) \begin{pmatrix} 0 & E(\mu) \\ Z(\mu) & Z(\mu)E(\mu) \end{pmatrix}.$$

By using Lemma 2.7.2, we have

$$(\mu-\lambda)H_1 W + (\lambda-\mu)I = (\mu-\lambda)\mathscr{I}_1(\mu).$$

Also, we get

$$\lambda I - \mathscr{L} = H_1 V(\lambda) W - (\lambda-\mu)\mathscr{I}_1(\mu) + \mathscr{H}(\mu).$$

This achieves the proof. Q.E.D.

Theorem 5.1.3 *Let $\mu \in \rho(A)$. If $(H2)$–$(H8)$ are satisfied, then*

(i)

$$\mu I - \mathscr{L}_0 = H_1 \begin{pmatrix} \mu I - A & 0 \\ 0 & \mu I - M(\mu) \end{pmatrix} J_1 + \mathscr{J}(\mu),$$

where J_1 is the upper-triangular operator matrix, given by

$$J_1 = \begin{pmatrix} I & W(\mu) \\ 0 & I \end{pmatrix}$$

and

$$\mathscr{J}(\mu) = \begin{pmatrix} A - A & B - B \\ (F(\mu) - F(\mu))(\mu I - A) + Z(\mu)(A - A) & (F(\mu) - F(\mu))B + Z(\mu)(B - B) \end{pmatrix}.$$

(ii) *If $\mathscr{J}(\mu)$ is continuous, then $\mathscr{L}_0$ is closable and its closure $\mathscr{L}$ is given by the relation:*

$$\mathscr{L} = \overline{\mathscr{L}_0} = \mu I + H_1 V(\mu) J_1 + \mathscr{J}(\mu).$$

(iii) *Let $\mu \in \rho(A)$, then for $\lambda \in \mathbb{C}$, we have*

$$\lambda I - \mathscr{L} = H_1 V(\lambda) J_1 - (\lambda - \mu)\mathscr{I}_2(\mu) + \mathscr{J}(\mu),$$

where

$$\mathscr{I}_2(\mu) = \begin{pmatrix} 0 & W(\mu) \\ Z(\mu) & Z(\mu)W(\mu) \end{pmatrix}.$$

◇

Proof (i) If we take $W(\mu)$ be a selection of $E(\mu)$, hence

$$E(\mu) = W(\mu) + E(\mu) - E(\mu).$$

Therefore,

$$\begin{aligned} \mu I - \mathscr{L}_0 &= H_1 \begin{pmatrix} \mu I - A & 0 \\ 0 & \mu I - M(\mu) \end{pmatrix} \begin{pmatrix} I & W(\mu) + E(\mu) - E(\mu) \\ 0 & I \end{pmatrix} + \mathscr{H}(\mu), \\ &= H_1 \begin{pmatrix} \mu I - A & 0 \\ 0 & \mu I - M(\mu) \end{pmatrix} \left[\begin{pmatrix} I & W(\mu) \\ 0 & I \end{pmatrix} + \begin{pmatrix} 0 & E(\mu) - E(\mu) \\ 0 & 0 \end{pmatrix} \right] + \mathscr{H}(\mu). \end{aligned}$$

Since $\mathscr{D}\left(\begin{pmatrix} \mu I - A & 0 \\ 0 & \mu I - M(\mu) \end{pmatrix}\right)$ contains the ranges of both $\begin{pmatrix} I & W(\mu) \\ 0 & I \end{pmatrix}$ and $\begin{pmatrix} 0 & E(\mu) - E(\mu) \\ 0 & 0 \end{pmatrix}$, then

$$\mu I - \mathscr{L}_0 = H_1 \begin{pmatrix} \mu I - A & 0 \\ 0 & \mu I - M(\mu) \end{pmatrix} \begin{pmatrix} I & W(\mu) \\ 0 & I \end{pmatrix} + \\ H_1 \begin{pmatrix} \mu I - A & 0 \\ 0 & \mu I - M(\mu) \end{pmatrix} \begin{pmatrix} 0 & E(\mu) - E(\mu) \\ 0 & 0 \end{pmatrix} + \mathscr{H}(\mu).$$

Let $\begin{pmatrix} x_1 \\ x_2 \end{pmatrix} \in X \times Y$. Then,

$$\begin{aligned} (\mu I - \mathscr{L}_0)\begin{pmatrix} x_1 \\ x_2 \end{pmatrix} &= \left(H_1 \begin{pmatrix} \mu I - A & 0 \\ 0 & \mu I - M(\mu) \end{pmatrix} J_1 + \mathscr{H}(\mu)\right)\begin{pmatrix} x_1 \\ x_2 \end{pmatrix} + \\ &\quad H_1 \begin{pmatrix} \mu I - A & 0 \\ 0 & \mu I - M(\mu) \end{pmatrix} \begin{pmatrix} 0 & E(\mu) - E(\mu) \\ 0 & 0 \end{pmatrix}\begin{pmatrix} x_1 \\ x_2 \end{pmatrix}, \\ &= \left(H_1 \begin{pmatrix} \mu I - A & 0 \\ 0 & \mu I - M(\mu) \end{pmatrix} J_1 + \mathscr{H}(\mu)\right)\begin{pmatrix} x_1 \\ x_2 \end{pmatrix} + \\ &\quad H_1 \begin{pmatrix} \mu I - A & 0 \\ 0 & \mu I - M(\mu) \end{pmatrix} \begin{pmatrix} (E(\mu) - E(\mu))x_2 \\ 0 \end{pmatrix}, \\ &= \left(H_1 \begin{pmatrix} \mu I - A & 0 \\ 0 & \mu I - M(\mu) \end{pmatrix} J_1 + \mathscr{H}(\mu)\right)\begin{pmatrix} x_1 \\ x_2 \end{pmatrix} + \\ &\quad H_1 \begin{pmatrix} (\mu I - A)(E(\mu) - E(\mu))x_1 \\ 0 \end{pmatrix}. \end{aligned}$$

Since

$$\begin{aligned} (\mu I - A)(E(\mu) - E(\mu))x_2 &= (\mu I - A)E(\mu)(0) \\ &= \overline{B}(0) + (A - \lambda I)(0) \\ &= B(0) + (A - \lambda I)(0) \\ &= B(0) + A(0) \\ &= (A - A)x_1 + (B - B)x_2, \end{aligned}$$

then we get

$$\begin{aligned} (\mu I - \mathscr{L}_0)\begin{pmatrix} x_1 \\ x_2 \end{pmatrix} &= \left(H_1 \begin{pmatrix} \mu I - A & 0 \\ 0 & \mu I - M(\mu) \end{pmatrix} J_1 + \mathscr{H}(\mu)\right)\begin{pmatrix} x_1 \\ x_2 \end{pmatrix} + \\ &\quad \begin{pmatrix} I & 0 \\ Z(\mu) & I \end{pmatrix} \begin{pmatrix} (A - A)x_1 + (B - B)x_2 \\ 0 \end{pmatrix} \\ &= \left[H_1 \begin{pmatrix} \mu I - A & 0 \\ 0 & \mu I - M(\mu) \end{pmatrix} J_1 + \right. \\ &\quad \left. \mathscr{H}(\mu) + \begin{pmatrix} A - A & B - B \\ Z(\mu)(A - A) & Z(\mu)(B - B) \end{pmatrix}\right]\begin{pmatrix} x_1 \\ x_2 \end{pmatrix}. \end{aligned}$$

Therefore,

$$(\mu I - \mathscr{L}_0)\begin{pmatrix} x_1 \\ x_2 \end{pmatrix} = (H_1 \begin{pmatrix} \mu I - A & 0 \\ 0 & \mu I - M(\mu) \end{pmatrix} J_1 + \mathscr{J}(\mu))\begin{pmatrix} x_1 \\ x_2 \end{pmatrix}. \quad (5.1.4)$$

Since

$$\begin{aligned}\mathscr{D}(\mathscr{J}(\mu)) &= \mathscr{D}\left[\mathscr{H}(\mu) + \begin{pmatrix} A - A & B - B \\ Z(\mu)(A - A) & Z(\mu)(B - B) \end{pmatrix}\right] \\ &= \mathscr{D}(\mathscr{H}(\mu)) \bigcap \mathscr{D}\left[\begin{pmatrix} A - A & B - B \\ Z(\mu)(A - A) & Z(\mu)(B - B) \end{pmatrix}\right] \\ &= \mathscr{D}(A) \times \mathscr{D}(B),\end{aligned}$$

then in this way

$$\begin{aligned}&\mathscr{D}\left(H_1 \begin{pmatrix} \mu I - A & 0 \\ 0 & \mu I - M(\mu) \end{pmatrix} J_1 + \mathscr{J}(\mu)\right) \\ &= \mathscr{D}\left(H_1 \begin{pmatrix} \mu I - A & 0 \\ 0 & \mu I - M(\mu) \end{pmatrix} J_1\right) \bigcap \mathscr{D}(\mathscr{J}(\mu)) \\ &= \mathscr{D}(A) \times \mathscr{D}(B). \quad (5.1.5)\end{aligned}$$

By using Eqs. (5.1.4) and (5.1.5), we have

$$\mu I - \mathscr{L}_0 = H_1 \begin{pmatrix} \mu I - A & 0 \\ 0 & \mu I - M(\mu) \end{pmatrix} J_1 + \mathscr{J}(\mu).$$

(ii) We start by determining the properties of the matrices

$$H_1 = \begin{pmatrix} I & 0 \\ Z(\mu) & I \end{pmatrix}$$

and

$$J_1 = \begin{pmatrix} I & W(\mu) \\ 0 & I \end{pmatrix}.$$

Indeed, J_1 is single-valued bounded, J_1 is bijective, and

$$J_1^{-1} = \begin{pmatrix} I & -W(\mu) \\ 0 & I \end{pmatrix}.$$

Further, H_1 is single-valued bounded with $\mathscr{D}(H) = \mathscr{D}(E(\mu)) \times Y = X \times Y$, H is bijective, and

$$H_1^{-1} = \begin{pmatrix} I & 0 \\ -Z(\mu) & I \end{pmatrix}.$$

On the one hand, $A - \mu I$ is closed. On the other hand, $M(\mu)$ is closable. By using Proposition 2.12.2, we have $M(\mu) - \mu I$ is closable. So,

$$\begin{pmatrix} \mu I - A & 0 \\ 0 & \mu I - M(\mu) \end{pmatrix}$$

is closable. Thus, from Theorem 2.12.3, we conclude that

$$H_1 \begin{pmatrix} \mu I - A & 0 \\ 0 & \mu I - M(\mu) \end{pmatrix} J_1$$

is closable and

$$\overline{H_1 \begin{pmatrix} \mu I - A & 0 \\ 0 & \mu I - M(\mu) \end{pmatrix} J_1} = H_1 V(\mu) J_1.$$

Also, we have

$$\mathscr{D}\left(H_1 \begin{pmatrix} \mu I - A & 0 \\ 0 & \mu I - M(\mu) \end{pmatrix} J_1 \right) \subset \mathscr{D}(\mathscr{J}(\mu))$$

and

$$\mathscr{J}(\mu)(0) \subset H_1 \begin{pmatrix} \mu I - A & 0 \\ 0 & \mu I - M(\mu) \end{pmatrix} J_1(0).$$

So, by using Proposition 2.12.2, we deduce that

$$H_1 \begin{pmatrix} \mu I - A & 0 \\ 0 & \mu I - M(\mu) \end{pmatrix} J_1 + \mathscr{J}(\mu)$$

is closable and

$$\begin{aligned} \overline{H_1 \begin{pmatrix} \mu I - A & 0 \\ 0 & \mu I - M(\mu) \end{pmatrix} J_1 + \mathscr{J}(\mu)} &= \overline{H_1 \begin{pmatrix} \mu I - A & 0 \\ 0 & \mu I - M(\mu) \end{pmatrix} J_1} + \mathscr{J}(\mu) \\ &= H_1 V(\mu) J_1 + \mathscr{J}(\mu). \end{aligned}$$

By applying Proposition 2.12.2, we have

$$\begin{aligned} \overline{\mathscr{L}_0 - \mu I} &= \overline{\mathscr{L}_0} - \mu I \\ &= H_1 V(\mu) J_1 + \mathscr{J}(\mu). \end{aligned}$$

So,

$$\overline{\mathscr{L}_0} = \mu I + H_1 V(\mu) J_1 + \mathscr{J}(\mu).$$

It follows that the relation $\mathscr{L}_0$ is closable.
(iii) Let $\lambda \in \mathbb{C}$. Then,

$$\begin{aligned}
\lambda I - \mathscr{L} &= \begin{pmatrix} I & 0 \\ Z(\mu) & I \end{pmatrix} \begin{pmatrix} \mu I - A & 0 \\ 0 & \mu I - \overline{M(\mu)} \end{pmatrix} \begin{pmatrix} I & W(\mu) \\ 0 & I \end{pmatrix} + \mathscr{J}(\mu) + (\lambda - \mu) I \\
&= H_1 \begin{pmatrix} \mu I - A - \lambda I + \lambda I & 0 \\ 0 & \mu I - \overline{M(\mu)} - \lambda I + \lambda I \end{pmatrix} J_1 + \mathscr{J}(\mu) + (\lambda - \mu) I \\
&= H_1 \left[\begin{pmatrix} \lambda I - A & 0 \\ 0 & \lambda I - \overline{M(\mu)} \end{pmatrix} + \begin{pmatrix} (\mu - \lambda) I & 0 \\ 0 & (\mu - \lambda) I \end{pmatrix} \right] J_1 + \mathscr{J}(\mu) + (\lambda - \mu) I.
\end{aligned}$$

Since $\mathscr{D}(H_1) = X \times Y$ and J_1 is single-valued, then we obtain

$$\lambda I - \mathscr{L} = H_1 V(\lambda) J_1 + H_1 \begin{pmatrix} (\mu - \lambda) I & 0 \\ 0 & (\mu - \lambda) I \end{pmatrix} J_1 + \mathscr{J}(\mu) + (\lambda - \mu) I.$$

The use of Lemma 2.8.4 (iii) leads to

$$\begin{aligned}
\lambda I - \mathscr{L} &= H_1 V(\lambda) J_1 + (\mu - \lambda) \begin{pmatrix} I & 0 \\ Z(\mu) & I \end{pmatrix} \begin{pmatrix} I & W(\mu) \\ 0 & I \end{pmatrix} + (\lambda - \mu) I + \mathscr{J}(\mu) \\
&= H_1 V(\lambda) J_1 + (\mu - \lambda) \begin{pmatrix} I & W(\mu) \\ Z(\mu) & Z(\mu) W(\mu) + I \end{pmatrix} + (\lambda - \mu) I + \mathscr{J}(\mu).
\end{aligned}$$

By using Lemma 2.8.4 (i), we get

$$\lambda I - \mathscr{L} = H_1 V(\lambda) J_1 + (\mu - \lambda) \begin{pmatrix} 0 & W(\mu) \\ Z(\mu) & Z(\mu) W(\mu) \end{pmatrix} + \mathscr{J}(\mu).$$

Consequently,

$$\lambda I - \mathscr{L} = H_1 V(\lambda) J_1 + (\mu - \lambda) \mathscr{I}(\mu) + \mathscr{J}(\mu),$$

where

$$\mathscr{I}(\mu) = \begin{pmatrix} 0 & W(\mu) \\ Z(\mu) & Z(\mu) W(\mu) \end{pmatrix}.$$

This achieves the proof. Q.E.D.

We assume the following hypothesis.
$(H9)$ A_1 and $M_1(\mu)$ are selections of A and $M(\mu)$, respectively, such that A_1 is closed and $M_1(\mu)$ is closable.

Theorem 5.1.4 *Let $\mu \in \rho(A)$. We suppose that the conditions $(H2)$–$(H9)$ are satisfied, then*

(i)

$$\mu I - \mathscr{L}_0 = H_1 K_1 J_1 + \mathscr{K}(\mu),$$

where K_1 is the diagonal operator matrix, given by

$$K_1 = \begin{pmatrix} \mu I - A_1 & 0 \\ 0 & \mu I - M_1(\mu) \end{pmatrix}$$

and

$$\mathscr{K}(\mu) = \begin{pmatrix} A - A & (A - A)W(\mu) \\ Z(\mu)(A - A) & Z(\mu)(A - A)W(\mu) + M(\mu) - M(\mu) \end{pmatrix}.$$

(ii) If $\mathscr{K}(\mu)$ is continuous, then $\mathscr{L}_0$ is closable and its closure $\mathscr{L}$ is identified with the relation:

$$\mathscr{L} = \overline{\mathscr{L}_0} = \mu I + H_1 V_1(\mu) J_1 + \mathscr{J}(\mu),$$

where $V_1(\mu)$ is the diagonal operator matrix, given by

$$V_1(\mu) = \begin{pmatrix} \mu I - A_1 & 0 \\ 0 & \mu I - \overline{M_1(\mu)} \end{pmatrix}.$$

(iii) Let $\mu \in \rho(A)$, then for $\lambda \in \mathbb{C}$, we have

$$\lambda I - \mathscr{L} = H_1 V_1(\lambda) J_1 - (\lambda - \mu)\mathscr{I}(\mu) + \mathscr{J}(\mu),$$

where

$$\mathscr{I}(\mu) = \begin{pmatrix} 0 & W(\mu) \\ Z(\mu) & Z(\mu)W(\mu) \end{pmatrix}.$$

◇

Proof Note first that

$$\begin{aligned} \mu I - \mathscr{L}_0 &= H_1 \begin{pmatrix} \mu I - A_1 + A - A & 0 \\ 0 & \mu I - M_1(\mu) + M(\mu) - M(\mu) \end{pmatrix} J_1 + \mathscr{J}(\mu) \\ &= H_1 \left[\begin{pmatrix} \mu I - A_1 & 0 \\ 0 & \mu I - M_1(\mu) \end{pmatrix} + \begin{pmatrix} A - A & 0 \\ 0 & M(\mu) - M(\mu) \end{pmatrix} \right] J_1 + \mathscr{J}(\mu). \end{aligned}$$

Since H_1 is bounded and J_1 is single-valued, then

$$\mu I - \mathscr{L}_0 = H_1 \begin{pmatrix} \mu I - A_1 & 0 \\ 0 & \mu I - M_1(\mu) \end{pmatrix} J_1 + H_1 \begin{pmatrix} A - A & 0 \\ 0 & M(\mu) - M(\mu) \end{pmatrix} J_1 + \mathscr{J}(\mu).$$

Let $\begin{pmatrix} x_1 \\ x_2 \end{pmatrix} \in \mathscr{D}(A) \times \mathscr{D}(B)$. Then,

$$\begin{aligned}
(\mu I - \mathscr{L}_0)\begin{pmatrix} x_1 \\ x_2 \end{pmatrix} &= H_1 \begin{pmatrix} \mu I - A_1 & 0 \\ 0 & \mu I - M_1(\mu) \end{pmatrix} J_1 \begin{pmatrix} x_1 \\ x_2 \end{pmatrix} + \\
&\quad H_1 \begin{pmatrix} A - A & 0 \\ 0 & M(\mu) - M(\mu) \end{pmatrix} J_1 \begin{pmatrix} x_1 \\ x_2 \end{pmatrix} + \mathscr{J}(\mu)\begin{pmatrix} x_1 \\ x_2 \end{pmatrix} \\
&= \left(H_1 \begin{pmatrix} \mu I - A_1 & 0 \\ 0 & \mu I - M_1(\mu) \end{pmatrix} J_1 + \mathscr{J}(\mu) \right) \begin{pmatrix} x_1 \\ x_2 \end{pmatrix} + \\
&\quad H_1 \begin{pmatrix} A - A & 0 \\ 0 & M(\mu) - M(\mu) \end{pmatrix} \begin{pmatrix} x_1 + W(\mu)x_2 \\ x_2 \end{pmatrix} \\
&= \left(H_1 \begin{pmatrix} \mu I - A_1 & 0 \\ 0 & \mu I - M_1(\mu) \end{pmatrix} J_1 + \mathscr{J}(\mu) \right) \begin{pmatrix} x_1 \\ x_2 \end{pmatrix} + \\
&\quad H_1 \begin{pmatrix} (A - A)(x_1 + W(\mu)x_2) \\ (M(\mu) - M(\mu))x_2 \end{pmatrix} \\
&= \left(H_1 \begin{pmatrix} \mu I - A_1 & 0 \\ 0 & \mu I - M_1(\mu) \end{pmatrix} J_1 + \mathscr{J}(\mu) \right) \begin{pmatrix} x_1 \\ x_2 \end{pmatrix} + \\
&\quad \begin{pmatrix} (A - A)(x_1 + W(\mu)x_2) \\ Z(\mu)(A - A)(x_1 + W(\mu)x_2) + (M(\mu) - M(\mu))x_2 \end{pmatrix}.
\end{aligned}$$

Since $B(0) \subset M(\mu)$, then

$$\begin{aligned}
(B - B)x_2 + (M(\mu) - M(\mu))x_2 &= B(0) + M(\mu)(0) \\
&= M(\mu)(0) \\
&= (M(\mu) - M(\mu))x_2.
\end{aligned}$$

We already know that

$$\begin{aligned}
(A - A)x_1 + (A - A)x_1 &= A(0) \\
&= (A - A)x_1.
\end{aligned}$$

By using Lemma 2.7.2, we have

$$\mathscr{J}(\mu)\begin{pmatrix} x_1 \\ x_2 \end{pmatrix} + \begin{pmatrix} (A - A)(x_1 + W(\mu)x_2) \\ Z(\mu)(A - A)(x_1 + W(\mu)x_2) + (M(\mu) - M(\mu))x_2 \end{pmatrix} = \mathscr{K}(\mu)\begin{pmatrix} x_1 \\ x_2 \end{pmatrix}.$$

Therefore,

$$(\mu I - \mathscr{L}_0)\begin{pmatrix} x_1 \\ x_2 \end{pmatrix} = \left(H_1 \begin{pmatrix} \mu I - A_1 & 0 \\ 0 & \mu I - M_1(\mu) \end{pmatrix} J_1 + \mathscr{K}(\mu) \right) \begin{pmatrix} x_1 \\ x_2 \end{pmatrix}.$$

Since

$$\begin{aligned}&\mathscr{D}(\mathscr{K}(\mu))\\&=\mathscr{D}\left(\mathscr{J}(\mu)\right)\bigcap\left(\mathscr{D}(A)\bigcap\mathscr{D}(Z(\mu)(A-A))\times\mathscr{D}(Z(\mu)(A-A)W(\mu))\bigcap\mathscr{D}(M(\mu))\right)\\&=\mathscr{D}(A)\times\mathscr{D}(B),\end{aligned}$$

then

$$\begin{aligned}\mathscr{D}(\mu I-\mathscr{L}_0)&=\mathscr{D}\left(H_1\begin{pmatrix}\mu I-A_1&0\\0&\mu I-M_1(\mu)\end{pmatrix}J_1+\mathscr{K}(\mu)\right)\\&=\mathscr{D}\left(H_1\begin{pmatrix}\mu I-A_1&0\\0&\mu I-M_1(\mu)\end{pmatrix}J_1\right)\bigcap\mathscr{D}(\mathscr{K}(\mu))\\&=\mathscr{D}(A)\times\mathscr{D}(B).\end{aligned}$$

The assertions (ii) and (iii) can be proved similar to (i). Q.E.D.

Theorem 5.1.5 *Let $\mathscr{L}_0$ be the matrix operator defined in* (5.1.1) *which satisfies* $(H2)$-$(H8)$ *and let $\mathscr{L}$ be its closure. Suppose that there is $\mu\neq 0$ such that $\frac{1}{\mu}\in\rho(A)$. If $B(0)\subset A(0)\subset N(F(\mu))$, then*

$$Q_{\mathscr{L}}(I-\mu\mathscr{L})=Q_{\mathscr{L}}\left(H_1V\left(\frac{1}{\mu}\right)J_1\right),$$

where H_1 and J_1 are the lower- and upper-triangular linear relations matrices defined by

$$H_1=\begin{pmatrix}I&0\\Z\left(\frac{1}{\mu}\right)&I\end{pmatrix}$$

,

$$J_1=\begin{pmatrix}I&W\left(\frac{1}{\mu}\right)\\0&I\end{pmatrix},$$

and $V(\frac{1}{\mu})$ is the diagonal multivalued linear relation matrix, given by

$$V\left(\frac{1}{\mu}\right)=\begin{pmatrix}I-\mu A&0\\0&\overline{I-\mu M\left(\frac{1}{\mu}\right)}\end{pmatrix}.$$ ◇

Proof By using Theorem 5.1.3, we have

$$I-\mu\mathscr{L}=H_1V\left(\frac{1}{\mu}\right)J_1+\mathscr{J}\left(\frac{1}{\mu}\right),$$

where

$$\mathscr{J}\left(\frac{1}{\mu}\right)=\begin{pmatrix} A-A & B-B \\ \mathscr{F}(\mu)\left(I-\mu A\right)+Z\left(\frac{1}{\mu}\right)(A-A) & \mathscr{F}(\mu)B+Z\left(\frac{1}{\mu}\right)(B-B) \end{pmatrix},$$

and

$$\mathscr{F}(\mu)=F\left(\frac{1}{\mu}\right)-F\left(\frac{1}{\mu}\right).$$

Then,

$$\begin{aligned} Q_{\mathscr{L}}\left(I-\mu\mathscr{L}\right) &= Q_{\mathscr{L}}\left(H_1 V\left(\frac{1}{\mu}\right)J_1+\mathscr{J}\left(\frac{1}{\mu}\right)\right) \\ &= Q_{\mathscr{L}}\left(H_1 V\left(\frac{1}{\mu}\right)J_1\right)+Q_{\mathscr{L}}\left(\mathscr{J}\left(\frac{1}{\mu}\right)\right). \end{aligned}$$

Since

$$\mathscr{L}\begin{pmatrix}0\\0\end{pmatrix}=\begin{pmatrix}A(0)\\C(0)\end{pmatrix},$$

we can write the following relation:

$$Q_{\mathscr{L}}\left(\mathscr{J}\left(\frac{1}{\mu}\right)\right)=\begin{pmatrix} Q_A(A-A) & Q_A(B-B) \\ Q_C\left(\mathscr{F}(\mu)(I-\mu A)+Z\left(\frac{1}{\mu}\right)(A-A)\right) & Q_C\left(\mathscr{F}(\mu)B+Z\left(\frac{1}{\mu}\right)(B-B)\right) \end{pmatrix}.$$

Let $x \in \mathscr{D}(A)$, then

$$Q_A(B-B)x = Q_A B(0) \subset Q_A A(0) = \{0\}. \tag{5.1.6}$$

Moreover, by making some simple calculations, we have

$$\begin{aligned} & Q_C\left(\mathscr{F}(\mu)\left(I-\mu A\right)+Z\left(\frac{1}{\mu}\right)(A-A)\right)x \\ & = Q_C\left(\mathscr{F}(\mu)\left(I-\mu A\right)\right)+Q_C\left(Z\left(\frac{1}{\mu}\right)(A-A)\right)x. \end{aligned}$$

Since $A(0) \subset N\left(F\left(\frac{1}{\mu}\right)\right)$, then

$$Q_C\left(\mathscr{F}(\mu)\left(I-\mu A\right)\right)x = Q_C\left(F\left(\frac{1}{\mu}\right)(I-\mu A)-F\left(\frac{1}{\mu}\right)(I-\mu A)\right)x.$$

Let $\frac{1}{\mu} \in \rho(A)$, then

$$F\left(\frac{1}{\mu}\right)(I-\mu A)=C.$$

We obtain that

$$\begin{aligned} Q_C(\mathscr{F}(\mu)(I-\mu A))x &= Q_C\left((F\left(\frac{1}{\mu}\right)(I-\mu A)\right)(0) \\ &= Q_C C(0). \end{aligned}$$

Thus,

$$Q_C(\mathscr{F}(\mu)(I-\mu A))x = 0.$$

Since $B(0) \subset N\left(F\left(\frac{1}{\mu}\right)\right)$, then

$$Q_C(\mathscr{F}(\mu)B)x = Q_C\left(F\left(\frac{1}{\mu}\right)B\right)(0).$$

Using the fact that $F\left(\frac{1}{\mu}\right)B(0) \subset C(0)$, we conclude that

$$Q_C(\mathscr{F}(\mu)B)x = 0.$$

Now, notice that

$$Q_C(\mathscr{F}(\mu)(B-B))x = Q_C(\mathscr{F}(\mu)B(0)).$$

Since $B(0) \subset N\left(F\left(\frac{1}{\mu}\right)\right)$, then

$$Q_C(\mathscr{F}(\mu)B(0)) = Q_C\left(F\left(\frac{1}{\mu}\right)B(0) - F\left(\frac{1}{\mu}\right)B(0)\right).$$

Therefore,

$$Q_C(\mathscr{F}(\mu)(B-B))x = 0. \tag{5.1.7}$$

By using Eq. (5.1.7), we have

$$\begin{aligned} Q_C\left(Z\left(\frac{1}{\mu}\right)(B-B)\right)x &= Q_C\left(Z\left(\frac{1}{\mu}\right)(B-B)\right)x + Q_C(\mathscr{F}(\mu)(B-B))x \\ &= Q_C\left(Z\left(\frac{1}{\mu}\right) + \mathscr{F}(\mu)\right)(B-B)x. \end{aligned}$$

Since $Z\left(\frac{1}{\mu}\right)$ is a selection of $F\left(\frac{1}{\mu}\right)$, then

$$\begin{aligned} Q_C\left(Z\left(\frac{1}{\mu}\right)(B-B)\right)x &= Q_C\left(F\left(\frac{1}{\mu}\right)(B-B)\right)x \\ &= Q_C\left(F\left(\frac{1}{\mu}\right)B(0)\right). \end{aligned}$$

We deduce that

$$Q_C\left(Z\left(\frac{1}{\mu}\right)(B-B)\right)x = 0. \tag{5.1.8}$$

This together with the relation (5.1.8) leads to

$$Q_C\left(Z\left(\frac{1}{\mu}\right)(A-A)\right)x = Q_C\left(Z\left(\frac{1}{\mu}\right)A(0)\right) \subset Q_C\left(Z\left(\frac{1}{\mu}\right)B(0)\right) = 0. \tag{5.1.9}$$

By applying Eqs. (5.1.6), (5.1.8) and (5.1.9), we get

$$Q_{\mathscr{L}}\left(\mathscr{I}\left(\frac{1}{\mu}\right)\right) = \begin{pmatrix} 0 & 0 \\ 0 & 0 \end{pmatrix}. \tag{5.1.10}$$

This achieves the proof. Q.E.D.

5.1.2 *Demicompact of the Block Multivalued Matrix*

Theorem 5.1.6 *Let $\mathscr{L}_0$ be the matrix operator defined in* (5.1.1) *satisfies* $(H2)$–$(H8)$ *and let $\mathscr{L}$ be its closure. Suppose that there is $\mu \neq 0$ such that $\frac{1}{\mu} \in \rho(A)$ and*

$$B(0) \subset A(0) \subset N\left(F\left(\frac{1}{\mu}\right)\right)$$

with $Q_A\left((I-\mu A)W\left(\frac{1}{\mu}\right)\right)$ is bounded. If the linear relations μA and $\mu\overline{M(\frac{1}{\mu})}$ are demicompact, then the linear relation $\mu\mathscr{L}$ is demicompact. ◇

Proof Giving $\begin{pmatrix} x_n \\ y_n \end{pmatrix}_n \in \mathscr{D}(\mathscr{L})$ be a bounded sequence such that

$$Q_{\mathscr{L}}\,(I-\mu\mathscr{L})\begin{pmatrix} x_n \\ y_n \end{pmatrix} \to \begin{pmatrix} x_0 \\ y_0 \end{pmatrix}.$$

By using Theorem 5.1.1, we have

$$Q_{\mathscr{L}}(I-\mu\mathscr{L})\begin{pmatrix}x_n\\ y_n\end{pmatrix}=Q_{\mathscr{L}}\left(H_1V\left(\frac{1}{\mu}\right)J_1\right)\begin{pmatrix}x_n\\ y_n\end{pmatrix}\to\begin{pmatrix}x_0\\ y_0\end{pmatrix}.$$

Since H_1 is bounded with

$$V\left(\frac{1}{\mu}\right)J_1\begin{pmatrix}0\\ 0\end{pmatrix}=V\left(\frac{1}{\mu}\right)\begin{pmatrix}0\\ 0\end{pmatrix},$$

then

$$Q_{\mathscr{L}}\left(H_1V\left(\frac{1}{\mu}\right)J_1\right)=Q_{\mathscr{L}}H_1Q^{-1}_{V\left(\frac{1}{\mu}\right)}Q_{V\left(\frac{1}{\mu}\right)}V\left(\frac{1}{\mu}\right)J_1. \tag{5.1.11}$$

We claim that $Q_{\mathscr{L}}H_1Q^{-1}_{V\left(\frac{1}{\mu}\right)}$ is operator. Indeed, since

$$Q_{\mathscr{L}}H_1Q^{-1}_{V\left(\frac{1}{\mu}\right)}\begin{pmatrix}0\\ 0\end{pmatrix}=Q_{\mathscr{L}}H_1V\left(\frac{1}{\mu}\right)\begin{pmatrix}0\\ 0\end{pmatrix},$$

then

$$H_1V\left(\frac{1}{\mu}\right)\begin{pmatrix}0\\ 0\end{pmatrix}=\begin{pmatrix}A(0)\\ Z\left(\frac{1}{\mu}\right)A(0)+\overline{M\left(\frac{1}{\mu}\right)}(0)\end{pmatrix}.$$

Hence,

$$\begin{aligned}Z\left(\frac{1}{\mu}\right)A(0)+\overline{M\left(\frac{1}{\mu}\right)}(0)&=Z\left(\frac{1}{\mu}\right)A(0)+C(0)\\ &=Z\left(\frac{1}{\mu}\right)A(0)+\mathscr{F}(\mu)A(0)\\ &=\left(Z\left(\frac{1}{\mu}\right)+\mathscr{F}(\mu)\right)A(0)\\ &=\mathscr{F}(\mu)A(0)\\ &=C(0).\end{aligned}$$

Therefore,

$$\begin{aligned}H_1V\left(\frac{1}{\mu}\right)\begin{pmatrix}0\\ 0\end{pmatrix}&=\begin{pmatrix}A(0)\\ C(0)\end{pmatrix}\\ &=\mathscr{L}\begin{pmatrix}0\\ 0\end{pmatrix}.\end{aligned}$$

It is obvious that

$$Q_{\mathscr{L}}H_1Q^{-1}_{V\left(\frac{1}{\mu}\right)}\begin{pmatrix}0\\ 0\end{pmatrix}=\begin{pmatrix}0\\ 0\end{pmatrix}.$$

Since $Q_{\mathscr{L}} H_1 Q^{-1}_{V\left(\frac{1}{\mu}\right)}$ is bounded operator inverse, it follows that

$$\left(Q_{\mathscr{L}} H_1 Q^{-1}_{V\left(\frac{1}{\mu}\right)}\right)^{-1} Q_{\mathscr{L}} (I - \mu\mathscr{L}) \begin{pmatrix} x_n \\ y_n \end{pmatrix} = Q_{V\left(\frac{1}{\mu}\right)} V\left(\frac{1}{\mu}\right) J_1 \begin{pmatrix} x_n \\ y_n \end{pmatrix}.$$

Moreover, by making some simple calculations, we may show that

$$V\left(\frac{1}{\mu}\right)\begin{pmatrix} 0 \\ 0 \end{pmatrix} = \begin{pmatrix} A(0) \\ C(0) \end{pmatrix}.$$

Hence,

$$\begin{aligned} V\left(\frac{1}{\mu}\right) J_1 &= V\left(\frac{1}{\mu}\right) \\ &= \begin{pmatrix} I - \mu A & 0 \\ 0 & \overline{I - \mu M\left(\frac{1}{\mu}\right)} \end{pmatrix} \begin{pmatrix} I & W\left(\frac{1}{\mu}\right) \\ 0 & I \end{pmatrix}. \end{aligned}$$

By using Lemma 2.7.2, we get

$$V\left(\frac{1}{\mu}\right) J_1 = \begin{pmatrix} I - \mu A & (I - \mu A) W\left(\frac{1}{\mu}\right) \\ 0 & \overline{I - \mu M\left(\frac{1}{\mu}\right)} \end{pmatrix}.$$

Then, by Lemma 2.9.2, we infer that

$$Q_{V\left(\frac{1}{\mu}\right)} V\left(\frac{1}{\mu}\right) J_1 = \begin{pmatrix} Q_A(I - \mu A) & Q_A\left((I - \mu A) W\left(\frac{1}{\mu}\right)\right) \\ 0 & Q_C\left(\overline{I - \mu M\left(\frac{1}{\mu}\right)}\right) \end{pmatrix}.$$

Therefore,

$$\begin{aligned} Q_{V\left(\frac{1}{\mu}\right)} V\left(\frac{1}{\mu}\right) J_1 \begin{pmatrix} x_n \\ y_n \end{pmatrix} &= \begin{pmatrix} Q_A(I - \mu A) & Q_A\left((I - \mu A) W\left(\frac{1}{\mu}\right)\right) \\ 0 & Q_C\left(\overline{I - \mu M\left(\frac{1}{\mu}\right)}\right) \end{pmatrix} \begin{pmatrix} x_n \\ \\ y_n \end{pmatrix} \\ &= \begin{pmatrix} Q_A(I - \mu A) x_n + Q_A\left((I - \mu A) W\left(\frac{1}{\mu}\right)\right) y_n \\ Q_C\left(\overline{I - \mu M\left(\frac{1}{\mu}\right)}\right) y_n \end{pmatrix}. \end{aligned}$$

We conclude that

$$Q_C\left(I - \mu\overline{M\left(\frac{1}{\mu}\right)}\right) y_n$$

is convergent. This together with the demicompact of $\frac{1}{\mu}\overline{M\left(\frac{1}{\mu}\right)}$ shows that $(y_n)_n$ has a convergent subsequence. Since

$$Q_A\left((I - \mu A)W\left(\frac{1}{\mu}\right)\right)$$

is bounded operator and μA is demicompact, we infer that $(x_n)_n$ has a convergent subsequence, which proves the demicompact of $\mu\mathcal{L}$. Q.E.D.

Theorem 5.1.7 *Let $\mathcal{L}_0$ be the operator defined in* (5.1.1) *and let $\mathcal{L}$ be its closure. Suppose that for a certain $\mu \in \rho(A)$, there is $\lambda \in \mathbb{C}\backslash\{0\}$ such that $Q_C Z(\mu) \in \mathcal{K}(X)$ and*

$$B(0) \subset A(0) \subset N\left(F\left(\frac{1}{\mu}\right)\right)$$

with $Q_A\left((I - \mu A)W\left(\frac{1}{\mu}\right)\right)$ is bounded. If λA and $\lambda S(\mu)$ are demicompact, then $\lambda\mathcal{L}$ is demicompact. ◇

Proof Take the following bounded sequence $\begin{pmatrix} x_n \\ y_n \end{pmatrix}_n \in \mathcal{D}(\mathcal{L})$ such that

$$(I - \lambda\mathcal{L})\begin{pmatrix} x_n \\ y_n \end{pmatrix} \to \begin{pmatrix} x_0 \\ y_0 \end{pmatrix}.$$

Recalling the factorization given in Theorem 5.1.3, one has

$$\frac{1}{\lambda}I - \mathcal{L} = HV\left(\frac{1}{\lambda}\right)J - \left(\frac{1}{\lambda} - \mu\right)\mathcal{I}(\mu) + \mathcal{J}(\mu).$$

From Eq. (5.1.10), it follows that

$$Q_{\mathcal{L}}(I - \lambda\mathcal{L}) = Q_{\mathcal{L}}(HV(\lambda)J) - (1 - \lambda\mu)Q_{\mathcal{L}}\mathcal{I}(\mu).$$

Since $Q_{\mathscr{L}} H Q_{V(\lambda)}^{-1}$ is bounded invertible operator and by using Eq. (5.1.11), we get

$$\begin{aligned}\left(Q_{\mathscr{L}} H Q_{V(\lambda)}^{-1}\right)^{-1} Q_{\mathscr{L}} (I-\mu\mathscr{L}) &= Q_{V(\lambda)} V(\lambda) J - (1-\lambda\mu)\left(Q_{\mathscr{L}} H Q_{V(\lambda)}^{-1}\right)^{-1} Q_{\mathscr{L}} \mathscr{I}(\mu)\\ &= Q_{V(\lambda)} V(\lambda) J - (1-\lambda\mu) Q_{V(\lambda)} H^{-1} Q_{\mathscr{L}}^{-1} Q_{\mathscr{L}} \mathscr{I}(\mu)\\ &= Q_{V(\lambda)} V(\lambda) J - (1-\lambda\mu) Q_{V(\lambda)} H^{-1}\left[\mathscr{I}(\mu) + Q_{\mathscr{L}}^{-1}(0)\right]\\ &= Q_{V(\lambda)} V(\lambda) J - (1-\lambda\mu) Q_{V(\lambda)} H^{-1}\mathscr{I}(\mu) +\\ &\quad (1-\lambda\mu) Q_{V(\lambda)} H^{-1}\mathscr{L}(0).\end{aligned}$$

Since

$$\begin{aligned}H^{-1}\mathscr{L}(0) &= \begin{pmatrix} A(0) \\ Z(\mu)A(0) + C(0)\end{pmatrix}\\ &= \begin{pmatrix} A(0) \\ C(0)\end{pmatrix}\\ &= V(\lambda)(0),\end{aligned}$$

then

$$(1-\lambda\mu) Q_{V(\lambda)} H^{-1}\mathscr{L}(0) = 0.$$

Consequently,

$$\left(Q_{\mathscr{L}} H Q_{V(\lambda)}^{-1}\right)^{-1} Q_{\mathscr{L}} (I-\mu\mathscr{L}) = Q_{V(\lambda)}\left[V(\lambda) J - (1-\lambda\mu) H^{-1}\mathscr{I}(\mu)\right].$$

By using Lemma 2.7.2, we get

$$H^{-1}\mathscr{I}(\mu) = \begin{pmatrix} 0 & W(\mu) \\ Z(\mu) & 0 \end{pmatrix}$$

and

$$V(\lambda) J = \begin{pmatrix} I-\lambda A & (I-\lambda) A W(\mu) \\ 0 & I - \lambda\overline{M(\mu)}\end{pmatrix}.$$

It is easy to conclude that

$$\left(Q_{\mathscr{L}} H Q_{V(\lambda)}^{-1}\right)^{-1} Q_{\mathscr{L}} (I-\mu\mathscr{L}) = \begin{pmatrix} Q_A(I-\lambda A) & (Q_A(I-\lambda A) - (1-\lambda\mu)Q_A) W(\mu) \\ (1-\lambda\mu) Q_C Z(\mu) & Q_C(I-\lambda\overline{M(\mu)})\end{pmatrix}.$$

Moreover, by making some simple calculations, we may show that

$$\begin{pmatrix} Q_A(I-\lambda A)x_n + (Q_A(I-\lambda A) - (1-\lambda\mu)Q_A) W(\mu) y_n \\ (1-\lambda\mu) Q_C Z(\mu) x_n + Q_C(I-\lambda\overline{M(\mu)}) y_n\end{pmatrix} \to Q_{V(\lambda)} H^{-1} Q_{\mathscr{L}}^{-1}\begin{pmatrix} x_n \\ y_n\end{pmatrix}. \tag{5.1.12}$$

Since $Q_C Z(\mu) \in \mathscr{K}(X)$ and $(x_n)_n$ is bounded, then $(I - \lambda\mu)Q_C Z(\mu)x_n$ has a convergent subsequence. Hence, from the second equation of (5.1.12), we infer that $Q_C(I - \lambda\overline{M(\mu)})y_n$ has a convergent subsequence. Using the demicompactness of $\lambda\overline{M(\mu)}$, we deduce that there exists a convergent subsequence of $(y_n)_n$. Now, since $Q_A(I - \lambda A)W(\mu) - (1 - \lambda\mu)Q_A W(\mu)$ is bounded, we conclude from the first equation of (5.1.12) that $(I - \lambda A)x_n$ has a convergent subsequence. Since λA is demicompact then $(x_n)_n$ has a convergent subsequence. Therefore, there exists a subsequence of $\begin{pmatrix} x_n \\ y_n \end{pmatrix}_n$ which converges in $\mathscr{D}(\mathscr{L})$. Thus, $\lambda - \mathscr{L}$ is demicompact. Q.E.D.

5.1.3 Essential Spectra of $\mathscr{L}$

Theorem 5.1.8 *Let $\mu \in \rho(A)$. We suppose that the conditions $(H2)$–$(H5)$ are satisfied. If $F(\mu)$, $E(\mu)$ are single-valued and the linear relation $\mathscr{I}(\mu) \in \mathscr{PR}(\Phi(X \times Y))$, then*

$$\sigma_{e5}(\mathscr{L}) \subset \sigma_{e5}(A) \bigcup \sigma_{e5}(\overline{M(\mu)}).$$

Furthermore, if the sets Φ_A and $\Phi_{\overline{M(\mu)}}$ are connected, then

$$\sigma_{e5}(\mathscr{L}) = \sigma_{e5}(A) \bigcup \sigma_{e5}(\overline{M(\mu)}).$$

Moreover, if the set Φ_A is connected and the sets $\rho(\mathscr{L})$ and $\rho(M(\mu))$ are not empty sets, then

$$\sigma_{eap}(\mathscr{L}) = \sigma_{eap}(A) \bigcup \sigma_{eap}(\overline{M(\mu)}).$$

(ii) If the linear relation $\mathscr{I}(\mu) \in \mathscr{PR}(\Phi_-(X \times Y))$, then

$$\sigma_{e\delta}(\mathscr{L}) \subset \sigma_{e\delta}(A) \bigcup \sigma_{e\delta}(\overline{M(\mu)}).$$

Additionally, if the set Φ_A is connected and the sets $\rho(\mathscr{L})$ and $\rho(M(\mu))$ are not empty, then

$$\sigma_{e\delta}(\mathscr{L}) = \sigma_{e\delta}(A) \bigcup \sigma_{e\delta}(\overline{M(\mu)}).$$ ◇

Proof Let $\mu \in \rho(A)$ and $\lambda \in \mathbb{C}$. Then, we find that

$$\lambda I - \mathscr{L} = \mu I - \mathscr{L} + (\lambda - \mu)I.$$

Now, by applying Theorem 5.1.1 (iii), we have

$$\lambda I - \mathscr{L} = UV(\lambda)W - (\lambda - \mu)\mathscr{I}(\mu). \qquad (5.1.13)$$

As though $\mathscr{I}(\mu) \in \mathscr{PR}(\Phi(X \times Y))$, then $\lambda I - \mathscr{L}$ is a Fredholm relation if, and only if, $UV(\lambda)W$ is a Fredholm relation. Now, observe that U and W are Fredholm relations. We deduce from Theorem 3.6.4 that $UV(\lambda)W$ is a Fredholm relation if, and only if, $V(\lambda)$ is a Fredholm relation if, and only if, $\lambda I - A$ (resp. $\lambda I - \overline{M(\mu)}$) is a Fredholm relation on X (resp. on Y). Hence,

$$\{\lambda \in \mathbb{C} : \lambda \in \Phi_{+\mathscr{L}}\} = \{\lambda \in \mathbb{C} : \lambda \in \Phi_{+A}\} \bigcup \left\{\lambda \in \mathbb{C} : \lambda \in \Phi_{+\overline{M(\mu)}}\right\}. \quad (5.1.14)$$

From Eq. (5.1.13), we get

$$i(\lambda - \mathscr{L}) = i(UV(\lambda)W - (\lambda - \mu)\mathscr{I}(\mu)).$$

By using Lemma 2.7.2, we have

$$\begin{aligned} i(\lambda I - \mathscr{L}) &= i(UV(\lambda)W), \\ &= i(U) + i(V(\lambda)W) - \dim(U(0) \bigcap (V(\lambda)W)^{-1}(0)), \\ &= i(U) + i(V(\lambda)) + i(W) - \dim(V(\lambda)(0) \bigcap W^{-1}(0)), \\ &= i(U) + i(V(\lambda)) + i(W). \end{aligned}$$

Since U and W are bijective, then

$$\begin{aligned} i(\lambda I - \mathscr{L}) &= i(V(\lambda)) \\ &= i(\lambda I - A) + i(\lambda I - \overline{M(\mu)}). \end{aligned} \quad (5.1.15)$$

It follows immediately from Eqs. (5.1.14) and (5.1.15) that

$$\sigma_{e5}(\mathscr{L}) \subset \sigma_{e5}(A) \bigcup \sigma_{e5}(\overline{M(\mu)}).$$

Suppose now that Φ_A is connected. It follows from $(H2)$, $\rho(A)$ is nonempty. Now, if we take $\mu_0 \in \rho(A)$, then $\mu_0 I - A \in \Phi(A)$ and $i(\mu_0 I - A) = 0$ is constant on any component of Φ_A. This implies that $i(\lambda I - A) = 0$ for all $\lambda \in \Phi_A$. Likewise $i(\lambda I - \overline{M(\mu)}) = 0$ for all $\lambda \in \Phi_{\overline{M(\mu)}}$. It follows that

$$\sigma_{e5}(\mathscr{L}) = \sigma_{e5}(A) \bigcup \sigma_{e5}(\overline{M(\mu)}).$$

This achieves the proof. Q.E.D.

Theorem 5.1.9 *Let $\mu \in \rho(A)$. Then,*

(i) If the conditions $(H2)$–$(H6)$ are satisfied and $E(\mu)$ is single-valued and if the linear relations $\mathscr{I}_1(\mu)$, $\mathscr{H}(\mu) \in \mathscr{PR}(\Phi(X \times Y))$, then

$$\sigma_{e5}(\mathscr{L}) \subset \sigma_{e5}(A) \bigcup \sigma_{e5}(\overline{M(\mu)}).$$

As well, if we suppose that the sets Φ_A *and* $\Phi_{\overline{M(\mu)}}$ *are connected, then*

$$\sigma_{e5}(\mathscr{L}) = \sigma_{e5}(A) \bigcup \sigma_{e5}(\overline{M(\mu)}).$$

(ii) If the conditions $(H2)$–$(H8)$ *are satisfied and the linear relations* $\mathscr{I}_2(\mu)$, $\mathscr{J}(\mu) \in \mathscr{PR}(\Phi(X \times Y))$, *then*

$$\sigma_{e5}(\mathscr{L}) \subset \sigma_{e5}(A) \bigcup \sigma_{e5}(\overline{M(\mu)}).$$

Additionally, if we suppose that the sets Φ_A *and* $\Phi_{\overline{M(\mu)}}$ *are connected, then*

$$\sigma_{e5}(\mathscr{L}) = \sigma_{e5}(A) \bigcup \sigma_{e5}(\overline{M(\mu)}).$$

(iii) We suppose that the conditions $(H2)$–$(H9)$ *are satisfied. If the linear relation* $\mathscr{I}_3(\mu)$, $\mathscr{K}(\mu) \in \mathscr{PR}(\Phi(X \times Y))$, *then*

$$\sigma_{e5}(\mathscr{L}) \subset \sigma_{e5}(A_1) \bigcup \sigma_{e5}(\overline{M_1(\mu)}).$$

In addition, if we suppose that the sets Φ_{A_1} *and* $\Phi_{\overline{M(\mu)}}$ *are connected, then*

$$\sigma_{e5}(\mathscr{L}) = \sigma_{e5}(A) \bigcup \sigma_{e5}(\overline{M(\mu)}). \qquad \diamond$$

Proof The proof of this theorem is a straightforward adoption of the proof of Theorem 5.1.8. Q.E.D.

5.2 Case Where the Domains Consisting of Vectors Satisfying Certain Relations Between Their Components

5.2.1 *Frobenius–Schur's Decomposition*

Let $A, C \in L\mathscr{R}(X, Y)$ and $B, D \in L\mathscr{R}(Y, X)$. We notice that the linear relation Γ_X goes from X into Y and the linear relation Γ_Y goes from Y into Z. In the product of the Banach spaces $X \times Y$, we define the linear relation $\mathscr{L}_0$ as follows:

$$\mathscr{L}_0 = \begin{pmatrix} A & B \\ C & D \end{pmatrix}, \tag{5.2.1}$$

and

$$\mathscr{D}(\mathscr{L}_0)=\left\{\begin{pmatrix} x\\ y\end{pmatrix} \text{ such that } x\in\mathscr{D}(A)\bigcap\mathscr{D}(C),\ y\in\mathscr{D}(B)\bigcap\mathscr{D}(D) \text{ and } \Gamma_X x=\Gamma_Y y\right\}.$$

In this subsection, we investigate the closedness of matrix multivalued linear operators in the product spaces. For this reason, we display some hypotheses, introduced in [1], which will be required in the sequel.

$(H10)$ A is a closable and densely defined linear relation.
$(H11)$ $B(0)\subset A(0)$ and Γ_X is bounded.
$(H12)$ The resolvent set of the restriction $A_1:=A|_{\mathscr{D}(A)\bigcap N(\Gamma_X)}$ is not empty and A_1 is closed and densely defined.
$(H13)$ C is bounded.

Remark 5.2.1 (i) Departing from the assumption $(H2)$ and by using Proposition 3.6.1 (ii), we infer that $A-\mu I$ is closable.
(ii) For $\mu\in\rho(A_1)$ and by using assumption $(H5)$, we deduce that the linear relation $W(\mu)=C(A_1-\mu I)^{-1}$ is bounded. ◇

Lemma 5.2.1 *For any $\mu\in\rho(A_1)$, the restriction*

$$\Gamma_\mu=\Gamma_X|_{N(A-\mu I)}$$

is one-to-one and $R(\Gamma_\mu)=\Gamma_X(N(A-\mu I))=\Gamma_X(\mathscr{D}(A))=Z_1$. ◇

Proof For $\mu\in\rho(A_1)$, then $(A_1-\mu I)^{-1}$ is bounded single-valued. Now,

$$\begin{aligned} x\in N(\Gamma_\mu) &\text{if, and only if, } \Gamma_X x=0 \textit{ and } x\in N(A-\mu I)\\ &\text{if, and only if, } x\in N(\Gamma_X) \textit{ and } x\in N(A-\mu I)\\ &\text{if, and only if, } x\in N(A-\mu I)\bigcap N(\Gamma_X)\\ &\text{if, and only if, } x\in N(A_1-\mu I)=(A_1-\mu I)^{-1}(0)=\{0\}.\end{aligned}$$

This implies that Γ_μ is one-to-one and

$$\begin{aligned} R(\Gamma_\mu)&=\Gamma_X(N(A-\mu I))\\ &=\Gamma_X(\mathscr{D}(A))\\ &=Z_1,\end{aligned}$$

which completes the proof. Q.E.D.

We can define the inverse of Γ_μ as follows:

$$K_\mu:=(\Gamma_X|_{N(A-\mu I)})^{-1}:Z_1\longrightarrow N(A-\mu I)\subset X.$$

It is easy to demonstrate that K_μ is single-valued and $K_\mu\Gamma_X(0)=0$ for $\mu\in\rho(A_1)$.

Lemma 5.2.2 *If* $\mu \in \rho(A_1)$ *and* $x \in \mathcal{D}(A)$, *then*

$$(A - \mu I)x = (A_1 - \mu I)(I - K_\mu \Gamma_X)x. \qquad \diamond$$

Proof Since $(A - \mu I)K_\mu \Gamma_X(0) = (A - \mu I)(0)$, we have

$$\begin{aligned} (A - \mu I)K_\mu \Gamma_X(0) &= (A - \mu I)K_\mu \Gamma_\mu(0) \\ &= (A - \mu I)\Gamma_\mu(0) \\ &= (A - \mu I)(0) \text{ (as } \Gamma_\mu \text{ is one-to-one).} \end{aligned}$$

From Proposition 2.7.2 (i), we get

$$(A - \mu I)x = (A - \mu I)(I - K_\mu \Gamma_X)x + (A - \mu I)K_\mu \Gamma_X x. \tag{5.2.2}$$

Since $x \in \mathcal{D}(A)$, we have $K_\mu \Gamma_X x \in N(A - \mu I)$. Then,

$$(A - \mu I)K_\mu \Gamma_X x = (A - \mu I)(0).$$

Let $x_1 = (I - K_\mu \Gamma_X)x \in \mathcal{D}(A)$, then

$$\begin{aligned} \Gamma_X x_1 &= \Gamma_X x - \Gamma_X K_\mu \Gamma_X x \\ &= \Gamma_X(0). \end{aligned}$$

This implies that $x_1 \in N(\Gamma_X)$. Thus,

$$x_1 \in N(\Gamma_X) \bigcap \mathcal{D}(A) = \mathcal{D}(A_1).$$

It follows from Eq. (5.2.2) that

$$(A - \mu I)x = (A_1 - \mu I)(I - K_\mu \Gamma_X)x.$$

This achieves the proof. Q.E.D.

In what follows, we suppose that the linear relations K_μ, D, Γ_Y, and B satisfy the following hypotheses:
($H14$) For some $\mu \in \rho(A_1)$, the linear relations $F(\mu) = (A_1 - \mu I)^{-1}B$, Γ_Y, and K_μ are bounded.
($H15$) The linear relation D is closable and $C(0) \subset C(A_1 - \mu I)^{-1}B(0) \subset D(0)$.
($H16$) $\mathcal{D}(B) \bigcap \mathcal{D}(D) \subset \mathcal{D}(\Gamma_Y)$. The linear relation $M(\mu) = D + CK_\mu \Gamma_Y - W(\mu)B$ is defined on the set Y_1, where

$$Y_1 = \left\{ y \in \mathcal{D}(B) \bigcap \mathcal{D}(D) \text{ such that } \Gamma_Y \in Z_1 \right\}.$$

Lemma 5.2.3 *Under assumptions* $(H14)$–$(H15)$, *for* $\mu \in \rho(A_1)$, *we have*

(i) $M(\mu) - CK_\mu\Gamma_Y + W(\mu)B = D$.
(ii) $M(\mu)$ *is closable.* ◇

Proof (i) We can write

$$M(\mu) - CK_\mu\Gamma_Y + W(\mu)B = D + T - T,$$

where $T = CK_\mu\Gamma_Y - W(\mu)B$. Thus,

$$\begin{aligned} T(0) &= (CK_\mu\Gamma_Y - W(\mu)B)(0) \\ &= CK_\mu\Gamma_X(0) - W(\mu)B(0) \\ &= C(0) - C(A_1 - \mu I)^{-1}B(0)(0) \\ &= C(0) \\ &\subset D(0). \end{aligned}$$

Moreover, we have

$$\begin{aligned} \mathscr{D}(T) &= \mathscr{D}(CK_\mu\Gamma_Y) \\ &\subset \mathscr{D}(W(\mu)B) \\ &= Y. \end{aligned}$$

By using Proposition 2.7.2 (i), we deduce that

$$D + T - T = D.$$

(ii) In the first step, we will prove that $D - W(\mu)B$ is closable. For this reason, we have D is closable, $W(\mu)B = C(A_1 - \mu I)^{-1}B$ is bounded and $C(A_1 - \mu I)^{-1}B(0) \subset D(0)$. In addition by using Proposition 2.11.2, we have $D - G(\mu)B$ is closable. On the other side, we have $CK_\mu\Gamma_Y$ is bounded and

$$\begin{aligned} CK_\mu\Gamma_Y(0) &= C(0) \\ &\subset D(0) \\ &\subset (D - W(\mu)B)(0). \end{aligned}$$

Therefore, combining this fact together with Proposition 2.11.2 allows us to conclude that $M(\mu) = D + CK_\mu\Gamma_Y - G(\mu)B$ is closable. Q.E.D.

In the next theorem, we shall determine the Frobenius–Schur decomposition for 2×2 block matrix multivalued linear operator $\mathscr{L}_0$ given in Eq. (5.2.1).

Theorem 5.2.1 *Let* $\mu \in \rho(A_1)$. *Assume that assumptions* $(H10)$–$(H16)$ *are satisfied. Then,*

$$\begin{aligned}\mathscr{L}_0 - \mu I &= \begin{pmatrix} I & 0 \\ W(\mu) & I \end{pmatrix} \begin{pmatrix} A_1 - \mu I & 0 \\ 0 & M(\mu) - \mu I \end{pmatrix} \begin{pmatrix} I & -K_\mu \Gamma_Y + F(\mu) \\ 0 & I \end{pmatrix} \\ &= RST,\end{aligned}$$

where

$$R = \begin{pmatrix} I & 0 \\ W(\mu) & I \end{pmatrix}, \; S = \begin{pmatrix} A_1 - \mu I & 0 \\ 0 & M(\mu) - \mu I \end{pmatrix} \text{ and } T = \begin{pmatrix} I & -K_\mu \Gamma_Y + F(\mu) \\ 0 & I \end{pmatrix}.$$

◇

Proof Let

$$\left[\begin{pmatrix} x_1 \\ x_2 \end{pmatrix}, \begin{pmatrix} y_1 \\ y_2 \end{pmatrix}\right] \in G\left(\mathscr{L}_0 - \mu I)\right).$$

Then,

$$\begin{pmatrix} x_1 \\ x_2 \end{pmatrix} \in \mathscr{D}(\mathscr{L}_0 - \mu I)$$

and

$$\begin{pmatrix} y_1 \\ y_2 \end{pmatrix} \in (\mathscr{L}_0 - \mu I) \begin{pmatrix} x_1 \\ x_2 \end{pmatrix}.$$

Hence, $x_1 \in \mathscr{D}(A) \bigcap \mathscr{D}(C) = \mathscr{D}(A)$, $x_2 \in \mathscr{D}(B) \bigcap \mathscr{D}(D)$ and $\Gamma_X x_1 = \Gamma_Y x_2$. Furthermore, we have

$$\begin{pmatrix} y_1 \\ y_2 \end{pmatrix} \in (\mathscr{L}_0 - \mu I) \begin{pmatrix} x_1 \\ x_2 \end{pmatrix} = \begin{pmatrix} (A - \mu I)x_1 + Bx_2 \\ Cx_1 + (D - \mu I)x_2 \end{pmatrix}.$$

It follows that

$$\begin{cases} y_1 \in (A - \mu I)x_1 + Bx_2 \\ y_2 \in Cx_1 + (D - \mu I)x_2. \end{cases}$$

Using Theorem 2.8.1, we obtain

$$y_1 \in (A_1 - \mu I)(x_1 - K_\mu \Gamma_X)x_1 + Bx_2,$$

which implies that

$$y_1 \in (A_1 - \mu I)x_1 - (A_1 - \mu I)K_\mu \Gamma_Y x_2 + Bx_2 + (A_1 - \mu I)(0).$$

Moreover, we have

$$F(\mu)x_2 = (A_1 - \mu I)^{-1} Bx_2.$$

Since $x_2 \in \mathscr{D}(B)$ and by using Theorem 2.8.1, we get

$$(A_1 - \mu I)F(\mu)x_2 = Bx_2 + (A_1 - \mu I)(0).$$

Then,

$$y_1 \in (A_1 - \mu I)x_1 + (A_1 - \mu I)\left[-K_\mu \Gamma_Y + F(\mu)\right] x_2.$$

Now, from Theorem 2.8.1, we obtain

$$y_2 \in Cx_1 + M(\mu)x_2 - CK_\mu \Gamma_Y x_2 + W(\mu)Bx_2 - \mu I.$$

Additionally, from Theorem 2.8.1, we get

$$\begin{aligned} W(\mu)(A_1 - \mu I)x_1 &= C(A_1 - \mu I)^{-1}(A_1 - \mu I)x_1 \\ &= Cx_1 \end{aligned}$$

and

$$\begin{aligned} W(\mu)(A_1 - \mu I)K_\mu \Gamma_Y x_2 &= CK_\mu \Gamma_Y x_2 + C(0) \\ &= CK_\mu \Gamma_Y x_2. \end{aligned}$$

Thus,

$$\begin{aligned} y_2 \in\, & W(\mu)(A_1 - \mu I)x_1 + M(\mu)x_2 - W(\mu)(A_1 - \mu I)K_\mu \Gamma_Y x_2 \\ & + W(\mu)(A_1 - \mu I)F(\mu)x_2 - \mu I. \end{aligned}$$

This implies that

$$y_2 \in W(\mu)\left[(A_1 - \mu I)x_1 + (A_1 - \mu I)(-K_\mu \Gamma_Y + F(\mu))x_2\right] + (M(\mu) - \mu I)x_2.$$

Let $t_1 = x_1 - K_\mu \Gamma_Y x_2 + F(\mu)x_2$, $t_2 = x_2$, $z_1 = (A_1 - \mu I)t_1$ and $z_2 = (M(\mu) - \mu I)t_2$, be which allows us to get $y_1 \in z_1$ and $y_2 \in W(\mu)z_1 + z_2$. This ensures that

$$\left[\begin{pmatrix} x_1 \\ x_2 \end{pmatrix}, \begin{pmatrix} t_1 \\ t_2 \end{pmatrix}\right] \in G\left[\begin{pmatrix} I & -K_\mu \Gamma_Y + F(\mu) \\ 0 & I \end{pmatrix}\right],$$

$$\left[\begin{pmatrix} t_1 \\ t_2 \end{pmatrix}, \begin{pmatrix} z_1 \\ z_2 \end{pmatrix}\right] \in G\left[\begin{pmatrix} (A_1 - \mu I) & 0 \\ 0 & M(\mu) - \mu I \end{pmatrix}\right],$$

and

$$\left[\begin{pmatrix} z_1 \\ z_2 \end{pmatrix}, \begin{pmatrix} y_1 \\ y_2 \end{pmatrix}\right] \in G\left[\begin{pmatrix} I & 0 \\ W(\mu) & I \end{pmatrix}\right].$$

Therefore,

$$\left[\begin{pmatrix} x_1 \\ x_2 \end{pmatrix}, \begin{pmatrix} y_1 \\ y_2 \end{pmatrix}\right] \in G\left[\begin{pmatrix} I & 0 \\ W(\mu) & I \end{pmatrix}\begin{pmatrix} A_1-\mu I & 0 \\ 0 & M(\mu)-\mu I \end{pmatrix}\begin{pmatrix} I & -K_\mu\Gamma_Y+F(\mu) \\ 0 & I \end{pmatrix}\right].$$

Then, we assert the following inclusion:

$$G(\mathscr{L}_0-\mu I)\subset G(RST). \tag{5.2.3}$$

Now, we shall prove that $\mathscr{D}(RST)=\mathscr{D}(\mathscr{L}_0-\mu I)$. Referring to Eq. (5.2.3), we get

$$\mathscr{D}(\mathscr{L}_0-\mu I)\subset \mathscr{D}(RST).$$

Moreover, we have

$$\begin{aligned}\mathscr{D}(RST) &= \left\{\begin{pmatrix} x \\ y \end{pmatrix} \in X\times Y \text{ such that } RST\begin{pmatrix} x \\ y \end{pmatrix}\neq \emptyset\right\} \\ &= \left\{\begin{pmatrix} x \\ y \end{pmatrix} \in X\times Y \text{ such that } ST\begin{pmatrix} x \\ y \end{pmatrix}\bigcap \mathscr{D}(R)\neq \emptyset\right\} \\ &= \left\{\begin{pmatrix} x \\ y \end{pmatrix} \in X\times Y \text{ such that } ST\begin{pmatrix} x \\ y \end{pmatrix}\neq \emptyset\right\} \quad (\text{as } \mathscr{D}(R)=X\times Y) \\ &= \mathscr{D}(ST).\end{aligned}$$

Let $\begin{pmatrix} x \\ y \end{pmatrix}\in \mathscr{D}(RST)=\mathscr{D}(ST)$. This means that there exists $\begin{pmatrix} x' \\ y' \end{pmatrix}\in \mathscr{D}(S)=\mathscr{D}(A_1)\times\mathscr{D}(M(\mu))$ such that $\begin{pmatrix} x' \\ y' \end{pmatrix}\in T\begin{pmatrix} x \\ y \end{pmatrix}$. This is equivalent to

$$\begin{aligned}T\begin{pmatrix} x \\ y \end{pmatrix} &= \begin{pmatrix} x' \\ y' \end{pmatrix}+T\begin{pmatrix} 0 \\ 0 \end{pmatrix} \\ &= \begin{pmatrix} x'-K_\mu\Gamma_Y(0)+F(\mu)(0) \\ y' \end{pmatrix},\end{aligned}$$

which implies that

$$\begin{pmatrix} x-K_\mu\Gamma_Y y+F(\mu)y \\ y \end{pmatrix}=\begin{pmatrix} x'-K_\mu\Gamma_Y(0)+F(\mu)(0) \\ y' \end{pmatrix}.$$

It follows that $y=y'\in\mathscr{D}(M(\mu))=Y_1$ and $x-K_\mu\Gamma_Y y+F(\mu)y\in\mathscr{D}(A_1)=\mathscr{D}(A)\bigcap N(\Gamma_X)$. Thus, $x\in\mathscr{D}(A)$ and $y\in\mathscr{D}(B)\bigcap\mathscr{D}(D)$. It is enough to demonstrate that $\Gamma_X x=\Gamma_Y y$. Indeed, we have $x-K_\mu\Gamma_Y y+F(\mu)y\in\mathscr{D}(A_1)$. Then,

$$\begin{aligned}\Gamma_X(x-K_\mu\Gamma_Y y+F(\mu)y) &= \Gamma_X(x)-\Gamma_X K_\mu\Gamma_Y y+\Gamma_X(F(\mu)y) \\ &= \Gamma_X(0).\end{aligned}$$

Moreover, we have $y\in Y_1$, which implies that $\Gamma_Y y\in Z_1$. Then,

$$\Gamma_X K_\mu \Gamma_Y y = \Gamma_Y y + \Gamma_X(0).$$

Besides, we have $F(\mu)y \in \mathscr{D}(A_1)$. Then,

$$\Gamma_X(F(\mu)y) = \Gamma_X(0).$$

Hence, we get

$$\Gamma_X(x) - \Gamma_Y y = \Gamma_Y(0).$$

Thus, we infer that

$$\Gamma_X x = \Gamma_Y y.$$

Therefore,

$$\mathscr{D}(\mathscr{L}_0 - \mu I) = \mathscr{D}(RST). \tag{5.2.4}$$

It remains to show that

$$(\mathscr{L}_0 - \mu I)\begin{pmatrix} 0 \\ 0 \end{pmatrix} = (RST)\begin{pmatrix} 0 \\ 0 \end{pmatrix}.$$

Indeed, we have

$$\begin{aligned} RST\begin{pmatrix} 0 \\ 0 \end{pmatrix} &= \left[\begin{pmatrix} I & 0 \\ W(\mu) & I \end{pmatrix}\begin{pmatrix} (A_1 - \mu I) & 0 \\ 0 & M(\mu) - \mu I \end{pmatrix}\begin{pmatrix} I & -K_\mu\Gamma_Y + F(\mu) \\ 0 & I \end{pmatrix}\right]\begin{pmatrix} 0 \\ 0 \end{pmatrix} \\ &= \begin{pmatrix} I & 0 \\ W(\mu) & I \end{pmatrix}\begin{pmatrix} (A_1 - \mu I) & 0 \\ 0 & M(\mu) - \mu I \end{pmatrix}\begin{pmatrix} -K_\mu\Gamma_Y(0) + F(\mu)(0) \\ 0 \end{pmatrix} \\ &= \begin{pmatrix} I & 0 \\ W(\mu) & I \end{pmatrix}\begin{pmatrix} (A_1 - \mu I)(-K_\mu\Gamma_Y(0) + F(\mu)(0)) \\ M(\mu)(0) \end{pmatrix} \\ &= \begin{pmatrix} (A_1 - \mu I)(-K_\mu\Gamma_Y(0) + F(\mu)(0)) \\ W(\mu)(A_1 - \mu I)(-K_\mu\Gamma_Y(0) + F(\mu)(0)) + M(\mu)(0) \end{pmatrix} \\ &= \begin{pmatrix} A(0) + B(0) \\ C(0) + D(0) \end{pmatrix}. \end{aligned}$$

Thus,

$$RST\begin{pmatrix} 0 \\ 0 \end{pmatrix} = (\mathscr{L}_0 - \mu I)\begin{pmatrix} 0 \\ 0 \end{pmatrix}.$$

Finally, applying Proposition 2.7.3, we get

$$\mathscr{L}_0 - \mu I = RST.$$ Q.E.D.

Theorem 5.2.2 *Let $\mu \in \rho(A_1)$. Suppose that the assumptions $(H10)$–$(H15)$ are satisfied. Then,*

(i) If $P(\mu)$ is a linear selection of $W(\mu)$, then

$$\mathscr{L}_0 - \mu I = \begin{pmatrix} I & 0 \\ P(\mu) & I \end{pmatrix} \begin{pmatrix} A_1 - \mu I & 0 \\ 0 & M(\mu) - \mu I \end{pmatrix} \begin{pmatrix} I & -K_\mu \Gamma_Y + F(\mu) \\ 0 & I \end{pmatrix} + \mathscr{H}(\mu)$$

$$= R_1 S T + \mathscr{H}(\mu),$$

where

$$R_1 = \begin{pmatrix} I & 0 \\ P(\mu) & I \end{pmatrix} \text{ and } \mathscr{H}(\mu) = \begin{pmatrix} 0 & 0 \\ (W(\mu) - W(\mu))(A - \mu I) & (W(\mu) - W(\mu))B \end{pmatrix}.$$

(ii) If $P(\mu)$ and $Z(\mu)$ are linear selections of $W(\mu)$ and $-K_\mu \Gamma_Y + F(\mu)$, respectively, then

$$\mathscr{L}_0 - \mu I = \begin{pmatrix} I & 0 \\ P(\mu) & I \end{pmatrix} \begin{pmatrix} A_1 - \mu I & 0 \\ 0 & M(\mu) - \mu I \end{pmatrix} \begin{pmatrix} I & Z(\mu) \\ 0 & I \end{pmatrix} + \mathscr{I}(\mu)$$

$$= R_1 S T_1 + \mathscr{I}(\mu),$$

where

$$R_1 = \begin{pmatrix} I & 0 \\ P(\mu) & I \end{pmatrix},$$

$$T_1 = \begin{pmatrix} I & Z(\mu) \\ 0 & I \end{pmatrix}$$

and

$$\mathscr{I}(\mu) = \begin{pmatrix} A - A & B - B \\ P(\mu)(A - A) + (W(\mu) - W(\mu))(A - \mu I) & (W(\mu) - W(\mu))B + P(\mu)(B - B) \end{pmatrix}. \quad \diamond$$

Proof *(i)* Let $P(\mu)$ be a linear selection of $W(\mu)$. Then,

$$W(\mu) = P(\mu) + W(\mu) - W(\mu).$$

Hence,

$$\begin{aligned} \mathscr{L}_0 - \mu I &= \begin{pmatrix} I & 0 \\ P(\mu) + W(\mu) - W(\mu) & I \end{pmatrix} \begin{pmatrix} A_1 - \mu I & 0 \\ 0 & M(\mu) - \mu I \end{pmatrix} \begin{pmatrix} I & -K_\mu \Gamma_Y + F(\mu) \\ 0 & I \end{pmatrix} \\ &= \left[\begin{pmatrix} I & 0 \\ P(\mu) & I \end{pmatrix} + \begin{pmatrix} 0 & 0 \\ W(\mu) - W(\mu) & 0 \end{pmatrix} \right] \begin{pmatrix} A_1 - \mu I & 0 \\ 0 & M(\mu) - \mu I \end{pmatrix} \begin{pmatrix} I & -K_\mu \Gamma_Y + F(\mu) \\ 0 & I \end{pmatrix}. \end{aligned}$$

Since

$$\begin{pmatrix} A_1 - \mu I & 0 \\ 0 & M(\mu) - \mu I \end{pmatrix} \begin{pmatrix} I & -K_\mu \Gamma_Y + F(\mu) \\ 0 & I \end{pmatrix} \begin{pmatrix} 0 \\ 0 \end{pmatrix} = \begin{pmatrix} A(0) \\ D(0) \end{pmatrix},$$

it yields that

$$\begin{pmatrix} A_1 - \mu I & 0 \\ 0 & M(\mu) - \mu I \end{pmatrix} \begin{pmatrix} I & -K_\mu \Gamma_Y + F(\mu) \\ 0 & I \end{pmatrix} \begin{pmatrix} 0 \\ 0 \end{pmatrix} \subset N \begin{pmatrix} 0 & 0 \\ F(\mu) - F(\mu) & 0 \end{pmatrix} = X \times Y.$$

Applying [2, P. 11], we obtain

$$\begin{aligned} \mathscr{L}_0 - \mu I &= \begin{pmatrix} I & 0 \\ P(\mu) & I \end{pmatrix} \begin{pmatrix} A_1 - \mu I & 0 \\ 0 & M(\mu) - \mu I \end{pmatrix} \begin{pmatrix} I & -K_\mu \Gamma_Y + F(\mu) \\ 0 & I \end{pmatrix} \\ &+ \begin{pmatrix} 0 & 0 \\ W(\mu) - W(\mu) & 0 \end{pmatrix} \begin{pmatrix} A_1 - \mu I & 0 \\ 0 & M(\mu) - \mu I \end{pmatrix} \begin{pmatrix} I & -K_\mu \Gamma_Y + F(\mu) \\ 0 & I \end{pmatrix}. \end{aligned}$$

Now, suppose that

$$T = \begin{pmatrix} 0 & 0 \\ W(\mu) - W(\mu) & 0 \end{pmatrix} \begin{pmatrix} A_1 - \mu I & 0 \\ 0 & M(\mu) - \mu I \end{pmatrix} \begin{pmatrix} I & -K_\mu \Gamma_Y + F(\mu) \\ 0 & I \end{pmatrix}.$$

Let $\begin{pmatrix} x_1 \\ x_2 \end{pmatrix} \in X \times Y$. Then,

$$\begin{aligned} T \begin{pmatrix} x_1 \\ x_2 \end{pmatrix} &= \begin{pmatrix} 0 & 0 \\ W(\mu) - W(\mu) & 0 \end{pmatrix} \begin{pmatrix} A_1 - \mu I & 0 \\ 0 & M(\mu) - \mu I \end{pmatrix} \begin{pmatrix} I & -K_\mu \Gamma_Y + F(\mu) \\ 0 & I \end{pmatrix} \begin{pmatrix} x_1 \\ x_2 \end{pmatrix} \\ &= \begin{pmatrix} 0 & 0 \\ W(\mu) - W(\mu) & 0 \end{pmatrix} \begin{pmatrix} A_1 - \mu I & 0 \\ 0 & M(\mu) - \mu I \end{pmatrix} \begin{pmatrix} x_1 + (-K_\mu \Gamma_Y + F(\mu))x_2 \\ x_2 \end{pmatrix} \\ &= \begin{pmatrix} 0 & 0 \\ W(\mu) - W(\mu) & 0 \end{pmatrix} \begin{pmatrix} (A_1 - \mu I)(I - K_\mu \Gamma_X)x_1 + (A_1 - \mu I)F(\mu))x_2 \\ (M(\mu) - \mu I)x_2 \end{pmatrix} \\ &= \begin{pmatrix} 0 & 0 \\ W(\mu) - W(\mu) & 0 \end{pmatrix} \begin{pmatrix} (A - \mu I)x_1 + Bx_2 \\ (M(\mu) - \mu I)x_2 \end{pmatrix} \\ &= \begin{pmatrix} 0 \\ (W(\mu) - W(\mu))(A - \mu I)x_1 + (W(\mu) - W(\mu))Bx_2 \end{pmatrix}. \end{aligned}$$

Hence,

$$\begin{aligned} T \begin{pmatrix} x_1 \\ x_2 \end{pmatrix} &= \begin{pmatrix} 0 \\ (W(\mu) - W(\mu))(A - \mu I)x_1 + (W(\mu) - W(\mu))Bx_2 \end{pmatrix} \\ &= \mathscr{H}(\mu) \begin{pmatrix} x_1 \\ x_2 \end{pmatrix}. \end{aligned}$$

Therefore,

$$(\mathscr{L}_0 - \mu I) \begin{pmatrix} x_1 \\ x_2 \end{pmatrix} = R_1 S T \begin{pmatrix} x_1 \\ x_2 \end{pmatrix} + \mathscr{H}(\mu) \begin{pmatrix} x_1 \\ x_2 \end{pmatrix}. \tag{5.2.5}$$

Now, we have

$$\mathscr{D}(R_1 ST + \mathscr{H}(\mu)) = \mathscr{D}(R_1 ST) \bigcap \mathscr{D}(\mathscr{H}(\mu)),$$

and

$$\mathscr{D}(\mathscr{H}(\mu)) = \left\{ \begin{pmatrix} x_1 \\ x_2 \end{pmatrix} \in X \times Y \text{ such that } (A - \mu I)x_1 \in \mathscr{D}(W(\mu)) \text{ and } Bx_2 \in \mathscr{D}(W(\mu)) \right\}.$$

Departing from the fact that $\mathscr{D}(W(\mu)) = X$, we get $\mathscr{D}(\mathscr{H}(\mu)) = \mathscr{D}(A) \times \mathscr{D}(B)$. Moreover, we have $P(\mu)$ which is a selection of $W(\mu)$ implying that $\mathscr{D}(W(\mu)) = \mathscr{D}(P(\mu))$. Hence, $\mathscr{D}(R_1 ST) = \mathscr{D}(RST)$. Therefore,

$$\mathscr{D}(\mathscr{L}_0 - \mu I) = \mathscr{D}(R_1 ST + \mathscr{H}(\mu)). \tag{5.2.6}$$

According to Eqs. (5.2.5) and (5.2.6), we deduce that

$$\mathscr{L}_0 - \mu I = R_1 ST + \mathscr{H}(\mu).$$

(ii) Let $Z(\mu)$ be a linear selection of $E(\mu) = -K_\mu \Gamma_Y + F(\mu)$. Now, using the decomposition determined in (i), we obtain

$$\begin{aligned} \mathscr{L}_0 - \mu I &= R_1 S \begin{pmatrix} I & Z(\mu) + E(\mu) - E(\mu) \\ 0 & I \end{pmatrix} + \mathscr{H}(\mu) \\ &= R_1 S \left[\begin{pmatrix} I & Z(\mu) \\ 0 & I \end{pmatrix} + \begin{pmatrix} 0 & E(\mu) - E(\mu) \\ 0 & 0 \end{pmatrix} \right] + \mathscr{H}(\mu). \end{aligned}$$

Since $\mathscr{D}(S)$ contains the ranges of both $T_1 = \begin{pmatrix} I & Z(\mu) \\ 0 & I \end{pmatrix}$ and $\begin{pmatrix} 0 & E(\mu) - E(\mu) \\ 0 & 0 \end{pmatrix}$, then by using Proposition 2.8.2, we get

$$\mathscr{L}_0 - \mu I = R_1 ST_1 + R_1 S \begin{pmatrix} 0 & E(\mu) - E(\mu) \\ 0 & 0 \end{pmatrix} + \mathscr{H}(\mu).$$

Let $\begin{pmatrix} x_1 \\ x_2 \end{pmatrix} \in X \times Y$. Then,

$$\begin{aligned} R_1 S \begin{pmatrix} 0 & E(\mu) - E(\mu) \\ 0 & 0 \end{pmatrix} \begin{pmatrix} x_1 \\ x_2 \end{pmatrix} &= R_1 S \begin{pmatrix} (E(\mu) - E(\mu)x_1 \\ 0 \end{pmatrix} \\ &= R_1 \begin{pmatrix} (A_1 - \mu I)(E(\mu) - E(\mu))x_1 \\ 0 \end{pmatrix}. \end{aligned}$$

Since

$$\begin{aligned}(A_1 - \mu I)(E(\mu) - E(\mu))x_1 &= (A_1 - \mu I)E(\mu)(0) \\ &= (A_1 - \mu I)(-K_\mu \Gamma_Y + F(\mu))(0),\end{aligned}$$

it follows that

$$\begin{aligned}(A_1 - \mu I)(E(\mu) - E(\mu))x_1 &= (A_1 - \mu I)F(\mu)(0) = A(0) + B(0) \\ &= (A - A)x_1 + (B - B)x_2.\end{aligned}$$

Hence,

$$(A_1 - \mu I)(E(\mu) - E(\mu))x_1 = (A - A)x_1 + (B - B)x_2.$$

$$\begin{aligned}R_1 S \begin{pmatrix} 0 & E(\mu) - E(\mu) \\ 0 & 0 \end{pmatrix} \begin{pmatrix} x_1 \\ x_2 \end{pmatrix} &= R_1 \begin{pmatrix} (A - A)x_1 + (B - B)x_2 \\ 0 \end{pmatrix} \\ &= \begin{pmatrix} (A - A)x_1 + (B - B)x_2 \\ P(\mu)(A - A)x_1 + P(\mu)(B - B)x_2 \end{pmatrix}.\end{aligned}$$

Thus,

$$\begin{aligned}(\mathscr{L}_0 - \mu I)\begin{pmatrix} x_1 \\ x_2 \end{pmatrix} &= \left[R_1 S T_1 + R_1 S \begin{pmatrix} 0 & E(\mu) - E(\mu) \\ 0 & 0 \end{pmatrix} + \mathscr{H}(\mu)\right]\begin{pmatrix} x_1 \\ x_2 \end{pmatrix} \\ &= R_1 S T_1 \begin{pmatrix} x_1 \\ x_2 \end{pmatrix} + \begin{pmatrix} (A - A)x_1 + (B - B)x_2 \\ P(\mu)(A - A)x_1 + P(\mu)(B - B)x_2 \end{pmatrix} \\ &\quad + \mathscr{H}(\mu)\begin{pmatrix} x_1 \\ x_2 \end{pmatrix}.\end{aligned}$$

Therefore,

$$(\mathscr{L}_0 - \mu I)\begin{pmatrix} x_1 \\ x_2 \end{pmatrix} = R_1 S T_1 \begin{pmatrix} x_1 \\ x_2 \end{pmatrix} + \mathscr{J}(\mu)\begin{pmatrix} x_1 \\ x_2 \end{pmatrix},$$

where

$$\begin{aligned}&\mathscr{J}(\mu)\begin{pmatrix} x_1 \\ x_2 \end{pmatrix} \\ &= \begin{pmatrix} (A - A)x_1 + (B - B)x_2 \\ (W(\mu) - W(\mu))(A - \mu I)x_1 + P(\mu)(A - A)x_1 + (W(\mu) - W(\mu))Bx_2 + P(\mu)(B - B)x_2 \end{pmatrix}.\end{aligned}$$

Now, we have

$$\mathscr{D}\left(R_1 S T_1 + \mathscr{J}(\mu)\right) = \mathscr{D}(R_1 S T_1) \bigcap \mathscr{D}(\mathscr{J}(\mu))$$

and

$$\mathcal{D}(\mathcal{J}(\mu)) = \mathcal{D}(\mathcal{H}(\mu)) \bigcap \mathcal{D}\begin{pmatrix} A - A & B - B \\ P(\mu)(A - A) & P(\mu)(B - B) \end{pmatrix}$$
$$= \mathcal{D}(A) \times \mathcal{D}(B).$$

Considering the fact that $P(\mu)$ and $Z(\mu)$ are linear selections of $W(\mu)$ and $E(\mu)$, respectively, we get $\mathcal{D}(P(\mu)) = \mathcal{D}(W(\mu))$ and $\mathcal{D}(Z(\mu)) = \mathcal{D}(E(\mu))$, which implies that $\mathcal{D}(R_1 S T_1) = \mathcal{D}(RST))$. Therefore,

$$\mathcal{D}(\mathcal{L}_0 - \mu I) = \mathcal{D}\left(R_1 S T_1 + \mathcal{J}(\mu)\right). \tag{5.2.7}$$

By combining Eqs. (5.2.4) and (5.2.7), we deduce that

$$\mathcal{L}_0 - \mu I = R_1 S T_1 + \mathcal{J}(\mu).$$

This achieves the proof. Q.E.D.

The next theorem is based on the decomposition adapted from Theorems 5.2.1 and 5.2.2.

Theorem 5.2.3 *Assume that the conditions* $(H10)$–$(H15)$ *are satisfied.*

(i) If C and $E(\mu)$ are single-valued, then $\mathcal{L}_0$ is closable and its closure $\mathcal{L}$ is provided by

$$\mathcal{L} = \overline{\mathcal{L}_0} = \mu I + \begin{pmatrix} I & 0 \\ W(\mu) & I \end{pmatrix} \begin{pmatrix} A_1 - \mu I & 0 \\ 0 & \overline{M}(\mu) - \mu I \end{pmatrix} \begin{pmatrix} I & E(\mu) \\ 0 & I \end{pmatrix}. \tag{5.2.8}$$

(ii) If $P(\mu)$ and $Z(\mu)$ are linear selections, respectively, of $W(\mu)$, $E(\mu)$ and $\mathcal{J}(\mu) \in B\mathcal{R}(X)$, then $\mathcal{L}_0$ is closable and its closure $\mathcal{L}$ is given by

$$\mathcal{L} = \overline{\mathcal{L}_0} = \mu I + \begin{pmatrix} I & 0 \\ P(\mu) & I \end{pmatrix} \begin{pmatrix} A_1 - \mu I & 0 \\ 0 & \overline{M}(\mu) - \mu I \end{pmatrix} \begin{pmatrix} I & Z(\mu) \\ 0 & I \end{pmatrix} + \mathcal{J}(\mu). \tag{5.2.9}$$

◇

Proof (i) We first determine the properties of the matrix

$$R = \begin{pmatrix} I & 0 \\ -W(\mu) & I \end{pmatrix}$$

and

$$T = \begin{pmatrix} I & E(\mu) \\ 0 & I \end{pmatrix}.$$

(a) R and T are single-valued bijective and

$$R^{-1} = \begin{pmatrix} I & 0 \\ -W(\mu) & I \end{pmatrix}$$

and

$$T^{-1} = \begin{pmatrix} I & -E(\mu) \\ 0 & I \end{pmatrix}.$$

(b) $R\begin{pmatrix} 0 \\ 0 \end{pmatrix}$ is closed and $\dim\left(T\begin{pmatrix} 0 \\ 0 \end{pmatrix}\right) < \infty$.

(c) The fact that $A_1 - \mu I$ is closed and $M(\mu) - \mu I$ is closable allows us to get

$$S = \begin{pmatrix} A_1 - \mu I & 0 \\ 0 & M(\mu) - \mu I \end{pmatrix}$$

which is closable.

(d), we have

$$\begin{aligned} \overline{S}T\begin{pmatrix} 0 \\ 0 \end{pmatrix} &= \overline{S}\begin{pmatrix} 0 \\ 0 \end{pmatrix} \quad \text{(as } T \text{ is single-valued)} \\ &= S\begin{pmatrix} 0 \\ 0 \end{pmatrix} \quad \text{(as } S \text{ is closable)} \\ &= ST\begin{pmatrix} 0 \\ 0 \end{pmatrix}. \end{aligned}$$

Now, applying Theorem 2.12.3, we conclude that RST is closable and $\overline{RST} = R\overline{S}T$. Thus, referring to Proposition 2.11.2, we deduce that

$$\begin{aligned} \overline{\mathscr{L}_0 - \mu I} &= \overline{\mathscr{L}_0} - \mu I \\ &= R\overline{S}T. \end{aligned}$$

Therefore,

$$\overline{\mathscr{L}} = \mu I + R\overline{S}T.$$

(ii) From Theorem 5.2.2 (ii), we have

$$\mathscr{L}_0 - \mu I = R_1 S T_1 + \mathscr{J}(\mu). \tag{5.2.10}$$

According to (i), we obtain $R_1 S T_1$ which is closable and

$$\overline{R_1 S T_1} = R_1 \overline{S} T_1.$$

Furthermore, we have $\mathscr{J}(\mu)$ is bounded and

$$\mathscr{I}(\mu)\begin{pmatrix}0\\0\end{pmatrix} \subset R_1ST_1\begin{pmatrix}0\\0\end{pmatrix}.$$

Indeed, we have

$$\begin{aligned} R_1ST_1\begin{pmatrix}0\\0\end{pmatrix} &= R_1S\begin{pmatrix}0\\0\end{pmatrix} \\ &= \begin{pmatrix} A(0) \\ P(\mu)A(0) + M(\mu)(0) \end{pmatrix}. \end{aligned}$$

On the other side,

$$\mathscr{I}\begin{pmatrix}0\\0\end{pmatrix} = \begin{pmatrix} A(0) + B(0) \\ P(\mu)A(0) + (W(\mu) - W(\mu))A(0) + (W(\mu) - W(\mu))B(0) + P(\mu)B(0) \end{pmatrix}.$$

Since $B(0) \subset A(0)$, we get

$$A(0) + B(0) = A(0)$$

and

$$P(\mu)A(0) + P(\mu)B(0) \subset P(\mu)A(0).$$

Moreover, we have

$$(W(\mu) - W(\mu))A(0) + (W(\mu) - W(\mu))B(0) = C(0) - C(0) + W(\mu)B(0) - W(\mu)B(0).$$

As $C(0) \subset W(\mu)B(0) \subset D(0)$, we get

$$-C(0) + W(\mu)B(0) \subset D(0).$$

Then,

$$\begin{aligned} (W(\mu) - W(\mu))A(0) + (W(\mu) - W(\mu))B(0) &\subset C(0) + D(0) - W(\mu)B(0) \\ &= M(\mu)(0). \end{aligned}$$

Thus, when applying Proposition 2.11.2 in Eq. (5.2.10), our claim holds. Q.E.D.

Proposition 5.2.1 *Assume that the conditions* $(H5)$–$(H14)$ *are satisfied.*
(i) *If* C *and* $E(\mu)$ *are single-valued, then*

$$\lambda I - \mathscr{L} = UV(\lambda)J - (\lambda - \mu)\mathscr{I}(\mu),$$

where

$$U = \begin{pmatrix} I & 0 \\ W(\mu) & I \end{pmatrix},$$

$$J = \begin{pmatrix} I & -K_\mu \Gamma_Y + F(\mu) \\ 0 & I \end{pmatrix}$$

$$V(\lambda) = \begin{pmatrix} \lambda - A_1 & 0 \\ 0 & \lambda - \overline{M(\mu)} \end{pmatrix}$$

and

$$\mathscr{I}(\mu) = \begin{pmatrix} 0 & E(\mu) \\ W(\mu) & W(\mu)E(\mu) \end{pmatrix}.$$

(ii) If $P(\mu)$ and $Z(\mu)$ are linear selections of $W(\mu)$ and $E(\mu)$, respectively, and having $\mathscr{J}(\mu) \in B\mathscr{R}(X)$, then

$$\lambda I - \mathscr{L} = U_1 V(\lambda) J_1 - (\lambda - \mu)\mathscr{I}_1(\mu) + \mathscr{J}(\mu),$$

where $U_1 = \begin{pmatrix} I & 0 \\ P(\mu) & I \end{pmatrix}$, $J_1 = \begin{pmatrix} I & Z(\mu) \\ 0 & I \end{pmatrix}$ *and* $\mathscr{I}_1(\mu) = \begin{pmatrix} 0 & Z(\mu) \\ P(\mu) & P(\mu)Z(\mu) \end{pmatrix}$. ◇

Proof (i) For $\lambda \in \mathbb{C}$, we write

$$\lambda I - \mathscr{L} = (\mu I - \mathscr{L}) + (\lambda - \mu)I$$

and by using Eq. (5.2.8), we have

$$\begin{aligned} \lambda I - \mathscr{L} &= \begin{pmatrix} I & 0 \\ W(\mu) & I \end{pmatrix} \begin{pmatrix} \mu I - A_1 & 0 \\ 0 & \mu I - \overline{M(\mu)} \end{pmatrix} \begin{pmatrix} I & E(\mu) \\ 0 & I \end{pmatrix} + (\lambda - \mu)I \\ &= \begin{pmatrix} I & 0 \\ W(\mu) & I \end{pmatrix} \begin{pmatrix} \mu - \lambda I + \lambda I - A_1 & 0 \\ 0 & \mu - \lambda I + \lambda I - \overline{M(\mu)} \end{pmatrix} \begin{pmatrix} I & E(\mu) \\ 0 & I \end{pmatrix} + (\lambda - \mu)I \\ &= \begin{pmatrix} I & 0 \\ W(\mu) & I \end{pmatrix} \left[\begin{pmatrix} \lambda - A_1 & 0 \\ 0 & \lambda - \overline{M(\mu)} \end{pmatrix} + \begin{pmatrix} (\mu - \lambda)I & 0 \\ 0 & (\mu - \lambda) \end{pmatrix} \right] \begin{pmatrix} I & E(\mu) \\ 0 & I \end{pmatrix} + (\lambda - \mu)I. \end{aligned}$$

Using Proposition 2.8.2, we get

$$\begin{aligned} \lambda I - \mathscr{L} = &\begin{pmatrix} I & 0 \\ W(\mu) & I \end{pmatrix} \begin{pmatrix} \lambda - A_1 & 0 \\ 0 & \lambda - \overline{M(\mu)} \end{pmatrix} \begin{pmatrix} I & E(\mu) \\ 0 & I \end{pmatrix} \\ &+ \begin{pmatrix} I & 0 \\ W(\mu) & I \end{pmatrix} \begin{pmatrix} (\mu - \lambda)I & 0 \\ 0 & (\mu - \lambda)I \end{pmatrix} \begin{pmatrix} I & E(\mu) \\ 0 & I \end{pmatrix} + (\lambda - \mu)I \\ = &\begin{pmatrix} I & 0 \\ W(\mu) & I \end{pmatrix} \begin{pmatrix} \lambda - A_1 & 0 \\ 0 & \lambda - \overline{M(\mu)} \end{pmatrix} \begin{pmatrix} I & E(\mu) \\ 0 & I \end{pmatrix} \\ &+ (\mu - \lambda) \begin{pmatrix} I & 0 \\ W(\mu) & I \end{pmatrix} \begin{pmatrix} I & E(\mu) \\ 0 & I \end{pmatrix} + (\lambda - \mu)I \\ = &\begin{pmatrix} I & 0 \\ W(\mu) & I \end{pmatrix} \begin{pmatrix} \lambda - A_1 & 0 \\ 0 & \lambda - \overline{M(\mu)} \end{pmatrix} \begin{pmatrix} I & E(\mu) \\ 0 & I \end{pmatrix} \end{aligned}$$

$$-(\lambda-\mu)\begin{pmatrix} 0 & E(\mu) \\ W(\mu) & W(\mu)E(\mu) \end{pmatrix}.$$

Thus, we deduce that

$$\lambda I - \mathscr{L} = UV(\lambda)J - (\lambda-\mu)\mathscr{I}(\mu).$$

(ii) Let $\mu \in \rho(A_1)$ and $\lambda \in \mathbb{C}$. We write

$$\lambda I - \mathscr{L} = \mu I - \mathscr{L} + (\lambda-\mu)I$$

and by using Eq. (5.2.9), we obtain

$$\lambda I - \mathscr{L} = \begin{pmatrix} I & 0 \\ P(\mu) & I \end{pmatrix}\begin{pmatrix} \mu I - A_1 & 0 \\ 0 & \mu I - \overline{M(\mu)}) \end{pmatrix}\begin{pmatrix} I & Z(\mu) \\ 0 & I \end{pmatrix} + (\lambda-\mu)I + \mathscr{J}(\mu).$$

Now, proceeding in the same way as (i) allows us to conclude that

$$\begin{aligned}\lambda I - \mathscr{L} &= \begin{pmatrix} I & 0 \\ P(\mu) & I \end{pmatrix}\begin{pmatrix} \mu I - A_1 & 0 \\ 0 & \mu I - \overline{M(\mu)}) \end{pmatrix}\begin{pmatrix} I & Z(\mu) \\ 0 & I \end{pmatrix} \\ &\quad -(\lambda-\mu)\begin{pmatrix} 0 & Z(\mu) \\ P(\mu) & P(\mu)Z(\mu) \end{pmatrix} + \mathscr{J}(\mu) \\ &= U_1V(\lambda)J_1 - (\lambda-\mu)\mathscr{I}_1(\mu) + \mathscr{J}(\mu).\end{aligned}$$

This achieves the proof. Q.E.D.

5.2.2 *Essential Spectra*

In the next theorems, based on the last results, we determine the stability of the essential spectra.

Theorem 5.2.4 *Let $\mu \in \rho(A_1)$. We suppose that the conditions $(H5)$–$(H14)$ are satisfied. If C and $E(\mu)$ are single-valued, then*

(i)

$$\sigma_{ei}(\mathscr{L}) = \sigma_{ei}(A_1)\bigcup\sigma_{ei}(\overline{M(\mu)} \quad i = 1,\ 2,\ 3,\ 4.$$

(ii) Furthermore, if we suppose that the sets Φ_{A_1} and $\Phi_{M(\mu)}$ are connected and $\rho(M(\mu)) \neq \emptyset$, then

$$\sigma_{e5}(\mathscr{L}) = \sigma_{e5}(A_1)\bigcup\sigma_{e5}(\overline{M(\mu)}).$$

◇

Proof Let $\mu \in \rho(A_1)$ and $\lambda \in \mathbb{C}$. Then, we can write

$$\lambda I - \mathscr{L} = \mu I - \mathscr{L} + (\lambda - \mu)I.$$

Relying on Proposition 5.2.3 (i), we have

$$\lambda - \mathscr{L} = UV(\lambda)J + (\mu - \lambda)\mathscr{I}(\mu).$$

Since $W(\mu)$ and $E(\mu)$ are bounded operators, then $\mathscr{I}(\mu) \in B\mathscr{R}(X \times Y)$ and trivially

$$\mathscr{I}(\mu)\begin{pmatrix}0\\0\end{pmatrix} \subset UV(\lambda)J\begin{pmatrix}0\\0\end{pmatrix}.$$

This implies $\mathscr{I}(\mu) \in \mathscr{K}_{UV(\lambda)J}(X)$. Since U and W are bounded and have bounded inverses, then U and J are a Fredholm relations. So, $\lambda I - \mathscr{L} \in \Phi(X \times X)$ if, and only if, $UV(\lambda)J \in \Phi(X \times Y)$ if, and only if, $V(\lambda) \in \Phi(X \times Y)$. This implies

$$\lambda I - A_1 \in \Phi(X) \text{ and } \lambda I - \overline{M(\mu)} \in \Phi(Y).$$

Thus, we obtain

$$\sigma_{e_4}(\mathscr{L}) = \sigma_{e_4}(A_1)\bigcup \sigma_{e_4}(\overline{M}(\mu)).$$

(ii), we have Φ_{A_1} and $\Phi_{M(\mu)}$ are connected. Then, we deduce that $\sigma_{e5}(\mathscr{L}) = \sigma_{e5}(A_1)\bigcup \sigma_{e5}(\overline{M(\mu)})$. Q.E.D.

Theorem 5.2.5 *Let $\mu \in \rho(A_1)$. We suppose that the conditions $(H5)$–$(H13)$ are satisfied. If $P(\mu)$ and $Z(\mu)$ are linear selections of $W(\mu)$ and $E(\mu)$, respectively, and assuming $\mathscr{J}(\mu) \in B\mathscr{R}(X)$, then*

(i)

$$\sigma_{ei}(\mathscr{L}) = \sigma_{ei}(A_1)\bigcup \sigma_{ei}(\overline{M(\mu)}) \quad i = 1,\ 2,\ 3,\ 4.$$

(ii) *In addition, if we suppose that the sets Φ_{A_1} and $\Phi_{M(\mu)}$ are connected and $\rho(M(\mu)) \neq \emptyset$, then*

$$\sigma_{e5}(\mathscr{L}) = \sigma_{e5}(A_1)\bigcup \sigma_{e5}(\overline{M(\mu)}).$$ ◇

Proof Let $\mu \in \rho(A_1)$ and $\lambda \in \mathbb{C}$. Then, we can write $\lambda I - \mathscr{L} = \mu I - \mathscr{L} + (\lambda - \mu)I$. Using Proposition 5.2.3 (ii), we have

$$\lambda - \mathscr{L} = U_1V(\lambda)J_1 + (\mu - \lambda)\mathscr{I}_1(\mu) + \mathscr{J}(\mu).$$

Then, using a similar reasoning as before, we obtain the desired result. Q.E.D.

5.3 Essential Spectra of a 2 × 2 Matrix of Linear Relations

5.3.1 Essential Spectra of an 2 × 2 Upper-Triangular Matrix of Linear Relations in Hilbert Space

Theorem 5.3.1 *Let $A \in CB\mathscr{R}(H)$ and $B \in CB\mathscr{R}(K)$. If $\sigma_{eap}(A) = \sigma_{e1}(B)$ or $\sigma_{e1}(B) \bigcap \sigma_{e2}(A) = \emptyset$, then for all $C \in \mathscr{L}(K, H)$,*

$$\sigma_{eap}(\mathscr{M}_C) = \sigma_{eap}(\mathscr{M}_0). \qquad \diamondsuit$$

Proof Let $C \in \mathscr{L}(K, H)$ and $\lambda \in \mathbb{C}$. Write

$$\mathscr{M}_C - \lambda I = \begin{pmatrix} I & 0 \\ 0 & B - \lambda I \end{pmatrix} \begin{pmatrix} I & C \\ 0 & I \end{pmatrix} \begin{pmatrix} A - \lambda I & 0 \\ 0 & I \end{pmatrix}.$$

We denote

$$D_1 = \begin{pmatrix} I & 0 \\ 0 & B - \lambda I \end{pmatrix},$$

$$D_2 = \begin{pmatrix} I & C \\ 0 & I \end{pmatrix}$$

and

$$D_3 = \begin{pmatrix} A - \lambda I & 0 \\ 0 & I \end{pmatrix}.$$

We have

$$\begin{aligned} \mathscr{M}_0 - \lambda I &= \begin{pmatrix} A - \lambda I & 0 \\ 0 & B - \lambda I \end{pmatrix} \\ &= D_1 D_3. \end{aligned}$$

For the first inclusion, let $\lambda_0 \notin \sigma_{eap}(\mathscr{M}_0)$. Then, $\mathscr{M}_0 - \lambda_0 I \in \Phi_{\pm}(H \bigoplus K)$. Thus, $D_3 \in \Phi_+(H \bigoplus K)$. We can also prove that $D_1 \in \Phi_+(H \bigoplus K)$. Indeed, let $\begin{pmatrix} x \\ y \end{pmatrix} \in N(D_1)$. Then, $x = 0$ and $0 \in (B - \lambda_0 I)y$. Therefore,

$$\begin{pmatrix} 0 \\ 0 \end{pmatrix} \in (\mathscr{M}_0 - \lambda_0 I) \begin{pmatrix} x \\ y \end{pmatrix}.$$

Hence, $N(D_1) \subseteq N(\mathscr{M}_0)$. Thus, $\alpha(D_1) \leq \alpha(\mathscr{M}_0) < +\infty$. Since $R(\mathscr{M}_0 - \lambda_0 I)^*) \subseteq H \bigoplus R((B - \lambda_0 I)^*) = R(D_1^*)$, then

$$\begin{aligned} \beta(D_1^*) &\leq \beta((\mathscr{M}_0 - \lambda_0 I)^*) \\ &= \alpha(M_0 - \lambda_0 I) < +\infty, \end{aligned}$$

which implies that $D_1 \in \Phi_+(H \bigoplus K)$. As D_2 is invertible, then $\mathcal{M}_C - \lambda_0 I \in \Phi_+(H \bigoplus K)$ with

$$i(\mathcal{M}_C - \lambda_0 I) = i(\mathcal{M}_0 - \lambda_0 I) \leq 0.$$

Now, we will prove that $\sigma_{eap}(\mathcal{M}_0) \subseteq \sigma_{eap}(\mathcal{M}_C)$. Suppose that $\sigma_{eap}(A) = \sigma_{e1}(B)$. If $\mathcal{M}_C - \lambda_0 I \in \Phi_\pm(H \bigoplus K)$, then $A - \lambda_0 I \in \Phi_+(H)$. Using Theorem 3.6.13, we know that if $\lambda_0 \in \sigma_{e1}(B)$, then $\beta(A - \lambda_0 I) = +\infty$ and hence $A - \lambda_0 I \in \Phi_\pm(H)$. It follows that $\lambda_0 \notin \sigma_{eap}(A)$, which is a contradiction. Then, $B - \lambda_0 I \in \Phi_+(K)$. Therefore, $\mathcal{M}_0 - \lambda_0 I \in \Phi_+(H \bigoplus K)$ with $i(\mathcal{M}_0 - \lambda_0 I) = i(\mathcal{M}_C - \lambda_0 I) \leq 0$ which means that $\lambda_0 \notin \sigma_{eap}(\mathcal{M}_0)$. Similarly, if $\sigma_{e2}(A) \bigcap \sigma_{e1}(B) = \emptyset$, the result is true also. Q.E.D.

Lemma 5.3.1 *Let $A \in L\mathcal{R}(X_1)$, $B \in L\mathcal{R}(X_2)$ be closed and everywhere defined (so A and B are bounded (see Theorem 2.11.1) and $A(0)$ and $B(0)$ are closed). We consider the 2×2 matrix of linear relation:*

$$\mathcal{M}_C := \begin{pmatrix} A & C \\ 0 & B \end{pmatrix},$$

where C is a bounded single-valued. Then,

(i) If $A \in \Phi_+(X_1)$ and $B \in \Phi_+(X_2)$, then $\mathcal{M}_C \in \Phi_+(X_1 \times X_2)$. Hence, in particular, we have

$$\sigma_{e1}(A) \subset \sigma_{e1}(\mathcal{M}_C) \subset \sigma_{e1}(A) \bigcup \sigma_{e1}(B).$$

Moreover, if $\rho(A)$, $\rho(B)$, and $\rho(\mathcal{M}_C)$ are nonempty sets and Φ_A, Φ_B, and $\Phi_{\mathcal{M}_C}$ are connected, then

$$\sigma_{eap}(A) \subset \sigma_{eap}(\mathcal{M}_C) \subset \sigma_{eap}(A) \bigcup \sigma_{eap}(B).$$

(ii) If $A \in \Phi_-(X_1)$ and $B \in \Phi_-(X_2)$, then $\mathcal{M}_C \in \Phi_-(X_1 \times X_2)$. In particular, we have

$$\sigma_{e2}(A) \subset \sigma_{e2}(\mathcal{M}_C) \subset \sigma_{e2}(A) \bigcup \sigma_{e2}(B).$$

Moreover, if $\rho(A)$, $\rho(B)$, and $\rho(\mathcal{M}_C)$ are nonempty and Φ_A, Φ_B, and $\Phi_{\mathcal{M}_C}$ are connected, then

$$\sigma_{e\delta}(A) \subset \sigma_{e\delta}(\mathcal{M}_C) \subset \sigma_{e\delta}(A) \bigcup \sigma_{e\delta}(B).$$

(iii) If $A \in \Phi(X_1)$ and $B \in \Phi(X_2)$, then $\mathcal{M}_C \in \Phi(X_1 \times X_2)$. Hence, in particular, we have

$$\sigma_{e4}(A) \subset \sigma_{e4}(\mathcal{M}_C) \subset \sigma_{e4}(A) \bigcup \sigma_{e4}(B).$$

Moreover, if $\rho(A)$, $\rho(B)$, and $\rho(\mathcal{M}_C)$ are nonempty sets and Φ_A, Φ_B, and $\Phi_{\mathcal{M}_C}$ are connected, then

$$\sigma_{e5}(A) \subset \sigma_{e5}(\mathcal{M}_C) \subset \sigma_{e5}(A) \bigcup \sigma_{e5}(B).$$

◇

Proof (*i*) Write $\mathcal{M}_C$ in the form

$$\mathcal{M}_C = \begin{pmatrix} I & 0 \\ 0 & B \end{pmatrix} \begin{pmatrix} I & C \\ 0 & I \end{pmatrix} \begin{pmatrix} A & 0 \\ 0 & I \end{pmatrix}.$$

Since $A \in \Phi_+(X_1)$ and $B \in \Phi_+(X_2)$, then

$$\begin{pmatrix} A & 0 \\ 0 & I \end{pmatrix}$$

and

$$\begin{pmatrix} I & 0 \\ 0 & B \end{pmatrix}$$

are both upper semi-Fredholm relations. On the other hand,

$$\begin{pmatrix} I & C \\ 0 & I \end{pmatrix}$$

is invertible. Indeed, let $\begin{pmatrix} x_1 \\ x_2 \end{pmatrix} \in N\left[\begin{pmatrix} I & C \\ 0 & I \end{pmatrix}\right]$ if, and only if,

$$\begin{cases} 0 \in x_1 + Cx_2 \\ 0 \in x_2 \end{cases}$$

if, and only if,

$$\begin{cases} x_1 = 0 \\ x_2 \in C(0) \ \ \text{(which is subspace)}. \end{cases}$$

Since C is single-valued ($C(0) = \{0\}$), then

$$N\left[\begin{pmatrix} I & C \\ 0 & I \end{pmatrix}\right] = \begin{pmatrix} 0 \\ 0 \end{pmatrix}.$$

On the other hand,

$$R\left[\begin{pmatrix} I & C \\ 0 & I \end{pmatrix}\right] = X_1 \times X_2.$$

So, $\mathcal{M}_C \in \Phi_+(X_1 \times X_2)$. For every $\lambda \in \mathbb{C}$, we have

$$\mathcal{M}_C - \lambda = \begin{pmatrix} I & 0 \\ 0 & B - \lambda \end{pmatrix} \begin{pmatrix} I & C \\ 0 & I \end{pmatrix} \begin{pmatrix} A - \lambda & 0 \\ 0 & I \end{pmatrix}. \quad (5.3.1)$$

If $\lambda \notin \sigma_{e1}(A) \bigcup \sigma_{e1}(B)$, then

$$\begin{pmatrix} A-\lambda & 0 \\ 0 & I \end{pmatrix} \in \Phi_+(X_1 \times X_2)$$

and

$$\begin{pmatrix} I & 0 \\ 0 & B-\lambda \end{pmatrix} \in \Phi_+(X_1 \times X_2).$$

Hence, by using Lemma 3.6.4 (i), we infer that $\mathcal{M}_C - \lambda \in \Phi_+(X_1 \times X_2)$ implies that $\lambda \notin \sigma_{e1}(\mathcal{M}_C)$. Therefore,

$$\sigma_{e1}(\mathcal{M}_C) \subset \sigma_{e1}(A) \bigcup \sigma_{e1}(B).$$

Now, if $\lambda \notin \sigma_{e1}(\mathcal{M}_C)$, then by using Eq. (5.3.1) and Theorem 3.6.4 (iv), we have

$$\begin{pmatrix} A-\lambda & 0 \\ 0 & I \end{pmatrix} \in \Phi_+(X_1 \times X_2).$$

So, $A - \lambda \in \Phi_+(X_1)$. Therefore, $\lambda \notin \sigma_{e1}(A)$. We derive that

$$\sigma_{e1}(A) \subset \sigma_{e1}(\mathcal{M}_C).$$

(ii) and (iii) can be checked in the same way as (i). Q.E.D.

Remark 5.3.1 (a) Using Lemma 5.3.1 (i) and (ii), we have

$$\sigma_{e3}(\mathcal{M}_C) \subset \left[\sigma_{e3}(A) \bigcup \sigma_{e3}(B)\right] \bigcup \left[\sigma_{e2}(A) \bigcap \sigma_{e1}(B)\right] \bigcup \left[\sigma_{e1}(A) \bigcap \sigma_{e2}(B)\right].$$

(b) The same analysis as the proof of Lemma 5.3.1, we can show

(i) If $A \in \Phi_+(X_1)$ and $B \in \Phi_+(X_2)$, then $\mathcal{M}_D \in \Phi_+(X_1 \times X_2)$ for every single-valued D, where

$$\mathcal{M}_D = \begin{pmatrix} A & 0 \\ D & B \end{pmatrix}.$$

(ii) If $A \in \Phi_-(X_1)$ and $B \in \Phi_-(X_2)$, then $\mathcal{M}_D \in \Phi_-(X_1 \times X_2)$ for every single-valued D. ◇

Lemma 5.3.2 *Let A and B be everywhere defined linear relations on X with $0 \in \rho(A) \bigcap \rho(B)$. Suppose that $C, D \in B\mathcal{R}(X)$ satisfies $C(0) \subset A(0)$, $D(0) \subset B(0)$,*

$$\|C\|(1 + \|D\|) < 1 \text{ and } 1 + \|D\| < \min(\gamma(A), \gamma(B)). \tag{5.3.2}$$

Then,

$$\sigma_{e5}\left[\begin{pmatrix} A & C \\ D & B \end{pmatrix}\right] \subset \sigma_{e5}(A) \bigcup \sigma_{e5}(B).$$ ◇

Proof Let us decompose the block matrix linear relation

$$\mathcal{M} = \begin{pmatrix} A & C \\ D & B \end{pmatrix}$$

in the following form $\mathcal{M} = \mathcal{T} + \mathcal{S}$, where

$$\mathcal{T} = \begin{pmatrix} A & 0 \\ 0 & B \end{pmatrix}$$

and

$$\mathcal{S} = \begin{pmatrix} 0 & C \\ D & 0 \end{pmatrix}.$$

Let $\lambda \notin \sigma_{e5}(A) \bigcup \sigma_{e5}(B)$. This is equivalent to saying that $\lambda - A$ and $\lambda - B$ are Fredholm linear relations with

$$i(\lambda - A) = i(\lambda - B) = 0.$$

By (iii) of Lemma 5.3.1, we have $\lambda - \mathcal{T}$ is a Fredholm linear relation and

$$i(\lambda - \mathcal{T}) = i(\lambda - A) + i(\lambda - B) = 0.$$

It is easy to see from (5.3.2) that $\|\mathcal{S}\| < \gamma(\mathcal{T})$. Now, let us assume that $0 < |\lambda| < \gamma(\mathcal{T}) - \|\mathcal{S}\|$. Then, the fact that $0 \in \rho(A) \bigcap \rho(B)$ implies from Lemma 2.10.10 that

$$\begin{aligned} \|\mathcal{S}\| &< \gamma(\mathcal{T}) - |\lambda| \\ &\leq \gamma(\lambda - \mathcal{T}). \end{aligned}$$

Hence, by using Corollary 3.6.3 (vi), we conclude that $\lambda - \mathcal{M}$ is a Fredholm linear relation and

$$\begin{aligned} i(\lambda - \mathcal{M}) &= i(\lambda - (\mathcal{T} + \mathcal{S})) \\ &= i(\lambda - \mathcal{T}) \\ &= 0. \end{aligned}$$

So, $\lambda \notin \sigma_{e5}(\mathcal{M})$. As a result,

$$\sigma_{e5}(\mathcal{M}) \subset \sigma_{e5}(A) \bigcup \sigma_{e5}(B),$$

as desired. Q.E.D.

Theorem 5.3.2 *Let X_1, X_2 be Banach spaces and*

$$\mathscr{F} = \begin{pmatrix} F_{11} & F_{12} \\ F_{21} & F_{22} \end{pmatrix},$$

where $F_{11} \in L\mathscr{R}(X_1)$, $F_{12} \in L\mathscr{R}(X_2, X_1)$, $F_{21} \in L\mathscr{R}(X_1, X_2)$ and $F_{22} \in L\mathscr{R}(X_2)$, F_{11} is single-valued and $\dim F_{ij}(0) < \infty$ *for $i, j = 1, 2$. If $F_{11} \in K\mathscr{R}(X_1)$, $F_{12} \in K\mathscr{R}(X_2, X_1)$, $F_{21} \in K\mathscr{R}(X_1, X_2)$, and $F_{22} \in K\mathscr{R}(X_2)$, then $\mathscr{F} \in K\mathscr{R}(X_1 \times X_2)$, where $K\mathscr{R}(X_i, X_j)$ will denote the class of compact linear relation from X_i into X_j.* ◇

Proof We consider the following decomposition:

$$\mathscr{F} = \begin{pmatrix} F_{11} & 0 \\ 0 & 0 \end{pmatrix} + \begin{pmatrix} 0 & F_{12} \\ 0 & 0 \end{pmatrix} + \begin{pmatrix} 0 & 0 \\ F_{21} & 0 \end{pmatrix} + \begin{pmatrix} 0 & 0 \\ 0 & F_{22} \end{pmatrix}. \tag{5.3.3}$$

It is sufficient to show that if F_{ij} is a compact relation $i, j = 1, 2$, then each relation linear in the right-hand side of Eq. (5.3.3) is a compact relation on $X_1 \times X_2$. We will prove the result, for example, for the first linear relation. The proof for the other linear relations will be obtained in the same way. Let

$$\widetilde{\mathscr{F}} = \begin{pmatrix} F_{11} & 0 \\ 0 & 0 \end{pmatrix}$$

and

$$\mathscr{L} = \begin{pmatrix} A & B \\ C & D \end{pmatrix}$$

with $\widetilde{\mathscr{F}}(0) \subset \mathscr{L}(0)$ and $\mathscr{D}(\mathscr{L}) \subset \mathscr{D}(\widetilde{F})$. Assume that $\mathscr{L} \in \Phi(X_1 \times X_2)$. Then, in view of Lemma 3.6.5, there exist

$$\mathscr{L}_0 = \begin{pmatrix} A_0 & B_0 \\ C_0 & D_0 \end{pmatrix}$$

bounded matrix operator and

$$K = \begin{pmatrix} K_{11} & K_{12} \\ K_{21} & K_{22} \end{pmatrix}$$

bounded finite rank matrix operator such that

$$\mathscr{L}\mathscr{L}_0 = I - K + \mathscr{L}\mathscr{L}_0 - \mathscr{L}\mathscr{L}_0.$$

Hence,

$$
\begin{aligned}
(\mathscr{L}+\widetilde{\mathscr{F}})\mathscr{L}_0 - \begin{pmatrix} 0 & 0 \\ -K_{21} & 0 \end{pmatrix} &= \begin{pmatrix} I - K_{11} + F_{11}A_0 & -K_{12} + F_{11}B_0 \\ 0 & I - K_{22} \end{pmatrix} + \mathscr{L}\mathscr{L}_0 - \mathscr{L}\mathscr{L}_0 \\
&= \mathscr{P} + \mathscr{L}\mathscr{L}_0 - \mathscr{L}\mathscr{L}_0
\end{aligned}
$$

with

$$
\mathscr{P} = \begin{pmatrix} I - K_{11} + F_{11}A_0 & -K_{12} + F_{11}B_0 \\ 0 & I - K_{22} \end{pmatrix}.
$$

Since K_{12}, F_{11}, and B_0 are single-valued implies that

$$
(-K_{12} + F_{11}B_0)(0) = \{0\}.
$$

So, $-K_{12} + F_{11}B_0$ is single-valued. Since $-K_{11} + F_{11}A_0$ is compact, then by using Lemma 3.6.11, we have

$$
I - K_{11} + F_{11}A_0 \in \Phi(X_1).
$$

Now, we assert that $I - K_{11} + F_{11}A_0$ is bounded. Clearly, it is continuous and further

$$
\begin{aligned}
\mathscr{D}(I - K_{11} + F_{11}A_0) &:= \mathscr{D}(I - K_{11}) \bigcap \mathscr{D}(F_{11}A_0) \\
&:= \mathscr{D}(F_{11}A_0) \\
&:= \left\{x \in \mathscr{D}(A_0) \text{ such that } A_0x \bigcap \mathscr{D}(F_{11}) \neq \emptyset\right\} \\
&:= \left\{x \in \mathscr{D}(A_0) \text{ such that } A_0x \bigcap \mathscr{D}(F_{11}) \neq \emptyset\right\}, \text{ (as } F_{11} \text{ everywhere defined)} \\
&:= \mathscr{D}(A_0) \\
&:= X_1.
\end{aligned}
$$

On the other hand, $I - K_{22} \in \Phi(X_2)$ and $\mathscr{D}(I - K_{22}) = X_2$, and thus we infer from Lemma 5.3.1 that

$$
\mathscr{P} \in \Phi(X_1 \times X_2).
$$

Hence,

$$
(\mathscr{L}+\widetilde{\mathscr{F}})\mathscr{L}_0 - \begin{pmatrix} 0 & 0 \\ -K_{21} & 0 \end{pmatrix} \in \Phi(X_1 \times X_2).
$$

As K_{21} is a bounded compact operator, we have

$$
\begin{pmatrix} 0 & 0 \\ -K_{21} & 0 \end{pmatrix}
$$

is a compact matrix operator. Therefore, $(\mathscr{L}+\widetilde{\mathscr{F}})\mathscr{L}_0 \in \Phi(X_1 \times X_2)$. Hence, $\mathscr{L} + \widetilde{\mathscr{F}} \in \Phi(X_1 \times X_2)$. Q.E.D.

Remark 5.3.2 If $F_{ij}(0) = \{0\}$ for $i, j = 1, 2$, then $\mathscr{F} \in \mathscr{PR}(\Phi(X_1 \times X_2))$ if, and only if, $F_{11} \in \mathscr{PR}(\Phi(X_1))$, $F_{12} \in \mathscr{PR}(\Phi(X_2, X_1))$, $F_{21} \in \mathscr{PR}(\phi(X_1, X_2))$, and $F_{22} \in \mathscr{PR}(\Phi(X_2))$. ◇

In the case of upper-triangular 2 × 2-block matrices of linear relations, we will have the following description of the spectra and the essential spectra.

Theorem 5.3.3 *Let $A \in C\mathscr{R}(H)$, $B \in B\mathscr{R}(K, H)$, and $D \in C\mathscr{R}(K)$ be such that $B(0) \subset A(0)$ and $\mathscr{D} = K$. Then,*

(i) $\sigma(\mathscr{M}_B) \subset \sigma(A) \bigcup \sigma(D)$, where $\mathscr{M}_B$ is given in (2.13.2).
(ii) $\sigma(A) \bigcup \sigma(D) \subset \sigma(\mathscr{M}_B) \bigcup \Xi(D) \bigcup \{\lambda \in \mathbb{C}$ such that $\overline{\lambda} \in \Xi(A^*)\}$.
(iii) $\sigma_{e4}(\mathscr{M}_B) \subset \sigma_{e4}(A) \bigcup \sigma_{e4}(D)$.
(iv) $\sigma_{e4}(A) \bigcup \sigma_{e4}(D) \subset \sigma_{e4}(\mathscr{M}_B) \bigcup \Xi(D) \bigcup \{\lambda \in \mathbb{C}$ such that $\overline{\lambda} \in \Xi(A^*)\}$. ◇

Proof It follows from Remark 2.13.2 that $\mathscr{M}_B$, given in (2.13.2), is closed. For $\lambda \in \mathbb{C}$, we infer from [3, Proposition 2.1] that

$$\begin{aligned} Q_{(\lambda - \mathscr{M}_B)}(\lambda - \mathscr{M}_B) &= \begin{pmatrix} Q_{\lambda - A}(\lambda - A) & Q_A(B) \\ 0 & Q_{\lambda - D}(\lambda - D) \end{pmatrix} \\ &= \begin{pmatrix} Q_{\lambda - A}(\lambda - A) & Q_A(B) \\ 0 & Q_{\lambda - D}(\lambda - D) \end{pmatrix}. \end{aligned}$$

Since $Q_A B(0) \subset Q_A A(0) = \{0\}$, then $Q_A B$ is an operator. For all $x \in \mathscr{D}(Q_A B) = \mathscr{D}(B) = K$, we have

$$\begin{aligned} \|Q_A Bx\| &= \operatorname{dist}(Bx, A(0)) \\ &\leq \operatorname{dist}(Bx, B(0)), \quad (\text{as } B(0) \subset A(0)) \\ &= \|Q_B Bx\| \\ &= \|Bx\| < \infty. \end{aligned}$$

Hence, we can conclude that $Q_A B$ is a bounded operator. Now, suppose that

$$Q_{\lambda - \mathscr{M}_B}(\lambda - \mathscr{M}_B) = \mathscr{I}_1 \mathscr{J}_1 \mathscr{K}_1, \tag{5.3.4}$$

where

$$\mathscr{I}_1 = \begin{pmatrix} I & 0 \\ 0 & Q_{\lambda - D}(\lambda - D) \end{pmatrix},$$

$$\mathscr{J}_1 = \begin{pmatrix} I & Q_A B \\ 0 & I \end{pmatrix}$$

and

$$\mathcal{K}_1 = \begin{pmatrix} Q_{\lambda-A}(\lambda - A) & 0 \\ 0 & I \end{pmatrix}.$$

Moreover, $\mathcal{J}_1$ is an invertible operator.

(i) Let $\lambda \notin \sigma(A)\bigcup\sigma(D)$. Then, $(\lambda - A)^{-1}$ and $(\lambda - D)^{-1}$ are operators everywhere defined. Consequently, $\lambda - A$ and $\lambda - D$ are one-to-one and onto. The fact that $\lambda - A$ and $\lambda - D$ are closed implies from Proposition 2.10.7 (i) that $Q_{\lambda-A}(\lambda - A)$ and $Q_{\lambda-D}(\lambda - D)$ are one-to-one. It follows from Lemma 2.9.1 that

$$\beta\left(Q_{\lambda-A}(\lambda - A)\right) = \beta\left(\lambda - A\right) = 0$$

and

$$\beta\left(Q_{\lambda-D}(\lambda - D)\right) = \beta\left(\lambda - D\right) = 0,$$

which yields $Q_{\lambda-A}(\lambda - A)$ and $Q_{\lambda-D}(\lambda - D)$ are onto. Hence, we conclude that $Q_{\lambda-A}(\lambda - A)$ and $Q_{\lambda-D}(\lambda - D)$ are invertible. This implies that $\mathcal{I}_1$ and $\mathcal{K}_1$ are invertible. Now, by referring to (5.3.4), we have $Q_{\lambda-\mathcal{M}_B}(\lambda - \mathcal{M}_B)$ is invertible. The fact that $\lambda - \mathcal{M}_B$ is closed implies from Lemma 2.9.1 and Proposition 2.10.7 (i) that $\lambda - \mathcal{M}_B$ is one-to-one and onto. Thus, $\lambda \in \rho(\mathcal{M}_B)$. This is equivalent to saying that $\sigma(\mathcal{M}_B) \subset \sigma(A)\bigcup\sigma(D)$.
(ii) Let $\lambda \notin \sigma(\mathcal{M}_B)\bigcup\Xi(D)\bigcup\left\{\lambda \in \mathbb{C} \text{ such that } \overline{\lambda} \in \Xi(A^*)\right\}$. Then, $(\lambda - \mathcal{M}_B)^{-1}$is an operator everywhere defined which yields that $\lambda - \mathcal{M}_B$ is one-to-one and onto. This implies from Lemma 2.9.1 and Proposition 2.10.7 (i) that $Q_{\lambda-\mathcal{M}_B}(\lambda - \mathcal{M}_B)$ is invertible. It is easy to see that

$$R(\mathcal{I}_1\mathcal{J}_1\mathcal{K}_1) \subset R(\mathcal{I}_1)$$

and

$$N(\mathcal{K}_1) \subset N(\mathcal{I}_1\mathcal{J}_1\mathcal{K}_1).$$

Hence,

$$\beta(\mathcal{I}_1) \leq \beta(\mathcal{I}_1\mathcal{J}_1\mathcal{K}_1) \text{ and } \alpha(\mathcal{K}_1) \leq \alpha(\mathcal{I}_1\mathcal{J}_1\mathcal{K}_1). \tag{5.3.5}$$

This leads from (5.3.4) to

$$\alpha(\mathcal{K}_1) = 0 \text{ and } \beta(\mathcal{I}_1) = 0. \tag{5.3.6}$$

Our purpose is to prove that $\lambda \notin \sigma(A)\bigcup\sigma(D)$. Then, we divide this part of the proof into two steps.
Step 1. We propose to show that $\lambda - D$ is one-to-one and onto. We have

$$\begin{aligned} H \times K &= R(\mathcal{I}_1) \\ &= R(I) \times R(Q_{\lambda-D}(\lambda - D)) \end{aligned}$$

$$= H \times R(Q_{\lambda-D}(\lambda - D)).$$

Consequently,

$$R(Q_{\lambda-D}(\lambda - D)) = K$$

which yields from Lemma 2.9.1 that

$$\beta(\lambda - D) = 0.$$

This leads to $R(\lambda - D) = K$. Moreover, we have $\lambda \notin \Xi(D)$. Hence, by Proposition 3.8.7, we conclude that $\lambda - D$ is one-to-one.

Step 2. Our purpose is to show that $\lambda - A$ one-to-one and onto. We have

$$\begin{aligned}\left\{\begin{pmatrix}0\\0\end{pmatrix}\right\} &= N(\mathscr{K}_1) \quad (\text{by } (4.3.6))\\ &= N(Q_{\lambda-A}(\lambda - A)) \times N(I)\\ &= N(\lambda - A) \times \{0\} \quad (\text{by Proposition 1.10.7 (i)}).\end{aligned}$$

Then, $\lambda - A$ is one-to-one. Now, it remains to prove that $\lambda - A$ is onto. Hence, from (3.1.4), we must show that $N(\overline{\lambda} - A^*) = \{0\}$.

The fact that $\lambda - A$ is one-to-one implies from (3.1.3) and (3.1.4) that $R(\overline{\lambda} - A^*)^{\perp} = \{0\}$. This implies that

$$\overline{R(\overline{\lambda} - A^*)} = H.$$

At this level, we will prove that $R(\overline{\lambda} - A^*)$ is closed. By referring to Theorem 2.15.1, it is sufficient to show that $R(\lambda - A)$ is closed. Since $(\lambda - A)(0)$ is closed and

$$R(Q_{\lambda-A}(\lambda - A)) = R(\lambda - A)/_{(\lambda - A)(0)}, \tag{5.3.7}$$

then we shall show that $R(Q_{\lambda-A}(\lambda - A))$ is closed. As we have

$$\begin{aligned}R(\mathscr{K}_1) &= R(Q_{\lambda-A}(\lambda - A)) \times R(I)\\ &= R(Q_{\lambda-A}(\lambda - A)) \times K,\end{aligned}$$

it is sufficient to prove that $R(\mathscr{K}_1)$ is closed. Let $\begin{pmatrix}z\\t\end{pmatrix} \in \overline{R(\mathscr{K}_1)}$, then there exists a sequence $\begin{pmatrix}x_n\\y_n\end{pmatrix} \subset \mathscr{D}(\mathscr{K}_1)$ such that $\mathscr{K}_1\begin{pmatrix}x_n\\y_n\end{pmatrix} \to \begin{pmatrix}z\\t\end{pmatrix}$ as $n \to +\infty$. Since $\mathscr{I}_1\mathscr{J}_1$ is a bounded operator, then we get

$$\mathscr{I}_1\mathscr{J}_1\mathscr{K}_1\begin{pmatrix}x_n\\y_n\end{pmatrix} \to \mathscr{I}_1\mathscr{J}_1\begin{pmatrix}z\\t\end{pmatrix} \quad \text{as } n \to +\infty.$$

The fact that $R(\mathscr{I}_1 \mathscr{J}_1 \mathscr{K}_1)$ is closed implies that $\mathscr{I}_1 \mathscr{J}_1 \begin{pmatrix} z \\ t \end{pmatrix} \in R(\mathscr{I}_1 \mathscr{J}_1 \mathscr{K}_1)$. Hence, by using (2.8.3), we obtain $\mathscr{I}_1 \mathscr{J}_1 \begin{pmatrix} z \\ t \end{pmatrix} \in \mathscr{I}_1 \mathscr{J}_1 (R(\mathscr{K}_1))$. This implies that

$$(\mathscr{I}_1 \mathscr{J}_1)^{-1} \mathscr{I}_1 \mathscr{J}_1 \begin{pmatrix} z \\ t \end{pmatrix} \in (\mathscr{I}_1 \mathscr{J}_1)^{-1} \mathscr{I}_1 \mathscr{J}_1 (R(\mathscr{K}_1)). \tag{5.3.8}$$

Now, the fact that $\lambda - D$ is one-to-one implies from Proposition 2.10.7 (i) that $N(Q_{\lambda - D}(\lambda - D)) = \{0\}$. Thus, $\mathscr{I}_1$ is one-to-one. In addition, we have $\mathscr{J}_1$ is one-to-one. Hence, we get $\mathscr{I}_1 \mathscr{J}_1$ is one-to-one. Consequently, by referring to Lemma 2.6.3, we deduce from (5.3.8) that $\begin{pmatrix} z \\ t \end{pmatrix} \in R(\mathscr{K}_1)$. This shows that $R(\mathscr{K}_1)$ is closed. Thus, we infer from (5.3.7) that $R(\lambda - A)$ is closed. Therefore,

$$R(\overline{\lambda} - A^*) = \overline{R(\overline{\lambda} - A^*)} = H.$$

The fact that $\overline{\lambda} \notin \Xi(A^*)$ implies from Proposition 3.8.7 that $\overline{\lambda} - A^*$ is one-to-one. Hence, the use of (3.1.3) and (3.1.4) allows us to conclude that

$$N(\overline{\lambda} - A^*) = R(\lambda - A)^{\perp} = \{0\},$$

and so $R(\lambda - A) = H$. As a result, $\lambda \notin \sigma(A) \bigcup \sigma(D)$.

(iii) Let us assume that $\lambda \notin \left[\sigma_{e4}(A) \bigcup \sigma_{e4}(D)\right]$. Then, $\lambda - A$ and $\lambda - D$ are Fredholm. Since $\lambda - A$ and $\lambda - D$ are closed, then by using [4, Lemma 4.2], we obtain $Q_{\lambda - A}(\lambda - A)$ and $Q_{\lambda - D}(\lambda - D)$ are Fredholm. This implies that $\mathscr{I}_1$ and $\mathscr{K}_1$ are Fredholm. Moreover, since $\mathscr{J}_1$ is invertible, then it is Fredholm. Hence, by using (5.3.4) and [5, Theorem 7.3], we infer that $Q_{\lambda - \mathscr{M}_B}(\lambda - \mathscr{M}_B)$ is a Fredholm. The use of [4, Lemma 4.2] makes us conclude that $\lambda - \mathscr{M}_B$ is a Fredholm. Thus, $\lambda \notin \sigma_{e4}(\mathscr{M}_B)$.

(iv) Let $\lambda \notin \sigma_{e4}(\mathscr{M}_B) \bigcup \Xi(D) \bigcup \left\{\lambda \in \mathbb{C} \text{ such that } \overline{\lambda} \in \Xi(A^*)\right\}$. This implies from (5.3.4) that $\mathscr{I}_1 \mathscr{J}_1 \mathscr{K}_1$ is a Fredholm relation. It follows from (5.3.5) that

$$\beta(\mathscr{I}_1) < \infty \text{ and } \alpha(\mathscr{K}_1) < \infty. \tag{5.3.9}$$

Our purpose is to prove that $\lambda \notin \sigma_{e4}(A) \bigcup \sigma_{e4}(D)$. Then, we divide this part of the proof into two steps.

Step 1. We propose to show that $\lambda - D$ is a Fredholm relation. First, the fact that $Q_{\lambda - D}(\lambda - D)$ and I are two closed operators implies from [6, Corollary 3.4 (i)] that $\mathscr{I}_1$ is closed. Hence, it follows from (5.3.9) and [7, Lemma 332] that $R(\mathscr{I}_1)$ is closed. This yields that $\mathscr{I}_1$ is lower semi-Fredholm, which implies that $Q_{\lambda - D}(\lambda - D)$ is so. Therefore, by referring to [4, Lemma 4.2], we get that $\lambda - D$ is lower semi-Fredholm. Moreover, the fact that $\lambda \notin \Xi(D)$ implies from Proposition 3.8.7 (i) and [8, Theorem 3.4] that

$$\alpha(\lambda - D) = 0.$$

Consequently, $\lambda - D$ is Fredholm.

Step 2. We will show that $\lambda - A$ is a Fredholm relation. Again from the same reasoning as in the proof of (ii), we get $R(\mathscr{K}_1)$ is closed. This implies from (5.3.9) that $\mathscr{K}_1$ is upper semi-Fredholm. It makes us conclude that $Q_{\lambda-A}(\lambda - A)$ is upper semi-Fredholm. As a result, we infer from [4, Lemma 4.2] that $\lambda - A$ is so. At this level, it remains to prove that

$$\beta(\lambda - A) < \infty.$$

The fact that $R(\lambda - A)$ is closed implies from Theorem 2.15.1 that $R((\lambda - A)^*)$ is so. Therefore, it follows from (3.1.3) that $R(\overline{\lambda} - A^*)$ is closed. Hence, based on the hypotheses $\overline{\lambda} \notin \Xi(A^*)$, we conclude from Proposition 3.8.7 (i) and [8, Theorem 3.4] that $N(\overline{\lambda} - A^*) = \{0\}$. Finally, we have

$$\begin{aligned}\{0\} &= N(\overline{\lambda} - A^*) \\ &= N((\lambda - A)^*) \quad \text{(by (2.1.3))} \\ &= R(\lambda - A)^{\perp} \quad \text{(by (2.1.4))}.\end{aligned}$$

This allows us to conclude that $\beta(\lambda - A) < \infty$. This leads us to $\lambda - A$ is Fredholm. As a result, $\lambda \notin \left[\sigma_{e4}(A) \bigcup \sigma_{e4}(D)\right]$, as desired. Q.E.D.

As a direct consequence of Theorem 5.3.3, we infer the following result.

Corollary 5.3.1 *Let $A \in C\mathscr{R}(H)$, $D \in C\mathscr{R}(K)$ and $B \in B\mathscr{R}(K, H)$ be such that $B(0) \subset A(0)$ and $\mathscr{D}(D) = K$. If A^* and D have a SVEP, then*

(i) $\sigma(\mathscr{M}_B) = \sigma(A) \bigcup \sigma(D)$, where $\mathscr{M}_B$ is given in (2.13.2).
(ii) $\sigma_{e4}(\mathscr{M}_B) = \sigma_{e4}(A) \bigcup \sigma_{e4}(D)$. ◇

Theorem 5.3.4 *Let $A \in C\mathscr{R}(H)$, $D \in C\mathscr{R}(K)$ and let B, C be continuous linear relations from K into H and from H into K, respectively. Suppose that $\mathscr{D}(A) \subset N(C)$, $\mathscr{D}(D) \subset N(B)$, $B(0) \subset A(0)$ and $C(0) \subset D(0)$. Then,*

(i) $\sigma_{e1}(A) \bigcup \sigma_{e1}(D) \subset \Xi_+(\mathscr{L}_0)$, where $\mathscr{L}_0$ is given in (2.13.1).
(ii) $\sigma_{e2}(A) \bigcup \sigma_{e2}(D) \subset \sigma_{e2}(\mathscr{L}_0) \subset \Xi_-(\mathscr{L}_0)$.
(iii) $\sigma_{e3}(A) \bigcup \sigma_{e3}(D) \subset \Xi_\pm(\mathscr{L}_0)$.
(iv) $\sigma_{e4}(A) \bigcup \sigma_{e4}(D) \subset \Xi_-(\mathscr{L}_0) \subset \Xi_0(\mathscr{L}_0)$.
(v) $\sigma(A) \bigcup \sigma(D) \subset \sigma_r(\mathscr{L}_0) \bigcup \Xi_-(\mathscr{L}_0)$. ◇

Proof It follows from Remark 2.13.2 that $\mathscr{L}_0$ is a closed linear relation.

(i) Let us assume that $\lambda \notin \Xi_+(\mathscr{L}_0)$. Then, $\lambda - \mathscr{L}_0 \in \Phi_+(H \times K)$. The fact that $R(\lambda - \mathscr{L}_0)$ is closed implies from (ii) of Lemma 2.8.5 that $R(\lambda - A)$ and $R(\lambda - D)$ are closed. Since $\mathscr{L}_0$ have SVEP at λ, then by Proposition 3.8.7 (i) and Theorem 3.9.7 that $\lambda - A$ and $\lambda - D$ have SVEP at 0. Based on the assumptions that $\lambda - A$ and

$\lambda - D$ are closed relations, we infer from [8, Theorem 3.4] that $\lambda - A$ and $\lambda - D$ are one-to-one. Consequently,

$$\alpha(\lambda - A) = \alpha(\lambda - D) = 0.$$

This shows that $\lambda \notin \sigma_{e1}(A) \bigcup \sigma_{e1}(D)$.

(ii) It follows from Remark 3.8.5 (i) that $\sigma_{e2}(\mathscr{L}_0) \subset \Xi_-(\mathscr{L}_0)$. Now, we will prove that

$$\sigma_{e2}(A) \bigcup \sigma_{e2}(D) \subset \sigma_{e2}(\mathscr{L}_0).$$

Let $\lambda \notin \sigma_{e2}(\mathscr{L}_0)$. Then, $\lambda - \mathscr{L}_0 \in \Phi_-(H \times K)$. This implies from Lemma 2.8.5 (ii) that $R(\lambda - A)$ and $R(\lambda - D)$ are closed. It remains to prove that

$$\beta(\lambda - A) < \infty \text{ and } \beta(\lambda - D) < \infty. \tag{5.3.10}$$

Using the fact that $\beta(\lambda - \mathscr{L}_0) < \infty$, together with [5, Lemma 5.3], we infer that there is a subspace $M \times N$ of $H \times K$ such that $\dim(M \times N) < \infty$ and

$$\begin{aligned} H \times K &= R(\lambda - \mathscr{L}_0) \bigoplus (M \times N) \\ &= [R(\lambda - A) \times R(\lambda - D)] \bigoplus [M \times N] \text{ (by Lemma 1.8.5 } (ii)) \\ &= [R(\lambda - A) \bigoplus M] \times [R(\lambda - D) \bigoplus N] \text{ (by Lemma 1.1.2 (i)).} \end{aligned}$$

This leads us to

$$H = R(\lambda - A) \bigoplus M$$

and

$$K = R(\lambda - D) \bigoplus N.$$

Since $R(\lambda - A)$ and $R(\lambda - D)$ are closed, then we can expressed H and K in the form

$$H = R(\lambda - A) \bigoplus R(\lambda - A)^{\perp}$$

and

$$K = R(\lambda - D) \bigoplus R(\lambda - D)^{\perp}.$$

The fact that the decomposition is unique implies from the above that

$$M = R(\lambda - A)^{\perp}$$

and

$$N = R(\lambda - D)^{\perp}.$$

This implies that $\dim(R(\lambda - A)^{\perp}) < \infty$ and $\dim(R(\lambda - D)^{\perp}) < \infty$. Consequently, (5.3.10) holds. Hence, we conclude that $\lambda \notin \left[\sigma_{e2}(A) \bigcup \sigma_{e2}(D)\right]$.

(iii) It follows immediately from (i), (ii), and Remark 3.8.5 (ii) that

$$\sigma_{e3}(A) \bigcup \sigma_{e3}(D) \subset \Xi_{\pm}(\mathcal{L}_0).$$

(iv) First, we shall prove the first inclusion. Let $\lambda \notin \Xi_{-}(\mathcal{L}_0)$. Then, by using (ii), we infer that

$$\lambda \notin \sigma_{e2}(A) \bigcup \sigma_{e2}(D),$$

which yields that $\lambda - A \in \Phi_{-}(H)$ and $\lambda - D \in \Phi_{-}(K)$. The fact that A and D are closed relations and have SVEP at λ implies from Proposition 3.8.7 (i) and [8, Theorem 3.4] that $\lambda - A$ and $\lambda - D$ are one-to-one. Consequently, $\lambda - A$ and $\lambda - D$ are Fredholm. Hence,

$$\sigma_{e4}(A) \bigcup \sigma_{e4}(D) \subset \Xi_{-}(\mathcal{L}_0).$$

Finally, to show the second inclusion, it suffices to use Remark 3.8.5 (ii).

(v) Let $\lambda \notin \Xi_{-}(\mathcal{L}_0) \bigcup \sigma_r(\mathcal{L}_0)$. Then, $\lambda - \mathcal{L}_0 \in \Phi_{-}(H \times K)$, $\mathcal{L}_0$ have SVEP at λ and $\lambda \notin \sigma_r(\mathcal{L}_0)$. The fact that $R(\lambda - \mathcal{L}_0)$ is closed and $\mathcal{L}_0$ have SVEP at λ implies from Proposition 3.8.7 (i) and [8, Theorem 3.4] that $\lambda - \mathcal{L}_0$ is one-to-one. This implies from the assumption $\lambda \notin \sigma_r(\mathcal{L}_0)$ and $R(\lambda - \mathcal{L}_0)$ is closed that $\lambda - \mathcal{L}_0$ is onto. Hence, by Lemma 2.8.5, we infer that $\lambda - A$, $\lambda - D$ are onto and one-to-one. This shows that $\lambda \notin \sigma(A) \bigcup \sigma(D)$. Q.E.D.

Theorem 5.3.5 *Let $A \in C\mathcal{R}(H)$, $D \in C\mathcal{R}(K)$ and let B, C be continuous linear relations from K into H and from H into K, respectively. Suppose that $\mathcal{D}(A) \subset N(C)$, $\mathcal{D}(D) \subset N(B)$, $B(0) \subset A(0)$, $C(0) \subset D(0)$ and*

$$\|C\|(1 + \|B\|) < 1. \tag{5.3.11}$$

Then,

$$\sigma_{e4}(\mathcal{L}_0) \subset \{\lambda \in \sigma_{e4}(A) : \gamma(\lambda - A) \leq 1 + \|B\|\} \bigcup \{\lambda \in \sigma_{e4}(D) : \gamma(\lambda - D) \leq 1 + \|B\|\},$$

where $\mathcal{L}_0$ is given in (2.13.1) and $\gamma(\cdot)$ is the minimum modulus. ◇

Proof On the one hand, we have to prove that

$$\gamma(\lambda - \mathcal{T}) \geq \min\{\gamma(\lambda - A), \gamma(\lambda - D)\}, \quad \text{for } \lambda \in \mathbb{C}, \tag{5.3.12}$$

where $\mathcal{T} = \begin{pmatrix} A & 0 \\ 0 & D \end{pmatrix}$. For any $\begin{pmatrix} x \\ y \end{pmatrix} \in \mathcal{D}(\lambda - \mathcal{T}) = \mathcal{D}(\lambda - A) \times \mathcal{D}(\lambda - D)$, we have

$$\left\|(\lambda-\mathscr{T})\begin{pmatrix}x\\y\end{pmatrix}\right\| = \left\|\begin{pmatrix}(\lambda-A)x\\(\lambda-B)y\end{pmatrix}\right\|$$
$$= \|(\lambda-A)x\| + \|(\lambda-D)y\| \quad \text{(by Remark 2.8.4).}$$

Hence, it follows from [2, Definition II.2.1] that

$$\begin{aligned}\left\|(\lambda-\mathscr{T})\begin{pmatrix}x\\y\end{pmatrix}\right\| &\geq \gamma(\lambda-A)\,\mathrm{dist}(x,N(\lambda-A)) + \gamma(\lambda-D)\,\mathrm{dist}(y,N(\lambda-D))\\ &\geq \min\{\gamma(\lambda-A),\gamma(\lambda-D)\}\,(\mathrm{dist}(x,N(\lambda-A)) + \mathrm{dist}(y,N(\lambda-D)))\\ &\geq \min\{\gamma(\lambda-A),\gamma(\lambda-D)\}\left(\inf_{z\in N(\lambda-A)}\|x-z\| + \inf_{t\in N(\lambda-D)}\|y-t\|\right)\\ &\geq \min\{\gamma(\lambda-A),\gamma(\lambda-D)\}\inf_{\substack{z\in N(\lambda-A)\\ t\in N(\lambda-D)}}(\|x-z\|+\|y-t\|).\end{aligned}$$

By referring to Remark 3.8.4 and Lemma 2.8.5 (i), we can conclude that

$$\begin{aligned}\left\|(\lambda-\mathscr{T})\begin{pmatrix}x\\y\end{pmatrix}\right\| &\geq \min\{\gamma(\lambda-A),\gamma(\lambda-D)\}\inf_{\begin{pmatrix}z\\t\end{pmatrix}\in N(\lambda-\mathscr{T})}\left\|\begin{pmatrix}x-z\\y-t\end{pmatrix}\right\|\\ &\geq \min\{\gamma(\lambda-A),\gamma(\lambda-D)\}\inf_{\begin{pmatrix}z\\t\end{pmatrix}\in N(\lambda-\mathscr{T}))}\left\|\begin{pmatrix}x\\y\end{pmatrix}-\begin{pmatrix}z\\t\end{pmatrix}\right\|\\ &\geq \min\{\gamma(\lambda-A),\gamma(\lambda-D)\}\,\mathrm{dist}\left(\begin{pmatrix}x\\y\end{pmatrix},N(\lambda-\mathscr{T})\right).\end{aligned}$$

This implies from [2, Definition II.2.1] that (5.3.12) holds. On the other hand, let

$$\lambda \notin \{\lambda\in\sigma_{e4}(A) : \gamma(\lambda-A)\leq 1+\|B\|\}\bigcup\{\lambda\in\sigma_{e4}(D) : \gamma(\lambda-D)\leq 1+\|B\|\}.$$

This is equivalent to saying that $\lambda - A \in \Phi(H)$, $\lambda - D \in \Phi(K)$ such that

$$\gamma(\lambda-A) > 1+\|B\| \text{ and } \gamma(\lambda-D) > 1+\|B\|. \tag{5.3.13}$$

Hence, we infer from [9, Lemma 4.1 (ii)] that $\lambda - \mathscr{T} \in \Phi(H\times K)$.

At this level, based on the assumptions $C(0)\subset D(0)$ and $B(0)\subset A(0)$, we obtain

$$\mathscr{S}(0)\subset(\lambda-\mathscr{T})(0),$$

where $\mathscr{S} = \begin{pmatrix}0 & B\\ C & 0\end{pmatrix}$. The fact that $\mathscr{D}(A)\subset\mathscr{D}(C)$ and $\mathscr{D}(D)\subset\mathscr{D}(B)$ implies that

$$\mathscr{D}(\lambda-\mathscr{T}) = \mathscr{D}(A)\times\mathscr{D}(D)$$

$$
\begin{aligned}
&\subset \mathscr{D}(C) \times \mathscr{D}(B) \\
&\subset \mathscr{D}(\mathscr{S}).
\end{aligned}
$$

In addition, by using (2.13.4), we obtain

$$
\begin{aligned}
\|\mathscr{S}\| &\leq \|B\| + \|C\|(1 + \|B\|) \\
&\leq \|B\| + 1 \quad (\text{by } (4.3.11)) \\
&< \min\{\gamma(\lambda - A), \gamma(\lambda - D)\} \quad (\text{by } (4.3.13)) \\
&< \gamma(\lambda - \mathscr{T}) \quad (\text{by } (4.3.12)).
\end{aligned}
$$

This show from Corollary 3.6.3 and Remark 2.13.2 that $\lambda - \mathscr{T} + \mathscr{S} = \lambda - \mathscr{L}_0$ is a Fredholm relation on $H \times K$. Q.E.D.

The following result is a direct consequence of Theorems 5.3.4 and 5.3.5.

Corollary 5.3.2 *Let $A \in C\mathscr{R}(H)$, $D \in C\mathscr{R}(K)$ and let B, C be continuous linear relations from K into H and from H into K, respectively. Suppose that A, D have SVEP, $\mathscr{D}(A) \subset N(C)$, $\mathscr{D}(D) \subset N(B)$, $B(0) \subset A(0)$, $C(0) \subset D(0)$ with*

$$\|C\|(1 + \|B\|) < 1$$

and

$$1 + \|B\| < \min\{\gamma(\lambda - A), \gamma(\lambda - D)\}.$$

Then, $\lambda \in \sigma_{e4}(\mathscr{L}_0)$ if, and only if, $\lambda \in \sigma_{e4}(A) \bigcup \sigma_{e4}(D)$, where $\mathscr{L}_0$ is given in (2.13.1). ◇

5.3.2 Essential Spectra of 2 × 2-Block Matrices Linear Relations

In the rest of this chapter, we extend in the first place the main results of Jeribi in [10] and Tretter in [11] to linear relations. Moreover, we set forward a matrix linear relation. For diagonally dominant and off-diagonally dominant of block matrix linear relations $\mathscr{L}_0$, given in (5.1.1), we provide a necessary and sufficient condition for $\mathscr{L}_0$ to become closed and closable and we try to determine the conditions under which a block matrix linear relations becomes a demicompact block matrix linear relation.

5.3.2.1 Closedness and Closability of the Diagonally and Off-Diagonally Dominant Block Matrix Linear Relations $\mathscr{L}_0$

Theorem 5.3.6 *The block matrix linear relation $\mathscr{L}_0$ is closable if either (i) or (ii) holds.*

(i) $\mathscr{L}_0$ is diagonally dominant, A and D are closable, $B(0) \subset A(0)$, $C(0) \subset D(0)$ and the relative bounds δ_C and δ_B of C and B, respectively, satisfy

$$\delta_C^2(1+\delta_B^2) < 1$$

or

$$\delta_B^2(1+\delta_C^2) < 1.$$

(ii) $\mathscr{L}_0$ is off-diagonally dominant, B and C are closable, $A(0) \subset B(0)$, $D(0) \subset C(0)$ and the relative bounds δ_A and δ_D of A and D, respectively, satisfy

$$\delta_A^2(1+\delta_D^2) < 1$$

or

$$\delta_D^2(1+\delta_A^2) < 1. \qquad \diamond$$

Proof We prove (i), the proof of (ii) is similar. Let us decompose the linear relation $\mathscr{L}_0$ in the following form:

$$\mathscr{L}_0 = \mathscr{T} + \mathscr{S}$$

with

$$\mathscr{T} = \begin{pmatrix} A & B \\ 0 & D \end{pmatrix}$$

and

$$\mathscr{S} = \begin{pmatrix} 0 & 0 \\ C & 0 \end{pmatrix}.$$

Firstly,

$$\begin{aligned} \mathscr{T}(0) &= \begin{pmatrix} A & B \\ 0 & D \end{pmatrix} \begin{pmatrix} 0 \\ 0 \end{pmatrix} \\ &= \begin{pmatrix} A(0) + B(0) \\ D(0) \end{pmatrix} \\ &= \begin{pmatrix} A(0) \\ D(0) \end{pmatrix}. \end{aligned}$$

Since A and D are closable, then $A(0)$ and $D(0)$ are closed. Then, $\mathcal{T}(0)$ is closed. Let $((x_n, y_n)^t)_n \subset \mathcal{D}(A) \times \mathcal{D}(D)$ be such that $((x_n, y_n))_n$ converges to $\begin{pmatrix} 0 \\ 0 \end{pmatrix}$. Then,

$$\begin{aligned} Q_{\mathcal{T}} \mathcal{T} \begin{pmatrix} x_n \\ y_n \end{pmatrix} &= \begin{pmatrix} Q_A A & Q_A B \\ 0 & Q_D D \end{pmatrix} \begin{pmatrix} x_n \\ y_n \end{pmatrix} \\ &= \begin{pmatrix} Q_A A x_n + Q_A B y_n \\ Q_D D y_n \end{pmatrix} \to \begin{pmatrix} x \\ y \end{pmatrix}. \end{aligned}$$

Since D is closable, then $Q_D D$ is closable, so $y = 0$. Besides, B is D-bounded, then there exist $a \geq 0$, $b \geq 0$ such that

$$\begin{aligned} \|Q_A B y_n\| &\leq \|B y_n\| \\ &\leq a\|y_n\| + b\|D y_n\| \\ &\leq a\|y_n\| + b\|Q_D D y_n\|. \end{aligned}$$

Hence $(Q_A B y_n)_n$ converges to 0 and $(Q_A A x_n)_n$ converges to x. The fact that A is closable implies that $Q_A A$ is closable and $(x_n)_n$ converges to 0. Then, $x = 0$. So, $Q_{\mathcal{T}} \mathcal{T}$ is closable. Hence, by using Lemma 2.12.2, we have $\mathcal{T}$ is closable. Using Lemma 2.13.1, we prove that $\mathcal{S}$ is $\mathcal{T}$-bounded with $\mathcal{T}$-bound $\delta < 1$. Finally, using Theorem 2.13.1 (i), we obtain the desired result. Q.E.D.

Theorem 5.3.7 *Let us assume that the entries of the relation matrix $\mathcal{L}_0$, given in (5.1.1), are closable such that C and B are relatively bounded with respect to A and D with A-bound and D-bound δ_C and δ_B, respectively. If $B(0) \subset A(0)$, $C(0) \subset D(0)$ and $\delta_C \delta_B < 1$, then the linear relation $\mathcal{L}_0$ is also closable and its closure has the form:*

$$\overline{\mathcal{L}_0} = \begin{pmatrix} \overline{A} & \overline{B} \\ \overline{C} & \overline{D} \end{pmatrix}. \qquad \diamond$$

Proof Let us decompose the linear relation $\mathcal{L}_0$ in the following form $\mathcal{L}_0 = \mathcal{T} + \mathcal{S}$, where

$$\mathcal{T} = \begin{pmatrix} A & B \\ 0 & D \end{pmatrix}$$

and

$$\mathcal{S} = \begin{pmatrix} 0 & 0 \\ C & 0 \end{pmatrix}.$$

Since the property of closability does not depend on the choice of an equivalent norm, then by using Lemma 2.10.1, it is sufficient to prove that $\mathcal{L}_0$ is closable in the norm defined in (2.10.2). First, in order to show that $\mathcal{T}$ is closable, we will prove that $Q_{\mathcal{T}} \mathcal{T}$ is closable and $\mathcal{T}(0)$ is closed. Let $\begin{pmatrix} x_n \\ y_n \end{pmatrix}$ be a sequence in $\mathcal{D}(Q_{\mathcal{T}} \mathcal{T}) = \mathcal{D}(A) \times \mathcal{D}(D)$ such that

$$\begin{pmatrix} x_n \\ y_n \end{pmatrix} \to \begin{pmatrix} 0 \\ 0 \end{pmatrix} \quad \text{as } n \to \infty,$$

and let

$$\begin{pmatrix} u_n \\ v_n \end{pmatrix} = Q_{\mathcal{T}} \mathcal{T} \begin{pmatrix} x_n \\ y_n \end{pmatrix} \to \begin{pmatrix} u \\ v \end{pmatrix} \quad \text{as } n \to \infty.$$

It follows from Lemma 2.9.2 that

$$\begin{pmatrix} Q_A A x_n + Q_A B y_n \\ Q_D D y_n \end{pmatrix} \to \begin{pmatrix} u \\ v \end{pmatrix} \quad \text{as } n \to \infty. \tag{5.3.14}$$

Since A and D are closable, then $Q_A A$ and $Q_D D$ are closable and $A(0)$, $D(0)$ are closed. It follows from (5.3.14) that

$$Q_D D y_n \to v \text{ as } n \to \infty.$$

This implies that $v = 0$. Now, using the fact that $B(0) \subset A(0)$, we get

$$\begin{aligned} \|Q_A B y_n\| &= \operatorname{dist}(B y_n, \overline{A(0)}) \\ &= \operatorname{dist}(B y_n, A(0)) \\ &\leq \operatorname{dist}(B y_n, B(0)) \\ &\leq \|Q_B B y_n\|. \end{aligned}$$

Therefore, since B is D-bounded, it is easy to see that there exists $a_B, b_B \geq 0$ such that

$$\begin{aligned} \|Q_A B y_n\| &\leq \|Q_B B y_n\| \\ &\leq a_B \|y_n\| + b_B \|Q_D D y_n\|, \text{ for all } (y_n)_n \subset \mathcal{D}(D). \end{aligned} \tag{5.3.15}$$

Hence, by using (5.3.15), we deduce that $Q_A B y_n \to 0$, when $n \to \infty$. This implies from (5.3.14) that

$$Q_A A x_n \to u, \quad \text{as } n \to \infty.$$

Thus, the fact that $x_n \to 0$, when $n \to \infty$ and $Q_A A$ is closable implies that $u = 0$. This enables us to conclude that $Q_{\mathcal{T}} \mathcal{T}$ is closable. Finally, we have to prove that $\mathcal{T}(0)$ is closed. For this purpose, let us assume that $B(0) \subset A(0)$. Then,

$$A(0) + B(0) = A(0).$$

This implies that

$$\mathcal{T}(0) = \begin{pmatrix} A & B \\ 0 & D \end{pmatrix} \begin{pmatrix} 0 \\ 0 \end{pmatrix}$$

$$= \begin{pmatrix} A(0) + B(0) \\ \operatorname{dist}(0) \end{pmatrix}$$
$$= \begin{pmatrix} A(0) \\ D(0) \end{pmatrix}.$$

Hence, the fact that $A(0)$ and $D(0)$ are closed implies that $\mathscr{T}(0)$ is closed. As result, $\mathscr{T}$ is closable, as desired. At this level, let us show that the linear relation $\mathscr{S}$ is $\mathscr{T}$-bounded in the norm defined in (2.10.2), and let us find its $\mathscr{T}$-bound. According to the assumptions, there exist constants $a_C, b_C, a_B, b_B \geq 0$ such that

$$\|Cx\| \leq a_C\|x\| + b_C\|Ax\|, \text{ for all } x \in \mathscr{D}(A). \tag{5.3.16}$$

$$\|By\| \leq a_B\|y\| + b_B\|Dy\|, \text{ for all } y \in \mathscr{D}(D). \tag{5.3.17}$$

Let the condition $\delta_B\delta_C < 1$ be fulfilled. Assume that $\delta_C < 1$. Later, we will say about the changes in the proof in the case $\delta_B < 1$. Obviously, we can choose numbers b_C and b_B in the estimates (5.3.16) and (5.3.17) to be close to the numbers δ_C and δ_B. So, the inequalities $b_C < 1$ and $b_C b_B < 1$ are preserved. Let us assume that $z = \begin{pmatrix} x \\ y \end{pmatrix} \in \mathscr{D}(A) \times \mathscr{D}(D)$. Then, by using (5.3.16), we can write

$$\begin{aligned}
\|\mathscr{S}z\| &= \|Q_{\mathscr{S}}\mathscr{S}z\| \\
&\leq \left\| \begin{pmatrix} 0 & 0 \\ Q_C C & 0 \end{pmatrix} \begin{pmatrix} x \\ y \end{pmatrix} \right\| \\
&= \|Q_C Cx\| \\
&\leq a_C\|x\| + b_C\|Q_A Ax\| \\
&\leq a_C\|x\| + b_C \ \left[\|Q_A Ax + Q_A By\| + \|Q_A By\|\right].
\end{aligned}$$

Since B is D-bounded, then it follows from (5.3.15) that

$$\begin{aligned}
\|\mathscr{S}z\| &\leq a_C\|x\| + b_C \ \left[\|Q_A Ax + Q_A By\| + a_B\|y\| + b_B\|Q_D Dy\|\right] \\
&\leq a_C\|x\| + a_B b_C\|y\| + b_C\|Q_A Ax + Q_A By\| + b_C b_B\|Q_D Dy\| \\
&\leq \max(a_C, a_B b_C)\left[\|x\| + \|y\|\right] + \max(b_C, b_C b_B)\left[\|Q_A Ax + Q_A By\| + \|Q_D Dy\|\right] \\
&\leq \max(a_C, a_B b_C)\left\| \begin{pmatrix} x \\ y \end{pmatrix} \right\| + \max(b_C, b_C b_B)\left\| \begin{pmatrix} Q_A A & Q_A B \\ 0 & Q_D D \end{pmatrix} \begin{pmatrix} x \\ y \end{pmatrix} \right\|.
\end{aligned}$$

Thus, we have

$$\begin{aligned}
\|\mathscr{S}z\| &\leq \max(a_C, a_B b_C)\|z\| + \max(b_C, b_C b_B)\|Q_{\mathscr{T}}\mathscr{T}z\| \\
&\leq \max(a_C, a_B b_C)\|z\| + \max(b_C, b_C b_B)\|\mathscr{T}z\|.
\end{aligned}$$

Consequently, the $\mathscr{T}$-bound of the linear relation $\mathscr{S}$ is equal to $\max(b_C, b_C b_B) < 1$. The fact that $\mathscr{T}$ is closable, $\mathscr{S}(0) \subset \mathscr{T}(0)$, and $\mathscr{D}(\mathscr{T}) \subset \mathscr{D}(\mathscr{S})$ implies from Lemma 2.12.3 that

$$\mathscr{L}_0 = \mathscr{T} + \mathscr{S}$$

is closable and

$$\begin{aligned}\overline{\mathscr{L}_0} &= \overline{\mathscr{T}} + \overline{\mathscr{S}} \\ &= \begin{pmatrix} \overline{A} & \overline{B} \\ \overline{C} & \overline{D} \end{pmatrix}.\end{aligned}$$

The proof of theorem is carried out analogously if the inequality $\delta_B < 1$ holds instead of $\delta_C < 1$. In this case, the operator $\mathscr{L}_0$ has to be presented in the form

$$\mathscr{L}_0 = \mathscr{T}_1 + \mathscr{S}_1,$$

where

$$\mathscr{T}_1 = \begin{pmatrix} A & 0 \\ C & D \end{pmatrix}$$

and

$$\mathscr{S}_1 = \begin{pmatrix} 0 & B \\ 0 & 0 \end{pmatrix}.$$

The proof of closability of $\mathscr{T}_1$ and $\mathscr{T}_1$-boundedness of the operator $\mathscr{S}_1$ with $\mathscr{T}_1$-bound < 1 is carried out as before. Q.E.D.

Remark 5.3.3 The condition $\delta_C \delta_B < 1$ of Theorem 5.3.7 is weaker than that of Theorem 5.3.6, in which it is required that $\delta_C^2(1 + \delta_B^2) < 1$ or $\delta_B^2(1 + \delta_C^2) < 1$. ◇

Corollary 5.3.3 *The block matrix linear relation $\mathscr{L}_0$, given in (5.1.1), is closed if one of the following conditions holds*

(i) $\mathscr{L}_0$ is diagonally dominant, A and D are closed, $B(0) \subset A(0)$, $C(0) \subset D(0)$ and the relative bounds δ_C and δ_B of C and B, respectively, satisfy

$$\delta_C^2(1 + \delta_B^2) < 1$$

or

$$\delta_B^2(1 + \delta_C^2) < 1.$$

(ii) $\mathscr{L}_0$ is off-diagonally dominant, B and C are closed, $A(0) \subset B(0)$, $D(0) \subset C(0)$ and the relative bounds δ_A and δ_D of A and D, respectively, satisfy

$$\delta_A^2(1 + \delta_D^2) < 1$$

or

$$\delta_D^2(1+\delta_A^2) < 1.$$

◇

Corollary 5.3.4 *The block matrix linear relation $\mathscr{L}_0$, given in (5.1.1), is closed if one of the following conditions holds:*

(i) C is A-bounded, B is D-bounded, A and D are closed, $B(0) \subset A(0)$, $C(0) \subset D(0)$ and at least one relative bound is 0.
(ii) A is C-bounded, D is B-bounded, B and C are closed, $A(0) \subset B(0)$, $D(0) \subset C(0)$ and at least one relative bound is 0. ◇

In the next two corollaries, we consider particular cases of linear relations with relative bound 0. As in Corollary 5.3.4, we distinguish the diagonally dominant case (i) and the off-diagonally dominant case (ii). Clearly, if X, Y, and Z are three Banach spaces and S is bounded from X into Z, then S is T-bounded with T-bound 0 for every linear relation T from X into Y. Thus, the following corollary is verified.

Corollary 5.3.5 *The block matrix linear relation $\mathscr{L}_0$, given in (5.1.1), is closed if one of the following conditions holds:*

(i) A and D are closed, $B(0) \subset A(0)$, $C(0) \subset D(0)$ and either (a) or (b) holds

(a) $\mathscr{D}(A) \subset \mathscr{D}(C)$ and B is bounded,
(b) $\mathscr{D}(D) \subset \mathscr{D}(B)$ and C is bounded.

(ii) B and C are closed, $A(0) \subset B(0)$, $D(0) \subset C(0)$ and either (a) or (b) holds

(a) $\mathscr{D}(C) \subset \mathscr{D}(A)$ and D is bounded,
(b) $\mathscr{D}(B) \subset \mathscr{D}(D)$ and A is bounded. ◇

Remark 5.3.4 If X and Y are reflexive, the linear relations T and the operator S are closable, and S is T-compact, then S is T-bounded with T-bound 0. ◇

This immediately implies the next corollary.

Corollary 5.3.6 *The block matrix linear relation $\mathscr{L}_0$, given in (5.1.1), is closed if X and Y are reflexive Banach spaces and either (i) or (ii) holds*

(i) A and D are closed, $B(0) \subset A(0)$, $C(0) \subset D(0)$ and either (a) or (b) holds

(a) $\mathscr{D}(A) \subset \mathscr{D}(C)$ and B is D-compact,
(b) $\mathscr{D}(D) \subset \mathscr{D}(B)$ and C is A-compact.

(ii) B and C are closed, $A(0) \subset B(0)$, $D(0) \subset C(0)$ and either (a) or (b) holds

(a) $\mathscr{D}(C) \subset \mathscr{D}(A)$ and D is B-compact,
(b) $\mathscr{D}(B) \subset \mathscr{D}(D)$ and A is C-compact. ◇

5.3.2.2 Stability of the Essential Spectra with the Diagonally and Off-Diagonally Dominant Block Matrix Linear Relations $\mathscr{L}_0$

In this subsection, we investigate the stability of the essential spectra with the diagonally and off-diagonally dominant block matrix linear relation $\mathscr{L}_0$. We present the following two theorems and two corollaries, useful to have results about the stability of $\sigma_{e1}(\mathscr{L}_0)$.

Theorem 5.3.8 *Let A, B, C, and $D \in L\mathscr{R}(X)$, we define the block matrix of linear relation*

$$\mathscr{T} = \begin{pmatrix} A & B \\ 0 & D \end{pmatrix}$$

and

$$\mathscr{S} = \begin{pmatrix} 0 & 0 \\ C & 0 \end{pmatrix}.$$

Let $\mathscr{L}_0$, given in (5.1.1), be diagonally dominant, A and D be closed, $B(0) \subset A(0)$ and $C(0) \subset D(0)$ and the relative bounds δ_C and δ_B of C and B, respectively, satisfy

$$\delta_C^2(1 + \delta_B^2) < 1$$

or

$$\delta_B^2(1 + \delta_C^2) < 1.$$

If $G(\mathscr{S}) \subset G(\mathscr{T})$ and $\dim \mathscr{D}(\mathscr{S}) = \infty$, then

$$\sigma_{e1}(\mathscr{L}_0) \subset \sigma_{e1}(\mathscr{T}).$$

If, in addition, $\dim \mathscr{S}(0) < \infty$, then

$$\sigma_{e1}(\mathscr{L}_0) = \sigma_{e1}(\mathscr{T}). \qquad \diamond$$

Proof By using Lemma 2.13.1, we prove that $\mathscr{S}$ is $\mathscr{T}$-bounded with $\mathscr{T}$-bound $\delta < 1$. Applying Theorem 2.13.1, we get the result. Q.E.D.

Corollary 5.3.7 *Define the block matrix of linear relation*

$$\mathscr{T} = \begin{pmatrix} A & B \\ 0 & D \end{pmatrix}$$

and

$$\mathscr{S} = \begin{pmatrix} 0 & 0 \\ C & 0 \end{pmatrix}.$$

Let $\mathscr{L}_0$, given in (5.1.1), be off-diagonally dominant, B and C be closed, $A(0) \subset B(0)$ and $D(0) \subset C(0)$ and the relative bounds δ_A and δ_D of A and D, respectively, satisfy

$$\delta_A^2(1+\delta_D^2) < 1$$

or

$$\delta_D^2(1+\delta_A^2) < 1.$$

If $G(\mathcal{T}) \subset G(\mathcal{S})$ *and* $\dim \mathcal{D}(\mathcal{T}) = \infty$, *then*

$$\sigma_{e1}(\mathcal{L}_0) \subset \sigma_{e1}(\mathcal{S}) = \sigma_{e1}(C).$$

If, in addition, $\dim \mathcal{T}(0) < \infty$, *then*

$$\sigma_{e1}(\mathcal{L}_0) = \sigma_{e1}(C). \qquad \diamond$$

Corollary 5.3.8 *Define the block matrix of linear relation*

$$\mathcal{L} = \begin{pmatrix} A_1 & B_1 \\ C_1 & D_1 \end{pmatrix},$$

$$\mathcal{M} = \begin{pmatrix} A_2 & B_2 \\ C_2 & D_2 \end{pmatrix}$$

,

$$\mathcal{T} = \begin{pmatrix} A_1 & 0 \\ 0 & D_1 \end{pmatrix},$$

and

$$\mathcal{S} = \begin{pmatrix} 0 & B_1 \\ C_1 & 0 \end{pmatrix}.$$

Let $\mathcal{L}$ *and* $\mathcal{M}$ *be diagonally dominant of the order* < 1, A_1, A_2, D_1, *and* D_2 *be closed,* $B_1(0) \subset A_1(0)$, $C_1(0) \subset D_1(0)$, $\mathcal{M}(0) \subset \mathcal{L}(0)$, $\mathcal{D}(\mathcal{L}) \subset \mathcal{D}(\mathcal{M})$ *and* $Q_{\mathcal{L}}(\mathcal{M} - I)$ *is compact. If* $G(\mathcal{S}) \subset G(\mathcal{T})$ *and* $\dim \mathcal{D}(\mathcal{S}) = \infty$, *then*

$$\sigma_l(\mathcal{L}) = \sigma_r(\mathcal{L}) \subset \sigma_r(A_1) \bigcup \sigma_r(D_1) = \sigma_l(A_1) \bigcup \sigma_l(D_1).$$

If, in addition, $\dim \mathcal{S}(0) < \infty$, *then*

$$\sigma_l(\mathcal{L}) = \sigma_r(\mathcal{L}) \subset \sigma_r(A_1) = \sigma_r(D_1) = \sigma_l(A_1) = \sigma_l(D_1). \qquad \diamond$$

Referring to this Corollary 5.3.8, we infer the next corollary.

Corollary 5.3.9 *Define the block matrix of linear relation*

$$\mathcal{L} = \begin{pmatrix} A_1 & B_1 \\ C_1 & D_1 \end{pmatrix},$$

$$\mathcal{M} = \begin{pmatrix} A_2 & B_2 \\ C_2 & D_2 \end{pmatrix},$$

$$\mathscr{T} = \begin{pmatrix} A_1 & 0 \\ 0 & D_1 \end{pmatrix},$$

and

$$\mathscr{S} = \begin{pmatrix} 0 & B_1 \\ C_1 & 0 \end{pmatrix}.$$

Let $\mathscr{L}$ and $\mathscr{M}$ be off-diagonally dominant of the order < 1, B_1, B_2, C_1 and C_2 be closed, $A_1(0) \subset B_1(0)$, $D_1(0) \subset C_1(0)$, $\mathscr{M}(0) \subset \mathscr{L}(0)$, $\mathscr{D}(\mathscr{L}) \subset \mathscr{D}(\mathscr{M})$ and $Q_{\mathscr{L}}(\mathscr{M} - I)$ is compact. If $G(\mathscr{T}) \subset G(\mathscr{S})$ and $\dim \mathscr{D}(\mathscr{T}) = \infty$, then

$$\sigma_l(\mathscr{L}) = \sigma_r(\mathscr{L}) \subset \sigma_r(B_1) \bigcup \sigma_r(C_1) = \sigma_l(B_1) \bigcup \sigma_l(C_1).$$

If, in addition, $\dim \mathscr{S}(0) < \infty$, then

$$\sigma_l(\mathscr{L}) = \sigma_r(\mathscr{L}) \subset \sigma_r(B_1) \cup \sigma_r(C_1) = \sigma_l(B_1) \cup \sigma_l(C_1). \qquad \diamond$$

Theorem 5.3.9 *Define the block matrix of linear relation*

$$\mathscr{L} = \begin{pmatrix} A_1 & B_1 \\ C_1 & D_1 \end{pmatrix},$$

$$\mathscr{M} = \begin{pmatrix} A_2 & B_2 \\ C_2 & D_2 \end{pmatrix},$$

$$\mathscr{T} = \begin{pmatrix} A_1 & B_1 \\ 0 & D_1 \end{pmatrix},$$

and

$$\mathscr{S} = \begin{pmatrix} 0 & 0 \\ C_1 & 0 \end{pmatrix}.$$

Let $\mathscr{L}$ and $\mathscr{M}$ be diagonally dominant such that the relative bounds δ_{C_1},δ_{C_2} and δ_{B_1} δ_{B_2} of C_1, C_2, B_1, and B_2, respectively, satisfy $\delta^2_{C_1}(1+\delta^2_{B_1}) < 1$ or $\delta^2_{B_1}(1+\delta^2_{C_1}) < 1$ and $\delta^2_{C_2}(1+\delta^2_{B_2}) < 1$ or $\delta^2_{B_2}(1+\delta^2_{C_2}) < 1$. Let A_1, A_2, D_1, and D_2 be closed, $B_1(0) \subset A_1(0)$, $C_1(0) \subset D_1(0)$, $\mathscr{M}(0) \subset \mathscr{L}(0)$, $\mathscr{D}(\mathscr{L}) \subset \mathscr{D}(\mathscr{M})$ and $Q_{\mathscr{L}}(\mathscr{M} - I)$ is compact. If $G(\mathscr{S}) \subset G(\mathscr{T})$ and $\dim \mathscr{D}(\mathscr{S}) = \infty$, then

$$\sigma_l(\mathscr{L}) = \sigma_r(\mathscr{L}) \subset \sigma_r(A_1) \bigcup \sigma_r(D_1) = \sigma_l(A_1) \bigcup \sigma_l(D_1).$$

If, in addition, $\dim \mathscr{S}(0) < \infty$, then

$$\sigma_l(\mathscr{L}) = \sigma_r(\mathscr{L}) \subset \sigma_r(A_1) \cup \sigma_r(D_1) = \sigma_l(A_1) = \sigma_l(D_1). \qquad \diamond$$

Hence, Theorem 5.3.9 immediately implies the next corollary.

Corollary 5.3.10 *Define the block matrix of linear relation*

$$\mathscr{L} = \begin{pmatrix} A_1 & B_1 \\ C_1 & D_1 \end{pmatrix},$$

$$\mathscr{M} = \begin{pmatrix} A_2 & B_2 \\ C_2 & D_2 \end{pmatrix},$$

$$\mathscr{T} = \begin{pmatrix} A_1 & B_1 \\ 0 & D_1 \end{pmatrix},$$

and

$$\mathscr{S} = \begin{pmatrix} 0 & 0 \\ C_1 & 0 \end{pmatrix}.$$

Let $\mathscr{L}$ *and* $\mathscr{M}$ *be diagonally dominant such that the relative bounds* δ_{D_1}, δ_{D_2} *and* δ_{A_1} δ_{A_2} *of* D_1, D_2, A_1 *and* A_2, *respectively, satisfy* $\delta^2_{D_1}(1+\delta^2_{A_1}) < 1$ *or* $\delta^2_{A_1}(1+\delta^2_{D_1}) < 1$ *and* $\delta^2_{D_2}(1+\delta^2_{A_2}) < 1$ *or* $\delta^2_{A_2}(1+\delta^2_{D_2}) < 1$. *Let* B_1, B_2, C_1 *and* C_2 *be closed,* $A_1(0) \subset B_1(0)$, $D_1(0) \subset C_1(0)$, $\mathscr{M}(0) \subset \mathscr{L}(0)$, $\mathscr{D}(\mathscr{L}) \subset \mathscr{D}(\mathscr{M})$ *and* $Q_{\mathscr{L}}(\mathscr{M} - I)$ *is compact. If* $G(\mathscr{T}) \subset G(\mathscr{S})$ *and* $\dim \mathscr{D}(\mathscr{T}) = \infty$, *then*

$$\sigma_l(\mathscr{L}) = \sigma_r(\mathscr{L}) \subset \sigma_r(\mathscr{S}) = \sigma_l(\mathscr{S}).$$

If, in addition, $\dim \mathscr{S}(0) < \infty$, *then*

$$\sigma_l(\mathscr{L}) = \sigma_r(\mathscr{L}) = \sigma_r(\mathscr{S}) = \sigma_l(\mathscr{S}).$$ ◇

5.4 The Pencils Essential Spectra of Block Matrices of Multivalued Linear Operators

The purpose of this section is to discuss the pencils essential spectra of the 2×2 matrix linear relation $\mathscr{L}_0$, given in (5.1.1), on the Banach space $X \times Y$, where S is a continuous linear relation formally defined on the product space $X \times Y$, by a matrix

$$S = \begin{pmatrix} S_1 & S_2 \\ S_3 & S_4 \end{pmatrix},$$

the linear relation S_1 acts on the Banach space X and has the domain $\mathscr{D}(S_1)$, S_4 is defined on $\mathscr{D}(S_4)$ and acts on the Banach space Y, and the intertwining linear relation S_2 (resp. S_3) is defined on the domain $\mathscr{D}(S_2)$ (resp. $\mathscr{D}(S_3)$) and acts between these spaces.

5.4.1 *Closedness and Closability of Pencils Matrix of Multivalued Linear Operator*

The aim of this section is to discuss the closure of $\mathscr{L}_0$. We will consider the following hypotheses:
$(H17)$ The linear relation A is continuous, $\mathscr{D}(A)$ is closed, $\dim(A(0)) < \infty$, S_1 is continuous with $S_1(0) \subset A(0)$, $\mathscr{D}(A) \subset \mathscr{D}(S_1)$ and $\rho_{S_1}(A) \neq \emptyset$.
$(H18)$ The linear relation C is closable satisfies $\mathscr{D}(A) \subset \mathscr{D}(C)$, S_3 is continuous with $S_3(0) \subset C(0)$ and $\overline{\mathscr{D}(C)} \subset \mathscr{D}(S_3)$.
$(H19)$ The linear relation B is closable, S_2 is continuous with $S_2(0) \subset B(0)$ and $\overline{\mathscr{D}(B)} = \mathscr{D}(S_2) = Y$.
$(H20)$ The linear relation D is bounded and, $D(0) \subset (C - \lambda S_3)(A - \lambda S_1)^{-1}B(0) \subset C(0)$.

The following lemma is introduced in order to justify the choices of the preceding hypotheses.

Lemma 5.4.1 (i) *Let C be closable satisfies $\mathscr{D}(A) \subset \mathscr{D}(C)$, S_3 be continuous with $S_3(0) \subset C(0)$ and $\overline{\mathscr{D}(C)} \subset \mathscr{D}(S_3)$. Then, for all $\lambda \in \rho_{S_1}(A)$, we have $F(\lambda) = (C - \lambda S_3)(A - \lambda S_1)^{-1}$ is a bounded linear relation.*
(ii) *If B is closable and S_2 is continuous with $S_2(0) \subset B(0)$ and $\overline{\mathscr{D}(B)} = \mathscr{D}(S_2) = Y$, then for all $\lambda \in \rho_{S_1}(A)$, the operator $\Psi(\lambda) = (A - \lambda S_1)^{-1}(B - \lambda S_2)$ is closable. It follows that the operator $E(\lambda) = \overline{\Psi(\lambda)}$ is bounded on Y.*
(iii) *If $D \in B\mathscr{R}(Y)$ with $D(0) \subset C(0)$, then $M(\lambda) = D - (C - \lambda S_3)(A - \lambda S_1)^{-1}(B - \lambda S_2)$ is closable for all $\lambda \in \rho_{S_1}(A)$ and*

$$\overline{M(\lambda)} = D - \overline{(C - \lambda S_3)(A - \lambda S_1)^{-1}(B - \lambda S_2)}$$

. ◇

Proof (i) Using Proposition 2.12.2, we get $C - \lambda S_3$ is closable. We know that if $\dim(A(0)) < \infty$, then $A(0)$ is closed and we have A is continuous and $\mathscr{D}(A)$ is closed. Therefore, by using Proposition 2.15.3, we get A is closed. Thus, by using Proposition 2.11.2 (i) for all $\lambda \in \rho_{S_1}(A)$, we obtain $A - \lambda S_1$ is closed. For all $\lambda \in \rho_{S_1}(A)$, we have $A - \lambda S_1$ is closed. Then, by using Proposition 2.11.1, we get $(A - \lambda S_1)^{-1}$ is closed single-valued and

$$\mathscr{D}((A - \lambda S_1)^{-1}) = R((A - \lambda S_1)) = Y.$$

From Theorem 2.11.1, we have $(A - \lambda S_1)^{-1}$ is bounded. Using Lemma 2.12.5, we obtain $(C - \lambda S_3)(A - \lambda S_1)^{-1}$ is closable. Moreover,

$$\begin{aligned} X &= R(A - \lambda S_1) \\ &= (A - \lambda S_1)\mathscr{D}(A - \lambda S_1) \\ &= (A - \lambda S_1)\mathscr{D}(A) \end{aligned}$$

$$\subset (A-\lambda S_1)\mathscr{D}(C).$$

Since

$$\begin{aligned}(A-\lambda S_1)\mathscr{D}(C) &= (A-\lambda S_1)\mathscr{D}(C-\lambda S_3) \\ &= \mathscr{D}\left((C-\lambda S_3)(A-\lambda S_1)^{-1}\right) \\ &= \mathscr{D}(F(\lambda)),\end{aligned}$$

and consequently $\mathscr{D}(F(\lambda)) = X$. According to Theorem 2.11.1, we have $F(\lambda)$ is continuous. Hence, $F(\lambda)$ is bounded.

(ii) Using Proposition 2.12.2, we have $B - \lambda S_2$ is closable. It is clear that

$$\alpha((A-\lambda S_1)^{-1}) = \dim(A(0)) < \infty$$

and $R((A-\lambda S_1)^{-1}) = \mathscr{D}(A)$ is closed. Using Lemma 2.11.4, we have $(A-\lambda S_1)^{-1}\overline{(B-\lambda S_2)}$ is closed. Hence, by Remark 2.11.5, we infer that $(A-\lambda S_1)^{-1}\overline{(B-\lambda S_2)}$ is closed. But

$$(A-\lambda S_1)^{-1}(B-\lambda S_2)(0) \subset \overline{(A-\lambda S_1)^{-1}(B-\lambda S_2)(0)},$$

then

$$\overline{(A-\lambda S_1)^{-1}(B-\lambda S_2)(0)} \subset \overline{(A-\lambda S_1)^{-1}(B-\lambda S_2)}(0).$$

Moreover,

$$(A-\lambda S_1)^{-1}(B-\lambda S_2) \subset (A-\lambda S_1)^{-1}\overline{(B-\lambda S_2)}.$$

So,

$$\overline{(A-\lambda S_1)^{-1}(B-\lambda S_2)(0)} \subset \overline{(A-\lambda S_1)^{-1}\overline{(B-\lambda S_2)}}(0).$$

Since $(A-\lambda S_1)^{-1}\overline{(B-\lambda S_2)}$ is closed, then

$$\overline{(A-\lambda S_1)^{-1}(B-\lambda S_2)(0)} \subset (A-\lambda S_1)^{-1}\overline{(B-\lambda S_2)}(0).$$

Assume that $B-\lambda S_2$ is closable, consequently,

$$(A-\lambda S_1)^{-1}\overline{(B-\lambda S_2)}(0) = (A-\lambda S_1)^{-1}(B-\lambda S_2)(0).$$

We conclude that,

$$(A-\lambda S_1)^{-1}(B-\lambda S_2)(0) = \overline{(A-\lambda S_1)^{-1}(B-\lambda S_2)(0)}.$$

So,

$$(A-\lambda S_1)^{-1}(B-\lambda S_2)(0) \subset \overline{(A-\lambda S_1)^{-1}(B-\lambda S_2)(0)}$$

$$\begin{aligned}&\subset \overline{(A-\lambda S_1)^{-1}(B-\lambda S_2)(0)}\\&\subset \overline{(A-\lambda S_1)^{-1}}\,\overline{(B-\lambda S_2)}(0)\\&= (A-\lambda S_1)^{-1}\overline{(B-\lambda S_2)}(0)\\&= (A-\lambda S_1)^{-1}(B-\lambda S_2)(0),\ \text{(since } B-\lambda S_2 \text{ is closable).}\end{aligned}$$

Consequently,

$$(A-\lambda S_1)^{-1}(B-\lambda S_2)(0) = \overline{(A-\lambda S_1)^{-1}(B-\lambda S_2)}(0).$$

Therefore,

$$\Psi(\lambda) = (A-\lambda S_1)^{-1}(B-\lambda S_2)$$

is closable. Now, we show that $E(\lambda)$ is bounded. We have

$$\mathcal{D}(\Psi(\lambda)) \times Y = G(\Psi(\lambda)) + (\{0\} \times Y),$$

and

$$\begin{aligned}\overline{G(\Psi(\lambda))} + (\{0\} \times Y) &\subset \overline{G(\Psi(\lambda)) + (\{0\} \times Y)}\\&= \overline{\mathcal{D}(\Psi(\lambda))} \times Y.\end{aligned}$$

Clearly,

$$\begin{aligned}\overline{G(\Psi(\lambda))} + (\{0\} \times Y) &= G(\overline{\Psi(\lambda)}) + (\{0\} \times Y)\\&= \mathcal{D}(\overline{\Psi(\lambda)}) \times Y.\end{aligned}$$

Hence,

$$\mathcal{D}(\overline{\Psi(\lambda)}) \subset \overline{\mathcal{D}(\Psi(\lambda))}.$$

We can extend $\overline{\Psi(\lambda)}$ by $\widetilde{\overline{\Psi(\lambda)}}$ which is continuous over all space Y such that $\widetilde{\overline{\Psi(\lambda)}} = \overline{\Psi(\lambda)} = \Psi(\lambda)$ on $\mathcal{D}(\Psi(\lambda)) = \mathcal{D}(B)$. Let $x \in \overline{\mathcal{D}(\Psi(\lambda))}$, then there exists $x_n \in \mathcal{D}(B)$ such that $x_n \to x$. Since $\widetilde{\overline{\Psi(\lambda)}}$ is continuous single-valued, then $\widetilde{\overline{\Psi(\lambda)}}x_n \to \widetilde{\overline{\Psi(\lambda)}}x$, which implies that

$$\Psi(\lambda)x_n \to \widetilde{\overline{\Psi(\lambda)}}x$$

because $x_n \in \mathcal{D}(B)$. Let $((x_n, \Psi(\lambda)x_n))_n \in G(\Psi(\lambda))$ be a sequence which converges to $(x, \widetilde{\overline{\Psi(\lambda)}}x) \in \overline{G(\Psi(\lambda))} = G(\overline{\Psi(\lambda)})$. Then, $x \in \mathcal{D}(\overline{\Psi(\lambda)})$. Therefore,

$$\begin{aligned}\mathcal{D}(\overline{\Psi(\lambda)}) &= \overline{\mathcal{D}(\Psi(\lambda))}\\&= \overline{\mathcal{D}(B)}\\&= Y.\end{aligned}$$

So, according to Theorem 2.11.1, $E(\lambda) = \overline{\Psi(\lambda)}$ is bounded.

(iii), we have

$$\mathscr{D}((C - \lambda S_3)(A - \lambda S_1)^{-1}(B - \lambda S_2)) = \mathscr{D}(B)$$

and $C - \lambda S_3$ is closable. Moreover, we get by (ii) that the operator $\Psi(\lambda)$ is bounded on $\mathscr{D}(B)$. Then, by virtue of Lemma 2.12.5 (i), we have $(C - \lambda S_3)\Psi(\lambda)$ is closable and D is continuous with $\overline{\mathscr{D}(B)} = \mathscr{D}(D) = Y$ and

$$D(0) \subset D - (C - \lambda S_3)(A - \lambda S_1)^{-1} B(0).$$

By using Corollary 2.12.1 (ii), we deduce that $M(\lambda)$ is closable and

$$\overline{M(\lambda)} = D - \overline{(C - \lambda S_3)(A - \lambda S_1)^{-1}(B - \lambda S_2)}. \qquad \text{Q.E.D.}$$

Remark 5.4.1 Let's take the following assumptions: $F(\lambda)$ and $E(\lambda)$ are bounded linear relations and $M(\lambda)$ is a closable linear relation. Particularly, when $E(\lambda)$ is operator, we find that the preceding assumptions are still valid. This was already proven to be true throughout Lemma 5.4.1. ◇

5.4.2 *Frobenius–Schur's Decomposition of the Pencils* 2×2 *Matrix of Multivalued Linear Operator*

Theorem 5.4.1 *Let $\mu \in \rho_{S_1}(A)$ and $S_4 \in B\mathscr{R}(Y)$ be with $S_4(0) \subset C(0)$. We suppose that the conditions $(H2)$–$(H5)$ are satisfied. Then,*

$$\mu S - \mathscr{L}_0 = HKJ,$$

where

$$H = \begin{pmatrix} I & 0 \\ F(\mu) & I \end{pmatrix},$$

$$J = \begin{pmatrix} I & E(\mu) \\ 0 & I \end{pmatrix},$$

and K is the diagonal multivalued linear relation matrix, given by

$$K = \begin{pmatrix} \mu S_1 - A & 0 \\ 0 & \mu S_4 - M(\mu) \end{pmatrix}. \qquad ◇$$

Proof Let $\begin{pmatrix} x_1 \\ x_2 \end{pmatrix} \in \mathscr{D}(HKJ)$ be such that

$$(HKJ)\begin{pmatrix} x_1 \\ x_2 \end{pmatrix} = H\begin{pmatrix} A - \mu S_1 & 0 \\ 0 & M(\mu) - \mu S_4 \end{pmatrix}\begin{pmatrix} I & E(\mu) \\ 0 & I \end{pmatrix}\begin{pmatrix} x_1 \\ x_2 \end{pmatrix}$$

$$= \begin{pmatrix} I & 0 \\ F(\mu) & I \end{pmatrix} \begin{pmatrix} (A - \mu S_1)x_1 + (A - \mu S_1)E(\mu)x_2 \\ (M(\mu) - \mu S_4)x_2 \end{pmatrix}.$$

Since $x_2 \in \mathscr{D}(B)$, then

$$(A - \mu S_1)E(\mu)x_2 = (B - \mu S_2)x_2 + (A - \mu S_1)(0).$$

It follows that

$$\begin{aligned} (HKJ)\begin{pmatrix} x_1 \\ x_2 \end{pmatrix} &= \begin{pmatrix} I & 0 \\ F(\mu) & I \end{pmatrix} \begin{pmatrix} (A - \mu S_1)x_1 + (B - \mu S_2)x_2 + (A - \mu S_1)(0) \\ (M(\mu) - \mu S_4)x_2 \end{pmatrix} \\ &= \begin{pmatrix} (A - \mu S_1)x_1 + (B - \mu S_2)x_2 \\ F(\mu)((A - \mu S_1)x_1 + (B - \mu S_2)x_2) + M(\mu)x_2 - \mu S_4 x_2 \end{pmatrix}. \end{aligned}$$

First of all, it should be mentioned that $A - \mu S_1$ is one-to-one, then we obtain

$$F(\mu)(A - \mu S_1)x_1 = (C - \mu S_3)x_1.$$

Also, we have $\mathscr{D}(M(\mu)) \subset \mathscr{D}(D) = Y$ and $D(0) \subset C(0) = M(\mu)(0)$. So, we get

$$M(\mu) = M(\mu) - D + D.$$

Thus,

$$\begin{aligned} (HKJ)\begin{pmatrix} x_1 \\ x_2 \end{pmatrix} &= \begin{pmatrix} (A - \mu S_1)x_1 + (B - \mu S_2)x_2 \\ (C - \mu S_3)x_1 - M(\mu)x_2 + M(\mu)x_2 + Dx_2 - \mu S_4 x_2 \end{pmatrix} \\ &= \begin{pmatrix} (A - \mu S_1)x_1 + (B - \mu S_2)x_2 \\ (C - \mu S_3)x_1 + (C - \mu S_3)(0) + (D - \mu S_4)x_2 \end{pmatrix} \\ &= \left[\begin{pmatrix} A & B \\ C & D \end{pmatrix} - \mu \begin{pmatrix} S_1 & S_2 \\ S_3 & S_4 \end{pmatrix} \right] \begin{pmatrix} x_1 \\ x_2 \end{pmatrix}. \end{aligned}$$

However, we have

$$(HKJ)\begin{pmatrix} x_1 \\ x_2 \end{pmatrix} = (\mathscr{L}_0 - \mu S)\begin{pmatrix} x_1 \\ x_2 \end{pmatrix}.$$

From other angle,

$$\begin{aligned} \mathscr{D}(HKJ) &= \left\{ \begin{pmatrix} x_1 \\ x_2 \end{pmatrix} \in X \times Y \text{ such that } x_1 + F(\mu)x_2 \in \mathscr{D}(A) \text{ and } x_2 \in \mathscr{D}(B) \right\} \\ &= \mathscr{D}(A) \times \mathscr{D}(B) \\ &= \left[\mathscr{D}(A) \bigcap \mathscr{D}(C) \right] \times \left[\mathscr{D}(B) \bigcap \mathscr{D}(D) \right] \\ &= \mathscr{D}(\mathscr{L}_0). \end{aligned}$$

We infer that $\mathscr{D}(HKJ) = \mathscr{D}(\mathscr{L}_0 - \mu S)$. Q.E.D.

Theorem 5.4.2 *Let $\mu \in \rho_{S_1}(A)$ and $S_4 \in B\mathscr{R}(Y)$ be with $S_4(0) \subset C(0)$. We suppose that the conditions $(H17)$–$(H20)$ are satisfied. If $F(\mu)$ and $E(\mu)$ are single-valued, then*

(i) $\mathscr{L}_0$ is closable and its closure $\mathscr{L}$ is given by

$$\mathscr{L} = \overline{\mathscr{L}_0} = \mu S + H\overline{K}J,$$

where

$$\overline{K} = \begin{pmatrix} A - \mu S_1 & 0 \\ 0 & \overline{M(\mu)} - \mu S_4 \end{pmatrix}.$$

(ii) Let $\mu \in \rho_{S_1}(A)$, then for $\lambda \in \mathbb{C}$, we have

$$\lambda S - \mathscr{L} = H\overline{V(\lambda)}J - (\lambda - \mu)\mathscr{I}(\mu),$$

where

$$\overline{V(\lambda)} = \begin{pmatrix} \lambda S_1 - A & 0 \\ 0 & \lambda S_4 - \overline{M(\mu)} \end{pmatrix},$$

and

$$\mathscr{I}(\mu) = \begin{pmatrix} S_1 - S_1 & S_1 E(\mu) - S_2 \\ F(\mu)S_1 - S_3 & F(\mu)S_1 E(\mu) + S_4 - S_4 \end{pmatrix}.$$ ◇

Proof (i) J is single-valued bounded bijective and

$$J^{-1} := \begin{pmatrix} I & -E(\mu) \\ 0 & I \end{pmatrix}.$$

H is single-valued bounded bijective with $\mathscr{D}(H) = \mathscr{D}(E(\mu)) \times Y = X \times Y$ and

$$H^{-1} := \begin{pmatrix} I & 0 \\ -F(\mu) & I \end{pmatrix}.$$

On the one hand, we have $A - \mu S_1$ is closed. On the other hand, we have $M(\mu)$ is closable with $S_4(0) \subset C(0)$ and $\overline{\mathscr{D}(M(\mu))} = \overline{\mathscr{D}(B)} = \mathscr{D}(S_4)$. Then, by using Proposition 2.12.2, we have $M(\mu) - \mu S_4$ is closable and

$$\overline{M(\mu) - \mu S_4} = \overline{M(\mu)} - \mu S_4.$$

So,

$$K = \begin{pmatrix} A - \mu S_1 & 0 \\ 0 & M(\mu) - \mu S_4 \end{pmatrix}$$

is closable. Theorem 2.12.3 allowed us to conclude that HKJ is closable and $\overline{HKJ} = H\overline{K}J$. Since S is continuous, $\mathscr{D}(H\overline{K}J) = \mathscr{D}(A) \times \mathscr{D}(B) \subset \mathscr{D}(S)$ and

$$S\begin{pmatrix}0\\0\end{pmatrix}=\begin{pmatrix}S_1(0)+S_2(0)\\S_3(0)+S_4(0)\end{pmatrix}\\\subset\begin{pmatrix}A(0)+B(0)\\C(0)\end{pmatrix}\\=H\overline{K}J\begin{pmatrix}0\\0\end{pmatrix},$$

then by using Corollary 2.12.1 (ii), we get $\mu S+H\overline{K}J$ is closable. By applying Proposition 2.12.2, we have

$$\overline{\mathscr{L}_0-\mu S}=\overline{\mathscr{L}_0}-\mu S\\=H\overline{K}J.$$

Hence,

$$\overline{\mathscr{L}_0}=\mu S+H\overline{K}J.$$

It follows that the relation $\mathscr{L}_0$ is closable.
(ii) Since $S\begin{pmatrix}0\\0\end{pmatrix}\subset\mathscr{L}\begin{pmatrix}0\\0\end{pmatrix}$ and $\overline{\mathscr{D}(\mathscr{L})}\subset\mathscr{D}(S)$, then

$$\lambda S-\mathscr{L}=\mu S-\mathscr{L}+(\lambda-\mu)S.$$

Therefore,

$$\begin{aligned}\lambda S-\mathscr{L}&=H\begin{pmatrix}\mu S_1-A&0\\0&\mu S_4-\overline{M(\mu)}\end{pmatrix}J+(\lambda-\mu)\,S,\\&=H\begin{pmatrix}\mu S_1-A-\lambda S_1+\lambda S_1&0\\0&\mu S_4-\overline{M(\mu)}-\lambda S_4+\lambda S_4\end{pmatrix}J+(\lambda-\mu)\,S,\\&=H\left[\begin{pmatrix}\lambda S_1-A&0\\0&\lambda S_4-\overline{M(\mu)}\end{pmatrix}+\begin{pmatrix}(\mu-\lambda)S_1&0\\0&(\mu-\lambda)S_4\end{pmatrix}\right]J+(\lambda-\mu)S.\end{aligned}$$

As $\mathscr{D}(H)=X\times Y$ and J is single-valued, we have

$$\lambda S-\mathscr{L}=HV(\lambda)J+H\begin{pmatrix}(\mu-\lambda)S_1&0\\0&(\mu-\lambda)S_4\end{pmatrix}J+(\lambda-\mu)\,S.$$

The use of Lemma 2.8.4 (iii) leads to

$$\begin{aligned}\lambda S-\mathscr{L}&=HV(\lambda)J+(\mu-\lambda)\begin{pmatrix}I&0\\F(\mu)&I\end{pmatrix}\begin{pmatrix}S_1&S_1E(\mu)\\0&S_4\end{pmatrix}+(\lambda-\mu)\,S,\\&=HV(\lambda)J+(\mu-\lambda)\begin{pmatrix}S_1&S_1E(\mu)\\F(\mu)S_1&F(\mu)S_1E(\mu)+S_4\end{pmatrix}+(\lambda-\mu)\,S.\end{aligned}$$

By Lemma 2.8.4 (i), we can deduce that

$$\lambda S - \mathscr{L} = HV(\lambda)J - (\lambda - \mu)\begin{pmatrix} S_1 - S_1 & S_1E(\mu) - S_2 \\ F(\mu)S_1 - S_3 & F(\mu)S_1E(\mu) + S_4 - S_4 \end{pmatrix}.$$

Consequently, $\lambda S - \mathscr{L} = H\overline{V(\lambda)}J - (\lambda - \mu)\mathscr{I}(\mu)$. Q.E.D.

Now, we assume that
$(H21)$ $Z(\mu)$ is a continuous selection of $F(\mu)$.
$(H22)$ $W(\mu)$ is a continuous selection of $E(\mu)$.
$(H23)$ For some $\mu \in \rho_{S_1}(A)$, $\mathscr{D}(K)$ contains the ranges of both $\begin{pmatrix} I & W(\mu) \\ 0 & I \end{pmatrix}$ and $\begin{pmatrix} 0 & E(\mu) - E(\mu) \\ 0 & 0 \end{pmatrix}$.
Now, we suppose that A_1, T_1, T_2, and $M_1(\mu)$ are selections, respectively, of A, S_1, S_4, and $M(\mu)$ such that A_1 is closed, S_1, S_4 are continuous, and $M_1(\mu)$ is closable.

These conditions are mainly based on the Frobenius–Schur decomposition of $\mathscr{L}_0$, where A, B, C, and D are the linear relations.

Theorem 5.4.3 *Let $\mu \in \rho_{S_1}(A)$ and $S_4 \in B\mathscr{R}(Y)$ be with $S_4(0) \subset C(0)$. We suppose that the conditions $(H17)$–$(H21)$ are satisfied. If $E(\mu)$ is single-valued, then*

(i)

$$\mu S - \mathscr{L}_0 = H_1KJ + \mathscr{H}(\mu),$$

where

$$H_1 = \begin{pmatrix} I & 0 \\ Z(\mu) & I \end{pmatrix},$$

and

$$\mathscr{H}(\mu) = \begin{pmatrix} 0 & 0 \\ (F(\mu) - F(\mu))(\mu S_1 - A) & (F(\mu) - F(\mu))(\mu S_2 - B) \end{pmatrix}.$$

(ii) If $\mathscr{H}(\mu)$ is continuous, then $\mathscr{L}_0$ is closable and its closure $\mathscr{L}$ is given by the relation:

$$\mathscr{L} = \overline{\mathscr{L}_0} = \mu S + H_1\overline{K}J + \mathscr{H}(\mu).$$

(iii) Let $\mu \in \rho_{S_1}(A)$, then for $\lambda \in \mathbb{C}$, we have

$$\lambda S - \mathscr{L} = H_1\overline{V(\lambda)}J - (\lambda - \mu)\mathscr{I}_1(\mu) + \mathscr{H}(\mu),$$

where

$$\mathscr{I}_1(\mu) = \begin{pmatrix} S_1 - S_1 & S_1E(\mu) - S_2 \\ Z(\mu)S_1 - S_3 & Z(\mu)S_1E(\mu) + S_4 - S_4 \end{pmatrix}.$$ ◇

Proof (*i*) It is clear that

$$\mu S - \mathcal{L}_0 = \begin{pmatrix} I & 0 \\ Z(\mu) + F(\mu) - F(\mu) & I \end{pmatrix} KJ.$$

We infer from Lemma 2.8.4 (*i*) that

$$\mu S - \mathcal{L}_0 = \left[\begin{pmatrix} I & 0 \\ Z(\mu) & I \end{pmatrix} + \begin{pmatrix} 0 & 0 \\ F(\mu) - F(\mu) & 0 \end{pmatrix} \right] KJ.$$

Since $\begin{pmatrix} A(0) \\ C(0) \end{pmatrix} \subset N \begin{pmatrix} 0 & 0 \\ F(\mu) - F(\mu) & 0 \end{pmatrix} = X \times Y$, then

$$\mu S - \mathcal{L}_0 = H_1 KJ + \begin{pmatrix} 0 & 0 \\ F(\mu) - F(\mu) & 0 \end{pmatrix} KJ.$$

Let $\begin{pmatrix} x_1 \\ x_2 \end{pmatrix} \in X \times Y$. Then,

$$\begin{aligned} (\mu S - \mathcal{L}_0) \begin{pmatrix} x_1 \\ x_2 \end{pmatrix} &= H_1 KJ \begin{pmatrix} x_1 \\ x_2 \end{pmatrix} + \begin{pmatrix} 0 & 0 \\ F(\mu) - F(\mu) & 0 \end{pmatrix} KJ \begin{pmatrix} x_1 \\ x_2 \end{pmatrix}, \\ &= H_1 KJ \begin{pmatrix} x_1 \\ x_2 \end{pmatrix} + \begin{pmatrix} 0 & 0 \\ F(\mu) - F(\mu) & 0 \end{pmatrix} K \begin{pmatrix} x_1 + E(\mu) x_2 \\ x_2 \end{pmatrix}, \\ &= H_1 KJ \begin{pmatrix} x_1 \\ x_2 \end{pmatrix} + \begin{pmatrix} 0 & 0 \\ F(\mu) - F(\mu) & 0 \end{pmatrix} \begin{pmatrix} (\mu S_1 - A)x_1 + (\mu S_2 - B)x_2 \\ (\mu S_4 - M(\mu))x_2 \end{pmatrix}, \\ &= H_1 KJ \begin{pmatrix} x_1 \\ x_2 \end{pmatrix} + \begin{pmatrix} 0 \\ (F(\mu) - F(\mu)) ((\mu S_1 - A)x_1 + (\mu S_2 - B)x_2) \end{pmatrix}. \end{aligned}$$

It follows that

$$(\mu S - \mathcal{L}_0) \begin{pmatrix} x_1 \\ x_2 \end{pmatrix} = H_1 KJ \begin{pmatrix} x_1 \\ x_2 \end{pmatrix} + \mathcal{H}(\mu) \begin{pmatrix} x_1 \\ x_2 \end{pmatrix}. \tag{5.4.1}$$

On the one hand,

$$\mathcal{D}(H_1 KJ + \mathcal{H}(\mu)) = \mathcal{D}(H_1 KJ) \bigcap \mathcal{D}(\mathcal{H}(\mu)).$$

On the other hand, we have

$$\mathcal{D}(\mathcal{H}(\mu)) = \left\{ \begin{pmatrix} x_1 \\ x_2 \end{pmatrix} \in X \times Y : (\mu S_1 - A)x_1 \in \mathcal{D}(F(\mu)) \text{ and } (\mu S_2 - B)x_2 \in \mathcal{D}(F(\mu)) \right\}.$$

Since $\mathcal{D}(F(\mu)) = X$, then $\mathcal{D}(\mathcal{H}(\mu)) = \mathcal{D}(A) \times \mathcal{D}(B)$. In addition $\mathcal{D}(Z(\mu)) = \mathcal{D}(F(\mu))$, whereas $\mathcal{D}(HKJ) = \mathcal{D}(H_1 KJ)$ and

$$\mathcal{D}(H_1 KJ + \mathcal{H}(\mu)) = \mathcal{D}(A) \times \mathcal{D}(B).$$

Therefore,

$$\begin{aligned}\mathscr{D}(\mu S-\mathscr{L}_0)&=\mathscr{D}(A)\times\mathscr{D}(B)\\&=\mathscr{D}(H_1KJ+\mathscr{H}(\mu)).\end{aligned}\tag{5.4.2}$$

By Eqs. (5.4.1) and (5.4.2), we have

$$\mu S-\mathscr{L}_0=H_1KJ+\mathscr{H}(\mu).$$

(ii) We start by determining the properties of the matrices H_1 and J. It is easy to check that J and H_1 are single-valued bounded bijective with

$$J^{-1}:=\begin{pmatrix} I & -E(\mu)\\ 0 & I\end{pmatrix}$$

and

$$H_1^{-1}:=\begin{pmatrix} I & 0\\ -Z(\mu) & I\end{pmatrix}.$$

We have μS_1-A is closed. Also, $M(\mu)$ is closable with $S_4(0)\subset C(0)$ and $\overline{\mathscr{D}(M(\mu))}=\overline{\mathscr{D}(B)}=\mathscr{D}(S_4)$. Then, by using Proposition 2.12.2, we get $\mu S_4-M(\mu)$ is closable. So,

$$K=\begin{pmatrix}\mu S_1-A & 0\\ 0 & \mu S_4-M(\mu)\end{pmatrix}$$

is closable and

$$\overline{K}=\begin{pmatrix}\mu S_1-A & 0\\ 0 & \mu S_4-\overline{M(\mu)}\end{pmatrix}$$

denotes its closure. By using Theorem 2.12.3, we conclude that H_1KJ is closable and $\overline{H_1KJ}=H_1\overline{K}J$. Moreover, $\mathscr{D}(H_1KJ)=\mathscr{D}(A)\times\mathscr{D}(B)=\mathscr{D}(\mathscr{H}(\mu))$ and

$$\mathscr{H}(\mu)\begin{pmatrix}0\\0\end{pmatrix}=\begin{pmatrix}0\\ F(\mu)(\mu S_1-A)(0)+F(\mu)(\mu S_2-B)(0)\end{pmatrix}.$$

This implies that

$$\begin{aligned}F(\mu)(\mu S_1-A)(0)&=(\mu S_1-A)A(0)+(C-\lambda S_3)(0)\\&=A(0)+C(0)\end{aligned}$$

and

$$F(\mu)B(0)\subset C(0).$$

This shows that

$$\mathscr{H}(\mu)\begin{pmatrix}0\\0\end{pmatrix} \subset \begin{pmatrix}0\\A(0)+C(0)\end{pmatrix}.$$

Obviously,

$$\mathscr{H}(\mu)\begin{pmatrix}0\\0\end{pmatrix} \subset H_1KJ\begin{pmatrix}0\\0\end{pmatrix}.$$

By using Corollary 2.12.1 (ii), we get $H_1KJ + \mathscr{H}(\mu)$ is closable and

$$\begin{aligned}\overline{H_1KJ+\mathscr{H}(\mu)} &= \overline{H_1KJ} + \mathscr{H}(\mu)\\ &= H_1\overline{K}J + \mathscr{H}(\mu).\end{aligned}$$

Now, applying Proposition 2.12.2, we have

$$\begin{aligned}\overline{\mathscr{L}_0 - \mu S} &= \overline{\mathscr{L}_0} - \mu S\\ &= H_1\overline{K}J + \mathscr{H}(\mu).\end{aligned}$$

Hence,

$$\overline{\mathscr{L}_0} = \mu S + H_1\overline{K}J + \mathscr{H}(\mu).$$

This leads us to the conclusion that $\mathscr{L}_0$ is closable.

(iii) Since $S\begin{pmatrix}0\\0\end{pmatrix} \subset \mathscr{L}\begin{pmatrix}0\\0\end{pmatrix}$ and $\overline{\mathscr{D}(\mathscr{L})} \subset \mathscr{D}(S)$, then

$$\begin{aligned}\lambda S - \mathscr{L} &= \mu S - \mathscr{L} + (\lambda-\mu)S\\ &= H_1\overline{K}J + \mathscr{H}(\mu) + (\lambda-\mu)S.\end{aligned}$$

Hence,

$$\begin{aligned}\lambda S - \mathscr{L} &= \begin{pmatrix}I & 0\\ Z(\mu) & I\end{pmatrix}\begin{pmatrix}\mu S_1 - A & 0\\ 0 & \mu S_4 - \overline{M(\mu)}\end{pmatrix}\begin{pmatrix}I & E(\mu)\\ 0 & I\end{pmatrix} + \mathscr{H}(\mu) + (\lambda-\mu)\,S,\\ &= H_1\begin{pmatrix}\mu S_1 - A - \lambda S_1 + \lambda S_1 & 0\\ 0 & \mu S_4 - \overline{M(\mu)} - \lambda S_4 + \lambda S_4\end{pmatrix}J + \mathscr{H}(\mu) + (\lambda-\mu)\,S,\\ &= H_1\left[V(\lambda) + \begin{pmatrix}(\mu-\lambda)S_1 & 0\\ 0 & (\mu-\lambda)S_4\end{pmatrix}\right]J + \mathscr{H}(\mu) + (\lambda-\mu)\,S.\end{aligned}$$

Since $\mathscr{D}(H_1) = X \times Y$ and J is single-valued, then

$$\lambda S - \mathscr{L} = H_1\overline{V(\lambda)}J + H_1\begin{pmatrix}(\mu-\lambda)S_1 & 0\\ 0 & (\mu-\lambda)S_4\end{pmatrix}J + \mathscr{H}(\mu) + (\lambda-\mu)\,S.$$

We shall show that

$$\mathscr{I}_1(\mu) = H_1\begin{pmatrix}(\mu-\lambda)S_1 & 0\\ 0 & (\mu-\lambda)S_4\end{pmatrix}J + (\lambda-\mu)\,S.$$

Let $\begin{pmatrix} x_1 \\ x_2 \end{pmatrix} \in X \times Y$. Then,

$$\begin{aligned}
H_1 &\begin{pmatrix} (\mu-\lambda)S_1 & 0 \\ 0 & (\mu-\lambda)S_4 \end{pmatrix} J \begin{pmatrix} x_1 \\ x_2 \end{pmatrix} \\
&= (\mu-\lambda)H_1 \begin{pmatrix} S_1 & 0 \\ 0 & S_4 \end{pmatrix} \begin{pmatrix} x_1 + E(\mu)x_2 \\ x_2 \end{pmatrix}, \\
&= H_1 \begin{pmatrix} (\mu-\lambda)S_1(x_1 + E(\mu)x_2) \\ (\mu-\lambda)S_4 x_2 \end{pmatrix}, \\
&= \begin{pmatrix} (\mu-\lambda)S_1x_1 + (\mu-\lambda)S_1E(\mu)x_2 \\ Z(\mu)(\mu-\lambda)S_1x_1 + (Z(\mu)(\mu-\lambda)S_1E(\mu) + (\mu-\lambda)S_4)x_2 \end{pmatrix}, \\
&= \begin{pmatrix} (\mu-\lambda)S_1 & (\mu-\lambda)S_1E(\mu) \\ Z(\mu)(\mu-\lambda)S_1 & Z(\mu)(\mu-\lambda)S_1E(\mu) + (\mu-\lambda)S_4 \end{pmatrix} \begin{pmatrix} x_1 \\ x_2 \end{pmatrix}.
\end{aligned}$$

Since

$$\begin{aligned}
\mathscr{D}\left[(\mu-\lambda) H_1 \begin{pmatrix} S_1 & 0 \\ 0 & S_4 \end{pmatrix} J\right] &= \left\{\begin{pmatrix} x_1 \\ x_2 \end{pmatrix} \in X \times Y \text{ such that } \begin{matrix} x_1 + E(\mu)x_2 \in \mathscr{D}(S_1) \\ \text{and } x_2 \in \mathscr{D}(S_4) \end{matrix}\right\}, \\
&= \mathscr{D}(S_1) \times \mathscr{D}(S_2), \\
&= \mathscr{D}\left[(\mu-\lambda)\begin{pmatrix} S_1 & S_1E(\mu) \\ Z(\mu)S_1 & Z(\mu)S_1E(\mu) + S_4 \end{pmatrix}\right],
\end{aligned}$$

then

$$(\mu-\lambda)H_1 \begin{pmatrix} S_1 & 0 \\ 0 & S_4 \end{pmatrix} J + (\lambda-\mu)S = (\mu-\lambda)\begin{pmatrix} S_1 & S_1E(\mu) \\ Z(\mu)S_1 & Z(\mu)S_1E(\mu) + S_4 \end{pmatrix} - (\mu-\lambda)S.$$

By using Lemma 2.8.4 (i), we have

$$(\mu-\lambda)\, H_1 \begin{pmatrix} S_1 & 0 \\ 0 & S_4 \end{pmatrix} J + (\lambda-\mu)\, S = (\mu-\lambda)\, \mathscr{I}_1(\mu).$$

We conclude that $\lambda S - \mathscr{L} = H_1\overline{V(\lambda)}J - (\lambda-\mu)\mathscr{I}_1(\mu) + \mathscr{H}(\mu)$. Q.E.D.

In the remainder of this section, we shall tacitly assume that $\mathfrak{A} = A - A$ and $\mathscr{B} = B - B$.

Theorem 5.4.4 *Let $\mu \in \rho_{S_1}(A)$ and $S_4 \in B\mathscr{R}(Y)$ be with $S_4(0) \subset C(0)$. We suppose that the conditions $(H16)$–$(H21)$ are satisfied, then*

(i)

$$\mu S - \mathscr{L}_0 = H_1 K J_1 + \mathscr{J}(\mu),$$

where J_1 is the upper-triangular operator matrix $J_1 = \begin{pmatrix} I & W(\mu) \\ 0 & I \end{pmatrix}$ and

$$\mathscr{J}(\mu) = \begin{pmatrix} 0 & \mathfrak{A} + \mathscr{B} \\ (F(\mu) - F(\mu))(\mu S_1 - A) & (F(\mu) - F(\mu))(\mu S_2 - B) + Z(\mu)(\mathfrak{A} + \mathscr{B}) \end{pmatrix}.$$

(ii) If $\mathscr{J}(\mu)$ is continuous and $B(0) \subset A(0)$, then $\mathscr{L}_0$ is closable and its closure $\mathscr{L}$ is given by the relation:

$$\mathscr{L} = \overline{\mathscr{L}_0} = \mu S + H_1 \overline{K} J_1 + \mathscr{J}(\mu).$$

(iii) Let $\mu \in \rho_{S_1}(A)$, then for $\lambda \in \mathbb{C}$, we have

$$\lambda S - \mathscr{L} = H_1 \overline{V(\lambda)} J_1 - (\lambda - \mu)\mathscr{I}_2(\mu) + \mathscr{J}(\mu),$$

where

$$\mathscr{I}_2(\mu) = \begin{pmatrix} S_1 - S_1 & S_1 W(\mu) - S_2 \\ Z(\mu)S_1 - S_3 & Z(\mu)S_1 W(\mu) + S_4 - S_4 \end{pmatrix}. \qquad \diamond$$

Proof (i) It is clear that

$$\begin{aligned} \mu S - \mathscr{L}_0 &= H_1 K \begin{pmatrix} I & W(\mu) + E(\mu) - E(\mu) \\ 0 & I \end{pmatrix} + \mathscr{H}(\mu) \\ &= H_1 K \left[\begin{pmatrix} I & W(\mu) \\ 0 & I \end{pmatrix} + \begin{pmatrix} 0 & E(\mu) - E(\mu) \\ 0 & 0 \end{pmatrix} \right] + \mathscr{H}(\mu). \end{aligned}$$

Since $\mathscr{D}(H_1 K)$ contains the ranges of both $\begin{pmatrix} I & W(\mu) \\ 0 & I \end{pmatrix}$ and $\begin{pmatrix} 0 & E(\mu) - E(\mu) \\ 0 & 0 \end{pmatrix}$. In order to get this equality

$$\mu S - \mathscr{L}_0 = H_1 K \begin{pmatrix} I & W(\mu) \\ 0 & I \end{pmatrix} + H_1 K \begin{pmatrix} 0 & E(\mu) - E(\mu) \\ 0 & 0 \end{pmatrix} + \mathscr{H}(\mu).$$

Let $\begin{pmatrix} x_1 \\ x_2 \end{pmatrix} \in X \times Y$ be such that

$$\begin{aligned} (\mu S - \mathscr{L}_0)\begin{pmatrix} x_1 \\ x_2 \end{pmatrix} &= \left[H_1 K J_1 + \mathscr{H}(\mu) + H_1 K \begin{pmatrix} 0 & E(\mu) - E(\mu) \\ 0 & 0 \end{pmatrix} \right] \begin{pmatrix} x_1 \\ x_2 \end{pmatrix} \\ &= (H_1 K J_1 + \mathscr{H}(\mu)) \begin{pmatrix} x_1 \\ x_2 \end{pmatrix} + H_1 K \begin{pmatrix} (E(\mu) - E(\mu))x_2 \\ 0 \end{pmatrix} \\ &= (H_1 K J_1 + \mathscr{H}(\mu)) \begin{pmatrix} x_1 \\ x_2 \end{pmatrix} + H_1 \begin{pmatrix} (\mu S_1 - A)(E(\mu) - E(\mu))x_1 \\ 0 \end{pmatrix}. \end{aligned}$$

Since

$$\begin{aligned} (\mu S_1 - A)(E(\mu) - E(\mu))x_2 &= (\mu S_1 - A)E(\mu)(0) \\ &= \overline{(B - \lambda S_2)}(0) + (A - \lambda S_1)(0) \\ &= (B - \lambda S_2)(0) + (A - \lambda S_1)(0) \\ &= (\mathfrak{A} + \mathscr{B})x_2, \end{aligned}$$

then

$$\begin{aligned}(\mu S-\mathscr{L}_0)\begin{pmatrix}x_1\\x_2\end{pmatrix}&=(H_1KJ_1+\mathscr{H}(\mu))\begin{pmatrix}x_1\\x_2\end{pmatrix}+H_1\begin{pmatrix}(\mathfrak{A}+\mathscr{B})x_2\\0\end{pmatrix}\\&=(H_1KJ_1+\mathscr{H}(\mu))\begin{pmatrix}x_1\\x_2\end{pmatrix}+\begin{pmatrix}(\mathfrak{A}+\mathscr{B})x_2\\Z(\mu)(\mathfrak{A}+\mathscr{B})x_2\end{pmatrix}\\&=\left[H_1KJ_1+\mathscr{H}(\mu)+\begin{pmatrix}0&\mathscr{L}+\mathscr{B}\\0&Z(\mu)(\mathfrak{A}+\mathscr{B})\end{pmatrix}\right]\begin{pmatrix}x_1\\x_2\end{pmatrix}.\end{aligned}$$

Subsequently,

$$(\mu S-\mathscr{L}_0)\begin{pmatrix}x_1\\x_2\end{pmatrix}=\big(H_1KJ_1+\mathscr{J}(\mu)\big)\begin{pmatrix}x_1\\x_2\end{pmatrix}. \tag{5.4.3}$$

We have

$$\mathscr{D}\big(H_1KJ_1+\mathscr{J}(\mu)\big)=\mathscr{D}(H_1KJ_1)\bigcap\mathscr{D}\big(\mathscr{J}(\mu)\big).$$

Thus, we can write

$$\begin{aligned}\mathscr{D}\big(\mathscr{J}(\mu)\big)&=\mathscr{D}\left[\mathscr{H}(\mu)+\begin{pmatrix}0&\mathfrak{A}+\mathscr{B}\\0&Z(\mu)(\mathfrak{A}+\mathscr{B})\end{pmatrix}\right]\\&=\mathscr{D}[\mathscr{H}(\mu)]\bigcap\mathscr{D}\left[\begin{pmatrix}0&\mathfrak{A}+\mathscr{B}\\0&Z(\mu)(\mathfrak{A}+\mathscr{B})\end{pmatrix}\right]\\&=\mathscr{D}(A)\times\mathscr{D}(B).\end{aligned}$$

It follows that

$$\mathscr{D}\big(H_1KJ_1+\mathscr{J}(\mu)\big)=\mathscr{D}(A)\times\mathscr{D}(B). \tag{5.4.4}$$

By Eqs. (5.4.3) and (5.4.4), we have $\mu S-\mathscr{L}_0=H_1KJ_1+\mathscr{J}(\mu)$.
(ii) We start by determining the properties of the matrix

$$H_1:=\begin{pmatrix}I&0\\Z(\mu)&I\end{pmatrix}$$

and

$$J_1:=\begin{pmatrix}I&W(\mu)\\0&I\end{pmatrix}.$$

(a) J_1 is single-valued bounded bijective and $J_1^{-1}:=\begin{pmatrix}I&-W(\mu)\\0&I\end{pmatrix}$.
(b) H_1 is single-valued bounded bijective with $\mathscr{D}(H)=\mathscr{D}(E(\mu))\times Y=X\times Y$ and

$$H_1^{-1}:=\begin{pmatrix}I&0\\-Z(\mu)&I\end{pmatrix}.$$

(c), we have $\overline{A - \mu S_1}$ is closed. In addition that $M(\mu)$ is closable with $S_4(0) \subset C(0)$ and $\overline{\mathscr{D}(M(\mu))} = \overline{\mathscr{D}(B)} = \mathscr{D}(S_4)$. Then, by Proposition 2.12.2, we have $M(\mu) - \mu S_4$ is closable. So,

$$K = \begin{pmatrix} A - \mu S_1 & 0 \\ 0 & M(\mu) - \mu S_4 \end{pmatrix}$$

is closable. The use of (a), (b), (c) and Theorem 2.12.3 allowed us to conclude that $H_1 K J_1$ is closable and $\overline{H_1 K J_1} = H_1 \overline{K} J_1$. Having

$$\begin{aligned} \mathscr{D}(H_1 K J_1) &= \mathscr{D}(\mathscr{J}(\mu)) \\ &= \mathscr{D}(A) \times \mathscr{D}(B) \end{aligned}$$

and

$$\mathscr{J}(\mu) \begin{pmatrix} 0 \\ 0 \end{pmatrix} = \begin{pmatrix} A(0) + B(0) \\ F(\mu)(\mu S_1 - A)(0) + F(\mu)(\mu S_2 - B)(0) + Z(\mu)(\mathscr{L} + \mathscr{B})(0) \end{pmatrix}.$$

Since

$$\begin{aligned} F(\mu)(\mu S_1 - A)(0) &= (\mu S_3 - C)(0) + (\mu S_3 - C)(\mu S_1 - A)^{-1}(0) \\ &= C(0), \end{aligned}$$

then by using the hypotheses $(H5)$, we deduce that $F(\mu)(\mu S_2 - B)(0) \subset C(0)$. Hence,

$$\begin{aligned} \mathscr{J}(\mu) \begin{pmatrix} 0 \\ 0 \end{pmatrix} &= \begin{pmatrix} A(0) \\ C(0) + Z(\mu) A(0) \end{pmatrix} \\ &= H_1 K J_1 \begin{pmatrix} 0 \\ 0 \end{pmatrix}. \end{aligned}$$

So, we conclude $H_1 K J_1 + \mathscr{J}(\mu)$ is closable and

$$\begin{aligned} \overline{H_1 K J_1 + \mathscr{J}(\mu)} &= \overline{H_1 K J_1} + \mathscr{J}(\mu) \\ &= H_1 \overline{K} J_1 + \mathscr{J}(\mu). \end{aligned}$$

Thus, applying Proposition 2.12.2, we have

$$\begin{aligned} \overline{\mathscr{L}_0 - \mu S} &= \overline{\mathscr{L}_0} - \mu S \\ &= H_1 \overline{K} J_1 + \mathscr{J}(\mu). \end{aligned}$$

Hence,

$$\overline{\mathscr{L}_0} = \mu S + H_1 \overline{K} J_1 + \mathscr{J}(\mu).$$

So, the relation $\mathscr{L}_0$ is closable.

(iii) Since $S\begin{pmatrix}0\\0\end{pmatrix} \subset \mathscr{L}\begin{pmatrix}0\\0\end{pmatrix}$ and $\overline{\mathscr{D}(\mathscr{L})} \subset \mathscr{D}(S)$, then

$$\begin{aligned}\lambda S - \mathscr{L} &= \mu S - \mathscr{L} + (\lambda - \mu)S \\ &= H_1 \overline{K} J_1 + \mathscr{J}(\mu) + (\lambda - \mu)S.\end{aligned}$$

Thus,

$$\begin{aligned}\lambda S - \mathscr{L} &= H_1 \begin{pmatrix} \mu S_1 - A & 0 \\ 0 & \mu S_4 - \overline{M(\mu)} \end{pmatrix} J_1 + \mathscr{J}(\mu) + (\lambda - \mu)\, S \\ &= H_1 \begin{pmatrix} \mu S_1 - A - \lambda S_1 + \lambda S_1 & 0 \\ 0 & \mu S_4 - \overline{M(\mu)} - \lambda S_4 + \lambda S_4 \end{pmatrix} J_1 \\ &\quad + \mathscr{J}(\mu) + (\lambda - \mu)\, S \\ &= H_1 \left[\begin{pmatrix} \lambda S_1 - A & 0 \\ 0 & \lambda S_4 - \overline{M(\mu)} \end{pmatrix} + \begin{pmatrix} (\mu - \lambda) S_1 & 0 \\ 0 & (\mu - \lambda) S_4 \end{pmatrix} \right] J_1 \\ &\quad + \mathscr{J}(\mu) + (\lambda - \mu)\, S.\end{aligned}$$

Now, by using the fact that $\mathscr{D}(H_1) = X \times Y$ and J_1 is single-valued, then

$$\lambda S - \mathscr{L} = H_1 \overline{V(\lambda)} J_1 + H_1 \begin{pmatrix} (\mu - \lambda) S_1 & 0 \\ 0 & (\mu - \lambda) S_4 \end{pmatrix} J_1 + \mathscr{J}(\mu) + (\lambda - \mu)\, S.$$

The use of Lemma 2.8.4 (iii) leads to

$$\begin{aligned}\lambda S - \mathscr{L} &= H_1 \overline{V(\lambda)} J_1 + (\mu - \lambda)\, H_1 \begin{pmatrix} S_1 & S_1 W(\mu) \\ 0 & S_4 \end{pmatrix} + (\lambda - \mu)\, S + \mathscr{J}(\mu) \\ &= H_1 \overline{V(\lambda)} J_1 + (\mu - \lambda) \begin{pmatrix} S_1 & S_1 W(\mu) \\ Z(\mu) S_1 & Z(\mu) S_1 W(\mu) + S_4 \end{pmatrix} + (\lambda - \mu)\, S + \mathscr{J}(\mu).\end{aligned}$$

By applying Lemma 2.8.4 (i), we deduce that

$$\lambda S - \mathscr{L} = H_1 \overline{V(\lambda)} J_1 + (\mu - \lambda) \mathscr{I}_2(\mu) + \mathscr{J}(\mu).$$

This achieves the proof. Q.E.D.

Theorem 5.4.5 *Let $\mu \in \rho_{S_1}(A)$ and $S_4 \in B\mathscr{R}(Y)$ be with $S_4(0) \subset C(0)$. We suppose that the conditions $(H15)$–$(H24)$ are satisfied, then*

(i)

$$\mu S - \mathscr{L}_0 = H_1 K_1 J_1 + \mathscr{K}(\mu),$$

where K_1 is the diagonal operator matrix

$$K_1 = \begin{pmatrix} \mu T_1 - A_1 & 0 \\ 0 & \mu T_2 - M_1(\mu) \end{pmatrix}$$

and

$$\mathscr{K}(\mu) = \begin{pmatrix} \mathfrak{A} & \mathfrak{A}(I + W(\mu)) + \mathscr{B} \\ F(\mu) - F(\mu)(\mu S_1 - A) + Z(\mu)\mathfrak{A} & Z(\mu)\mathscr{L}(I + W(\mu)) + \mathscr{B} + M(\mu) - M(\mu) \end{pmatrix}.$$

(ii) If $\mathscr{K}(\mu)$ is continuous, then $\mathscr{L}_0$ is closable and its closure $\mathscr{L}$ is given by the relation:

$$\mathscr{L} = \overline{\mathscr{L}_0} = \mu S + H_1 \overline{K_1} J_1 + \mathscr{J}(\mu),$$

where $\overline{K_1}$ is the diagonal operator matrix, given by

$$\overline{K_1} = \begin{pmatrix} \mu T_1 - A_1 & 0 \\ 0 & \mu T_2 - \overline{M_1(\mu)} \end{pmatrix}.$$

(iii) Let $\mu \in \rho_{S_1}(A)$, then for $\lambda \in \mathbb{C}$, we have

$$\lambda S - \mathscr{L} = H_1 V_1(\lambda) J_1 - (\lambda - \mu)\mathscr{I}_3(\mu) + \mathscr{J}(\mu),$$

where

$$\overline{V_1(\lambda)} = \begin{pmatrix} \lambda T_1 - A_1 & 0 \\ 0 & \lambda T_2 - \overline{M_1(\mu)} \end{pmatrix},$$

and

$$\mathscr{I}_3(\mu) = \begin{pmatrix} T_1 - S_1 & T_1 W(\mu) - S_2 \\ Z(\mu) T_1 - S_3 & Z(\mu) S_1 W(\mu) + T_2 - S_4 \end{pmatrix}. \qquad \diamond$$

Proof We assume that $Q = \mu S_1 - \mu S_1 + \mathfrak{A}$ and $P = \mu S_4 - \mu S_4 + M(\mu) - M(\mu)$. Then,

$$\begin{aligned} \mu S - \mathscr{L}_0 &= H_1 \begin{pmatrix} \mu T_1 - A_1 + Q & 0 \\ 0 & \mu T_2 - M_1(\mu) + P \end{pmatrix} J_1 + \mathscr{J}(\mu) \\ &= H_1 \left[\begin{pmatrix} \mu T_1 - A_1 & 0 \\ 0 & \mu T_2 - M_1(\mu) \end{pmatrix} + \begin{pmatrix} Q & 0 \\ 0 & P \end{pmatrix} \right] J_1 + \mathscr{J}(\mu). \\ &= H_1 K_1 J_1 + H_1 \begin{pmatrix} Q & 0 \\ 0 & P \end{pmatrix} J_1 + \mathscr{J}(\mu). \end{aligned}$$

Let $\begin{pmatrix} x_1 \\ x_2 \end{pmatrix} \in \mathscr{D}(A) \times \mathscr{D}(B)$. Then,

$$\begin{aligned} (\mu S - \mathscr{L}_0)\begin{pmatrix} x_1 \\ x_2 \end{pmatrix} &= H_1 K_1 J_1 \begin{pmatrix} x_1 \\ x_2 \end{pmatrix} + H_1 \begin{pmatrix} Q & 0 \\ 0 & P \end{pmatrix} J_1 \begin{pmatrix} x_1 \\ x_2 \end{pmatrix} + \mathscr{J}(\mu)\begin{pmatrix} x_1 \\ x_2 \end{pmatrix} \\ &= (H_1 K_1 J_1 + \mathscr{J}(\mu))\begin{pmatrix} x_1 \\ x_2 \end{pmatrix} + H_1 \begin{pmatrix} Q & 0 \\ 0 & P \end{pmatrix} \begin{pmatrix} x_1 + W(\mu) x_2 \\ x_2 \end{pmatrix} \end{aligned}$$

$$= \left(H_1K_1J_1 + \mathscr{J}(\mu)\right)\begin{pmatrix} x_1 \\ x_2 \end{pmatrix} + H_1\begin{pmatrix} Q(x_1 + W(\mu)x_2) \\ Px_2 \end{pmatrix}$$

$$= \left(H_1K_1J_1 + \mathscr{J}(\mu)\right)\begin{pmatrix} x_1 \\ x_2 \end{pmatrix} + \begin{pmatrix} Q(x_1 + W(\mu)x_2) \\ Z(\mu)Q(x_1 + W(\mu)x_2) + Px_2 \end{pmatrix}.$$

Since

$$\begin{aligned} Q(x_1 + W(\mu)x_2) &= (S_1 + A)(0) \\ &= A(0) \\ &= \mathfrak{A}(x_1 + W(\mu)x_2) \end{aligned}$$

and

$$\begin{aligned} Px_2 &= S_4(0) + M(\mu)(0) \\ &= M(\mu)(0) \\ &= (M(\mu) - M(\mu))x_2, \end{aligned}$$

then

$$\begin{aligned} \begin{pmatrix} Q(x_1 + W(\mu)x_2) \\ Z(\mu)Q(x_1 + W(\mu)x_2) + Px_2 \end{pmatrix} &= \begin{pmatrix} \mathfrak{A}(x_1 + W(\mu)x_2) \\ Z(\mu)\mathfrak{A}(x_1 + W(\mu)x_2) + (M(\mu) - M(\mu))x_2 \end{pmatrix} \\ &= \begin{pmatrix} \mathfrak{A} & \mathfrak{A}W(\mu) \\ Z(\mu)\mathfrak{A} & Z(\mu)\mathfrak{A}W(\mu) + M(\mu) - M(\mu) \end{pmatrix}\begin{pmatrix} x_1 \\ x_2 \end{pmatrix} \\ &= \mathscr{K}(\mu)\begin{pmatrix} x_1 \\ x_2 \end{pmatrix}. \end{aligned}$$

We conclude that,

$$(\mu S - \mathscr{L}_0)\begin{pmatrix} x_1 \\ x_2 \end{pmatrix} = \left(H_1K_1J_1 + \mathscr{K}(\mu)\right)\begin{pmatrix} x_1 \\ x_2 \end{pmatrix}.$$

Since

$$\begin{aligned} \mathscr{D}(\mathscr{K}(\mu)) &= \mathscr{D}(A) \times \mathscr{D}(B) \\ &= \mathscr{D}(\mathscr{J}(\mu))\bigcap\left[\mathscr{D}(A)\bigcap\mathscr{D}(Z(\mu)(A - A)) \times \mathscr{D}(Z(\mu)(A - A)(I + W(\mu)))\bigcap\mathscr{D}(M(\mu))\right] \\ &= \mathscr{D}(A) \times \mathscr{D}(B), \end{aligned}$$

then

$$\begin{aligned} \mathscr{D}(\mu S - \mathscr{L}_0) &= \mathscr{D}(H_1K_1J_1 + \mathscr{K}(\mu)) \\ &= \mathscr{D}(H_1K_1J_1)\bigcap\mathscr{D}(\mathscr{K}(\mu)) \\ &= \mathscr{D}(A) \times \mathscr{D}(B). \end{aligned}$$

The assertions (ii) and (iii) can be proved in the same way as the evidence of Theorem 5.4.4. Q.E.D.

5.4.3 The Pencils Essential Spectra of $\mathscr{L}$

Theorem 5.4.6 *Assume the conditions of Theorem 5.4.2 hold, then if for some* $\mu \in \rho_{S_1}(A)$, *we have*

(i) *If the linear relation* $\mathscr{I}(\mu) \in \mathscr{PR}(\Phi_+(X \times Y))$, *then*

$$\sigma_{eap,S}(\mathscr{L}) \subset \sigma_{eap,S_1}(A) \bigcup \sigma_{eap,S_4}(\overline{M(\mu)}).$$

If, in addition, we suppose that the sets Φ_{A,S_1} *and* Φ_{A,S_4} *are connected and the sets* $\rho_S(\mathscr{L})$ *and* $\rho_{S_4}(\overline{M(\mu)})$ *are not empty, then*

$$\sigma_{eap,S}(\mathscr{L}) = \sigma_{eap,S_1}(A) \bigcup \sigma_{eap,S_4}(\overline{M(\mu)}).$$

(ii) *If the linear relation* $\mathscr{I}(\mu) \in \mathscr{PR}(\Phi_-(X \times Y))$, *then*

$$\sigma_{e\delta,S}(\mathscr{L}) \subset \sigma_{e\delta,S_1}(A) \bigcup \sigma_{e\delta,S_4}(\overline{M(\mu)}).$$

If, in addition, we suppose that the sets Φ_{A,S_1} *and* Φ_{A,S_4} *are connected and the sets* $\rho_S(\mathscr{L})$ *and* $\rho_{S_4}(\overline{M(\mu)})$ *are not empty, then*

$$\sigma_{e\delta,S}(\mathscr{L}) = \sigma_{e\delta,S_1}(A) \bigcup \sigma_{e\delta,S_4}(\overline{M(\mu)}). \qquad \diamond$$

Proof (i) Let $\mu \in \rho_{S_1}(A)$, set $\lambda \in \mathbb{C}$, and writing

$$\lambda S - \mathscr{L} = \mu S - \mathscr{L} + (\lambda - \mu)S.$$

Now, applying Theorem 5.4.2 (iii), we have

$$\lambda S - \mathscr{L} = U\overline{V(\lambda)}W - (\lambda - \mu)\mathscr{I}(\mu). \tag{5.4.5}$$

As though $\mathscr{I}(\mu) \in \mathscr{PR}(\Phi_+(X \times Y))$, then $\lambda S - \mathscr{L}$ is an upper semi-Fredholm relation if, and only if, $U\overline{V(\lambda)}W$ is upper semi-Fredholm relation. Now, observe that the relations U and W are Fredholm relations, we deduce from Lemma 3.6.9 that $U\overline{V(\lambda)}W$ is upper semi-Fredholm relation if, and only if, $\overline{V(\lambda)}$ is an upper semi-Fredholm relation if, and only if, $\lambda S_1 - A$ (resp. $\lambda S_4 - \overline{M(\mu)}$) is an upper semi-Fredholm relation on X (resp. on Y). Hence

$$\left\{\lambda \in \mathbb{C} \text{ such that } \lambda \in \Phi_{+\mathscr{L},S}\right\} = \left\{\lambda \in \mathbb{C} \text{ such that } \lambda \in \Phi_{+A,S_1}\right\} \bigcup \left\{\lambda \in \mathbb{C} \text{ such that } \lambda \in \Phi_{+\overline{M(\mu)},S_4}\right\}.$$

From Eq. (5.4.5), we get

$$i(\lambda S - \mathscr{L}) = i(U\overline{V(\lambda)}W - (\lambda - \mu)\mathscr{I}(\mu)).$$

By using Lemma 2.7.2, we have

$$\begin{aligned} i(\lambda S - \mathscr{L}) &= i(U\overline{V(\lambda)}W) \\ &= i(U) + i(\overline{V(\lambda)}W) - \dim\left[U(0)\bigcap(\overline{V(\lambda)}W)^{-1}(0)\right] \\ &= i(U) + i(\overline{V(\lambda)}) + i(W) - \dim\left[\overline{V(\lambda)}(0)\bigcap W^{-1}(0)\right] \\ &= i(U) + i(\overline{V(\lambda)}) + i(W). \end{aligned}$$

Since U and W are bijective, then

$$\begin{aligned} i(\lambda S - \mathscr{L}) &= i(\overline{V(\lambda)}) \\ &= i(\lambda S_1 - A) + i(\lambda S_4 - \overline{M(\mu)}). \end{aligned}$$

Now, let $\lambda \notin \sigma_{eap,S_1}(A)\bigcup\sigma_{eap,S_4}(\overline{M(\mu)})$, then $\lambda \in \Phi_{+A,S_1}$, $\lambda \in \Phi_{+\overline{M(\mu)},S_4}$,

$$i(\lambda S_1 - A) \leq 0$$

and

$$i(\lambda S_4 - \overline{M(\mu)}) \leq 0.$$

Hence,

$$i(\overline{V(\lambda)}) \leq 0$$

and $\overline{V(\lambda)}$ is an upper semi-Fredholm relation. This is equivalent to

$$i(\lambda S - \mathscr{L}) \leq 0$$

and $\lambda S - \mathscr{L}$ is an upper semi-Fredholm relation. So, $\lambda \notin \sigma_{eap,S}(\mathscr{L})$, and we infer that

$$\sigma_{eap,S}(\mathscr{L}) \subset \sigma_{eap,S_1}(A)\bigcup\sigma_{eap,S_4}(\overline{M(\mu)}).$$

Now, suppose that Φ_{A,S_1} and $\Phi_{\overline{M(\mu)},S_4}$ are connected. It is easy to check that

$$\{\lambda \in \mathbb{C} \text{ such that } \lambda \notin \Phi_{+A,S_1}\} \subset \sigma_{eap,S_1}(A).$$

Conversely, assume that $\lambda \notin \Phi_{+A,S_1}$, then $\lambda \in \Phi_{A,S_1}\bigcup\Phi_{+A,S_1}\backslash\Phi_{A,S_1}$. Hence, we will discuss the following two cases:
<u>Case 1</u>: If $\lambda \in \Phi_{A,S_1}$, then

$$i(\lambda S_1 - A) = 0.$$

Indeed, let $\lambda_0 \in \rho_{S_1}(A)$, then

$$i(\lambda_0 S_1 - A) = 0.$$

Since $i(\lambda S_1 - A)$ is constant for all $\lambda \in \Phi_{A,S_1}$, then

$$i(\lambda S_1 - A) = 0$$

for all $\lambda \in \Phi_{A,S_1}$. Therefore, $\lambda \notin \sigma_{eap,S_1}(A)$.
Case 2: If $\lambda \in \Phi_{+A,S_1} \backslash \Phi_{A,S_1}$, then

$$\alpha(\lambda S_1 - A) < +\infty$$

and

$$\beta(\lambda S_1 - A) = +\infty.$$

So, $i(\lambda S_1 - A) = -\infty < 0$ for this reason, we have $\lambda \notin \sigma_{eap,S_1}(A)$. Then,

$$\sigma_{eap,S_1}(A) = \{\lambda \in \mathbb{C} \text{ such that } \lambda \notin \Phi_{+A,S_1}\}$$

and in the same way we find that

$$\sigma_{eap,S_4}(\overline{M(\mu)}) = \left\{\lambda \in \mathbb{C} \text{ such that } \lambda \notin \Phi_{+\overline{M(\mu)},S_4}\right\}.$$

We deduce that

$$\sigma_{eap,S}(\mathscr{L}) = \sigma_{eap,S_1}(A) \bigcup \sigma_{eap,S_4}(\overline{M(\mu)}).$$

(ii) The proof of this assertion may be checked in the same way as the proof of (i). Q.E.D.

Theorem 5.4.7 *Assume the conditions of Theorem 5.4.3 hold, then if for some* $\mu \in \rho_{S_1}(A)$, *we have*

(i) *If the linear relation* $\mathscr{I}_1(\mu)$, $\mathscr{H}(\mu) \in \mathscr{PR}(\Phi_+(X \times Y))$, *then*

$$\sigma_{eap,S}(\mathscr{L}) \subset \sigma_{eap,S_1}(A) \bigcup \sigma_{eap,S_4}(\overline{M(\mu)}).$$

Besides, if we suppose that the sets Φ_{A,S_1} *and* Φ_{A,S_4} *are connected and the sets* $\rho_S(\mathscr{L})$ *and* $\rho_{S_4}(\overline{M(\mu)})$ *are not empty, then*

$$\sigma_{eap,S}(\mathscr{L}) = \sigma_{eap,S_1}(A) \bigcup \sigma_{eap,S_4}(\overline{M(\mu)}).$$

(ii) *If the linear relation* $\mathscr{I}_1(\mu)$, $\mathscr{H}(\mu) \in \mathscr{PR}(\Phi_-(X \times Y))$, *then*

$$\sigma_{e\delta,S}(\mathscr{L}) \subset \sigma_{e\delta,S_1}(A) \bigcup \sigma_{e\delta,S_4}(\overline{M(\mu)}).$$

Moreover, if we suppose that the sets Φ_{A,S_1} *and* Φ_{A,S_4} *are connected and the sets* $\rho_S(\mathscr{L})$ *and* $\rho_{S_4}(\overline{M(\mu)})$ *are not empty, then*

$$\sigma_{e\delta,S}(\mathscr{L}) = \sigma_{e\delta,S_1}(A) \bigcup \sigma_{e\delta,S_4}(\overline{M(\mu)}).$$ ◇

Proof The proof is a straightforward adoption of the proof of Theorem 5.4.6. Q.E.D.

Theorem 5.4.8 *Assume the conditions of Theorem 5.4.4 hold, then if for some $\mu \in \rho_{S_1}(A)$, we have*

(i) If the linear relation $\mathscr{I}_2(\mu)$ and $\mathscr{J}(\mu) \in \mathscr{PR}(\Phi_+(X \times Y))$, then

$$\sigma_{eap,S}(\mathscr{L}) \subset \sigma_{eap,S_1}(A) \bigcup \sigma_{eap,S_4}(\overline{M(\mu)}).$$

If, in addition, we suppose that the sets Φ_{A,S_1} and Φ_{A,S_4} are connected and the sets $\rho_S(\mathscr{L})$ and $\rho_{S_4}(\overline{M(\mu)})$ are not empty, then

$$\sigma_{eap,S}(\mathscr{L}) = \sigma_{eap,S_1}(A) \bigcup \sigma_{eap,S_4}(\overline{M(\mu)}).$$

(ii) If the linear relation $\mathscr{I}_2(\mu)$, $\mathscr{J}(\mu) \in \mathscr{PR}(\Phi_-(X \times Y))$, then

$$\sigma_{e\delta,S}(\mathscr{L}) \subset \sigma_{e\delta,S_1}(A) \bigcup \sigma_{e\delta,S_4}(\overline{M(\mu)}).$$

Additionally, if we suppose that the sets Φ_{A,S_1} and Φ_{A,S_4} are connected and the sets $\rho_S(\mathscr{L})$ and $\rho_{S_4}(\overline{M(\mu)})$ are not empty, then

$$\sigma_{e\delta,S}(\mathscr{L}) = \sigma_{e\delta,S_1}(A) \bigcup \sigma_{e\delta,S_4}(\overline{M(\mu)}).$$ ◇

Proof The proof is a straightforward adoption of the proof of Theorem 5.4.6. Q.E.D.

Theorem 5.4.9 *Assume the conditions of Theorem 5.4.5 hold, then if for some $\mu \in \rho_{S_1}(A)$, we have*

(i) If the linear relation $\mathscr{I}_3(\mu)$ and $\mathscr{K}(\mu) \in \mathscr{PR}(\Phi_+(X \times Y))$, then

$$\sigma_{eap,S}(\mathscr{L}) \subset \sigma_{eap,T_1}(A_1) \bigcup \sigma_{eap,T_2}(\overline{M_1(\mu)}).$$

Additionally, if we suppose that the sets Φ_{A_1,T_1} and Φ_{A_1,T_2} are connected and the sets $\rho_S(\mathscr{L})$ and $\rho_{T_2}(\overline{M_1(\mu)})$ are not empty, then

$$\sigma_{eap,S}(\mathscr{L}) = \sigma_{eap,T_1}(A_1) \bigcup \sigma_{eap,T_2}(\overline{M_1(\mu)}).$$

(ii) If the linear relation $\mathscr{I}_3(\mu)$ and $\mathscr{K}(\mu) \in \mathscr{PR}(\Phi_-(X \times Y))$, then

$$\sigma_{e\delta,S}(\mathscr{L}) \subset \sigma_{e\delta,T_1}(A_1) \bigcup \sigma_{e\delta,T_2}(\overline{M_1(\mu)}).$$

Moreover, if we suppose that the sets Φ_{A_1,T_1} and Φ_{A_1,T_2} are connected and the sets $\rho_S(\mathscr{L})$ and $\rho_{T_2}(\overline{M_1(\mu)})$ are not empty, then

$$\sigma_{e\delta,S}(\mathscr{L}) = \sigma_{e\delta,T_1}(A_1) \bigcup \sigma_{e\delta,T_2}(\overline{M_1(\mu)}).$$ ◇

Proof The proof of this theorem is a straightforward adoption of the proof of Theorem 5.4.6. Q.E.D.

5.5 Essential Spectra of 3 × 3-Block Matrices Linear Relations

Let X, Y, and Z be three Banach spaces. In this section, we are concerned with the essential spectra of linear relations defined by a 3 × 3-block matrices linear relations:

$$\mathscr{A}_0 := \begin{pmatrix} A & B & C \\ D & E & F \\ G & H & K \end{pmatrix}, \tag{5.5.1}$$

where the entries of the matrix are, in general, unbounded linear relations. The relation A acts on X with domain $\mathscr{D}(A)$, the relation E acts on Y with domain $\mathscr{D}(E)$, and the relation K acts on Z with domain $\mathscr{D}(K)$. The intertwining relation B is defined on the domain $\mathscr{D}(B) \subset Y$ into X, the relation H is defined on the domain $\mathscr{D}(H) \subset Y$ into Z, the relation C is defined on the domain $\mathscr{D}(C) \subset Z$ into X, the relation F is defined on the domain $\mathscr{D}(F) \subset Z$ into Y, the relation D is defined on the domain $\mathscr{D}(D) \subset X$ into Y, and the relation G is defined on the domain $\mathscr{D}(G) \subset X$ into Z. The relation $\mathscr{A}_0$ is defined on

$$\begin{cases} G(\mathscr{A}_0) = \left\{ \left(\begin{pmatrix} u_1 \\ u_2 \\ u_3 \end{pmatrix}, \begin{pmatrix} v_1 \\ v_2 \\ v_3 \end{pmatrix} \right) \in (X \times Y \times Z)^2 \text{ such that } \begin{matrix} v_1 \in Au_1 + Bu_2 + Cu_3 \\ v_2 \in Du_1 + Eu_2 + Fu_3 \\ v_3 \in Gu_1 + Hu_2 + Ku_3 \end{matrix} \right\} \\ \mathscr{D}(\mathscr{A}_0) = \left(\mathscr{D}(A) \bigcap \mathscr{D}(D) \bigcap \mathscr{D}(G)\right) \times \left(\mathscr{D}(B) \bigcap \mathscr{D}(E) \bigcap \mathscr{D}(H)\right) \times \left(\mathscr{D}(C) \bigcap \mathscr{D}(F) \bigcap \mathscr{D}(K)\right). \end{cases}$$

5.5.1 *The Closedness and Closability of 3 × 3 Matrices Linear Relations*

Lemma 5.5.1 *Let*

$$\mathscr{T} = \begin{pmatrix} A & 0 & 0 \\ 0 & E & 0 \\ 0 & 0 & K \end{pmatrix},$$

$$\mathscr{S} = \begin{pmatrix} 0 & B & 0 \\ 0 & 0 & F \\ G & 0 & 0 \end{pmatrix}$$

and

$$\mathscr{K} = \begin{pmatrix} 0 & 0 & C \\ D & 0 & 0 \\ 0 & H & 0 \end{pmatrix}.$$

(a) If the following conditions are satisfied

(i) B is E-bounded with E-bound δ_1,
(ii) F is K-bounded with K-bound δ_2,
(ii) G is A-bounded with A-bound δ_3.

Then, $\mathscr{S}$ is $\mathscr{T}$-bounded with $\mathscr{T}$-bound $\delta = \max\{\delta_1, \delta_2, \delta_3\}$.

(b) If the following conditions are satisfied

(i) C is F-bounded with F-bound δ_1,
(ii) D is G-bounded with G-bound δ_2,
(ii) H is A-bounded with A-bound δ_3.

Then, $\mathscr{K}$ is $\mathscr{S}$-bounded with $\mathscr{S}$-bound $\delta = \max\{\delta_1, \delta_2, \delta_3\}$. ◇

Proof Let $\varepsilon > 0$. By the above assumptions there exist constants $a_1, a_2, a_3, b_1, b_2, b_3 \geq 0$ such that $\delta_1 \leq b_1 < \delta_1 + \varepsilon$, $\delta_2 \leq b_2 < \delta_2 + \varepsilon$, $\delta_3 \leq b_3 < \delta_3 + \varepsilon$,

$$\|Bg\|^2 \leq a_1^2\|g\|^2 + b_1^2\|Eg\|^2 \text{ for all } g \in \mathscr{D}(E) \subset \mathscr{D}(B),$$

$$\|Fh\|^2 \leq a_2^2\|h\|^2 + b_2^2\|Kh\|^2 \text{ for all } h \in \mathscr{D}(K) \subset \mathscr{D}(F),$$

and

$$\|Gf\|^2 \leq a_3^2\|f\|^2 + b_3^2\|Af\|^2 \text{ for all } f \in \mathscr{D}(A) \subset \mathscr{D}(G).$$

Hence, for $(f, g, h) \in \mathscr{D}(A) \times \mathscr{D}(E) \times \mathscr{D}(K)$, we have

$$\begin{aligned} \left\| \begin{pmatrix} 0 & B & 0 \\ 0 & 0 & F \\ G & 0 & 0 \end{pmatrix} \begin{pmatrix} f \\ g \\ h \end{pmatrix} \right\|^2 &= \left\| \begin{pmatrix} 0 & Q_B B & 0 \\ 0 & 0 & Q_F F \\ Q_G G & 0 & 0 \end{pmatrix} \begin{pmatrix} f \\ g \\ h \end{pmatrix} \right\|^2 \\ &= \|Q_B Bg\|^2 + \|Q_F Fh\|^2 + \|Q_G Gf\|^2 \\ &= \|Bg\|^2 + \|Fh\|^2 + \|Gf\|^2 \end{aligned}$$

and

$$\begin{aligned} \left\| \begin{pmatrix} A & 0 & 0 \\ 0 & E & 0 \\ 0 & 0 & K \end{pmatrix} \begin{pmatrix} f \\ g \\ h \end{pmatrix} \right\|^2 &= \left\| \begin{pmatrix} Q_A A & 0 & 0 \\ 0 & Q_E E & 0 \\ 0 & 0 & Q_K K \end{pmatrix} \begin{pmatrix} f \\ g \\ h \end{pmatrix} \right\|^2. \\ &= \|Q_A Af\|^2 + \|Q_E Fg\|^2 + \|Q_K Kh\|^2 \\ &= \|Af\|^2 + \|Eg\|^2 + \|Kh\|^2. \end{aligned}$$

Then,

$$\begin{aligned}
\left\| \begin{pmatrix} 0 & B & 0 \\ 0 & 0 & F \\ G & 0 & 0 \end{pmatrix} \begin{pmatrix} f \\ g \\ h \end{pmatrix} \right\|^2 &= \|Bg\|^2 + \|Fh\|^2 + \|Gf\|^2 \\
&\le a_3^2\|f\|^2 + b_3^2\|Gf\|^2 + a_1^2\|g\|^2 \\
&\le \eta \left\| \begin{pmatrix} f \\ g \\ h \end{pmatrix} \right\|^2 + \chi \left\| \begin{pmatrix} A & 0 & 0 \\ 0 & B & 0 \\ 0 & 0 & C \end{pmatrix} \begin{pmatrix} f \\ g \\ h \end{pmatrix} \right\|^2,
\end{aligned}$$

where $\eta = \max\{a_1, a_2, a_3\}^2$ and $\chi = \max\{b_1, b_2, b_3\}^2$ as

$$\max\{b_1, b_2, b_3\} = \{\delta_1 + \varepsilon, \delta_2 + \varepsilon, \delta_3 + \varepsilon\} = \delta + \varepsilon.$$

This shows that $\mathscr{S}$ is $\mathscr{T}$-bounded with T-bound $< \delta$. The proof of (ii) can be done in a similar fashion as that of (i). Q.E.D.

5.5.2 *Frobenius–Schur Factorization for Multivalued* 3 × 3 *Matrices Linear Operator*

In what follows, we will consider the following hypotheses:

$(H25)$ The multivalued linear operator A is a closed, densely defined on X, with a nonempty resolvent set $\rho(A)$.

$(H26)$ The multivalued linear operator D (resp. G) verifies that $\mathscr{D}(A) \subset \mathscr{D}(D)$ (resp. $\mathscr{D}(A) \subset \mathscr{D}(G)$) and, for some (hence for all) $\mu \in \rho(A)$, the multivalued linear operator $D(A-\mu)^{-1}$ (resp. $G(A-\mu)^{-1}$) is bounded.

Now, let

$$\begin{aligned} F_1(\mu) &= D(A-\mu)^{-1}, \\ F_2(\mu) &= G(A-\mu)^{-1}. \end{aligned}$$

$(H27)$ The multivalued linear operator B (resp. C) is densely defined on Y (resp. Z) and, for some (hence for all) $\mu \in \rho(A)$ the multivalued linear operator $(A-\mu)^{-1}B$ (resp. $(A-\mu)_{-1}C$) is bounded on its domain. Now, let

$$\begin{aligned} G_1(\mu) &:= \overline{(A-\mu)^{-1}B}, \\ G_2(\mu) &:= \overline{(A-\mu)^{-1}C}. \end{aligned}$$

$(H28)$ The linear $\mathscr{D}(B) \bigcap \mathscr{D}(E)$ is dense in Y and, for some (hence for all) $\mu \in \rho(A)$, the multivalued linear operator

$$S_1(\mu) := E - D(A-\mu)^{-1}B$$

is closed.

$(H29)$ $\mathscr{D}(C) \subset \mathscr{D}(F)$, and the multivalued linear operator $F - D(A-\mu)^{-1}C$ is bounded on its domain, for some $\mu \in \rho(A) \bigcap \rho(S_1(\mu))$ we will denote $G_3(\mu)$ by

$$G_3(\mu) := \overline{(S_1(\mu) - \mu)^{-1}(F - D(A-\mu)^{-1}C)}.$$

$(H30)$ The multivalued linear operator H satisfies the fact that $\mathscr{D}(B) \subset \mathscr{D}(H)$ and, for some (hence for all) $\mu \in \rho(A) \bigcap \rho(S_1(\mu))$ the multivalued linear operator $H - G(A-\mu)^{-1}B)(S_1(\mu) - \mu)^{-1}$ is bounded. Set

$$F_3(\mu) := (H - G(A-\mu)^{-1}B)(S_1(\mu) - \mu)^{-1}.$$

$(H31)$ For the multivalued linear operator K, we will assume that $\mathscr{D}(C) \subset \mathscr{D}(K)$ and, for some (hence for all) $\mu \in \rho(A) \bigcap \rho(S_1(\mu))$, the operator

$$K - G(A-\mu)^{-1}C - [H - G(A-\mu)^{-1})B](S_1(\mu) - \mu)^{-1}[F - D(A-\mu)^{-1}C]$$

is closable. Let us denote by $S_2(\mu)$ this multivalued linear operator, and by $\overline{S_2(\mu)}$ its closure.

Theorem 5.5.1 *We suppose that the conditions* $(H25)$–$(H31)$ *are satisfied. Then,*

$$\mathscr{A}_0 - \mu I + \begin{pmatrix} 0 & 0 & 0 \\ 0 & \mathbb{E}(\mu) - \mathbb{E}(\mu) & \mathbb{F}(\mu) - \mathbb{F}(\mu) \\ 0 & \mathbb{H}(\mu) - \mathbb{H}(\mu) & \mathbb{K}(\mu) - \mathbb{K}(\mu) \end{pmatrix} = UVW, \tag{5.5.2}$$

where $\mathscr{A}_0$ *is given in (5.5.1),* $\mathbb{E}(\mu) = D(A-\mu)^{-1}B$, $\mathbb{F}(\mu) = D(A-\mu)^{-1}C$, $\mathbb{H}(\mu) = G(A-\mu)^{-1}B$, $\mathbb{K}(\mu) = G(A-\mu)^{-1}C + HG_3(\mu) + G(A-\mu)^{-1} BG_3(\mu)$,

$$U = \begin{pmatrix} I_X & 0 & 0 \\ F_1(\mu) & I_Y & 0 \\ F_2(\mu) & F_3(\mu) & I_Z \end{pmatrix}, \tag{5.5.3}$$

$$V = \begin{pmatrix} A-\mu & 0 & 0 \\ 0 & S_1(\mu) - \mu & 0 \\ 0 & 0 & S_2(\mu) - \mu \end{pmatrix} \tag{5.5.4}$$

and

$$W = \begin{pmatrix} I_X & G_1(\mu) & G_2(\mu) \\ 0 & I_Y & G_3(\mu) \\ 0 & 0 & I_Z \end{pmatrix}. \tag{5.5.5}$$

◇

Proof Let $\begin{pmatrix} x \\ y \\ z \end{pmatrix} \in \mathscr{D}(UVW)$. Then,

$$UVW \begin{pmatrix} x \\ y \\ z \end{pmatrix}$$

$$= UV \begin{pmatrix} x + G_1(\mu)y + G_2(\mu)z \\ y + G_3(\mu)z \\ z \end{pmatrix}$$

$$= U \begin{pmatrix} (A-\mu)(x + G_1(\mu)y + G_2(\mu)z) \\ (S_1(\mu) - \mu)(y + G_3(\mu)z) \\ (S_2(\mu) - \mu)z \end{pmatrix}$$

$$= \begin{pmatrix} (A-\mu)(x + G_1(\mu)y + G_2(\mu)z) \\ (F_1(\mu))(A-\mu)(x + G_1(\mu)y + G_2(\mu)z) + (S_1(\mu) - \mu)(y + G_3(\mu)z) \\ F_2(\mu)(A-\mu)(x + G_1(\mu)y + G_2(\mu)z) + F_3(\mu)(S_1(\mu) - \mu)(y + G_3(\mu)z) + (S_2(\mu) - \mu)z \end{pmatrix}.$$

Since $\mathscr{D}(A)$ containing the ranges of both $R(G_1(\mu))$ and $R(G_2(\mu))$, then

$$(A-\mu)(x + G_1(\mu)y + G_2(\mu)z) = (A-\mu)x + (A-\mu)G_1(\mu)y + (A-\mu)G_2(\mu)z.$$

Hence, $y \in \mathscr{D}(B)$ and $z \in \mathscr{D}(C)$. So,

$$G_1(\mu)y = (A-\mu)^{-1}By$$

and

$$G_2(\mu)y = (A-\mu)^{-1}Cz.$$

Therefore,

$$(A-\mu)(A-\mu)^{-1}By = By + A(0)$$

and

$$(A-\mu)(A-\mu)^{-1}Cz = Cz + A(0).$$

This leads to

$$\begin{aligned}(A-\mu)x + (A-\mu)G_1(\mu)y + (A-\mu)G_2(\mu)z &= (A-\mu)x + By + A(0) + Cz + A(0) \\ &= (A-\mu)x + By + Cz.\end{aligned}$$

Since $F_1(\mu)$ is bounded, then

$$\begin{aligned}(F_1(\mu))(A-\mu)(x + G_1(\mu)y + G_2(\mu)z) &= (F_1(\mu))((A-\mu)x + By + Cz) \\ &= (F_1(\mu)(A-\mu)x + F_1(\mu)By + F_1(\mu)Cz) \\ &= D(A-\mu)^{-1}(A-\mu)x + \mathbb{E}(\mu)y + \mathbb{F}(\mu)z.\end{aligned}$$

Since

$$\begin{aligned}D(A-\mu)^{-1}(A-\mu)x &= D(x + (A-\mu)^{-1}(0)) \\ &= Dx,\end{aligned}$$

then

$$(F_1(\mu))(A-\mu)(x+G_1(\mu)y+G_2(\mu)z) = Dx+\mathbb{E}(\mu)y+\mathbb{F}(\mu)z. \qquad (5.5.6)$$

Since $y \in \mathscr{D}(B)\bigcap\mathscr{D}(E) = \mathscr{D}(S_1(\mu))$ and $R(G_3(\mu)) \subset \mathscr{D}((S_1(\mu)-\mu)^{-1})$, then

$$\begin{aligned}(S_1(\mu)-\mu)(y+G_3(\mu)z) &= (S_1(\mu)-\mu)y+(S_1(\mu)-\mu)G_3(\mu)z\\ &= (S_1(\mu)-\mu)y+\\ &\quad (S_1(\mu)-\mu)(S_1(\mu)-\mu)^{-1}(F-D(A-\mu)^{-1}C)z\\ &= (S_1(\mu)-\mu)y+(F-D(A-\mu)^{-1}Cz+(S_1(\mu)-\mu)(0)\\ &= (S_1(\mu)-\mu)y+(F-D(A-\mu)^{-1}C)z\\ &= (E-\mu)y+Fz-\mathbb{E}(\mu)y-D(A-\mu)^{-1}Cz.\end{aligned}$$

We can easily check that

$$\begin{aligned}&(F_1(\mu))(A-\mu)(x+G_1(\mu)y+G_2(\mu)z)+(S_1(\mu)-\mu)(y+G_3(\mu)z)\\ &\qquad = Dx+(E-\mu)y+Fz+\mathbb{E}(\mu)(0)+\mathbb{F}(\mu)(0).\end{aligned}$$

It is clear that,

$$\begin{aligned}F_2(\mu)(A-\mu)\,(x+G_1(\mu)y+G_2(\mu)z) &= F_2(\mu)\,((A-\mu)x+By+Cz)\\ &= G(A-\mu)^{-1}\,((A-\mu)x+By+Cz)\,.\end{aligned}$$

Since $F_2(\mu)$ is bounded, then

$$\begin{aligned}F_2(\mu)(A-\mu)\,(x+G_1(\mu)y+G_2(\mu)z) &= G(A-\mu)^{-1}(A-\mu)x\\ &\quad +G(A-\mu)^{-1}By+G(A-\mu)^{-1}Cz.\end{aligned}$$

Hence, $\mu \in \rho(A)$ and

$$(A-\mu)^{-1}(A-\mu)x = x.$$

Consequently,

$$F_2(\mu)(A-\mu)\,(x+G_1(\mu)y+G_2(\mu)z) = Gx+G(A-\mu)^{-1}By+G(A-\mu)^{-1}Cz.$$

On the one hand,

$$\begin{aligned}F_3(\mu)(S_1(\mu)-\mu)\left(y+G_(\mu)z\right) &= \left(H-G(A-\mu)^{-1}B\right)(S_1(\mu)-\mu)^{-1}\\ &\quad (S_1(\mu)-\mu)\left(y+G_(\mu)z\right).\end{aligned}$$

Since $\mu \in \rho(S_1(\mu))$, we will check that

$$\begin{aligned}F_3(\mu)(S_1(\mu)-\mu)\left(y+G_(\mu)z\right) &= \left(H-G(A-\mu)^{-1}B\right)\\ &\quad (S_1(\mu)-\mu)^{-1}(S_1(\mu)-\mu)\left(y+G_(\mu)z\right)\\ &= \left(H-G(A-\mu)^{-1}B\right)\left(y+G_(\mu)z\right)\end{aligned}$$

$$= Hy + HG_3(\mu)z - G(A-\mu)^{-1}By - G(A-\mu)^{-1}$$
$$BG_3(\mu)z.$$

On the other hand,

$$(S_2(\mu)-\mu)\,z = (K-\mu)z - G(A-\mu)^{-1}Cz - HG_3(\mu)z + G(A-\mu)^{-1}BG_{(}\mu)z.$$

Finally, we get

$$\begin{aligned}
&F_2(\mu)(A-\mu)\,(x + G_1(\mu)y + G_2(\mu)z)\\
&+ F_3(\mu)(S_1(\mu)-\mu)\,\big(y + G_{(}\mu)z\big) + (S_2(\mu)-\mu)\,z\\
&= Gx + G(A-\mu)^{-1}By + G(A-\mu)^{-1}Cz + Hy + HG_3(\mu)z - G(A-\mu)^{-1}By\\
&- G(A-\mu)^{-1}BG_3(\mu)z + (K-\mu)z - G(A-\mu)^{-1}Cz\\
&- HG_3(\mu)z + G(A-\mu)^{-1}BG_3(\mu)z\\
&= Gx + Hy + (K-\mu)z + \mathbb{K}(\mu)(0).
\end{aligned}$$

This implies that

$$\begin{pmatrix} (A-\mu)x + By + Cz \\ Dx + (E-\mu)y + Fz + \mathbb{E}(\mu)(0) + \mathbb{F}(\mu)(0) \\ Gx + Hy + (K-\mu)z + \mathbb{K}(\mu)(0) \end{pmatrix}$$

$$= \begin{pmatrix} A-\mu & B & C \\ D & E-\mu & F \\ G & H & K-\mu \end{pmatrix}\begin{pmatrix} x \\ y \\ z \end{pmatrix} + \begin{pmatrix} 0 & 0 & 0 \\ 0 & \mathbb{E}(\mu)-\mathbb{E}(\mu) & \mathbb{F}(\mu)-\mathbb{F}(\mu) \\ 0 & \mathbb{H}(\mu)-\mathbb{H}(\mu) & \mathbb{K}(\mu)-\mathbb{K}(\mu) \end{pmatrix}\begin{pmatrix} x \\ y \\ z \end{pmatrix}.$$

Therefore,

$$\left[\mathscr{A}_0 - \mu I + \begin{pmatrix} 0 & 0 & 0 \\ 0 & \mathbb{E}(\mu)-\mathbb{E}(\mu) & \mathbb{F}(\mu)-\mathbb{F}(\mu) \\ 0 & \mathbb{H}(\mu)-\mathbb{H}(\mu) & \mathbb{K}(\mu)-\mathbb{K}(\mu) \end{pmatrix}\right]\begin{pmatrix} x \\ y \\ z \end{pmatrix} = UVW\begin{pmatrix} x \\ y \\ z \end{pmatrix}.$$

Observe that,

$$\begin{aligned}
&\mathscr{D}\left[\mathscr{A}_0 - \mu I + \begin{pmatrix} 0 & 0 & 0 \\ 0 & \mathbb{E}(\mu)-\mathbb{E}(\mu) & \mathbb{F}(\mu)-\mathbb{F}(\mu) \\ 0 & \mathbb{H}(\mu)-\mathbb{H}(\mu) & \mathbb{K}(\mu)-\mathbb{K}(\mu) \end{pmatrix}\right]\\
&= \mathscr{D}(\mathscr{A}_0)\bigcap \mathscr{D}\left[\begin{pmatrix} 0 & 0 & 0 \\ 0 & \mathbb{E}(\mu)-\mathbb{E}(\mu) & \mathbb{F}(\mu)-\mathbb{F}(\mu) \\ 0 & \mathbb{H}(\mu)-\mathbb{H}(\mu) & \mathbb{K}(\mu)-\mathbb{K}(\mu) \end{pmatrix}\right]\\
&= \Big(\mathscr{D}(A) \times (\mathscr{D}(B)\bigcap \mathscr{D}(E)) \times \mathscr{D}(C)\Big)\bigcap \Big(X \times (\mathscr{D}(\mathbb{E}(\mu))\bigcap \mathscr{D}(\mathbb{H}(\mu))) \times (\mathscr{D}(\mathbb{F}(\mu))\bigcap \mathscr{D}(\mathbb{K}(\mu)))\Big).
\end{aligned}$$

Since $\mathscr{D}(\mathbb{E}(\mu)) = \mathscr{D}(\mathbb{H}(\mu)) = \mathscr{D}(B)$ and $\mathscr{D}(\mathbb{F}(\mu)) = \mathscr{D}(\mathbb{K}(\mu)) = \mathscr{D}(C)$, then

$$\mathscr{D}\left[\mathscr{A}_0 - \mu I + \begin{pmatrix} 0 & 0 & 0 \\ 0 & \mathbb{E}(\mu) - \mathbb{E}(\mu) & \mathbb{F}(\mu) - \mathbb{F}(\mu) \\ 0 & \mathbb{H}(\mu) - \mathbb{H}(\mu) & \mathbb{K}(\mu) - \mathbb{K}(\mu) \end{pmatrix}\right]$$

$$= \mathscr{D}(\mathscr{A}_0) \bigcap \mathscr{D}\left[\begin{pmatrix} 0 & 0 & 0 \\ 0 & \mathbb{E}(\mu) - \mathbb{E}(\mu) & \mathbb{F}(\mu) - \mathbb{F}(\mu) \\ 0 & \mathbb{H}(\mu) - \mathbb{H}(\mu) & \mathbb{K}(\mu) - \mathbb{K}(\mu) \end{pmatrix}\right]$$

$$= \mathscr{D}(A) \times (\mathscr{D}(B) \bigcap \mathscr{D}(E)) \times \mathscr{D}(C).$$

In fact,

$$\mathscr{D}(UVW) = \left\{ \begin{pmatrix} x \\ y \\ z \end{pmatrix} \in X \times Y \times Z \text{ such that } \begin{matrix} x + G_1(\mu)y + G_2(\mu)z \in \mathscr{D}(A), \\ y + G_3(\mu)z \in \mathscr{D}(S_1(\mu)) \\ \text{and } z \in \mathscr{D}(S_2(\mu)) \end{matrix} \right\}$$

$$= \left\{ \begin{pmatrix} x \\ y \\ z \end{pmatrix} \in X \times Y \times Z \text{ such that } \begin{matrix} x + G_1(\mu)y + G_2(\mu)z \in \mathscr{D}(A), \\ y + G_3(\mu)z \in \mathscr{D}(B) \bigcap \mathscr{D}(E) \\ \text{and } z \in \mathscr{D}(C) \end{matrix} \right\}$$

$$= \mathscr{D}(A) \times (\mathscr{D}(B) \bigcap \mathscr{D}(E)) \times \mathscr{D}(C).$$

It follows that

$$\mathscr{D}\left[\mathscr{A}_0 - \mu I + \begin{pmatrix} 0 & 0 & 0 \\ 0 & \mathbb{E}(\mu) - \mathbb{E}(\mu) & \mathbb{F}(\mu) - \mathbb{F}(\mu) \\ 0 & \mathbb{H}(\mu) - \mathbb{H}(\mu) & \mathbb{K}(\mu) - \mathbb{K}(\mu) \end{pmatrix}\right] = \mathscr{D}(UVW).$$

So, we get Eq. (5.5.2). Q.E.D.

We can find another decomposition which will be used in the sequel. For this reason, we add the following hypothesis:

$(H32)$ We will assume that $\mathbb{E}(\mu)(0) \subset E(0)$, $\mathbb{H}(\mu)(0) \subset H(0)$, $\mathbb{F}(\mu)(0) \subset F(0)$, $\mathbb{K}(\mu)(0) \subset K(0)$, $\mathscr{D}(E) \subset \mathscr{D}(B)$, $\mathscr{D}(H) = \mathscr{D}(B)$, and $\mathscr{D}(F) = \mathscr{D}(K) = \mathscr{D}(C)$.

Theorem 5.5.2 *We suppose that the conditions* $(H25)$–$(H30)$ *are satisfied. Then,*

$$\mathscr{A}_0 - \mu I = UVW,$$

where $\mathscr{A}_0$ *is given in (5.5.1).* ◇

Proof By using Theorem 5.5.1, we get

$$\mathscr{A}_0 - \mu I + \begin{pmatrix} 0 & 0 & 0 \\ 0 & \mathbb{E}(\mu) - \mathbb{E}(\mu) & \mathbb{F}(\mu) - \mathbb{F}(\mu) \\ 0 & \mathbb{H}(\mu) - \mathbb{H}(\mu) & \mathbb{K}(\mu) - \mathbb{K}(\mu) \end{pmatrix}$$
$$= \begin{pmatrix} A - \mu & B & C \\ D & E - \mu + \mathbb{E}(\mu) - \mathbb{E}(\mu) & F + \mathbb{F}(\mu) - \mathbb{F}(\mu) \\ G & H + \mathbb{H}(\mu) - \mathbb{H}(\mu) & K - \mu + \mathbb{K}(\mu) - \mathbb{K}(\mu) \end{pmatrix}.$$

Assume that $\mathbb{E}(\mu) \subset E(0)$ and $\mathscr{D}(E) \subset \mathscr{D}(\mathbb{E}(\mu)) = \mathscr{D}(B)$ are satisfied. Then,

$$E - \mu + \mathbb{E}(\mu) - \mathbb{E}(\mu) = E - \mu.$$

In the same way, we prove that

$$H + \mathbb{H}(\mu) - \mathbb{H}(\mu) = H,$$

$$F + \mathbb{F}(\mu) - \mathbb{F}(\mu) = F,$$

and

$$K - \mu + \mathbb{K}(\mu) - \mathbb{K}(\mu) = K - \mu.$$

This implies that

$$\mathscr{A}_0 - \mu I = UVW. \qquad \text{Q.E.D.}$$

Theorem 5.5.3 *We suppose that the conditions* $(H25)$*-*$(H30)$ *are satisfied. If* U *and* W *are single-valued, then the multivalued linear operator* $\mathscr{A}_0$*, given in (5.5.1), is closable if, and only if, the multivalued linear operator* $S_2(\mu)$ *is closable. Moreover, the closure* $\mathscr{A}$ *of* $\mathscr{A}_0$ *is given by*

$$\mathscr{A} = \overline{\mathscr{A}_0} = \mu I + U\overline{V}W,$$

where U*,* V*, and* W *are given in (5.5.3), (5.5.4), and (5.5.5), respectively.* ◇

Proof We start by determining the properties of the matrices

$$U = \begin{pmatrix} I_X & 0 & 0 \\ F_1(\mu) & I_Y & 0 \\ F_2(\mu) & F_3(\mu) & I_Z \end{pmatrix}$$

and

$$W = \begin{pmatrix} I_X & G_1(\mu) & G_2(\mu) \\ 0 & I_Y & G_3(\mu) \\ 0 & 0 & I_Z \end{pmatrix}.$$

The operator matrix U is bounded and bijective with

$$U^{-1} = \begin{pmatrix} I_X & 0 & 0 \\ -F_1(\mu) & I_Y & 0 \\ F_1(\mu)F_3(\mu) - F_2(\mu) & -F_3(\mu) & I_Z \end{pmatrix}.$$

The operator matrix W is bounded and bijective with

$$W^{-1} = \begin{pmatrix} I_X & -G_1(\mu) & G_1(\mu)G_3(\mu) - G_2(\mu) \\ 0 & I_Y & -G_3(\mu) \\ 0 & 0 & I_Z \end{pmatrix}.$$

We have $A - \mu I$ is closed. On the one hand, $S_1(\mu)$ is closed. Then, $S_1(\mu) - \mu$ is closed. On the other hand, $S_2(\mu)$ is closable, then by Proposition 2.12.2, we have $S_2(\mu) - \mu$ is closable, where $\overline{S_2(\mu) - \mu} = \overline{S_2(\mu)} - \mu$. So,

$$V = \begin{pmatrix} A - \mu & 0 & 0 \\ 0 & S_1(\mu) - \mu & 0 \\ 0 & 0 & S_2(\mu) - \mu \end{pmatrix}$$

is closable and

$$\overline{V} = \begin{pmatrix} A - \mu & 0 & 0 \\ 0 & S_1(\mu) - \mu & 0 \\ 0 & 0 & \overline{S_2(\mu)} - \mu \end{pmatrix}.$$

Therefore, by using Theorem 2.12.3, we conclude that UVW is closable and $\overline{UVW} = U\overline{V}W$. Hence,

$$\begin{aligned} \overline{\mathscr{A}_0 - \mu I} &= \overline{\mathscr{A}_0} - \mu I \\ &= U\overline{V}W. \end{aligned}$$

So,

$$\overline{\mathscr{A}_0} = \mu I + U\overline{V}W.$$

It follows that the relation $\mathscr{A}_0$ is closable. Q.E.D.

Theorem 5.5.4 *We suppose that the conditions $(H25)$–$(H32)$ are satisfied. If U and W are single-valued, then the multivalued linear operator $\mathscr{A}_0$, given in (5.5.1), is closable if, and only if, the multivalued linear operator $S_2(\mu)$ is closable. Moreover, for all $\lambda \in \mathbb{C}$ the closure $\mathscr{A}$ of $\mathscr{A}_0$ is given by*

$$\mathscr{A} - \lambda = \overline{\mathscr{A}_0} - \lambda = U\overline{V(\lambda)}W + \mathbb{M}(\mu),$$

where

$$V(\lambda) = \begin{pmatrix} A - \lambda & 0 & 0 \\ 0 & S_1(\mu) - \lambda & 0 \\ 0 & 0 & \overline{S_2(\mu)} - \lambda \end{pmatrix}$$

and

$$\mathbb{M}(\mu) = \begin{pmatrix} 0 & G_1(\mu) & G_2(\mu) \\ F_1(\mu) & F_1(\mu)G_1(\mu) & F_1(\mu)G_2(\mu) + G_3(\mu) \\ F_2(\mu) & F_2(\mu)G_1(\mu) + F_3(\mu) & F_2(\mu)G_2(\mu) + F_3(\mu)G_3(\mu) \end{pmatrix}. \quad \diamond$$

Proof Since I is single-valued, then

$$\mathscr{A} - \lambda I = U\overline{V}W - \mu I + (\mu - \lambda)I$$

for $\lambda \in \mathbb{C}$. We can deduce that

$$\begin{aligned} \mathscr{A} - \lambda I &= U \begin{pmatrix} A - \mu - \lambda + \lambda & 0 & 0 \\ 0 & S_1(\mu) - \mu - \lambda + \lambda & 0 \\ 0 & 0 & \overline{S_2(\mu)} - \mu - \lambda + \lambda \end{pmatrix} W + (\mu - \lambda)I \\ &= U \left[\begin{pmatrix} A - \lambda & 0 & 0 \\ 0 & S_1(\mu) - \lambda & 0 \\ 0 & 0 & \overline{S_2(\mu)} - \lambda \end{pmatrix} + (\lambda - \mu) \begin{pmatrix} I_X & 0 & 0 \\ 0 & I_Y & 0 \\ 0 & 0 & I_Z \end{pmatrix} \right] W + (\mu - \lambda)I. \end{aligned}$$

Since U is bounded and W is single-valued, then

$$\begin{aligned} \mathscr{A} - \lambda I = U &\begin{pmatrix} A - \lambda & 0 & 0 \\ 0 & S_1(\mu) - \lambda & 0 \\ 0 & 0 & \overline{S_2(\mu)} - \lambda \end{pmatrix} W + \\ U &\begin{pmatrix} (\lambda - \mu)I_X & 0 & 0 \\ 0 & (\lambda - \mu)I_Y & 0 \\ 0 & 0 & (\lambda - \mu)I_Z \end{pmatrix} W + (\mu - \lambda)I. \end{aligned}$$

Since U and W are single-valued, then

$$\begin{aligned} \mathbb{M}(\mu) &= U \begin{pmatrix} (\lambda - \mu)I_X & 0 & 0 \\ 0 & (\lambda - \mu)I_Y & 0 \\ 0 & 0 & (\lambda - \mu)I_Z \end{pmatrix} W + (\mu - \lambda)I \\ &= \begin{pmatrix} 0 & G_1(\mu) & G_2(\mu) \\ F_1(\mu) & F_1(\mu)G_1(\mu) & F_1(\mu)G_2(\mu) + G_3(\mu) \\ F_2(\mu) & F_2(\mu)G_1(\mu) + F_3(\mu) & F_2(\mu)G_2(\mu) + F_3(\mu)G_3(\mu) \end{pmatrix}. \end{aligned}$$

This achieves the proof. Q.E.D.

Concerning the multivalued linear operator $F_i(\mu)$, and $G_i(\mu)$ with $i = \{1, 2, 3\}$ we impose the following conditions:

$(H33)$ The operator $Z_i(\mu)$ is a bounded selection of $F_i(\mu)$ with $i = \{1, 2, 3\}$.
$(H34)$ The operator $Y_i(\mu)$ is a bounded selection of $G_i(\mu)$ with $i = \{1, 2, 3\}$.
$(H35)$ For some $\mu \in \rho(A)$, $\mathscr{D}(V)$ contains the ranges of both

$$\begin{pmatrix} I_X & Y_1(\mu) & Y_2(\mu) \\ 0 & I_Y & Y_3(\mu) \\ 0 & 0 & I_Z \end{pmatrix}$$

and

$$\begin{pmatrix} 0 & G_1(\mu) - G_1(\mu) & G_2(\mu) - G_2(\mu) \\ 0 & 0 & G_3(\mu) - G_3(\mu) \\ 0 & 0 & 0 \end{pmatrix}.$$

Theorem 5.5.5 *We suppose that the conditions* $(H25)$*–*$(H33)$ *are satisfied. If* $G_i(\mu)$ *with* $i = \{1, 2, 3\}$ *are single-valued, then*

(i)

$$\mathscr{A}_0 - \mu I = U_1 V W + \begin{pmatrix} 0 & 0 & 0 \\ D - D & 0 & 0 \\ G - G & H - H & 0 \end{pmatrix},$$

where $\mathscr{A}_0$ *is given in (5.5.1), and*

$$U_1 = \begin{pmatrix} I_X & 0 & 0 \\ Z_1(\mu) & I_Y & 0 \\ Z_2(\mu) & Z_3(\mu) & I_Z \end{pmatrix}.$$

(ii) *If*

$$\begin{pmatrix} 0 & 0 & 0 \\ D - D & 0 & 0 \\ G - G & H - H & 0 \end{pmatrix}$$

is continuous, then the multivalued linear operator $\mathscr{A}_0$ *is closable if, and only if, the multivalued linear operator* $S_2(\mu)$ *is closable. Moreover, the closure* $\mathscr{A}$ *of* $\mathscr{A}_0$ *is given by*

$$\mathscr{A} - \mu I = \overline{\mathscr{A}_0} - \mu I = U_1 \overline{V} W + \begin{pmatrix} 0 & 0 & 0 \\ D - D & 0 & 0 \\ G - G & H - H & 0 \end{pmatrix}.$$

(iii) *If*

$$\begin{pmatrix} 0 & 0 & 0 \\ D - D & 0 & 0 \\ G - G & H - H & 0 \end{pmatrix}$$

is continuous, then for $\lambda \in \mathbb{C}$*, we have*

$$\mathscr{A} - \lambda = \overline{\mathscr{A}_0} - \lambda = U_1 \overline{V(\lambda)} W + \mathbb{M}_1(\mu),$$

where

$$\mathbb{M}_1(\mu) = \begin{pmatrix} 0 & G_1(\mu) & G_2(\mu) \\ Z_1(\mu) + D - D & Z_1(\mu)G_1(\mu) & Z_1(\mu)G_2(\mu) + G_3(\mu) \\ Z_2(\mu) + G - G & Z_2(\mu)G_1(\mu) + Z_3(\mu) + H - H & Z_2(\mu)G_2(\mu) + Z_3(\mu)G_3(\mu) \end{pmatrix}.$$

◇

Proof (i) We can write

$$\begin{aligned} \mathscr{A}_0 - \mu I &= \begin{pmatrix} I_X & 0 & 0 \\ Z_1(\mu) + F_1(\mu) - F_1(\mu) & I_Y & 0 \\ Z_2(\mu) + F_2(\mu) - F_2(\mu) & Z_3(\mu) + F_3(\mu) - F_3(\mu) & I_Z \end{pmatrix} VW \\ &= \left[\begin{pmatrix} I_X & 0 & 0 \\ Z_1(\mu) & I_Y & 0 \\ Z_2(\mu) & Z_3(\mu) & I_Z \end{pmatrix} + \begin{pmatrix} 0 & 0 & 0 \\ F_1(\mu) - F_1(\mu) & 0 & 0 \\ F_2(\mu) - F_2(\mu) & F_3(\mu) - F_3(\mu) & 0 \end{pmatrix} \right] VW. \end{aligned}$$

Since

$$\begin{aligned} VW(0) &= V(0) \\ &\subset N\left[\begin{pmatrix} 0 & 0 & 0 \\ F_1(\mu) - F_1(\mu) & 0 & 0 \\ F_2(\mu) - F_2(\mu) & F_3(\mu) - F_3(\mu) & 0 \end{pmatrix} \right] \\ &= X \times Y \times Z, \end{aligned}$$

then

$$\mathscr{A}_0 - \mu I = \begin{pmatrix} I_X & 0 & 0 \\ Z_1(\mu) & I_Y & 0 \\ Z_2(\mu) & Z_3(\mu) & I_Z \end{pmatrix} V + \begin{pmatrix} 0 & 0 & 0 \\ F_1(\mu) - F_1(\mu) & 0 & 0 \\ F_2(\mu) - F_2(\mu) & F_3(\mu) - F_3(\mu) & 0 \end{pmatrix} VW.$$

Let $\left[\begin{pmatrix} x_1 \\ y_1 \\ z_1 \end{pmatrix}, \begin{pmatrix} x_2 \\ y_2 \\ z_2 \end{pmatrix} \right] \in G\left[\begin{pmatrix} 0 & 0 & 0 \\ F_1(\mu) - F_1(\mu) & 0 & 0 \\ F_2(\mu) - F_2(\mu) & F_3(\mu) - F_3(\mu) & 0 \end{pmatrix} VW \right].$

Then, $\begin{pmatrix} x_2 \\ y_2 \\ z_2 \end{pmatrix} \in \begin{pmatrix} 0 & 0 & 0 \\ F_1(\mu) - F_1(\mu) & 0 & 0 \\ F_2(\mu) - F_2(\mu) & F_3(\mu) - F_3(\mu) & 0 \end{pmatrix} \begin{pmatrix} (A - \mu)(x_1 + G_1(\mu)y_1 + G_2(\mu)z_1) \\ (S_1(\mu) - \mu)(y_1 + G_3(\mu)z_1) \\ (S_2(\mu) - \mu)z_1 \end{pmatrix}.$
This is equivalent to

$$\begin{aligned} & x_2 = 0 \\ & y_2 \in (F_1(\mu) - F_1(\mu))(A - \mu)(x_1 + G_1(\mu)y_1 + G_2(\mu)z_1) \\ & z_2 \in (F_2(\mu) - F_2(\mu))(A - \mu)(x_1 + G_1(\mu)y_1 + G_2(\mu)z_1) \\ & \quad + (F_3(\mu) - F_3(\mu))(S_1(\mu) - \mu)(y_1 + G_3(\mu)z_1). \end{aligned}$$

Hence,

$$\begin{aligned} y_2 \in (F_1(\mu) - F_1(\mu))(A - \mu)(x_1 + G_1(\mu)y_1 + G_2(\mu)z_1 \\ \subset F_1(\mu)(A - \mu)(x_1 + G_1(\mu)y_1 + G_2(\mu)z_1) - F_1(\mu)(A - \mu). \end{aligned}$$

From the first equation of (5.5.6), we can deduce that

$$y_2 \in D(0) + \mathbb{E}(\mu)(0) + \mathbb{F}(\mu)(0).$$

Hence, $G_1(\mu)$, $G_2(\mu)$ are single-valued and $\mathbb{E}(\mu)(0) = \mathbb{F}(\mu)(0) = D(0)$. Observe that $y_2 \in D(0) = (D - D)x_1$. Using the same reasoning, we get

$$z_2 \in (G - G)x_1 + (H - H)y_1.$$

Consequently,

$$\begin{aligned} x_2 &= 0 \\ y_2 &\in D(0) \\ z_2 &\in G(0) + H(0). \end{aligned}$$

Therefore,

$$\begin{pmatrix} x_2 \\ y_2 \\ z_2 \end{pmatrix} \in \begin{pmatrix} 0 & 0 & 0 \\ D - D & 0 & 0 \\ G - G & H - H & 0 \end{pmatrix} \begin{pmatrix} x_1 \\ y_1 \\ z_1 \end{pmatrix}.$$

It follows that

$$G\left[\begin{pmatrix} 0 & 0 & 0 \\ F_1(\mu) - F_1(\mu) & 0 & 0 \\ F_2(\mu) - F_2(\mu) & F_3(\mu) - F_3(\mu) & 0 \end{pmatrix} VW\right] \subset G\left[\begin{pmatrix} 0 & 0 & 0 \\ D - D & 0 & 0 \\ G - G & H - H & 0 \end{pmatrix}\right]. \tag{5.5.7}$$

Now,

$$\begin{aligned}
&\begin{pmatrix} 0 & 0 & 0 \\ F_1(\mu) - F_1(\mu) & 0 & 0 \\ F_2(\mu) - F_2(\mu) & F_3(\mu) - F_3(\mu) & 0 \end{pmatrix} VW \begin{pmatrix} 0 \\ 0 \\ 0 \end{pmatrix} \\
&= \begin{pmatrix} 0 & 0 & 0 \\ F_1(\mu) - F_1(\mu) & 0 & 0 \\ F_2(\mu) - F_2(\mu) & F_3(\mu) - F_3(\mu) & 0 \end{pmatrix} V \begin{pmatrix} 0 \\ 0 \\ 0 \end{pmatrix}, \\
&= \begin{pmatrix} 0 & 0 & 0 \\ F_1(\mu) - F_1(\mu) & 0 & 0 \\ F_2(\mu) - F_2(\mu) & F_3(\mu) - F_3(\mu) & 0 \end{pmatrix} \begin{pmatrix} (A - \mu)(0) \\ (S_1(\mu) - \mu)(0) \\ (S_2(\mu) - \mu)(0) \end{pmatrix}, \\
&= \begin{pmatrix} 0 \\ (F_1(\mu) - F_1(\mu))(A - \mu)(0) \\ (F_2(\mu) - F_2(\mu))(A - \mu)(0) + (F_3(\mu) - F_3(\mu))(S_1(\mu) - \mu)(0) \end{pmatrix}, \\
&= \begin{pmatrix} 0 \\ (D - D)(A - \mu)^{-1}(A - \mu)(0) \\ (G - G)(A - \mu)^{-1}(A - \mu)(0) + ((H - G(A - \mu)^{-1}B) - (H - G(A - \mu)^{-1}B))(0) \end{pmatrix}, \\
&= \begin{pmatrix} 0 \\ D(0) \\ G(0) + H(0) - G(A - \mu)^{-1}B(0) \end{pmatrix},
\end{aligned}$$

$$= \begin{pmatrix} 0 & 0 & 0 \\ D-D & 0 & 0 \\ G-G & H-H+\mathbb{H}(\mu)-\mathbb{H}(\mu) & 0 \end{pmatrix} \begin{pmatrix} 0 \\ 0 \\ 0 \end{pmatrix}.$$

Since $\mathbb{H}(\mu)(0) \subset H(0)$, then

$$\begin{pmatrix} 0 & 0 & 0 \\ F_1(\mu)-F_1(\mu) & 0 & 0 \\ F_2(\mu)-F_2(\mu) & F_3(\mu)-F_3(\mu) & 0 \end{pmatrix} VW \begin{pmatrix} 0 \\ 0 \\ 0 \end{pmatrix} = \begin{pmatrix} 0 & 0 & 0 \\ D-D & 0 & 0 \\ G-G & H-H & 0 \end{pmatrix} \begin{pmatrix} 0 \\ 0 \\ 0 \end{pmatrix}. \quad (5.5.8)$$

Recall that

$$\mathscr{D}\left[\begin{pmatrix} 0 & 0 & 0 \\ D-D & 0 & 0 \\ G-G & H-H & 0 \end{pmatrix}\right] = \left(\mathscr{D}(D)\bigcap\mathscr{D}(G)\right) \times \mathscr{D}(H) \times Y$$

and

$$\begin{aligned} &\mathscr{D}\left[\begin{pmatrix} 0 & 0 & 0 \\ F_1(\mu)-F_1(\mu) & 0 & 0 \\ F_2(\mu)-F_2(\mu) & F_3(\mu)-F_3(\mu) & 0 \end{pmatrix} VW\right] \\ &= \mathscr{D}(F_1(\mu))\bigcap\mathscr{D}(F_2(\mu)) \times \mathscr{D}(F_3(\mu)) \times Y \\ &= \left(\mathscr{D}(D)\bigcap\mathscr{D}(G)\right) \times \mathscr{D}(H) \times Y. \end{aligned}$$

We infer that

$$\mathscr{D}\left[\begin{pmatrix} 0 & 0 & 0 \\ F_1(\mu)-F_1(\mu) & 0 & 0 \\ F_2(\mu)-F_2(\mu) & F_3(\mu)-F_3(\mu) & 0 \end{pmatrix} VW\right] = \mathscr{D}\left[\begin{pmatrix} 0 & 0 & 0 \\ D-D & 0 & 0 \\ G-G & H-H & 0 \end{pmatrix}\right]. \quad (5.5.9)$$

By using estimation (5.5.7), (5.5.8), and (5.5.9), we have

$$\mathscr{A}_0 - \mu I = \begin{pmatrix} I_X & 0 & 0 \\ Z_1(\mu) & I_Y & 0 \\ Z_2(\mu) & Z_3(\mu) & I_Z \end{pmatrix} VW + \begin{pmatrix} 0 & 0 & 0 \\ D-D & 0 & 0 \\ G-G & H-H & 0 \end{pmatrix}.$$

(ii) We start by determining the properties of the matrices U_1 and W. It is easy to notice that U_1 and W are single-valued bounded bijective, where

$$U_1^{-1} = \begin{pmatrix} I_X & 0 & 0 \\ -Z_1(\mu) & I_Y & 0 \\ Z_1(\mu)Z_3(\mu)-Z_2(\mu) & -Z_3(\mu) & I_Z \end{pmatrix}.$$

We have

$$V = \begin{pmatrix} A-\mu & 0 & 0 \\ 0 & S_1(\mu)-\mu & 0 \\ 0 & 0 & S_2(\mu)-\mu \end{pmatrix}$$

is closable. Therefore, by using Theorem 2.12.3, we conclude that U_1VW is closable and $\overline{U_1VW} = U_1\overline{V}W$. Then, applying Proposition 2.12.1 (ii), we have

$$\begin{aligned}\overline{\mathscr{A}_0 - \mu I} &= \overline{\mathscr{A}_0} - \mu I \\ &= U_1\overline{V}W.\end{aligned}$$

So,

$$\overline{\mathscr{A}_0} = \mu I + U_1\overline{V}W.$$

It follows that the relation $\mathscr{A}_0$ is closable and we have

$$\mathscr{D}(U_1VW) \subset \mathscr{D}\left[\begin{pmatrix} 0 & 0 & 0 \\ D-D & 0 & 0 \\ G-G & H-H & 0 \end{pmatrix}\right]$$

and

$$\begin{pmatrix} 0 & 0 & 0 \\ D-D & 0 & 0 \\ G-G & H-H & 0 \end{pmatrix}\begin{pmatrix} 0 \\ 0 \\ 0 \end{pmatrix} \subset U_1VW(0).$$

Then,

$$U_1VW + \begin{pmatrix} 0 & 0 & 0 \\ D-D & 0 & 0 \\ G-G & H-H & 0 \end{pmatrix}$$

is closable and

$$\begin{aligned}\overline{U_1VW + \begin{pmatrix} 0 & 0 & 0 \\ D-D & 0 & 0 \\ G-G & H-H & 0 \end{pmatrix}} &= \overline{U_1VW} + \begin{pmatrix} 0 & 0 & 0 \\ D-D & 0 & 0 \\ G-G & H-H & 0 \end{pmatrix} \\ &= U_1\overline{V}W + \begin{pmatrix} 0 & 0 & 0 \\ D-D & 0 & 0 \\ G-G & H-H & 0 \end{pmatrix}.\end{aligned}$$

Then,

$$\begin{aligned}\overline{\mathscr{A}_0 - \mu} &= \overline{\mathscr{A}_0} - \mu I \\ &= U_1\overline{V}W + \begin{pmatrix} 0 & 0 & 0 \\ D-D & 0 & 0 \\ G-G & H-H & 0 \end{pmatrix}.\end{aligned}$$

Hence,

$$\overline{\mathscr{A}_0} = \mu I + U_1 \overline{V} W + \begin{pmatrix} 0 & 0 & 0 \\ D - D & 0 & 0 \\ G - G & H - H & 0 \end{pmatrix}.$$

It follows that the relation $\mathscr{A}_0$ is closable.

(iii) Since I is single-valued, then we get

$$\mathscr{A} - \lambda I = U_1 \overline{V} W - \mu I + (\mu - \lambda) I$$

for $\lambda \in \mathbb{C}$. We can deduce that

$$\begin{aligned}
\mathscr{A} - \lambda I &= \overline{\mathscr{A}_0} - \lambda I \\
&= U_1 \begin{pmatrix} A - \mu - \lambda + \lambda & 0 & 0 \\ 0 & S_1(\mu) - \mu - \lambda + \lambda & 0 \\ 0 & 0 & \overline{S_2(\mu)} - \mu - \lambda + \lambda \end{pmatrix} W \\
&+ \begin{pmatrix} 0 & 0 & 0 \\ D - D & 0 & 0 \\ G - G & H - H & 0 \end{pmatrix} + (\mu - \lambda) I \\
&= U_1 \left[\begin{pmatrix} A - \lambda & 0 & 0 \\ 0 & S_1(\mu) - \lambda & 0 \\ 0 & 0 & \overline{S_2(\mu)} - \lambda \end{pmatrix} + (\lambda - \mu) \begin{pmatrix} I_X & 0 & 0 \\ 0 & I_Y & 0 \\ 0 & 0 & I_Z \end{pmatrix} \right] W \\
&+ (\mu - \lambda) I + \begin{pmatrix} 0 & 0 & 0 \\ D - D & 0 & 0 \\ G - G & H - H & 0 \end{pmatrix}.
\end{aligned}$$

Since U_1 is bounded and W is single-valued, then

$$\begin{aligned}
\mathscr{A} - \lambda I = U_1 &\begin{pmatrix} A - \lambda & 0 & 0 \\ 0 & S_1(\mu) - \lambda & 0 \\ 0 & 0 & \overline{S_2(\mu)} - \lambda \end{pmatrix} W + U_1 \begin{pmatrix} (\lambda - \mu) I_X & 0 & 0 \\ 0 & (\lambda - \mu) I_Y & 0 \\ 0 & 0 & (\lambda - \mu) I_Z \end{pmatrix} W \\
&+ (\mu - \lambda) I + \begin{pmatrix} 0 & 0 & 0 \\ D - D & 0 & 0 \\ G - G & H - H & 0 \end{pmatrix}.
\end{aligned}$$

Since U_1 and W are single-valued, then

$$\begin{aligned}
\mathbb{M}_1(\mu) = &\begin{pmatrix} 0 & G_1(\mu) & G_2(\mu) \\ Z_1(\mu) & Z_1(\mu) G_1(\mu) & Z_1(\mu) G_2(\mu) + G_3(\mu) \\ Z_2(\mu) & Z_2(\mu) G_1(\mu) + Z_3(\mu) & Z_2(\mu) G_2(\mu) + Z_3(\mu) G_3(\mu) \end{pmatrix} \\
&+ \begin{pmatrix} 0 & 0 & 0 \\ D - D & 0 & 0 \\ G - G & H - H & 0 \end{pmatrix}.
\end{aligned}$$

This achieves the proof. Q.E.D.

Theorem 5.5.6 *We suppose that the conditions* $(H25)$*–*$(H35)$ *are satisfied. Then,*

(i)

$$\mathscr{A}_0 - \mu I = U_1 V W_1 + \mathscr{I}(\mu),$$

where $\mathscr{A}_0$ *is given in (5.5.1),*

$$W_1 = \begin{pmatrix} I_X & Y_1(\mu) & Y_2(\mu) \\ 0 & I_Y & Y_3(\mu) \\ 0 & 0 & I_Z \end{pmatrix}$$

and

$$\mathscr{I}(\mu) = \begin{pmatrix} A - A & B - B & C - C \\ Z_1(\mu)A - Z_1(\mu)A + D - D & T_1 & T_2 \\ Z_2(\mu)A - Z_2(\mu)A + G - G & T_3 + H - H & T_4 \end{pmatrix}$$

with

$$\begin{aligned}
T_1 &= Z_1(\mu)B + S_1(\mu) - Z_1(\mu)B - S_1(\mu), \\
T_2 &= Z_1(\mu)C - Z_1(\mu)C + F - D(A-\mu)^{-1}C - F + D(A-\mu)^{-1}C, \\
T_3 &= Z_2(\mu)B - Z_2(\mu)B + Z_3(\mu)S_1(\mu) - Z_3(\mu)S_1(\mu), \\
T_4 &= Z_2(\mu)C - Z_2(\mu)C + Z_3(\mu)(F - D(A-\mu)^{-1}C) - Z_3(\mu)(F - D(A-\mu)^{-1}C).
\end{aligned}$$

(ii) If $\mathscr{I}(\mu)$ *is continuous and* $\mathscr{I}(\mu)(0) \subset U_1V(0)$*, then the multivalued linear operator* $\mathscr{A}_0$ *is closable if, and only if, the multivalued linear operator* $S_2(\mu)$ *is closable. Moreover, the closure* $\mathscr{A}$ *of* $\mathscr{A}_0$ *is given by*

$$\mathscr{A} - \mu I = \overline{\mathscr{A}_0} - \mu I = U_1 \overline{V} W_1 + \mathscr{I}(\mu).$$

(iii) If $\mathscr{I}(\mu)$ *is continuous and* $\mathscr{I}(\mu)(0) \subset U_1V(0)$*, then for* $\lambda \in \mathbb{C}$*, we have*

$$\mathscr{A} - \lambda = \overline{\mathscr{A}_0} - \lambda = U_1 \overline{V(\lambda)} W_1 + \mathbb{M}_2(\mu),$$

where

$$\mathbb{M}_2(\mu) = \begin{pmatrix} 0 & Y_1(\mu) & Y_2(\mu) \\ Z_1(\mu) & Z_1(\mu)Y_1(\mu) & Z_1(\mu)Y_2(\mu) + Y_3(\mu) \\ Z_2(\mu) & Z_2(\mu)Y_1(\mu) + Z_3(\mu) & Z_2(\mu)Y_2(\mu) + Z_3(\mu)Y_3(\mu) \end{pmatrix} + \mathscr{I}(\mu).$$

◇

Proof (i) We can write

$$\begin{aligned}
\mathscr{A}_0 - \mu I &= (U_1 + U - U)\, V \begin{pmatrix} I_X & Y_1(\mu) + G_1(\mu) - G_1(\mu) & Y_2(\mu) + G_2(\mu) - G_2(\mu) \\ 0 & I_Y & Y_3(\mu) + G_3(\mu) - G_3(\mu) \\ 0 & 0 & I_Z \end{pmatrix}, \\
&= (U_1 + U - U)\, V \left[\begin{pmatrix} I_X & Y_1(\mu) & Y_2(\mu) \\ 0 & I_Y & Y_3(\mu) \\ 0 & 0 & I_Z \end{pmatrix} + W - W \right], \\
&= (U_1 + U - U)\, V\, (W_1 + W - W)\,.
\end{aligned}$$

Since $V\,(W_1 + W - W)\,(0) = V W(0) = V(0) \subset N(U - U) = X \times Y \times Z$, then

$$\mathscr{A}_0 - \mu I = U_1 V \left[W_1 + \begin{pmatrix} 0 & G_1(\mu) - G_1(\mu) & G_2(\mu) - G_2(\mu) \\ 0 & 0 & G_3(\mu) - G_3(\mu) \\ 0 & 0 & 0 \end{pmatrix} \right] + (U - U)\, V W.$$

If $(H35)$ is satisfied, then

$$\mathscr{A}_0 - \mu I = U_1 V W_1 + U_1 V \begin{pmatrix} 0 & G_1(\mu) - G_1(\mu) & G_2(\mu) - G_2(\mu) \\ 0 & 0 & G_3(\mu) - G_3(\mu) \\ 0 & 0 & 0 \end{pmatrix} + \begin{pmatrix} 0 & 0 & 0 \\ D - D & 0 & 0 \\ G - G & H - H & 0 \end{pmatrix}.$$

Let $\left[\begin{pmatrix} x_1 \\ y_1 \\ z_1 \end{pmatrix}, \begin{pmatrix} x_2 \\ y_2 \\ z_2 \end{pmatrix} \right] \in G \left[U_1 V \begin{pmatrix} 0 & G_1(\mu) - G_1(\mu) & G_2(\mu) - G_2(\mu) \\ 0 & 0 & G_3(\mu) - G_3(\mu) \\ 0 & 0 & 0 \end{pmatrix} \right].$

Then,

$$\begin{pmatrix} x_2 \\ y_2 \\ z_2 \end{pmatrix} \in U_1 \begin{pmatrix} A - \mu & 0 & 0 \\ 0 & S_1(\mu) - \mu & 0 \\ 0 & 0 & S_2(\mu) - \mu \end{pmatrix} \begin{pmatrix} G_1(\mu)(0) + G_2(\mu)(0) \\ G_3(\mu)(0) \\ 0 \end{pmatrix}$$

if, and only if,

$$\begin{pmatrix} x_2 \\ y_2 \\ z_2 \end{pmatrix} \in \begin{pmatrix} I_X & 0 & 0 \\ Z_1(\mu) & I_Y & 0 \\ Z_2(\mu) & Z_3(\mu) & I_Z \end{pmatrix} \begin{pmatrix} (A - \mu)(G_1(\mu)(0) + G_2(\mu)(0)) \\ (S_1(\mu) - \mu) G_3(\mu)(0) \\ 0 \end{pmatrix}$$

if, and only if,

$$\begin{pmatrix} x_2 \\ y_2 \\ z_2 \end{pmatrix} \in \begin{pmatrix} (A - \mu)(G_1(\mu)(0) + G_2(\mu)(0)) \\ Z_1(\mu)(A - \mu)(G_1(\mu)(0) + G_2(\mu)(0)) + (S_1(\mu) - \mu) G_3(\mu)(0) \\ Z_2(\mu)(A - \mu)(G_1(\mu)(0) + G_2(\mu)(0)) + Z_3(\mu)(S_1(\mu) - \mu) G_3(\mu)(0) \end{pmatrix}.$$

Hence, $\mathscr{D}(A - \mu)$ contains the ranges of both $G_1(\mu)$ and $G_2(\mu)$. Hence, we conclude that

$$(A - \mu)(G_1(\mu)(0) + G_2(\mu)(0)) = (A - \mu) G_1(\mu)(0) + (A - \mu) G_2(\mu)(0).$$

Moreover, we get the following result:

$$\begin{cases} (A-\mu)(A-\mu)^{-1}B(0)+(A-\mu)(A-\mu)^{-1}C(0)=B(0)+C(0)+A(0) \\ (S_1(\mu)-\mu)(S_1(\mu)-\mu)^{-1}(F-D(A-\mu)^{-1}C)(0)=(F-D(A-\mu)^{-1}C)(0)+S_1(\mu)(0). \end{cases}$$

Therefore,

$$\begin{aligned} x_2 &\in A(0)+B(0)+C(0) \\ y_2 &\in Z_1(\mu)(A(0)+B(0)+C(0))+(F-D(A-\mu)^{-1}C)(0)+S_1(\mu)(0) \\ z_2 &\in Z_2(\mu)(A(0)+B(0)+C(0))+Z_3(\mu)\left((F-D(A-\mu)^{-1}C)(0)+S_1(\mu)(0)\right). \end{aligned}$$

Since $Z_1(\mu)$, $Z_2(\mu)$ and $Z_3(\mu)$ are bounded, then

$$\begin{aligned} x_2 &\in A(0)+B(0)+C(0) \\ y_2 &\in Z_1(\mu)A(0)+Z_1(\mu)B(0)+Z_1(\mu)C(0)+(F-D(A-\mu)^{-1}C)(0)+S_1(\mu)(0) \\ z_2 &\in Z_2(\mu)A(0)+Z_2(\mu)B(0)+Z_2(\mu)C(0)+Z_3(\mu)(F-D(A-\mu)^{-1}C)(0)+Z_3(\mu)S_1(\mu)(0). \end{aligned}$$

This is equivalent to

$$\begin{pmatrix} x_2 \\ y_2 \\ z_2 \end{pmatrix} \in \begin{pmatrix} A-A & B-B & C-C \\ Z_1(\mu)A-Z_1(\mu)A & T_1 & T_2 \\ Z_2(\mu)A-Z_2(\mu)A & T_3 & T_4 \end{pmatrix} \begin{pmatrix} x_1 \\ y_1 \\ z_1 \end{pmatrix}.$$

Then,

$$\left[\begin{pmatrix} x_1 \\ y_1 \\ z_1 \end{pmatrix}, \begin{pmatrix} x_2 \\ y_2 \\ z_2 \end{pmatrix}\right] \in G\left[\begin{pmatrix} A-A & B-B & C-C \\ Z_1(\mu)A-Z_1(\mu)A & T_1 & T_2 \\ Z_2(\mu)A-Z_2(\mu)A & T_3 & T_4 \end{pmatrix}\right].$$

We deduce that

$$\begin{aligned} G&\left[U_1V\begin{pmatrix} 0 & G_1(\mu)-G_1(\mu) & G_2(\mu)-G_2(\mu) \\ 0 & 0 & G_3(\mu)-G_3(\mu) \\ 0 & 0 & 0 \end{pmatrix}\right] \\ &\subset G\left[\begin{pmatrix} A-A & B-B & C-C \\ Z_1(\mu)A-Z_1(\mu)A & T_1 & T_2 \\ Z_2(\mu)A-Z_2(\mu)A & T_3 & T_4 \end{pmatrix}\right]. \end{aligned} \tag{5.5.10}$$

We will show that $\mathscr{A}=\mathscr{B}$, where

$$\mathscr{A}=U_1V\begin{pmatrix} 0 & G_1(\mu)-G_1(\mu) & G_2(\mu)-G_2(\mu) \\ 0 & 0 & G_3(\mu)-G_3(\mu) \\ 0 & 0 & 0 \end{pmatrix}$$

and

$$\mathscr{B} = \begin{pmatrix} A - A & B - B & C - C \\ Z_1(\mu)A - Z_1(\mu)A & T_1 & T_2 \\ Z_2(\mu)A - Z_2(\mu)A & T_3 & T_4 \end{pmatrix}.$$

In order to get this equality, we will prove that $\mathscr{A}(0) = \mathscr{B}(0)$ and $\mathscr{D}(\mathscr{A}) = \mathscr{D}(\mathscr{B})$. First,

$$\begin{aligned}
\mathscr{A}(0) &= U_1 V \begin{pmatrix} 0 & G_1(\mu) - G_1(\mu) & G_2(\mu) - G_2(\mu) \\ 0 & 0 & G_3(\mu) - G_3(\mu) \\ 0 & 0 & 0 \end{pmatrix} \begin{pmatrix} 0 \\ 0 \\ 0 \end{pmatrix}, \\
&= U_1 \begin{pmatrix} A - \mu & 0 & 0 \\ 0 & S_1(\mu) - \mu & 0 \\ 0 & 0 & S_2(\mu) - \mu \end{pmatrix} \begin{pmatrix} G_1(\mu)(0) + G_2(\mu)(0) \\ G_3(\mu)(0) \\ 0 \end{pmatrix}, \\
&= U_1 \begin{pmatrix} (A - \mu)(G_1(\mu)(0) + G_2(\mu)(0)) \\ (S_1(\mu) - \mu)G_3(\mu)(0) \\ 0 \end{pmatrix}, \\
&= \begin{pmatrix} I_X & 0 & 0 \\ Z_1(\mu) & I_Y & 0 \\ Z_2(\mu) & Z_3(\mu) & I_Z \end{pmatrix} \begin{pmatrix} A(0) + B(0) + C(0) \\ (F - D(A - \mu)^{-1}C)(0) + S_1(\mu)(0) \\ 0 \end{pmatrix}, \\
&= \begin{pmatrix} A(0) + B(0) + C(0) \\ Z_1(\mu)(A(0) + B(0) + C(0)) + (F - D(A - \mu)^{-1}C)(0) + S_1(\mu)(0) \\ Z_2(\mu)(A(0) + B(0) + C(0)) + Z_3(\mu)(F - D(A - \mu)^{-1}C)(0) + S_1(\mu)(0) \end{pmatrix},
\end{aligned}$$

then

$$\mathscr{A}(0) = \mathscr{B}(0). \tag{5.5.11}$$

Second,

$$\begin{aligned}
\mathscr{D}(\mathscr{B}) &= \mathscr{D}\left[\begin{pmatrix} A - A & B - B & C - C \\ Z_1(\mu)A - Z_1(\mu)A & T_1 & T_2 \\ Z_2(\mu)A - Z_2(\mu)A & T_3 & T_4 \end{pmatrix}\right], \\
&= \mathscr{D}(A) \bigotimes \mathscr{D}(B) \bigcap \mathscr{D}(T_1) \bigcap \mathscr{D}(T_3) \bigotimes \mathscr{D}(C) \bigcap \mathscr{D}(T_2) \bigcap \mathscr{D}(T_4), \\
&= \mathscr{D}(A) \bigotimes \mathscr{D}(B) \times \mathscr{D}(C),
\end{aligned}$$

and

$$\begin{aligned}
\mathscr{D}(\mathscr{A}) &= \mathscr{D}\left[U_1 V \begin{pmatrix} 0 & G_1(\mu) - G_1(\mu) & G_2(\mu) - G_2(\mu) \\ 0 & 0 & G_3(\mu) - G_3(\mu) \\ 0 & 0 & 0 \end{pmatrix}\right], \\
&= \left\{ \begin{pmatrix} x \\ y \\ z \end{pmatrix} \in X \times Y \times Z \text{ such that } \begin{array}{c} G_1(\mu)(0) + G_2(\mu)(0) \in \mathscr{D}(A), \\ G_3(\mu)(0) \in \mathscr{D}(B) \\ \text{and } z \in \mathscr{D}(C) \end{array} \right\}, \\
&= \mathscr{D}(A) \times \mathscr{D}(B) \times \mathscr{D}(C).
\end{aligned}$$

Finally, we get

$$\mathscr{D}(\mathscr{A}) = \mathscr{D}(\mathscr{B}). \tag{5.5.12}$$

By using estimation (5.5.10), (5.5.11), and (5.5.12), we have

$$U_1 V \begin{pmatrix} 0 & G_1(\mu) - G_1(\mu) & G_2(\mu) - G_2(\mu) \\ 0 & 0 & G_3(\mu) - G_3(\mu) \\ 0 & 0 & 0 \end{pmatrix} = \begin{pmatrix} A - A & B - B & C - C \\ Z_1(\mu)A - Z_1(\mu)A & T_1 & T_2 \\ Z_2(\mu)A - Z_2(\mu)A & T_3 & T_4 \end{pmatrix}.$$

(ii) We start by determining the properties of the matrices U_1 and W_1. It is easy to notice that U_1 and W_1 are single-valued bounded bijective, where

$$W_1^{-1} = \begin{pmatrix} I_X & -Y_1(\mu) & Y_1(\mu)Y_3(\mu) - Y_2(\mu) \\ 0 & I_Y & -Y_3(\mu) \\ 0 & 0 & I_Z \end{pmatrix}.$$

We have V is closable. Therefore, by using Theorem 2.12.3, we conclude that $U_1 V W_1$ is closable and $\overline{U_1 V W_1} = U_1 \overline{V} W_1$. Hence,

$$\begin{aligned} \overline{\mathscr{A}_0 - \mu I} &= \overline{\mathscr{A}_0} - \mu I \\ &= U_1 \overline{V} W_1. \end{aligned}$$

So, we have

$$\overline{\mathscr{A}_0} = \mu I + U_1 \overline{V} W_1.$$

It follows that the relation $\mathscr{A}_0$ is closable,

$$\begin{aligned} \mathscr{D}(U_1 V W_1) &= \mathscr{D}(\mathscr{I}(\mu)) \\ &= \mathscr{D}(A) \times \mathscr{D}(B) \times \mathscr{D}(C) \end{aligned}$$

and

$$\mathscr{I}(\mu)(0) \subset U_1 V W_1(0).$$

Then, we get $U_1 V W_1 + \mathscr{I}(\mu)$ is closable and

$$\begin{aligned} \overline{U_1 V W_1 + \mathscr{I}(\mu)} &= \overline{U_1 V W_1} + \mathscr{I}(\mu) \\ &= U_1 \overline{V} W_1 + \mathscr{I}(\mu). \end{aligned}$$

Thus,

$$\begin{aligned} \overline{\mathscr{A}_0 - \mu I} &= \overline{\mathscr{A}_0} - \mu I \\ &= U_1 \overline{V} W_1 + \mathscr{I}(\mu). \end{aligned}$$

Hence,

$$\overline{\mathscr{A}_0} = \mu I + U_1 \overline{V} W_1 + \mathscr{I}(\mu).$$

It follows that the relation $\mathscr{A}_0$ is closable.
(iii) Since I is single-valued, then we get

$$\mathscr{A} - \lambda I = U_1 \overline{V} W_1 - \mu I + (\mu - \lambda) I$$

for $\lambda \in \mathbb{C}$. We can deduce that

$$\begin{aligned}
\mathscr{A} - \lambda I &= \overline{\mathscr{A}_0} - \lambda I \\
&= U_1 \begin{pmatrix} A - \mu - \lambda + \lambda & 0 & 0 \\ 0 & S_1(\mu) - \mu - \lambda + \lambda & 0 \\ 0 & 0 & \overline{S_2(\mu)} - \mu - \lambda + \lambda \end{pmatrix} W_1 \mathscr{I}(\mu) + (\mu - \lambda) I \\
&= U_1 \left[\begin{pmatrix} A - \lambda & 0 & 0 \\ 0 & S_1(\mu) - \lambda & 0 \\ 0 & 0 & \overline{S_2(\mu)} - \lambda \end{pmatrix} + (\lambda - \mu) \begin{pmatrix} I_X & 0 & 0 \\ 0 & I_Y & 0 \\ 0 & 0 & I_Z \end{pmatrix} \right] W_1 + (\mu - \lambda) I + \mathscr{I}(\mu).
\end{aligned}$$

Since U_1 is bounded and W_1 is single-valued, then

$$\begin{aligned}
\mathscr{A} - \lambda I = U_1 &\begin{pmatrix} A - \lambda & 0 & 0 \\ 0 & S_1(\mu) - \lambda & 0 \\ 0 & 0 & \overline{S_2(\mu)} - \lambda \end{pmatrix} W_1 \\
+ U_1 &\begin{pmatrix} (\lambda - \mu) I_X & 0 & 0 \\ 0 & (\lambda - \mu) I_Y & 0 \\ 0 & 0 & (\lambda - \mu) I_Z \end{pmatrix} W_1 + (\mu - \lambda) I + \mathscr{I}(\mu).
\end{aligned}$$

Since U_1 and W_1 are single-valued, then

$$\mathbb{M}_2(\mu) = \begin{pmatrix} 0 & Y_1(\mu) & Y_2(\mu) \\ Z_1(\mu) & Z_1(\mu) Y_1(\mu) & Z_1(\mu) Y_2(\mu) + Y_3(\mu) \\ Z_2(\mu) & Z_2(\mu) Y_1(\mu) + Z_3(\mu) & Z_2(\mu) Y_2(\mu) + Z_3(\mu) Y_3(\mu) \end{pmatrix} + \mathscr{I}(\mu).$$

This achieves the proof. Q.E.D.

5.5.3 *Essential Spectra* 3 × 3 *Matrices Linear Operator*

In this section, we investigate the essential spectra of a 3 × 3 matrix linear relation on a Banach space.

5.5.3.1 Essential Spectra of the Multivalued Linear Operator $\mathscr{A}$

Theorem 5.5.7 *We suppose that the conditions* $(H25)$*–*$(H30)$ *are satisfied. Let* U*,* W *be single-valued, we have*

(i) *If* $\mathbb{M}(\mu) \in \mathscr{P}\mathscr{R}(X \times Y \times Z)$*, then*

$$\sigma_{e4}(\mathscr{A}) = \sigma_{e4}(A) \bigcup \sigma_{e4}(S_1(\mu)) \bigcup \sigma_{e4}(\overline{S_2(\mu)}).$$

(ii) *If* $\mathbb{M}(\mu) \in \mathscr{P}\mathscr{R}_+(X \times Y \times Z)$*, then*

$$\sigma_{e1}(\mathscr{A}) = \sigma_{e1}(A) \bigcup \sigma_{e1}(S_1(\mu)) \bigcup \sigma_{e1}(\overline{S_2(\mu)}).$$

(iii) *If* $\mathbb{M}(\mu) \in \mathscr{P}\mathscr{R}_-(X \times Y \times Z)$*, then*

$$\sigma_{e2}(\mathscr{A}) = \sigma_{e2}(A) \bigcup \sigma_{e2}(S_1(\mu)) \bigcup \sigma_{e2}(\overline{S_2(\mu)}).$$

(iv) *If* $\mathbb{M}(\mu) \in \mathscr{P}\mathscr{R}_\pm(X \times Y \times Z)$*, then*

$$\begin{aligned}\sigma_{e3}(\mathscr{A}) = \sigma_{e3}(A) &\bigcup \sigma_{e3}(S_1(\mu)) \bigcup \sigma_{e3}(\overline{S_2(\mu)}) \\ &\bigcup \left[\sigma_{e1}(A) \bigcup \sigma_{e2}(S_1(\mu)) \bigcup \sigma_{e2}(\overline{S_2(\mu)})\right] \\ &\bigcup \left[\sigma_{e2}(A) \bigcup \sigma_{e1}(S_1(\mu)) \bigcup \sigma_{e2}(\overline{S_2(\mu)})\right] \\ &\bigcup \left[\sigma_{e2}(A) \bigcup \sigma_{e2}(S_1(\mu)) \bigcup \sigma_{e1}(\overline{S_2(\mu)})\right].\end{aligned}$$

◇

Proof Let $\mu \in \rho(A) \bigcap \rho(S_1(\mu))$ and $\lambda \in \mathbb{C}$, then by using Theorem 5.5.4, we can rewrite

$$\begin{aligned}\lambda - \mathscr{A} &= \overline{\mathscr{A}_0} \\ &= U\overline{V(\lambda)}W + \mathbb{M}(\mu).\end{aligned}$$

As though $\mathbb{M}(\mu) \in \mathscr{P}\mathscr{R}(X \times Y \times Z)$, then $\lambda - \mathscr{A}$ is a Fredholm relation if, and only if, $U\overline{V(\lambda)}W$ is a Fredholm relation. Now, let us notice that the operators U and W are bounded and have bounded inverses. Then, U and W are Fredholm relations. We deduce from Theorem 3.6.4 that $UV(\lambda)W$ is a Fredholm relation if, and only if, $V(\lambda)$ is a Fredholm relation if, and only if, $\lambda - A$, $\lambda - S_1(\mu)$, and $\lambda - S_2(\mu)$ are a Fredholm relation. Therefore

$$\sigma_{e4}(\mathscr{A}) = \sigma_{e4}(A) \bigcup \sigma_{e4}(S_1(\mu)) \bigcup \sigma_{e4}(\overline{S_2(\mu)}).$$

The proof of (ii) and (iii) may be checked in the same way as in the proof of (i).
(iv) This assertion is an immediate consequence of (ii) and (iii). Q.E.D.

Theorem 5.5.8 *We suppose that the conditions* $(H25)$–$(H33)$ *are satisfied. If* $G_i(\mu)$ *with* $i = \{1, 2, 3\}$ *are single-valued and*

$$\begin{pmatrix} 0 & 0 & 0 \\ D - D & 0 & 0 \\ G - G & H - H & 0 \end{pmatrix}$$

is continuous, then

(i) *If* $\mathbb{M}_1(\mu) \in \mathcal{PR}(X \times Y \times Z)$*, then*

$$\sigma_{e4}(\mathcal{A}) = \sigma_{e4}(A) \bigcup \sigma_{e4}(S_1(\mu)) \bigcup \sigma_{e4}(\overline{S_2(\mu)}).$$

(ii) *If* $\mathbb{M}_1(\mu) \in \mathcal{PR}_+(X \times Y \times Z)$*, then*

$$\sigma_{e1}(\mathcal{A}) = \sigma_{e1}(A) \bigcup \sigma_{e1}(S_1(\mu)) \bigcup \sigma_{e1}(\overline{S_2(\mu)}).$$

(iii) *If* $\mathbb{M}_1(\mu) \in \mathcal{PR}_-(X \times Y \times Z)$*, then*

$$\sigma_{e2}(\mathcal{A}) = \sigma_{e2}(A) \bigcup \sigma_{e2}(S_1(\mu)) \bigcup \sigma_{e2}(\overline{S_2(\mu)}).$$

(iv) *If* $\mathbb{M}_1(\mu) \in \mathcal{PR}_\pm(X \times Y \times Z)$*, then*

$$\begin{aligned} \sigma_{e3}(\mathcal{A}) = \sigma_{e3}(A) \bigcup \sigma_{e3}(S_1(\mu)) \bigcup \sigma_{e3}(\overline{S_2(\mu)}) \\ \bigcup \left[\sigma_{e1}(A) \bigcup \sigma_{e2}(S_1(\mu)) \bigcup \sigma_{e2}(\overline{S_2(\mu)})\right] \\ \bigcup \left[\sigma_{e2}(A) \bigcup \sigma_{e1}(S_1(\mu)) \bigcup \sigma_{e2}(\overline{S_2(\mu)})\right] \\ \bigcup \left[\sigma_{e2}(A) \bigcup \sigma_{e2}(S_1(\mu)) \bigcup \sigma_{e1}(\overline{S_2(\mu)})\right]. \end{aligned}$$

◇

Proof The proof is a straightforward adoption of the proof of Theorem 5.5.7.Q.E.D.

Theorem 5.5.9 *We suppose that the conditions* $(H25)$–$(H35)$ *are satisfied. If* $\mathcal{I}(\mu)$ *is continuous and* $\mathcal{I}(\mu)(0) \subset U_1 V(0)$*, then*

(i) *If* $\mathbb{M}_2(\mu) \in \mathcal{PR}(\Phi(X \times Y \times Z))$*, then*

$$\sigma_{e4}(\mathcal{A}) = \sigma_{e4}(A) \bigcup \sigma_{e4}(S_1(\mu)) \bigcup \sigma_{e4}(\overline{S_2(\mu)}).$$

(ii) If $\mathbb{M}_2(\mu) \in \mathscr{PR}(\Phi_+(X \times Y \times Z))$, *then*

$$\sigma_{e1}(\mathscr{A}) = \sigma_{e1}(A) \bigcup \sigma_{e1}(S_1(\mu)) \bigcup \sigma_{e1}(\overline{S_2(\mu)}).$$

(iii) If $\mathbb{M}_2(\mu) \in \mathscr{PR}(\Phi_-(X \times Y \times Z))$, *then*

$$\sigma_{e2}(\mathscr{A}) = \sigma_{e2}(A) \bigcup \sigma_{e2}(S_1(\mu)) \bigcup \sigma_{e2}(\overline{S_2(\mu)}).$$

(iv) If $\mathbb{M}_2(\mu) \in \mathscr{PR}(\Phi_\pm(X \times Y \times Z))$, *then*

$$\begin{aligned}\sigma_{e3}(\mathscr{A}) = \sigma_{e3}(A) \bigcup \sigma_{e3}(S_1(\mu)) \bigcup \sigma_{e3}(\overline{S_2(\mu)}) \\ \bigcup \left[\sigma_{e1}(A) \bigcup \sigma_{e2}(S_1(\mu)) \bigcup \sigma_{e2}(\overline{S_2(\mu)})\right] \\ \bigcup \left[\sigma_{e2}(A) \bigcup \sigma_{e1}(S_1(\mu)) \bigcup \sigma_{e2}(\overline{S_2(\mu)})\right] \\ \bigcup \left[\sigma_{e2}(A) \bigcup \sigma_{e2}(S_1(\mu)) \bigcup \sigma_{e1}(\overline{S_2(\mu)})\right].\end{aligned}$$

◇

Proof The proof is a straightforward adoption of the proof of Theorem 5.5.7.Q.E.D.

5.5.4 Essential Spectra of Upper-Triangular Matrix

5.5.4.1 Essential Spectra of $\mathscr{M}_C$

Theorem 5.5.10 *Let* $A_1 \in L\mathscr{R}(X_1)$, $A_2 \in L\mathscr{R}(X_2)$, *and* $A_3 \in L\mathscr{R}(X_3)$ *be closed and everywhere defined. We consider the* 3×3 *matrix linear relation:*

$$\mathscr{M}_C = \begin{pmatrix} A_1 & B & C \\ 0 & A_2 & D \\ 0 & 0 & A_3 \end{pmatrix},$$

where B, C, *and* D *are bounded single-valued. Then,*

(i) If $A_1 \in \Phi_+(X_1)$, $A_2 \in \Phi_+(X_2)$ *and* $A_3 \in \Phi_+(X_3)$, *then* $\mathscr{M}_C \in \Phi_+(X_1 \times X_2 \times X_3)$. *In particular, we have*

$$\sigma_{e1}(A_1) \subset \sigma_{e1}(\mathscr{M}_C) \subset \sigma_{e1}(A_1) \bigcup \sigma_{e1}(A_2) \bigcup \sigma_{e1}(A_3).$$

Moreover, if $\rho(A_1)$, $\rho(A_2)$, $\rho(A_3)$, *and* $\rho(\mathscr{M}_C)$ *are nonempty and* Φ_{A_1}, Φ_{A_2}, Φ_{A_3}, *and* $\Phi_{\mathscr{M}_C}$ *are connected, then*

$$\sigma_{eap}(A_1) \subset \sigma_{eap}(\mathscr{M}_C) \subset \sigma_{eap}(A_1) \bigcup \sigma_{eap}(A_2) \bigcup \sigma_{eap}(A_3).$$

(ii) If $A_1 \in \Phi_-(X_1)$, $A_2 \in \Phi_-(X_2)$ and $A_3 \in \Phi_-(X_3)$, then $\mathscr{M}_C \in \Phi_-(X_1 \times X_2 \times X_3)$. In particular, we have

$$\sigma_{e2}(A_1) \subset \sigma_{e2}(\mathscr{M}_C) \subset \sigma_{e2}(A_1) \bigcup \sigma_{e2}(A_2) \bigcup \sigma_{e2}(A_3).$$

Moreover, if $\rho(A_1)$, $\rho(A_2)$, $\rho(A_3)$ and $\rho(\mathscr{M}_C)$ are nonempty and Φ_{A_1}, Φ_{A_2}, Φ_{A_3}, and $\Phi_{\mathscr{M}_C}$ are connected. Then,

$$\sigma_{e\delta}(A_1) \subset \sigma_{e\delta}(\mathscr{M}_C) \subset \sigma_{e\delta}(A_1) \bigcup \sigma_{e\delta}(A_2) \bigcup \sigma_{e\delta}(A_3).$$

(iii) If $A_1 \in \Phi(X_1)$, $A_2 \in \Phi(X_2)$ and $A_3 \in \Phi(X_3)$, then $\mathscr{M}_C \in \Phi(X_1 \times X_2 \times X_3)$. In particular, we have

$$\sigma_{e4}(A_1) \subset \sigma_{e4}(\mathscr{M}_C) \subset \sigma_{e4}(A_1) \bigcup \sigma_{e4}(A_2) \bigcup \sigma_{e4}(A_3).$$

Moreover, if $\rho(A_1)$, $\rho(A_2)$, $\rho(A_3)$, and $\rho(\mathscr{M}_C)$ are nonempty and Φ_{A_1}, Φ_{A_2}, Φ_{A_3}, and $\Phi_{\mathscr{M}_C}$ are connected. Then,

$$\sigma_{e5}(A_1) \subset \sigma_{e5}(\mathscr{M}_C) \subset \sigma_{e5}(A_1) \bigcup \sigma_{e5}(A_2) \bigcup \sigma_{e5}(A_3). \qquad \diamond$$

Proof (i) Write $\mathscr{M}_C$ in the from

$$\mathscr{M}_C = \begin{pmatrix} I & 0 & 0 \\ 0 & I & 0 \\ 0 & 0 & A_3 \end{pmatrix} \begin{pmatrix} I & 0 & C \\ 0 & I & D \\ 0 & 0 & I \end{pmatrix} \begin{pmatrix} I & 0 & 0 \\ 0 & A_2 & 0 \\ 0 & 0 & I \end{pmatrix} \begin{pmatrix} I & B & 0 \\ 0 & I & 0 \\ 0 & 0 & I \end{pmatrix} \begin{pmatrix} A_1 & 0 & 0 \\ 0 & I & 0 \\ 0 & 0 & I \end{pmatrix}.$$

Since $A_1 \in \Phi_+(X_1)$, $A_2 \in \Phi_+(X_2)$, and $A_3 \in \Phi_+(X_3)$, then

$$\begin{pmatrix} A_1 & 0 & 0 \\ 0 & I & 0 \\ 0 & 0 & I \end{pmatrix},$$

$$\begin{pmatrix} I & 0 & 0 \\ 0 & A_2 & 0 \\ 0 & 0 & I \end{pmatrix},$$

and

$$\begin{pmatrix} I & 0 & 0 \\ 0 & I & 0 \\ 0 & 0 & A_3 \end{pmatrix}$$

are both an upper semi-Fredholm relations. On the other hand

$$\begin{pmatrix} I & 0 & C \\ 0 & I & D \\ 0 & 0 & I \end{pmatrix}$$

and

$$\begin{pmatrix} I & B & 0 \\ 0 & I & 0 \\ 0 & 0 & I \end{pmatrix}$$

are invertible. Indeed, let

$$\begin{pmatrix} x_1 \\ x_2 \\ x_3 \end{pmatrix} \in N \begin{pmatrix} I & 0 & C \\ 0 & I & D \\ 0 & 0 & I \end{pmatrix}$$

if, and only if,

$$\begin{cases} 0 \in x_1 + Cx_3, \\ 0 \in x_2 + Dx_3, \\ 0 \in x_3 \end{cases}$$

if, and only if,

$$\begin{cases} x_1 = x_2 = 0 \\ x_2 \in C(0) \\ x_3 \in D(0). \end{cases}$$

Since C and D are single-valued, then $C(0) = D(0) = \{0\}$. Hence,

$$N\left[\begin{pmatrix} I & 0 & C \\ 0 & I & D \\ 0 & 0 & I \end{pmatrix}\right] = \begin{pmatrix} 0 \\ 0 \\ 0 \end{pmatrix}.$$

On the other hand,

$$R\left[\begin{pmatrix} I & 0 & C \\ 0 & I & D \\ 0 & 0 & I \end{pmatrix}\right] = X_1 \times X_2 \times X_3.$$

We can show that

$$\begin{pmatrix} I & B & 0 \\ 0 & I & 0 \\ 0 & 0 & I \end{pmatrix}$$

is invertible. So, $\mathscr{M}_C \in \Phi_+(X_1 \times X_2 \times X_3)$. For every $\lambda \in \mathbb{C}$, we have

$$\mathscr{M}_C = \begin{pmatrix} I & 0 & 0 \\ 0 & I & 0 \\ 0 & 0 & A_3 - \lambda \end{pmatrix} \begin{pmatrix} I & 0 & C \\ 0 & I & D \\ 0 & 0 & I \end{pmatrix} \begin{pmatrix} I & 0 & 0 \\ 0 & A_2 - \lambda & 0 \\ 0 & 0 & I \end{pmatrix} \begin{pmatrix} I & B & 0 \\ 0 & I & 0 \\ 0 & 0 & I \end{pmatrix} \begin{pmatrix} A_1 - \lambda & 0 & 0 \\ 0 & I & 0 \\ 0 & 0 & I \end{pmatrix}. \tag{5.5.13}$$

If $\lambda \notin \sigma_{e1}(A_1) \bigcup \sigma_{e1}(A_2) \bigcup \sigma_{e1}(A_3)$, then

$$\begin{pmatrix} A_1 - \lambda & 0 & 0 \\ 0 & I & 0 \\ 0 & 0 & I \end{pmatrix} \in \Phi_+(X_1 \times X_2 \times X_3),$$

$$\begin{pmatrix} I & 0 & 0 \\ 0 & A_2 - \lambda & 0 \\ 0 & 0 & I \end{pmatrix} \in \Phi_+(X_1 \times X_2 \times X_3)$$

and

$$\begin{pmatrix} I & 0 & 0 \\ 0 & I & 0 \\ 0 & 0 & A_3 - \lambda \end{pmatrix} \in \Phi_+(X_1 \times X_2 \times X_3).$$

Hence,

$$\mathscr{M}_C - \lambda \in \Phi_+(X_1 \times X_2 \times X_3).$$

This implies that $\lambda \notin \sigma_1(\mathscr{M}_C)$. Therefore,

$$\sigma_{e1}(\mathscr{M}_C) \subset \sigma_{e1}(A_1) \bigcup \sigma_{e1}(A_2) \bigcup \sigma_{e1}(A_3).$$

Now, if $\lambda \notin \sigma_1(\mathscr{M}_C)$, then by using Eq. (5.5.13), we have

$$\begin{pmatrix} A_1 - \lambda & 0 & 0 \\ 0 & I & 0 \\ 0 & 0 & I \end{pmatrix} \in \Phi_+(X_1 \times X_2 \times X_3).$$

Then, $A_1 - \lambda \in \Phi_+(X_1)$. Therefore, $\lambda \notin \sigma_1(A_1)$. We derive that

$$\sigma_{e1}(A_1) \subset \sigma_{e1}(\mathscr{M}_C).$$

Now, if $\rho(A_1)$, $\rho(A_2)$, $\rho(A_3)$, and $\rho(\mathcal{M}_C)$ are nonempty and Φ_{A_1}, Φ_{A_2}, Φ_{A_3}, and $\Phi_{\mathcal{M}_C}$ are connected, then

$$\begin{aligned}\sigma_{eap}(A_1) \subset & \sigma_{eap}(\mathcal{M}_C)\\ & \subset \sigma_{eap}(A_1)\bigcup \sigma_{eap}(A_2)\bigcup \sigma_{eap}(A_3).\end{aligned}$$

(ii) and (iii) can be checked in the same way as (i). Q.E.D.

Corollary 5.5.1 *Let $A_1 \in L\mathcal{R}(X_1)$, $A_2 \in L\mathcal{R}(X_2)$ and $A_3 \in L\mathcal{R}(X_3)$ are closed and everywhere defined. We consider the 3×3 matrix linear relation:*

$$\mathcal{M}_C = \begin{pmatrix} A_1 & B & C \\ 0 & A_2 & D \\ 0 & 0 & A_3 \end{pmatrix}$$

where B, C, and D are single-valued. Then,

(i) If $\mathcal{M}_C \in \Phi_+(X_1 \times X_2 \times X_3)$, then $A_1 \in \Phi_+(X_1)$.
(ii) If $\mathcal{M}_C \in \Phi_-(X_1 \times X_2 \times X_3)$, then $A_3 \in \Phi_-(X_3)$.
(iii) If $\mathcal{M}_C \in \Phi(X_1 \times X_2 \times X_3)$, then $A_1 \in \Phi(X_1)$, $A_2 \in \Phi(X_2)$ and $A_3 \in \Phi(X_3)$. ◇

5.5.4.2 Essential Spectra of a 3 × 3 Upper-Triangular Matrix of Linear Relations

The next theorem shows that an inclusion between the essential spectrum of $\mathcal{T}$ and the essential spectrum of its entries by using the analytic residuum, where the matrix $\mathcal{T}$ is given in (3.9.19).

Theorem 5.5.11 *Let $T_{ii} \in B\mathcal{R}(X_i)$ and $T_{ij} \in B\mathcal{R}(X_j, X_i)$, for all $i \in \{1, 2, 3\}$ and $j \in \{2, 3\}$ such that $T_{ii}(0) \subset T_{ij}(0)$ for $j \neq i \in \{1, 2\}$ and $\mathcal{D}(T_{33}) \subset N(T_{23})$. If T_{11} (resp. T_{22}) has a selection E_1 (resp. E_2) which satisfies $T_{11}(0) \subset R(E_1)$ (resp. $T_{22}(0) \subset R(E_2)$), then*

$$\bigcup_{i\in\{1,2,3\}} \sigma_{e_4}(T_{ii}) \subset \sigma_{e_4}(\mathcal{T})\bigcup\left(S_{T_{11}^*}\bigcup S_{T_{22}}\bigcup S_{T_{33}}\right),$$

where the matrix $\mathcal{T}$ is given in (3.9.19). ◇

Proof Let $\lambda \in \mathbb{C}$, then

$$\mathcal{T}-\lambda = \begin{pmatrix} I & 0 & 0 \\ 0 & I & 0 \\ 0 & 0 & T_{33}-\lambda \end{pmatrix}\begin{pmatrix} I & 0 & T_{13} \\ 0 & I & T_{23} \\ 0 & 0 & I \end{pmatrix}\begin{pmatrix} I & 0 & 0 \\ 0 & T_{22}-\lambda & 0 \\ 0 & 0 & I \end{pmatrix}\begin{pmatrix} I & T_{12} & 0 \\ 0 & I & 0 \\ 0 & 0 & I \end{pmatrix}\begin{pmatrix} T_{11}-\lambda & 0 & 0 \\ 0 & I & 0 \\ 0 & 0 & I \end{pmatrix}. \tag{5.5.14}$$

Now, let us suppose that $\lambda \notin \sigma_{e_4}(\mathscr{T}) \bigcup \left(S_{T_{11}^*} \bigcup S_{T_{22}} \bigcup S_{T_{33}}\right)$. Then, $\mathscr{T} - \lambda$ is Fredholm. Our purpose is to prove that $\lambda \notin \bigcup\limits_{i \in \{1,2,3\}} \sigma_{e_4}(T_{ii})$. By using (5.5.14), we have

$$\begin{pmatrix} I & 0 & 0 \\ 0 & I & 0 \\ 0 & 0 & T_{33} - \lambda \end{pmatrix}$$

is lower semi-Fredholm, then $T_{33} - \lambda$ is so. Now, to prove that $T_{33} - \lambda$ is Fredholm, it remains to show that $\dim(N(T_{33} - \lambda))$ is finite. Since $\lambda \notin S_{T_{33}}$, then by using Proposition 3.8.7 and Theorem 3.8.5, we obtain that $\dim(N(T_{33} - \lambda)) = 0$. Thus, $\lambda \notin \sigma_{e_4}(T_{33})$. The same reasoning as before, we can show that $\lambda \notin \sigma_{e_4}(T_{22})$. Let us prove that $T_{11} - \lambda$ is a Fredholm linear relation. Using the fact that $\mathscr{T} - \lambda$ is Fredholm and by (5.5.14), we infer that

$$\begin{pmatrix} T_{11} - \lambda & 0 & 0 \\ 0 & I & 0 \\ 0 & 0 & I \end{pmatrix}$$

is upper semi-Fredholm. Therefore, $T_{11} - \lambda$ is upper semi-Fredholm. It follows from the fact $\lambda \notin S_{T_{11}^*}$ and again by using Proposition 3.8.7 and Theorem 3.8.5 that $N(T_{11}^* - \lambda) = \{0\}$. Since T_{11} is a bounded linear relation, then

$$\begin{aligned} \{0\} &= N(T_{11}^* - \lambda) \\ &= N((T_{11} - \lambda)^*) \\ &= R(T_{11} - \lambda)^{\perp}. \end{aligned}$$

This implies that $\operatorname{codim}(T) < \infty$. Accordingly, $T_{11} - \lambda$ is a Fredholm linear relation. This achieves the proof. Q.E.D.

As a direct consequence of Theorem 5.5.11, we get

Corollary 5.5.2 *Let $T_{ii} \in B\mathscr{R}(X_i)$ and $T_{ij} \in B\mathscr{R}(X_j, X_i)$, for all $i \in \{1, 2, 3\}$ and $j \in \{2, 3\}$ be such that $T_{ii}(0) \subset T_{ij}(0)$ for $j \neq i \in \{1, 2\}$ and $\mathscr{D}(T_{33}) \subset N(T_{23})$. If T_{11} (resp. T_{22}) has a selection E_1(resp. E_2) which satisfies $T_{11}(0) \subset R(E_1)$ (resp. $T_{22}(0) \subset R(E_2)$), and T_{11}^*, T_{22}, and T_{33} have a SVEP, then*

$$\sigma_{e_4}(\mathscr{T}) = \bigcup_{i \in \{1,2,3\}} \sigma_{e_4}(T_{ii}),$$

where the matrix $\mathscr{T}$ is given in (3.9.19). ◇

References

1. Bátkai, A., Binding, P., Dijksma, A., Hryniv, R., Langer, H.: Spectral problems for operator matrices. Math. Nachr. **278**, 1408–1429 (2005)
2. Cross, R.: Multivalued linear operators. In: Monographs and Textbooks in Pure and Applied Mathematics, vol. 213, pp. x+335. Marcel Dekker, Inc., New York (1998). ISBN: 0-8247-0219-0
3. Elleuch, S., Mnif, M.: Essential approximate point spectra for upper triangular matrix of linear relations. Acta Math. Sci. Ser. B (Engl. Ed.) **33**(4), 1187–1201 (2013)
4. lvarez, T., Ammar, A., Jeribi, A.: On the essential spectra of some matrix of linear relations. Math. Methods Appl. Sci. **37**(5), 620–644 (2014)
5. Schechter, M.: Principles of Functional Analysis, 2nd edn., Graduate Studies in Mathematics, vol. 36. American Mathematical Society, Providence, RI (2002)
6. Ammar, A., Fakhfakh, S., Jeribi, A.: Stability of the essential spectrum of the diagonally and off-diagonally dominant block matrix linear relations. J. Pseudo-Differ. Oper. Appl. **7**(4), 493–509 (2016)
7. Kato, T.: Perturbation theory for nullity, deficiency and other quantities of linear operators. J. Analyse Math. **6**, 261–322 (1958)
8. Ammar, A., Bouchekoua, A., Jeribi, A.: The ε-pseudospectra and the essential ε-pseudospectra of linear relations. J. Pseudo-Differ. Oper. Appl. **11**(2), 879–915 (2020)
9. Ammar, A., Diagana, T., Jeribi, A.: Perturbations of Fredholm linear relations in Banach spaces with application to 3×3-block matrices of linear relations. Arab J. Math. Sci. **22**(1), 59–76 (2016)
10. Jeribi, A.: Spectral Theory and Applications of Linear Operators and Block Operator Matrices. Springer, New York (2015)
11. Tretter, C.: Spectral Theory of Block Operator Matrices and Applications. Imperial College Press, London (2008)

Index

Symbols
$B\mathscr{R}(X)$, 75
$B\mathscr{R}(X, Y)$, 75
$K\mathscr{R}(X)$, 210
$K\mathscr{R}(X, Y)$, 210
$X = H \bigoplus G$, 15
diam (U), 17
dist(M, N), 13
$\rho_{e1,V}(U)$, 274
$\rho_{e1}(V)$, 269
$\rho_{e2,V}(U)$, 274
$\rho_{e2}(V)$, 269
$\rho_{e3,V}(U)$, 274
$\rho_{e3}(V)$, 269
$\rho_{e4,V}(U)$, 274
$\rho_{e4}(V)$, 269
$\rho_{e5,V}(U)$, 274
$\rho_{e5}(V)$, 269
$\rho_{e6,V}(U)$, 274
$\rho_{e6}(V)$, 269
$\rho_{e8}(V)$, 269
$\rho_{eap}(V)$, 269
$2^N \backslash \emptyset$, 22
$BC\mathscr{R}(X)$, 88
$BC\mathscr{R}(X, Y)$, 88
$C\mathscr{R}(X)$, 87
$C\mathscr{R}(X, Y)$, 87
$C\sigma(U)$, 186
$G(I_K)$, 22
$G(U)$, 22
$G\,(U_\infty)$, 100
$G\,(U_s)$, 100
$G_U(x)$, 88
$H(O, X)$, 21
I_Ξ, 22
J_M, 212
J_M^X, 212
$K^\perp$, 2
$L\mathscr{R}(X, Y)$, 27
$L^\top$, 2
$N(U)$, 27
$P\sigma(U)$, 186
Q, 63
$Q^Y_{\overline{T(0)}}$, 63
$Q_T T(B_X)$, 210
Q_U, 63
$R(U)$, 22, 27
$R\sigma(U)$, 186
R^*, 226
$R_V(\lambda, U)$, 199
S_U, 207
T-compact, 210
$T_{|M}$, 23
U is a continuous, 75
U is bounded, 75
U-bound, 106
U-bounded, 106
$U(\Xi)$, 22
U^Q, 103
U^*, 2, 137
$U^{-1}(0)$, 27
U_G, 124
$U_{|M}$, 34
$W(A)$, 445, 446
$W\mathscr{R}(X)$, 210
X/E, 14
X^*, 15
X_U, 88
$\Gamma(T)$, 211
$\Gamma_0(T)$, 211
$\Sigma_\varepsilon(T)$, 375
$\Sigma_{e1,\varepsilon}(T)$, 419
$\Sigma_{e2,\varepsilon}(T)$, 419
$\Sigma_{e3,\varepsilon}(T)$, 419

A. Ammar and A. Jeribi, *Spectral Theory of Block Multivalued Operator Matrices*, Infosys Science Foundation Series, https://doi.org/10.1007/978-981-95-0397-1

$\Sigma_{e4,\varepsilon}(T)$, 419
$\Sigma_{e5,\varepsilon}(T)$, 419
$\|Q_u U\|_{L\mathscr{R}(X,Y)}$, 72
$\|U\|_{L\mathscr{R}(X,Y)}$, 72
$\Xi(T)$, 306
$\Xi_+(T)$, 308
$\Xi_-(T)$, 308
$\Xi_0(T)$, 308
$\Xi_\pm(T)$, 308
$\alpha(U)$, 30
$\beta(U)$, 30
$\delta(E)$, 18
$\delta(U, V)$, 19
$\frac{X}{E}$, 14
co(U), 18
dist$(\lambda, W(A))$, 446
$\mathbb{B}(0, 1)$, 20
$\mathbb{D}_\varepsilon$, 21
$\mathbb{S}_X$, 445
$\mathscr{C}(X)$, 211
$\mathscr{DC}(X)$, 213
$\mathscr{DC}(X, Y)$, 213
$\mathscr{DC}_V(X)$, 217
$\mathscr{DC}_V(X, Y)$, 217
$\mathscr{DC}_\mu(X, Y)$, 219
$\mathscr{D}(U)$, 21
$\mathscr{EPRK}(X)$, 210
$\mathscr{EPRK}(X, Y)$, 210
$\mathscr{F}_+(X, Y)$, 226
$\mathscr{F}_-(X, Y)$, 226
$\mathscr{GR}_r(X, Y)$, 185
$\mathscr{GR}(X, Y)$, 185
$\mathscr{GR}_l(X, Y)$, 185
$\mathscr{I}(X)$, 211
$\mathscr{J}_T(X)$, 401, 421
$\mathscr{K}(T)$, 310
$\mathscr{K}_V(X)$, 270
$\mathscr{L}(X)$, 75
$\mathscr{L}(X, Y)$, 75
$\mathscr{PRK}(X)$, 210
$\mathscr{PRK}(X, Y)$, 210
$\mathscr{P}(X)$, 211
$\mathscr{R}(M, N)$, 22
$\mathscr{SRC}(X)$, 212
$\mathscr{SSR}(X)$, 211
$\mathscr{SSR}(X, Y)$, 211
$\overline{U}$, 87, 104
$\overline{\Gamma}_0(T)$, 211
$\overline{\beta}(U)$, 30
$\rho_V(T)$, 198
$\sigma_{\varepsilon,S}(T)$, 429
$\rho_{e1,S}(U)$, 274
$\rho_{e2,S}(U)$, 274
$\rho_{e3,S}(U)$, 274
$\rho_{e4,S}(U)$, 274
$\rho_{e5,S}(U)$, 274
$\rho_{e5,\varepsilon,S}(T)$, 438
$\rho_{e6,S}(U)$, 274
$\rho_{ei,S}(U)$, 274
$\rho_{ei}(V)$, 269
$\sigma_V(T)$, 198
$\sigma_\varepsilon(U)$, 365
$\sigma_{\delta,V}(U)$, 198
$\sigma_{\varepsilon,S}(T)$, 429
$\sigma_{ap,V}(U)$, 198
$\sigma_{dis}(V)$, 192
$\sigma_{e1,V}(U)$, 274
$\sigma_{e1,\varepsilon}(T)$, 401
$\sigma_{e1}(T)$, 270
$\sigma_{e1}(V)$, 270
$\sigma_{e2,V}(U)$, 274
$\sigma_{e2,\varepsilon}(T)$, 401
$\sigma_{e2}(T)$, 270
$\sigma_{e3,V}(U)$, 274
$\sigma_{e3,\varepsilon}(T)$, 401
$\sigma_{e3}(V)$, 270
$\sigma_{e4,V}(U)$, 274
$\sigma_{e4,\varepsilon}(T)$, 401
$\sigma_{e4}(V)$, 270
$\sigma_{e5,V}(U)$, 274
$\sigma_{e5,\varepsilon,S}(T)$, 438
$\sigma_{e5,\varepsilon}(T)$, 401
$\sigma_{e5}(V)$, 270
$\sigma_{e6,V}(U)$, 274
$\sigma_{e6}(V)$, 270
$\sigma_{e\alpha}(V)$, 270
$\sigma_{e\delta,V}(U)$, 274
$\sigma_{e\delta}(V)$, 270
$\sigma_{eap,V}(U)$, 274
$\sigma_{eap}(V)$, 270
$\sigma_{ei}(T)$, 269
$\sigma_j(A)$, 357
$\sigma_{p,A}(T)$, 471
ε-pseudospectra of linear relations, 375
$\widehat{U}$, 229
$\widehat{\delta}(U, V)$, 19
$\widetilde{\Delta}(T)$, 211
$\widetilde{(\cdot)}$, 88
$\widetilde{M}$, 88
$\widetilde{U}$, 88
$ci(U)$, 30
$i(U)$, 30
i_U, 88
$G(U^{-1})$, 27
$\mathscr{W}_*(X)$, 357
ε-essential pseudospectra, 419

A
Adjoint, 2, 137
Analytical core of, 310
Analytic function, 20
Approximate point spectrum, 188

B
Banach space, 20
Bijective, 23, 27
Bounded below, 185
Bounded linear operators, 75

C
Cantor's intersection theorem, 17
Chain of kernels, 44
Chain of range, 45
Closable, 104
Closed linear relation, 87
Closure and completion of a linear relation T, 87
Co-index, 30
Compact, 210
Completion, 88
Completion of a linear relation, 88
Continuous selections, 127
Continuous spectrum, 186
Corner of the numerical range, 446

D
Defect spectrum, 188
Demicompact, 213
Diameter of a set X, 17
Domain, 21, 27

E
Essential pseudospectra, 401
Essential resolvent sets, 269
Essential spectra, 269
Extension, 34

F
Function, 22

G
Gap between two linear subspaces, 19
Graph, 22
Graph operator, 88

H
Hermitian relation, 141

I
Identity relation, 22
Index of U, 30
Infimum δ, 106
Injective, 23
Inverse, 27
Involution, 141
Involutive, 141, 143, 144

J
Jeribi essential spectrum, 357

K
Kernel, 27
Kuratowski measure of noncompactness, 18

L
Left invertible, 185
Left spectrum, 186
Lower semi-Fredholm, 226

M
Maximal gap, 19
Measure of compactness, 211
Metrizable topology, 20
Minimum modulus, 81
Multivalued linear operator, 26

N
Neighbourhood $\mathcal{O}$ in $R(U)$, 75
Neighbourhood in $\mathcal{D}(U)$, 75
Norm, 73
Norm dual, 15
Normed vector space, 14, 20
Norm of linear relation, 72
Null space, 27
Numerical range, 445, 446

O
One-to-one, 27
Open, 75
Open and one-to-one with dense range, 86

P
Pencil essential pseudoresolvent set, 438
Pencil essential pseudospectra, 438
Pencil essential resolvent sets, 274
Pencil point spectrum, 471
Pencil resolvent, 198
Pencil resolvent of U at λ, 199
Pencil spectrum, 198
Point spectrum, 186
Precompact, 210
Pseudospectra of linear relation, 365

Q
Quadratic form, 141
Quasi-Banach space, 20
Quasi-norm, 20
Quotient space, 14

R
Range, 27
Range of relation, 22
Relation, 21
Relative bound of S with respect to T, 106
Relatively bounded, 106
Residual spectrum, 186
Resolvent, 177
Restriction, 34
Restriction of a relation, 23
Riesz integral, 191
Right invertible, 185
Right spectrum, 186

S
Selection, 123
Semi-Fredholm linear relation, 226
Single-valued, 27
Single-Valued Extension Property (SVEP), 298
Single-valued operator, 22
Spectrum of T, 186
Surjective, 23, 27
Symmetric gap, 19

T
Topologically complemented, 15

U
Upper semi-Fredholm, 226

MIX
Papier aus verantwortungsvollen Quellen
Paper from responsible sources
FSC® C105338

If you have any concerns about our products,
you can contact us on
ProductSafety@springernature.com

In case Publisher is established outside the EU,
the EU authorized representative is:
Springer Nature Customer Service Center GmbH
Europaplatz 3, 69115 Heidelberg, Germany

Printed by Libri Plureos GmbH
in Hamburg, Germany